Microbial Technology for Sustainable Environment

Pankaj Bhatt • Saurabh Gangola •
Dhanushka Udayanga • Govind Kumar
Editors

Microbial Technology for Sustainable Environment

Springer

Editors
Pankaj Bhatt
State Key Laboratory for Conservation and Utilization of Subtropical Agro-bioresources, Guangdong Laboratory for Lingnan Modern Agriculture Integrative Microbiology Research Centre
South China Agricultural University
Guangzhou, China

Saurabh Gangola
School of Agriculture
Graphic Era Hill University
Bhimtal, Uttarakhand, India

Dhanushka Udayanga
Department of Biosystems Technology
Faculty of Technology,
University of Sri Jayewardenepura,
Pitipana
Homagama, Sri Lanka

Govind Kumar
Division of Crop Production
ICAR- Central Institute For Subtropical Horticulture
Lucknow, India

ISBN 978-981-16-3839-8 ISBN 978-981-16-3840-4 (eBook)
https://doi.org/10.1007/978-981-16-3840-4

This Springer imprint is published by the registered company Springer Nature Singapore Pte Ltd.
The registered company address is: 152 Beach Road, #21-01/04 Gateway East, Singapore 189721, Singapore

This book is dedicated to all the researchers working throughout the world to make life easier and the authors who have contributed to each of the chapters.

Foreword

It gives me immense pleasure to write a foreword for an upcoming book on microbiology entitled *Microbial Technology for Sustainable Environment*, edited by a learned group of young microbiologists. This book provides some interesting and straightforward approaches to study complex microbiological themes and theories related to microbial life governing the environment and their role in sustenance of life under stressed ecosystems in a simple manner. A careful balance of basic and advanced concepts of microbial systems undertaken in the book provides an insight of the subject to understand precise networking of microbial life for environmental benefit. Streamlined updated information on burning topics related to microbial biotechnology encourages not only students to understand microbial systems and their physiology like signals, metabolism and metagenomics but also researchers of various streams to visualize concepts on microbial metabolism/biodegradation of different pollutants, rhizosphere biology, modern tools for biofertilizer production and microbiome analysis to address stress environment posed by rapid industrialization. The content of the book related to the theory and practical approaches of microbiology would also help to inculcate scientific temperament among new aspirants. To the best of my knowledge, the editors of the book (Dr(s) Pankaj Bhatt, Saurabh Gangola, Dhanuska and Govind Kumar) have contributed significantly to the area of Environmental Microbiology as evident from their quality research papers. I sincerely congratulate them for their attempt to work on such a critical subject which would certainly bring a new height in the area of microbial technology in a simple manner.

Department of Microbiology, GB Pant University of Ag. and Technology, Pantnagar, Uttarakhand, India

Anita Sharma

Preface

Microorganisms are highly diverse and ubiquitous on earth. These microorganisms are able to perform various biological functions in the environment. Microbial applications are used as biofertilizers, bioremediation, biofortification and other sustainable approaches of environmental development. Indigenous microbial strains have the potential to perform various functions that are beneficial to achieve sustainable goals.

To date, different strains of microorganisms have been commercialized globally, for industrial and other applications towards the achievement of sustainable development. Microbial strains and their consortia have been widely utilized in crop improvement and protection, bioremediation of xenobiotics and other sustainable applications in food, agriculture and environment technologies.

In this book, we have compiled a collection of chapters about implementation of microbes in various sectors in sustainable environment. Therefore, the chapters of this book cover a vast area of research in the field of microbial biotechnology, including both traditional and emerging applications. Therefore, the book can be used as an essential reference source for students, aspiring researchers, industrialists, entrepreneurs and policy makers in the field of agriculture, food security, environmental engineering and management.

Authors from various fields of expertise have provided their valuable inputs as chapters complied in this book. Without their expertise contribution, commitment and dedication, this book could not have ever been accomplished. We would like to extend our gratitude to Kripa Guruprasad of Springer Nature who worked hard towards the publication of this book, as well as the families who supported us.

Guangzhou, China — Pankaj Bhatt
Bhimtal, Uttarakhand, India — Saurabh Gangola
Pitipana, Sri Lanka — Dhanushka Udayanga
Lucknow, Uttar Pradesh, India — Govind Kumar

Acknowledgements

In the year 2020, the COVID-19 pandemic affected each of the life throughout the world. We have made a team of four editors from three different countries for the theme microbial technology for sustainable environment. We all were excited to complete a book on Microbial Technologies in sustainable environment. The proposal was initially drafted by Pankaj Bhatt and send for revisions and suggestions to Saurabh Gangola, Dhanushka Udayanga and Govind Kumar. All the editors have revised the proposal. Furthermore, Pankaj Bhatt has submitted it for review to Springer Nature. The reviewer reports were positive and we got acceptance for our book proposal. We all then embarked on this wonderful journey together. This has been a journey of learning and full excitement. We followed the tireless efforts of each of the authors who contributed chapters to this book, and their unwavering determination has made this project from a dream to reality.

We also want to thank the publisher Springer Nature and specially the projector coordinator Ms. Kripa Guruprasad for their support and guidance. She helped us in providing a valuable response to each of the query that was raised by the editors and authors from time to time during the preparation of the manuscript.

We are also very much thankful to the contributing authors from various corners of the globe to make our project meaningful. The authors shared rich information, expert insights and collections of informative scientific figures.

The editors are thankful to Dr. Kalpana Bhatt, researcher at Gurukul Kangri University, Haridwar; Professor Anita Sharma, Department of Microbiology, G.B Pant University of Agriculture and Technology, Pantnagar, U.S Nagar, India; Professor Shaohua Chen, Integrative Microbiology Research Centre, South China Agriculture University, Guangzhou, China; and Mr. Rakesh Bhatt, Department of

Civil Engineering, Indian Institute of Technology, Kanpur, India, for the inspiration to complete the book on time.

We are also grateful to each of the members of Springer's technical staff who supported us in the completion of the project even during the COVID-19 pandemic.

Pankaj Bhatt
Saurabh Gangola
Dhanushka Udayanga
Govind Kumar

Contents

1 **Microbial World for Sustainable Development** 1
Shubhangi Sharma, Raja Singh Rawal, Deepa Pandey, and Neha Pandey

2 **Insights into the Rhizospheric Microbes and Their Application for Sustainable Agriculture** 13
Ankit Negi, Anchal Giri, Pooja Pant, and Rishendra Kumar

3 **Different Biofertilizers and Their Application for Sustainable Development** 31
Dharmendra Kumar, Som Dutt, Pinky Raigond, Sushil Sudhakar Changan, Milan Kumar Lal, Rahul Kumar Tiwari, Kumar Nishant Chourasia, and Brajesh Singh

4 **Microbial Mediated Natural Farming for Sustainable Environment** 49
Asha Rani and Beenam Saxena

5 **Rhizosphere Manipulations for Sustainable Plant Growth Promotion** 61
Pooja Pant, Ankit Negi, Anchal Giri, Pankaj Bhatt, and Rishendra Kumar

6 **Rhizospheric Microbes and Their Mechanism** 79
Anuj Chaudhary, Heena Parveen, Parul Chaudhary, Hina Khatoon, and Pankaj Bhatt

7 **Endophytes and Their Applications as Biofertilizers** 95
Gaurav Yadav, Rishita Srivastva, and Preeti Gupta

8 **Microbial Action on Degradation of Pesticides** 125
Hira Singh Gariya and Arun Bhatt

9 Biofortification of Plants by Using Microbes 141
Ankur Adhikari, Kamal Pandey, Vinita Pant, Tara Singh Bisht, and Himanshu Punetha

10 Microbial Biopesticides: Development and Application 167
H. T. Mandakini and Dimuthu S. Manamgoda

11 Microbial Consortia and Their Application for the Development of a Sustainable Environment 191
Sneha Trivedi, Naresh Butani, Helina Patel, Manoj Nath, and Deepesh Bhatt

12 Microbial Engineering and Applications for the Development of Value-Added Products 203
Ashutosh Paliwal, Abhishek Verma, Ashwini Kumar Nigam, Jalaj Kumar Gour, Manoj Kumar Singh, and Rohit Kumar

13 Plant Growth-Promoting Rhizobacteria and Their Application in Sustainable Crop Production 217
Parul Chaudhary, Heena Parveen, Saurabh Gangola, Govind Kumar, Pankaj Bhatt, and Anuj Chaudhary

14 Reinstating Microbial Diversity in Degraded Ecosystems for Enhancing Their Functioning and Sustainability 235
Sachini Wayanthimali Meepegamage, Ambalangodage Thilini Dhanushka Rathnathilake, Mahesh Premarathna, and Gamini Seneviratne

15 Recent Advancements and Mechanism of Microbial Enzymes in Sustainable Agriculture 247
Pankaj Bhatt, Saurabh Gangola, Charu Joshi, Parul Chaudhary, Govind Kumar, Geeta Bhandari, Saurabh Kumar, Samiksha Joshi, Avikal Kumar, Narendra Singh Bhandari, and Samarth Tewari

16 Application of Microbial Technology for Waste Removal 261
Ravi Ranjan Kumar, Chitra Bhattacharya, and Nutan Prakash Vishwakarma

17 Metagenomics: Insights into Microbial Removal of the Contaminants 293
Dipti Singh, Shruti Bhasin, Anshi Mehra, Manali Singh, Neha Suyal, Nasib Singh, Ravindra Soni, and Deep Chandra Suyal

18 Methods of Strain Improvement for Crop Improvement 307
Jyoti Rawat and Veena Pande

19 Microbial Technologies in Pest and Disease Management of Tea (*Camellia sinensis* (L.) O. Kuntze) 325
Ganga Devi Sinniah and Padmini Dharmalatha Senanayake

20 Field Application of the Microbial Technology and Its Importance in Sustainable Development 347
Saloni Kunwar, Shristi Bhatt, Deepa Pandey, and Neha Pandey

21 Solubilization of Micronutrients Using Indigenous Microorganisms 365
A. D. Sarangi N. P. Athukorala

22 Synergistic Interaction of Methanotrophs and Methylotrophs in Regulating Methane Emission 419
Vijaya Rani, Rajeev Kaushik, Sujan Majumder, A. T. Rani, Asha Arambam Devi, Pratap Divekar, Priyanka Khati, K. K. Pandey, and Jagdish Singh

23 Biopesticides: An Alternative to Synthetic Insecticides 439
A. T. Rani, Vasudev Kammar, M. C. Keerthi, Vijaya Rani, Sujan Majumder, K. K. Pandey, and Jagdish Singh

24 Impact of Pesticides on Microbial Population 467
Sujan Majumder, Anindita Paul, Anup Kumar, Chandan K. Verma, Pratap A. Divekar, Vijaya Rani, A. T. Rani, Jaydeep Halder, K. K. Pandey, and Jagdish Singh

25 Microbe-Mediated Removal of Xenobiotics for Sustainable Environment 483
Helina Patel, Sneha Trivedi, Deepesh Bhatt, Manoj Nath, and Naresh Butani

26 Harnessing the Rhizomicrobiome Interactions for Plant Growth Promotion and Sustainable Agriculture: Mechanisms, Applications and Recent Advances 499
Geeta Bhandari and Niki Nautiyal

27 Fungal Mycelium-Based Biocomposites: An Emerging Source of Renewable Materials 529
Dhanushka Udayanga and Shaneya Devmini Miriyagalla

28 An Endophytic Bacterial Approach: A Key Regulator of Drought Stress Tolerance in Plants 551
Sudha Bind, Sandhya Bind, and Dinesh Chandra

Chapter 1
Microbial World for Sustainable Development

Shubhangi Sharma, Raja Singh Rawal, Deepa Pandey, and Neha Pandey

Abstract Increasing population and decreasing sustainability of natural resources is a global concern; indiscriminate use of natural resources has led to a large-scale exploitation of nature. Change in lifestyle and urbanisation is also a major cause for various conditions such as pollution, greenhouse effect etc. It needs an immediate measure with regard to curb the damage being caused to nature. Sustainability of natural resources is a major concern. A wise and applicable step at this time could provide the privilege to upcoming generations to live an efficient life. Microorganisms being ubiquitous have both harmful and beneficial role. Though microbes are a cause of major pathogenic ailments, efficiently harnessing microbes towards a developing role could help in achieving the major sustainable development goal (SDGs). The presence and usefulness of microbes in almost every field like agricultural, industry, health, education, pharmaceutical and environment is undeniable which can positively regulate nation's economy, whereas a single outbreak of pathogenic microbes could destroy the economy. A microscopic creature is potent enough to cause global disaster, but the misbalance spread by mankind in nature could be balanced by efficient use of these microscopic creatures. Thus, it depends on mankind how these microbes need to be handled with efficiency, in order to attain the best results and help fulfil the goals adopted by United Nations member state to make this planet a better place for us and upcoming generations.

Keywords Sustainability · Bioenergy · Education · Bioremediation · Ecosystem · Economy

---

S. Sharma · R. S. Rawal · N. Pandey (✉)
Department of Biotechnology and Life Sciences, Graphic Era Deemed to be University, Dehradun, Uttarakhand, India
e-mail: neha.pandey@geu.ac.in

D. Pandey
Department of Zoology, Government Post Graduate College, Ranikhet, Uttarakhand, India

P. Bhatt et al. (eds.), *Microbial Technology for Sustainable Environment*,
https://doi.org/10.1007/978-981-16-3840-4_1

1.1 Introduction

Life without higher organisms is feasible, but without microbes is not. It is not exaggeration to mention that life originated from microbes and every one life springs from microbes (Kuhad 2012; Bhatt et al. 2021a, b; Bhandari and Bhatt 2020; Kumar et al. 2017). Microbes play an integral role in various aspects of life. One can consider microbes beyond any imagination altogether the possible regions (Khati et al. 2018). Microbes if exploited judicially can mark a major effect in overall development, i.e. sustainable development (Kuhad 2012). Brundtland in 1987 stated that sustainable development generally meets the needs of the present without compromising the ability of future generations to meet their own needs (Brundtland 1987; Bhatt and Nailwal 2018).

To collaboratively make an endeavour during sustainable development, around 193 countries agreed to different sustainable development goals (SDG), which is a UN's sponsored effort for a sustainable economic development of the planet. These goals are classified into five (5) subgroups: People, Planet, Prosperity, Peace and Partnerships (Bhatt and Bhatt 2020). The SDGs goals are the answer that could permit financial and societal development, however now no longer on the fee of environmental damage (Bhatt and Maheshwari 2020). Rather, those efforts emphasise at the environmental safety with the aid of using stopping and controlling the illegal exploitation of herbal resources (Akinsemolu 2018).

The World Health Organisation (WHO) has stated certain areas for sustainable development goals (SDGs) as shown in Fig. 1.1.

- No poverty.
- No hunger.
- Good health and wellbeing.
- Education.
- Clean water and sanitation.
- Affordable clean energy.
- Economic growth.
- Industrial innovation.
- Reduce inequality amongst countries.
- Sustainable cities and community.
- Climate change.
- Life below water.
- Life on land.
- Peace and justice.
- Global partnership for development.

Microbes are capable of fulfilling all the above stated goals of SDGs. Microbes are omnipresent and also the predominant forms of life on the earth (Goel et al. 2020). Microbes are the backbone of the ecosystem, with many applications that can contribute in sustainable development. Microbes manifest spectrum of evolutionary,

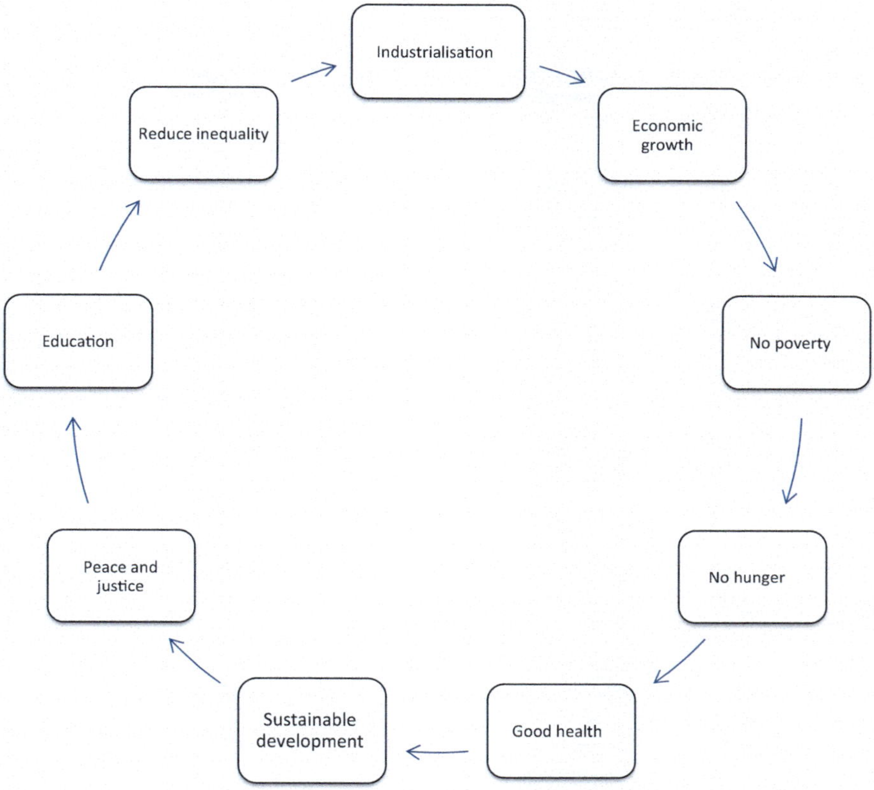

Fig. 1.1 Interlinked sustainable development goal (SDGs): The SDGs are somewhat interlinked; fulfilment of one will lead to attainment of many other SDGs

functional and metabolic diversity (Kumar et al. 2020; Suyal et al. 2019a, b; Bhatt and Maheshwari 2019).

Microbes omnipresence all over the environment, and therefore, their diverse and versatile nature makes them vital agents of planetary system. They have the tendency to facilitate and regulate biogeochemical cycles and consequently use biological materials and waste products. Microbes are also responsible for producing greenhouse gases, viz. carbon dioxide and methane, and are, therefore, necessary determinants of global climate change. In addition to this, they perform essential roles in soil structure and fertility and within the quality and productivity of land, seas, lakes and rivers. Microbes, therefore, are also key members of the committee of stewards of planetary health and property (Timmis et al. 2017).

1.2 Microbes and the Sustainable Development Goals

1.2.1 No Poverty, Economic Growth and Industrial Innovation

Eradication of poverty may also help in attaining various other SDGs directly or indirectly. Mass educating the economically backward class for generating income utilising microbes may also play a pivotal role in eradication of poverty. Various techniques such as using microbes to produce fermented food products may help in raising income, economic development and eradicating hunger. However, the role of microbes in SDGs is shown in Table 1.1. The economic growth can be efficiently made via microbes by medicine, vaccine production and lowering disease rate thereby improving economy (Drexler 2010). Various industries such as food and beverages and chemical synthesis can efficiently exploit microbial population for development (Adesulu and Awojobi 2014).

Table 1.1 Role of microorganisms in accomplishing SDGs

SDGs	Microorganism	Role	Reference
Life below water	*Aspergillus niger*	Decolourisation of pulp and paper industry water	Ahmad et al. (2018)
	Bacillus and *Pseudomonas*	Reduce metal toxicity	Ahmad et al. (2018)
Life on land	*Bordetella avium*	Degrade naphthalene	Abo-State et al. (2018)
No hunger, no poverty, economic growth	*Pseudomonas fluorescence*	Enhance root and shoot growth	Johansson et al. (2004)
	Bradyrhizobium	Enhance soil nutrient content (N,P,S)	Johansson et al. (2004)
	Vigna radiata	ABA production (plant growth under stress)	Ahmad et al. (2013)
No hunger, industrial innovation	*Lactobacilli*	Dairy production	Pereg and McMillan (2015)
Clean energy	*Chlorella vulgaris*	Biobutanol and biohydrogen	Srivastava (2019)
	Shewanella oneidensis	Produce electricity	Lal (2013)
Human health, economic growth	*Streptomyces*	Aminoglycoside antibiotic production	Finkelstein et al. (1996)
Human health, industry application	*Serratia marcescens*	Biotin production	Shimizu (2008)
	Propionibacterium shermanii	Vitamin B12 production	Shimizu (2008)

1.2.2 Good Health Wellbeing, Clean Water and Sanitation

Humans harbour growth of various microorganisms known as human microbiome. These microbes play various essential metabolic and physiological roles such as the intestinal microbiome helps in digestion and absorption of food. Various essential nutrients that are not synthesised in the body and not included in diet are also provided by the microbes. Various recent researches also suggest that microbiota may also influence brain function (Sampson and Mazmanian 2015). By various activities such as decomposition, environmental cleanup can be efficiently maintained, and thereby enhancing good health and wellbeing can be strengthened by intake of probiotics, antibiotics and vaccine.

1.2.3 No Hunger

Microbes play an inseparable part in agriculture by enhancing yield by *Bacillus thuringiensis* (Bt) crops. Various fermented products are produced via microbial activity. Microbes play an exceptionally important role in eradicating hunger positively regulating agricultural practices. Microbes enhance crop yield and soil fertility and play vital role in controlling plant pathogens. Soil fertility and in turn crop productivity can be enhanced by using arbuscular mycorrhizal symbiotic fungi and phosphate-solubilising and nitrogen-fixing microbes (Johansson et al. 2004).

Various microbes specifically effect various aspects of plant growth, e.g. Strains of *Pseudomonas aeruginosa* increases accumulation of dry matter, nodule formation, grain yield and protein content. Various strains of *Azospirillum* increase drought tolerance and enhance root and shoot growth in maize seedlings. *Pseudomonas fluorescens* provides good root and shoot growth and increases tolerance to salinity for cucumber plant. *Bradyrhizobium* species enhances nitrogen, phosphate, sulphur and yield of soybean grain. Microbes also enhance the plant growth by increasing phytohormone productivity and plant growth regulators by 60 times (Camerini et al. 2008). Microbial synthesised phytohormone can regulate physiological plant processes both under normal and stress condition. Auxin synthesised by Pseudomonas and Rhizobium strain helps in tolerating osmotic stress in *Vigna radiate (*Ahmad et al. 2013*)*. Abscisic acid (ABA) helps in growth under stress induction of photoperiodic flowering. Plant growth can also be enhanced by biological control of plant pathogens by competition for nutrients, producing antibiotic, hydrolytic enzymes, siderophores etc. (Glick 2012).

1.2.4 Education

By teaching, research and innovation microbes even play an important role in education field. Education forms the basis of various other SDGs. Mass educating people in turn generates growth opportunities, economic development, improvement in living conditions, good health and research. Education can provide growth opportunities to economically backward classes and find employment. Steps are now being taken to establish new teaching methods for numerous technologies such as environmental technology fermentation technology, food biotechnology and immunology, so that students can easily understand the present and potential use of microbiology and biotechnology for better livelihoods and environmental security (Simonneaux 2000).

1.2.5 Affordable Clean Energy

Bioenergy and biofuel are turning to be good alternate sources of energy, for example *Shewanella oneidensis* exploits organic matter to produce utilisable electricity (Lal 2013). Various wastes such as sewage sludge and municipal solid are being utilised by numerous fungal species including *Trichoderma* and *Aspergillus* to produce bioenergy (Elshahed 2010).

Fossil fuel burning possesses a great threat to environment and mankind. In order to curb this, inefficiency biogas and biomass-based energy are good alternatives that are both cost-effective and environment friendly. The third-generation biofuels can be developed by using microalgal population and curbing the environmental hazards, e.g. *Chlamydomonas reinhardtii* produce ethanol. *Chlorella vulgaris* produce biobutanol and biohydrogen (Srivastava 2019).

1.2.6 Reduced Inequality

Women, who make up half of the world's population, often have half the ability to work and account for more than half of the workforce in fields such as health care (Kaushik and Kapila 2009). Restricted access to education is a big setback for women across the globe. Women need sexual and reproductive health and hygiene knowledge as insufficient education on sexually transmitted diseases such as chlamydia, herpes, gonorrhoea, AIDS and syphilis possesses a higher risk of contracting them (Dehne and Riedner 2001). Therefore, empowerment of girls and women is urgently needed. Indeed, encouraging the completion of formal education, encouraging women to engage in higher education or to learn new skills and raising women's awareness of their rights can contribute to their growth (Penner 2015). It is possible to manage gender disparity by supporting women's education and making women qualified enough to earn a living. This can be managed through the advancement of agricultural, food and dairy and land management activities by women from rural areas. The general importance of various microorganisms both in pathogenic and non-pathogenic aspect is to be provided to women, in particular, for their involvement in agriculture, dairy and medical fields. The knowledge of microorganisms such as *Pseudomonas*, *Rhizobium*, *Trichoderma*, *Bradyrhizobium*, *Azospirillum*, *Lactobacillus*, yeasts etc. should be imparted to women in agriculture in order to increase crop productivity as well as to make various food and dairy products (Pereg and McMillan 2015).

1.2.7 Sustainable Cities and Communities

Proper solid waste disposal system to avoid clogged drains, floods and the spread of waterborne diseases is a primary necessity for the durable and sustainable growth of society. An expensive method is the disposal of agricultural waste. Proper waste management and dispersal is efficiently maintained by using microbial population. The development of green concrete wall and bioremediation curl pollution. The bioconversion of solid waste into useful products such as biofuel, biogas and animal feedstock, as well as its agricultural uses, is a resourceful, green and sustainable way to handle waste products. The composting of solid waste is an efficient and economically viable process in which different microorganisms such as *Pseudomonas*, *Bacillus*, *Microbispora*, *Actinobifida* and *Thermoactinomycetes* are being used to convert their organic constituents into usable end products. Compost can be used as crop manure, thereby improving its productivity and contributing to green growth (Finstein and Morris 1975).

1.2.8 Global Climate

Global climate can be efficiently controlled by microbes by controlling pollution and various biogeochemical cycles like nitrogen, carbon and phosphorus cycles. Various poisonous gases, e.g. released by various human interventions and processes such as fossil fuels burning and the processes of industrial development, are the main global climate change players. With different biotic and abiotic variables, microorganisms are involved in the recycling of elements. Many natural and engineered systems, such as wastewater treatment, agriculture, remediation, production of biofuels and metabolite production and mineralisation, are important (Bodelier 2011).

The marine microbial populations are one of the key regulators of carbon dioxide concentration in the environment. They are even responsible for recycling nutrients that are further used in marine food webs. Microbes are majorly responsible for decomposition of organic matter which in turn releases carbon dioxide (CO_2), methane (CH_4) and other gases into the atmosphere, thereby indirectly regulating global climate. Methanogens such as certain archaea produce large amount of CH_4 into environment (Cavicchioli 2019).

1.2.9 Life Below Water

Various industrial effluents discharged into the water bodies and surface run off from agricultural lands contain harmful chemicals which when reach water bodies may cause harmful effects to both life on land and life under water. Oil spills are one of the most common issues prevailing in oceans. Crude oil contains potential carcinogen products. Microbes play an efficient role in bioremediation and removal of harmful effluents and clearing oil spills, e.g. *Aspergillus niger* is used for decomposition of pulp and paper and wastewater. Microbes such as *Bacillus* and *Pseudomonas* are used at metal contaminated site to reduce toxicity and concentration of pollutants. Microbes control marine population by controlling pathogenic outburst, producing oxygen. As the most important contributor to global climate change, the combustion of fossil fuels may also be controlled by the use of microorganisms as a source of biofuels or as part of biofuel processing technologies (Ahmad et al. 2018).

1.2.10 Life on Land

Microbes have ubiquitous role on land in almost every field. Microbes stabilise the soil structure, permit nutrient uptake via way of means of plants, manage pests and diseases, decompose natural cloth and degrade dangerous chemicals, in addition to being a hallmark of the soil health. Increasing population and demands of humans has led to increase in destruction of forest, loss of biodiversity and increased pollution. Microbes could play an important role in limiting these effects such as microbes could increase agricultural yield. Microbes increase the soil content or quality by nitrogen fixation and phosphate solubilisation.

Microbes also help in bioremediation by degrading polyaromatic hydrocarbon, e.g. *Bordetella avium* MAM-P22 can degrade naphthalene (Abo-State et al. 2018).

1.2.11 Peace and Justice

In general, one cannot think about the connection of microbes with peace and justice; however, microbes contribute significantly to the preservation of a stable society. The occurrence of poverty, insufficient access to food and illiteracy have an adverse impact on children's growth. A significant contributing factor to the emotional wellbeing of children has been implicated in food insecurity (Chilton et al. 2007).

By countering bioterrorism, improving sources of nutrition, improving environmental conditions, introducing green technologies and improving national and international infrastructure would eventually lead to the growth of society and the prevalence of peace and justice (Bhatt and Maheshwari 2020).

1.2.12 Global Partnership for Development

Policy mechanisms must contribute to the social, economic and environmental needs of microbiology in order to achieve a sustainable future. The most important areas which need urgent attention are the use of microbes in agriculture, pharmaceutical science, biofuels and fermented food. Without active cooperation and partnership between nations, sustainable development is not viable. In order to make globalisation more efficient, more distinct, wider and intercontinental agreements are needed (Bhatt and Maheshwari 2020). Only by globalisation and breaking land barriers can the advantages of microbes and microbial technology reach the masses (Chambers et al. 2004; Finkelstein et al. 1996; Shimizu 2008; Rawat et al. 2019; Suyal et al. 2018; Mishra et al. 2020; Zhang et al. 2020). This green technology must be used to improve the ideals of equality and social justice.

1.3 Conclusion

It is not an exaggeration to state that the SDGs laid if fulfilled at this point of time slowly and steadily would definitely make this planet worth living for the upcoming countless generation. It is not an individual or national concern; it needs a global effort to curb the gap created in global sustainable development. A carefree approach towards nature would definitely end the upcoming generations sooner or later. Though microbes appear to be very insignificant with regard to their size, they are potent enough to be both a boon and a curse. It just needs a constructive approach, and microbes alone proves to be a great factor in achieving all sustainable development goals. Significantly, microbes contribute to enhance green production technologies, improve crop productivity and provide earning livelihood to needy people. However, it is now believed that these perspectives and better knowledge might help young people to make efforts in achieving sustainable development goals.

References

Abo-State MAM, Riad BY, Bakr AA, Abdel Aziz MF (2018) Biodegradation of naphthalene by Bordetella avium isolated from petroleum refinery wastewater in Egypt and its pathway. J Radiat Res Appl Sci 11:1–9. https://doi.org/10.1016/j.jrras.2017.10.001

Adesulu AT, Awojobi KO (2014) Enhancing sustainable development through indigenous fermented food products in Nigeria. Afr J Microbiol Res 8:1338–1343. https://doi.org/10.5897/AJMR2013.5439

Ahmad M, Zahir ZA, Nazli F, Akram F, Arshad M, Khalid M (2013) Effectiveness of halo-tolerant, auxin producing Pseudomonas and rhizobium strains to improve osmotic stress tolerance in mung bean (Vigna radiata L.). Braz J Microbiol 44:1341–1348. https://doi.org/10.1590/S1517-83822013000400045

Ahmad M, Pataczek L, Hilger T, Zahir Z, Hussain A, Rasche F, Schafleitner R, Solberg S (2018) Perspectives of microbial inoculation for sustainable development and environmental management. Front Microbiol 9:9–11

Akinsemolu A (2018) The role of microorganisms in achieving the sustainable development goals. J Clean Prod 182:140–146

Bhandari G, Bhatt P (2020) Concepts and application of plant microbe interaction in remediation of heavy metals. In: Sharma A (ed) Microbes and signaling biomolecules against plant stress. Rhizosphere biology. Springer, Singapore. https://doi.org/10.1007/978-981-15-7094-0_4

Bhatt K, Bhatt P (2020) Rhizosphere biology: alternate tactics for enhancing sustainable agriculture. Chapter 9. https://doi.org/10.1002/9781119644798.ch9

Bhatt K, Maheshwari DK (2019) Decoding multifarious role of cow dung bacteria in mobilization of zinc fractions along with growth promotion of *C. annuum* L. Sci Rep 9:14232. https://doi.org/10.1038/s41598-019-50788-8

Bhatt K, Maheshwari DK (2020) Zinc solubilizing bacteria (*Bacillus megaterium*) with multifarious plant growth promoting activities alleviates growth in *Capsicum annuum* L. 3 Biotech 10:36. https://doi.org/10.1007/s13205-019-2033-9

Bhatt P, Nailwal TK (2018) In: Prasad R, Gill SS, Tuteja N (eds) Crop improvement through microbial technology: a step towards sustainable agriculture. Elsevier, Amsterdam, pp 245–253

Bhatt P, Zhou X, Huang Y, Zhang W, Chen S (2021a) Characterization of the role of esterases in the biodegradation of organophosphate, carbamate and pyrethroid group pesticides. J Hazard Mater 411:125026

Bhatt P, Joshi T, Bhatt K, Zhang W, Huang Y, Chen S (2021b) Binding interaction of glyphosate oxidoreductase and C-P lyase: molecular docking and molecular dynamics simulation studies. J Hazard Mater 409:124927

Bodelier PLE (2011) Toward understanding, managing, and protecting microbial ecosystems. Front Microbiol 2:80. https://doi.org/10.3389/fmicb.2011.00080

Brundtland G (1987) Our common future—call for action. *Environ Conserv* 14(4):291

Camerini S, Senatore B, Lonardo E, Imperlini E, Bianco C, Moschetti G et al (2008) Introduction of a novel pathway for IAA biosynthesis to rhizobia alters vetch root nodule development. Arch Microbiol 190:67–77. https://doi.org/10.1007/s00203-008-0365-7

Cavicchioli R (2019) Scientists' warning to humanity: microorganisms and climate change. In: eScholarship. University of California, Los Angeles, CA, pp 572–575

Chambers WB, Green JF, Kambu A (2004) Trade, biotechnology and sustainable development: a report on the Southeast Asia Workshop for Policymakers. In: Global Environmental Change, pp 185–188. https://doi.org/10.1016/j.gloenvcha.2004.05.001

Chilton M, Chyatte M, Breaux J (2007) The negative effects of poverty & food insecurity on child development. Indian J Med Res 126(4):262–272

Dehne KL, Riedner G (2001) Sexually transmitted infections among adolescents: the need for adequate health services. Reprod Health Matters 9:170–183. https://doi.org/10.1016/S0968-8080(01)90021-7

Drexler M (2010) What you need to know about infectious disease. National Academies Press (US), Washington, DC

Elshahed MS (2010) Microbiological aspects of biofuel production: current status. J Adv Res 1 (2):103–111. https://doi.org/10.1016/j.jare.2010.03.001

Finkelstein E, Amichai B, Grunwald MH (1996) Griseofulvin and its uses. Int J Antimicrob Agents 6:189–194. https://doi.org/10.1016/0924-8579(95)00037-2

Finstein MS, Morris ML (1975) Microbiology of municipal solid waste composting. Adv Appl Microbiol 19:113–151. https://doi.org/10.1016/S0065-2164(08)70427-1
Glick BR (2012) Plant growth promoting bacteria: mechanisms and applications. Scientifica 2012:1–15. https://doi.org/10.6064/2012/963401
Goel R, Debbarma P, Kumar P, Suyal DC, Kumar S, Mahapatra BS (2020) Assessment of soil chemical quality, soil microbial population and plant growth parameters under organic and conventional rice–wheat cropping system. Agric Res. https://doi.org/10.1007/s40003-020-00499-8
Johansson JF, Paul LR, Finlay RD (2004) Microbial interactions in the mycorrhizosphere and their significance for sustainable agriculture. FEMS Microbiol Ecol. https://doi.org/10.1016/j.femsec.2003.11.012
Kaushik K, Kapila K (2009) Women in medical microbiology: reflections on contributions. Indian J Med Microbiol 27:285–288. https://doi.org/10.4103/0255-0857.55435
Khati P, Parul, Bhatt P, Nisha, Kumar R, Sharma A (2018) Effect of nanozeolite and plant growth promoting rhizobacteria on maize. 3 Biotech 8:141
Kuhad R (2012) Microbes and their role in sustainable development. Ind J Microbiol 52(2):310
Kumar G, Arya K, Verma A, Pankaj P, Khati P, Gangola S, Kumar R, Sharma A, Singh H (2017) Bioremediation of petrol engine oil polluted soil using microbial consortium and wheat crop. J Pure Appl Microbiol 11(3):1583–1588
Kumar S, Suyal DC, Yadav A, Shouche Y, Goel R (2020) Psychrophilic *Pseudomonas helmanticensis* proteome under simulated cold stress. Cell Stress Chaperones 25 (6):1025–1032. https://doi.org/10.1007/s12192-020-01139-4
Lal D (2013) Microbes to generate electricity. Indian J Microbiol 53:120–122. https://doi.org/10.1007/s12088-012-0343-2
Mishra S, Zhang W, Lin Z, Pang S, Huang Y, Bhatt P, Chen S (2020) Carbofuran toxicity and its microbial degradation in contaminated environments. Chemosphere 259:127429. https://doi.org/10.1016/j.chemosphere.2020.127419
Penner AM (2015) Gender inequality in science. Science (-80) 347:234–235. https://doi.org/10.1126/science.aaa3781
Pereg L, McMillan M (2015) Scoping the potential uses of beneficial microorganisms for increasing productivity in cotton cropping systems. Soil Biol Biochem 80:349–358. https://doi.org/10.1016/j.soilbio.2014.10.020
Rawat N, Sharma M, Suyal DC, Singh DK, Joshi D, Singh P, Goel R (2019) Psyhcrotolerant bio-inoculants and their co-inoculation to improve *Cicer arietinum* growth and soil nutrient status for sustainable mountain agriculture. J Soil Sci Plant Nutr 19(3):639–647
Sampson TR, Mazmanian SK (2015) Control of brain development, function, and behavior by the microbiome. Cell Host Microbe 17:565–576
Shimizu S (2008) Vitamins and related compounds: microbial production. In: Biotechnology, 2nd Completely Revised edn, pp 318–340. https://doi.org/10.1002/9783527620999.ch11k
Simonneaux L (2000) A study of pupils' conceptions and reasoning in connection with "microbes", as a contribution to research in biotechnology education. Int J Sci Edu 22(6):619–644. https://doi.org/10.1080/095006900289705
Srivastava R (2019) Bio-energy production by contribution of effective and suitable microbial system. Mat Sci Energy Technol 2(2):309
Suyal DC, Kumar S, Joshi D, Soni R, Goel R (2018) Quantitative proteomics of psychotrophic diazotroph in response to nitrogen deficiency and cold stress. J Proteomics 187:235–242. https://doi.org/10.1016/j.jprot.2018.08.005

Suyal DC, Joshi D, Kumar S, Soni R, Goel R (2019a) Differential protein profiling of soil diazotroph *Rhodococcus qingshengii* S10107 towards low-temperature and nitrogen deficiency. Sci Rep 9(20378):1–9. https://doi.org/10.1038/s41598-019-56592-8

Suyal DC, Kumar S, Joshi D, Yadav A, Shouche Y, Goel R (2019b) Comparative overview of red kidney bean (*Phaseolus vulgaris*) rhizospheric bacterial diversity in perspective of altitudinal variations. Biologia 74(10):1405–1413. https://doi.org/10.2478/s11756-019-00292-1

Timmis K, de Vos W, Ramos J, Vlaeminck S, Prieto A, Danchin A, Verstraete W, de Lorenzo V, Lee S, Brüssow H, Timmis J, Singh B (2017) The contribution of microbial biotechnology to sustainable development goals. Microb Biotechnol J 10(5):985

Zhang W, Lin Z, Pang S, Bhatt P, Che S (2020) Insights into the biodegradation of Lindane (γ-Hexachlorocyclohexane) using a microbial system. Front Microbiol 11:522. https://doi.org/10.3389/fmicb.2020.00522

Chapter 2
Insights into the Rhizospheric Microbes and Their Application for Sustainable Agriculture

Ankit Negi, Anchal Giri, Pooja Pant, and Rishendra Kumar

Abstract The rhizosphere soil of plant consists of diverse microorganisms. The study of plant-microbe interactions is necessary for plant health and promotion. The rhizosphere microbes not only provide important nutrients to plants but also prevent the growth of harmful pathogenic microbes. The processes that occur in biome include quorum sensing, nitrogen fixation, nutrients solubilization, volatiles, mobilization, and immobilization of nutrients. The plant allures microbes according to their requirements by releasing certain chemical compounds needed for microbes. In case of any pathogen attack, they recruit microbes to suppress pathogens. Some of the evidences revealed that the recruitment of microbes in rhizosphere shows effects on plant at physiological and molecular level. This increases the possibility that the soil microbiota can stimulate the ability of plant to tackle different biotic and abiotic stresses. In this chapter, we will discuss various mechanisms by which the microbial communities work for plants and how the plants recruit them for their development in detail. Altering the microbial population in the rhizosphere either by removing or adding new species can modify the plant and rhizosphere microbiome. The rhizosphere microbes used in making of agricultural products like biopesticides and biofertilizers for sustainable agriculture practices. The plant-microbe interaction can be further studied with the help of bioinformatics, molecular, and modeling tools.

Keywords Rhizosphere microbes · Rhizosphere plant growth mechanisms · Beneficial PGPR · Plant-microbe interactions

---

A. Negi · A. Giri · P. Pant · R. Kumar (✉)
Department of Biotechnology, Sir J.C. Bose Technical Campus, Bhimtal, Kumaun University, Nainital, Uttarakhand, India

P. Bhatt et al. (eds.), *Microbial Technology for Sustainable Environment*,
https://doi.org/10.1007/978-981-16-3840-4_2

2.1 Introduction

The rhizosphere is a closed narrow zone around the plant root that consists of large number of microbial communities. A single plant inhabits a large number of microbes in their rhizosphere soil and can contain >30,000 prokaryotic species (Mendes et al. 2011) and up to 1011 microbial cells per gram of root (Egamberdieva et al. 2008). In fact, the soil microbial communities are the greatest reservoir of biological diversity till now (Buee et al. 2009). The complete genome of the plant rhizosphere microbial communities is much more than that of the host plants itself, thus termed as the second genome of the plant. The functions performed by the rhizosphere microbial communities are very much similar to that of human gut (Bron et al. 2012). The microbial species present in the soil changes with time according to the plant requirements from thousands to millions (Nihorimbere et al. 2011; Sulbhi et al. 2021; Bhandari et al. 2021; Bhatt et al. 2021a, b, c).

2.2 Rhizosphere Microbes: Role in Plant Health and Growth Promotion

The plant-microbe interactions play vital roles in plant life like carbon and nitrogen sequestration and nutrient cycling (Singh et al. 2004). These interactions are mostly positive because plant attracts selectively only those microbes which are needed for their growth promotion such as plant growth-promoting rhizobacteria (PGPR), mycorrhizal fungi, and epiphytes. They can be beneficial or pathogenic and may live freely in the soil or in mutual or commensal associations (Philippot et al. 2013). Plants provide the microbes the root exudates which they use as a substrate, and in turn, they help plants in disease suppression (Haas and Défago 2005), increasing immunity to biotic (Badri et al. 2013) and abiotic stresses (Zolla et al. 2013). The plants release fixed carbon as amino acids, soluble sugars, and secondary metabolites (Badri et al. 2013; Chaparro et al. 2013) which are then used by microbes in the rhizosphere. The root exudate composition is determined by plant species, plant developmental stage, and environmental factors like soil pH, temperature, and the microorganisms present in the soil (Badri and Vivanco 2009). These factors cause the specificity of microbes to each plant species. The changing numbers and diversity of microbes' affects crop yield and soil fertility for developing better varieties. The root exudates of plant are primary factors for attracting and inhibiting the growth of certain microorganisms in the rhizosphere. They selectively permit those microbes which are essential for plant growth among the bulk soil population of microbes (Grayston et al. 1998). These consist of polysaccharides, amino acids, mucilage, and many secondary metabolites such as flavonoids, terpenes, and glucosinolates (Moore et al. 2014). Even the minor alteration in the quantity of substances released by plant can change the microbial community structure of the plant rhizosphere (Jones et al. 2004). For good plant productivity, healthy soil is necessary, which is affected by various biotic and abiotic factors among which the soil microorganisms are among the most dominant biotic components (Bhatt et al. 2020a, b, 2021d, e, f, g). Thus, plant-microbe interactions are not only important for growth, health, and biocontrol of plants but also influence the chemical, biological, and physical properties of soil (Bhatt et al. 2019a, 2020c, d, e, f).

2.3 Plant-Microbe Interactions

The plant and microbes interacted with each other by mutual or commensal associations, in which either both host and microbe get benefitted or only one of them gets benefits from associations. Sometimes microbes transit from pathogenic nature to symbiont nature depending on the environment conditions (Newton et al. 2010). The rhizobia, is a symbiont nitrogen-fixing bacteria changes from symbiont to neutral interaction with plant with change in soil nitrogen level (Zahran 1999; Bhatt et al. 2015a, b, 2016a, b, 2019b, c).

2.3.1 Plant Growth-Promoting Rhizobacteria (PGPR)

PGPR is a group of bacteria that are found in the rhizosphere colonizing the root of monocot and dicot plants to enhance the plant growth by various mechanisms (Ahemad and Khan 2012; Huang et al. 2021). PGPR are classified into two types: intracellular PGPR which are present in nodules of plant root and extracellular PGPR which are free-living bacteria (Martínez-Viveros et al. 2010). Common examples of intracellular PGPR are *Frankia*, *Rhizobium*, *Bradyrhizobium*, *Allorhizobium*, etc. (Bhattacharyya and Jha 2012), and extracellular PGPR genera are *Erwinia*, *Azospirillum*, *Burkholderia*, *Caulobacter*, *Pseudomonas*, *Serratia*, *Bacillus*, and *Chromobacterium*, etc. (Ahemad and Kibret 2014). They increase nutrient, like micronutrients, phosphorous, and potassium solubilization, uptake in plants (Singh et al. 2007) or release chemical substances like ethylene, IAA, GA, and cytokinins (Kloepper 1992) and improve plant growth under stress conditions (Egamberdieva and Kucharova 2009). PGPR along with other bacterial and fungal partners are inoculated in soil to enhance fertility (Kumar et al. 2013; Prasanna et al. 2011, 2014, 2015). The increase in salt tolerance and leaf water content of *Zea mays* and decrease in its electrolyte leakage occur with the co-inoculation of *Pseudomonas* and *Rhizobium* (Bano and Fatima 2009). Induced systemic resistance (ISR) is acquired by plant against pathogens when the PGPR are inoculated; they resist a broad spectrum of pathogens (van Hulten et al. 2006). Many PGPBs such as *Paenibacillus alvei*, *Azospirillum brasilense*, *Bacillus pumilus*, *Pseudomonas fluorescens*, etc. form a colony on roots and protect vegetables, crops, and trees from foliar diseases in field trials and greenhouse (Van Loon 2007).

2.3.2 Mycorrhizae

Mycorrhizal symbiotic relationship is a non-disease-producing mutual association between higher plants and fungi in which both plant and the fungus benefit (Morgan et al. 2005; Bhatt et al. 2019d). The fungus invades the plant root to absorb nutrients, while 90% of land plants depend on mycorrhizal fungi for minerals mainly phosphorus (Bhatt et al. 2019d; Sharma and Bhatt 2016). In winter season, the plants do not get sufficient exposure to sunlight which they get dependent on fungi for nitrogenous compounds, sugar, and other nutrients that fungi absorb from soil. In lowland forests, mycelia networks are formed that connect various trees; together with the help of them, trees and their seedlings use that network to exchange chemical messages and nutrients (Bhatt et al. 2019d; Sharma et al. 2016; Bhatt and Nailwal 2018; Khati et al. 2018a; Gangola et al. 2018a). Mycorrhizae are of two types: ectomycorrhiza and endomycorrhiza. Ectomycorrhiza are found in trees, while endomycorrhiza are mostly found on agricultural crops (Bhatt et al. 2019d, e; Bhatt 2018; Bhatt and Barh 2018; Bhandari and Bhatt 2020; Bhatt and Bhatt 2020). The examples of mycorrhizae fungi are *Scutellospora*, *Glomus*, *Entrophospora*, and *Gigaspora* along with providing minerals to plant mycorrhizae provide tolerance from drought, heavy metals, and pathogens to plants (Bago et al. 2003).

2.3.3 Nitrogen-Fixing Microbes

Nitrogen-fixing microbes play a great role in plant life by providing them soil nitrogen through nitrogen fixation process. They may live freely or in symbiotic association to plant depending on the type of host plant (Deaker et al. 2004). Some of the examples of symbiotic nitrogen fixers are *Photorhizobium*, *Rhizobium*, *Sinorhizobium*, and *Bradyrhizobium*; they form nodules on plant root or stem (Moreira 2008). Rhizobia represent Proteobacteria; they are not only the one that form nodules on plant, but also some other examples of α-Proteobacteria such as Ochrobactrum and Phyllobacterium and β-Proteobacteria such as *Devosia* and *Burkholderia* form nodules (Daniel et al. 2007). Actinobacteria, Frankia (Prasanna et al. 2009), and few Cyanobacteria like Nostoc, Anabaena, etc. also take part in the fixation process (Unkovich and Baldock 2008). Common free-living nitrogen-fixing bacteria are *Pantoea*, *Bacillus*, *Azotobacter*, *Burkholderia*, etc. *Gluconacetobacter* and *Herbaspirillum* are endophytes that also fix nitrogen. *Clostridium*, *Enterobacter*, *Klebsiella*, and *Desulfovibrio* are facultative and obligate anaerobes that fix nitrogen only in the absence of oxygen (Arnold 2007).

2.3.4 Endophytes

Endophytes are the fungus and bacteria that reside inside the tissue of plant cells without harming them. They intake nutrition from their host plant and provide them various benefits. Their main functions include the plant growth promotion, yield enhancement, proper nutrient cycling, and pathogen suppression. They form colonies in different parts of plant like stem, root, leaves, bark, etc. They also produce phytohormones that protect plant from abiotic stresses (Ganley et al. 2004). A single plant microbiome consists of diverse endophytes which help plant including the modulation in gene expression (Coombs and Franco 2003). Common bacterial endophytes include *Azoarcus*, *Azospirillum*, *Pseudomonas*, *Gluconacetobacter*, *Achromobacter*, etc. (Jumpponen 2001). Common fungal endophytes basidiomycetes, Harpophora, Rhizoctonia/Ceratobasidium complex, *Periconia macrospinosa*, and Exophyla are involved in plant growth and promotion (Fierer and Jackson 2006).

2.4 Effect of Rhizosphere Microbes on Root Development

Beneficial rhizosphere microbes induce plant growth by modifying root development. Soil consists of huge microbial diversity like fungi, bacteria, archaea, etc. (Ferguson and Mathesius 2014). The microbes interact with the host plants and improve their health by direct and indirect mechanisms (Verbon and Liberman 2016). PGPR are most common among them; they modulate the significant determinant of crop yield, i.e., root system architecture. Though it is reviewed that it modulates the root system, the mechanism by which it influences the cell division, proliferation, and differentiation in root initiation sites is unknown. The recent findings also suggest that PGPR play role in root hair formation and lateral root development (Zamioudis et al. 2013). For studying the effect of bacterial and fungal species on root development, Arabidopsis root is taken as host. *Pseudomonas simiae WCS417* or *Bacillus megaterium UMCV1* colonizes around Arabidopsis root and causes the transition from proliferation to differentiation in the root (Ortíz-Castro et al. 2008; Zamioudis and Pieterse 2012). Hence, their effects are different, *P. simiae* increases the cell division, whereas *B. megaterium* decreases it, in the meristematic zone. These bacterial species decrease cell elongation, thus decreasing primary root length (Ortíz-Castro et al. 2008). The density of root hairs gets increased due to their colonization and grows longer. The volatile compounds released by *P. simiae WCS417* enhance lateral root formation without inhibiting the primary growth (Zamioudis and Pieterse 2012).

2.5 Impact of Rhizosphere Microbes on Host Immune System

The beneficial rhizosphere microbes enhance the defensive activity of plants by activating the induced systemic resistance (ISR) in which the host immune system gets activated for defense (De Vleesschauwer and Höfte 2009; Millet et al. 2010). In case of Arabidopsis, the induction of ISR is well studied by PGPR *Pseudomonas fluorescens* WCS417. WCS417 secretes low molecular weight molecules that suppress the flagellin-triggered immune response (Pineda et al. 2010), while an immune signaling cascade gets initiated that provides resistance to a broad spectrum of pathogens and insects (Van Oosten et al. 2008; Van der Ent et al. 2008). This initiation in Arabidopsis root is regulated by transcription factor MYB72 (Van der Ent et al. 2009). ISR response causes then the increased deposition of callose at the pathogen entry site (Pozo et al. 2008). Jasmonic acid, ethylene hormones, and transcriptional co-activators NPR1 and MYC2 help in the process of WCS417-ISR activation (Pozo et al. 2008; Segarra et al. 2009). Not only bacteria but also some fungus like *Trichoderma* sp. (Trivedi et al. 2012) and mycorrhizal fungi (Segarra et al. 2009) induce ISR. The rhizosphere microbiome structure also gets changed when a plant gets infected by a pathogen. For example, the infection caused by *Candidatus Liberibacter asiaticus* on citrus plant results in the alteration in the composition of citrus rhizosphere community (Zhang et al. 2011). Similarly, the infection caused by *Verticillium dahlia* alters the rhizospheric microbial composition of cotton (Bais et al. 2002). These changes may occur due to the release of some type of antimicrobial compounds by infected roots. Like in the case of *Ocimum basilicum*, infection by Pythium causes the release of rosmarinic acid, a caffeic acid ester with antimicrobial activity (Lanoue et al. 2010). *Fusarium graminum* infection on Barley causes the release of phenol compounds with antifungal activity (Meyer 2000).

2.6 Mechanisms Involved in Plant-Microbe Interactions

The microbes interact with plant through various mechanisms including quorum sensing, volatile production, microbial signaling, plant hormones, and siderophore synthesis.

2.6.1 Siderophore Production

Iron is a highly needed nutrient for all living being except for certain like *Legionella*, *Neisseria*, and *S. cerevisiae* (Atzorn et al. 1988). To satisfy the need of iron, microorganisms have evolved many pathways that include the low molecular weight iron chelators called siderophores. They are secreted to solubilize the iron from surrounding environments and form a complex of ferric-siderophore that move by diffusion and return to the cell surface (Rajkumar et al. 2010). Based on their structural features, iron-coordinating functional groups, and types of ligands, bacterial siderophores have been classified into four main classes: pyoverdines, carboxylate, phenol catecholates, and hydroxamates (Rodríguez et al. 2006). Hundreds of siderophores have been identified and reported for cultivable microorganisms; some of which are widely recognized and used by different microorganisms, while others are species-specific (Misra et al. 2012). Siderophore production confers competitive advantages to PGPR that can colonize roots and exclude other microorganisms from this ecological niche (Bhandari and Bhatt 2020). Under highly competitive conditions, the ability to acquire iron via siderophores may determine the outcome of competition for different carbon sources that are available as a result of root exudation or rhizodeposition (Rodríguez et al. 2006). Most organisms require iron as an essential element; it serves as a cofactor for a wide variety of cellular processes, such as electron transport chain, oxygen transport, cellular respiration, chlorophyll biosynthesis, and thylakoid biogenesis and chloroplast development (Neilands 1995). More than 100 enzymes involve in primary and secondary metabolic reactions contain ferric residues such as iron-sulfur clusters (Oves et al. 2013). Although iron is abundantly present in the environment, the low solubility and slow dissolution rates of iron-containing minerals often limit the bioavailability of iron. The rhizoremediation of soils by PGP microorganisms is believed to reduce chemical fertilizers in agriculture practices (Philippot et al. 2010). Plant growth promotion by siderophore-producing rhizobacterial inoculations have been reported in various studies. Siderophore-producing bacteria *Pseudomonas* strain GRP3 has been shown to enhance chlorophyll content and iron nutrition in *Vigna radiata* plants. Fe-siderophore complex, which is produced by rhizosphere microorganisms, can deliver iron to plant through specific transporter channels under iron starvation (Bhatt et al. 2016a). Moreover, chelation of trace elements by bacterial siderophores in the rhizosphere have employed as natural biodegradable chelators (Moreira 2008). Some siderophores, e.g., desferal, desferrioxamine B, dexrazoxane, O-trensox, desferri-exochelins, desferrithiocin, and tachpyridine, are found useful in sickle cell disease, thalassemia, malaria, haemochromatosis, and cancer therapy.

2.6.2 Quorum Sensing

The small signal molecules that can diffuse easily are termed as autoinducers and mediate quorum sensing (QS) by regulating the gene expression of the population (Hooshangi and Bentley 2008; Von Bodman et al. 1998). In case of Gram-negative bacteria, the N-acyl homoserine lactones (AHL) act as the signal molecules and regulate the density of population (Rinaudi and González 2009). QS plays a very important role in legume symbiosis (Fray 2002). Bacterial QS is a type of cell density-dependent population behavior, in which the production, detection, and response to a molecule regulate gene expression. Only proteobacteria among all bacteria exhibit QS by signaling molecule N-acyl homoserine lactone (AHL) (Elasri et al. 2001). A huge number of AHL-producing proteobacteria are found in rhizosphere (DeAngelis et al. 2007). Two studies show direct evidence of QS in natural soil (Burmølle et al. 2005) and compost soil (Passador et al. 1993), although the role of QS in control of soil processes has not been investigated in biologically intact soils. QS control of extracellular enzyme activity has been studied almost entirely within the context of pathogenesis and is known mostly in pathogenic Gammaproteobacteria like *Pseudomonas aeruginosa* PAO1 (Worm et al. 2000), *Pseudomonas fluorescens* (Rasch et al. 2005), Enterobacteria spp. (Swift et al. 1999), *Aeromonas hydrophila* (Jones et al. 1993), *Erwinia carotovora* (Pirhonen et al. 1993; Eberl et al. 1996), *Serratia* spp. (Croxatto et al. 2002), and *Vibrio* spp. (Aguilar et al. 2003) and Betaproteobacteria *Burkholderia cepacia* (Chernin et al. 1998) and *Chromobacterium violaceum* (Von Bodman et al. 2003). The prevalence of QS-controlled secretion of enzymes among pathogens (Berg et al. 2005) and the prevalence of pathogens in soil (Dweck et al. 2015) suggest that bacterial QS may also be important in the context of soil nitrogen cycling.

2.6.3 Volatile Organic Compounds

The chemicals that carry out the communication across all kingdoms of life are termed as volatile metabolites (Schmidt et al. 2015). They are able to alter the physiological processes in other bacteria, fungi, and plants (Blom et al. 2011). In some studies, it has been shown that the bacterium-bacterium and bacterium-host interactions also get facilitated by bacterial volatiles (Lowery et al. 2008). The biofilm formation and the motility are some of the bacterial processes that are regulated by producing quantitative and qualitative differences in the volatiles (Groenhagen et al. 2013). In *Pseudomonas aeruginosa*, *Burkholderia ambifaria*, and *Streptomyces* sp., the synthesis of QS-regulated volatile, i.e., 2-amino acetophenone, has been reported (Kai et al. 2016), where the pattern of volatile emission is influenced by plant-microbe interactions (Rosier et al. 2016).

2.6.4 Plant-Mediated Signaling

Plants are greatly responsible for assembling the rhizosphere microorganisms (Walling 2000). When plant get affected by biotic stress, i.e., pathogens, the defense system called induced systemic response (ISR) gets activated by plants (Glazebrook 2005). The most dominating signaling pathways in plants are based on salicylic acid-dependent systemic acquired resistance (SAR) (Giron et al. 2013) and jasmonic acid/ethylene-dependent ISR (Glazebrook 2005). Other hormones like abscisic acid, cytokinin, auxin, and gibberellins also take part in this signaling network (Stam et al. 2014). The activation of these pathways depends on the stress type (Doornbos et al. 2011) and influences the rhizospheric microbial community differently (Lee et al. 2012). In sweet pepper Rhizosphere, *Bacillus subtilis* GB03 population in soil are attracted by plants when they get feed by aphids (Lebeis et al. 2015). Salicylic acid promotes the colonization of very selective bacterial groups in the root (Lakshmanan et al. 2012). The root exudate composition gets affected by foliar infection in plant that facilitates the colonization of beneficial rhizobacteria in the roots (Lakshmanan et al. 2014; Singh et al. 2021; Zhang et al. 2020a, b; Mishra et al. 2020). From the previous research, it was concluded that rhizospheric and indigenous soil microbial strains perform various important role in sustainable environmental development (Zhang et al. 2020b; Feng et al. 2020; Lin et al. 2020; Zhan et al. 2020; Ye et al. 2019; Huang et al. 2019). These strains can be applied for greater agriculture production and bioremediation of xenobiotics from the soil and water environments (Huang et al. 2019, 2020; Fan et al. 2020; Pang et al. 2020; Gangola et al. 2018b). Furthermore, more validation of processes of these microorganism could be more beneficial for the resource recovery and sustainable agricultural environments (Bhatt et al. 2021f; Gangola et al. 2018b; Gupta et al. 2018; Khati et al. 2017a, b, 2018b; Kumar et al. 2017).

2.7 Conclusion

The world is facing a lot of issues regarding food quality, soil deterioration, crop production, soil fertility, and many more. These issues could be overcome with the help of beneficial microbes that include bacteria, fungi, and archaea. The plant-microbe interactions pay a great role in promoting plant growth and health. Still the understanding on this field is in its infancy. As discussed above, the root microbiome boosts the defense system of plant and also facilitates nutrient solubilization. It is also suggested that the plants assemble the group of microorganisms in their rhizosphere according to their requirements and necessity by alluring microbes with chemical compounds. Due to this selective interaction with beneficial microbes, the specificity between host plant and microbes has been increased. The beneficial microbes present in the soil for a long period make the soil disease suppressive and healthy for crop growth and yield. In future, it is expected that many more mechanisms would be revealed by which the plant and microbes interact. Thus, it will ultimately helpful in increasing crop productivity and quality.

References

Aguilar C, Friscina A, Devescovi G, Kojic M, Venturi V (2003) Identification of quorum-sensing-regulated genes of *Burkholderia cepacia*. J Bacteriol 185(21):6456–6462

Ahemad M, Khan MS (2012) Productivity of greengram in tebuconazole-stressed soil, by using a tolerant and plant growth-promoting Bradyrhizobium sp. MRM6 strain. Acta Physiol Plant 34 (1):245–254

Ahemad M, Kibret M (2014) Mechanisms and applications of plant growth promoting rhizobacteria: current perspective. J King Saud Univ Sci 26(1):1–20

Arnold AE (2007) Understanding the diversity of foliar endophytic fungi: progress, challenges, and frontiers. Fungal Biol Rev 21(2–3):51–66

Atzorn R, Crozier A, Wheeler CT, Sandberg G (1988) Production of gibberellins and indole-3-acetic acid by rhizobium phaseoli in relation to nodulation of *Phaseolus vulgaris* roots. Planta 175(4):532–538

Badri DV, Vivanco JM (2009) Regulation and function of root exudates. Plant Cell Environ 32 (6):666–681

Badri DV, Zolla G, Bakker MG, Manter DK, Vivanco JM (2013) Potential impact of soil microbiomes on the leaf metabolome and on herbivore feeding behavior. New Phytol 198 (1):264–273

Bago B, Pfeffer PE, Abubaker J, Jun J, Allen JW, Brouillette J, Douds DD, Lammers PJ, Shachar-Hill Y (2003) Carbon export from arbuscular mycorrhizal roots involves the translocation of carbohydrate as well as lipid. Plant Physiol 131(3):1496–1507

Bais HP, Walker TS, Schweizer HP, Vivanco JM (2002) Root specific elicitation and antimicrobial activity of rosmarinic acid in hairy root cultures of *Ocimum basilicum*. Plant Physiol Biochem 40(11):983–995

Bano A, Fatima M (2009) Salt tolerance in Zea mays (L). Following inoculation with rhizobium and Pseudomonas. Biol Fertil Soils 45(4):405–413

Berg G, Eberl L, Hartmann A (2005) The rhizosphere as a reservoir for opportunistic human pathogenic bacteria. Environ Microbiol 7(11):1673–1685

Bhandari G, Bhatt P (2020) Concepts and application of plant microbe interaction in remediation of heavy metals. In: Sharma A (ed) Microbes and signaling biomolecules against plant stress. rhizosphere biology. Springer, Singapore. https://doi.org/10.1007/978-981-15-7094-0_4

Bhandari G, Bagheri AR, Bhatt P, Bilal M (2021) Occurrence, potential ecological risks, and degradation of endocrine disruptor, nonylphenol from the aqueous environment. Chemosphere 275:130013. https://doi.org/10.1016/j.chemosphere.2021.130013

Bhatt P (2018) *Insilico* tools to study the bioremediation in microorganisms. In: Pathak V, Navneet (eds) Handbook of research on microbial tools for environmental waste management. IGI Global, Hershey, PA, pp 389–395. https://doi.org/10.4018/978-1-5225-3540-9.ch018

Bhatt P, Barh A (2018) Bioinformatic tools to study the soil microorganisms: an *in silico* approach for sustainable agriculture. In: Choudhary D, Kumar M, Prasad R, Kumar V (eds) In silico approach for sustainable agriculture. Springer, Singapore. https://doi.org/10.1007/978-981-13-0347-0_10

Bhatt K, Bhatt P (2020) Rhizosphere biology: alternate tactics for enhancing sustainable agriculture. ISBN: 9781119644620

Bhatt P, Nailwal TK (2018) In: Prasad R, Gill SS, Tuteja N (eds) Crop improvement through microbial technology: a step towards sustainable agriculture. Chapter 11. Elsevier, Amsterdam, pp 245–253. https://doi.org/10.1016/B978-0-444-63987-5.00011-6

Bhatt P, Bisht TS, Pathak VM, Barh A, Chandra D (2015a) Optimization of amylase production from the fungal isolates of Himalayan region Uttarakhand. Ecol Environ Conserv 21 (3):1517–1521

Bhatt P, Negi G, Gangola S, Khati P, Srivastava A, Sharma A (2015b) Optimization of sulfosulfuron biodegradation through response surface methodology using indigenous bacterial strain isolated from contaminated agriculture field. Int J Curr Microbiol App Sci 4(8):105–112

Bhatt P, Negi G, Gangola S, Khati P, Kumar G, Srivastava A, Sharma A (2016a) Differential expression and characterization of cypermethrin degrading potential proteins in *Bacillus thuringiensis* strain, SG4. 3 Biotech 6:225

Bhatt P, Sharma A, Gangola S, Khati P, Kumar G, Srivastava A (2016b) Novel pathway of cypermethrin biodegradation in *a Bacillus* sp. strain SG2 isolated from cypermethrin-contaminated agriculture field. 3 Biotech 6:45

Bhatt P, Huang Y, Zhan H, Chen S (2019a) Insight into microbial applications for the biodegradation of Pyrethroid insecticides. Front Microbiol 10:1778

Bhatt P, Pal K, Bhandari G, Barh A (2019b) Modeling of methyl halide biodegradation on bacteria and its effect on other environmental systems. Pestic Biochem Physiol 158:88–100

Bhatt P, Gangola S, Chaudhary P, Khati P, Kumar G, Sharma A, Srivastava A (2019c) Pesticide induced up-regulation of esterase and aldehyde dehydrogenase in indigenous *Bacillus* spp. Biorem J 23(1):42–52

Bhatt P, Joshi D, Kumar N, Kumar N (2019d) Recent trends to study the functional analysis of Mycorrhizosphere. In: Varma A, Choudhary D (eds) Mycorrhizosphere and pedogenesis. Springer, Singapore. https://doi.org/10.1007/978-981-13-6480-8_11

Bhatt P, Pathak VM, Joshi S, Bisht TS, Singh K, Chandra D (2019e) Major metabolites after degradation of xenobiotics and enzymes involved in these pathways. Chapter 12. In: Smart bioremediation technologies: microbial enzymes, pp 205–215. https://doi.org/10.1016/B978-0-12-818307-6.00012-3

Bhatt P, Rene ER, Kumar AJ, Kumar AJ, Zhang W, Chen S (2020a) Binding interaction of allethrin with esterase: bioremediation potential and mechanism. Bioresour Technol 315:13845. https://doi.org/10.1016/j.biortech.2020.123845

Bhatt P, Verma A, Verma S, Anwar MS, Prasher P, Mudila H, Chen S (2020b) Understanding phytomicrobiome: a potential reservoir for better crop management. Sustainability 12:5446

Bhatt P, Huang Y, Rene ER, Kumar AJ, Chen S (2020c) Mechanism of allethrin biodegradation by a newly isolated *Sphingomonas trueperi* strain CW3 from wastewater sludge. Bioresour Technol 305:123074

Bhatt P, Zhang W, Lin Z, Pang S, Huang Y, Chen S (2020d) Biodegradation of allethrin by a novel fungus *Fusarium proliferatum* strain CF2, isolated from contaminated. Soils Microorganisms 8:593

Bhatt P, Huang Y, Zhang W, Sharma A, Chen S (2020e) Enhanced Cypermethrin degradation kinetics and metabolic pathway in *Bacillus* thuringiensis strain, SG4. Microorganisms 8:223

Bhatt P, Bhatt K, Huang Y, Ziqiu L, Chen S (2020f) Esterase is a powerful tool for the biodegradation of pyrethroid insecticides. Chemosphere 244:125507

Bhatt P, Zhou X, Huang Y, Zhang W, Chen S (2021a) Characterization of the role of esterases in the biodegradation of organophosphate, carbamate and pyrethroid group pesticides. J Hazard Mater 411:125026

Bhatt P, Joshi T, Bhatt K, Zhang W, Huang Y, Chen S (2021b) Binding interaction of glyphosate oxidoreductase and C-P lyase: molecular docking and molecular dynamics simulation studies. J Hazard Mater 409:124927

Bhatt P, Bhatt K, Sharma A, Zhang W, Mishra S, Chen S (2021c) Biotechnological basis of microbial consortia for the removal of pesticides from the environment. Crit Rev Biotechnol 41(3):317–338. https://doi.org/10.1080/07388551.2020.1853032

Bhatt P, Sethi K, Gangola S, Bhandari G, Verma A, Adnan M, Singh Y, Chaube S (2021d) Modeling and simulation of atrazine biodegradation in bacteria and its effect in other living systems. J Biomol Struct Dyn:1–11. https://doi.org/10.1080/07391102.2020.1846623

Bhatt P, Gangola S, Bhandari G, Zhang W, Maithani D, Mishra S, Chen S (2021e) New insights into the degradation of synthetic pollutants in contaminated environments. Chemosphere 268:128827. https://doi.org/10.1016/j.chemosphere.2020.128827

Bhatt P, Sharma A, Rene ER, Kumar AJ, Zhang W, Chen S (2021f) Bioremediation mechanism, kinetics of fipronil degradation using *Bacillus* sp. FA3 and resource recovery potential from contaminated environments. J Water Process Eng 39:101712

Bhatt P, Bhandari G, Bhatt K, Maithani D, Mishra S, Gangola S, Bhatt R, Huang Y, Chen S (2021g) Plasmid-mediated catabolism for the removal of xenobiotics from the environment. J Haz Mat 420:126618

Bhattacharyya PN, Jha DK (2012) Plant growth-promoting rhizobacteria (PGPR): emergence in agriculture. World J Microbiol Biotechnol 28(4):1327–1350

Blom D, Fabbri C, Connor EC, Schiestl FP, Klauser DR, Boller T, Eberl L, Weisskopf L (2011) Production of plant growth modulating volatiles is widespread among rhizosphere bacteria and strongly depends on culture conditions. Environ Microbiol 13(11):3047–3058

Bron PA, Van Baarlen P, Kleerebezem M (2012) Emerging molecular insights into the interaction between probiotics and the host intestinal mucosa. Nat Rev Microbiol 10(1):66–78

Buee M, Reich M, Murat C, Morin E, Nilsson RH, Uroz S, Martin F (2009) 454 pyrosequencing analyses of forest soils reveal an unexpectedly high fungal diversity. New Phytol 184 (2):449–456

Burmølle M, Hansen LH, Sørensen SJ (2005) Use of a whole-cell biosensor and flow cytometry to detect AHL production by an indigenous soil community during decomposition of litter. Microb Ecol 50(2):221–229

Chaparro JM, Badri DV, Bakker MG, Sugiyama A, Manter DK, Vivanco JM (2013) Root exudation of phytochemicals in Arabidopsis follows specific patterns that are developmentally programmed and correlate with soil microbial functions. PLoS One 8(2):e55731

Chernin LS, Winson MK, Thompson JM, Haran S, Bycroft BW, Chet I, Williams P, Stewart GS (1998) Chitinolytic activity in *Chromobacterium violaceum*: substrate analysis and regulation by quorum sensing. J Bacteriol 180(17):4435–4441

Coombs JT, Franco CM (2003) Isolation and identification of actinobacteria from surface-sterilized wheat roots. Appl Environ Microbiol 69(9):5603–5608

Croxatto A, Chalker VJ, Lauritz J, Jass J, Hardman A, Williams P, Cámara M, Milton DL (2002) VanT, a homologue of Vibrio harveyi LuxR, regulates serine, metalloprotease, pigment, and biofilm production in *Vibrio anguillarum*. J Bacteriol 184(6):1617–1629

Daniel G, Jaffre T, Prin Y (2007) Abundance of Frankia from Gymnostoma spp. in the rhizosphere of *Alphitonia neocaledonica*, a nonnodulated Rhamnaceae endemicto New Caledonia. Eur J Soil Biol 36:169–175

De Vleesschauwer D, Höfte M (2009) Rhizobacteria-induced systemic resistance. Adv Bot Res 51:223–281

Deaker R, Roughley RJ, Kennedy IR (2004) Legume seed inoculation technology—a review. Soil Biol Biochem 36(8):1275–1288

DeAngelis KM, Firestone MK, Lindow SE (2007) Sensitive whole-cell biosensor suitable for detecting a variety of N-acyl homoserine lactones in intact rhizosphere microbial communities. Appl Environ Microbiol 73(11):3724–3727

Doornbos RF, Geraats BP, Kuramae EE, Van Loon LC, Bakker PA (2011) Effects of jasmonic acid, ethylene, and salicylic acid signaling on the rhizosphere bacterial community of *Arabidopsis thaliana*. Mol Plant Microbe Interact 24(4):395–407

Dweck HK, Ebrahim SA, Thoma M, Mohamed AA, Keesey IW, Trona F, Lavista-Llanos S, Svatoš A, Sachse S, Knaden M, Hansson BS (2015) Pheromones mediating copulation and attraction in drosophila. Proc Natl Acad Sci 112(21):E2829–E2835

Eberl L, Winson MK, Sternberg C, Stewart GS, Christiansen G, Chhabra SR, Bycroft B, Williams P, Molin S, Givskov M (1996) Involvement of N-acyl-l-homoserine lactone autoinducers in controlling the multicellular behaviour of *Serratia liquefaciens*. Mol Microbiol 20(1):127–136

Egamberdieva D, Kucharova Z (2009) Selection for root colonising bacteria stimulating wheat growth in saline soils. Biol Fertil Soils 45(6):563–571

Egamberdieva D, Kamilova F, Validov S, Gafurova L, Kucharova Z, Lugtenberg B (2008) High incidence of plant growth-stimulating bacteria associated with the rhizosphere of wheat grown on salinated soil in Uzbekistan. Environ Microbiol 10(1):1–9

Elasri M, Delorme S, Lemanceau P, Stewart G, Laue B, Glickmann E, Oger PM, Dessaux Y (2001) Acyl-homoserine lactone production is more common among plant-associated Pseudomonas spp. than among soilborne Pseudomonas spp. Appl Environ Microbiol 67(3):1198–1209

Fan X, Ye T, Li Q, Bhatt P, Zhang L, Chen S (2020) Potential of a quorum quenching bacteria isolate Ochrobactrum intermedium D-2 against soft rot pathogen *Pectobacterium carotovorum* subsp. Carotovora. Front Microbiol 11:898

Feng Y, Huang Y, Zhan H, Bhatt P, Chen S (2020) An overview of strobilurin fungicide degradation: current status and future perspective. Front Microbiol 11:389

Ferguson BJ, Mathesius U (2014) Phytohormone regulation of legume-rhizobia interactions. J Chem Ecol 40(7):770–790

Fierer N, Jackson RB (2006) The diversity and biogeography of soil bacterial communities. Proc Natl Acad Sci 103(3):626–631

Fray RG (2002) Altering plant–microbe interaction through artificially manipulating bacterial quorum sensing. Ann Bot 89(3):245–253

Gangola S, Bhatt P, Chaudhary P, Khati P, Kumar N, Sharma A (2018a) Bioremediation of industrial waste using microbial metabolic diversity. In: Pankaj, Sharma A (eds) Microbial biotechnology in environmental monitoring and cleanup. IGI Global, Hershey, PA, pp 1–27. https://doi.org/10.4018/978-1-5225-3126-5.ch001

Gangola S, Sharma A, Bhatt P, Khati P, Chaudhary P (2018b) Presence of esterase and laccase in *Bacillus subtilis* facilitates biodegradation and detoxification of cypermethrin. Sci Rep 8:12755. https://doi.org/10.1038/s41598-018-31082-5

Ganley RJ, Brunsfeld SJ, Newcombe G (2004) A community of unknown, endophytic fungi in western white pine. Proc Natl Acad Sci 101(27):10107–10112

Giron D, Frago E, Glevarec G, Pieterse CM, Dicke M (2013) Cytokinins as key regulators in plant–microbe–insect interactions: connecting plant growth and defence. Funct Ecol 27(3):599–609

Glazebrook J (2005) Contrasting mechanisms of defense against biotrophic and necrotrophic pathogens. Annu Rev Phytopathol 43:205–227

Grayston SJ, Wang S, Campbell CD, Edwards AC (1998) Selective influence of plant species on microbial diversity in the rhizosphere. Soil Biol Biochem 30(3):369–378

Groenhagen U, Baumgartner R, Bailly A, Gardiner A, Eberl L, Schulz S, Weisskopf L (2013) Production of bioactive volatiles by different *Burkholderia ambifaria* strains. J Chem Ecol 39 (7):892–906

Gupta S, Bhatt P, Chaturvedi P (2018) Determination and quantification of asiaticoside in endophytic fungus from Centella asiatica (L.) Urban. World J Microbiol Biotechnol 34:111

Haas D, Défago G (2005) Biological control of soil-borne pathogens by fluorescent pseudomonads. Nat Rev Microbiol 3(4):307–319

Hooshangi S, Bentley WE (2008) From unicellular properties to multicellular behavior: bacteria quorum sensing circuitry and applications. Curr Opin Biotechnol 19(6):550–555

Huang Y, Zhan H, Bhatt P, Chen S (2019) Paraquat degradation from contaminated environments: current achievements and perspectives. Front Microbiol 10:1754

Huang Y, Lin Z, Zhang W, Pang S, Bhatt P, Rene ER, Kumar AJ, Chen S (2020) New insights into the microbial degradation of D-cyphenothrin in contaminated water/soil environments. Microorganisms 8:473

Huang Y, Zhang W, Pang S, Chen J, Bhatt P, Mishra S, Chen S (2021) Insights into the microbial degradation and catalytic mechanism of chlorpyrifos. Environ Res 194:110660

Jones S, Yu B, Bainton NA, Birdsall M, Bycroft BW, Chhabra SR, Cox AJ, Golby P, Reeves PJ, Stephens S (1993) The lux autoinducer regulates the production of exoenzyme virulence determinants in *Erwinia carotovora* and *Pseudomonas aeruginosa*. EMBO J 12(6):2477–2482

Jones DL, Hodge A, Kuzyakov Y (2004) Plant and mycorrhizal regulation of rhizodeposition. New Phytol 163(3):459–480

Jumpponen A (2001) Dark septate endophytes—are they mycorrhizal? Mycorrhiza 11(4):207–211

Kai M, Effmert U, Piechulla B (2016) Bacterial-plant-interactions: approaches to unravel the biological function of bacterial volatiles in the rhizosphere. Front Microbiol 7:108

Khati P, Sharma A, Gangola S, Kumar R, Bhatt P, Kumar G (2017a) Impact of agriusable nanocompounds on soil microbial activity: an indicator of soil health. Soil Air Water 45 (5):1600458

Khati P, Parul, Gangola S, Bhatt P, Sharma A (2017b) Nanochitosan induced growth of Zea Mays with soil health maintenance. 3 Biotech 7(1):81

Khati P, Gangola S, Bhatt P, Kumar R, Sharma A (2018a) Application of nanocompounds for sustainable agriculture system. In: Pankaj, Sharma A (eds) Microbial biotechnology in environmental monitoring and cleanup. IGI Global, Hershey, PA, pp 194–211. https://doi.org/10.4018/978-1-5225-3126-5.ch012

Khati P, Parul, Bhatt P, Nisha Kumar R, Sharma A (2018b) Effect of nanozeolite and plant growth promoting rhizobacteria on maize. 3 Biotech 8:141

Kloepper JW (1992) Plant growth-promoting rhizobacteria as biological control agents. In: Soil microbial ecology: applications in agricultural and environmental management, pp 255–274

Kumar M, Prasanna R, Bidyarani N, Babu S, Mishra BK, Kumar A, Adak A, Jauhari S, Yadav K, Singh R, Saxena AK (2013) Evaluating the plant growth promoting ability of thermotolerant bacteria and cyanobacteria and their interactions with seed spice crops. Sci Hortic 164:94–101

Kumar G, Arya K, Verma A, Pankaj, Khati P, Gangola S, Kumar R, Sharma A, Singh H (2017) Bioremediation of petrol engine oil polluted soil using microbial consortium and wheat crop. J Pure Appl Microbiol 11(3):1583–1588

Lakshmanan V, Kitto SL, Caplan JL, Hsueh YH, Kearns DB, Wu YS, Bais HP (2012) Microbe-associated molecular patterns-triggered root responses mediate beneficial rhizobacterial recruitment in Arabidopsis. Plant Physiol 160(3):1642–1661

Lakshmanan V, Selvaraj G, Bais HP (2014) Functional soil microbiome: below ground solutions to an above ground problem. Plant Physiol 166(2):689–700

Lanoue A, Burlat V, Henkes GJ, Koch I, Schurr U, Röse US (2010) De novo biosynthesis of defense root exudates in response to Fusarium attack in barley. New Phytol 185(2):577–588

Lebeis SL, Paredes SH, Lundberg DS, Breakfield N, Gehring J, McDonald M, Malfatti S, Del Rio TG, Jones CD, Tringe SG, Dangl JL (2015) Salicylic acid modulates colonization of the root microbiome by specific bacterial taxa. Science 349(6250):860–864

Lee B, Lee S, Ryu CM (2012) Foliar aphid feeding recruits rhizosphere bacteria and primes plant immunity against pathogenic and non-pathogenic bacteria in pepper. Ann Bot 110(2):281–290

Lin Z, Zhang W, Pang S, Huang Y, Mishra S, Bhatt P, Chen S (2020) Current approaches to and future perspectives on methomyl degradation in contaminated soil/water environments. Molecules 25:738

Lowery CA, Dickerson TJ, Janda KD (2008) Interspecies and interkingdom communication mediated by bacterial quorum sensing. Chem Soc Rev 37(7):1337–1346

Martínez-Viveros O, Jorquera MA, Crowley DE, Gajardo GM, Mora ML (2010) Mechanisms and practical considerations involved in plant growth promotion by rhizobacteria. J Soil Sci Plant Nutr 10(3):293–319

Mendes R, Kruijt M, De Bruijn I, Dekkers E, van der Voort M, Schneider JH, Piceno YM, DeSantis TZ, Andersen GL, Bakker PA, Raaijmakers JM (2011) Deciphering the rhizosphere microbiome for disease-suppressive bacteria. Science 332(6033):1097–1100

Meyer JM (2000) Pyoverdines: pigments, siderophores and potential taxonomic markers of fluorescent Pseudomonas species. Arch Microbiol 174(3):135–142

Millet YA, Danna CH, Clay NK, Songnuan W, Simon MD, Werck-Reichhart D, Ausubel FM (2010) Innate immune responses activated in Arabidopsis roots by microbe-associated molecular patterns. Plant Cell 22(3):973–990

Mishra S, Zhang W, Lin Z, Pang S, Huang Y, Bhatt P, Chen S (2020) Carbofuran toxicity and its microbial degradation in contaminated environments. Chemosphere 259:127429

Misra N, Gupta G, Jha PN (2012) Assessment of mineral phosphate-solubilizing properties and molecular characterization of zinc-tolerant bacteria. J Basic Microbiol 52(5):549–558

Moore BD, Andrew RL, Külheim C, Foley WJ (2014) Explaining intraspecific diversity in plant secondary metabolites in an ecological context. New Phytol 201(3):733–750

Moreira FMS (2008) Nitrogen fixing Leguminosae-nodulating bacteria. In: Hand book of tropical soil biology. Earthscan, London, pp 107–130

Morgan JA, Bending GD, White PJ (2005) Biological costs and benefits to plant–microbe interactions in the rhizosphere. J Exp Bot 56(417):1729–1739

Neilands JB (1995) Siderophores: structure and function of microbial iron transport compounds. J Biol Chem 270(45):26723–26726

Newton AC, Fitt BD, Atkins SD, Walters DR, Daniell TJ (2010) Pathogenesis, parasitism and mutualism in the trophic space of microbe–plant interactions. Trends Microbiol 18(8):365–373

Nihorimbere V, Ongena M, Smargiassi M, Thonart P (2011) Beneficial effect of the rhizosphere microbial community for plant growth and health. Biotechnol Agron Soc Environ 15 (2):327–337

Ortíz-Castro R, Valencia-Cantero E, López-Bucio J (2008) Plant growth promotion by *Bacillus megaterium* involves cytokinin signaling. Plant Signal Behav 3(4):263–265

Oves M, Khan MS, Zaidi A (2013) Chromium reducing and plant growth promoting novel strain *Pseudomonas aeruginosa* OSG41 enhance chickpea growth in chromium amended soils. Eur J Soil Biol 56:72–83

Pang S, Lin Z, Zhang W, Mishra S, Bhatt P, Chen S (2020) Insights into the microbial degradation and biochemical mechanisms of neonicotinoids. Front Microbiol 11:868

Passador L, Cook JM, Gambello MJ, Rust L, Iglewski BH (1993) Expression of *Pseudomonas aeruginosa* virulence genes requires cell-to-cell communication. Science 260(5111):1127–1130

Philippot L, Andersson SG, Battin TJ, Prosser JI, Schimel JP, Whitman WB, Hallin S (2010) The ecological coherence of high bacterial taxonomic ranks. Nat Rev Microbiol 8(7):523–529

Philippot L, Raaijmakers JM, Lemanceau P, Van Der Putten WH (2013) Going back to the roots: the microbial ecology of the rhizosphere. Nat Rev Microbiol 11(11):789–799

Pineda A, Zheng SJ, van Loon JJ, Pieterse CM, Dicke M (2010) Helping plants to deal with insects: the role of beneficial soil-borne microbes. Trends Plant Sci 15(9):507–514

Pirhonen M, Flego D, Heikinheimo R, Palva ET (1993) A small diffusible signal molecule is responsible for the global control of virulence and exoenzyme production in the plant pathogen *Erwinia carotovora*. EMBO J 12(6):2467–2476

Pozo MJ, Van Der Ent S, Van Loon LC, Pieterse CM (2008) Transcription factor MYC2 is involved in priming for enhanced defense during rhizobacteria-induced systemic resistance in *Arabidopsis thaliana*. New Phytol 180(2):511–523

Prasanna R, Jaiswal P, Nayak S, Sood A, Kaushik BD (2009) Cyanobacterial diversity in the rhizosphere of rice and its ecological significance. Indian J Microbiol 49(1):89–97

Prasanna R, Pattnaik S, Sugitha TK, Nain L, Saxena AK (2011) Development of cyanobacterium-based biofilms and their in vitro evaluation for agriculturally useful traits. Folia Microbiol 56 (1):49–58

Prasanna R, Triveni S, Bidyarani N, Babu S, Yadav K, Adak A, Khetarpal S, Pal M, Shivay YS, Saxena AK (2014) Evaluating the efficacy of cyanobacterial formulations and biofilmed inoculants for leguminous crops. Arch Agron Soil Sci 60(3):349–366

Prasanna R, Hossain F, Babu S, Bidyarani N, Adak A, Verma S, Shivay YS, Nain L (2015) Prospecting cyanobacterial formulations as plant-growth-promoting agents for maize hybrids. South Afr J Plant Soil 32(4):199–207

Rajkumar M, Ae N, Prasad MN, Freitas H (2010) Potential of siderophore-producing bacteria for improving heavy metal phytoextraction. Trends Biotechnol 28(3):142–149

Rasch M, Andersen JB, Nielsen KF, Flodgaard LR, Christensen H, Givskov M, Gram L (2005) Involvement of bacterial quorum-sensing signals in spoilage of bean sprouts. Appl Environ Microbiol 71(6):3321–3330

Rinaudi LV, González JE (2009) The low-molecular-weight fraction of exopolysaccharide II from *Sinorhizobium meliloti* is a crucial determinant of biofilm formation. J Bacteriol 191 (23):7216–7224

Rodríguez H, Fraga R, Gonzalez T, Bashan Y (2006) Genetics of phosphate solubilization and its potential applications for improving plant growth-promoting bacteria. Plant and Soil 287 (1):15–21

Rosier A, Bishnoi U, Lakshmanan V, Sherrier DJ, Bais HP (2016) A perspective on inter-kingdom signaling in plant–beneficial microbe interactions. Plant Mol Biol 90(6):537–548
Schmidt R, Cordovez V, De Boer W, Raaijmakers J, Garbeva P (2015) Volatile affairs in microbial interactions. ISME J 9(11):2329–2335
Segarra G, Van der Ent S, Trillas I, Pieterse CM (2009) MYB72, a node of convergence in induced systemic resistance triggered by a fungal and a bacterial beneficial microbe. Plant Biol 11 (1):90–96
Sharma A, Bhatt P (2016) Bioremediation: a microbial technology for improvising wildlife. Chapter 3. In: Wildlife management concept analysis and conservation, pp 29–40
Sharma A, Pankaj, Khati P, Gangola S, Kumar G (2016) Microbial degradation of pesticides for environmental cleanup. Chapter 6. In: Bioremediation of industrial pollutants. Write and print publication, pp 178–205
Singh BK, Millard P, Whiteley AS, Murrell JC (2004) Unravelling rhizosphere–microbial interactions: opportunities and limitations. Trends Microbiol 12(8):386–393
Singh R, Behl R, Jain P, Narula N, Singh K (2007) Performance and gene effects for root characters and micronutrient uptake in wheat inoculated with arbuscular mycorrhizal fungi and *Azotobacter chroococcum*. Acta Agron Hungarica 55(3):325–330
Singh K, Gera R, Sharma R, Maithani D, Chandra D, Bhat AM, Kumar R, Bhatt P (2021) Mechanism and application of *Sesbania* root nodulating bacteria: an alternative for chemical fertilizers and sustainable development. Arch Microbiol 203:1259–1270. https://doi.org/10.1007/s00203-020-02137-x
Stam JM, Kroes A, Li Y, Gols R, van Loon JJ, Poelman EH, Dicke M (2014) Plant interactions with multiple insect herbivores: from community to genes. Annu Rev Plant Biol 65:689–713
Sulbhi V, Bhatt P, Verma A, Mudila H, Prasher P, Rene ER (2021) Microbial technologies for heavy metal remediation: effect of process conditions and current practices. Clean Technol Environ Policy. https://doi.org/10.1007/s10098-021-02029-8
Swift S, Lynch MJ, Fish L, Kirke DF, Tomás JM, Stewart GS, Williams P (1999) Quorum sensing-dependent regulation and blockade of exoprotease production in Aeromonas hydrophila. Infect Immun 67(10):5192–5199
Trivedi P, He Z, Van Nostrand JD, Albrigo G, Zhou J, Wang N (2012) Huanglongbing alters the structure and functional diversity of microbial communities associated with citrus rhizosphere. ISME J 6(2):363–383
Unkovich M, Baldock J (2008) Measurement of asymbiotic N2 fixation in Australian agriculture. Soil Biol Biochem 40(12):2915–2921
Van der Ent S, Verhagen BW, Van Doorn R, Bakker D, Verlaan MG, Pel MJ, Joosten RG, Proveniers MC, Van Loon LC, Ton J, Pieterse CM (2008) MYB72 is required in early signaling steps of rhizobacteria-induced systemic resistance in Arabidopsis. Plant Physiol 146 (3):1293–1304
Van der Ent S, Van Hulten M, Pozo MJ, Czechowski T, Udvardi MK, Pieterse CM, Ton J (2009) Priming of plant innate immunity by rhizobacteria and β-aminobutyric acid: differences and similarities in regulation. New Phytol 183(2):419–431
van Hulten M, Pelser M, Van Loon LC, Pieterse CM, Ton J (2006) Costs and benefits of priming for defense in Arabidopsis. Proc Natl Acad Sci 103(14):5602–5607
Van Loon LC (2007) Plant responses to plant growth-promoting rhizobacteria. In: New perspectives and approaches in plant growth-promoting Rhizobacteria research. Springer, Dordrecht, pp 243–254
Van Oosten VR, Bodenhausen N, Reymond P, Van Pelt JA, Van Loon LC, Dicke M, Pieterse CM (2008) Differential effectiveness of microbially induced resistance against herbivorous insects in Arabidopsis. Mol Plant Microbe Interact 21(7):919–930
Verbon EH, Liberman LM (2016) Beneficial microbes affect endogenous mechanisms controlling root development. Trends Plant Sci 21(3):218–229

Von Bodman SB, Majerczak DR, Coplin DL (1998) A negative regulator mediates quorum-sensing control of exopolysaccharide production in *Pantoea stewartii* subsp. stewartii. Proc Natl Acad Sci 95(13):7687–7692

Von Bodman SB, Bauer WD, Coplin DL (2003) Quorum sensing in plant-pathogenic bacteria. Annu Rev Phytopathol 41(1):455–482

Walling LL (2000) The myriad plant responses to herbivores. J Plant Growth Regul 19(2):195–216

Worm J, Jensen LE, Hansen TS, Søndergaard M, Nybroe O (2000) Interactions between proteolytic and non-proteolytic *Pseudomonas fluorescens* affect protein degradation in a model community. FEMS Microbiol Ecol 32(2):103–109

Ye T, Zhou T, Fan X, Bhatt P, Zhang L, Chen S (2019) *Acinetobacter lactucae* strain QL-1, a novel quorum quenching candidate against bacterial pathogen *Xanthomonas campestris* pv. Campestris. Front Microbiol 10:2867

Zahran HH (1999) Rhizobium-legume symbiosis and nitrogen fixation under severe conditions and in an arid climate. Microbiol Mol Biol Rev 63(4):968–989

Zamioudis C, Pieterse CM (2012) Modulation of host immunity by beneficial microbes. Mol Plant Microbe Interact 25(2):139–150

Zamioudis C, Mastranesti P, Dhonukshe P, Blilou I, Pieterse CM (2013) Unraveling root developmental programs initiated by beneficial Pseudomonas spp. bacteria. Plant Physiol 162 (1):304–318

Zhan H, Huang Y, Lin Z, Bhatt P, Chen S (2020) New insights into the microbial degradation and catalytic mechanism of synthetic pyrethroids. Environ Res 182:109138

Zhang Y, Du BH, Jin ZG, Li ZH, Song HN, Ding YQ (2011) Analysis of bacterial communities in rhizosphere soil of healthy and diseased cotton (Gossypium sp.) at different plant growth stages. Plant and Soil 339(1):447–455

Zhang W, Pang S, Lin Z, Mishra S, Bhatt P, Chen S (2020a) Biotransformation of perfluoroalkyl acid precursors from various environmental system: advances and perspectives. Environ Pollut 272:115908

Zhang W, Lin Z, Pang S, Bhatt P, Chen S (2020b) Insights into the biodegradation of Lindane (γ-Hexachlorocyclohexane) using a microbial system. Front Microbiol 11:522

Zolla G, Badri DV, Bakker MG, Manter DK, Vivanco JM (2013) Soil microbiomes vary in their ability to confer drought tolerance to Arabidopsis. Appl Soil Ecol 68:1–9

Chapter 3
Different Biofertilizers and Their Application for Sustainable Development

Dharmendra Kumar, Som Dutt, Pinky Raigond, Sushil Sudhakar Changan, Milan Kumar Lal, Rahul Kumar Tiwari, Kumar Nishant Chourasia, and Brajesh Singh

Abstract The excessive use of chemical fertilizers causes many serious negative impacts on the agriculture production system and natural resources. Therefore, we need to have alternative options of chemical fertilizers sustainably. Biofertilizer has become increasingly important in agriculture due to its potential role in food security and environmentally friendly methods. Organic farming is not possible without the use of biofertilizers. The biofertilizers are living and latent cells of microbes that supply nutrients for crop production. The present book chapter highlighted different biofertilizers such as nitrogen-fixing microbes, phosphorus-solubilizing and phosphorus-mobilizing microbes, potassium solubilizer microbes, blue-green algae and *Azolla*, etc., uses in crop production, their method of production, and illustration of beneficial microbes which are used in biofertilizer industries.

Keywords Biofertilizers · Sustainable agriculture · Rhizobium · Organic farming

3.1 Introduction

Present day's world human population is rapidly increasing and is now 7.7 billion. India's population is rapidly growing, putting pressure on the agricultural production system and our natural resources, both of which are required to feed this huge population on limited land (FAO 2020). The agricultural sector is under pressure to meet the demand for food security as a result of the increasing human population, which forces farmers to use modern intensive farming methods with intensive application of chemical fertilizers, insecticides, fungicides, nematicides, and

D. Kumar (✉) · S. Dutt · P. Raigond · S. S. Changan · M. K. Lal · R. K. Tiwari · K. N. Chourasia · B. Singh
Division of Crop Physiology, Biochemistry and PHT, ICAR-Central Potato Research Institute, Shimla, Himachal Pradesh, India
e-mail: sushil.changan@icar.gov.in; milankumarlal@naarm.org.in; rahul.tiwari@icar.gov.in; kumar.chourasia@icar.gov.in

P. Bhatt et al. (eds.), *Microbial Technology for Sustainable Environment*,
https://doi.org/10.1007/978-981-16-3840-4_3

Table 3.1 Major biofertilizers and target crops

Different biofertiliser	Biofertilizer application crops
Rhizobium sp.	Leguminous crops (specific for pulses)
Azotobacter sp.	Wheat, barley, oats, cotton, mustard, and tomatoes, also some of the most common crops (potato, chili, okra, cucumber, onion, tomato, brinjal, and others)
Azospirillum sp.	Maize, wheat, barley, oat, Kodo, finger millets, bajra and other millets, sorghum, and sugarcane are all cereal crops
Cyanobacteria/BGA	Rice
PSM	All crops
Arbuscular mycorrhiza	Crops cultivated in nurseries and orchard trees
K solubilizing microbes	All crops
Micronutrient-solubilizing microbes	All crops
PGPRs	All crops
Azolla	Rice

pesticides for increasing food production and productivity. As a result, the continued use of agrochemicals for increased soil fertility and plant growth often has negative environmental effects, such as soil, groundwater, and aquifer pollution. In this context, environmentally friendly and sustainable crop production that makes efficient use of natural resources is becoming increasingly common as a means of meeting food security demands. Biofertilizers play a vital role in delivering better agricultural production in modern crop production systems (FAO 2020; Sharma et al. 2020a; Bhattacharyya and Jha 2012; Dal Cortivo et al. 2020; Bargaz et al. 2018).

The nutritional content of soil enriched with biofertilizers is improved by products containing living cells of various types of microbes. Bacteria, fungi, and cyanobacteria are the most important sources of biofertilizers (blue-green algae). Biofertilizers that fall into one of the categories mentioned above include nitrogen-fixing microbes, phosphate-solubilizing and phosphate-mobilizing microbes, potassium-solubilizing microbes, micronutrient-solubilizing microbes, and plant growth-promoting rhizobacteria (Chittora et al. 2020; Kour et al. 2020a; Mahanty et al. 2017).

Some major biofertilizers are given in Table 3.1. Rhizobium fixes nitrogen in symbiotic association with the root nodules of legume plants, and other microbes may fix nitrogen associatively or independently (Bennett et al. 2020). By dissolving rock phosphate and tricalcium phosphate, phosphorus-solubilizing microorganisms that secrete organic acids assist plants in consuming more phosphorus (Alori et al. 2017a).

The most common phosphorus-mobilizing microbes are arbuscular mycorrhizal fungi types that are omnipresent (Berruti et al. 2016). Plant growth-promoting rhizobacteria (PGPR) are a type of bacteria that aid plant growth by fixing nitrogen, solubilizing phosphorus, or producing plant growth-promoting metabolites, as well as producing antibiosis or antibiotics compounds for disease control. The crop productivity enhanced by using potential PGPR is used as microbial inoculants (Glick 2012).

The chemical fertilizer application in agriculture crops in the last three decades has been drastically increasing in the use of production, and it is causing serious concern. The quality of soil and groundwater is adversely affected by the effect of excessive fertilizers. The amount of of chemical fertilizers used can be decreased by the application of environmentally friendly biofertilizers that can definite advantage over chemical fertilizers. Biofertilizers are a cheap source of nutrients in economical use compared to synthetic chemical fertilizers. Chemical fertilizers are harmful to life at higher concentrations, while biofertilizers have no toxic effects (Singh et al. 2016; Kong et al. 2018; Vessey 2003).

3.2 Types of Biofertilizers

Integrated plant nutrient management has the combination of different sources of nutrients; one of the most important is the biofertilizers. Biofertilizers are important for crop growth, soil productivity, and water sustainability, as well as ecosystem protection. The renewable source of plant nutrients is biofertilizers which are sustainable to the agricultural ecosystem.

Biofertilizers are items that contain living latent cells of microbes when applied to seed, nursery, soil, or plant canopy and are found in the interior or rhizospheric zone of the plant. Therefore, agriculture production improves through the biological process such as BNF, solubilization of unavailable form of phosphorus, or mobilization of phosphorus or solubilization of other elements which are important for crops (Kour et al. 2020b; Thomas and Singh 2019).

The additional protection of crops and plants from pest and pathogens will improve plant growth and development. Many researchers across the world have been reviewed for the sustainable development of agriculture by the application of different soil microorganisms. Various biofertilizer are described below.

3.2.1 Nitrogen-Fixing Bacteria

The plants cannot utilize the atmosphere nitrogen which is nearly 78% in the environment in free form because of its inert nature. The ammonia or nitrate forms of N are uptake by plants.

However, the source of an available form of nitrogen is many such as lightening small relatively amount of ammonia produced and by the Haber-Bosch process, industrially ammonia produced on very high pressures and fairly high temperature by using the iron-based catalyst. However, through nitrogenase enzyme activities, the major conversion of atmospheric nitrogen into ammonia which is the available form of N is achieved using the BNF process which is performed by microorganisms. The nitrogenase enzyme converts nitrogen to ammonia and further by transamination process into proteins; this process is known as nitrogen fixation or dinitrogen fixation (Islam et al. 2016; Laguerre et al. 2007; Raklami et al. 2019).

The prokaryote group only belongs to nitrogen-fixing microbes. The nitrogen fixing of diazotroph microorganisms belongs to the different groups of prokaryotes in nature.

The diazotrophs are classified mainly into the following categories:

1. Mutualistic microorganism: *Rhizobium* sp.—legume symbiosis *Rhizobium-Parasponia* (non-legume) symbiosis *Frankia*—Trees (e.g., *Alder*, *Casuarina*) *Azolla-Anabaena*.
2. Free-living microbes: *Azotobacter paspali—Paspalum notatum*.
   (a) Aerobic: *Azotobacter*, *Beijerinckia*, Cyanobacteria (e.g., *Nostoc*, *Anabaena*, *Tolypothrix*, *Aulosira*).
   (b) Facultative: *Klebsiella pneumonia*, *Bacillus polymyxa*.
   (c) Anaerobic: *Clostridium*, *Desulfovibrio*, *Rhodospirillum*, *Rhodopseudomonas*, *Desulfotomaculum*, *Desulfovibrio*, *Chromatium*, *Chlorobium*.
3. Symbiosis in associative microbes (Kour et al. 2020a; Laguerre et al. 2007; Govindarajulu et al. 2005): *Azospirillum*, *Herbaspirillum*.
   (a) *Gluconobacter diazotrophicus*, *Azoarcus*.

Some important types of nitrogen-fixing related biofertilizers can be considered for the agriculture production system.

3.2.1.1 *Rhizobium*

Rhizobia is a category of microbes that includes *Rhizobium*, *Bradyrhizobium*, *Mesorhizobium*, *Sinorhizobium*, and *Azorhizobium*, among others. These rhizobia genera are symbiotic associations with leguminous plants for fixation nitrogen to ammoniacal form (Mutch and Young 2004; Keet et al. 2017). The rhizobial colonies are cultivated on the yeast extract mannitol agar (YEMA) medium and appear raised, wet, shining, translucent, or opaque with a smooth margin. The nitrogen-fixing rhizobia are found in the root or stem nodules of rhizobium-legume symbiosis. Therefore, these alternatives to the nitrogen-fixing process are agronomically structural unique for reducing energy-expensive ammonium and nitrate biofertilizers. All legume crops cannot fix atmospheric nitrogen. The mainly three families of legumes fix nitrogen such as Caesalpiniaceae, Mimosaceae, and Fabaceae. However, while all members of the Caesalpiniaceae family cannot form nodules, all members of the Mimosaceae and Fabaceae families do. The legume-rhizobia symbiosis relations repair nearly 75 million metric tons of nitrogen per year. The nitrogen fixation depends on the species of rhizobia, legume, soil, and environmental conditions (Dal Cortivo et al. 2020; Laguerre et al. 2007; Andrews and Andrews 2017; Boivin et al. 2020; Cavassim et al. 2019). Therefore, the legume inoculation in the legumes is a very important aspect for the manipulation of rhizobial microflora in the agriculture field for improving soil fertility and crop productivity. Due to high levels of mineral nitrogen and the presence of adequate native rhizobia, rhizobial inoculation often fails in tropical areas. Therefore, it is very necessary to identify the situation where rhizobia inoculation can do. Many researchers suggested different diagnostic tools to measure the rhizobia inoculation in legumes (Missbah El Idrissi et al. 2020; Rocha et al. 2019).

If rhizobial inoculation is needed, the following criteria should be followed;

- Species-specific rhizobia should have a low population density.
- The previous crop should be not grown as the same legume fields.
- The unproductive or waste soil should be reclaimed before inoculation of rhizobia.
- The crop rotation should be followed in legume and non-legume manners.
- The nitrogen status of soil should not be high level.
- The soils should not be acidic, alkaline, and saline (Rocha et al. 2019; John et al. 2011).

3.2.1.1.1 Rhizobial Strain Selection for Inoculant Development

The ideal inoculant strain for different legume crops should be screening, identifying, and classifying at a large scale. To differentiate legume inoculants, use acidic, sodic, saline, nitrate-rich, or heavy metal-polluted soil types (John et al. 2011; Vassilev et al. 2020; Mosa et al. 2016a).

The following are some of the characteristics that an inoculant strain should have in order to be suitable for commercial inoculants:

- It should have the ability to shape nitrogen-fixing modules in target legumes.
- The inoculant strain can compete with native rhizobia populations in the soil for nodule formation.
- They must be able to fix nitrogen in a number of environmental conditions.
- In artificial media, inoculant carriers and, in the soil, the inoculant strain must be able to expand.
- The inoculant strain should be able to persist in soil, particularly for perennial legumes such as alfalfa.
- The inoculant strain should be able to migrate away from the initial inoculation site.
- In the absence of a legume host, the inoculant strain should be able to colonize the soil.
- The strain used for inoculation should be able to withstand environmental stresses.
- With a wide variety of host genotypes, the inoculant strain should be able to fix nitrogen.

- The inoculant strain should have genetic stability.
- The inoculant strain should have compatibility with agrochemicals (Rocha et al. 2019; Vassilev et al. 2020).

3.2.1.2 *Azotobacter*

Azotobacter is a nitrogen-fixing bacteria that lives in the rhizosphere of a wide range of plants and forms cysts. Examples are *A. niger*, *A. chroococcum*, *A. vinelandii*, *A. beijerinckii*, *A. nigricans*, *A. armeniacus*, and *A. paspali*. The other members are mostly soil-borne and rhizospheric, except for a few last rhizoplane bacterium. *A. chroococcum* and *A. paspali* are well-known for their ability to serve as a biofertilizer for a number of non-legume crops (Thomas and Singh 2019; Noar and Bruno-B Arcena 2018). Azotobacter is a Gram-negative aerobic rod-shaped bacteria that can be found singly, in chains, or in clumps. It does not produce endospores, but it does contain cysts with thick walls. Desiccation and some harmful chemical and physical agents do not affect these cysts. They, on the other hand, are unable to resist extreme temperatures. They do not fix nitrogen and are optically refractile during the cyst stage of their life cycle. It may be motile or non-mobile due to peritrichous flagella. It can be used to produce a water-soluble pigment that is yellow-green, fluorescent, or red-violet/brownish-black. It thrives in a temperature range of 20 to 30 °C and a pH of 7.0 to 7.5. They may expand on a variety of carbohydrates, alcohols, and organic acids, among other aspects (Das 2019; Khosravi and Dolatabad 2020). *Azotobacter* can be found in the rhizosphere and phyllosphere of neutral and alkaline soils. Plant growth is influenced by antifungal metabolites and phosphate solubilization's ability to fix molecular nitrogen; and plant growth is influenced by growth-promoting substances like IAA, gibberellin, or gibberellin-like compounds, and vitamins' ability to fix molecular nitrogen (Das 2019; Van Oosten et al. 2018). Seed inoculation and carrier-based Azotobacter inoculant Azotobacter mass multiplication are equivalent to rhizobial inoculation. The Jensen's N-free medium is widely used in mass multiplication. Seed inoculation with *A. niger* increased field crop yields by about 10% and cereal yields by about 15–20%. *A. chroococcum* is a type of chroococcum. Azotobacter co-inoculation with other bio-inoculants including *Rhizobium*, *Azospirillum*, P-solubilizers, and vesicular-arbuscular mycorrhizae has been shown to increase legume, cereal, and vegetable crop growth and yield (Das 2019; Mosa et al. 2016b). Many experiments have shown that *Azotobacter chroococcum* inoculation has beneficial effects on a variety of cereal, vegetable, oilseed, legume, and others (Table 3.2). Higher nitrogen levels (0 to 30 kg N ha^{-1}) only improved yield efficiency in Azotobacter inoculation experiments. These diazotrophic bacteria need a lot of available carbon to survive in the soil. When farmyard manure (FYM), compost, and other organic additives are applied to agricultural soils, Azotobacter production increases, and the plant grows and yields more (Thomas and Singh 2019; Das 2019; Mosa et al. 2016b; Wu et al. 2005).

Table 3.2 Effect of *Azotobacter* on crop yield

Agriculturally some important crop	Yield (%) increases	Other crops	Yield (%) increases
Food grains		Other	
Wheat	10–15	Potato	10–18
Rice	10–25	Carrot	5–20
Maize	10–20	Cauliflower	20–30
Sorghum	15–20	Tomato	9–24
Sugarcane	11–24	Cotton	11–27

3.2.1.3 *Azospirillum*

Beijerinck described the nitrogen-fixing bacterium *Spirillum lipoferum* for the first time in 1925. *Azospirillum* isolation necessitates the surface sterilization of roots with 70% alcohol and the formation of microaerophilic (low oxygen requirements) conditions in the medium. The *Azospirillum* is a soil organism that can colonize both the roots and the above-ground sections of a wide variety of plants, forming an associative symbiosis (Cassán and Diaz-Zorita 2016). *Azospirillum* is a type of bacteria that lives in the soil. Isolation from half-centimeter-long root fragments surface-sterilized for 3–5 s in 70% alcohol. Phosphate buffer is used to wash roots (pH 7). A semi-solid, nitrogen-free sodium malate medium is used to extract *Azospirillum* from roots. Before being plated in, the root fragments are washed several times. The tubes are incubated for 48 h at 35 °C. The formation of a white pellicle 2–4 mm beneath the medium's surface is associated with *Azospirillum* growth (Alexandre 2017).

3.2.1.3.1 Characterization of *Azospirillum* Strain

Azospirillum is a Gram-negative, motile, and usually vibroid bacterium that contains poly-hydroxybutyrate granules. The *Azospirillum* genus has peritrichous flagella that allow it to swarm on surfaces, is extremely motile, and has a long, polar flagellum that needs swimming. Additionally, cysts are formed when the shape and size of the cells change with time in the culture. The *Azospirillum* can develop in microaerophilic (N_2 or NH_3 as nitrogen sources), anaerobic (NO_3- as the electron acceptor, denitrification), and totally aerobic conditions (NH_3, NO_3-, amino acids) and needs the cultivation of organic acids like malate, succinate, lactate, and pyruvate. On Rojo-Congo red medium, *Azospirillum* forms distinct scarlet red, dry, and wrinkled colonies (Mosa et al. 2016a; Alexandre 2017; Lin et al. 2016). *Azospirillum* is a type of alpha Proteobacterium. Five species of *Azospirillum* group include *A. brasilense*, *A. lipoferum*, *A. amazonense*, *A. halopraeferens*, and *A. irakense* (Ayyaz et al. 2016).

3.2.1.3.2 Crop Response to *Azospirillum* Inoculation

Wheat, sorghum, pearl millet, finger millet, barley, and corn are among the Gramineae plants that *Azospirillum* inoculants use to fix nitrogen. Sorghum (*Sorghum bicolor*), pearl millet (*Pennisetum americanum*), and finger millet (*Eleusine coracana*) were all consistently responsive to *Azospirillum* in many places across India (Veresoglou and Menexes 2010). *Azospirillum* strains increase the yield of agriculturally important crops in a variety of soil and climate conditions. The seed inoculants with *Azospirillum* sp. saved 20–30 kg N/ha equivalents. However, *Azospirillum*'s main impacts extend far beyond providing nitrogen to host plants (Fukami et al. 2018; Steenhoudt and Vanderleyden 2000a). When *Azospirillum* was inoculated onto plant roots, the entire root system's morphology and behavior changed drastically. The average density and length of the root system increases as hairs at the root tip become more distinct. Root hairs are root epidermal cells that help bind the root to its surroundings while also transporting water and nutrients. The plant roots increase the diameter and length of both lateral and adventitious roots when inoculated with *Azospirillum* sp., which typically contributes to further branching of the lateral roots.

These root system changes are important because they increase the plant's absorptive area and the volume of soil substrate available to it (Cassán and Diaz-Zorita 2016; Steenhoudt and Vanderleyden 2000b). Azos strains contain siderophores as well. The plant roots increase the diameter and length of both lateral and adventitious roots after being inoculated with *Azospirillum* sp., which normally leads to more lateral root branching. Microorganisms excrete siderophores, which are low molecular weight iron-binding compounds produced and excreted into the culture medium when iron levels are low.

The siderophores are transported through the bacterial envelope from the culture medium, where they are bound with metal ions. By extracting iron, plants can improve their iron uptake and protect themselves from pathogens by *Azospirillum* (Fukami et al. 2018; Raffi and Charyulu 2021). *Azospirillum* has also been implicated in the biosynthesis of antifungal, antibacterial, and growth-promoting compounds such as phytohormones and vitamins. The most well-known growth promoters are IAA, gibberellins, cytokinin-like compounds, and vitamins (Alexandre 2017). *Azospirillum* strains differ in their ability to produce antibiotic substances. *Azospirillum* controls a wide range of phytopathogenic fungi through fungistatic activity; for example, certain *Azospirillum* protect cotton plants from *Thielaviopsis basicola* and *Fusarium oxysporum*. By inoculating crops like wheat, sorghum, pearl millet, and corn with *Azospirillum*, they can increase root numbers while also boosting yields. Sorghum plants used moisture accumulated in soils from winter precipitation better than uninoculated plants by inoculated *Azospirillum* in a field experiment in Israel. *Azospirillum*-inoculated plants are more successful than uninoculated plants at absorbing nitrogen, phosphorus, potassium, and other microelements from the soil in both field and greenhouse experiments (Cassán and Diaz-Zorita 2016; Alexandre 2017; Steenhoudt and Vanderleyden 2000b; Bashan and Levanony 1990; Venkateswarlu and Rao 1985). In recent years, plant-microbe-

microbe interactions have changed from plant-microbe interactions. Many studies have shown that when *Azospirillum* is co-inoculated with other microorganisms like *Rhizobium*, *Azotobacter*, and PSB, the beneficial effects on plants are improved. The stimulation of lateral roots and root hair branching has resulted in a rise in lateral root and root hair branching development. *Azospirillum* is predominantly responsible for the production of phytohormones and other growth factors. At low application rates of mineral N, the researchers discovered that inoculating non-legumes with *Azospirillum brasilense* and *Azotobacter chroococcum* had a positive effect on associative nitrogen fixation and crop yield (Mosa et al. 2016a; Cassán and Diaz-Zorita 2016; Venkateswarlu and Rao 1985).

3.2.1.4 *Gluconoacetobacter diazotrophicus*

On a semi-solid N-free sugar medium acidified with acetic acid to pH 4.5, the *Gluconoacetobacter diazotrophicus* was isolated from washed roots and stems of sugarcane. This nitrogen-fixing microorganism is Gram-negative, microaerobic, and motile by lateral or peritrichous flagella and grows best at temperatures about 30 °C. Azolla forms a thick mat on the water's surface in 15–20 days under ideal conditions. Just about two-thirds of it is harvested, leaving the remainder to grow. It multiplies again in 2–3 weeks and forms a dense mat. A 100 m^2 nursery can produce about 100 kg of fresh Azolla inoculum per week. This bacterium uses sucrose as a carbon source, but it also uses glucose, fructose, galactose, mannitol, and other sugars. In the pH range of 3.8 to 5.8, Gluconacetobacter diazotrophicus has excellent nitrogenase activity and development. Sugar concentrations of 10 to 30% are favorable for this organism's nitrogen fixation in sugarcane (Suman et al. 2005; Muñoz-Rojas and Caballero-Mellado 2003). Endophytic diazotrophic bacteria like *Gluconoacetobacter diazotrophicus* can be found in sugar-rich plants like sugarcane, sweet potato, Cameroon grass, sweet sorghum, and coffee. This bacterium can colonize the stem, roots, and leaves of the host plant, among other places. Gluconoacetobacter diazotrophicus contributes more than half of the biologically fixed nitrogen in sugarcane for the first time in Brazil.

By fixing atmospheric nitrogen, this bacterium excreted ammonia into the medium. It was also reported that *Gluconoacetobacter* strains produce a substantial amount of IAA. When sugarcane, sweet potato, and sweet sorghum plants were inoculated with *Gluconobacter diazotrophicus* and AM fungi, they showed synergistic effects on plant growth and yield (Suman et al. 2005; De Oliveira et al. 2016; Muthukumarasamy et al. 2006; Saravanan et al. 2007).

3.2.1.5 Blue-Green Algae and *Azolla*

Blue-green algae are found all over the world (cyanobacteria). BGA cells can be single-celled or can be made up of branched or unbranched filaments. BGA is a free-living group of microorganisms that fix biological nitrogen in paddy fields in large amounts. The BGA that fix nitrogen from the atmosphere biologically have a special structure known as a "heterocyst:, and all heterocystous forms are capable of biological nitrogen fixation. However, due to newly identified special conditions in low oxygen environments, some BGA can fix atmospheric nitrogen without heterocyst. The BGA such as *Scytonema*, *Aulosira*, *Tolypothrix*, *Nostoc*, *Anabaena* and *Plectonema* as a mixture applied in the field for fixation of atmospheric nitrogen (Dhar et al. 2007; Lumpkin and Plucknett 1980). The BGA is the only organism on the biosphere with both photosynthesis and biological nitrogen fixation abilities. Cyanobacteria contain a wide range of plant growth-promoting compounds, including growth hormones, amino acids, small proteins and peptides, carbohydrates, complex polysaccharides, and vitamins, all of which have played an important role in crop development.

In the paddy field, the BGA fix biological nitrogen in the range of 25–40 kg nitrogen per hectare per season and supplied 5–20 tons of fresh weight of cyanobacteria. This is due to a large amount of organic matter that is commonly accessible, which greatly benefits succeeding crops (Dhar et al. 2007; Prasanna et al. 2003).

3.2.1.5.1 Production of Algae for Field Application

To grow these algae in large quantities, a simple rural-oriented open-air method based on the natural ecology of these algae has been developed. The underlying assumption is to grow them in natural sunshine in rice fields with promoting conditions. Various agricultural universities and private dealers backed a starter culture consisting of a soil-based mixture of productive BGA strains (Chatterjee et al. 2017; Mahapatra et al. 2018).

BGA biofertilizer is made of shallow stays (15 cm × 7.5 cm × 22.5 cm) made of galvanized iron sheet, brick and mortar, or pits lined with polythene sheets. The size of the pit for processing algal biofertilizer may also be increased. First, about 10 kg of soil is inserted in the trays and mixed with 0.2 kg of superphosphate; then, the trays are filled with water based on local conditions such as evaporation rate; the pH of the soil should be about neutral (5–20 cm).

After the soil has settled, the surface of the standing water is sprayed with sawdust and the starter culture. The whole setup should be in full light. The algae will continue to grow, and a thick algal mat will form on the soil's surface in about a week, and it may even float up in the hot summer months.

More water is applied intermittently if the regular rate of evaporation is high due to the high temperature during the summer season. The water should be drained and allowed to dry out in the sun until the algal growth has thickened enough. The dried algal flakes are collected on the surface and stored in bags to be used later in the fields. Inoculum is added in the form of a small amount of dry algal flakes, and the trays are refilled with water.

The operation is repeated in the same manner as before. When the soil in the tray is depleted (usually after three to four harvests), new soil is added, combined with superphosphate, and the process is repeated. To prevent insect reproduction, malathion (0.00075 ppm) or carbofuran (3% granule) should be used (Chatterjee et al. 2017; Win et al. 2018).

3.2.1.5.2 Method of Algae in Field Application

One week after paddy transplantation, cyanobacteria is applied at a rate of 10–15 kg/ha over standing water in the region. After algal treatment, the field is kept damp for at least a couple of days. The establishment and operation of these algae in the fields are unaffected by recommended pest control measures and other management activities. The sun-dried algal material can be processed and used in the field for a long time. Chemical fertilizers or other agricultural chemicals should not be processed in close proximity to algal material (Mahapatra et al. 2018).

3.2.1.5.3 *Azolla-Anabaena* Symbiosis

Azolla is a small aquatic fern that grows in almost any climate. Azolla has a thick aerial dorsal lobe, two lobes on each leaf, and a thin ventral lobe that is sometimes larger. The dorsal lobe is green, with a central cavity containing a blue-green algal symbiont (Anabaena azollae). The site of nitrogen fixation in the symbiont Anabaena's heterocyst. Azolla is a nitrogen-fixing plant that also provides nutrients. Azolla fixes nitrogen, feeds Anabaena, and adds a protective leaf cavity for the fern (Singh et al. 1984, 1988; Prasanna et al. 2008).

Azolla can be found in temperate and tropical wetlands, streams, swamps, ditches, and paddy fields where there is still water. It has been used as a rice fertilizer due to its rapid growth, high nitrogen content, and ability to grow in still water. This has been used as a biofertilizer in Vietnam and China for decades, but it is a relatively new invention in India. Azolla has seven living species: *Azolla sp.*, *A. pinnata*, *A. caroliniana*, *A. filiculoides*, *A. rubra*, *A. nilotica*, and *A. mexicana*, and mexicana microphylla is a type of microphylla. *A. pinnata* is a native of India, but many of these species have been imported to the United States (Singh et al. 1984; Yadav et al. 2014). Azolla's potential as a rice biofertilizer is determined by its high N_2 fixation ability, rapid growth, high biomass accumulation, and N content. Via biological nitrogen fixation, the Azolla-Anabaena complex is thought to be a promising biological system for increasing rice yield at a low cost. Azolla's N_2 repair ability is approximately 1.1 kg N/ha/day. The gross biomass ranged from 0.8 to 5.2 t dry matter/ha, with an average of 2.1 t/ha and a 2 to 10 day doubling period (Prasanna et al. 2003; Singh et al. 1988).

3.2.1.5.4 On the Large-Scale Production of *Azolla*

The concrete tanks that keep the soil flooded are the potential Azolla listed. Due to the high rate of evaporation, partial shade is required during the summer months. Azolla is harvested from these and used as an inoculum in rice-growing villages' larger plots or small ponds (Lumpkin and Plucknett 1980). It is cultivated on a large scale in a well-kept field divided into small subplots with sufficient irrigation (4–50 sq.m. plot with 5–10 cm water depth). Azolla is inoculated at a rate of 0.5 to 1.0 t/ha. Superphosphate, at a rate of 4–8 kg/ha, promotes fern growth. Furadan (2.5–3.0 kg/ha) and other insecticides are also used. Under optimal conditions, Azolla forms a dense mat on the water's surface in 15–20 days. Only about two-thirds of it is harvested, leaving the rest to multiply. In 2–3 weeks, it multiplies once more and forms a dense mat. A 100 m^2 nursery produces approximately 100 kg of fresh Azolla inoculum per week. Superphosphate, at a rate of 60 kg/ha, can be administered in two or three doses, or at weekly intervals, to boost performance. There is no need to repeat Azolla multiplication if the results are satisfactory without adding P (Dhar et al. 2007; Singh et al. 1984; Yadav et al. 2014).

3.2.1.6 Phosphate-Solubilizing and Phosphate-Mobilizing Microbes

Phosphorus is an essential nutrient for plant growth. While soils contain large amounts of phosphorus, only a small amount is available to plants. Microorganisms in the soil can solubilize phosphorus that is unavailable to plants and make it available to them. Phosphate-solubilizing microorganisms (PSM) are what they are called. A fungus colonizes the roots of higher plants and transports phosphorus from the soil to the plant system (Khan et al. 2014a).

3.2.1.6.1 Phosphate-Solubilizing Microorganisms (PSM)

Regular applications of phosphatic fertilizer in crops lead to significant phosphorus reserves in agricultural soils, which are constantly accumulated in the ecosystem. Many factors contain significant quantities, fixation, and precipitation of phosphorus in the soil, including pH sensitivity, which results in low-soluble phosphate fertilizer efficacy. Phosphate precipitates as aluminum and iron phosphates in acidic soils, while phosphate precipitates as a result of a high calcium concentration in calcareous soils. The PSM release soluble phosphorus from insoluble phosphates into the soil habitat through a series of solubilization reactions. These PSM have been described as bio-inoculants for crops in soils that have been amended with rock phosphate or tricalcium phosphate and have low available phosphate (Nautiyal 1999; Abd-Alla 1994). Bacteria, fungi, and actinomycetes are phosphate-solubilizing microorganisms that help in the transformation of insoluble inorganic phosphate into simple and soluble forms. Pikovaskaya medium is now being used to isolate, identify, and classify phosphorus-solubilizing bacteria (PSB). The suspended phosphorus source is insoluble phosphates such as tricalcium phosphate (Alori et al. 2017a, b; Khan et al. 2007). Phosphate-solubilizing species are being identified by the formation of clearing zones around their colonies. Microorganisms have been found to mineralize organic phosphorus to a soluble state due to enzymatic activity (Kour et al. 2020a, b; Sharma et al. 2013; Kumar et al. 2018). The ability of the effective isolates to generate organic acids has shown that they can solubilize insoluble inorganic phosphates including rock phosphate, tricalcium phosphate, iron, and aluminum phosphates. Seed or seedling inoculation with PSM improves crop grain yield. They are said to contribute 30–35 kg P_2O_5 per hectare (Alori et al. 2017b; Khan et al. 2014b). Acidification, chelation, exchange reactions in the growth medium, and proton transfer during ammonium assimilation all require inorganic phosphate solubilization by microbes (Sharma et al. 2020b).

3.2.1.6.2 Phosphate Mobilizing Microbes: Mycorrhizae

Fungi and plant roots form symbiotic relationships called mycorrhizae. Mycorrhiza is a Greek term that combines the words myces and rhizo, with myceas referring to fungi and rhizo referring to roots. Under natural conditions, mycorrhizae can be found in virtually all soils, from mine spoils to agricultural soils, as well as the soil under horticultural or field crops. Mycorrhizal associations are recognized by over 95% of plant taxa. The fungi receive carbon from the host, while the latter benefits from increased nutrient absorption as a result of the fungi's soil nutrient transfer. Endomycorrhiza, ectomycorrhiza, and ectoendomycorrhiza are the three morphologically distinct types of mycorrhizae, depending on whether or not fungal penetration of root cells occurs. Of the three classes, endomycorrhizal fungi are the most effective biofertilizers (Berruti et al. 2016; Hodge and Storer 2014). Nearly 90% of land plants grow endomycorrhizae.

The fungi form external hyphal networks in the soil and thrive extensively within the cells of the root cortex in this relationship. The Hartig net is a network of fungal hyphae found within the root cortex. Different forms of endomycorrhizal interactions include basidiomycetes, ascomycetes, and zygomycetes fungi.

The Ericaceae (Ericoid mycorrhizae) and Orchidaceae (Orchidaceous mycorrhizae) families include endomycorrhizae, but arbuscular mycorrhizae are the most common (earlier referred to as vesicular-arbuscular mycorrhizae). It is made up of 120 different zygomycetes species, all of which are members of the Glomales order (Glomus, Acaulospora, Gigaspora, Sclerocystis, Entrophospora, and Scutellospora).

Mycorrhiza are yet to be successfully axenic or pure cultured (Wu et al. 2005; Ritika and Utpal 2014). Numerous researchers have reported the role of mycorrhizae in promoting plant growth in a variety of plants. The increase in nutrient uptake, especially phosphorus, has been attributed to mycorrhizae's beneficial effect on plant development.

Mycorrhizal fungi improve soil phosphorus supply by solubilizing inorganic sources of phosphorus or mineralizing organic phosphorus. By solubilizing inorganic sources of phosphorus or mineralizing organic phosphorus, mycorrhizal fungi significantly increase phosphorus supply.

Other nutrients, such as NH_4^+, NO_3^-, K^+, Ca^{2+}, $SO4^{2-}$, Cu^+, Zn^{2+}, and Fe^{+3}, can be taken up and delivered to plants by mycorrhizal external hyphae. AM's external hyphae have been shown to produce up to 80% of plant P, 25% of plant N, 10% of plant K, 25% of plant Zn, and 60% of plant Cu in experimental chambers. Mycorrhizal fungi produce ectoenzymes that allow host plants to access organic nitrogen and phosphorus that would otherwise be unavailable to AM fungi or non-mycorrhizal roots (Bargaz et al. 2018; Kumar et al. 2018; Hodge and Storer 2014; Ritika and Utpal 2014; Wahbi et al. 2016).

3.3 Conclusion

Biofertilizers are important part of integrated nutrients management and renewable source of plant nutrients as an alternative to chemical fertilizers in a sustainable agricultural system. As environmentally friendly and cost-effective inputs for farmers, biofertilizers are important in organic crop production and ecosystem protection, as well as protecting the environment. Therefore, we need to understand the direct interactions between different microbes and crops which will ultimately benefit the development and growth of the plants and soil health. First of all, we will access affordable potential biofertilizers from laboratory and greenhouse to the farmer's field. Therefore, we will require some novel approaches such as the application method of biofertilizers, storage of different biofertilizer, transportation and creation of awareness among farmers for particular biofertilizers for specific crops, etc. So we need more focus on interdisciplinary research such as microbial genetics, biotechnology, and agriculture extension in order to finally flourish the biofertilizer industries.

References

Abd-Alla MH (1994) Solubilization of rock phosphates by rhizobium and Bradyrhizobium. Folia Microbiol 39:53–56

Alexandre G (2017) Azospirillum brasilense, a beneficial soil bacterium: isolation and cultivation. Curr Protoc Microbiol 2017:3E.1.1–3E.1.10

Alori ET, Glick BR, Babalola OO (2017a) Microbial phosphorus solubilization and its potential for use in sustainable agriculture. Front Microbiol 8:971. https://doi.org/10.3389/fmicb.2017.00971

Alori ET, Glick BR, Babalola OO (2017b) Microbial phosphorus solubilization and it's potential for use in sustainable agriculture. Front Microbiol 8:971

Andrews M, Andrews ME (2017) Specificity in legume-rhizobia symbioses. Int J Mol Sci 18:705. https://doi.org/10.3390/ijms18040705

Ayyaz K, Zaheer A, Rasul G, Mirza MS (2016) Isolation and identification by 16S rRNA sequence analysis of plant growth-promoting azospirilla from the rhizosphere of wheat. Braz J Microbiol 47:542–550

Bargaz A, Lyamlouli K, Chtouki M, Zeroual Y, Dhiba D (2018) Soil microbial resources for improving fertilizers efficiency in an integrated plant nutrient management system. Front Microbiol 9:1606. https://doi.org/10.3389/fmicb.2018.01606

Bashan Y, Levanony H (1990) Current status of Azospirillum inoculation technology: Azospirillum as a challenge for agriculture. Can J Microbiol 36:591–608

Bennett AB, Pankievicz VCS, Ané JM (2020) A model for nitrogen fixation in cereal crops. Trends Plant Sci 25:226–235

Berruti A, Lumini E, Balestrini R, Bianciotto V (2016) Arbuscular mycorrhizal fungi as natural biofertilizers: let's benefit from past successes. Front Microbiol 6:1559. https://doi.org/10.3389/fmicb.2015.01559

Bhattacharyya PN, Jha DK (2012) Plant growth-promoting rhizobacteria (PGPR): emergence in agriculture. World J Microbiol Biotechnol 28:1327–1350

Boivin S, Ait Lahmidi N, Sherlock D et al (2020) Host-specific competitiveness to form nodules in *Rhizobium leguminosarum* symbiovar *viciae*. New Phytol 226:555–568

Cassán F, Diaz-Zorita M (2016) Azospirillum sp. in current agriculture: from the laboratory to the field. Soil Biol Biochem 103:117–130

Cavassim MIA, Moeskjær S, Moslemi C et al (2019) Symbiosis genes show a unique pattern of introgression and selection within a Rhizobium leguminosarum species complex. BioRxiv. https://doi.org/10.1101/526707

Chatterjee A, Singh S, Agrawal C, Yadav S, Rai R, Rai LC (2017) Role of algae as a biofertilizer. In: Algal Green Chemistry recent progress in biotechnology. Elsevier, Amsterdam, pp 189–200

Chittora D, Meena M, Barupal T, Swapnil P (2020) Cyanobacteria as a source of biofertilizers for sustainable agriculture. Biochem Biophys Rep 22:100737

Dal Cortivo C, Ferrari M, Visioli G et al (2020) Effects of seed-applied biofertilizers on rhizosphere biodiversity and growth of common wheat (Triticum aestivum L.) in the field. Front Plant Sci 11:72

Das HK (2019) Azotobacters as biofertilizer. Adv Appl Microbiol 108:1–43

De Oliveira MVV, Intorne AC, de Vespoli L et al (2016) Differential effects of salinity and osmotic stress on the plant growth-promoting bacterium Gluconacetobacter diazotrophicus PAL5. Arch Microbiol 198:287–294

Dhar DW, Prasanna R, Singh BV (2007) Comparative performance of three carrier based blue green algal biofertilizers for sustainable rice cultivation. J Sustain Agric 30:41–50

FAO (2020) Food security and nutrition in the world. FAO, Rome

Fukami J, Cerezini P, Hungria M (2018) Azospirillum: benefits that go far beyond biological nitrogen fixation. AMB Express 8:1–12

Glick BR (2012) Plant growth-promoting bacteria: mechanisms and applications. Scientifica (Cairo) 2012:1–15

Govindarajulu M, Pfeffer PE, Jin H et al (2005) Nitrogen transfer in the arbuscular mycorrhizal symbiosis. Nature 435:819–823

Hodge A, Storer K (2014) Arbuscular mycorrhiza and nitrogen: implications for individual plants through to ecosystems. Plant and Soil 386:1–19. https://doi.org/10.1007/s11104-014-2162-1

Islam S, Akanda AM, Prova A, Islam MT, Hossain MM (2016) Isolation and identification of plant growth promoting rhizobacteria from cucumber rhizosphere and their effect on plant growth promotion and disease suppression. Front Microbiol 6:1360. https://doi.org/10.3389/fmicb.2015.01360

John RP, Tyagi RD, Brar SK, Surampalli RY, Prévost D (2011) Bio-encapsulation of microbial cells for targeted agricultural delivery. Crit Rev Biotechnol 31:211–226

Keet JH, Ellis AG, Hui C, Le Roux JJ (2017) Legume-rhizobium symbiotic promiscuity and effectiveness do not affect plant invasiveness. Ann Bot 119:1319–1331

Khan MS, Zaidi A, Wani PA (2007) Role of phosphate-solubilizing microorganisms in sustainable agriculture—a review. Agron Sustain Dev 27:29–43

Khan MS, Zaidi A, Musarrat J (2014a) Phosphate solubilizing microorganisms: principles and application of microphos technology. Springer International, Cham

Khan MS, Zaidi A, Ahmad E (2014b) Mechanism of phosphate solubilization and physiological functions of phosphate-solubilizing microorganisms. In: Phosphate solubilizing microorganisms: principles and application of microphos technology. Springer International, Cham, pp 31–62
Khosravi H, Dolatabad HK (2020) Identification and molecular characterization of Azotobacter chroococcum and Azotobacter salinestris using ARDRA, REP, ERIC, and BOX. Mol Biol Rep 47:307–316
Kong Z, Hart M, Liu H (2018) Paving the way from the lab to the field: using synthetic microbial consortia to produce high-quality crops. Front Plant Sci 9:1467. https://doi.org/10.3389/fpls.2018.01467
Kour D, Rana KL, Yadav AN et al (2020a) Microbial biofertilizers: bioresources and eco-friendly technologies for agricultural and environmental sustainability. Biocatal Agric Biotechnol 23:101487
Kour D, Rana KL, Yadav AN et al (2020b) Microbial biofertilizers: bioresources and eco-friendly technologies for agricultural and environmental sustainability. Biocatal Agric Biotechnol 23:101487
Kumar SM, Reddy CG, Phogat M, Korav S (2018) Role of bio-fertilizers towards sustainable agricultural development: a review. J Pharmacogn Phytochem 7:1915–1921
Laguerre G, Depret G, Bourion V, Duc G (2007) Rhizobium leguminosarum bv. Viciae genotypes interact with pea plants in developmental responses of nodules, roots and shoots. New Phytol 176:680–690
Lin SY, Liu YC, Hameed A et al (2016) Azospirillum agricola sp. nov., a nitrogen-fixing species isolated from cultivated soil. Int J Syst Evol Microbiol 66:1453–1458
Lumpkin TA, Plucknett DL (1980) Azolla: botany, physiology, and use as a green manure. Econ Bot 34:111–153
Mahanty T, Bhattacharjee S, Goswami M et al (2017) Biofertilizers: a potential approach for sustainable agriculture development. Environ Sci Pollut Res 24:3315–3335
Mahapatra DM, Chanakya HN, Joshi NV, Ramachandra TV, Murthy GS (2018) Algae-based biofertilizers: a biorefinery approach. Dermatol Sin:177–196
Missbah El Idrissi M, Lamin H, Bouhnik O et al (2020) Characterization of Pisum sativum and Vicia faba microsymbionts in Morocco and definition of symbiovar viciae in Rhizobium acidisoli. Syst Appl Microbiol 43:126084. https://doi.org/10.1016/j.syapm.2020.126084
Mosa WFAEG, Sas-Paszt L, Frac M, Trzciński P (2016a) Microbial products and biofertilizers in improving growth and productivity of apple—a review. Pol J Microbiol 65:243–251
Mosa WFAEG, Sas-Paszt L, Frac M, Trzciński P (2016b) Microbial products and biofertilizers in improving growth and productivity of apple—a review. Pol J Microbiol 65:243–251
Muñoz-Rojas J, Caballero-Mellado J (2003) Population dynamics of Gluconacetobacter diazotrophicus in sugarcane cultivars and its effect on plant growth. Microb Ecol 46:454–464
Mutch LA, Young JPW (2004) Diversity and specificity of rhizobium leguminosarum biovar viciae on wild and cultivated legumes. Mol Ecol 13:2435–2444
Muthukumarasamy R, Govindarajan M, Vadivelu M, Revathi G (2006) N-fertilizer saving by the inoculation of Gluconacetobacter diazotrophicus and Herbaspirillum sp. in micropropagated sugarcane plants. Microbiol Res 161:238–245
Nautiyal CS (1999) An efficient microbiological growth medium for screening phosphate solubilizing microorganisms. FEMS Microbiol Lett 170:265–270
Noar JD, Bruno-B Arcena JM (2018) Azotobacter vinelandii: the source of 100 years of discoveries and many more to come. Microbiology 164(4):421–436. https://doi.org/10.1099/mic.0.000643
Prasanna R, Tripathi U, Dominic TK, Singh AK, Yadav AK, Singh PK (2003) An improvized technique for measurement of nitrogen fixation by blue green algae and Azolla using moist soil cores from rice fields. Exp Agric 39:145–150
Prasanna R, Jaiswal P, Singh YV, Singh PK (2008) Influence of biofertilizers and organic amendments on nitrogenase activity and phototrophic biomass of soil under wheat. Acta Agron Hungarica 56:149–159

Raffi MM, Charyulu PBBN (2021) Azospirillum-biofertilizer for sustainable cereal crop production: current status. Recent Dev Appl Microbiol Biochem 10:193–209

Raklami A, Bechtaoui N, Tahiri A, Anli M, Meddich A, Oufdou K (2019) Use of Rhizobacteria and mycorrhizae consortium in the open field as a strategy for improving crop nutrition, productivity and soil fertility. Front Microbiol 10:1106

Ritika B, Utpal D (2014) Biofertilizer, a way towards organic agriculture: a review. Afr J Microbiol Res 8:2332–2343

Rocha I, Ma Y, Souza-Alonso P, Vosátka M, Freitas H, Oliveira RS (2019) Seed coating: a tool for delivering beneficial microbes to agricultural crops. Front Plant Sci 10:1357. https://doi.org/10.3389/fpls.2019.01357

Saravanan VS, Madhaiyan M, Thangaraju M (2007) Solubilization of zinc compounds by the diazotrophic, plant growth promoting bacterium Gluconacetobacter diazotrophicus. Chemosphere 66:1794–1798

Sharma SB, Sayyed RZ, Trivedi MH, Gobi TA (2013) Phosphate solubilizing microbes: Sustainable approach for managing phosphorus deficiency in agricultural soils. SpringerPlus 2:587. https://doi.org/10.1186/2193-1801-2-587

Sharma D, Gahtyari NC, Chhabra R, Kumar D (2020a) Role of microbes in improving plant growth and soil health for sustainable agriculture, pp 207–256

Sharma D, Gahtyari NC, Chhabra R, Kumar D (2020b) Role of microbes in improving plant growth and soil health for sustainable agriculture. Dermatol Sin 19:207–256

Singh PK, Misra SP, Singh AL (1984) Azolla biofertilization to increase rice production with emphasis on dual cropping. In: Practical application of Azolla for rice production. Springer, Dordrecht, pp 132–144

Singh AL, Singh PK, Singh PL (1988) Effects of different herbicides on the Azolla and blue-green algal biofertilization of rice. J Agric Sci 111:451–458

Singh M, Dotaniya ML, Mishra A, Dotaniya CK, Regar KL, Lata M (2016) Role of biofertilizers in conservation agriculture. In: Conservation agriculture: an approach to combat climate change in Indian Himalaya. Springer, Singapore, pp 113–134

Steenhoudt O, Vanderleyden J (2000a) Azospirillum, a free-living nitrogen-fixing bacterium closely associated with grasses: genetic, biochemical and ecological aspects. FEMS Microbiol Rev 24:487–506

Steenhoudt O, Vanderleyden J (2000b) Azospirillum, a free-living nitrogen-fixing bacterium closely associated with grasses: genetic, biochemical and ecological aspects. FEMS Microbiol Rev 24:487–506

Suman A, Gaur A, Shrivastava AK, Yadav RL (2005) Improving sugarcane growth and nutrient uptake by inoculating Gluconacetobacter diazotrophicus. Plant Growth Regul 47:155–162

Thomas L, Singh I (2019) Microbial biofertilizers: types and applications, pp 1–19. https://doi.org/10.1007/978-3-030-18933-4_1:1-19

Van Oosten MJ, Di Stasio E, Cirillo V et al (2018) Root inoculation with Azotobacter chroococcum 76A enhances tomato plants adaptation to salt stress under low N conditions. BMC Plant Biol 18:205. https://doi.org/10.1186/s12870-018-1411-5

Vassilev N, Vassileva M, Martos V et al (2020) Formulation of microbial inoculants by encapsulation in natural polysaccharides: focus on beneficial properties of carrier additives and derivatives. Front Plant Sci 11:270. https://doi.org/10.3389/fpls.2020.00270

Venkateswarlu B, Rao AV (1985) Physiological studies on Azospirillum spp. isolated from the roots of diverse plants. Zentralbl Mikrobiol 140:521–526

Veresoglou SD, Menexes G (2010) Impact of inoculation with Azospirillum spp. on growth properties and seed yield of wheat: a meta-analysis of studies in the ISI web of science from 1981 to 2008. Plant and Soil 337:469–480

Vessey JK (2003) Plant growth promoting rhizobacteria as biofertilizers. Plant and Soil 255:571–586

Wahbi S, Prin Y, Thioulouse J et al (2016) Impact of wheat/faba bean mixed cropping or rotation systems on soil microbial functionalities. Front Plant Sci 7:13. https://doi.org/10.3389/fpls.2016.01364

Win TT, Barone GD, Secundo F, Fu P (2018) Algal biofertilizers and plant growth stimulants for sustainable agriculture. Ind Biotechnol 14:203–211

Wu SC, Cao ZH, Li ZG, Cheung KC, Wong MH (2005) Effects of biofertilizer containing N-fixer, P and K solubilizers and AM fungi on maize growth: a greenhouse trial. Geoderma 125:155–166

Yadav RK, Abraham G, Singh YV, Singh PK (2014) Advancements in the utilization of Azolla-anabaena system in relation to sustainable agricultural practices. Proc Indian Natl Sci Acad 80:301–316

Chapter 4
Microbial Mediated Natural Farming for Sustainable Environment

Asha Rani and Beenam Saxena

Abstract India is an agriculture-based country, and agriculture is the backbone of Indian economy. More than half of the population depends upon agriculture. The majority of the farmers rely on conventional farming in comparison to natural or organic farming. To fulfil the food requirement, it is necessary to increase yield and production of crops. Different types of chemical fertilizers are used to increase total yield. Due to the large use of these fertilizers, heavy metal ions increased in the soil which may be toxic to animals and humans. The heavy metals are also present in city waste water (CWW) in toxic amount, and when this polluted water reaches to adjoining areas of the city, it contaminates the soil. When these heavy metals are absorbed by the plants, it may lead to some adverse effect on different growth parameters which directly affects the total yield of the crop. The quantity of these chemicals can be reduced with the help of microbes present in soil or by use of biofertilizers. This book chapter describes the importance of organic farming to maintain sustainable agriculture.

Keywords Biofertilizers · Conventional farming · Heavy metal · Soil microbes

4.1 Introduction

Soil is very important and an essential factor for plant growth. However, by the use of enormous number of chemical fertilizers, it can be contaminated (Chao et al. 2014). Continuous use of chemical fertilizers and regular addition of heavy metals may cause the different other types of pollution in soil and water environment (Bhatt et al. 2019a, b; Pankaj et al. 2015a, b, 2016a, b). Soil can be contaminated by heavy metals which reache through city waste water and other industrial wastes to the

A. Rani (✉)
Department of Botany, Bareilly College Bareilly, Bareilly, Uttar Pradesh, India

B. Saxena
Department of Zoology, Bareilly College Bareilly, Bareilly, Uttar Pradesh, India

P. Bhatt et al. (eds.), *Microbial Technology for Sustainable Environment*,
https://doi.org/10.1007/978-981-16-3840-4_4

agricultural fields. The heavy metal ions and other chemicals may also cause decrease in nutrients in the soil. Heavy metal-contaminated soil cannot be remediated (Verma et al. 2021). Due to heavy metal pollution, size, composition, and activity of the microbial community are also adversely affected along with plant quality and yield (Wang et al. 2016). Heavy metals can interfere with the enzymatic activity of microbes so organic matter decreases in soil (Shun-hong et al. 2009). Human exposure to these metals occurs through ingestion of contaminated food or water. The high cost of chemical fertilizers and their adverse effect on environment have encouraged scientists to develop alternative method to increase soil productivity (Huang et al. 2021; Singh et al. 2021; Zhang et al. 2020, 2021; Mishra et al. 2020; Feng et al. 2020; Lin et al. 2020; Vaxevanidou et al. 2015). Microorganisms play very important role to increase soil fertility which is contaminated by heavy metals. Phytoremediation is another aspect for the treatment of polluted soil. In this, plants are used to reduce soil contamination. Some plants have the capacity to absorb heavy metals when they are planted at the boundary of fields. Highly resistant plants like sunflower (*Helianthus annuus*), Indian mustard (*Brassica juncea* L.), willow (*Salix alba*), popular tree (*Populus deltoids*), vetiver (*Chrysopogon zizanioides*), etc. can be used for a remediation of the pollution site. For phytoremediation, molecular mechanisms of resistance to heavy metals should be studied in different types of plants. It will be helpful in the near future to find out more plant species having heavy metal resistance. Effected bioremediation of heavy metal-polluted soils can be possible by using combination of both microorganisms and plants. However, success of this approach will depend on species of organisms involved in the process. Bioremediation is very economical in comparison to the other techniques for remediation of contaminated soil. However, it has been found that growth of different plant growth-promoting (PGP) microbes was proper in the organic soil. This was due to the frequent use of green manure. The soil health depends upon the diversity of microbes present in the soil. The productivity of the crop directly depends upon soil health (Bhatt and Maheshwari 2019, 2020a, b, c).

To reduce the amount of chemicals, conventional farming should be replaced by organic or natural farming with use of soil microbes. Indigenous microbial consortium inhabits the soil and has potential to improve soil fertility. By increasing natural farming and use of biofertilizers or organic fertilizer, the food quality can be improved. An "organic fertilizer" can be derived from non-synthetic or organic sources such as plant or animal, microbes, and rock powders; by different processes like drying, cooking, composting (Dadi et al. 2019), chopping, grinding, and fermenting (Mario et al. 2019); or other method (Thanaporn and Nuntavun 2019). The soil enriched with microbes is considered as healthy, and it helps in plant growth and makes them resistant against stress. Although the maintenance of organic soil quality is quite tough in Indian agricultural practices and expensive too. India has 1.94 million hectares of organic farmland in 2018–2019 (Fig. 4.1) accounted for 1.08% of total agricultural land, and certified organic production for all crop categories stood at 2.6 million metric tons (MT). According to World of Organic Agricultural Report 2018, India produces 30% of total organic production and has maximum number of organic producers in world, i.e., about 835,000. In the year

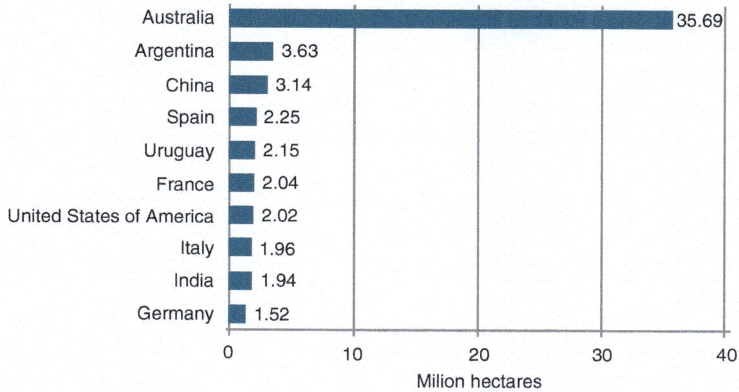

Fig. 4.1 Farmland under organic cultivation of ten top countries in 2018. (Source FiBL 2020)

2018–2019, India exported 6.389 lakh MT, and total earning was around INR 4686 crore (Li et al. 2019).

In India, approximately 3.67-million-hectare agriculture area are used for the organic farming. Among the Indian provinces, Madhya Pradesh represent the large land for organic farming followed by Rajasthan, Maharashtra, Gujarat, Karnataka, Odisha, Sikkim, and Uttar Pradesh. In 2016, Sikkim converts its entire land for organic farming production. Globally, the USA, Canada, Switzerland, Australia, Japan, UAE, New Zealand, etc., pay more attention for organic farming (Li et al. 2019). Despite of development in this field, organic farming has not yet so popular, and it is not an easy task for Indian farmers to switch to organic farming as there is no policy for encouraging the spirit of farmers to opt organic agriculture. Still, it is necessary to promote organic farming over the conventional farming as it is the need of the hour (Bhatt and Nailwal 2018; Khati et al. 2018; Gangola et al. 2018; Bhatt 2018; Bhatt and Barh 2018; Bhatt et al. 2019c; Bhandari and Bhatt 2020; Bhatt and Bhatt 2021).

4.2 Effect of Heavy Metals on Different Crops

Heavy metals are present in toxic amount in the city waste water which goes to the adjoining areas of different cities. It contains considerable quantities of toxic elements. Many unused electronic instruments and heavy metal containing batteries are discarded which also serve as a source of heavy metals in groundwater resources. The various elements Cd, Cu, Zn, and Pb are most likely to cause phytotoxicity when waste water is applied to agricultural field or land where different types of crops are growing (Bhatt and Maheshwari 2019, 2020b, c). However, heavy metals are required for growth and upkeep of plants, but their excessive amounts become toxic to plants. Accumulation of essential metals in plants enable them to acquire other nonessential metals (Zhou et al. 2008). Some heavy metals in the soil also have an effect on the growth of soil microbes (Gulser and Erdogan 2008).

Increased application of agrochemicals and inorganic fertilizers is more in practice which has caused agricultural pollution leading to degradation of the ecosystem and the environment (Malik et al. 2017). Industrial development also caused negative impact on the environment (Dhami et al. 2013); however, due to industrialization, there is rise in global economy over the last century, but it has led to a dramatic increase in production and release of hazardous metals to the environment (Gerhardt et al. 2009; Gallego et al. 2012; Burger 2008; Central Pollution Control Board [CPCB] 2007).

Among the heavy metals, zinc and copper are very essential for plant growth, but when present at elevated levels in soils, they become toxic and can ultimately cause the death of plants. When the effect of these heavy metals studied, it is found phytotoxic to mung bean and have adverse effect on different growth parameters such as seedling height (Narwal et al. 1992), chlorophyll content (Khandelwal 1993), and nitrogen content (Singh 1999; Rani 2011). Reduction in all these parameters ultimately affects the total grain yield of the crops. Except this, Pb and Cd are also found in very low concentration. These are not beneficial for the plants, but even their low concentration has adverse effects on plant growth. High concentration of arsenic showed inhibitory effect on seed germination and seedling growth of wheat (Zhang et al. 2010) as well as on length of plumule and radicle of *Helianthus annuus* (Imran et al. 2013). The vegetable crops production at the heavy metal-contaminated soil showed variability in heavy metal accumulation. The vegetables can be successfully grown into the zinc- and copper-contaminated soils, where some of them such as mustard, soybean, and spinach cannot be cultivated (Singh et al. 2012). The accumulation of the heavy metals into the vegetable crops affect the human health directly due to their entry via food chain (Fu et al. 2008; Bonanomi et al. 2016).

The occurrence of heavy metals in groundwater is reported from western Uttar Pradesh, India, and all four districts Shahjahanpur, Bareilly, Moradabad, and Rampur have excessive presence of cadmium (Idrees et al. 2018). Status of different heavy metals like As, Cd, Pb, and Hg has been investigated in most commonly used cereals and legumes of Bareilly district of Uttar Pradesh (India). Among cereals, rice contains the highest levels of all these heavy metals; however, As, Pb, and Hg accumulation is also found in wheat and maize at lower level. Cd level remains significantly higher in maize than wheat, and levels of arsenic remain similar among different legumes (Lipismita and Garg 2012). Growth reduction as a result of changes in physiological and biochemical processes in plants growing on heavy metal-polluted soils has been recorded (Chatterjee and Chatterjee 2000; Oncel et al. 2000; Oancea et al. 2005).

It has been clear that heavy metal contamination causes loss of bacterial species richness and a relative increase in soil actinomycetes or even decreases in both the biomass and diversity of the bacterial communities in soil (Karaca et al. 2010). By using microbial-based fertilizers, the soil health can be improved, and by doing so, sustainability of environment can be maintained.

4.3 Soil Health in Non-organic and Organic Farming Sites

Soil health is affected by the presence of microorganisms which play an important role for crop production and final yield. Soil bacteria and fungus increase soil fertility. 1gm. of fertile soil may have around 400,000 fungi (Griffiths et al. 1999). During the comparative study of organic and inorganic sites, it has been found that the organic soil has enormous amount of microorganism than inorganic sites. This is due to the frequent use of green manure in organic soil (Khanna et al. 2010). Due to the presence of richness of nutrients in the organic soil, growth of microorganism is directly affected. It has been consistently reported that soil organic matter favours the growth of bacteria present in soil. The studies have revealed that bacterial diversity in soil is approximately 100 times greater than the other microbial diversity (Barns et al. 1999). *Pseudomonas* and *Bacilli* however are found in both types of farming sites, but the richness is much higher in organic sites. Nitrogen (N) is a very essential element for the growth of leaves and stem which also plays an important role in the formation and proper functioning of chloroplast. The organic field has high nitrogen content as compared to non-organic farming site. The higher amount of nitrogen in organic site is due to addition of compost and green manure which increases soil fertility. Although chemical compounds as urea and nitrogen fertilizers are also use in non-organic site, they are not available for plants due to their precipitation (Barns et al. 1999; Sharma and Bhatt 2016).

Soil organic carbon (SOC) of organic farming site was found to be higher as compared to non-organic farming site. Soil acts as a main reservoir of carbon, and the higher SOC value is the direct indication of level of soil health. Soil organic matter (SOM) present in the soil adds more nutrients to the soil. Good soil fertility increases aeration, water holding capacity, proper root growth, and soil microflora which finally affects crop yield. According to a global review, the soils in organic cropping systems have significantly higher levels of SOC than those in conventional systems (Sharma et al. 2016).

Phytohormones also play an important role in plant growth and directly affect the final yield of any crop. Indole acetic acid (auxin) has many physiological roles in plant development. Low concentration of auxin induces the root growth which increases the water absorption. The environmental factors and soil microflora affect the auxin activity.

4.4 Role of Microbes in Treatment of Soil Polluted with Xenobiotics

Agriculture plays an important role in Indian economy. India holds the second largest position in growing wheat and rice, the staple food of the world. It is the need of the hour to increase the soil fertility and productivity of crops to fulfil the food requirements of the large population. Different types of fertilizers and

Table 4.1 Microbial diversity in soil

Microorganism	Plant	Plant growth regulation	References
Bacillus amyloliquefaciens 5113 and *Azospirillum brasilense* NO 40	Wheat	Enhance plant growth under drought condition and increase enzyme activity in wheat	Edwards and Lofty (1974)
Pseudomonas aeruginosa FP6	Chili	Siderophore produced by bio-control strain to reduce metal pollution	Amir and Fouzia (2011)
Bacillus and *Pseudomonas* spp.	*C. annum* L	Plant growth enhancement and bio-control management to control plant disease	Kasim et al. (2013)
Mesorhizobium spp.	Chickpea	Increase nodulation, enhance, and uptake of nutrient yield	Sasirekha and Srividya (2016)
Bacillus thuringiensis	Wheat	Decrease volatile emissions and increase photosynthesis	Kumar et al. (2014)
Trichoderma harzianum Tr6 and *Pseudomonas* sp. *Ps 14*	Cucumber and *Arabidopsis thaliana*	Induced systemic resistance	Verma et al. (2013)

pesticides are being used to increase production; thus further, intensive utilization of chemical fertilizers and pesticides for higher crop production may become destructive and detrimental for soil and food quality (Gattinger et al. 2012). Soil rich in microorganism directly affects the agricultural productivity. Number of microorganism can be increased by use of biofertilizers and biopesticides. With the help of microorganism, plants absorb nutrients at a promising speed. These microorganisms get food from the waste by products of plants.

Plant growth-promoting (PGP) microbes and PGPR play very important role to cope up with heavy metal pollution of soil. They increase soil fertility, bioremediation, and stress management for development of eco-friendly sustainable agriculture. Different types of bacteria such as *Bacillus*, *Pseudomonas*, *Azotobacter*, etc. are beneficial for plant growth (Table 4.1). The bacterial count remains higher in the organic farming site in comparison to non-organic farming site. Regular use of chemicals in fields decreases the C—compound in the soil which is necessary for microbial growth. High CFU counts in organic farming soil may be due to nutrient richness and absence of high concentration of heavy metal ions that are inhibitory for microbial growth (Kang et al. 2016). Organic manure increases the carbon source in the soil which is beneficial for the microbes as it increases the growth and activity. By increasing microbial count, bioremediation of the soil can be done as this is the way to treat heavy metal-polluted soil. Several comparisons of organic and conventional farming systems have indicated significant impact of soil microbial community on agricultural practices (Smith et al. 2012; Liao et al. 2018; Hartmann 2015; Li et al. 2012).

Naturally available technologies for enhancement of agriculture and management of agricultural waste are being aimed by scientist. Indigenous microorganism (IMO)-based technology is being applied in the eastern part of the world for the extraction of minerals, enhancement of agriculture, and waste management (Rajeshwari 2017). Bacteria are helpful in nitrogen fixation and many other biological processes. *Rhizobia* are found in symbiotic association in root nodules of legumes. Cyanobacteria are helpful in binding the soil molecules as they act as cementing material. *Pseudomonas* sp. are used for remediation. Secondary metabolites have very effective and vital role in plant growth. Microbes are also helpful in production of such metabolite which stimulates the growth and development in plants. Micro-organism may also be protective towards the plants, and rhizosphere soil microbes form a physical barrier around the roots of plants and reduce the invasion of pathogens and pests by providing healthy micro-ecological environment (Table 4.1) (Wu and Lin 2003).

Vermiculture is also a very important tool for organic farming. It is low input farming in comparison to conventional farming. Many researchers reported that vermiculture in organic farming sites is more benefited than in conventional farming site (Timmusk et al. 2014). It is reported that biodegradation process is enhanced when earthworms and microbes work together and produce vermicompost, which is worm fecal matter with worm casts. Vermicompost provided macro-elements such as N, P, K, Ca, and Mg and microelements such as Fe, Mo, Zn, and Cu (Lim and Kim 2013).

The indigenous microbial strains are able to remediate the xenobiotic compounds from the soil and water system (Bhandari et al. 2021). The bacterial and fungal strains are able to degrade the pesticides, antibiotics, endocrine disrupting chemicals, and other organic compounds from the environment (Bhatt et al. 2021a). These microbial strains accelerate the residual level of toxic chemicals from the environment and enhance the sustainable developments (Bhatt et al. 2021b). These potential microbial strains are used throughout the globe for the remediation of the toxic xenobiotics from the contaminated sites (Bhatt et al. 2021c).

4.5 Conclusion

It is the need of the hour to fulfil the food requirements of the huge Indian population. Due to the increasing population and industrialization, the discharge of polluted waste water and agricultural waste is also increasing. As a result, the heavy metals are adversely affecting soil health due to their toxic and non-biodegradable nature. An ideal agriculture system should be developed to improve soil health and for sustainable environment. To increase the yield of any crop, chemicals and pesticides are frequently used by Indian farmers; as a result, soil health is continuously deteriorating. There are many techniques to improve the soil health. Microorganisms play very important role for improving soil health contaminated by heavy metals and for the sustainable environment. It has been proved that they are beneficial for society and environment. By using them, we can get social, economic, and environmental benefits. It is well understood that by increasing microbial communities in the soil and by detection of heavy metals present in the soil, total yield can be enhanced. Organic green manure is well suited for the proper growth of PGPRs and other microbes. The frequent use of the chemicals and pesticides in inorganic fields is harmful. Although it is bitter truth that natural and organic farming is costly as compared to conventional farming and the farmers adopting organic farming face difficulty to survive and market the organic products, but to improve soil health and for development of sustainable environment, farmers should be motivated for organic farming as it can provide quality food without any harmful effect on soil health. Organic farming can be done with proper planning for the betterment of mankind and upcoming generations, and economically sustainable organic farming is the prerequisite for ensuring affordability of organic products at consumer's end.

References

Amir K, Fouzia I (2011) Chemical nutrient analysis of different composts (Vermi compost and Pitcompost) and their effect on the growth of a vegetative crop Pisum sativum. Asian J Plant Sci Res 1(1):116–130

Barns SM, Takala SL, Kurke CR (1999) Wide distribution and diversity of members of the bacterial kingdom acidobacterium in the environment. Appl Environ Microbiol 65:1731–1737

Bhandari G, Bhatt P (2020) Concepts and application of plant microbe interaction in remediation of heavy metals. In: Sharma A (ed) Microbes and signaling biomolecules against plant stress. Rhizosphere biology. Springer, Singapore. https://doi.org/10.1007/978-981-15-7094-0_4

Bhandari G, Bagheri AR, Bhatt P, Bilal M (2021) Occurrence, potential ecological risks, and degradation of endocrine disruptor, nonyphenol from the aqueous environment. Chemosphere 275:130013. https://doi.org/10.1016/j.chemosphere.2021.130013

Bhatt P (2018) *Insilico* tools to study the bioremediation in microorganisms. In: Pathak V, Navneet (eds) Handbook of research on microbial tools for environmental waste management. IGI Global, Hershey, PA, pp 389–395. https://doi.org/10.4018/978-1-5225-3540-9.ch018

Bhatt P, Barh A (2018) Bioinformatic tools to study the soil microorganisms: an *in silico* approach for sustainable agriculture. In: Choudhary D, Kumar M, Prasad R, Kumar V (eds) In silico approach for sustainable agriculture. Springer, Singapore. https://doi.org/10.1007/978-981-13-0347-0_10

Bhatt K, Bhatt P (2021) Rhizosphere biology: alternate tactics for enhancing sustainable agriculture. In: Plant microbiome interactions and sustainable agriculture. https://doi.org/10.1002/9781119644798.ch9

Bhatt K, Maheshwari DK (2019) Decoding multifarious role of cow dung bacteria in mobilization of zinc fractions along with growth promotion of *Capsicum annum* L. Sci Rep 9:14232

Bhatt K, Maheshwari DK (2020a) *Bacillus Megaterium* strain CDK25, a novel plant growth promoting bacterium enhances proximate chemical and nutritional composition of *Capsicum annum* L. Front Plant Sci 11:1147

Bhatt K, Maheshwari DK (2020b) Zinc solubilizing bacteria (*Bacillus Megaterium*) with multifarious plant growth promoting activities alleviates growth in *Capsicum annum* L. 3 Biotech 10:36

Bhatt K, Maheshwari DK (2020c) Insights into zinc-sensing metalloregulator 'Zur' deciphering mechanism of zinc transportation in *Bacillus* spp. by modeling, simulation and molecular docking. J Biomol Struct Dyn. https://doi.org/10.1080/07391102.2020.1818625

Bhatt P, Nailwal TK (2018) In: Prasad R, Gill SS, Tuteja N (eds) Crop improvement through microbial technology: a step towards sustainable agriculture. Elsevier, Amsterdam, pp 245–253. https://doi.org/10.1016/B978-0-444-63987-5.00011-6

Bhatt P, Pal K, Bhandari G, Barh A (2019a) Modeling of methyl halide biodegradation on bacteria and its effect on other environmental systems. Pest Biochem Physiol 158:88–100

Bhatt P, Gangola S, Chaudhary P, Khati P, Kumar G, Sharma A, Srivastava A (2019b) Pesticide induced up-regulation of esterase and aldehyde dehydrogenase in indigenous *Bacillus* spp. Biorem J 23(1):42–52

Bhatt P, Pathak VM, Joshi S, Bisht TS, Singh K, Chandra D (2019c) Major metabolites after degradation of xenobiotics and enzymes involved in these pathways. Chapter 12. In: Smart bioremediation technologies: microbial enzymes, pp 205–215. https://doi.org/10.1016/B978-0-12-818307-6.00012-3

Bhatt P, Zhou X, Huang Y, Zhang W, Chen S (2021a) Characterization of the role of esterases in the biodegradation of organophosphate, carbamate and pyrethroid group pesticides. J Hazard Mater 411:125026

Bhatt P, Joshi T, Bhatt K, Zhang W, Huang Y, Chen S (2021b) Binding interaction of glyphosate oxidoreductase and C-P lyase: molecular docking and molecular dynamics simulation studies. J Hazard Mater 409:124927

Bhatt P, Bhatt K, Sharma A, Zhang W, Mishra S, Chen S (2021c) Biotechnological basis of microbial consortia for the removal of pesticides from the environment. Crit Rev Biotechnol 41(3):317–338. https://doi.org/10.1080/07388551.2020.1853032

Bonanomi G, Filippis F, Cesarano G, Storia AL, Ercolini D, Scala F (2016) Organic farming induces changes in soil microbiota that affect agro–ecosystem functions. Soil Biol Biochem 103:327–336

Burger J (2008) Assessment and management of risk to wild life from cadmium. Sci Total Environ 389:37–45

Central Pollution Control Board [CPCB] (2007) Cadmium levels in environment. News Letter Central Pollution Control Board, New Delhi

Chao S, Liqin J, Zhang W (2014) A review on heavy metal contamination in the soil worldwide: situation, impact and remediation techniques. Environ Skeptics Critics 3(2):24–38

Chatterjee J, Chatterjee C (2000) Phytotoxicity of cobalt, chromium and copper in cauliflower. Environ Pollut 109(1):69–74

Dadi D, Daba G, Beyene A, Luis P, Van der Bruggen B (2019) Composting and co-composting of coffee husk and pulp with source-separated municipal solid waste: a breakthrough in valorization of coffee waste. Int J Recycl Organic Waste Agric 8(3):263–270

Dhami JK, Singh H, Gupta M (2013) Industrialization at the cost of environment degradation. A case of leather and iron and steel industry from Punjab economy. Innov J Bus Manag 2:19–21

Edwards CA, Lofty JR (1974) The invertebrate fauna of the park grassplots. I: soil fauna, Rothamsted report. eRAdoc 2:133–154

Feng Y, Huang Y, Zhan H, Bhatt P, Chen S (2020) An overview of strobilurin fungicide degradation: current status and future perspective. Front Microbiol 11:389

Fu J, Zhou Q, Liu J, Liu W, Wang T, Zhang Q (2008) Jiang G high levels of heavy metal in rice (*Oryza sativa* L.) from a typical E-waste recycling area in Southeast China and its potential risk to human health. Chemosphere 71:1269–1127

Gallego SM, Pena LB, Barcia RA, Azpilicueta CE, Iannone MF, Rosales EP (2012) Unravelling cadmium toxicity and tolerance in plants: insight into regulatory mechanism. Environ Exp Bot 83:33–46

Gangola S, Bhatt P, Chaudhary P, Khati P, Kumar N, Sharma A (2018) Bioremediation of industrial waste using microbial metabolic diversity. In: Pankaj, Sharma A (eds) Microbial biotechnology in environmental monitoring and cleanup. IGI Global, Hershey, PA, pp 1–27. https://doi.org/10.4018/978-1-5225-3126-5.ch001

Gattinger A, Muller A, Haeni M, Skinner C, Fliessbach A, Buchmann N, Paul M, Stolze M, Smith P, Scialabba NE, Niggli U (2012) Enhanced top soil carbon stock under organic farming. Proc Natl Acad Sci 109(44):18226–18231

Gerhardt KE, Huang XD, Glick BR, Greenberg BM (2009) Phytoremediation and rhizo remediation of organic soil contaminants: potential and challenges. Plant Sci J 176:20–30

Griffiths BS, Ritz K, White EN, Dobson G (1999) Soil microbial community structure and their effect on substrate loading rates. Soil Biol Biochem 31:145–153

Gulser F, Erdogan E (2008) The effects of heavy metal pollution on enzyme activities and basal soil respiration of roadside soils. Environ Monit Assess 145(1–3):127–133

Hartmann M (2015) Frey B, Mayer J, Mader P, Widmer F distinct soil microbial diversity under long–term organic and conventional farming. ISME J 9:1177–1194

Huang Y, Zhang W, Pang S, Chen J, Bhatt P, Mishra S, Chen S (2021) Insights into the microbial degradation and catalytic mechanism of chlorpyrifos. Environ Res 192:110660

Idrees N, Tabassum B, Allah EF, Hashem A, Sarah R, Hashim M (2018) Groundwater contamination with cadmium concentrations in some West U.P. Regions, India. Saudi J Biol Sci 25:1365–1368

Imran MA, Ch MN, Khan RM, Ali Z, Mahmood T (2013) Toxicity of arsenic (As) on seed germination of sunflower (*Helianthus annuus* L.). Int J Phys Sci 8:840–847

Kang Y, Hao Y, Shen M, Zhao Q, Li Q, Hu J (2016) Impacts of supplementing chemical fertilizers with organic fertilizers manufactured using pig manure as a substrate on the spread of tetracycline resistance genes in soil. Ecotoxicol Environ Saf 130:279–288

Karaca A, Cetin SC, Turgay OC, Kizikaya R (2010) Effects of heavy metals on soil enzyme activities. In: Sharameti I, Verma A (eds) Soil heavy metals, soil biology, vol 19. Springer, Heidelberg, pp 237–265

Kasim KA, Osman ME, Omar MN, Abd El-Daim IA, Bejai S, Meijer J (2013) Control of drought stress in wheat using plant-growth-promoting bacteria. J Plant Growth Regul 32:1–9

Khandelwal RK (1993) Arey NC effect of zinc tailing on the chlorophyll content of Glycine max and Lathyrus sativus. J Ecotoxicol Environ Monit 3(2):129–133

Khanna DR, Ishaq F, Matta G, Khan A, Semwal KC (2010) Comparison between organic and inorganic soil microbial diversity of different agronomic fields. Int J Environ Rehabil Conserv 1 (2):43–51

Khati P, Gangola S, Bhatt P, Kumar R, Sharma A (2018) Application of nanocompounds for sustainable agriculture system. In: Pankaj, Sharma A (eds) Microbial biotechnology in environmental monitoring and cleanup. IGI Global, Hershey, PA, pp 194–211. https://doi.org/10.4018/978-1-5225-3126-5.ch012

Kumar PG, Mir Hassan Ahmed SK, Desai S, Leo Daniel Amalraj E, Rasul A (2014) In vitro screening for abiotic stress tolerance in potent biocontrol and plant growth promoting strains of *Pseudomonas* and *Bacillus* spp. Int J Bacteriol 2014:195946. https://doi.org/10.1155/2014/195946

Li FR, Liu LL, Liu JL, Yang K (2019) Abiotic and biotic controls on dynamics of labile phosphorus fractions in calcareous soils under agricultural cultivation. Sci Total Environ 681:163–174

Li R, Khafipour E, Krause DO, Entz MH, de Kievit TR, Fernando WGD (2012) Pyrosequencing Reveals the Influence of Organic and Conventional Farming Systems on Bacterial Communities. PLoS One 7:e51897

Liao J, Liang Y, Huang D (2018) Organic farming improves soil microbial abundance and diversity under greenhouse condition: a case study in Shanghai (eastern China). Sustainability 10:3825

Lim JH, Kim SD (2013) Induction of drought stress resistance by multi-functional PGPR *Bacillus licheniformis* K11 in pepper. Plant Pathol J 29(2):201–208

Lin Z, Zhang W, Pang S, Huang Y, Mishra S, Bhatt P, Chen S (2020) Current approaches to and future perspectives on methomyl degradation in contaminated soil/water environments. Molecules 25:738

Lipismita S, Garg AK (2012) Status of toxic heavy metals in cereal grains and pulses in Bareilly district of Uttar Pradesh. Indian Vet J 89(3):25–27

Malik Z, Ahmad A, Abassi M, Dawood GH, Hussain A, Jamil M (2017) Agrochemicals and soil microbes: interaction for soil health. In: Xenobiotics in the soil environment. monitoring, toxicity and management. Springer International, Cham, pp 139–152

Mario AHC, Catalino JLC, Nereida RO, Joel VV, Ariadna LG, Gustavo LR (2019) Nutrient content of fermented fertilizers and its efficacy in combination with hydrogel in Zea mays L. Int J Recycl Organic Waste Agric 8(3):309–315

Mishra S, Zhang W, Lin Z, Pang S, Huang Y, Bhatt P, Chen S (2020) Carbofuran toxicity and its microbial degradation in contaminated environments. Chemosphere 259:127429

Narwal RP, Antil RS, Gupta AP (1992) Soil pollution through industrial effluent and its management. J Soil Contam 1:265–272

Oancea S, Foca N, Airinei A (2005) Effects of heavy metals on plant growth and photosynthetic activity. In: Analele Ştiinţifice ale Universităţii AL. I. CUZA1 IAŞI, Tomul I, s. Biofizică, Fizică medicală şi Fizica mediului, pp 107–110

Oncel I, Keleş Y, Ustun AS (2000) Interactive effects of temperature and heavy metal stress on the growth and some biochemical compounds in wheat seedlings. Environ Pollut 107(3):315–320

Pankaj BTS, Pathak VM, Barh A, Chandra D (2015a) Optimization of amylase production from the fungal isolates of Himalayan region Uttarakhand. Ecol Environ Cons 21(3):1517–1521

Pankaj, Negi G, Gangola S, Khati P, Srivastava A, Sharma A (2015b) Optimization of sulfosulfuron biodegradation through response surface methodology using indigenous bacterial strain isolated from contaminated agriculture field. Int J Curr Microbiol Appl Sci 4(8):105–112

Pankaj, Negi G, Gangola S, Khati P, Kumar G, Srivastava A, Sharma A (2016a) Differential expression and characterization of cypermethrin degrading potential proteins in *Bacillus thuringiensis* strain, SG4. 3 Biotech 6:225

Pankaj, Sharma A, Gangola S, Khati P, Kumar G, Srivastava A (2016b) Novel pathway of cypermethrin biodegradation in a *Bacillus* sp. strain SG2 isolated from cypermethrin-contaminated agriculture field. 3 Biotech 6:45

Rajeshwari G (2017) Role of microbes in environmental sustainability. Int J Sci Res 6(7):413–414

Rani A (2011) Phytotoxic and mutagenic effects of heavy metals present in City waste water of Bareilly in *Vigna radiata* (L) mill Sp. Indian J Appl Pure Biol 26(2):213–218

Sasirekha B, Srividya S (2016) Siderophore production by *Pseudomonas aeruginosa* FP6, a biocontrol strain for Rhizoctonia solani and Colletotrichum gloeosporioides causing diseases in chilli. Agric Nat Resour 50(4):250–256

Sharma A, Bhatt P (2016) Bioremediation: a microbial technology for improvising wildlife. Chapter 3. In: Wildlife management concept analysis and conservation, pp 29–40

Sharma A, Pankaj, Khati P, Gangola S, Kumar G (2016) Microbial degradation of pesticides for environmental cleanup. Chapter 6. In: Bioremediation of industrial pollutants, write and print publication, pp 178–205

Shun-hong H, Bing P, Zhi-hui Y, Li-yuan C, Li-cheng Z (2009) Chromium accumulation, micro-organism population and enzyme activities in soils around chromium-containing slag heap of steel alloy factory. Trans Nonferrous Met Soc Chin 19:241–248

Singh P (1999) Mutation Breeding in *Cajanas cajan* (L) Millisp. with special reference to the induction of effluent resistant plant a type. PhD Thesis, M.J.P. Rohilkhand University, Bareilly

Singh S, Zacharias M, Kalpana S, Mishra S (2012) Heavy metals accumulation and distribution pattern in different vegetable crops. J Environ Chem Ecotoxicol 4(10):170–177

Singh K, Gera R, Sharma R, Maithani D, Chandra D, Bhat AM, Kumar R, Bhatt P (2021) Mechanism and application of *Sesbania* root nodulating bacteria: an alternative for chemical fertilizers and sustainable development. Arch Microbiol 203(4):1259–1270. https://doi.org/10.1007/s00203-020-02137-x

Smith SC, Brandeau ML, Hunter GE, Bavinger JC, Pearson M, Eschbach PJ, Sundaram V, Liu H, Schirmer P, Stave C, Olkin I, Bravata DM (2012) Are organic foods safer or healthier than conventional alternatives? A systematic review. Ann Intern Med 157(5):348–366

Thanaporn P, Nuntavun R (2019) Liquid organic fertilizer production for growing vegetables under hydroponic condition. Int J Recycl Organic Waste Agric 8:1–12

Timmusk S, Islam A, Abd El D, Lucian C, Tanilas T, Kännaste A, Behers L, Nevo E, Seisenbaeva G, Stenstrom E, Niinemets (2014) Drought-tolerance of wheat improved by rhizosphere bacteria from harsh environments: enhanced biomass production and reduced emissions of stress volatiles. PLoS One 9(5):1–13

Vaxevanidou K, Christou C, Kremmydas GF, Georgakopoulos DG, Papassiopi N (2015) Role of indigenous arsenate and iron (III) respiring microorganisms in controlling the mobilization of arsenate in a contaminated soil sample. Bull Environ Contam Toxicol 94(3):282–288

Verma JP, Yadav J, Tiwari KN, Kumar A (2013) Effect of indigenous *Mesorhizobium* spp. and plant growth promoting rhizobacteria on yields and nutrients uptake of chickpea (Cicer arietinum L.) under sustainable agriculture. Ecol Eng 51:282–286

Verma S, Bhatt P, Verma A, Mudila H, Prasher P, Rene ER (2021) Microbial technologies for heavy metal remediation: Effect of process conditions and current practices. Clean Techn Environ Policy. https://doi.org/10.1007/s10098-021-02029-8

Wang W, Wang H, Feng Y, Wang L, Xiao X, Xi Y, Luo X, Sun R, Ye X, Huang Y (2016) Consistent responses of the microbial community structure to organic farming along the middle and lower reaches of the Yangtze River. Sci Rep 6:35046

Wu JF, Lin XG (2003) Effects of soil microbes on plant growth. Soil 35(1):18–21

Zhang ZC, Chen BX, Qiu BS (2010) Phytochelatin synthesis plays a similar role in shoots of the cadmium hyper accumulator *Sedum alfredii* as in non-resistant plants. Plant Cell Environ 33:1248–1255

Zhang W, Lin Z, Pang S, Bhatt P, Chen S (2020) Insights into the biodegradation of Lindane (γ-Hexachlorocyclohexane) using a microbial system. Front Microbiol 11:522

Zhang W, Pang S, Lin Z, Mishra S, Bhatt P, Chen S (2021) Biotransformation of perfluoroalkyl acid precursors from various environmental system: advances and perspectives. Environ Pollut 272:115908

Zhou FJ, Liu J, Liu W, Wang T, Zhang Q, Jiang G (2008) High levels of heavy metals in Rice (Oryza sativa L.) from a typical E-waste recycling area in Southeast China and its potential risk to human health. Chemosphere 71:1269–1275

Chapter 5
Rhizosphere Manipulations for Sustainable Plant Growth Promotion

Pooja Pant, Ankit Negi, Anchal Giri, Pankaj Bhatt, and Rishendra Kumar

Abstract A plant keeps a mesmerizing diversity of microbes inside and outside of its root surface, known as rhizosphere microbiome. This plant-associated microbial community is among the most complex communities found on earth. It is also termed as the second genome of plant. Plants modulate their microbiome by various mechanisms like modifying their structures and releasing the secondary metabolites in order to maximize their fitness. Manipulating the rhizosphere microbiome is an artificial method but eco-friendly. In this review, we have focused on the great role of beneficial microorganisms and their interactions with plants. The role of rhizobacteria on promoting plant growth and the effect of their inoculation over resident bacteria have been discussed with few examples. We have highlighted the contribution of biotic and abiotic factors over rhizosphere microbiome shape alteration. Finally, the process and significance of synthetic or artificial microbial community construction for sustainable development of plant and some of the techniques developed for harnessing the plant and the beneficial microbes with some example have been discussed.

Keywords Microbiome engineering · Plant growth promotion · PGPRs · Sustainable plant growth promotion · Rhizosphere microbiome

5.1 Introduction

The rhizosphere concept is not a newly generated one, but it has been defined over a century ago. The German agronomist and plant physiologist Lorenz Hiltner in 1904 was the first one who coined the term "rhizosphere" and centered the basic idea that

P. Pant · A. Negi · A. Giri · R. Kumar (✉)
Department of Biotechnology, Sir J. C. Bose Technical Campus, Bhimtal, Kumaun University, Nainital, Uttarakhand, India

P. Bhatt
State Key Laboratory for Conservation and Utilization of Subtropical Agro-bioresources, Guangdong Laboratory for Lingnan Modern Agriculture Integrative Microbiology Research Centre, South China Agricultural University, Guangzhou, China

P. Bhatt et al. (eds.), *Microbial Technology for Sustainable Environment*,
https://doi.org/10.1007/978-981-16-3840-4_5

somehow the plant health is related to the microbial composition on the root soil (Hartmann et al. 2008). The narrow zone of soil around the plant root is known as rhizosphere. It is consisting of a large number of microorganisms which include bacteria, fungi, nematodes, etc. Some of the microorganisms are the good ones which give a huge contribution toward the plant growth and development, while some of the them are the bad ones that are harmful not only for plants but also for the mankind. Rhizosphere is basically consisting of three components that interact with one another: rhizoplane, rhizosphere, and root. The rhizosphere is comprised of large number of microorganisms that are affected by the release of substrate from root, the rhizoplane is a closely adhering soil on the surface of the root, while the root is the main component responsible for assembling of microorganisms. Root also contains many endophytes inside the root tissues (Barea et al. 2005).

5.2 Plant Growth-Promoting Rhizobacteria (PGPR)

Sustainable agriculture for plant growth promotion is important and a much better practice as compared to the conventional methods for future agricultural demands. The rhizospheric soil contains a huge number of useful bacteria which are referred as "plant growth-promoting rhizobacteria (PGPR)" that gives plant a healthy and promoting environment. The basic characteristics required for being a good PGPR are it must compete, survive, and multiply with other microbes for plant growth promotion and must be able to form colony on root surface. These are present naturally in the soil and get attracted toward the plant roots and fulfill their requirements from the root excretion and secretions. The rhizobacteria lodged around the plant roots are more capable of solubilizing and mobilizing the nutrients as compared to the rest of the bacterial populations. Thus, they are the crucial ones needed to be focused for fertile soil (Glick 2012). Various mechanisms have been explained by scientists for plant development which involves rhizoremediation, biofertilization, disease suppression, etc. (Lugtenberg et al. 2001; Raaijmakers et al. 2009). The mechanisms of *Pseudomonas* and *Bacillus*, belonging to Proteobacteria and Firmicutes, and the fungi *Trichoderma* and *Gliocladium*, belonging to Sebacinales order, have already been documented (Qiang et al. 2012). PGPRs not only enhance immunity and growth of a plant but also provide "induced systemic tolerance" to certain abiotic stresses like salt and drought. They help in uptaking the nutrients through roots to the plant and help in fulfilling the deficiency of phosphorous, nitrogen, etc. It has already been reported that plants release certain compounds that selectively recruit microorganisms in rhizosphere soil, which are beneficial for plant health (Reinhold-Hurek et al. 2015). The beneficial PGPRs are deliberately inoculated in the rhizosphere for checking their effects on the host plant and on the already present indigenous microbial population. Certain rhizobacteria show a positive result and suppress the disease-causing pathogens. The effective functioning of these bio-inoculates can be obtained by exploring the large population pool of indigenous soil microbes (Hill et al. 2000). These rhizobacteria can be co-inoculated with other fungi or bacteria for their synergistic response. Generally, a bio-control is consisting of a mixture of antagonists rather than a single one for effective outcome. These are used as inoculants in biofertilizers, phytostimulators, and bioremediators. Their mechanism of suppression involves either releasing harmful chemicals against the pathogen to restrict their growth or by in taking all the nutrition available in their surrounding and letting the pathogens die with hunger. Some of the very common and abundant PGPRs in soil are *Pseudomonas*, *Flavobacterium*, *Rhizobium*, *Pseudomonas*, *Azotobacter*, *Bacillus*, etc. They are widely used as biocontrol agents for suppressing pathogenic diseases. The diseases like common leaf spot, *Fusarium* root rot, anthracnose, and rust caused by *Pseudomonas syringae pv. syringae*, *Fusarium oxysporum*, *Colletotrichum* sp., and *Uromyces striatus*, respectively, (Nyvall 2013) are certain examples of diseases which affect many crop plants such as legumes like horse gram (*Macrotyloma uniflorum*), tomato (*Solanum lycopersicum*), potato (*Solanum tuberosum*), and fruits like banana, etc. In evidence, activity of GBO3 (*Bacillus subtilis*) and PGPR strains INR 7 (*Bacillus pumilus*) were observed against *P. syringae*, *C. orbiculare*, and *E. tracheiphila* (causing cucurbit wilt disease) (Raupach and Kloepper 1998). *Bacillus velezensis* isolates (Y6 and F7) have potential antagonist activity against *Fusarium oxysporum* and *R. solanacearum* (Cao et al. 2018). *B. amyloliquefaciens* DGA14 showed antagonism against *Colletotrichum gloeosporioides* (causing anthracnose) in mango (Alvindia and Acda 2015), etc. This review is an effort to focus on the urgent need of bio-control practices and explains the rhizosphere manipulation benefits on plant growth with recent updates (Fig. 5.1).

5.3 The Rhizosphere Microbiome

The rhizosphere microbiome consists of large number of actively metabolizing soil microbial communities; they enhance plant health by either providing nutrition, playing as an antagonist against harmful pathogens, producing siderophores, HCN, or many other direct and indirect ways beyond our imagination. This is the reason why scientists are attracting toward this hotspot concept of plant and microbial interactions. The rhizosphere microbiome of each plant varies from species to species according to their requirement. The rooted plant soil contains a large number of microorganisms as compared to the non-rooted ones (Foster et al. 1983). This can be explained with the help of the concept of the "rhizospheric effect." According to this concept, the growth and the increase in the number of microorganisms within the rhizosphere depends on the excretions and the organic carbon released from the roots of a plant (Brahmaprakash et al. 2017). The microorganisms not only intake the nutrients from the plant roots but also help them in stimulating plant growth and health. For example, rhizobacteria help plant by fixing atmospheric nitrogen and improve soil nitrogen deficiency. Symbionts like mycorrhizal fungi translocate minerals and nutrients from soil to the plant (Johnson and Graham 2013) and suppress the soil-borne plant pathogens (Whipps 2001), and these are well

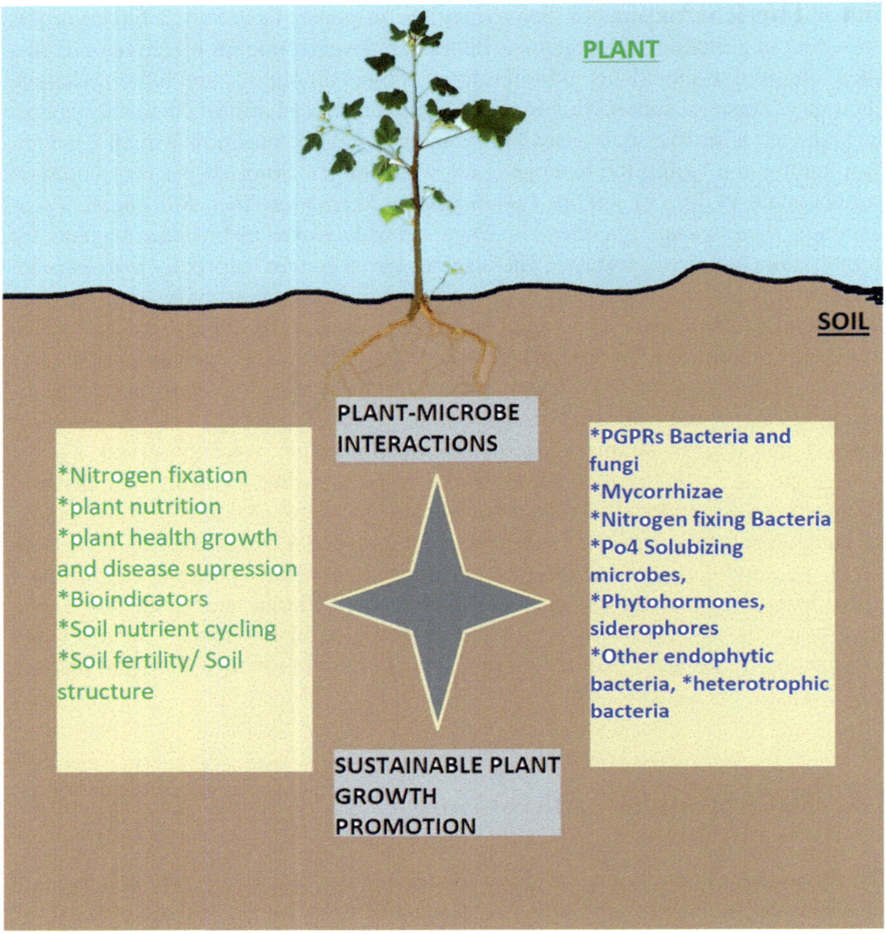

Fig. 5.1 Benefits of plant-microbe interactions to plant

documented and explained (Johnson and Nielsen 2012). Bacteria like *Burkholderia*, *Rhizobium*, *Bradyrhizobium*, and *Achromobacter* species can form a nodule on cowpea and perform effective nitrogen (Guimarães et al. 2012). Some bacteria regulate the concentrations of intracellular iron, in plants by releasing siderophores (Andrews et al. 2003; Hider and Kong 2010). *R. arrhizus* produces a siderophore called rhizoferrin that is a carrier of iron much efficient than synthetic chelates (Yehuda et al. 2000). Some rhizobacteria produce hormones for plant growth. These rhizobacteria are often termed as PGPR.

5.4 The Rhizosphere Microbiome-Mysterious Members?

Does complete rhizosphere microbiota have been identified? Do plants require only the known bacteria? Does removing the uncultured microbiota would affect plant growth adversely? These are certain questions which are still needed to be answered. However, some of the evidences from culture-independent methods have shown that the diverse microbial population of plant root soils has been highly underestimated for a long period. Till now only up to 5% of bacteria phyla have been cultured; still a large proportion is unculturable, and different approach is required for them to culture. In a soil metagenomic study, the rarefaction curve, prepared from 16SrRNA gene sequencing data, got failed to reach saturation. From that, the soil clone library was made by taking 150,000 sequencing reads, less than 1% among them exhibited overlapping with the sequencing reads of other independent clone libraries (Tringe et al. 2005). In a study, 1 g of boreal forest soil contains c. 10,000 number of bacteria sp. (Torsvik et al. 2002). Further, the researchers showed that 1 gram of soil contains >one million different bacteria, much higher compared to the previous researches (Gans et al. 2005). In a study from four distinct soils, around 9340 crenarchaeotal and 139,819 bacterial rRNA gene sequences were obtained. Alphaproteobacteria, Betaproteobacteria, and Bacteroidetes were abundantly present bacteria group among them (Roesch et al. 2007). Till now, in the field of rhizosphere studies, scientists work more on the diversity and number of bacteria populations and less on other rhizosphere inhabitants (Mendes et al. 2013).

5.5 Rhizosphere Microbiome Support Over Abiotic and Biotic Stresses

Plants are affected largely by many abiotic and biotic stresses in their complete life cycle. This directly or indirectly triggers the recruitment of specific microorganisms to the rhizosphere microbiome. For example, plants intake phosphorous only in the form of inorganic orthophosphate, present in the soil which activates phosphate starvation responses (PSRs) thus causing the synthesis of organic acids, glucosinolates (Castrillo et al. 2017), etc. which attracts the specific microbial population to fulfill the phosphorous deficiency. The bacteria belongs to *Bacillus decolorationis*, *Halobacillus* [uncultured], and *Cesiribacter* sp. JJ02 isolated from the atrazine contaminated soil (Xu et al. 2018). Similarly, glyphosate sprayed on maize leaves causes enrichment of *Fusarium* around the maize roots (Kremer and Means 2009). Diclofop-methyl treatment causes excess release of amino acids, fatty acids, and organic acids in rice seedlings ultimately enriching the population of *Massilia* and *Anderseniella* genera (Qian et al. 2018). There are many more environmental contaminants that alter the rhizosphere community either directly with their presence or causing root exudates to release certain molecules. Moreover, plant pathogen also sometimes causes the rhizosphere alteration, for example,

Pseudomonas syringae pv tomato enriches the presence of *Roseiflexus* genus in soil (Yuan et al. 2018). Similarly, *Fusarium culmorum* infection causes the enrichment of *Carex arenaria* in plant roots (Schulz-Bohm et al. 2018). The rhizosphere microbiome gives support to the plant against various types of stresses, and it is also the primary defense system to the plants against soil-borne pathogens. The important mechanisms against the biotic stresses involves competition for food, nutrients (Maloy and Murray 2001), antibiosis (Raaijmakers and Mazzola 2012), ISR (Pieterse 2012), and parasitism (Mela et al. 2011). A study revealed that *Paenibacillus lentimorbus* and *Paenibacillus polymyxa* have antifungal activities against *Meloidogyne incognita* and *Fusarium oxysporum* pathogens of tomato plants (Son et al. 2009). Rhizosphere microbiome members not only suppress the pathogens and enhance the plant growth but also can modulate the plant immune system (De Vleesschauwer and Höfte 2009; Berendsen et al. 2012). *Achromobacter piechaudii* ARV8, obtained from saline soil environment, enhanced the pepper biomass and tomato seedlings under drought stress (Mayak et al. 2004a, b). Rhizobacteria supports plant growth in flooding stress (Grichko and Glick 2001) (Upadhyay et al. 2009). revealed that about 24 rhizobacterial isolates out of 130 isolates were 8% tolerant to high salt concentration levels. Some of them produced siderophores, some produced gibberellins, and some were able to solubilize phosphorous. The dominant genus found was *Bacillus*. There are many more evidences to prove that rhizosphere microbiome members are primarily responsible for providing support to the plant from every day stresses.

5.6 Manipulating the Rhizosphere for Sustainable Plant Growth Promotion

The rhizosphere is among the most complex habitat on the earth. It is a large reservoir of plant roots, soil, and diverse microorganisms including bacteria, fungi, viruses, etc. Many studies have been done in this field are still much more needed to be explored. Plants have co-evolved with microbes. Plant domestication and crop breeding fertilization practices have created certain disturbances among the rhizosphere microbiome which has caused loss of important interactions. Thus, by re-engineering the beneficial rhizosphere microbiome to the agricultural cropping systems will work in getting back the lost microbe-plant interactions. Predicting and then controlling the function and structure of rhizosphere will help in understanding the plant microbial interactions and also improve the plant response to the climate change effects and various other environmental stresses. Enhancing the plant health and growth is not only the problem to be tackled, but another big problem that affects the plant development is the diseases that are caused by harmful pathogens in crop plants. Using the chemical fertilizers makes them worse. Thus, for the sustainable plant growth, alternative method called biological control is required. Microbiome manipulations are a new approach for this problem. Probably, the first changes that occurred in rhizosphere microbiome assemblage were involuntary while performing plant domestication and breeding practices. Genetic alterations in plant caused the reduction of many beneficial microbial interactions (Weese et al. 2015). By manipulating the genetic structure of plants in the process of creating improved varieties, many plants do not support the beneficial microbe as much as their ancestors (Philippot et al. 2013). In case of wheat, landraces assemble more complex microbiome as compared to the modern cultivars (Rossmann et al. 2020). However, this is not only the case in some cases, but the domestication causes less or negligible impacts on the microbial communities. For example, in *Helianthus annuus* (common sunflower), the effects were shown in fungus only (Leff et al. 2017) (Schuhegger et al. 2006). explained that *Serratia liquefaciens* and *P. putida* produce the [N-acyl homoserine] AHL signal molecules in plant against the pathogen *Alternaria alternata*, by inducing the systemic resistance. Similar approach was done for *A. thaliana (*Liu et al. 2012*)*. In the rhizosphere of *Lolium perenne*, the effects of Pseudomonas on the phosphate availability were investigated (Zyśko et al. 2012). This shows that the root exudates affect the bacterial gene expression. The gene expression of hydrogen cyanide (HCN) biosynthesis in *Pseudomonas* sp. LBUM300 got increased by the infection caused by *Verticillium dahlia* on strawberry plant rhizosphere (DeCoste et al. 2010). Thus, biocontrol was stimulated. There are many more evidences to prove that manipulating the rhizosphere microbiome structure can solve problems related to the plant growth. Till now a large proportion of soil microbiome is unculturable; yet metagenomics is a field for providing a complete picture to the rhizosphere microbiome. Exposing these microorganisms and their interactions with plant root will give a better insight toward the crop yield and health in adverse environmental conditions which are the major problems of today's world.

5.7 Induction of Soil Suppressive Ability Against Plant Pathogens Through Alterations

The organic fertilizers are widely used to suppress the plant pathogens and enhance plant health in present days (Bonanomi et al. 2007; Cao et al. 2011). The soil suppressiveness against soil-borne pathogens can be induced by introducing the beneficial microbes in rhizosphere soil that release certain compounds that inhibit the growth of pathogen. Some of the pathogenic fungi that affect plant growth are *Verticillium dahlia*, *Sclerotinia* spp., *R. solani*, and *Fusarium* spp., and a pathogenic bacterium is *R. solanacearum (*Bonanomi et al. 2007*)*. Another interesting technique that is being used nowadays is solarization technique. The organic matter along with solarization technique helps in changing the biological structure of plant root soil. This technique induces the suppressiveness of soil and reduces the growth of some of the pathogens (Klein et al. 2011). The solar energy is used to raise the temperature of soil at that extent at which the soil pathogens are unable to survive or become weak to show any harmful effects. This technique is beneficial because it shows effective results without removing the soil microorganisms, by modifying the soil environment in the support of useful microbial community (Raaijmakers et al. 2009). Next strategy involves the use of synthetic biology by applying probiotics in plants. The microbiome is manipulated in this strategy. Vegetal probiotics costs low and increases growth and production more (Kim and Anderson 2018; Picard et al. 2008). The synergistic characteristics of different microorganisms in the consortium have more potential as compared to the isolated microbe, thus increasing biocontrol efficiency in consortium. Different microbes in the consortium use different niches and thus prevent competition among them. For example, *Bradyrhizobium japonicum* when inoculated with *Azospirillum brasilense* enhance the stabilization of arsenic (Armendariz et al. 2019). Similarly, *R. tropici* along with *C. balustinum* enhance common bean growth in normal and saline conditions (Estévez et al. 2009). Many more evidences have been given in various studies that prove that the microbiome plays an important role in raising the soil disease suppressiveness (Srivastava et al. 2010). Many companies are using the products made based on mixture of *Trichoderma* and *Bacillus* species (Agrofit 2020). Another important strategy is based on controlling the microbiome functions in which the regulation of antibiotics, gene, and their expressions are manipulated through signaling introduction of required molecules (Bakker et al. 2012). The organisms' abilities are studied for developing selective enrichment techniques to be used for long-term manipulations (Bakker et al. 2012).

5.8 Effects of PGPR Inoculation on Resident Rhizosphere Microbes

The rhizosphere is a complex habitat consists of plant roots, biotic components like resident microorganisms, and abiotic soil components. These all function together for plant growth. Any disturbance can change the whole scenario. The PGPRs are exogenous bacteria that are inoculated deliberately to the new environment for better performance of plants. It is necessary to check their interactions with the plant and their effects on already present resident bacteria. They must be inoculated properly. Large amount of PGPR inoculation can affect the functioning of resident bacteria; some groups may be enhanced, some may be inhibited, while some may not be affected at all (Dobbelaere et al. 2003; Nacamulli et al. 1997). PGPR *Azospirillum brasilense* inoculation in maize caused root volume increase as compared to the uninoculated maize plant (Dobbelaere et al. 2003). PGPR *Pseudomonas fluorescens* F113Rif (producing 2,4-diacetylphoroglucinol) inoculation causes the reduction in rhizobium microbial diversity (Walsh et al. 2003) (Robleto et al. 1998). reported that *Rhizobium etli* (producing trifolitoxin) inoculation reduced the population of trifolitoxin-sensitive bacteria of the Alphaproteobacteria. The PGPR indirectly affects the resident microorganisms by releasing some chemicals or antimicrobial compounds. In evidence, *Bacillus subtilis* strain 330–2 produces some lytic enzymes like laminarase, cellulase, etc. that inhibit *Rhizoctonia solani Kuhn* growth by degrading the pathogenic fungi cell wall (Ahmad et al. 2017). *B. subtilis* strains PCL1612 secrete iturin A to resist *Rosellinia necatrix* and *Fusarium oxysporum* (Cazorla et al. 2007). *B. amyloliquefaciens* and *B. mojavensis* release antifungal compounds that inhibit the growth of *F. oxysporum* and *F. verticillioides*, respectively (Wu et al. 2019).

5.9 The Process of Using Synthetic Communities for Plant Promotion

The microbial consortia successfully inoculated to enhance plants health were firstly tested in labs according to their potentials and then were transferred to the field (Sessitsch et al. 2019). This is based on two strategies: either the closely related strains are tailored for increasing diverse resources that the strains would use (Wei et al. 2015) or the distant-related microbes combined to suppress pathogens, tolerate diverse plant genotypes, and promote plant growth (Compant et al. 2019). In evidence, when the *Pseudomonas consortia was introduced in tomato plant soil rhizosphere, it reduced the growth of R. solanacearum suppressing the disease in tomato plants* (Hu et al. 2018). *Synthetic community made up of Bacillus isolates enhances tomato growth* (Tsolakidou et al. 2019). *The complex inoculation of different bacteria has shown effectively increased plant health and growth as compared to the single strains* (Tsolakidou et al. 2019; Niu et al. 2017). Constructing a large synthetic community and managing each member of the mixture is quite a difficult task to perform. A machine learning computational approach has been developed recently, based on "cry for help" theory to design the bacterial synthetic community. This method is useful for creating a community with predictable plant phenotypes. Another method developed by (Berendsen et al. 2018) was more plant-dependent approach based on the same theory, in which plant attracts bacteria consortium to produce desirable plant phenotypes. This experiment showed that the microbes that show response on plant stress signals can be used as reliable predictors for exploring useful microbes (Pascale et al. 2020). Basically, the assembling of rhizospheric microbiota is a result of series of events that can produce stable and diversified microbial consortia (Philippot et al. 2013). The factors like environmental (abiotic) and interactions among microbes and plants (biotic) act as modulators in the process of assemblage. Competitive and cooperative interactions promote diversity and microbial co-occurrence and determine the invasive pathogen fate (Hacquard et al. 2017).

5.10 Framing the Artificial Rhizosphere Microbiome

The rhizosphere is a large pool of microorganisms which include a diverse number of bacterial and fungal populations. They may be beneficial or harmful for plants or may change their activities with the environment. Thus, it is required to develop the new strategies for framing the rhizosphere microbiome of a plant for growth promotion. Synthetic or artificial microbial community is a new approach in this perspective. It is a method in which the pure strains are transferred into an axenic system mimicking the natural system to assemble microbial communities (Vorholt et al. 2017). Researchers can study the rhizosphere environment in controlled conditions according to their needs, and they can add, remove, or replace organisms at the strain level according to the needs. A study showed that the phosphate starvation mutants and the wild-type *Arabidopsis thaliana* had a different root microbiome in natural soils as well as in synthetic community experiments (Castrillo et al. 2017). Plant-microbe interaction can be understood more easily in synthetic community experiments than the natural process. In evidence, under the same nitrogen condition, indica-enriched synthetic microbial community promotes the rice growth much more than the japonica-enriched ones (Zhang et al. 2019). Thus, there is a lot more work to explore in the field of artificial rhizosphere microbiome field. Firstly, for constructing a rhizosphere microbiome, it is highly necessary to have an expert knowledge about the members in the microbiome. The advanced molecular and "omics" tools can help in better understanding the plant-microbe interactions. Those microbes which have a high capability of promoting plant growth in every aspect must be isolated, characterized, and assembled for further use. On the other hand, the microbes that are harmful for plant as well as human welfare must be avoided for the reconstructed microbiome. There are many bacterial and fungal populations that have been included as the beneficial ones like *Bacillus*, *Pseudomonas*, *Azotobacter*, *Flavobacterium*, etc. and fungus like *Trichoderma*. Many studies have suggested that the components that decide the rhizosphere microbiome shape are the soil type and plant genotype (Bakker et al. 2012). The plant roots along with their general roles of support and nutrient uptake also perform some specialized roles like synthesis, accumulation, and secretion of diverse array of compounds (Flores et al. 1999). Microbes get attracted toward the chemicals or excretions that the plant roots secrete. In evidence, the beneficial microbe *Bacillus subtilis* got attracted toward the malic-acid secretion from a plant (Rudrappa et al. 2008). *Pseudomonas putida* is attracted to DIMBOA, the allelochemical released from young maize seedling roots (Neal et al. 2012). It has been reported that *Arabidopsis thaliana* produces a swathe of chemicals belonging to a family triterpene, and about half of the microbes attracted towards its root are due to the network of these chemicals (Huang et al. 2019). Certain cases have been reported where some of the pathogens also get attracted toward the chemical secretions along with the beneficial bacteria, e.g., isoflavones released from soybean roots attract symbionts *B. japonicum* along with the pathogen *Phytophthora sojae* (Morris et al. 1998; Zhalnina et al. 2018). showed in a recent study that the root exudates released during the growth cycle of *Avena barbata* showed variations like high sucrose content at early age and more defense molecules and amino acids at later developmental stages. Later on, using exometabolomics, it was studied that the aromatic organic acids like salicylic, cinnamic, IAA, shikimic, and nicotinic cause the inhibition and proliferation of desired microbes during plant life cycle. The secreted molecules, Arabinogalactan proteins, regulate Agrobacterium and Rhizobium presence on roots (Xie et al. 2012). Thus, one can construct a rhizosphere microbiome according to one's choice keeping the genetic appearance and the nature of a plant in mind. The plant's genetic makeup can be altered with the help of genetic engineering which is another interesting topic to discuss.

5.11 Techniques Involved in Harnessing Plants and Engineer Beneficial Microbiomes

The concept of engineering microbiomes is a promising approach toward the plant health and fitness for sustainable future; however, it is not easy to apply. For overcoming the limitations, certain mechanisms and considerations have been given by various scientists (Bhatt et al. 2021g). One of the recent reviews has been given by (Lawson et al. 2019). Metatranscriptomics, metagenomics, metabolomics, and plant transcriptomics are the approaches that extricate the complexity in the connections happening among the holobiont members (Stringlis et al. 2018) combined the metabolomics and shotgun metagenomics on an order of Arabidopsis mutants and said that the rhizosphere microbiome can be reshaped by the coumarin root exudation. Similarly, the combined analysis of amplicon-based metagenomics and metabolomics was performed on wild and benzoxazinoid precursor mutant of maize genotype by (Hu et al. 2018). He explained that benzoxazinoid metabolites can structure fungal and bacterial community of maize rhizosphere microbiome. Another approach in this field is system biology approach which is based on the discovery of microbial associations with the help of correlation networks. For example, the antagonist and synergist interactions can be analyzed from negative and positive correlations in system biology approach (Poudel et al. 2016). With the help of this, the most interactive nodes in the network called microbial hub taxa can be identified (Pascale et al. 2020) (Agler et al. 2016). identified plant pathogens *Dioszegia and Albugo in* Arabidopsis phyllosphere microbiome as microbiome hub. Metagenome-wide association study [MWAS] is another very important and powerful approach nowadays (Sulbhi et al. 2021; Bhandari et al. 2021; Bhatt et al. 2020a, 2021a, b, c, d, e, f). It is helpful in identifying the associations between the several plant pathogens and the rhizosphere microbiome (Bhatt et al. 2015a; b, 2016a, b, 2019a, b, c, 2020b, c, d, e, f, Ahemad and Khan 2012; Huang et al. 2021).

5.12 Concluding Remarks

With the increase in concern about the chemicals used as fertilizers and their harmful effects over mankind and environment, it is no doubt to promote the ecofriendly practices. The relationships between the plant and the microbiome are co-evolved. There are many possibilities and missing links that must be expected. Rhizosphere manipulation is one of the best ecofriendly approaches. The PGPR used as inoculants are one of the members of soil microbiome from another plant and thus are easy and harmless to experiment which have already been proved by effective results obtained from various researches discussed above. The understanding of plant and microbes is still needed to be explored for more innovative findings regarding plant health and fitness. Most of the microbes in the soil are still not explored maybe they are holding some secrets regarding the mechanisms involved in plant-microbe associations. The knowledge of various fields like metabolomics, plant genetics, plant transcriptomics, metagenomics, and metatranscriptomics must be gained more for further development of techniques regarding plant and microbes harnessing. The synthetic microbial community construction is a novel approach, and the results are showing that it has a bright future in the field of rhizosphere microbiome.

References

Agler MT, Ruhe J, Kroll S, Morhenn C, Kim ST, Weigel D, Kemen EM (2016) Microbial hub taxa link host and abiotic factors to plant microbiome variation. PLoS Biol 14(1):e1002352

Agrofit (2020). http://agrofit.agricultura.gov.br/agrofit_cons/principal_agrofit_cons. Accessed 18 Mar 2020

Ahemad M, Khan MS (2012) Productivity of greengram in tebuconazole-stressed soil, by using a tolerant and plant growth-promoting Bradyrhizobium sp. MRM6 strain. Acta Physiol Plant 34 (1):245–254

Ahmad Z, Wu J, Chen L, Dong W (2017) Isolated *Bacillus subtilis* strain 330-2 and its antagonistic genes identified by the removing PCR. Sci Rep 7(1):1–3

Alvindia DG, Acda MA (2015) The antagonistic effect and mechanisms of *Bacillus amyloliquefaciens* DGA14 against anthracnose in mango cv. 'Carabao'. Biocontrol Sci Tech 25(5):560–572

Andrews SC, Robinson AK, Rodríguez-Quiñones F (2003) Bacterial iron homeostasis. FEMS Microbiol Rev 27(2–3):215–237

Armendariz AL, Talano MA, Nicotra MF, Escudero L, Breser ML, Porporatto C, Agostini E (2019) Impact of double inoculation with *Bradyrhizobium japonicum* E109 and *Azospirillum brasilense* Az39 on soybean plants grown under arsenic stress. Plant Physiol Biochem 138:26–35

Bakker MG, Manter DK, Sheflin AM, Weir TL, Vivanco JM (2012) Harnessing the rhizosphere microbiome through plant breeding and agricultural management. Plant and Soil 360(1):1–3

Barea JM, Pozo MJ, Azcon R, Azcon-Aguilar C (2005) Microbial co-operation in the rhizosphere. J Exp Bot 56(417):1761–1778

Berendsen RL, Pieterse CM, Bakker PA (2012) The rhizosphere microbiome and plant health. Trends Plant Sci 17(8):478–486

Berendsen RL, Vismans G, Yu K, Song Y, de Jonge R, Burgman WP, Burmølle M, Herschend J, Bakker PA, Pieterse CM (2018) Disease-induced assemblage of a plant-beneficial bacterial consortium. ISME J 12(6):1496–1507

Bhandari G, Bagheri AR, Bhatt P, Bilal M (2021) Occurrence, potential ecological risks, and degradation of endocrine disruptor, nonylphenol from the aqueous environment. Chemosphere 275:130013. https://doi.org/10.1016/j.chemosphere.2021.130013

Bhatt P, Bisht TS, Pathak VM, Barh A, Chandra D (2015a) Optimization of amylase production from the fungal isolates of Himalayan region Uttarakhand. Ecol Environ Conserv 21 (3):1517–1521

Bhatt P, Negi G, Gangola S, Khati P, Srivastava A, Sharma A (2015b) Optimization of sulfosulfuron biodegradation through response surface methodology using indigenous bacterial strain isolated from contaminated agriculture field. Int J Curr Microbiol App Sci 4(8):105–112

Bhatt P, Negi G, Gangola S, Khati P, Kumar G, Srivastava A, Sharma A (2016a) Differential expression and characterization of cypermethrin degrading potential proteins in *Bacillus thuringiensis* strain, SG4. 3 Biotech 6:225

Bhatt P, Sharma A, Gangola S, Khati P, Kumar G, Srivastava A (2016b) Novel pathway of cypermethrin biodegradation in *a Bacillus* sp. strain SG2 isolated from cypermethrin-contaminated agriculture field. 3 Biotech 6:45

Bhatt P, Huang Y, Zhan H, Chen S (2019a) Insight into microbial applications for the biodegradation of Pyrethroid insecticides. Front Microbiol 10:1778

Bhatt P, Pal K, Bhandari G, Barh A (2019b) Modeling of methyl halide biodegradation on bacteria and its effect on other environmental systems. Pestic Biochem Physiol 158:88–100

Bhatt P, Gangola S, Chaudhary P, Khati P, Kumar G, Sharma A, Srivastava A (2019c) Pesticide induced up-regulation of esterase and aldehyde dehydrogenase in indigenous *Bacillus* spp. Biorem J 23(1):42–52

Bhatt P, Rene ER, Kumar AJ, Kumar AJ, Zhang W, Chen S (2020a) Binding interaction of allethrin with esterase: bioremediation potential and mechanism. Bioresour Technol 315:13845. https://doi.org/10.1016/j.biortech.2020.123845

Bhatt P, Verma A, Verma S, Anwar MS, Prasher P, Mudila H, Chen S (2020b) Understanding phytomicrobiome: a potential reservoir for better crop management. Sustainability 12:5446

Bhatt P, Huang Y, Rene ER, Kumar AJ, Chen S (2020c) Mechanism of allethrin biodegradation by a newly isolated *Sphingomonas trueperi* strain CW3 from wastewater sludge. Bioresour Technol 305:123074

Bhatt P, Zhang W, Lin Z, Pang S, Huang Y, Chen S (2020d) Biodegradation of allethrin by a novel fungus *Fusarium proliferatum* strain CF2, isolated from contaminated. Soils Microorganisms 8:593

Bhatt P, Huang Y, Zhang W, Sharma A, Chen S (2020e) Enhanced Cypermethrin degradation kinetics and metabolic pathway in *Bacillus* thuringiensis strain, SG4. Microorganisms 8:223

Bhatt P, Bhatt K, Huang Y, Ziqiu L, Chen S (2020f) Esterase is a powerful tool for the biodegradation of pyrethroid insecticides. Chemosphere 244:125507

Bhatt P, Zhou X, Huang Y, Zhang W, Chen S (2021a) Characterization of the role of esterases in the biodegradation of organophosphate, carbamate and pyrethroid group pesticides. J Hazard Mater 411:125026

Bhatt P, Joshi T, Bhatt K, Zhang W, Huang Y, Chen S (2021b) Binding interaction of glyphosate oxidoreductase and C-P lyase: molecular docking and molecular dynamics simulation studies. J Hazard Mater 409:124927

Bhatt P, Bhatt K, Sharma A, Zhang W, Mishra S, Chen S (2021c) Biotechnological basis of microbial consortia for the removal of pesticides from the environment. Crit Rev Biotechnol 41(3):317–338. https://doi.org/10.1080/07388551.2020.1853032

Bhatt P, Sethi K, Gangola S, Bhandari G, Verma A, Adnan M, Singh Y, Chaube S (2021d) Modeling and simulation of atrazine biodegradation in bacteria and its effect in other living systems. J Biomol Struct Dyn:1–11. https://doi.org/10.1080/07391102.2020.1846623

Bhatt P, Gangola S, Bhandari G, Zhang W, Maithani D, Mishra S, Chen S (2021e) New insights into the degradation of synthetic pollutants in contaminated environments. Chemosphere 268:128827. https://doi.org/10.1016/j.chemosphere.2020.128827

Bhatt P, Sharma A, Rene ER, Kumar AJ, Zhang W, Chen S (2021f) Bioremediation mechanism, kinetics of fipronil degradation using *Bacillus* sp. FA3 and resource recovery potential from contaminated environments. J Water Process Eng 39:101712

Bhatt P, Bhandari G, Bhatt K, Maithani D, Mishra S, Gangola S, Bhatt R, Huang Y, Chen S (2021g) Plasmid-mediated catabolism for the removal of xenobiotics from the environment. J Haz Mat 420:126618

Bonanomi G, Antignani V, Pane C, Scala F (2007) Suppression of soilborne fungal diseases with organic amendments. J Plant Pathol 1:311–324

Brahmaprakash GP, Sahu PK, Lavanya G, Nair SS, Gangaraddi VK, Gupta A (2017) Microbial functions of the rhizosphere. In: Plant-microbe interactions in agro-ecological perspectives. Springer, Berlin, pp 177–210

Cao Y, Zhang Z, Ling N, Yuan Y, Zheng X, Shen B, Shen Q (2011) *Bacillus subtilis* SQR 9 can control Fusarium wilt in cucumber by colonizing plant roots. Biol Fertil Soils 47(5):495–506

Cao Y, Pi H, Chandrangsu P, Li Y, Wang Y, Zhou H, Xiong H, Helmann JD, Cai Y (2018) Antagonism of two plant-growth promoting *Bacillus velezensis* isolates against *Ralstonia solanacearum* and *Fusarium oxysporum*. Sci Rep 8(1):1–4

Castrillo G, Teixeira PJ, Paredes SH, Law TF, de Lorenzo L, Feltcher ME, Finkel OM, Breakfield NW, Mieczkowski P, Jones CD, Paz-Ares J (2017) Root microbiota drive direct integration of phosphate stress and immunity. Nature 543(7646):513–518

Cazorla FM, Romero D, Pérez-García A, Lugtenberg BJ, Vicente AD, Bloemberg G (2007) Isolation and characterization of antagonistic *Bacillus subtilis* strains from the avocado rhizoplane displaying biocontrol activity. J Appl Microbiol 103(5):1950–1959

Compant S, Samad A, Faist H, Sessitsch A (2019) A review on the plant microbiome: ecology, functions, and emerging trends in microbial application. J Adv Res 19:29–37

De Vleesschauwer D, Höfte M (2009) Rhizobacteria-induced systemic resistance. Adv Bot Res 51:223–281

DeCoste NJ, Gadkar VJ, Filion M (2010) Verticillium dahliae alters Pseudomonas spp. populations and HCN gene expression in the rhizosphere of strawberry. Can J Microbiol 56(11):906–915

Dobbelaere S, Vanderleyden J, Okon Y (2003) Plant growth-promoting effects of diazotrophs in the rhizosphere. Crit Rev Plant Sci 22(2):107–149

Estévez J, Dardanelli MS, Megías M, Rodríguez-Navarro DN (2009) Symbiotic performance of common bean and soybean co-inoculated with rhizobia and *Chryseobacterium balustinum* Aur9 under moderate saline conditions. Symbiosis 49(1):29–36

Flores HE, Vivanco JM, Loyola-Vargas VM (1999) 'Radicle' biochemistry: the biology of root-specific metabolism. Trends Plant Sci 4(6):220–226

Foster RC, Rovira AD, Cock TW (1983) Ultrastructure of the root-soil interface. American Phytopathological Society, Saint Paul, MN

Gans J, Wolinsky M, Dunbar J (2005) Computational improvements reveal great bacterial diversity and high metal toxicity in soil. Science 309(5739):1387–1390

Glick BR (2012) Plant growth-promoting bacteria: mechanisms and applications. Scientifica 2012:963401

Grichko VP, Glick BR (2001) Amelioration of flooding stress by ACC deaminase-containing plant growth-promoting bacteria. Plant Physiol Biochem 39(1):11–17

Guimarães AA, Jaramillo PM, Nóbrega RS, Florentino LA, Silva KB, de Souza Moreira FM (2012) Genetic and symbiotic diversity of nitrogen-fixing bacteria isolated from agricultural soils in the western Amazon by using cowpea as the trap plant. Appl Environ Microbiol 78(18):6726–6733

Hacquard S, Spaepen S, Garrido-Oter R, Schulze-Lefert P (2017) Interplay between innate immunity and the plant microbiota. Annu Rev Phytopathol 55:565–589

Hartmann A, Rothballer M, Schmid M (2008) Lorenz Hiltner, a pioneer in rhizosphere microbial ecology and soil bacteriology research. Plant and Soil 312:1–2

Hider RC, Kong X (2010) Chemistry and biology of siderophores. Nat Prod Rep 27(5):637–657
Hill GT, Mitkowski NA, Aldrich-Wolfe L, Emele LR, Jurkonie DD, Ficke A, Maldonado-Ramirez-S, Lynch ST, Nelson EB (2000) Methods for assessing the composition and diversity of soil microbial communities. Appl Soil Ecol 15(1):25–36
Hu L, Robert CA, Cadot S, Zhang X, Ye M, Li B, Manzo D, Chervet N, Steinger T, Van Der Heijden MG, Schlaeppi K (2018) Root exudate metabolites drive plant-soil feedbacks on growth and defense by shaping the rhizosphere microbiota. Nat Commun 9(1):1–3
Huang AC, Jiang T, Liu YX, Bai YC, Reed J, Qu B, Goossens A, Nützmann HW, Bai Y, Osbourn A (2019) A specialized metabolic network selectively modulates Arabidopsis root microbiota. Science 364(6440)
Huang Y, Zhang W, Pang S, Chen J, Bhatt P, Mishra S, Chen S (2021) Insights into the microbial degradation and catalytic mechanism of chlorpyrifos. Environ Res 194:110660
Johnson NC, Graham JH (2013) The continuum concept remains a useful framework for studying mycorrhizal functioning. Plant and Soil 363(1):411–419
Johnson SN, Nielsen UN (2012) Foraging in the dark–chemically mediated host plant location by belowground insect herbivores. J Chem Ecol 38(6):604–614
Kim YC, Anderson AJ (2018) Rhizosphere pseudomonads as probiotics improving plant health. Mol Plant Pathol 19(10):2349–2359
Klein E, Katan J, Gamliel A (2011) Soil suppressiveness to Fusarium disease following organic amendments and solarization. Plant Dis 95(9):1116–1123
Kremer RJ, Means NE (2009) Glyphosate and glyphosate-resistant crop interactions with rhizosphere microorganisms. Eur J Agron 31(3):153–161
Lawson CE, Harcombe WR, Hatzenpichler R, Lindemann SR, Löffler FE, O'Malley MA, Martín HG, Pfleger BF, Raskin L, Venturelli OS, Weissbrodt DG (2019) Common principles and best practices for engineering microbiomes. Nat Rev Microbiol 17(12):725–741
Leff JW, Lynch RC, Kane NC, Fierer N (2017) Plant domestication and the assembly of bacterial and fungal communities associated with strains of the common sunflower, *Helianthus annuus*. New Phytol 214(1):412–423
Liu F, Bian Z, Jia Z, Zhao Q, Song S (2012) The GCR1 and GPA1 participate in promotion of Arabidopsis primary root elongation induced by N-acyl-homoserine lactones, the bacterial quorum-sensing signals. Mol Plant Microbe Interact 25(5):677–683
Lugtenberg BJ, Dekkers L, Bloemberg GV (2001) Molecular determinants of rhizosphere colonization by Pseudomonas. Annu Rev Phytopathol 39(1):461–490
Maloy OC, Murray TD (2001) Encyclopedia of plant pathology. Wiley, Hoboken, NJ
Mayak S, Tirosh T, Glick BR (2004a) Plant growth-promoting bacteria that confer resistance to water stress in tomatoes and peppers. Plant Sci 166(2):525–530
Mayak S, Tirosh T, Glick BR (2004b) Plant growth-promoting bacteria confer resistance in tomato plants to salt stress. Plant Physiol Biochem 42(6):565–572
Mela F, Fritsche K, De Boer W, Van Veen JA, De Graaff LH, Van Den Berg M, Leveau JH (2011) Dual transcriptional profiling of a bacterial/fungal confrontation: *Collimonas fungivorans* versus *Aspergillus Niger*. ISME J 5(9):1494–1504
Mendes R, Garbeva P, Raaijmakers JM (2013) The rhizosphere microbiome: significance of plant beneficial, plant pathogenic, and human pathogenic microorganisms. FEMS Microbiol Rev 37 (5):634–663
Morris PF, Bone E, Tyler BM (1998) Chemotropic and contact responses of *Phytophthora sojae* hyphae to soybean isoflavonoids and artificial substrates. Plant Physiol 117(4):1171–1178
Nacamulli C, Bevivino A, Dalmastri C, Tabacchioni S, Chiarini L (1997) Perturbation of maize rhizosphere microflora following seed bacterization with *Burkholderia cepacia* MCI 7. FEMS Microbiol Ecol 23(3):183–193
Neal AL, Ahmad S, Gordon-Weeks R, Ton J (2012) Benzoxazinoids in root exudates of maize attract *Pseudomonas putida* to the rhizosphere. PLoS One 7(4):e35498
Niu B, Paulson JN, Zheng X, Kolter R (2017) Simplified and representative bacterial community of maize roots. Proc Natl Acad Sci 114(12):E2450–E2459

Nyvall RF (ed) (2013) Field crop diseases handbook. Springer Science & Business Media, Berlin
Pascale A, Proietti S, Pantelides IS, Stringlis IA (2020) Modulation of the root microbiome by plant molecules: the basis for targeted disease suppression and plant growth promotion. Front Plant Sci 10:1741
Philippot L, Raaijmakers JM, Lemanceau P, Van Der Putten WH (2013) Going back to the roots: the microbial ecology of the rhizosphere. Nat Rev Microbiol 11(11):789–799
Picard C, Baruffa E, Bosco M (2008) Enrichment and diversity of plant-probiotic microorganisms in the rhizosphere of hybrid maize during four growth cycles. Soil Biol Biochem 40(1):106–115
Pieterse CMJ (2012) Prime time for transgenerational defense. Plant Physiol 158:545
Poudel R, Jumpponen A, Schlatter DC, Paulitz TC, Gardener BM, Kinkel LL, Garrett KA (2016) Microbiome networks: a systems framework for identifying candidate microbial assemblages for disease management. Phytopathology 106(10):1083–1096
Qian H, Zhu Y, Chen S, Jin Y, Lavoie M, Ke M, Fu Z (2018) Interacting effect of diclofop-methyl on the rice rhizosphere microbiome and denitrification. Pestic Biochem Physiol 146:90–96
Qiang X, Weiss M, KOGEL KH, Schäfer P (2012) *Piriformospora indica*—a mutualistic basidiomycete with an exceptionally large plant host range. Mol Plant Pathol 13(5):508–518
Raaijmakers JM, Mazzola M (2012) Diversity and natural functions of antibiotics produced by beneficial and plant pathogenic bacteria. Annu Rev Phytopathol 50:403–424
Raaijmakers JM, Paulitz TC, Steinberg C, Alabouvette C, Moënne-Loccoz Y (2009) The rhizosphere: a playground and battlefield for soil borne pathogens and beneficial microorganisms. Plant and Soil 321(1):341–361
Raupach GS, Kloepper JW (1998) Mixtures of plant growth-promoting rhizobacteria enhance biological control of multiple cucumber pathogens. Phytopathology 88(11):1158–1164
Reinhold-Hurek B, Bünger W, Burbano CS, Sabale M, Hurek T (2015) Roots shaping their microbiome: global hotspots for microbial activity. Annu Rev Phytopathol 53:403–424
Robleto EA, Borneman J, Triplett EW (1998) Effects of bacterial antibiotic production on rhizosphere microbial communities from a culture-independent perspective. Appl Environ Microbiol 64(12):5020–5022
Roesch LF, Fulthorpe RR, Riva A, Casella G, Hadwin AK, Kent AD, Daroub SH, Camargo FA, Farmerie WG, Triplett EW (2007) Pyrosequencing enumerates and contrasts soil microbial diversity. ISME J 1(4):283–290
Rossmann M, Pérez-Jaramillo JE, Kavamura VN, Chiaramonte JB, Dumack K, Fiore-Donno AM, Mendes LW, Ferreira MM, Bonkowski M, Raaijmakers JM, Mauchline TH (2020) Multitrophic interactions in the rhizosphere microbiome of wheat: from bacteria and fungi to protists. FEMS Microbiol Ecol 96(4):fiaa032
Rudrappa T, Czymmek KJ, Paré PW, Bais HP (2008) Root-secreted malic acid recruits beneficial soil bacteria. Plant Physiol 148(3):1547–1556
Schuhegger R, Ihring A, Gantner S, Bahnweg G, Knappe C, Vogg G, Hutzler P, Schmid M, Van Breusegem F, Eberl LE, Hartmann A (2006) Induction of systemic resistance in tomato by N-acyl-L-homoserine lactone-producing rhizosphere bacteria. Plant Cell Environ 29 (5):909–918
Schulz-Bohm K, Gerards S, Hundscheid M, Melenhorst J, de Boer W, Garbeva P (2018) Calling from distance: attraction of soil bacteria by plant root volatiles. ISME J 12(5):1252–1262
Sessitsch A, Pfaffenbichler N, Mitter B (2019) Microbiome applications from lab to field: facing complexity. Trends Plant Sci 24(3):194–198
Son SH, Khan Z, Kim SG, Kim YH (2009) Plant growth-promoting rhizobacteria, *Paenibacillus polymyxa* and *Paenibacillus lentimorbus* suppress disease complex caused by root-knot nematode and fusarium wilt fungus. J Appl Microbiol 107(2):524–532
Srivastava R, Khalid A, Singh US, Sharma AK (2010) Evaluation of arbuscular mycorrhizal fungus, fluorescent Pseudomonas and *Trichoderma harzianum* formulation against *Fusarium oxysporum* f. sp. lycopersici for the management of tomato wilt. Biol Control 53(1):24–31

Stringlis IA, Yu K, Feussner K, De Jonge R, Van Bentum S, Van Verk MC, Berendsen RL, Bakker PA, Feussner I, Pieterse CM (2018) MYB72-dependent coumarin exudation shapes root microbiome assembly to promote plant health. Proc Natl Acad Sci 115(22):E5213–E5222

Sulbhi V, Bhatt P, Verma A, Mudila H, Prasher P, Rene ER (2021) Microbial technologies for heavy metal remediation: effect of process conditions and current practices. Clean Technol Environ Policy. https://doi.org/10.1007/s10098-021-02029-8

Torsvik V, Øvreås L, Thingstad TF (2002) Prokaryotic diversity--magnitude, dynamics, and controlling factors. Science 296(5570):1064–1066

Tringe SG, Von Mering C, Kobayashi A, Salamov AA, Chen K, Chang HW, Podar M, Short JM, Mathur EJ, Detter JC, Bork P (2005) Comparative metagenomics of microbial communities. Science 308(5721):554–557

Tsolakidou MD, Stringlis IA, Fanega-Sleziak N, Papageorgiou S, Tsalakou A, Pantelides IS (2019) Rhizosphere-enriched microbes as a pool to design synthetic communities for reproducible beneficial outputs. FEMS Microbiol Ecol 95(10):fiz138

Upadhyay SK, Singh DP, Saikia R (2009) Genetic diversity of plant growth promoting rhizobacteria isolated from rhizospheric soil of wheat under saline condition. Curr Microbiol 59(5):489–496

Vorholt JA, Vogel C, Carlström CI, Müller DB (2017) Establishing causality: opportunities of synthetic communities for plant microbiome research. Cell Host Microbe 22(2):142–155

Walsh UF, Moënne-Loccoz Y, Tichy HV, Gardner A, Corkery DM, Lorkhe S, O'gara F (2003) Residual impact of the biocontrol inoculant *Pseudomonas fluorescens* F113 on the resident population of rhizobia nodulating a red clover rotation crop. Microb Ecol 45(2):145–155

Weese DJ, Heath KD, Dentinger BT, Lau JA (2015) Long-term nitrogen addition causes the evolution of less-cooperative mutualists. Evolution 69(3):631–642

Wei Z, Yang T, Friman VP, Xu Y, Shen Q, Jousset A (2015) Trophic network architecture of root-associated bacterial communities determines pathogen invasion and plant health. Nat Commun 6(1):1–9

Whipps JM (2001) Ecological and biotechnological considerations in enhancing disease biocontrol. In: Nato science series sub series I. Life and behavioural sciences, vol 339, pp 43–51

Wu Y, Zhou J, Li C, Ma Y (2019) Antifungal and plant growth promotion activity of volatile organic compounds produced by *Bacillus amyloliquefaciens*. Microbiol Open 8(8):e00813

Xie F, Williams A, Edwards A, Downie JA (2012) A plant arabinogalactan-like glycoprotein promotes a novel type of polar surface attachment by *rhizobium leguminosarum*. Mol Plant Microbe Interact 25(2):250–258

Xu L, Li T, Wu Z, Feng H, Yu M, Zhang X, Chen B (2018) Arbuscular mycorrhiza enhances drought tolerance of tomato plants by regulating the 14-3-3 genes in the ABA signaling pathway. Appl Soil Ecol 125:213–221

Yehuda Z, Shenker M, Hadar Y, Chen Y (2000) Remedy of chlorosis induced by iron deficiency in plants with the fungal siderophore rhizoferrin. J Plant Nutr 23(11–12):1991–2006

Yuan J, Zhao J, Wen T, Zhao M, Li R, Goossens P, Huang Q, Bai Y, Vivanco JM, Kowalchuk GA, Berendsen RL (2018) Root exudates drive the soil-borne legacy of above ground pathogen infection. Microbiome 6(1):1–2

Zhalnina K, Louie KB, Hao Z, Mansoori N, da Rocha UN, Shi S, Cho H, Karaoz U, Loqué D, Bowen BP, Firestone MK (2018) Dynamic root exudate chemistry and microbial substrate preferences drive patterns in rhizosphere microbial community assembly. Nat Microbiol 3 (4):470–480

Zhang J, Liu YX, Zhang N, Hu B, Jin T, Xu H, Qin Y, Yan P, Zhang X, Guo X, Hui J (2019) NRT1. 1B is associated with root microbiota composition and nitrogen use in field-grown rice. Nat Biotechnol 37(6):676–684

Zyśko A, Sanguin H, Hayes A, Wardleworth L, Zeef LA, Sim A, Paterson E, Singh BK, Kertesz MA (2012) Transcriptional response of *Pseudomonas aeruginosa* to a phosphate-deficient *Lolium perenne* rhizosphere. Plant and Soil 359(1):25–44

Chapter 6
Rhizospheric Microbes and Their Mechanism

Anuj Chaudhary, Heena Parveen, Parul Chaudhary, Hina Khatoon, and Pankaj Bhatt

Abstract With constantly rising global population, requirement for higher agronomic productivity is increasing. There is a need to adopt safe and effective strategies contributing toward precision farming and sustainable crop production. Rhizosphere is the region of soil where root of plants has large population of microbes and makes its environment highly complex. The pattern of root exudates affects microbial activity in the soil. Beneficial and harmful interactions occur in rhizosphere microbes and plants which ultimately affect root functions and plant development. A better understanding of this process is needed to manage the microbes and plant productivity, and maintaining the soil health is important. Plant growth-promoting activities of PGPRs help to overcome challenges such as low crop productivity, overuse of agrochemicals, and nutrient loss. This chapter discusses the rhizospheric microbes and how they facilitate the nutrient solubilization and uptake by plants and helps in agriculture productivity.

Keywords Rhizosphere · Microbes · Agriculture · Sustainable · Environment

---

A. Chaudhary
School of Agriculture and Environmental Sciences, Shobhit University, Gangoh, Uttar Pradesh, India

H. Parveen
Department of Dairy Microbiology, NDRI, Karnal, Haryana, India

P. Chaudhary (✉)
Department of Animal Biotechnology, NDRI, Karnal, Haryana, India

H. Khatoon
Uttarakhand Pollution Control Board, Roorkee, Uttarakhand, India

P. Bhatt
State Key Laboratory for Conservation and Utilization of Subtropical Agro-bioresources, Guangdong Laboratory for Lingnan Modern Agriculture Integrative Microbiology Research Centre, South China Agricultural University, Guangzhou, China

P. Bhatt et al. (eds.), *Microbial Technology for Sustainable Environment*,
https://doi.org/10.1007/978-981-16-3840-4_6

6.1 Introduction

The soil has the possible capacity of plant cultivation, in order to produce food crops for animal and human consumption (Sulbhi et al. 2021; Bhandari et al. 2021; Bhatt et al. 2021a, b, c). The health of human being and animals depends solely on consistency of the soil, since it serves as the primary means of fibrous crops and food production (Bhatt et al. 2020a, 2021d, e, f, 2020b, c, d, e, f, 2019a). The qualitative content of directly ingested air and water depends on the conditions of the soil, as it is an important relation to the overall climate (Bhatt et al. 2019b, c, 2016a, b, 2015a, b; Huang et al. 2021). The quality of the land is therefore directly proportional to the wellbeing of the living beings and the environment (Odelade and Babalola 2019). Soil is an area where various factors interact with each other either in a favorable or in a harmful way. However, the most important and highly focused area is called rhizospheric zone, which is a hot spot, the richest zone where plant roots interact with various rhizospheric microbes. This zone becomes a multiplex ecosystem on this planet where multitrophic interactions have been occurred (Mendes et al. 2013). Broadly three different interconnected participants documented in the rhizospheric soil, viz., rhizoplane, rhizosphere, and root itself. Rhizosphere is the soil region affected by the roots, frequently by releasing the substrates that influence soil microbial activity. Including the firmly adhering soil particles, the rhizoplane is the root surface. Root is the part of system itself as different microbes; endophytes are capable to inhabit root tissues (Kennedy 1998; Bowen and Rovira 1999).

Rhizosphere is affected by the activities of root like exudation of carbon compound, nutrient uptake, and water absorption (George et al. 2006). Rhizospheric microbial community responds to these plant root exudates through chemotaxis and affects the rhizospheric microbial population that eventually changes in response to plant age and seasonal change (Hartmann et al. 2009). These exudates serve as a basis of carbon for soil heterotrophic microbiota either as a growth substrate or as signal molecules that elicit plant-microbe signaling pathways. Microbial groups participated as the major player in plant functioning via affecting their metabolism and growth. Among the various associates of rhizospheric microbiota, some are valuable toward plant, and others are pathogenic ones, which inhabit the rhizospheric zone and cause disease to plant. The main rhizospheric microbial community comprises of bacteria, fungi, nematodes, viruses, algae, and protozoa. Majority of the members of microbiome are fraction of a dynamic food web and utilized the nutrients secreted by plants (Raaijmakers et al. 2009). These organisms are classified according to their effect on the plant system: the beneficial microbiota (nitrogen-fixing bacteria, arbuscular-mycorrhizal fungi, and the most prominent and active group called as plant growth-promoting rhizobacteria) and the harmful microbiota (pathogenic fungi, pathogenic bacteria, and nematodes) (Yang et al. 2009). These microorganisms are the key player in the improvement of sustainable agriculture and help in maintaining the growth of plants as well as the soil structure. With the emergence of molecular studies on the soil microbiota, it has resolved many queries with regard to their role and behavior in this most complex zone of the plant. Metagenomics approach, the most acceptable technique which has the ability to discover the unculturable microbiota with all the functions and genes with the purpose of improving agricultural productions. Hence, in this chapter, the rhizospheric microbiota has been discussed with their effect on plant as a crop producer or as a crop destructor. The interaction, role, and application are also reviewed for the development of a sustainable environment in terms of agriculture (Bhatt et al. 2019d, e; Sharma and Bhatt 2016; Sharma et al. 2016; Bhatt and Nailwal 2018; Gangola et al. 2018a; Bhatt 2018; Bhatt and Barh 2018; Bhandari and Bhatt 2020; Bhatt and Bhatt 2020).

6.2 Organisms in the Rhizosphere

Various microorganisms colonize plants, which can reach cell densities much higher than the amount of plant cells. This colonization is a kind of biochemical communication among plants and microorganisms, in which molecules are released by plant/microbe system and responsible for the harmful and beneficial activities in soil system. This underground communication is increasingly recognized in improving crop productivity as well as giving an understanding of the effect of microbe on plant health. For the betterment of farming environment health, stability, and sustainability, the interspecies or inter-kingdom communication in the rhizospheric zone that is created during plant-plant/plant-microbe interactions is important. The most considered microbial relationships for plant productivity are those produced by PGPR (plant growth-promoting rhizobacteria), NFB (nitrogen fixing bacteria), AMF (arbuscular- mycorrhizal fungi), pathogenic fungi/bacteria, nematodes, and invasive plant species.

6.3 Beneficial Microbial Community

6.3.1 Plant Growth-Promoting Rhizobacteria

A population of free-living bacteria, entering in the rhizospheric zone and supporting plant root growth. Roots of plants are capable to attract valuable soil bacteria from a large range of genera, called PGPRs, viz., *Bacillus*, *Acinetobacter*, *Pseudomonas*, and *Rhizobium* (Agri et al. 2021; Chaudhary et al. 2021c, d). These bacteria are important because of their position as producers of growth regulators and phosphorus solubilizers that help in abiotic and biotic stress tolerance (Bhattacharyya and Jha 2012; Bakker et al. 2007). Colonization of root entails the capacity of bacteria, in the presence of the indigenous microflora, to develop on or in the plant root and to reproduce, thrive, and spread along the growing root. Besides with the activity of plant growth promoters, the PGPR group also involved in controlling phytopathogens as BCAs (bio-control agents), enhancing fertilizer's efficiency and rhizoremediation process (Podile and Kishore 2007). As an off-shoot of biological control of soil-borne pathogens, the significance of PGPR was realized in other aspects of plant promotion which divided PGPRs into two groups, according to the mode of action, i.e., PGPBs and PGPBs as BCAs. Inseparable from the plant growth-promoting activity, it has been observed that the identified PGPRs have defense ability of bacterial-inoculated seedlings against soil-borne pathogens. As a matter of fact, PGPR were considered as protectants for soil-borne pathogens (Vessey 2003). With the growing value of organic culturing, the requirement for PGPR as biofertilizers has been re-discovered in recent years with minimum or no inputs. Biofertilizers are used to enhance the plant growth by providing essential nutrients to the plants (Kumari et al. 2021). There are different biofertilizers such as *Rhizobium*, *Allorhizobium* spp., *Trichoderma* spp., and *Pseudomonas fluorescens* which enhance the plant development (Badar and Qureshi 2012; Shen et al. 2013). It was observed that plants require optimal nutrient input and resilience under stress conditions in order to have the highest crop productivity (Koller et al. 2013). This enables plants to use soil nutrients effectively by maximizing root/rhizosphere efficiency in the mobilization and acquisition of nutrients. Such mobilization is carried out by rhizosphere bacteria at a high rate. Various previous and recent studies reported the plant growth promoter activity of PGPRs on different plant system. Pieterse and Van Loon, 1999 introduced *P. fluorescence* WCS417, a rhizobacterial strain under in vitro conditions in autoclaved soil, and observed an increase in growth promotion of *Arabidopsis* accession Col-O by 33%. Application of *Pseudomonas taiwanensis*, *Pantoea agglomerans*, and *Bacillus* spp. improved the plant health parameters in a pot experiment which shows the plant growth-promoting activities like IAA and siderophore production (Chaudhary and Sharma 2019; Khati et al. 2019a, 2017). Bioinoculation of *Bacillus* spp. also improved the plant health parameters, enhanced the yield of maize crop, and improved the beneficial population of soil (Khati et al. 2018, 2019b; Chaudhary et al. 2021a, b; Rudrappa et al. 2008).

PGPRs also favor plant development by releasing some signal molecules under infection/pathogen attacking condition that act as an alarming call for beneficial bacteria. In a report, it was observed that *Pseudomonas syringae* pv tomato causes infection to *Arabidopsis* and induced the expression of L-malic acid transporter and elevates the release of malic acid in roots. Once the concentration of malic acid was increased in the rhizosphere region at some instance, it recruits the beneficial *B. subtilis*, which endorse the biofilm formation on *Arabidopsis* roots (Lakshmanan et al. 2013; Asaka and Shoda 1996), developing a general resistance reaction toward the disease-causing microbes. A number of bacteria release antimicrobial metabolites, viz., surfactin and iturin A, that provide a protective guard to roots against pathogenic fungi and bacteria such as *Rhizoctonia* and *P. syringae* (Bais et al. 2004; Kumari et al. 2020). Kumari et al. (2020) found that application of *Bacillus* spp. increased the growth of Fenugreek plants and improved the soil health too. Kukreti et al. (2020) also observed the positive impact of bioinoculants on maize crop.

6.3.2 *Nitrogen-Fixing Bacteria*

Plant roots provide a position for the growth of soil bacteria that flourish on root exudates and lysates. The rhizospheric microbiota consume these exudates as nutrient factor for their development, and in response to the plant root exudation, these bacteria excrete some metabolites into the rhizosphere, which can act as signaling molecules that help in plant promotion. One of the best examples of this interaction is *Rhizobium* and *Legume* symbiosis. Plant secretes flavonoid which acts as signal molecule for *Rhizobium*, and in response to this, the bacterium starts secreting Nod factors which are received by plant roots and inducing root nodule formation. In these nodules, the *Rhizobium* initiates fixing atmospheric nitrogen and promoting plant growth under nitrogen-poor environments. Bacteria rise at the expense of the carbohydrates of the host and in exchange provide fixed nitrogen for the biosynthesis of amino acid (Brencic and Winans 2005; Gray and Smith 2005; Herridge et al. 2008). This mutual symbiotic association has resulted in fixing $50–70 \times 10^6$ tons of nitrogen annually into the soil which results in a great reduction on the application of chemical fertilizers in agriculture systems (Graham 2008). The atmospheric nitrogen-fixing rhizobacteria fixed nitrogen either in a symbiotic or non-symbiotic association and have been employed in agricultural practices for the development of growth and yield of many crops. The bacteria which develop root nodule formation in leguminous plants and convert atmospheric nitrogen into utilizable form of nitrogen are collectively called rhizobia. Nowadays, the rhizobia genera also include other genera, such as *Sinorhizobium*, *Rhizobium*, *Bradyrhizobium*, *Mesorhizobium*, *Azorhizobium*, and *Allorhizobium* (Kurrey et al. 2018). Other category of nitrogen-fixing bacteria belongs to free-living bacteria (*Azotobacter* and *Azospirillum*), and these free-living diazotrophs are recognized to utilize nitrogen from the atmosphere for their cellular protein biosynthesis which is mineralized in the soil, maintaining a large portion of nitrogen available from the soil supply to crop plants. Research into *Azotobacter chroococcum* in crop production has shown that it is important to improve plant nutrients and improve soil fertility (Gonzalez-Lopez et al. 2005). Numerous *Azotobacter* strains are found to be capable of producing amino acids supplemented with different sources of carbon and nitrogen when grown in culture media (Vikhe 2014). They have the ability to fix nitrogen and release some phytohormones, namely, GA3, IAA, and cytokinins that could improve plant growth and increase the supply of plant root nutrients (Brakel and Hilger 1965). In vitro studies showed that indole-3-acetic acid (IAA) is released by *Azotobacter* when tryptophan is added to the medium (Hennequin and Blachere 1966). Small amounts of IAA were present in old cultures of *Azotobacter* that have no supplement with tryptophan (Jeffries et al. 2003). Besides, auxin and gibberellins are also found in *A. chroococcum*.

6.3.3 *Arbuscular-Mycorrhizal Fungi (AMF)*

AMF also enhance plant growth by expanding their hyphal networks into the soil to obtain nutrients, such as phosphorous, which are then delivered to their hosts (Salam et al. 2017). They promote the vigorous growth of host plants under stressful conditions by mediating a variety of complex contact events between the plant and the fungus, resulting in increased photosynthetic rate and other characteristics associated with gas exchange, as well as increased water absorption (Kumar et al. 2021). Several studies identify improved tolerance to a range of stresses due to fungal symbiosis including drought, salinity, temperature, and metals (Redecker et al. 2013). Basically, AMF develops arbuscules, vesicles, and hyphal network with plant roots that significantly augment the accessibility of plant nutrient uptake, causing improvement of plant growth. Fungal hyphae can accelerate the degradation process of soil organic matter. Arbuscular-mycorrhizal fungi are classified into four different orders including Glomerales, Archaeosporales, Paraglomerales, and Diversisporales (Zeng et al. 2014). Better contents of flavonoids, sugars, organic acids, and minerals enhanced the quality of citrus fruit due to *Glomus versiforme* (Chen et al. 2017). AMF inoculation can significantly increase the absorption of various macro-/micronutrients, resulting in increased production of photosynthate and thus increased accumulation of biomass (Mitra et al. 2019; MacLean et al. 2017). The distribution of phosphate is one of the most significant advantages for the host in AM symbiosis, and collective knowledge indicates that the arbuscules are the position of phosphate transfer from the fungal partner to its host (Wu et al. 2008). Thus, the plant growth-promoting rhizospheric microorganisms are the key players in soil ecosystem that enhance the plant growth by protecting them from various phytopathogens and also providing essential nutrients to the plant, which augment the crop productivity by providing a sight of sustainable agriculture system.

6.4 Harmful Microbial Community

In the above section, the beneficial rhizospheric microbiota are reviewed, but as we know, soil is a dynamic ecosystem which interacts with the useful and deleterious subjects that directly or indirectly affect the plant growth. Besides good community, bad community of rhizospheric microbes also exists which is mainly responsible for the development of diseases and causes major reduction of food and fuel crops. Two foremost communities come under this category, namely, the pathogenic fungi and bacteria and nematode.

6.4.1 Pathogenic Fungi

Phytopathogenic fungi have multifaceted life cycles of pathogenesis in rhizospheric region for plant disease development. The initial infection process usually proceeds by fungal spores' development into conidia that further grows and starts penetrating into the host plant. Sections of the plants or the whole plant are destroyed by the pathogens, and the pathogenesis depends upon the type of plant tissue that is affected by the pathogen and the nature of the infection. Finally, in the contaminated plant tissue, the fungal pathogens develop spores and begin the disease cycle in their host system. For the dispersal of fungal spores, environmental factors participated in an active way which includes wind, rain, or insect vectors and finally settles on their new hosts. For the processing of pathogenesis cycle, various surrounding factors are involved for the germination, development, and establishment in rhizospheric part of the soil. Various reports have concluded that, for triggering dormancy of fungal spores, different factors such as soil pH and root exudation play pivotal role. In a report by Gonzalez et al. (2011), gallic, salicylic, and ferulic acid stimulates the conidial germination of pathogenic fungi. Major groups of rhizospheric pathogenic fungi include *Fusarium*, *Rhizoctonia*, *Pythium*, *Phytophthora*, and many others (Rasmann et al. 2012).

6.4.2 Nematodes

Nematodes are free-living soil creature that feed on the plant roots either as an ectoparasite or by penetrating inside the root cells and reproduce as a sedentary endoparasite. The plant exudates have a major role in chemotaxis of nematodes toward plant roots as these compounds have the ability to attract these organisms. Most of the recent studies have reports on different plant-derived exudates to attract nematodes (Johnson and Nielsen 2012). Among the various volatile root exudates, CO_2 is the predominant long-distance chemotaxis molecule that is released by the plant roots for positioning of parasitic nematode, with a potential of up to 1 m action radius for one root and for full root system around 2 m of action radius required (Turlings et al. 2012). CO_2 acts as "response activator" that alarms other species to the general existence of entomopathogenic nematodes and can increase their responsiveness to more specific indications (Ormeño-Orrillo et al. 2013). In addition to CO_2, chemotaxis in nematodes is induced by 2,4-dihydroxy-7-methoxy-1, 4-benzoxazine-3-one and ascorbic acid (Zhu et al. 2011).

6.5 Mechanism of PGPRs

Their mode of action can be categorized into two:

6.5.1 Direct Mechanisms

6.5.1.1 Nutrient Acquisition

This mechanism supports the plant growth by providing the essential nutrients like nitrogen fixation, zinc, potassium, iron, and phosphorus solubilization.

- **Fixation of Nitrogen:** Nitrogen plays an essential role for plant growth. Seventy eight percent of total nitrogen is available in the atmosphere and remains unavailable to plants. N_2-fixing bacteria and fungi are involved in fixing the atmospheric nitrogen with the help of nitrogenase enzyme. Thirty percent to 50% of biological nitrogen is fixed using bacteria, a major contribution to the nitrogen in agriculture crop (Hussain et al. 2015).
- **Phosphorus Solubilization:** P is also a necessary constituent for the growth of plants. It makes 0.2% dry weight of plants, and only 0.1% of P is available to plants and limits their growth (Sattar et al. 2019). Phosphorus-solubilizing bacteria and fungi mineralize the insoluble soil P into soluble form by releasing the organic acids along with decrease in pH.
- **Zinc Solubilization:** Zinc is the main micronutrient for plant development and present in bound form as insoluble complexes and minerals. Zinc solubilizing bacteria have the ability to solubilize the zinc using acidification process and by organic acid production and decrease the pH of soil (Hider and Kong 2010).
- **Potassium Solubilization:** Potassium functions as a cofactor to regulate the enzymatic reaction and helps in plant development. Lowering pH, acidolysis, and production of organic acids by soil microbes solubilize the K by conversion of insoluble form of potassium into soluble form which is easily up taken by plants (Cassán et al. 2014).
- **Iron Sequestration:** Iron is abundantly found in soil. Its assimilated form (Fe^{3+}) is less predominant in nature because it chelates with several compounds and its bioavailability becomes very low. Therefore, rhizospheric microbes that synthesize siderophores (low molecular weight compounds) can help in sequester (Fe^{3+}) with great affinity. Siderophores have binding affinity toward microbial membrane receptors which can hold iron-siderophore complex and help in the uptake of iron, favored in stress conditions (Lugtenberg and Kamilova 2009).
- **Production of Phytohormones:** They are known as plant growth regulators that help the plants to combat various abiotic stresses and increase the plant growth and productivity. These include the IAA, Gibberellins, ABA, and cytokinins (Waweru et al. 2014). Indole acetic acid is an auxin produced by microbes which promotes plant development such as cell division and elongation. It has been found that more than 80% of auxin are produced by *Rhizobium* sp. Many bacterial species and fungal species produced growth hormones such as *Bacillus megaterium*, *Pseudomonas fluorescens*, and *Trichoderma harzianum*.

6.5.2 Indirect Mechanism

- **Production of Lytic Enzymes:** PGPRs are well-known for antibiotics production and lytic enzymes which help in the suppression of the growth of phytopathogens. There are various antifungal compounds like phenazines, pyrrolnitrin, cyclic lipopeptides, glucanases, lipases, cellulases, and chitinases which directly inhibit the growth of phytopathogens (Gamalero and Glick 2015).

- **Competition:** PGPR and PGPF protect the plants from destructive phytopathogens and help in enhancing the growth of plants. Inoculation of *Fusarium oxysporum* significantly suppressed nematode pathogen in banana plants and increases their yields (Van Loon et al. 1998).
- **ACC Deaminase:** PGPR produces 1-aminocyclopropane-1-carboxylic acid deaminase. This enzyme cleaves ethylene precursor into ammonia and α-ketobutyrate and controls growth of plants by cleaving the ACC and minimizing the level of ethylene, when present in high amount. This enzyme is encoded by acdS gene in several bacterial and fungal species. ACC production was also observed in several fungi such as *Issatchenkia occidentalis* and *Penicillium citrinum* (Itelima et al. 2018).
- **ISRs (Induced Systemic Resistance):** Rhizobacteria suppress diseases via inducing the resistance mechanism in plants known as ISR (Singh et al. 2021; Zhang et al. 2020a; Mishra et al. 2020). It involves ethylene and jasmonate signaling which help in the stimulation of defense response in host plant against a large number of pathogens.

6.6 Challenges of Using PGPR as a Bioinoculant

The use of PGPRs as a bioinoculant is very old. PGPR strains selected in laboratory result in low performance of a variety of activities, while do not for all time results under field conditions. So, there is need to improve the methods. Developing bioinocula containing vastly efficient microbes with a lengthy shelf life and high-rhizospheric colonization rate poses a most important dispute for commercialization of bioinoculants. PGPR and PGPF are often used in an inappropriate carrier that does not allow the efficient colonization under field conditions due to competition with resident soil microbial flora. Creation of bioinoculants for specific soil type and to train farmers to well apply them to crops are very important in the improvement of more beneficial inocula (Zhang et al. 2020b; Feng et al. 2020; Lin et al. 2020; Zhan et al. 2020; Ye et al. 2019; Huang et al. 2019, 2020; Fan et al. 2020; Pang et al. 2020; Gangola et al. 2018b; Gupta et al. 2018).

6.7 Conclusion

Today, concern arises due to the harsh impact of agrochemicals; there is a rising attention in improving our knowledge to understand the function of rhizospheric microorganisms in agriculture. The plant rhizospheric microbes show outstanding prospective for wider application in sustainable agriculture as they improve plant health and productivity in an ecofriendly and commercial manner. Using bioinoculants is a potential way to enhance nutrient use efficiency in soil and can be a good alternative to agrochemicals used in the agricultural fields. However, more examination is needed in different disciplines to recognize the exact mechanisms of bioinoculants under different set of conditions.

References

Agri U, Chaudhary P, Sharma A (2021) In vitro compatibility evaluation of agriusable nanochitosan on beneficial plant growth-promoting rhizobacteria and maize plant. Natl Acad Sci Lett. https://doi.org/10.1007/s40009-021-01047-w

Asaka O, Shoda M (1996) Biocontrol of Rhizoctonia solani damping-off of tomato with *Bacillus subtilis* RB14. Appl Environ Microbiol 62:4081–4085

Badar R, Qureshi SA (2012) Use of *Trichoderma hamatum* alone and in combination with rhizobial isolates as bio-fertilizer for improving the growth and strength of sunflower. J Basic Appl Sci Res 2:6307–6314

Bais HP, Fall R, Vivanco JM (2004) Biocontrol of *Bacillus subtilis* against infection of Arabidopsis roots by *Pseudomonas syringae* is facilitated by biofilm formation and surfactin production. Plant Physiol 134:307–319

Bakker PAHM, Raaijmakers JM, Bloemberg G, Hdfte M, Lemanceau P, Cooke BM (2007) New perspectives and approaches in Plant Growth-Promoting Rhizobacteria research. Eur J Plant Pathol 119:241–242

Bhandari G, Bhatt P (2020) Concepts and application of plant microbe interaction in remediation of Heavy metals. In: Sharma A (ed) Microbes and signaling biomolecules against plant stress. Rhizosphere biology. Springer, Singapore. https://doi.org/10.1007/978-981-15-7094-0_4

Bhandari G, Bagheri AR, Bhatt P, Bilal M (2021) Occurrence, potential ecological risks, and degradation of endocrine disruptor, nonylphenol from the aqueous environment. Chemosphere 275:130013. https://doi.org/10.1016/j.chemosphere.2021.130013

Bhatt P (2018) *In silico* tools to study the bioremediation in microorganisms. In: Pathak V, Navneet (eds) Handbook of research on microbial tools for environmental waste management. IGI Global, Hershey, PA, pp 389–395. https://doi.org/10.4018/978-1-5225-3540-9.ch018

Bhatt P, Barh A (2018) Bioinformatic tools to study the soil microorganisms: an *in silico* approach for sustainable agriculture. In: Choudhary D, Kumar M, Prasad R, Kumar V (eds) In silico approach for sustainable agriculture. Springer, Singapore. https://doi.org/10.1007/978-981-13-0347-0_10

Bhatt K, Bhatt P (2020) Rhizosphere biology: alternate tactics for enhancing sustainable agriculture. In: Phytomicrobiome interactions and sustainable agriculture. John Wiley & Sons, New York, NY. ISBN: 9781119644620

Bhatt P, Nailwal TK (2018) Chapter 11: Crop improvement through microbial technology: a Step towards sustainable agriculture. In: Prasad R, Gill SS, Tuteja N (eds) Crop improvement through microbial biotechnology. Elsevier, Amsterdam, pp 245–253. https://doi.org/10.1016/B978-0-444-63987-5.00011-6

Bhatt P, Bisht TS, Pathak VM, Barh A, Chandra D (2015a) Optimization of amylase production from the fungal isolates of Himalayan region Uttarakhand. Ecol Environ Conserv 21 (3):1517–1521

Bhatt P, Negi G, Gangola S, Khati P, Srivastava A, Sharma A (2015b) Optimization of sulfosulfuron biodegradation through response surface methodology using indigenous bacterial strain isolated from contaminated agriculture field. Int J Curr Microbiol App Sci 4(8):105–112

Bhatt P, Negi G, Gangola S, Khati P, Kumar G, Srivastava A, Sharma A (2016a) Differential expression and characterization of cypermethrin degrading potential proteins in *Bacillus thuringiensis* strain, SG4. 3 Biotech 6:225

Bhatt P, Sharma A, Gangola S, Khati P, Kumar G, Srivastava A (2016b) Novel pathway of cypermethrin biodegradation in *a Bacillus* sp. strain SG2 isolated from cypermethrin-contaminated agriculture field. 3 Biotech 6:45

Bhatt P, Huang Y, Zhan H, Chen S (2019a) Insight into microbial applications for the biodegradation of pyrethroid insecticides. Front Microbiol 10:1778

Bhatt P, Pal K, Bhandari G, Barh A (2019b) Modeling of methyl halide biodegradation on bacteria and its effect on other environmental systems. Pestic Biochem Physiol 158:88–100

Bhatt P, Gangola S, Chaudhary P, Khati P, Kumar G, Sharma A, Srivastava A (2019c) Pesticide induced up-regulation of esterase and aldehyde dehydrogenase in indigenous *Bacillus* spp. Bioremed J 23(1):42–52

Bhatt P, Joshi D, Kumar N, Kumar N (2019d) Recent trends to study the functional analysis of mycorrhizosphere. In: Varma A, Choudhary D (eds) Mycorrhizosphere and pedogenesis. Springer, Singapore. https://doi.org/10.1007/978-981-13-6480-8_11

Bhatt P, Pathak VM, Joshi S, Bisht TS, Singh K, Chandra D (2019e) Chapter 12: Major metabolites after degradation of xenobiotics and enzymes involved in these pathways. In: Smart bioremediation technologies: microbial enzymes, pp 205–215. https://doi.org/10.1016/B978-0-12-818307-6.00012-3

Bhatt P, Rene ER, Kumar AJ, Kumar AJ, Zhang W, Chen S (2020a) Binding interaction of allethrin with esterase: bioremediation potential and mechanism. Bioresour Technol 315:13845. https://doi.org/10.1016/j.biortech.2020.123845

Bhatt P, Verma A, Verma S, Anwar MS, Prasher P, Mudila H, Chen S (2020b) Understanding phytomicrobiome: a potential reservoir for better crop management. Sustainability. 12:5446

Bhatt P, Huang Y, Rene ER, Kumar AJ, Chen S (2020c) Mechanism of allethrin biodegradation by a newly isolated *Sphingomonas trueperi* strain CW3 from wastewater sludge. Bioresour Technol 305:123074

Bhatt P, Zhang W, Lin Z, Pang S, Huang Y, Chen S (2020d) Biodegradation of allethrin by a novel fungus *Fusarium proliferatum* strain CF2, Isolated from Contaminated. Soils Microorg 8:593

Bhatt P, Huang Y, Zhang W, Sharma A, Chen S (2020e) Enhanced cypermethrin degradation kinetics and metabolic pathway in *Bacillus* thuringiensis strain, SG4. Microorganisms 8:223

Bhatt P, Bhatt K, Huang Y, Ziqiu L, Chen S (2020f) Esterase is a powerful tool for the biodegradation of pyrethroid insecticides. Chemosphere 244:125507

Bhatt P, Zhou X, Huang Y, Zhang W, Chen S (2021a) Characterization of the role of esterases in the biodegradation of organophosphate, carbamate and pyrethroid group pesticides. J Hazard Mater 411:125026

Bhatt P, Joshi T, Bhatt K, Zhang W, Huang Y, Chen S (2021b) Binding interaction of glyphosate oxidoreductase and C-P lyase: molecular docking and molecular dynamics simulation studies. J Hazard Mater 409:124927

Bhatt P, Bhatt K, Sharma A, Zhang W, Mishra S, Chen S (2021c) Biotechnological basis of microbial consortia for the removal of pesticides from the environment. Crit Rev Biotechnol 41:317. https://doi.org/10.1080/07388551.2020.1853032

Bhatt P, Sethi K, Gangola S, Bhandari G, Verma A, Adnan M, Singh Y, Chaube S (2021d) Modeling and simulation of atrazine biodegradation in bacteria and its effect in other living systems. J Biomol Struct Dyn. https://doi.org/10.1080/07391102.2020.1846623

Bhatt P, Gangola S, Bhandari G, Zhang W, Maithani D, Mishra S, Chen S (2021e) New insights into the degradation of synthetic pollutants in contaminated environments. Chemosphere 268:128827. https://doi.org/10.1016/j.chemosphere.2020.128827

Bhatt P, Sharma A, Rene ER, Kumar AJ, Zhang W, Chen S (2021f) Bioremediation mechanism, kinetics of fipronil degradation using *Bacillus* sp. FA3 and resource recovery potential from contaminated environments. J Water Process Eng 39:101712

Bhattacharyya PN, Jha DK (2012) Plant growth-promoting rhizobacteria (PGPR): emergence in agriculture. World J Microbiol Biotechnol 28:1327–1350

Bowen GD, Rovira AD (1999) The rhizosphere and its management to improve plant growth. Adv Agron 66:1–102

Brakel J, Hilger F (1965) Qualatitive et quantative de la synthese de substances de nature auxinique par *Azotobacter chrococcum* in vitro. Bull Inst AgroStns Res Gembloux 33:469–487

Brencic A, Winans SC (2005) Detection of and response to signals involved in host-microbe interactions by plant associated bacteria. Microbiol Mol Biol Rev 69:155–194

Cassán F, Vanderleyden J, Spaepen S (2014) Physiological and agronomical aspects of phytohormone production by model plant-growth-promoting rhizobacteria (PGPR) belonging to the genus *Azospirillum*. J Plant Growth Regul 33(2):440–459

Chaudhary P, Sharma A (2019) Response of nanogypsum on the performance of plant growth promoter bacteria recovered from nanocompound infested agriculture field. Environ Ecol 37:363–372

Chaudhary P, Khati P, Chaudhary A, Gangola S, Kumar R, Sharma A (2021a) Bioinoculation using indigenous *Bacillus* spp. improves growth and yield of *Zea mays* under the influence of nanozeolite. 3 Biotech 11:11

Chaudhary P, Sharma A, Chaudhary A, Sharma P, Gangola S, Maithani D (2021b) Illumina based high throughput analysis of microbial diversity of rhizospheric soil of maize infested with nanocompounds and *Bacillus* sp. Appl Soil Ecol 159:103836

Chaudhary P, Khati P, Gangola S, Kumar A, Kumar R, Sharma A (2021c) Impact of nanochitosan and Bacillus spp. on health, productivity and defence response in Zea mays under field condition. 3 Biotech 11:237. https://doi.org/10.1007/s13205-021-02790-z

Chaudhary P, Khati P, Chaudhary A, Maithani D, Kumar G, Sharma A (2021d) Cultivable and metagenomic approach to study the combined impact of nanogypsum and Pseudomonas taiwanensis on maize plant health and its rhizospheric microbiome. PLoS One 16(4): e0250574. https://doi.org/10.1371/journal.pone.0250574

Chen S, Zhao H, Zou C, Li Y, Chen Y, Wang Z (2017) Combined Inoculation with multiple arbuscular mycorrhizal fungi improves growth, nutrient uptake and photosynthesis in cucumber seedlings. Front Microbiol 8:25–16

Fan X, Ye T, Li Q, Bhatt P, Zhang L, Chen S (2020) Potential of a quorum quenching bacteria isolate Ochrobactrum intermedium D-2 against soft rot pathogen *Pectobacterium carotovorum* subsp. Carotovora. Front Microbiol 11:898

Feng Y, Huang Y, Zhan H, Bhatt P, Chen S (2020) An overview of strobilurin fungicide degradation: current status and future perspective. Front Microbiol 11:389

Gamalero E, Glick BR (2015) Bacterial modulation of plant ethylene levels. Plant Physiol 169:13–22

Gangola S, Bhatt P, Chaudhary P, Khati P, Kumar N, Sharma A (2018a) Bioremediation of industrial waste using microbial metabolic diversity. In: Pankaj, Sharma A (eds) Microbial biotechnology in environmental monitoring and cleanup. IGI Global, Hershey, PA, pp 1–27. https://doi.org/10.4018/978-1-5225-3126-5.ch001

Gangola S, Sharma A, Bhatt P, Khati P, Chaudhary P (2018b) Presence of esterase and laccase in *Bacillus subtilis* facilitates biodegradation and detoxification of cypermethrin. Sci Rep 8:12755. https://doi.org/10.1038/s41598-018-31082-5

George TS, Turner BL, Gregory PJ, Cade-Menun BJ, Richardson AE (2006) Depletion of organic phosphorus from Oxisols in relation to phosphatase activities in the rhizosphere. Eur J Soil Sci 57:47–57

Gonzalez M, Pujol M, Metraux JP, Gonzalez-Garcia V, Bolton MD, Borrás-Hidalgo O (2011) Tobacco leaf spot and root rot caused by *Rhizoctonia solani* Kuhn. Mol Plant Pathol 12:209–216

Gonzalez-Lopez L, Rodelas PC, Salmeron-Lopez V, Martinez-Toledo MV, Salmeron V (2005) Liberation of amino acids by heterotrophic nitrogen fixing bacteria. Amino Acids 28:363–367

Graham PH (2008) Ecology of the root-nodule bacteria of legumes. In: Dilworth MJ, James EK, Spent JI, Newton WE (eds) Nitrogen-fixing leguminous symbioses. Springer, Dordrecht, pp 23–58

Gray EJ, Smith DL (2005) Intracellular and extracellular PGPR: commonalities and distinctions in the plant bacterium signaling process. Soil Biol Biochem 37:395–412

Gupta S, Bhatt P, Chaturvedi P (2018) Determination and quantification of asiaticoside in endophytic fungus from Centella asiatica (L.) Urban. World J Microbiol Biotechnol 34:111

Hartmann A, Schmid M, van Tuinen D, Berg G (2009) Plant-driven selection of microbes. Plant Soil 321:235–257

Hennequin JR, Blachere H (1966) Research on the synthesis of phytohormones and phenolic compounds by Azotobacter and bacteria of the rhizosphere. Annales de IInstitut Pasteur 111:3

Herridge D, Peoples M, Boddey R (2008) Global inputs of biological nitrogen fixation in agricultural systems. Plant Soil 311:1–18

Hider RC, Kong X (2010) Chemistry and biology of siderophores. Nat Prod Rep 27:637–657

Huang Y, Zhan H, Bhatt P, Chen S (2019) Paraquat degradation from contaminated environments: current achievements and perspectives. Front Microbiol 10:1754

Huang Y, Lin Z, Zhang W, Pang S, Bhatt P, Rene ER, Kumar AJ, Chen S (2020) New insights into the microbial degradation of D-cyphenothrin in contaminated water/soil environments. Microorganisms 8:473

Huang Y, Zhang W, Pang S, Chen J, Bhatt P, Mishra S, Chen S (2021) Insights into the microbial degradation and catalytic mechanism of chlorpyrifos. Environ Res 194:110660

Hussain A, Arshad M, Zahir ZA, Asghar M (2015) Prospects of zinc solubilizing bacteria for enhancing growth of maize. Pak J Agric Sci 52:915–922

Itelima JU, Bang WJ, Onyimba IA, Egbere OJ (2018) A review: biofertilizer; a key player in enhancing soil fertility and crop productivity. J Microbiol Biotechnol Rep 2:22–28

Jeffries P, Gianinazzi S, Perotto S, Turnau K, Barea JM (2003) The contribution of arbuscular mycorrhizal fungi in sustainable maintenance of plant health and soil fertility. Biol Fertil Soils 37:1–16

Johnson SN, Nielsen UN (2012) Foraging in the dark – chemically mediated host plant location by belowground insect herbivores. J Chem Ecol 38:604–614

Kennedy AC (1998) The rhizosphere and spermosphere. In: Sylvia DM, Fuhrmann JJ, Hartel PG, Zuberer DA (eds) Principles and applications of soil microbiol. Prentice Hall, Upper Saddle River, NJ, pp 389–407

Khati P, Chaudhary P, Gangola S, Bhatt P, Sharma A (2017) Nanochitosan supports growth of *Zea mays* and also maintains soil health following growth. 3 Biotech 7:81

Khati P, Parul, Bhatt P, Nisha KR, Sharma A (2018) Effect of nanozeolite and plant growth promoting rhizobacteria on maize. 3 Biotech 8:141

Khati P, Chaudhary P, Gangola S, Sharma P (2019a) Influence of nanozeolite on plant growth promoter bacterial isolates recovered from nanocompound infested agriculture field. Environ Ecol 37:521–527

Khati P, Sharma A, Chaudhary P, Singh AK, Gangola S, Kumar R (2019b) High- throughput sequencing approach to access the impact of nanozeolite treatment on species richness and evens of soil metagenome. Biocatal Agric Biotechnol 20:101249

Koller R, Robin C, Bonkowski M, Ruess L, Scheu S (2013) Litter quality as driving factor for plant nutrition via grazing of protozoa on soil microorganisms. FEMS Microbiol Ecol 85:241–250

Kukreti B, Sharma A, Chaudhary P, Agri U, Maithani D (2020) Influence of nanosilicon dioxide along with bioinoculants on *Zea mays* and its rhizospheric soil. 3 Biotech 10:345

Kumar A, Sharma A, Chaudhary P, Gangola S (2021) Chlorpyrifos degradation using binary fungal strains isolated from industrial waste soil. Biologia. https://doi.org/10.1007/s11756-021-00816-8

Kumari S, Sharma A, Chaudhary P, Khati P (2020) Management of plant vigor and soil health using two agriusable nanocompounds and plant growth promoter rhizobacteria in Fenugreek. 3 Biotech 10:461

Kumari H, Khati P, Gangola S, Chaudhary P, Sharma A (2021) Performance of plant growth promotory rhizobacteria on maize and soil characteristics under the influence of TiO2 nanoparticles. Pantnagar J Res 19:28–39

Kurrey DK, Sharma R, Lahre MK, Kurrey RL (2018) Effect of *Azotobacter* on physio-chemical characteristics of soil in onion field. Pharma Innov J 7:108–113

Lakshmanan V, Castaneda R, Rudrappa T, Bais HP (2013) Root transcriptome analysis of *Arabidopsis thaliana* exposed to beneficial *Bacillus subtilis* FB17 rhizobacteria revealed genes for bacterial recruitment and plant defense independent of malate efflux. Planta 238:657–668

Lin Z, Zhang W, Pang S, Huang Y, Mishra S, Bhatt P, Chen S (2020) Current approaches to and future perspectives on methomyl degradation in contaminated soil/water environments. Molecules 25:738

Lugtenberg B, Kamilova F (2009) Plant-growth-promoting rhizobacteria. Annu Rev Microbiol 63:541–556

MacLean AM, Bravo A, Harrison MJ (2017) Plant signaling and metabolic pathways enabling arbuscular mycorrhizal symbiosis. Plant Cell 29:2319–2335

Mendes R, Garbeva P, Raaijmakers JM (2013) The rhizosphere microbiome: significance of plant beneficial, plant pathogenic, and human pathogenic microorganisms. FEMS Microbiol Rev 37:634–663

Mishra S, Zhang W, Lin Z, Pang S, Huang Y, Bhatt P, Chen S (2020) Carbofuran toxicity and its microbial degradation in contaminated environments. Chemosphere 259:127429

Mitra D, Navendra U, Panneerselvam U, Ansuman S, Ganeshamurthy AN, Divya J (2019) Role of mycorrhiza and its associated bacteria on plant growth promotion and nutrient management in sustainable agriculture. Int J Life Sci Appl Sci 1:1–10

Odelade KA, Babalola OO (2019) Bacteria, fungi and archaea domains in rhizospheric soil and their effects in enhancing agricultural productivity. Int J Environ Res Public Health 16:3873

Ormeño-Orrillo E, Hungria M, Martinez-Romero E (2013) Dinitrogen-fixing prokaryotes. In: The prokaryotes: prokaryotic physiology and biochemistry, pp 427–451

Pang S, Lin Z, Zhang W, Mishra S, Bhatt P, Chen S (2020) Insights into the microbial degradation and biochemical mechanisms of neonicotinoids. Front Microbiol 11:868

Pieterse CMJ, Van Loon LC (1999) Salicylic acid independent plant defence pathways. Trends Plant Sci 4:52–58

Podile AR, Kishore GK (2007) Plant growth-promoting rhizobacteria. In: Gnanamanickam SS (ed) Plant-associated bacteria. Springer, Dordrecht

Raaijmakers JM, Paulitz TC, Steinberg C (2009) The rhizosphere: a playground and battlefield for soilborne pathogens and beneficial microorganisms. Plant Soil 321:341–361

Rasmann S, Ali JG, Helder J, van der Putten W (2012) Ecology and evolution of soil nematode chemotaxis. J Chem Ecol 38:615–628

Redecker D, Schüssler A, Stockinger H, Stürmer SL, Morton JB, Walker C (2013) An evidence-based consensus for the classification of arbuscular mycorrhizal fungi (Glomeromycota). Mycorrhiza 23:515–531

Rudrappa T, Czymmek KJ, Paré PW, Bais HP (2008) Root-secreted malic acid recruits beneficial soil bacteria. Plant Physiol 148:1547–1556

Salam EA, Alatar A, El-Sheikh MA (2017) Inoculation with arbuscular mycorrhizal fungi alleviates harmful effects of drought stress on damask rose. Saudi J Biol Sci 25:1772–1780

Sattar A, Naveed M, Ali M, Zahir ZA, Nadeem SM, Yaseen M, Meena HN (2019) Perspectives of potassium solubilizing microbes in sustainable food production system: a review. Appl Soil Ecol 133:146

Sharma A, Bhatt P (2016) Chapter 3: Bioremediation: a microbial technology for improvising wildlife. In: Wildlife management concept analysis and conservation. DAYA Publishing House, New Delhi, pp 29–40

Sharma A, Pankaj, Khati P, Gangola S, Kumar G (2016) Chapter 6: Microbial degradation of pesticides for environmental cleanup. In: Bioremediation of industrial pollutants. Write and Print Publication, pp 178–205

Shen J, Li C, Mi G, Li L, Yuan L, Jiang R, Zhang F (2013) Maximizing root/rhizosphere efficiency to improve crop productivity and nutrient use efficiency in intensive agriculture of China. J Exp Bot 64:1181–1192

Singh K, Gera R, Sharma R, Maithani D, Chandra D, Bhat AM, Kumar R, Bhatt P (2021) Mechanism and application of *Sesbania* root nodulating bacteria: an alternative for chemical fertilizers and sustainable development. Arch Microbiol 203:1259. https://doi.org/10.1007/s00203-020-02137-x

Sulbhi V, Bhatt P, Verma A, Mudila H, Prasher P, Rene ER (2021) Microbial technologies for heavy metal remediation: effect of process conditions and current practices. Clean Techn Environ Policy. https://doi.org/10.1007/s10098-021-02029-8

Turlings TCJ, Hiltpold I, Rasmann S (2012) The importance of root-produced volatiles as foraging cues for entomopathogenic nematodes. Plant Soil 358:51–60

Van Loon LC, Bakker PAHM, Pieterse CMJ (1998) Systemic resistance induced by rhizosphere bacteria. Annu Rev Phytopathol 36:453–483

Vessey JK (2003) Plant growth promoting rhizobacteria as biofertilizers. Plant Soil 255:571–586

Vikhe PS (2014) *Azotobacter* species as a natural plant hormone synthesizer. Res J Recent Sci 3:59–63

Waweru B, Turoop L, Kahangi E, Coyne D, Dubois T (2014) Non-pathogenic *Fusarium oxysporum* endophytes provide field control of nematodes, improving yield of banana (Musa sp.). Biol Control 72:82–88

Wu HS, Liu DY, Ling N (2008) Effects of vanillic acid on the growth and development of *Fusarium oxysporum* f. sp *niveum*. Allelopath J 22:111–121

Yang J, Kloepper JW, Ryu CM (2009) Rhizosphere bacteria help plants tolerate abiotic stress. Trends Plant Sci 14:1–4

Ye T, Zhou T, Fan X, Bhatt P, Zhang L, Chen S (2019) *Acinetobacter lactucae* strain QL-1, a novel quorum quenching candidate against bacterial pathogen *Xanthomonas campestris* pv. Campestris. Front Microbiol 10:2867

Zeng L, JianFu L, JianFu L, MingYuan W (2014) Effects of arbuscular mycorrhizal (AM) fungi on citrus quality under nature conditions. Southwest China J Agric Sci 27:2101–2105

Zhan H, Huang Y, Lin Z, Bhatt P, Chen S (2020) New insights into the microbial degradation and catalytic mechanism of synthetic pyrethroids. Environ Res 182:109138

Zhang W, Pang S, Lin Z, Mishra S, Bhatt P, Chen S (2020a) Biotransformation of perfluoroalkyl acid precursors from various environmental system: advances and perspectives. Environ Pollut 272:115908

Zhang W, Lin Z, Pang S, Bhatt P, Chen S (2020b) Insights into the biodegradation of lindane (γ-Hexachlorocyclohexane) using a microbial system. Front Microbiol 11:522

Zhu F, Qu L, Hong X, Sun X (2011) Isolation and characterization of a phosphate solubilizing halophilic bacterium *Kushneria* sp. YCWA18 from Daqiao Saltern on the coast of yellow sea of China. Evid Based Complement Alternat Med 2011:615032

Chapter 7
Endophytes and Their Applications as Biofertilizers

Gaurav Yadav, Rishita Srivastva, and Preeti Gupta

Abstract Endophytes are microorganisms residing inside plant tissues. The endophytes are not harmful for the plant health, and besides this, they provide nutrients to plants as well as protect plants from stress conditions and from plant pathogens. The endophytic microbial community includes different genus of bacteria, fungi, algae, actinomycetes, and transgenic microbes like *Bacillus* sp. and *Piriformospora indica*. These microbial strains can be used as consortium or single species as biofertilizers. These biofertilizers are free from chemicals and have a number of benefits for agricultural crops such as they help in plant growth, act as biocontrol agents, protect plants from stress, and also help in the recovery of diseased plant, N_2 fixation, and IAA production, etc. This compendium will accentuate on the different types of endophytic microorganisms and their extensive role as a biofertilizer in the field of agriculture.

Keywords Endophytes · Agriculture · Biocontrol · Biofertilizers

7.1 Introduction

There are some microorganisms, especially bacteria and fungi, which can thrive on their plant host externally, and are commonly known as epiphytes, i.e., present on leaf surfaces, and some live internally and are known as endophytes (Afzal et al. 2019). Endophytes are the microorganism which resides in the inner part of the plants' body such as xylem vessels without vitiating or infecting the plant, and there are approximately 300,000 plant species present on earth which carry one or two endophytes (Afzal et al. 2019; Fadiji and Babalola 2020; Maela and Serepa 2019). Endophytes are of two types: obligate endophytes and facultative endophytes.

G. Yadav (✉)
Amity Institute of Microbial Technology, Noida, Uttar Pradesh, India

R. Srivastva · P. Gupta
Dolphin PG Institute of Biomedical and Natural Sciences, Dehradun, India

P. Bhatt et al. (eds.), *Microbial Technology for Sustainable Environment*,
https://doi.org/10.1007/978-981-16-3840-4_7

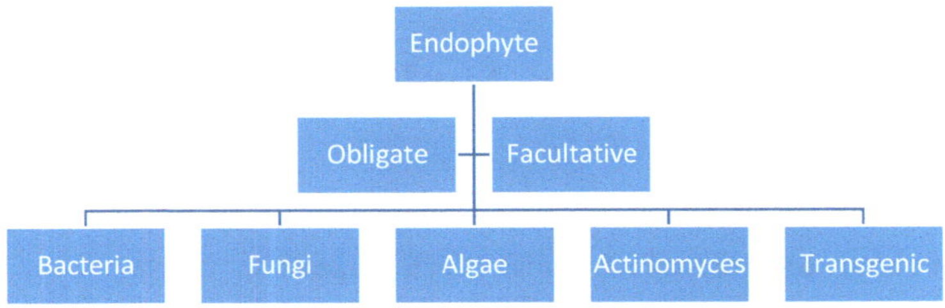

Fig. 7.1 Classification of an endophyte

Obligate endophytes are those which require more specific nutrients and strict conditions to grow in in vitro conditions; without them they cannot be cultivable, while facultative are those endophytes which do not need any specific nutrients or conditions to grow, i.e., non-fastidious in nature; they can harbour themselves in soil, artificial medium, and inside plants (Maela and Serepa 2019). These salutary microbes can be transmitted to various generation of plants through sexual units such as seeds and spores (Soldan et al. 2019). Wild species of dicotyledonous and monocotyledonous plants are also a hub of various endophytes. On the basis of various studies on endophytes, they are classified as bacteria, fungi, algae, actinomycetes, and transgenic endophytes as shown in Fig. 7.1. Archaebacterial and mycoplasma endophytes also exist in plants, but there are not enough studies on them (Maela and Serepa 2019; White et al. 2019).

These endophytes may perform various roles in plants such as adaptation and resistance against temperature stress and environmental stress and overcome the salinity, and they also help in stimulating plant growth, which may be direct and indirect. Direct roles are biological nitrogen fixation (BNF), phosphorus solubilization, potassium solubilization, siderophore production, IAA production, ACC utilization, ammonia excretion, lytic enzyme secretion, and phytohormone production, and indirect roles may be designated as production of metabolite, induction of plant resistance, and promotion of plant growth (Fadiji and Babalola 2020; Sansanwal et al. 2017). An endophyte can be a good candidate for biofertilizers. A biofertilizer is a consortium of viable microbes which when used on seeds, plants, and soil will colonize the rhizosphere and internal structures of plants, seed, etc. Furthermore, when these endophytes colonize the inner part of plants or its reproductive bodies, they exert direct beneficial effects. This exercise will obliterate the use of chemical fertilizers because of the adverse effect of chemical fertilizer on crops and on human health. They not only affect crops or humans but also deteriorate the quality of soil and damage the environment (Panpatte et al. 2017; Sansanwal et al. 2017). Endophytes also help in phytoremediation of heavy metals from soil, and they also produce gluconic acid (GA) (Oteino et al. 2015).

7.2 Mode of Transmission of Endophytes

Transmission is important for the survival of endophytes as well as for plant species because endophytes help plants in certain ways; thus, endophytes can be transmitted from parents to progeny and from one individual plant to another in the same community. Therefore, based on this view, the mode of transmission can be of two types (Kuzniar et al. 2019).

7.2.1 Vertical Mode of Transmission

It is a direct mode of transmission in which endophytes are transmitted from parents to offspring via seeds and through pollens.

7.2.1.1 Vertical Transmission Through Seeds

In the surface-sterilized seeds of various plant species such as alfalfa, maize, rice, coffee, tobacco, barley, pumpkin, and quinoa, various bacterial species are found to be more dominating than other microorganisms. These endophytic bacterial species have been also detected in a range of wild plant species, including *Pachycereus pringlei*, *Lolium rigidum*, *Eucalyptus* spp., and *Picea abies*. These endophytes are found to be present in different parts of seeds such as embryonic tissues, coat, and endosperm. *Bacillus*, *Pseudomonas*, *Staphylococcus*, *Acinetobacter*, and *Micrococcus* are examples of some common bacterial genera which are present as endophytes in seeds. Endophytes are comparable to biofertilizers as they are known to impart beneficial properties to plants. For example, in rye grass, they produce cytokines which help in releasing seed dormancy. Endophytes help in fixing nitrogen and help plants to survive in extreme environments. In vitro tests demonstrated that they also have antifungal properties against fungal plant pathogens. It was observed that in rice seed if endophytes are removed, seedling of the rice seed will become restricted (Frank et al. 2017). Those endophytes present inside the seeds protect plants against diseases and abiotic stress. They also help in better germination and preserve seeds for a longer period of time (Girsowicz et al. 2019).

7.2.1.2 Vertical Transfer Through Pollens

Male gametes are a possible way for transmission of endophytes, but pollens are also being a method of horizontal way of transmission because they can easily be colonized by the atmosphere, animals, and through pollinators. *Enterobacter* spp. were isolated from fertilized *P. brutia* ovules. 10^6–10^9 numbers of bacteria are found on per gram of pollen; they are present as clusters, biofilms, and single cells. There are different types of endophytes and it depends upon plant species, plant-specific endophytes, antimicrobial content of pollen or plants, and nutritional composition of plants.

7.2.2 *Horizontal Mode of Transmission*

It is an indirect mode of transmission which occurs between two individual plants in the same community. This mainly happens in fungal species in which fungal spores can be easily dispersed among plants through air, animals, and flying insects. Soil is the main source of microorganisms for both below and above the ground because soil contains most of the vital nutrients which are needed to survive in the environment. In the horizontal mode of transmission, endophytes can colonize seeds and roots via soil, and from soil, endophytes first colonize spermosphere, and in this endophytes have beneficial effects on the germination process. Secondly, they can also colonize through the root endosphere region via rhizosphere where vast numbers of microorganisms are present which promote plant growth and also protect against plant pathogens. Flowers and fruits and plant insects can also be a horizontal mode of transmission and colonization of leaves through stomata (Frank et al. 2017).

7.3 Bacteria

Previously isolated endophytic bacteria from Napier grass and characterized by molecular description include *Sphingomonas*, *Bacillus*, *Enterobacter* sp., and *Pantoea* sp. Endophytes were described after sequence analysis and after analysis of the phylogenetic relationship. In this study, representative isolates were selected for the purpose of observing their sequence-based plant developmental capacity and development relationship. Lately, four iso-member endophytic bacteria are capable of colonizing the plant root more rapidly. Plant growth-enhancing features such as production of IAA, production of siderophores, production of ammonia, solubility of phosphates, fixation of nitrogen, and ACC deaminase are the properties of endophytic bacteria and endophytic bacteria also produce some chemicals which have antagonist properties as shown in Table 7.1.

Endophytic bacteria have been able to cause an increase in shoot and length in hybrid pennisetum relative to inoculation controls in both saline and natural environments and to colonize host plant roots. Up to 200 mM NaCl were retained. These PGP attributes are therefore adorable to endophytic bacteria and favourable for growth of plants and higher yields in unfruitful soil, salt ground, and infertile areas. With this peculiarity of the endophytic population, the bioinoculants of agriculture will obviously be used as better biofertilizers. The growth of plants encourages endophytic bacteria and fuels plant growth by discrete mechanisms. They encourage phytohormone production, nutrient absorption (Bibi et al. 2012;

Table 7.1 Benefits and some chemicals produced by endophytes which help plants to grow and survive in stress condition, and these chemicals also have antagonist properties

Benefits of endophytes	Chemicals produced by endophytes	Bacterial spp.
Antibiotics	Antioxidants enzymes	*Acinetobacter*
Antiviral	Volatile compounds	*Cellulomonas*, *Azotobacter*
Immunosuppressive and anticancer compounds	Phytohormones	*Clavibacter*, *Azospirillum*
Antibacterial	IAA (indole acetic acid)	*Pseudomonas*, *Bacillus species*
Antifungal	Cytokinin	*Curtobacterium*, *Pseudomonas*
Antiprotozoal	GA (Gibberellic acid)	*Microbacterium*
Bio-insecticides		

Mei et al. 2014; Sturz et al. 2000), and biocontrol by reduction of phytopathogens. Throughout the plant life cycle, it is created at various times, through various mechanisms, to promote growth and produce endophytic bacteria (Glick 2003). The bioinoculant preparation can be used as a biofertilizer in the field in conformation of endophyte bacterial consortia by using these endophytic bacteria. The use of mixed useful endophytes in agriculture increases the consistency of the soil and eventually helps grow plants. Besides other PGP attributes, ACC deaminase is present in endophytic bacteria activity, and thus the plant root is capable of reducing ethylene levels compared to other endophytes which was recorded earlier with ACC deaminase activity. ACC (ethylene precursor) is converted to ammonia and alkaline butyrates using ACC deaminase enzymes in a stressed setting such as acidic soil conditions (Jha and Kumar 2009; Jha et al. 2012; Glick 1995; Alexander and Zuberer 1991; Burd et al. 2000; Nabti et al. 2010) and promotes the expansion of plants in adverse conditions. High deaminase activity (about 225.2–1106.6 nmol-KB/h/mg) in the endophytic bacterial population has been documented in contrast to non-endophytic deaminase activity (approximately 20 nmol-KB/h/mg) (Glick 2003). The use of plant-building endophytic bacteria as biofertilizer in a region, however, requires attention to maximize the gain within the host plant. In order to increase crop size, and growth use of different endophtes will be beneficial. Promoting low-dose chemical fertilizer and increased use of different biofertilizer has proven to be effective (Wolinska et al. 2017; Singh et al. 2021; Zhang et al. 2020a, b; Mishra et al. 2020; Feng et al. 2020; Lin et al. 2020; Zhan et al. 2020; Ye et al. 2019; Huang et al. 2019, 2020; Fan et al. 2020; Pang et al. 2020; Gangola et al. 2018; Gupta et al. 2018). By increasing the absorption of nutrients in plants and also by preserving soil microbial floral dynamics, biofertilizers are increasing soil fertility (Bhatt et al. 2020a, b, c, d, e, f, 2021a, b, c). Plants are immune to harmful conditions and pesticides by biofertilizers. By plant nutrient intake and also retaining the microbial dynamics of soil, biofertilizers improve soil fertility to enhance their health. Endophytic microbials are discussed in the current sense of how they can make better crop production. Studies show that PGPP (plant growth promotion) attributes are present in both Rhizobium and Pseudomonas and that they make plant nutrients accessible through metabolic modifications, phosphate solubilization, iron chelation, and many more attributes (Bhatt et al. 2015a, b, 2016a, b, 2019a, b, c; Huang et al. 2021).

7.4 Nitrogen-Fixing Endophytes

In the agricultural outlook, some important advantage of endophytes is nitrogen fixation, siderophore production and antagonism against phytopathogens (Franche et al. 2009). In soil, diverse kinds of microorganisms are found including bacteria, fungi, protozoa actinomycetes, and algae. Among all these spp., *Rhizobia* were studied extensively in various aspects like physiological, biochemical, and molecular level to facilitate plant growth and to increase plant productivity without fertilizers, and their applications are widely used in maize, wheat, etc. Diazotrophic endophytic bacteria not only promote plant growth; besides they produce a variety of compounds that have antagonistic properties and are used as a defense system as shown in Fig. 7.2. Endophytic microorganisms not only remove nitrogen from the environment but also provide nitrogen to plants for their stable growth.

The relation between nitrogen fixation and bacteria is very important but how endophytes fix nitogen is needed to understand and these are some organisms which help in fixing the nitrogen, *Azoarcus*, *Azospirillum brasilense*, *Burkholderia* spp., *Klebsiella*, *Gluconacetobacter*, *Herbaspirillum*, and *Serratia*, and many others

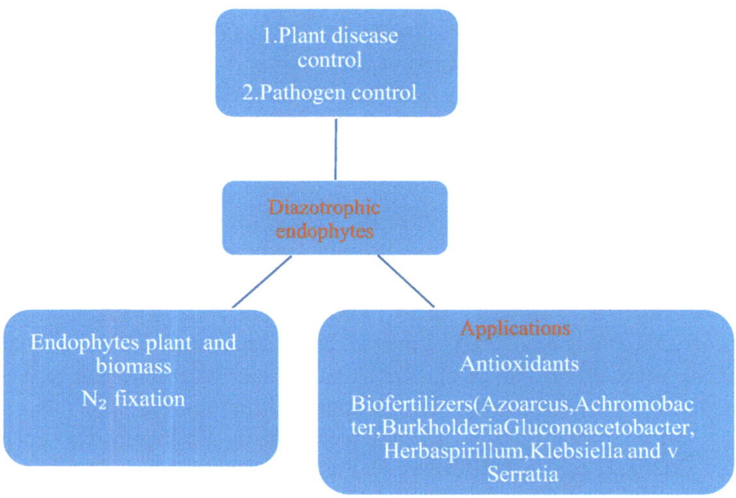

Fig. 7.2 Diazotrophic endophytes as biofertilizers

availability much easily (Rothballer et al. 2008; Ahmad et al. 2008). Various bacteria such as *Azoarcus* spp. produce siderophores to deplete the deficiency of iron (Newton 2000) by provided iron to plants by soil bacteria help in several situations like heavy metal pollution. However, the concentration of siderophores affects the growth of plants. Nitrogen attachment mechanism is performed in the *nif* genes, and a total of 16 ATP is consumed throughout the process. As compared to the free-living bacteria, symbiotic association fixes more nitrogen. Oxygen works as an inhibitor for enzyme nitrogenase and acts as a negative regulator of nif gene expression. The nitrogenase is responsible for catalysing reduction of (N_2) to ammonia (NO_3) (Gupta and Nath 2015). Nitrogenase enzyme is sensitive to oxygen that irreversibly inactivates diazotrophs (N_2) fixation which, on the other hand, permits supply of oxygen required for every regeneration and protects from deleterious effect of O_2. In this situation, bacterium *Rhizobium* spp., which populate internal plant tissue niche with or without minor complaint to host plants, switch on a process that is known as biological nitrogen fixation (Gupta et al. 2012). The ongoing research assuming that some endophytes such as diazotrophic *Gluconacetobacter*, *Serratia marcescens*, and *Azoarcus* sp. will be good as biofertilizers in agricultural areas (Annapurna et al. 2018). They are also suitable for use.

Information is also available on N fixation by non-legumes and bacteria other than rhizobia (Tanjung et al. 2017; Ohyama et al. 2014). *Sphingomonas azotifigens* were first identified in India as a novel bacterium when they were N fasteners (Naylor et al. 2017; Singh et al. 2017). The earlier documented or identified plant pathogens were *Stenotrophomonas maltophilia* and *Herbaspirillum rubrisubalbicans* that showed high N fixation and potentials for auxin production. Diazotrophic endophytes bacteria colonize the palm leaves and also found that *Bacillus cereus* do excessive nitrogen fixation (Aryantha and Hidiyah 2018; Sharma and Roy 2017). For *Exiguobacterium*, the ability to attach N has also confirmed N4 strong strain of *Amaranthus spinosus* colonization (Rodriguez et al. 2009).

7.5 Fungal Endophytes

Microorganisms have provided a large variety of biologically active compounds, which have broad applications in human health and diseases. Fungus is ubiquitous in nature, and it lives inside the host plant for being a part of their life cycle without causing visible proclamation. In 1928, antibiotic penicillin from *Penicillium notatum* was discovered by Alexander Fleming, which was used later during the Second World War. After that, various antibiotics were discovered, such as Chloramphenicol, Streptomycin, Tetracycline, Griseofulvin, Cyclosporine, and Taxol from different fungal species, and therefore, this period is known as the Antibiotics Era. After that in the twenty-first century, novel antibiotic compounds were isolated from different fungal species as it has the ability to produce potential pharmaceutical products which will be used for various purposes (Naylor et al. 2017; Singh et al. 2017; Aryantha and Hidiyah 2018; Sharma and Roy 2017; Rodriguez et al. 2009).

7.5.1 Classification of Endophytic Fungi

Fungi are classified into two major groups on the basis of their evolutionary relatedness, taxonomy, plant hosts, and ecological functions.

1. Clavicipitaceous endophytes
2. Non-clavicipitaceous endophytes

Apart from these two main groups, there are four classes of endophytic fungi:

- Class I endophytic fungi
- Class II endophytic fungi
- Class III endophytic fungi
- Class IV endophytic fungi

7.5.1.1 Class I Endophytic Fungi

Clavicipitaceous endophytes are the class I endophytes reported in the late nineteenth century by European investigators from seeds of *Lolium arvense*, *Lolium temulentum*, and *Lolium remotum* (Arnold et al. 2002).

7.5.1.2 Class II Endophytic Fungi

The class II endophytic fungi contain a great diversity of species belongs to Dikaryomycota, which consists of Ascomycota and Basidiomycota. In 1915, Rayner reported a phoma species called *Calluna vulgaris*. The crucial point is that they always colonized all parts of the plant including the seed coat. Now recently, they become a common root endophyte that confers benefit to plant.

7.5.1.3 Class III Endophytic Fungi

This class is differentiated on the basis of occurrence, that is, above-ground tissue; this class transmits horizontally and is extremely high in plant biodiversity. This class includes the hyper/high endophytic fungi that are associated with the leaves of tropical trees (Suryanarayanan et al. 2005). Class III endophytes span from above-ground tissues of non-vascular plants to the seedless vascular plants, conifers, and herbaceous angiosperms in biomes and differ from tropical forests to Boreal and Arctic/Antarctica communities (Arnold et al. 2002).

7.5.1.4 Class IV Endophytic Fungi

The discovery of class IV endophytes is concluded during the study of ectomycorrhizal fungi, where Naik Shankar et al. 2008 remarks a pigmented fungus, which is brown to blackish in appearance, and it is associated with terrestrial plant roots that are called as MRA. This species is found with mycorrhizal fungi and implied as pseudomycorrhizal fungi. Recently, these fungi are mentioned as DSE (dark septate endophytes) and come under class IV endophyte. A century has passed since it was discovered, but the role of these queer fungal symbionts is still unknown. Apart from diseases, fungi are beneficial to humans as they produce numerous alkaloids (ergot alkaloids from *Claviceps*), some enzymes (cellulose, lipase, ligninolytic enzymes), and also some pigments (anthraquinone, betalains, aroma, and flavours); besides these, they also play an important role in the control of nematodes. Some fungi are edible like mushrooms and they are useful to our health as they are abudant in minerals, vitamins and proteins like selenium, potassium, riboflavin, niacin, vitamin D, proteins. Endophytes are reported from plants that can be found in various environments including trophic, temperate, aquatic oceans, xerophytic deserts, Antarctic, geothermal soils, rainforests, mangrove swamps, and also coastal forests (Naik Shankar et al. 2008). Endophytic fungi from 15 shrubby medicinal plants grow in Malnad region and southern India (Raghukumar 2008). She also reported that the isolation of a greater number of endophytic fungi happens in the winter season rather than in monsoon and the summer season. In India less number of discoveries are available or we can say that few reports are available on endophytic fungi, blooming in mangrove plants. It is interesting that mangrove plants have shown to adapt to anaerobe soil, muddy saline waters, and brackish tidal activation. Narayanan et al. (2014) reported on cell wall enzymes such as pectinases, proteases, and pectate transeliminase from endophytic fungi and their leaf litter degradation activity when exposed to extreme conditions in terrestrial and marine environment such as high temperature at tropical areas, elevated hydrostatic pressure, low temperature in deep sea, and aloft hydrostatic pressure. Like all other endophytes, endophytic fungi with mangrove plants protect mangrove from adverse environment conditions (Jeffery 2008). Many species of fungi include *Aspergillus* (Narayanan et al. 2014; Jeffery 2008), *Beauveria bassiana*, *Trichoderma harzianum*, *Lecanicillium lecanii*, *Metarhizium anisopliae*, and *Fusarium* spp. According to reports, most common endophytes are *Aspergillus*, *Phomopsis*, *Wardomyces*, *Penicillium*, and many unidentified fungi. Species *Colletotrichum gloeosporioides*, *Fusarium solani*, *Pestalotia*, and *Phomopsis* were predominant endophytes though *Colletotrichum gloeosporioides* establish as endophyte and it is a pathogenic fungus for Cashew tree, which is reported by some authors during studying different hosts. A substantial research was go through using several other fungi isolated mainly a epiphytic in an attempt to Jurisdiction pathogen *C. gloeosporioides* being a thrichoderma strain, the most hopeful one. Fungal endophytes have the most unique adaptation in nature; with respect to future prospects that how endophytes communicate with each other in contrast to their pathogenicity for such innovative purposes, endophytic fungi should be an effective topic to study.

7.6 Actinomycetes Endophytes

Actinomycetes are one of the most abundant groups of microbes which is extensively distributed in nature. Predominantly, they are found in dry alkaline soils (Fouda et al. 2018). Endophytic population was greatly impacted by climatic conditions and locality where their host plant grows (Waheeda and Shyam 2017). Endophytes reside within plant tissue to finalize their life cycle, despite no deleterious effect, and play a noteworthy role in aggravate growth of host plant by producing phytohormone and build plant tolerant to various stresses, other growth-encouraging factors in order they are compensation with nutrients and shields within host plant as shown in Fig. 7.3.

Bacteria present in the rhizosphere is different from endophyte because endophytes are aunthetic and specific to its habitat as they resides within the plant tissue (Kumar and Jadeja 2016). Endophytic actinobacteria are examined to be a substitute to combat multidrug-resistant human pathogens as they serve as a latent source of novel antimicrobial compounds (Limaye et al. 2017). Plant diseases can be reduced by actinomycetes using distinct mechanisms. A method known as pyrosequencing is used to disclose a wide range of bacteria that live in and around roots of plants such as *Actinobacteria*, *Bacteroide*, *Verrucomicrobia*, and *Proteobacteria* (Limaye et al. 2017). Some actinomycetes secrete a range of enzymes that can entirely degrade all components of lignocelluloses such as lignin, hemicelluloses, and cellulose (Andreote et al. 2014). Because of their ability to secrete enzymes, they are virtual in attacking raw beer. Bioactive molecules from actinobacteria and their biosynthetic

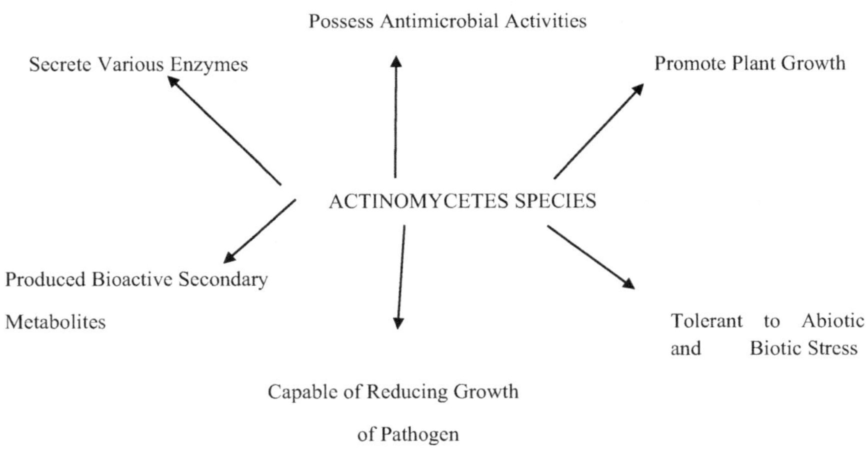

Fig. 7.3 Actinomycetes species and their various applications

genes have been less studied; bioprospecting of actinobacteria for bioactive molecules holds great promise.

There is an increasing necessity for discovery of new drugs, due to the increasing threats day by day which incorporate drug-resistant pathogens. In order to achieve this, novel strategies are needed to be applied to search for a new molecules to fight against resistan microorganism. One of which is "Renaissance" in antibacterial discovery from actinomycetes. Internally colonizing microbes play various roles in widening and plant fitness. A few of them also cause diseases (Yang et al. 2015) but the interaction is said to be beneficial. Antibacterial compound-producing strains belonging to genera *Streptomyces*, *Nocardiopsis*, *Pseudonocardia*, *Agrococcus*, and *Isoptericola* have been reported from mangrove plants in Beilun River, Beilum Estuary National Nature Reserve, China (Purushotham et al. 2018). Actinomycete strains have genetic ability to produce 10–20 secondary metabolites. *Pseudowintera colorata* (Horopito) is an indigenous medicinal plant of New Zealand. In the study of *Pseudowintera colorata endophytes* it was observed that microbial communities in roots and stems are not diverse but in leaves microbial communities were found to be spare diverse on the basis of DGGE pattern. Profusion of actinobacteria taxa was steeper in stems (39%) and roots (27%). However, three clones among them were identified as uncultured bacteria (Oskay et al. 2004). Compounds from the rare organisms include *Teicoplanin* and *Actinoplanes teichomyceticus* (Kaaria et al. 2012). In vitro actinomycetes isolated from Turkey's farming soil have shown the ability to inhibit *Erwinia amylovora* bacteria that cause fire blight to apple and *Agrobacterium tumefaciens*, a causal agent of crown gall diseases. In view of the increasing threats day by day which include drug-resistant pathogens, there is an increasing need for discovery of new drugs. In order to, this challenge, novel strategies are to be applied to searching new molecules.

7.7 Algae Endophytes

We all know that endophytes have a symbiotic association with plants, and sometimes they act as a biocontrol agent because they protect plants from animals to produce certain compounds which help plants to become safe. When we heard about the word endophyte, automatically in our mind bacterial endophytes strike. In various aspects of life, algal endophytes also play an important role (Bacon and White Jr. 2000). Marine algae is found in coastal regions including some cyanobacteria and other microbes. The marine microbes were isolated from red algae, green algae, and brown algae as shown in Fig. 7.4.

In red algae, brown algae and green algae, red algae has highest bioactivity among all. It is found that marine algae harbour some epiphytic and endophytic microbes that produce antimicrobial substances that can inhibit human pathogens. They also produce bioactive compounds that can be used by plants (Fremlin et al. 2009). Besides this, they produce some antiprotozoal, antiparasitic, antiviral, and

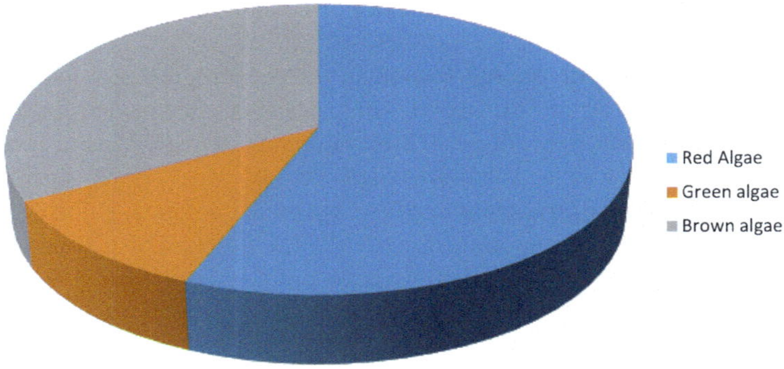

Fig. 7.4 Pie chart indicating the bioactivity (%) based on class of algae

antitumour activities (Vesterlund et al. 2011). It is shown that host plant tolerance to various stress, such as abiotic, is increased by endophytic infection (Gonzalez-Bashan et al. 2000) and biotic. There are some naturally grown microalgae which are found to be associated with bacteria (Suminto and Hirayama 1997). Currently it is seen that the plant growth-promoting bacterium (PGPB) *Flavobacterium* spp. was built to aid the growth of a marine alga (*diatom Chaetoceros gracilis*) which is used as a feed in oyster hatcheries (Correa et al. 1987). With the help of the above example, we understand how bacteria and microalgae are mutually beneficial for each other.

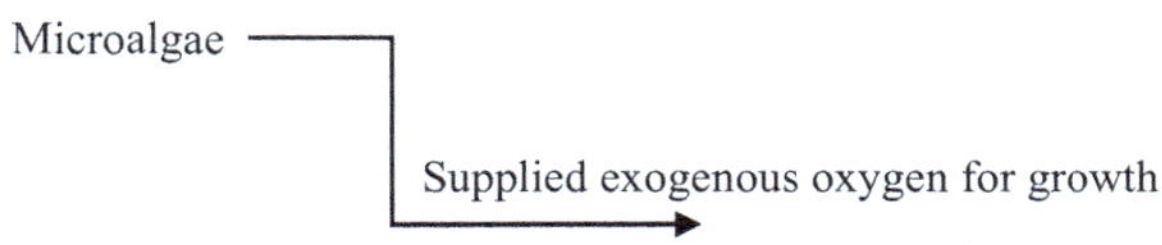

7.7.1 *Chlorella vulgaris*

There are several advantages of PGPB in which environmental application is most important, For example, *Azospirillum* species add in wastewater for its treatment with microalgae to enhancing the development microalgae metabolites. Microalgae have numerous uses such as water bioremediation. *Chlorella vulgaris* and *Scenedesmus dimorphus is* a unicellular microalgae, which is used for separating the phosphate and nearly all of the ammonium from dairy industry and pig farm wastewaters. Now in such studies, microalgae were in a shelving. Although these crisp populations of such algal endophytes are pigmented (Plumb 1999; Correa et al. 1988). Some endophytes have been capable of nutritional independence; this indicates that they have cultured separately from their host Acrochaete heteroclada (Nielsen 1979; Kumar et al. 2015) have previously been allocated to green algal *Chondrus crispus*-associated endophytes. A variety of endophytic strains are host to Stackhouse Red Algae *Chondrus Crispus*, including pigmented algae, Multicellular species (Chlorophyta) and pigmented brown (Heterokontophyta) respectively.

7.8 Methanogens as Endophytes

Methylotrophs are those, which can grow by using reduced carbon sources such as methanol or methane can grow by apply the diminish carbon substrates (Kumar et al. 2016; Meena et al. 2012) some are found advantageous and non-pathogenic for plants (Pirttila et al. 2005; Holland and Polacco 1992) during study of soybean seedling. Endophytic methylotrophic bacteria have been marked to increase seedling growth closely with root biomass growth (Dourado et al. 2012). Some methylotrophic endophytic reported are *Methylobacterium* sp., *Methylovorus mays*, mesophilic *methylobacterium extorquens* and *methanotropics*, directly or indirectly, plant growth plants directly or indirectly (Ferreira et al. 2008; Raghoebarsing et al. 2005). Some methanotrophs are described as biofertilizers to assembled the agricultural development up to the agricultural fields (Keerthi et al. 2015; Rekadwad 2014; Dourado et al. 2015). Ubiquity and worldwide appendage of genus *Methylobacterium* sp. is found as Epiphytic and Endophytic bacteria and PPFM of *Methylobacterium* sp. are found (pink pigmented facultative methylotroph) the biotechnology and agronomic future group is reported (Baldani et al. 2000). In the current conclusion, PPFM and the spinal pseudomonas Biofertilizers have differed and the intensification of plant growth has been shown in the field to be continuously forward with positive microbial soil production. Endophytic *Methylobacterium* species NPFM-SB3 was found to be remote from the Sesbania stem nodules, a symbiotic company with rice plant can be systemized (Gyaneshwar et al. 2005; James et al. 2000). Recently during Fe famine, *Methylobacterium phyllosphere* (insulated from the lowland rice phyllosphere) triggered the hydroxamate of the form siderophores, as well as tryptophan and tyrosine during Fe famine. The main forebear for methane producing by methanogenesis is frequently acetate, which is one of the largest products from anaerobic digestion of organic compound by bacterial metabolism (Karakashev et al. 2011; Smith and Mah 1978; Thauer 1998; Ardanov et al. 2011). Methylobacterium fertilizer and the priming ability of *Methylobacterium* sp. have been used in potato development. Also, IMBG290 was observed. The priming of plants with beneficial bacteria and advantage of using beneficial bacterial strain is that they induce host plant rescue energy and decrease growth and growth time. Plant priming the host saves energy and reduces the time taken by non-pathogenic bacteria during a pathogen attack for the development of the defense reaction (Doty 2008).

7.9 Transgenic Endophytes

Formerly research was conducted on plants to improvise its ability to reduce environmental pollution. Genes from many plants, microbes, and animals were inserted into plants to enhance their pollution degrading abilities, known as transgenic plants (Yang et al. 2017). For extensive use of endophytes in the field of agriculture and forestry, some useful genes can be inserted into endophytes to introduce new features into the microbes. In the study conducted by Glandorf et al. (2001) to control lepidoptera larvae, in which the insecticidal protein gene of *Bacillus thuringiensis* was inserted bacterial endophytes *Burkholderia pyrrocinia* JK-SH007 to control lepidopetra larvae. Cloning of cry218 gene was done using PCR, and PHKT2 expression vector was used for the introduction of gene into JK-SH007 (Glandorf et al. 2001). *Pseudomonas putida* WCS358r was genetically modified. Mini-Tn5 lacZ1 transposon was used as a delivery vector (size 6.8 kb), in this *Bgl*II-XbaI fragment was incorporated that carrying the *phzABCDEFG* genes from Pseudomonas fluorescens 2–79 and inserted into *Pseudomonas putida* to produce phenazine-I-carboxylic acid (PCA), which shown antifungal action, when released into wheat plant rhizosphere (Patra et al. 2017). Transgenic endophytes will have bright future in the field of agriculture because they can be used as biofertilizers, biocontrol agent and for plant growth promotion also by inserting the gene of interest into the endophytes.

7.10 Role of Endophytes as Biofertilizers

Biofertilizer is a consortium of living microorganisms together with minerals plus nutrients that help in plant growth promotion without changing soil properties; they also secrete secondary metabolites and bioactive compounds; biosynthesis of bactericides and fungicides protects plants from environmental stress (Kuzniar et al. 2019; Lobo et al. 2019). Biofertilizers provide plants a great amount of minerals such as potassium, phosphorous, nitrogen, and siderophores as well as exopolysaccharides (Lobo et al. 2019). Optimal condition must be followed to produce low cost-efficient inoculant consortium which can promote plant growth (Patle et al. 2018).

7.11 Plant Growth-Promoting Endophytes

Plant growth-promoting endophytes have vast functions like IAA production, phosphate solubilization, siderophores, N_2 fixation, and ACC utilization as shown in Fig. 7.5.

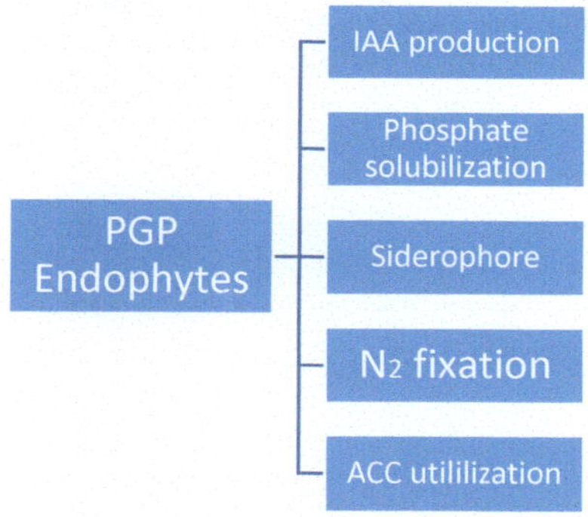

Fig. 7.5 Different roles of PGP endophytes

7.12 IAA Production

There are a lot of endophytic bacteria and fungi present, which can synthesize many phytohormones such as auxin, cytokines, and gibberellins, in which indole-3-acetic acid is very common and it contains the carboxyl group which is joined to the third carbon of indole group (Maela and Serepa 2019; Zahir et al. 2003; Valérie et al. 2007). IAA helps in cell division and enhances the root length and hair, thus increasing the number of sites for infection and nodulation. This change will help in better absorption of nutrients and in turn stimulate plant growth (Zahir et al. 2003; Susilowati et al. 2018). IAA helps in stimulation of tuber and germination of seed, it also increases the rate of root and xylem development, it stimulates pigment formation, and it also mediates the responses to gravity, florescence, and light and also affects photosynthesis (Sansanwal et al. 2017). It is the direct mechanism of PGPR that stimulated the growth and yield of plants but a high level of IAA can lead to abnormalities in plants during its development stage, and the low level of IAA stimulates root elongation (Maela and Serepa 2019). The production of IAA is stimulated by the amino acid known as L-tryptophan which acts as a physiological precursor for IAA production in plants and microorganisms. It was observed that the production of IAA can be increased up to 2.7 times in the presence of L-tryptophan amino acid (Audipudi et al. 2017). An increase in the amount of nitrogen in soil leads to the production of auxin that softens the cell wall, thus increasing the water retention capacity by increasing the cell size, and the addition of these cells can increase the weight of rice. Biofertilizers containing a mixture of *Azospirillum*, *Pseudomonas*, and *Bacillus* increase nutrient uptake and plant growth and also increase the size of the rice grain (Audipudi et al. 2017). Bacteria can convert tryptophan to IAA through four pathways, which include Indole-3-acetamide (IAM), Indole-3-pyruvate (IpyA), Tryptamine (TAM), and Indole-3-acetonitrile (IAN) pathways. The indole-3-acetamide pathway forms directly IAA without any intermediate compound. IAA-producing bacteria through the indole-3-pyruvate (IpyA) pathway are known, and the pathway is illustrated below in Fig. 7.6 (Audipudi et al. 2017).

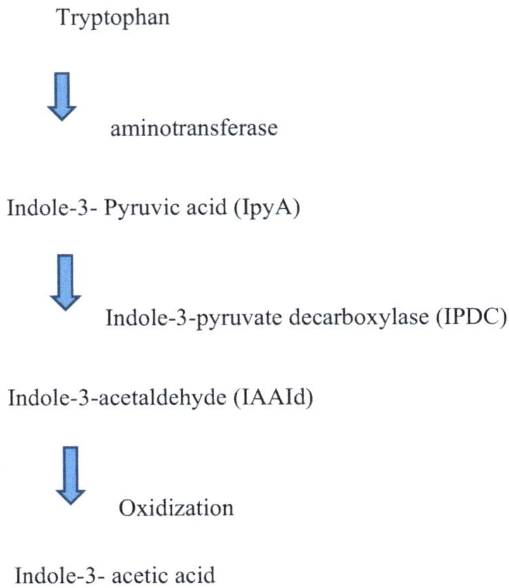

Fig. 7.6 IAA production through IpyA pathway

In one study the authors showed that 15 endophytes isolates were positive for IAA production out of 65 endophytes, and they produce 25 μg/mL of IAA. *Acetobacter diazotrophicus* and *Herbaspirillum seropedicae* produce IAA in chemically defined culture media. They tested ten isolates of *Typha australis*, of which seven isolates were positive for IAA production (Matos et al. 2017). In a study genera *Bacillus*, *Micrococcus*, *Escherichia*, *Staphylococcus*, and *Pseudomonas* were tested with wild herbaceous flora for the level of IAA production and growth of *Triticum aestivum*. Results were obtained using Gas Chromatography and Mass Spectrometry (GC-MS), which shows enhanced root length and seed weight by 16% and 70%, respectively, by bacterial endophytes (Maela and Serepa 2019).

7.13 Phosphate Solubilization

Phosphate is the second most essential macronutrient which plants need for their growth and development after nitrogen, and the average amount of phosphorous found in soil is 400–1200 mg/kg of soil. Soil types and its pH determine the precipitation and adsorption of phosphorous (Gull et al. 2004). It is hard to be utilized by plants itself because phosphate basically presents either as a mineral or in an insoluble form, which has poor mobility, and this is due to the reactivity of phosphate ion with other constituents present in the soil (Maela and Serepa 2019; Lin et al. 2013; Wei et al. 2018). Endophytic microorganisms secret some organic acids such as citric, acetic, succinic, and oxalic acid, and they also secret phosphate and also use mechanisms like acidification, chelation, and ion exchange, thus making phosphorous available to plants by converting it into soluble mono- and dibasic forms. Some environmental factors like pH, oxygen, humidity, and temperature can affect the whole process of phosphate solubilization (Afzal et al. 2019; Maela and Serepa 2019).

In one study, different bacterial endophytes were isolated from the two types of medicinal plants such as *Zingiber officinale* and *Azadirachta indica* on different culture media. After the isolation, isolates were screened for various activities like IAA, phosphate solubilization, siderophore production using biochemical characterization, morphological identification, and molecular ribotyping; fives isolates were characterized as *Bacillus tequilensis* (AAU K1), *Bacillus endophyticus* (AAU K2), *Beijerinckia fluminensis* (AAU K3), *Bacillus safensis* (AAU K4), and *Pseudomonas aeruginosa* (AAU K5). From the wheat root *Penicillium radicum* was isolated which is a phosphate solubilizing fungus; it shows plant promoting activity in vitro. Phosphate solubilizing endophytic bacteria can play a very important role in the field of agriculture as the demand for biofertilizers is increasing (Maela and Serepa 2019; Lacava et al. 2008).

Some of the compounds present in acidic soil such as weak aluminium, oxyhydroxides, and iron oxide can retain phosphorus and that results in a low amount of phosphorous. Phosphate fertilizers are of great importance, but in alkaline soil calcium leads to less efficient solubilization of phosphate fertilizers (Sansanwal et al. 2017; Panpatte et al. 2017). In one study, a total of 40 endophytic isolates were isolated from the root of banana tree and evaluated to check the phosphate solubilizing activity, in which 67.5% isolates solubilized phosphorus from tricalcium phosphate in solid medium and 7.5% isolates in soy lecithin solubilized phosphorus. In iron phosphate containing medium no isolates show P solubilization activity. Isolates *Aneurinibacillus* sp. and *Lysinibacillus* sp. showed best solubilization activity, and other genera and species that exhibited positive results were *Acetobacter* sp., *Agrobacterium tumefaciens*, *Bacillus amyloliquefaciens*, *Aneurinibacillus* sp., *Bacillus subtilis*, *Bacillus pumilus*, *Streptomyces* sp., *Micrococcus luteus*, and *Bacillus* sp. (Panpatte et al. 2017). Endophytes also prevent the absorption and fixation of phosphate when its amount is not sufficient (Afzal et al. 2019). Phosphate solubilizing bacteria (PSB) can solubilize inorganic phosphate, which is present in the form of $Ca_3(PO_4)^2$, $FePO_4$, and $AlPO_4$ by producing hydroxyl PGPB, organic acid, and siderophores in soil (Sansanwal et al. 2017).

7.14 Siderophore Production

Siderophores are the low molecular weight iron-containing complex present on the bacterial membrane in the form of Fe^{3+} and then get reduced to Fe^{2+} released into the cell from siderophore complex through the gating mechanism. When there is a high level of contamination of heavy metals and when plants starve from iron availability, bacterial endophytes make it available to plants (Fadiji and Babalola 2020; Sansanwal et al. 2017). Siderophores sometimes also serve as a great biocontrol agent such as phenolate, hydroxymate, and catecholate type which are produced by endophytes species, and they also help in fixing atmospheric nitrogen in diazotrophic organisms because for the functioning and biosynthesis of enzyme nitrogenase (key enzyme in N_2 fixation) diazotrophic organisms require Fe^{2+} and Mo factor (Fadiji and Babalola 2020). In the absence of heavy metal contamination, absorption of iron can be performed by other mechanisms like iron chelates directly lead to siderophores absorption or through ligand exchange. *Pseudomonas* strain GRP3 which is a siderophore-producing strain was tested for iron nutrition on *Vigna radiata*. Forty-five days of evaluation with endophytic organism, showed betterment in iron availability and in chlorotic symptoms and there is also an increment in the amount of chlorophyll a and b as compared to the control results in plants (Fadiji and Babalola 2020; Sansanwal et al. 2017). Actinomycetes endophytic species like *S. acidiscabies E*13, *Streptomyces* sp. Mhcr0816, *Nocardia* sp., *Streptomyces* sp. UKCW/8, and *Streptomyces* sp. GMKU 3100 are great producers of siderophores (1). In one study conducted on *Methylobacterium* sp. which belongs to the same niche as *X. fastidiosa* subsp. *pauca* (Xfp) isolated from citrus variegated chlorosis (CVC) showed that catechol-type siderophore is not produced by *Methylobacterium* sp. and it is the producer of hydroxamate-type siderophores (Tang et al. 2019).

7.15 N_2 Fixation

Endophytes perform various vital functions for plants, in which nitrogen fixation is one that abolishes the use of chemical fertilizer to prevent crops as well as the environment from harmful effects. Basically, nitrogen fixation is done by legume plant rhizobium. For the first time in India *Sphingomonas azotifigens* is a novel bacterium reported to be N_2 fixing bacteria. *Bacillus cereus* showed highest N_2 fixing ability in oil palm leaves. *Herbispirillum rubrisubalbicans* and *Stenotrophomonas maltophilia* were considered as plant pathogens but were efficient to do biological nitrogen fixation. *Exiguobacterium profundum* strain N4 which colonizes *Amaranthus spinosus*, *Azospirillum*, *Rhizobium*, *Agrobacterium*, and *Sphingomonas* in wild rye tissue also performs nitrogen fixation (Kuzniar et al. 2019). One non-rhizobial endophytic consortium which shows very promising results in BNF includes *Achromobacter*, *Burkholderia*, *Herbaspirillum*, *Azoarcus*, *Gluconacetobacter*, *Serratia*, and *Klebsiella*. In sugarcane *Gluconacetobacter diazotrophicus* is an endophytic organism which has the ability to fix nitrogen up to 150 kg N ha^{-1} $year^{-1}$. In kallar grass, obligate endophytic diazotrophs known as *Azoarcus* fix nitrogen (Sansanwal et al. 2017). A fungus known as *Phomopsis liquidambari* in rice promoted better functioning of nitrogen and phosphorus by inhibiting or promoting various indigenous soil microbes (Card et al. 2016).

7.16 1-Aminocyclopropane-1-Carboxylate (ACC) Utilization

Ethylene is the most essential plant hormone or metabolite which every plant has and that helps in the growth and development of the plant. It is affected by different abiotic and biotic stresses, and it also controls many activities in plants like cell elongation, leaf senescence, root nodulation, abscission, auxin transport, and fruit ripening (Afzal et al. 2019; Fadiji and Babalola 2020). Many stress conditions like drought, high salinity, pathogenicity, and the presence of extreme concentration of heavy metals cause the elevation of ethylene levels which lead to inhibition of root elongation, formation of root hairs, and alteration in cellular processes. In these stress conditions, endophytic organisms which reside inside the plant produce an enzyme known as 1-aminocyclopropane-1-carboxylate (ACC) deaminase; it is a pyridoxal phosphate-independent enzyme which breaks down the ACC (precursor of ethylene). ACC-degrading genera include *Agrobacterium*, *Enterobacter*, *Bacillus*, *Achromobacter*, *Acinetobacter*, *Pseudomonas*, *Alcaligenes*, *Ralstonia*, *Serratia*, *Rhizobium*, and *Burkholderia*. These organisms can bind to roots of plants and break down the ACC into α-ketobutyrate and ammonia and use ammonia as a nitrogen source, thus promoting plant growth under extreme conditions (Afzal et al. 2019; Fadiji and Babalola 2020). *Pseudomonas putida* HS-2, which has purified ACC enzyme, was used to enhance the growth of tobacco plants (Maela and Serepa 2019).

7.17 Biocontrol Activity by Endophytes

Protection of plants from plant pathogens or phytopathogens which can retard plant growth and can make plants diseased and sabotage the agricultural crop is known as biocontrol activity. Chemical fertilizers can prevent plants from phytopathogens but these chemicals have harmful effects on soil, environment, and microflora of plants and soil, so endophytic organisms are a great way to control phytopathogens. Some endophytes release bioactive compounds in their specific host plant species as mentioned in Table 7.2.

Endophytes that act as BCA have shown four mechanisms to control pathogens: (1) antibiosis, (2) competition, (3) host-induced resistance, and (4) direct parasitism. Some bacterial endophytes have shown plant protection from various diseases such as Fusarium, Verticillium, Eggplant, and Verticillium wilt, and these endophytes are *Erwinia persicina*, *Pantoea agglomerans*, *Achromobacter piechaudii*, *Enterobacter cloacae*, *P. fluorescens*, *Serratia plymuthica*, *S. marcescens*, *B. amyloliquefaciens*, *Paenibacillus*, sp., *Enterobacter* sp., *Bacillus subtilis*, etc. (Lobo et al. 2019; Agrillo et al. 2019). Siderophores play an important role to fight against phytopathogens by endophytes because endophytes release siderophores such as salicylic acid, chelate iron, and pyochelin which bind all available iron, so that other organisms starve for it and eventually die. 2,4-diacetylphloroglucinol (DAPG) is an antimicrobial

Table 7.2 Examples of endophytic strains and their bioactive compounds which help plants in different biotic and abiotic stress conditions (modified from Bhatt et al. 2019a)

S. No	Endophytes	Bioactive compounds	References
1	*Trichoderma harzianum*	Phytohormones and degradation of cell wall	Glick (2003)
2	*Bacillus amyloliquefaciens*	Antifungal compounds	Yang et al. (2015)
3	*Epichloe festucae* var. *lolii*	Alkaloids	Glick (2003)
4	*Arthrobacter endophyticus SYSU 33332*, *Nocardiopsis alba SYSU 333140*	Genes for water and potassium ion uptake, survival in stress condition	Yang et al. (2015)
5	*Aeromicrobium ponti*	1-Acetyl-β-carboline, indole-3-carbaldehyde,3-(Hydroxyacetyl)-Indole, Brevianamide F, and Cyclo-(L-Pro-L-Phe)	Burd et al. (2000)
6	*Serendipita vermifera*	Hydrolytic enzyme genes, protection from plant pathogen infection	Yang et al. (2015)
7	*Pseudomonas putida*	Trichloroethylene	Burd et al. (2000)
8	*Gibberella moniliformis*	Lawsone	Yang et al. (2015)

compound which is released by endophytes, which inhibit disease causing microorganisms. *Beauveria bassiana* is an entomopathogenic fungi for borer insects in coffee seedlings. Various fungi that have shown great biocontrol activity in the field of agriculture are *Trichoderma koningiopsis*, *Gibberella fujikuroi*, *Aspergillus tubingensis*, *A. flavus*, *Galactomyces geotrichum*, *P. simplicissimum*, *Eupenicillium javanicum*, and *P. ochrochloron* (Lobo et al. 2019). The most studied endophytic BCA are *Bacillus subtilis*, *Beauveria bassiana*, *Asexual Epichloe* spp., *Lecanicillium lecanii*, *Piriformospora indica*, *Rhizobia*, and *Trichoderma* spp. *Beauveria bassiana* works against a wide range of pests such as aphids, beetles, caterpillars, termites, thrips and whitefly; to use *Beauveria bassiana* fungi, regular spray of fungi should be done on infected plants. Asexual Epichloe belong to the family *clavicipitaceae* which are not found in the roots of plants. They have strong mutualism with grass species. Epichloe include *E. festucae* var. *lolii* and *E. coenophiala* which produce alkaloids—secondary metabolites during symbiosis and that work against invertebrate pests. *Lecanicillium lecanii* is a ubiquitous entomopathogenic fungi, which horizontally transmit from one plant to another plant such as from cotton plants to an insect known as *Aphis gossypii* and from insects to leaves. *L. lecanii* strain 41185 is pathogenic to aphides spp., such as *Myzus persicae*, *A. gossypii*, and *Aphis craccivora*. *L. lecanii* also has anti-fungal activity against *Sphaerotheca macularis* and *Hemileia vastatrix* and it also possesses anti-algae activity against soil-borne pathogens like *Pythium ultimum* by producing host-induced resistance and some structural barriers in the plant roots. *Piriformospora indica* is a root-colonizing fungus which is found in the root of xerophytic plants present in Indian Thar dessert. It also colonizes both mono- and dicotyledonous plants such as barley, tobacco, and *Arabidopsis thaliana*. *P. indica* has plant growth-promoting activity and function in both abiotic stress such as salinity, drought, water, cold, high temperature, and heavy metals and biotic stress like antagonism against root pathogens. *Trichoderma* is a fungal species which has a variety of advantages like anti-fungal properties, can survive in various conditions, simple nutritional requirement for growth in vivo and in vitro, fast growth, etc. *Trichoderma* genus involves many species like *T. atroviride*, *T. asperellum*, *T. harzianum*, *T. polysporum*, *T. viride*, and *T. hamatum*. These *Trichoderma* species can be used to target different soil fungal pathogens such as *Phytophthora*, *Rhizoctonia*, *Sclerotinia*, *Pythium*, and *Verticillium* and also work against foliar fungal pathogens like *Botrytis* and *Alternaria* (Valérie et al. 2007). *Erwinia carotovora* is a bacterial pathogen which causes food spoilage and is inhibited by the bacterial endophytes such as *Pseudomonas* sp., *Pantoea agglomerans* and *Curtobacterium luteum*. A fungal species *Cryphonectria parasitica* causing chestnut blight is inhibited by bacterial endophytes, i.e., *Bacillus subtilis*. *Clavibacter michiganensis* subsp. *Sepedonicus* was inhibited by bacterial endophytes isolated from potato stem tissues (Audipudi et al. 2017; Matos et al. 2017). Some bacterial endophytes have anti-nematode properties like *Bacillus megaterium* BP17 and *Curtobacterium luteum* TC10, which work against *Radopholus similis* Thorne. Bacterial species like *Bacillus thuringiensis* and *Serratia marcescens* produce toxin and enzyme chitinases which target the *Eldana saccharina* larvae (sugarcane borer). Some bacterial genus like *Pseudomonas*, *Bacillus*, and *Serratia* can protect plants by a mechanism known as induced systemic resistance (ISR). ISR can be initiated by three of the pathways: (1) salicylic acid [SA], (2) ethylene [ET], and (3) jasmonic acid [JA]. *Actinobacteria* was inoculated in a plant, i.e., *A. thaliana* showed that bacterial endophytes protected the plant from *Erwinia carotovora* and *Fusarium oxysporum* (Afzal et al. 2019). *Pseudomonas protegens* N has antifungal properties against the genus *Alternaria* and stops spore germination (Gull et al. 2004; Lin et al. 2013; Wei et al. 2018; Lacava et al. 2008; Tang et al. 2019; Card et al. 2016; Agrillo et al. 2019; Walitang et al. 2017).

7.18 Role of Endophytes in Overcoming Oxidative Stress, Salinity, Drought, and Temperature Stress

Endophytes play an important role in the prevention of crop loss by preventing different types of abiotic stress such as extreme heat, salinity, oxidative stress, drought, and temperature. Extreme conditions cause change in phenotype and genetic changes (Sansanwal et al. 2017). Endophytes produce osmoprotectant compounds such as proline, trehalose, exopolysaccharides, and volatile organic molecules (Kuzniar et al. 2019). *Pseudomonas* spp. help *Asparagus* spp. do better seedling and seed germination in extreme water stress. In coastal areas where high saline condition is present, *Pseudomonas fluorescens* MSP-393 has proven to act as PGPR for many crops in that area. *Pseudomonas putida* RS-198 has proven to promote cotton seedling in extreme salt presence by inhibiting the absorption of Na^+ and increasing the uptake of Mg^{2+}, K^+, and Ca^{2+}. Some species of bacterial endophytes such as *Bacillus polymyxa*, *Mycobacterium phlei*, *Alcaligenes* sp., and *Paenibacillus* sp., produce some compounds such as calcisol which has proven to promote the growth of maize plant under extreme temperature and saline conditions (Sansanwal et al. 2017). *Pseudomonas migulae* 8R6 and *Pseudomonas fluorescens* YsS6 can reduce salt stress in tomato plants by secreting ACC deaminase, and it was observed that ACC deaminase activity helps in the production of a greater amount of chlorophyll and increase in dry and fresh biomass. Plants infected with fungal endophytes, i.e., *Neotyphodium lolii* increase the survival chance of plants in drought conditions. Endophytic fungi belong to the group *Ascomycota* and *Basidiomycota*, which increase the tolerance of plants in drought and heat conditions. Endophytic fungi belongs to the genus *Penicillium*, *Trichoderma*, *Aspergillus sydowii*, *Myxotrichum stipitatum*, and *Acremonium variecolor*, which exhibit saline resistance (Kumar et al. 2016). During an environmental stress in plants that can trigger the production of reactive oxygen species such as superoxide, hydroperoxyl radicals, hydrogen peroxide, and hydroxyl radicals, these can cause damage to plant proteins, nucleic acids, and membranes. White et al. (2019) showed that colonization of bacteria at an early stage causes upregulated transcript level of ROS degrading genes which include the genes of SOD and glutathione reductase. *Festuca arundinacea* infected by the endophytic fungi, i.e., *Epichloe coenophiala* has greater concentration of mannitol. Osmoprotectants and other fungal carbohydrate which helps plant to survive in oxidative stress. Metagenomics analysis of rice crop endophytes has proven to have the presence of numerous genes which encode enzymes for protection from ROS (White et al. 2019). *Xanthomonas* sp., *Microbacterium* sp., and *Flavobacterium* sp. help plants to survive in osmotic and salinity stress conditions (Walitang et al. 2017).

7.19 Conclusion

A vast majority of endophytic microorganisms have been isolated from a variety of plants; they help plants in various aspects like plant growth, prevention to pathogen, phytoremediation, and nutrient availability; besides this, they also help plants to adapt to different biotic and abiotic stresses. The understanding of plant-endophyte mutual association has to be better, to give consistent results under field conditions. However, PGPR has an advantageous impact on the quality of plants by means of various action mechanisms. Endophytes synthesize a collection of bioactive metabolic compounds which are used to prevent viral infections as effective medicine. Secondary metabolites are also synthesized by some endophytes including flavonoids, alkaloids, steroids, tannin, quinones, benzopyrones, etc. divided according to their functional groups. Sometimes, it is difficult to identify some rare endophytes microorganisms that have beneficial characteristics, and these have a huge role to bring about less and minimal consumption of diverse types of agrochemical substances which could include pesticides, chemical fertilizers, and so on. If it is studied in more details, then it is very surprising, and interesting facts will come out, which will increase the interest of other people. The current compilation shows the numerous forms of advantageous endophytic microbes that are used to increase soil fertility as biofertilizers in the field and better crop yield and crop production.

References

Afzal I, Shinwari ZK, Sikandar S, Shahzad S (2019) Plant beneficial endophytic bacteria: mechanisms, diversity, host range and genetic determinants. Microbiol Res 221:36–49

Agrillo B, Mirino S, Tatè R, Gratino L, Gogliettino M et al (2019) An alternative biocontrol agent of soil-borne phytopathogens: a new antifungal compound produced by a plant growth promoting bacterium isolated from North Algeria. Microbiol Res 221:60–69

Ahmad F, Ahmad I, Khan M (2008) Screening of free-living rhizospheric bacteria for their multiple plant growth promoting activities. Microbiol Res 163:173–181

Alexander DB, Zuberer DA (1991) Use of chrome azurol S reagents to evaluate siderophore production by rhizosphere bacteria. Biol Fertil Soils 12:39–45

Andreote F, Gumiere T, Durrer A (2014) Exploring interactions of plant microbiomes. Sci Agric 71:528–539

Annapurna K, Govindasamy V, Sharma M, Ghosh A, Chikara SK (2018) Whole genome shotgun sequence of Bacillus paralicheniformis strain KMS 80, a rhizobacterial endophyte isolated from rice (Oryza sativa L.). 3 Biotech 8:223–226

Ardanov P, Ovcharenko L, Zaets I, Kozyrovska N, Pirttilä AM (2011) Endophytic bacteria enhancing growth and disease resistance of potato (*Solanum tuberosum* L.). Biol Control 56:43–49

Arnold AE, Maynard Z, Gilbert GS, Coley PD et al (2002) Are tropical fungal endophytes hyper diverse. Ecol Lett 3:267–274

Aryantha INP, Hidiyah ARM (2018) Colonization and performance of diazotroph endophytic bacteria on palm oil (Elaeis guineensis Jacq L.) leaves. IOP Conf Ser Earth Environ Sci 166:1–12

Audipudi AV, Chakicherla BV, Bhore SJ (2017) Bacterial endophytes as biofertilizers and biocontrol agents for sustainable agriculture. Biotech Sustain:223–247

Bacon CW, White JF Jr (2000) An overview of endophytic microbes: endophytism defined. Am J Plant Sci 6:3–29

Baldani JJ, Oliveira ALM, Guimmaraes SL, Baldani VLD, Reis FB, 2000 et al. Biological nitrogen fixation in non-leguminous plants: the role of endophytic diazotrophs. In: Nitrogen fixation: from molecules to crop productivity. Pedrosa, F. O., Hungria, M., Yates, G., and Newton, W. E., Eds., Kluwer Academic Publishers, Dordrecht. 397–400.

Bhatt P, Bisht TS, Pathak VM, Barh A, Chandra D (2015a) Optimization of amylase production from the fungal isolates of Himalayan region Uttarakhand. Ecol Environ Conserv 21 (3):1517–1521

Bhatt P, Negi G, Gangola S, Khati P, Srivastava A, Sharma A (2015b) Optimization of sulfosulfuron biodegradation through response surface methodology using indigenous bacterial strain isolated from contaminated agriculture field. Int J Curr Microbiol App Sci 4(8):105–112

Bhatt P, Negi G, Gangola S, Khati P, Kumar G, Srivastava A, Sharma A (2016a) Differential expression and characterization of cypermethrin degrading potential proteins in *Bacillus thuringiensis* strain, SG4. 3 Biotech 6:225

Bhatt P, Sharma A, Gangola S, Khati P, Kumar G, Srivastava A (2016b) Novel pathway of cypermethrin biodegradation in *a Bacillus* sp. strain SG2 isolated from cypermethrin-contaminated agriculture field. 3 Biotech 6:45

Bhatt P, Huang Y, Zhan H, Chen S (2019a) Insight into microbial applications for the biodegradation of pyrethroid insecticides. Front Microbiol 10:1778

Bhatt P, Pal K, Bhandari G, Barh A (2019b) Modeling of methyl halide biodegradation on bacteria and its effect on other environmental systems. Pestic Biochem Physiol 158:88–100

Bhatt P, Gangola S, Chaudhary P, Khati P, Kumar G, Sharma A, Srivastava A (2019c) Pesticide induced up-regulation of Esterase and Aldehyde dehydrogenase in Indigenous *Bacillus* spp. Bioremed J 23(1):42–52

Bhatt P, Rene ER, Kumar AJ, Kumar AJ, Zhang W, Chen S (2020a) Binding interaction of allethrin with esterase: bioremediation potential and mechanism. Bioresour Technol 315:13845. https://doi.org/10.1016/j.biortech.2020.123845

Bhatt P, Verma A, Verma S, Anwar MS, Prasher P, Mudila H, Chen S (2020b) Understanding phytomicrobiome: a potential reservoir for better crop management. Sustainability 12:5446

Bhatt P, Huang Y, Rene ER, Kumar AJ, Chen S (2020c) Mechanism of allethrin biodegradation by a newly isolated *Sphingomonas trueperi* strain CW3 from wastewater sludge. Bioresour Technol 305:123074

Bhatt P, Zhang W, Lin Z, Pang S, Huang Y, Chen S (2020d) Biodegradation of allethrin by a novel fungus *Fusarium proliferatum* strain CF2, isolated from contaminated Soils. Microorganisms 8:593

Bhatt P, Huang Y, Zhang W, Sharma A, Chen S (2020e) Enhanced Cypermethrin degradation kinetics and metabolic pathway in *Bacillus* thuringiensis strain, SG4. Microorganisms 8:223

Bhatt P, Bhatt K, Huang Y, Ziqiu L, Chen S (2020f) Esterase is a powerful tool for the biodegradation of pyrethroid insecticides. Chemosphere 244:125507

Bhatt P, Sethi K, Gangola S, Bhandari G, Verma A, Adnan M, Singh Y, Chaube S (2021a) Modeling and simulation of atrazine biodegradation in bacteria and its effect in other living systems. J Biomol Struct Dyn. https://doi.org/10.1080/07391102.2020.1846623

Bhatt P, Gangola S, Bhandari G, Zhang W, Maithani D, Mishra S, Chen S (2021b) New insights into the degradation of synthetic pollutants in contaminated environments. Chemosphere 268:128827. https://doi.org/10.1016/j.chemosphere.2020.128827

Bhatt P, Sharma A, Rene ER, Kumar AJ, Zhang W, Chen S (2021c) Bioremediation mechanism, kinetics of fipronil degradation using *Bacillus* sp. FA3 and resource recovery potential from contaminated environments. J Water Process Eng 39:101712

Bibi F, Yasir M, Song GC, Lee SY, Chung YR (2012) Diversity and characterization of endophytic bacteria associated with tidal flat plants and their antagonistic effects on oomycetous plant pathogens. Plant Pathol J 28:20–31

Burd GI, Dixon DG, Glick BR (2000) Plant growth promoting bacteria that decrease heavy metal toxicity in plants. Can J Microbiol 46:237–245

Card S, Johnson L, Teasdale S, Caradus J (2016) Deciphering endophyte behaviour - the link between endophyte biology and efficacious biological control agents. FEMS Microbiol Ecol 92:1–19

Correa J, Nielssen R, Grund DW, McLachla J (1987) Endophytic algae of Irish moss (Chondrus crispus Stackh.). Hydrobiologia 151/152:223–228

Correa J, Nielssen R, Grund DW (1988) Endophytic algae of Chondrus crispus (Rhodophyta). II. Acrochaete heteroclada sp. nov. A. operculata sp. nov. and Phaeophila dendroides (Chlorophyta). J Phycol 28:528–539

Doty SL (2008) Enhancing phytoremediation through the use of transgenics and endophytes. New Phytol 179:318–333

Dourado MN, Ferreira A, Araújo WL, Azevedo JL, Lacava PT (2012) The diversity of endophytic methylotrophic bacteria in an oil-contaminated and an oil-free mangrove ecosystem and their tolerance to heavy metals. Biotechnol Res Int 2012:1–8

Dourado MN, Neves AAC, Santos DS, Araújo WL (2015) Biotechnological and agronomic potential of endophytic pink-pigmented methylotrophic Methylobacterium spp. Bio Med Res 2015:1–19

Fadiji AE, Babalola OO (2020) Elucidating mechanisms of endophytes used in plant protection and other bioactivities with multifunctional prospects. Front Bioeng Biotechnol 8:467–971

Fan X, Ye T, Li Q, Bhatt P, Zhang L, Chen S (2020) Potential of a quorum quenching bacteria isolate Ochrobactrum intermedium D-2 against soft rot pathogen *Pectobacterium carotovorum* subsp. Carotovora. Front Microbiol 11:898

Feng Y, Huang Y, Zhan H, Bhatt P, Chen S (2020) An overview of strobilurin fungicide degradation: current status and future perspective. Front Microbiol 11:389

Ferreira A, Quecine MC, Lacava PT et al (2008) Diversity of endophytic bacteria from eucalyptus species seed and colonization of seedlings by Pantoea agglomerans. Microbiol Lett 287:8–14

Fouda M, Saad EL, Salem SS et al (2018) New approach for antimicrobial activity and biocontrol of various pathogens by biosynthesized copper nanoparticles using endophytic actinomycetes. J Radiat Res Appl Sci 11:262–270

Franche C, Lindstrom K, Elmerich C (2009) Nitrogen- fixing bacteria associated with leguminous and non-leguminous plants. Plant Soil 321:35–59

Frank A, Guzmán S, Shay J (2017) Transmission of bacterial endophytes. Microorganisms 5:70–91

Fremlin LJ, Piggott AM, Lacey E, Capon RJ (2009) Cottoquinazoline A and Cotteslosins A and B, metabolites from an Australian marine-derived strain of Aspergillus versicolor. J Nat Prod 72:666–670

Gangola S, Sharma A, Bhatt P, Khati P, Chaudhary P (2018) Presence of esterase and laccase in *Bacillus subtilis* facilitates biodegradation and detoxification of cypermethrin. Sci Rep 8:12755. https://doi.org/10.1038/s41598-018-31082-5

Girsowicz R, Moroenyane I, Steinberger Y (2019) Bacterial seed endophyte community of annual plants modulated by plant photosynthetic pathways. Microbiol Res 223–225:58–62

Glandorf DCM, Verheggen P, Jansen T, Jorritsma JW, Smit E et al (2001) Effect of genetically modified pseudomonas putida WCS358r on the fungal rhizosphere microflora of field-grown wheat. Appl Environ Microbiol 67:3371–3378

Glick BR (1995) The enhancement of plant growth by free living bacteria. Can J Microbiol 41:109–114

Glick BR (2003) Phytoremediation: synergistic use of plants and bacteria to clean up the environment. Biotechnol Adv 21:383–393

Gonzalez-Bashan LE, Lebsky V, Hernandez JP, Bustillos JJ, Bashan Y (2000) Changes in the metabolism of the microalgae Chlorella vulgaris when co-immobilized in alginate with the nitrogen-fixing Phyllobacterium myrsinacearum. Can J Microbiol 46:653–659

Gull M, Hafeez FY, Saleem M, Malik KA (2004) Phosphorus uptake and growth promotion of chickpea by co-inoculation of mineral phosphate solubilising bacteria and a mixed rhizobial culture. Aust J Exp Agric 44:623–628

Gupta G, Nath JP (2015) Screening of potential PGPR candidates as future biofertilizers-a strategic approach from lab to field. Res J Biotechnol 10:48–62

Gupta G, Panwar J, Akhtar MS, Jha PJ (2012) Endophytic nitrogen-fixing bacteria as biofertilizer. In: Lichtfouse E (ed) Sustainable agriculture reviews, vol 6, pp 183–221

Gupta S, Bhatt P, Chaturvedi P (2018) Determination and quantification of asiaticoside in endophytic fungus from Centella asiatica (L.) Urban. World J Microbiol Biotechnol 34:111

Gyaneshwar P, Reddy PM, Ladha JK (2005) Nutrient amendments influence endophytic colonization of rice by Serratia Marcescens IRBG500 and Herbaspirillum Z67. J Microbiol Biotechnol 10:694–699

Holland MA, Polacco JC (1992) Urease-null and hydrogenase- null phenotypes of a phylloplane bacterium reveal altered nickel metabolism in two soybean mutants. Plant Physiol 98:942–948

Huang Y, Zhan H, Bhatt P, Chen S (2019) Paraquat degradation from contaminated environments: current achievements and perspectives. Front Microbiol 10:1754

Huang Y, Lin Z, Zhang W, Pang S, Bhatt P, Rene ER, Kumar AJ, Chen S (2020) New insights into the microbial degradation of D-cyphenothrin in contaminated water/soil environments. Microorganisms 8:473

Huang Y, Zhang W, Pang S, Chen J, Bhatt P, Mishra S, Chen S (2021) Insights into the microbial degradation and catalytic mechanism of chlorpyrifos. Environ Res 194:110660

James EK, Gyaneshwar P, Barraquio WL, Mathan N, Ladha JK (2000) Endophytic diazotrophs associated with rice. In: Ladha JK, Reddy PM (eds) The quest for nitrogen fixation in Rice. International Rice Research Institute, Los Baños, pp 119–140

Jeffery LSH (2008) Isolation, characterization and identification of actinomycetes from agriculture soils at Semomgok, Sarawak. Afr J Biotechnol 7:3697–3702

Jha P, Kumar A (2009) Characterization of novel plant growth promoting endophytic bacterium Achromobacter xylosoxidans from wheat plant. Microb Ecol 58:179–188

Jha B, Gontia I, Hartmann A (2012) The roots of the halophyte Salicornia Brachiata are a source of new halotolerant diazotrophic bacteria with plant growth-promoting potential. Plant Soil 356:265–277

Kaaria P, Matiru V, Ndungu M (2012) Antimicrobial activities of secondary metabolites produced by endophytic bacteria from selected indigenous Kenyan plant. Afr J Microbiol Res 6:7253–7258

Karakashev D, Batstone DJ, Trably E, Angelidaki I (2011) Acetate oxidation is the dominant methanogenic pathway from acetate in the absence of Methanosaetaceae. Appl Environ Microbiol 27:5138–5141

Keerthi MM, Babu R, Jose M, Amutha R (2015) Optimizing plant geometry and nutrient management for grain yield and economics in irrigated greengram. Am J Plant Sci 6:1144–1150

Kumar RR, Jadeja JV (2016) Endophytic actinomycetes: a novel antibiotic source. Int J Pharm Sci Rev Res 45:32–36

Kumar M, Srivastava A, Pandey AK (2015) Biocontrol activity of some potent Methylotrophs isolated from Bhitarkanika mangrove sediment. Int J Curr Res Biosci Plant Biol 2:101–106

Kumar M, Tomar RS, Lade H, Diby P (2016) Methylotrophic bacteria in sustainable agriculture. World J Microbiol Biotechnol 32:120–129

Kuzniar A, Włodarczyk K, Wolinska A (2019) Agricultural and other biotechnological applications resulting from trophic plant-endophyte interactions. Agronomy 9:779–881

Lacava PT, Silva S, Maria E, Araújo WL, Simionato Ana VC et al (2008) Detection of siderophores in endophytic bacteria Methylobacterium spp. associated with Xylella fastidiosa subsp. pauca. Pesq Agropec Bras 43:521–528

Limaye L, Patil R, Ranadive P, Kamath G (2017) Application of potent actinomycete strains for bio-degradation of domestic agro-waste by composting and treatment of pulp-paper mill effluent. Adv Microbiol 7:94–108

Lin T, Zhao L, Yang Y, Guan Q, Gong M (2013) Potential of endophytic bacteria isolated from Sophora alopecuroides nodule in biological control against Verticillium wilt disease. Aust J Crop Sci 7:139–146

Lin Z, Zhang W, Pang S, Huang Y, Mishra S, Bhatt P, Chen S (2020) Current approaches to and future perspectives on methomyl degradation in contaminated soil/water environments. Molecules 25:738

Lobo CB, Juárez T, María S, Viruel E, Ferrero MA, Lucca ME (2019) Development of low-cost formulations of plant growth-promoting bacteria to be used as inoculants in beneficial agricultural technologies. Microbiol Res 219:12–25

Maela PM, Serepa DMH (2019) Current understanding of bacterial endophytes, their diversity, colonization and their roles in promoting plant growth. Appl Microbiol Open Access 5:1–157

Matos ADM, Gomes ICP, Nietsch S, Xavier AA, Gomes WS et al (2017) Phosphate solubilization by endophytic bacteria isolated from banana trees. An Acad Bras Cienc 89:2945–2954

Meena KK, Kumar M, Kalyuzhnaya MG, Yandigeri MS, Singh DP (2012) Epiphytic pink-pigmented methylotrophic bacteria enhance germination and seedling growth of wheat (Triticum aestivum) by producing phytohormone. Antonie Van Leeuwenhoek 101:777–786

Mei C, Lara-Chavez A, Lowman S, Flinn B (2014) The use of endophytes and mycorrhizae in switchgrass biomass production. In: Compendium of bioenergy plants, vol 4, pp 67–108

Mishra S, Zhang W, Lin Z, Pang S, Huang Y, Bhatt P, Chen S (2020) Carbofuran toxicity and its microbial degradation in contaminated environments. Chemosphere 259:127429

Nabti E, Sahnoune M, Ghoul M, Fischer D, Hofmann A et al (2010) Restoration of growth of durum wheat (Triticum Durum Var. waha) under saline conditions due to inoculation with the rhizosphere bacterium Azospirillum brasilense NH and extracts of the marine alga Ulva Lactuca. J Plant Growth Regul 29:6–22

Naik Shankar B, Shashikala J, Krishnamurthy YL (2008) Diversity of fungal endophytes in shrubby medicinal plants of Malnad region, Western Ghats, Southern India. Mycosphere 1:265–274

Narayanan K, Karthik A, Parameswaran B, Ashok P (2014) Production, Purification and properties of fungal chitinases-a review. Indian J Exp Biol 52:1025–1035

Naylor D, DeGraaf S, Purdom E, Coleman-Derr D (2017) Drought and host selection influence bacterial community dynamics in the grass root microbiome. ISME J 11:2691–2704

Newton WE (2000) Nitrogen fixation: from molecules to crop productivity. J Biol Chem 27:3–8

Nielsen R (1979) Culture studies on the type species of Acrochaete, Bolbocoleon and Entocladia (Chaetophoraceae, Chlorophyceae). Bot Notiser 132:441–449

Ohyama T, Momose A, Ohtake N, Sueyoshi K, Sato T et al (2014) Nitrogen fixation in sugarcane. Isolation and identification of a novel endophyte from a plant Amaranthus spinosus. Int J Curr Microbiol:50–70

Oskay M, Tamer UA, Azeri C (2004) Antibacterial activity of some Actinomycetes isolates from farming soils of turkey. Afr J Biotechnol 3:441–446

Oteino N, Lally RD, Kiwanuka S, Lloyd A, Ryan D et al (2015) Plant growth promotion induced by phosphate solubilizing endophytic Pseudomonas isolates. Front Microbiol 6:745–754

Pang S, Lin Z, Zhang W, Mishra S, Bhatt P, Chen S (2020) Insights into the microbial degradation and biochemical mechanisms of neonicotinoids. Front Microbiol 11:868

Panpatte DG, Jhala YK, Vyas RV, Shelat HN (2017) Endophytic microorganisms: promising candidate as biofertilizer, microorganisms for sustainability. In: Microorganisms for green revolution, vol 6, pp 77–85

Patle PN, Navnage NP, Ramteke PR (2018) Endophytes in plant system: roles in growth promotion, Mechanism and their potentiality in achieving agriculture sustainability. Int J Chem Stud 6:270–274

Patra JK, Vishnuprasad CN, Das G (2017) Microbial biotechnology bioprospecting of endophytes for agricultural and environmental sustainability. Microbiol Biotech:429–458. https://doi.org/10.1007/978-981-10-6847-8_19

Pirttila AM, Pospiech H, Laukkanen H et al (2005) Seasonal variation in location and population structure of endophytes in bus of scots pine. Tree Physiol 25:289–297

Plumb J (1999) Population dynamics and endophytic flora of Chondrus crispus Rhodophyta): a temporal study. PhD thesis, University of the West of England. Bristol, UK

Purushotham N, Jones E, Monk J, Ridgway H (2018) Community structure of endophytic actinobacteria in a New Zealand native medicinal plant Pseudowintera colorata (Horopito) and their influence on plant growth. Microb Ecol 76:729–740

Raghoebarsing AA, Smolders AJP, Schmid MC, Rijpstra WIC, Wolters-Arts M et al (2005) Methanotrophic symbionts provide carbon for photosynthesis in peat bogs. Nature 436:1153–1156

Raghukumar C (2008) Marine fungal biotechnology: an ecological perspective. Fungal Divers 31:19–35

Rekadwad BN (2014) Growth promotion of crop plants by Methylobacterium organophilum: efficient bio-inoculant and biofertilizer isolated from mud. Res Biotechnol 5:1–6

Rodriguez JR, White JF, Arnold AE, Redman RS (2009) Fungal endophytes: diversity and functional roles. New Phytol 182:314–330

Rothballer M, Eckert B, Schmid M, Fekete A, Schloter M et al (2008) Endophytic root colonization of gramineous plants by Herbaspirillum frisingense. FEMS Microbiol Ecol 66:85–95

Sansanwal R, Ahlawat U, Priyanka, Wati L (2017) Role of endophytes in agriculture. Chem Sci Rev Lett 6:2397–2407
Sharma S, Roy S (2017) Isolation and identification of a novel endophyte from a plant Amaranthus spinosus. Int J Curr Microbiol App Sci 4:785–798
Singh M, Kumar A, Singh R, Pandey KD (2017) Endophytic bacteria: a new source of bioactive compounds. 3. Biotech 7:315–329
Singh K, Gera R, Sharma R, Maithani D, Chandra D, Bhat AM, Kumar R, Bhatt P (2021) Mechanism and application of *Sesbania* root nodulating bacteria: an alternative for chemical fertilizers and sustainable development. Arch Microbiol. https://doi.org/10.1007/s00203-020-02137-x
Smith MR, Mah RA (1978) Growth and methanogenesis by Methanosarcina strain 227 on acetate and methanol. Appl Environ Microbiol 36:870–879
Soldan R, Mapelli F, Crotti E, Schnell S, Daffonchio D et al (2019) Bacterial endophytes of mangrove propagules elicit early establishment of the natural host and promote growth of cereal crops under salt stress. Microbiol Res 223-225:33–43
Sturz AV, Christie BR, Nowak J (2000) Bacterial endophytes: potential role in developing sustainable systems of crop production. Crit Rev Plant Sci 19:1–30
Suminto, Hirayama K (1997) Application of a growth-promoting bacteria for stable mass culture of three marine microalgae. Hydrobiologia 358:223–230
Suryanarayanan ST, Wittlinger SK, Faeth SH (2005) Endophytic fungi associated with cacti in Arizona. Mycol Res 109:635–639
Susilowati DN, Riyanti EI, Setyowati M, Mulya K (2018) Indole-3-acetic acid producing bacteria and its application on the growth of rice. In: AIP Conf Proc. 2002:020016-9
Tang MJ, Zhu Q, Zhang FM, Zhang W, Yuan J et al (2019) Enhanced nitrogen and phosphorus activation with an optimized bacterial community by endophytic fungus Phomopsis liquidambari in paddy soil. Microbiol Res 221:50–59
Tanjung ZA, Aditama R, Sudania WM, Utomo C, Liwang T (2017) Metagenomics workflow analysis of endophytic bacteria from oil palm fruits. J Phys Conf Ser 835:1–6
Thauer RK (1998) Biochemistry of methanogenesis: a tribute to Marjory Stephenson. Microbiology 144:2377–2406
Valérie G, Hani A, Russell JT (2007) Effect of indole-acetic acid (IAA) on the development of symptoms caused by Pythium ultimum on tomato plants. Eur J Plant Pathol 119:457–462
Vesterlund SR, Helander M, Faeth SH, Hyvonen T, Saikkonen K (2011) Environmental conditions and host plant origin override endophyte effects on invertebrate communities. Fungal Divers 47:109–118
Waheeda K, Shyam KV (2017) Formulation of novel surface sterilization method and culture media for the isolation of endophytic actinomycetes from medicinal plants and its antibacterial activity. J Plant Pathol Microbiol 8:399–408
Walitang DI, Kim K, Madhaiyan M, Kim YK, Kang Y et al (2017) Characterizing endophytic competence and plant growth promotion of bacterial endophytes inhabiting the seed endosphere of Rice. BMC Microbiol 17:1–19
Wei Y, Zhao Y, Shi M, Cao Z, Lu Q, Yang T, Fan Y, Wei Z (2018) Effect of organic acids production and bacterial community on the possible mechanism of phosphorus solubilization during composting with enriched phosphate-solubilizing bacteria inoculation. Bioresour Technol 247:190–199
White JF, Kingsley KL, Zhang Q, Verma R, Obi N et al (2019) Endophytic microbes and their potential applications in crop management. Pest Manag Sci 75:2558–2565
Wolinska A, Kuzniar A, Zielenkiewicz U, Banach A, Izak D et al (2017) Metagenomic analysis of some potential nitrogen-fixing bacteria in arable soils at different formation process. Soil Microbiol 73:162–176
Yang GL, Yang LF, Jiang MG, Wu JF, Gan GH et al (2015) Isolation, Identification and bioactivity of endophytic actinomycetes from mangrove plants in Beilun River. J Agric Biotech 23:894–904

Yang L, Choufei W, Zhong X, Bingli G, Liqin Z (2017) Engineering the bacterial endophyte Burkholderia pyrrocinia JK-SH007 for the control of lepidoptera larvae by introducing the cry218 genes of Bacillus thuringiensis. Biotechnol Biotechnol Equip 31:1167–1172

Ye T, Zhou T, Fan X, Bhatt P, Zhang L, Chen S (2019) *Acinetobacter lactucae* strain QL-1, a novel quorum quenching candidate against bacterial pathogen *Xanthomonas campestris* pv. Campestris. Front Microbiol 10:2867

Zahir Z, Arshad M, Frankenberger W (2003) Plant growth promoting rhizobacteria: applications and perspectives in agriculture. Adv Agron 81:97–168

Zhan H, Huang Y, Lin Z, Bhatt P, Chen S (2020) New insights into the microbial degradation and catalytic mechanism of synthetic pyrethroids. Environ Res 182:109138

Zhang W, Pang S, Lin Z, Mishra S, Bhatt P, Chen S (2020a) Biotransformation of perfluoroalkyl acid precursors from various environmental system: advances and perspectives. Environ Pollut 272:115908

Zhang W, Lin Z, Pang S, Bhatt P, Chen S (2020b) Insights into the biodegradation of lindane (γ-hexachlorocyclohexane) using a microbial system. Front Microbiol 11:522

Chapter 8
Microbial Action on Degradation of Pesticides

Hira Singh Gariya and Arun Bhatt

Abstract The ultimate end fate of chemical pesticide degradations in the environment depends on microbial activity. The degradation of any biomaterial depends on the decomposers for their necessary recycling. So, in the same way, pesticide degradation is one of the major tasks to prevent their accumulation in the food chain known as bioaccumulation or may be somewhat disastrous known as biomagnification at each trophic level of the ecosystem. The extensive implication of pesticides globally causes a serious imbalance in the soil, air, and finally the potable water. Pesticide degradation does not only safe as after degradation the resultant products are more harmful and noxious. Instead, it forms a new chemical product that may be more or less toxic than the original chemical compound. Generally, they are broken into resultant smaller and smaller pieces until only the formation of carbon dioxide, water, and minerals is left. Microbes often play a large role in this process of harmful product formation. Some of the pesticides also cause restriction in the growth of algae via the release of various biochemicals that are important for the growth of algae known as an algal bloom. Worldwide, a large fraction of pests causes loss, damage to crops, and subsequently productivity. Pesticides have been used extensively from ancient times, but due to increased application up to date, their release by a various chemical processes in the environment causes serious ecological problems. Use of pesticides in an unregulated way causes adverse effect to humans, animals, and non-targeted plants. The non-biodegradable and recalcitrant pesticides persist in the environment and cause serious health hazards. Despite of their restriction and ban by the government, their continuous use is ever increasing. So, it is mandatory to restrict their use to reduce risk related to the environment as well as to humans and animals. It means the indiscriminate application of pesticides causes an adverse effect on different life forms.

Keywords Bioaccumulation · Pesticides · Biodegradable · Hazards · Biomagnification · Toxic · Recalcitrant

---

H. S. Gariya (✉) · A. Bhatt
Biotechnology Department, G.B.P.I.E.T, Pauri Garhwal, Uttarakhand, India

P. Bhatt et al. (eds.), *Microbial Technology for Sustainable Environment*,
https://doi.org/10.1007/978-981-16-3840-4_8

8.1 Introduction

Pesticides are useful for controlling weeds, pests, and various diseases for the increased productivity of crops (Sulbhi et al. 2021). Nowadays, these pesticides have become one of the major tools of farmers to get rid out of various diseases and also the harmful pest that hampers the crop's productivity (Bhandari et al. 2021). Pesticides are useful for repelling and mitigating any kind of pest, but due to their huge application and non-biodegradable nature cause several harmful problems to the environment (Bhatt et al. 2021a, b, c). So, it is obvious to regularly monitor the effects of these harmful pesticides by the government and by the farmers. These pesticides provide various nutrient supplementation for soil microorganisms (Torres 2003; Bhatt et al. 2020a).

The adaptation of microorganisms to biodegradation is a very common phenomenon due to the rapid use of chemical pesticides (Bhatt et al. 2021d, e, f, 2020b, c). There is a lot of severe harmful cases of health hazard due to accumulation of pesticides in the agricultural system, and their quantification is very difficult to analyze and predict (Bhatt et al. 2020d). It is very important to know that the inbuilt soil fertility and the half-life is decreasing due to the persistence of the chemical compound in the soil is the major cause of reduction in soil texture, fertility, half-life, etc. (Bhatt et al. 2020e, f, 2019a, b, c, 2016a, b, 2015a, b; Huang et al. 2021). There are three pathways in which a chemical pesticide compound may undergo:

- Pesticides have no transformation, no chemical structure alternation, and direct compartmentalization
- Pesticides get transformed and from this may go physical transformation or chemical transformation and subsequently compartmentalization.
- Pesticides through the biotic transformation may go under co-metabolism or mineralization.

The ultimate fate of above all three pathways is compartmentalization, deposit in soil, and get sedimented. The exo-toxicological effect of several pesticides has been not yet analyzed with precise extent, so ultimately soil is the main reservoir of various pesticides chemical compounds and their resultant product that can be detectable (Bhatt et al. 2019d). It is a well-known fact that the uniformity of pesticides degrading microorganisms lacks at every polluted site so, with the combination of advancement in technologies and conventional procedures to get rid out of pesticides, the effect on different biological system can be severely decreased (Sharma and Bhatt 2016). Pesticides cause some of the species to restrict their growth in the environment. Pesticides are classified in two major classes depending upon microbial degradation: degradative and non-degradative known as recalcitrant or microbial resistant. Pesticides include the wide category that involves herbicides, insecticides, fungicides rodenticides, molluscicides, and nematicides (Sharma et al. 2016).

A diverse group of bacteria including the genera of *Flavobacterium*, *Pseudomonas*, and *Rhodococcus* can metabolize pesticides rapidly. Microbial metabolism to

Table 8.1 Classification of the major groups of pesticides

Groups	Examples
Organochlorine	DDT, Aldrin, 666, Chlordane, Mirex
Organophosphate	Malathion, diazinon
Carbamates	Sevin
Pyrethroids	Pyrethrins
Thiocarbamates	Ferbam
Organosulfur	CHEBI
Biological	Dispel, foray
Urea derivative	Diuron-desmethyl, diuron

degrade various classes of pesticides depends on environmental parameters like light, pH, moisture, etc. Microbial metabolic activity can also be enhanced via supplying the necessary supplements so that degradation rate can be enhanced (Bhatt and Nailwal 2018). Microorganisms have the capability to bio-transform the pesticide. A major and wide range of bacterial genera has the potential that has been reported to degrade the pesticides depending upon the pesticide's chemical nature, microbial action on pesticide, and pesticide composition (Khati et al. 2018a). Pesticides that are proven to be effective and widely used are DDT, BHC (commonly known as 666 or gammexane or lindane), aldrin, dieldrin, and 2–4D (Table 8.1).

Pesticides cause the contamination of ground and surface water. By the process of leaching, pesticides contaminate the ground water in a way that is not potable and cause serious health-related disorders (Gangola et al. 2018a; Bhatt 2018; Bhatt and Barh 2018; Bhatt et al. 2019e; Bhandari and Bhatt 2020; Bhatt and Bhatt 2020). The elimination of pesticides from the environment in a safe way is very important to reduce the risk mentioned earlier because of the rapid solubility of pesticides, bioaccumulation, and biomagnification in a non-targeted microorganisms (Agrawal et al. 2010). The precise concentration of pesticides is difficult to estimate in the environment as they have been used ancient before and also due to lack of documented information (Vischetti et al. 2008). So, the remediation of these chemical compounds is necessary for a way to maintain the kinetics of the ecosystem. Bioremediation is one of the emerging fields to cope up with this problem. Now in present time, biologist from different interdisciplinary field related to biology is continuously seeking and exploring to develop the organism via the help of genetic engineering with increased efficiency to degrade these contaminants rapidly with a high rate of degradation. The application of pesticides can kill a certain group of microorganisms that are useful for agro-farming industry (Hussain et al. 2009). Detection through voltammetry analysis of the electrode modification increases its ability to quantify uM or nM and up to pM concentrations of a specific analyte, for metals and metal oxide (Barbosa and Fernanda 2019).

There are several types of chemical-mediated reaction that is used to degrade the pesticides. Chemical compounds like atrazine are degraded by the elemental removal of chlorine with the help of the hydroxylation process, i.e., incorporating hydroxyl group (Singh et al. 2021; Zhang et al. 2020a; Mishra et al. 2020). The atrazine is finally converted into a nontoxic one (Zhang et al. 2020b). Enzymatic and nonenzymatic processes can be used for pesticide degradation depending upon the suitable one, which one is easier and relevant to non-harmful. The nonenzymatic process is also linked to some disadvantages like the chemical to treat pesticides can cause a biological hazard in the ecosystem (Feng et al. 2020). The nonenzymatic chemical component can leach out or their release into the water system is a disastrous event causing the Minamata due to mercury release. The second major sulfonylurea derivative herbicides are degraded by simply soil pH. Soil acidic conditions convert the compound into a hydroxylated intermediate product which is easier to cleave. Some of the cases are reported where the degradation process is slower as the acidity of soil increases (Lin et al. 2020; Zhan et al. 2020; Ye et al. 2019; Huang et al. 2019, 2020; Fan et al. 2020; Pang et al. 2020).

Photodecomposition method of degradation is efficient for cleaving the internal chemical bond. The molecules which are multicyclic and contain carbon nitrogen are vulnerable to be attacked by sunlight radiation. The electromagnetic radiation range between 290 and 450 nM is efficient for the reaction like oxidation, reduction, and hydroxylation. The main problem for photodecomposition is that if the toxic chemical compound is leached deep into the soil or released deep into the water system, then the sunlight radiation penetration becomes ineffective, and photodecomposition rate becomes almost negligible. Pesticides in the environment have several fates according to the prevailing condition. Some of them are following—run off by surface water, volatilization in the atmosphere, leaching by ground water, adsorption and desorption (surface phenomena) by soil particle, degradation by the microorganism, and photodecomposition (Gangola et al. 2018b; Gupta et al. 2018; Khati et al. 2018b).

The hot and humid prevailing conditions like the tropical regions, where sunlight has a direct influence, exist among the majority of countries. The hot and humid condition is favorable for the growth of microorganisms. The metabolic rate of microorganism is very high during these conditions and can efficiently degrade the pesticides when they encounter it. Pesticides reduce the waste of crops and food resources annually. Aerobic bacteria cleave the aromatic benzene ring and make it prone to be breakdown by further chemical processes. The anaerobic metabolism of the aromatic compound-cleaving mechanism is slightly different. In the absence of oxygen, benzene is disrupted by reduction. We can say these chemical processes are not exclusive for any aerobic or anaerobic bacteria (Kumar et al. 2017; Khati et al. 2017a, b; Van Eerd et al. 2003). Some of the techniques that biologist relies to tackle with toxic and harmful pesticides have been enlisted in Table 8.2.

Globally, the agriculture sector is the primary and major user of pesticides. On the positive aspect, pesticides have resulted in the enhanced crop productivity. Depending upon the chemical composition, the major classification of pesticides has been proposed by the scientist. Bioremediation is nowadays a novel technique and the least expensive to handle with these toxic pesticides. The conversion of harmful and bio-hazardous toxicants simply by breakdown into non-toxic compounds is known as biotransformation or biodegradation. This technique in modern times is globally hot spot for the researcher. Biodegradation has several advantages over the traditional one mentioned earlier (Van Eerd et al. 2003). The number of

Table 8.2 Techniques for the remediation of pesticides as well as for the xenobiotics

Treatment process	Examples
Physical treatment	Adsorption, percolation with variable size pore filters
Chemical treatment	Strong oxidizing agents
Photocatalysis	Titanium oxide
Incineration	High temperature
Photodegradation	Light
Acid hydrolysis	Strong acid
Alkaline hydrolysis	Strong base
Microorganism	Pseudomonas, flavobacterium, alcaligenes
GMO (genetically modified organisms)	Via RDT-modified organisms

pesticides is usually very large that is being used by farmers, and their consequences are almost proportional to their number. The pesticides are linked to various effects on humans as well as on the ecosystem. It can cause pulmonary and hematological morbidity as well. The various types of birth disorders are also related to these pesticides. The farmers are more prone and vulnerable to the special risk associated by coming in contact of pesticides and their byproducts. These contaminants can contaminate the food in the food chain. Fishes are also prone to be contaminated by the pesticides because pesticides are very harmful in the native as well as byproduct form. Ultimately, the consequences are disturbed food chain, reproductive failure, deformities in children, poor health, and teratogenic effects. Before going to the microbial mechanism of action, it is important to know about the isolation and screening of pesticides degrading bacteria from nature (Gangola et al. 2018b; Gupta et al. 2018; Khati et al. 2018b).

8.2 Isolation and Enrichment of Pesticide-Degrading Microbes

The bacterial and fungal strains have the potential to degrade the pesticides from the contaminated environment. The contaminated environment mainly includes the soil and water system. Pesticides acted as the sole source of carbon, nitrogen, and phosphorous for the growth and reproduction of bacteria. Media provide the nutrients to bacteria in optimal quantity. Both solid and liquid mediums in the laboratory can be used for the cultivation of pesticide-degrading microbes. Making any material contamination free is known as the sterilization. Sterilization can be done by heat or steam and by other methods also. But generally, autoclave is used to sterilize any material, e.g., media, petri plates, beaker, flask, etc. The media which is selected for the desired organism is allowed to grow and suppression of undesirable microorganisms with suitable modification in the composition of media (Zhang et al. 2020a, b; Mishra et al. 2020; Feng et al. 2020; Lin et al. 2020; Zhan et al. 2020). The soil sample is serially diluted to reduce over crowdedness, and isolation of CFU is possible. Dilution factor is also very important in the calculation of bacterial numbers. The sample is then poured on to the media with the specific concentration of the pesticides, enabling to support the growth of pesticide-resistant bacteria. The concentration of the pesticides can be varied to check the up-to-what concentration the target bacteria can degrade it and resist it (Gangola et al. 2018b; Gupta et al. 2018; Khati et al. 2018b, 2017a; Kumar et al. 2017).

8.3 Characterization of the Pesticide-Degrading Microbes

The microbiological, biochemical, and morphological methods have been used for the characterization of the pesticide-degrading microbes. The surviving bacteria can be then isolated in form of pure colonies and can be inoculated to the specific contaminant sites. Several microbial floras such as bacteria, fungi, actinomycetes, algae, and plants are useful in degradation process. Fungi and bacteria are involved in the biodegradation process that releases hydrolytic enzymes, peroxidases, and oxygenases (Van Eerd et al. 2003). Microbial action on the toxic chemical pesticide breaks down them and in reciprocation takes carbon source for their efficient metabolism. The microorganism can be isolated from the soil and inoculated to contaminated sites that help to get rid out of intoxicants. Various parameters are important for the microbial activity to degrade them. So, by providing the appropriate supplements and various appropriate influencing supplements, biodegradation can be enhanced up to such a desirable extent that is not harmful. Few microorganisms are specialized to degrade the specific toxic chemical compound; then microorganisms can be isolated, cultured in a suitable feeding media, and inoculated to the contaminated sites or in a place where there is high exposure to harmful chemicals pesticides. Microorganisms can be categorized according to their specific biodegradability to a compound so that biodegradation efficiency can be accelerated. Accelerating the rate of biodegradation is done by the following methods: addition of the surfactant, supplementation with appropriate inoculant nutrients, and controlling the environmental factor.

8.4 Microbial Activity Mechanism to Degrade Pesticides

8.4.1 Degradation of Parathion

In the aerobic pathway of degradation, parathion in the presence of phosphodiesterases (cleavage of phosphodiester bond) hydrolyzes to form the diethyl-thio-phosphoric acid and para-nitro-phenol. In other sets of aerobic pathways, the parathion undergoes the oxidation process in the presence of an oxidoreductase enzyme. This pathway results into the di-ethyl-phosphoric acid and para-nitro phenol as same as above pathway, but the intermediate compounds are different. In the anaerobic pathway, parathion undergoes the reduction process to produce amino parathion as an intermediate which after the process of hydrolysis results into the benzoquinone.

8.4.2 Atrazine Degradation

There are several methods to degrade the toxicant from the environment enlisted above. The *Pseudomonas* species used the compound atrazine as a carbon source and via the help of several enzymes for efficient degradation. Enzymes used by *Pseudomonas* ADP strain are AtzA, AtzB, and AtzC. AtzA helps in hydrolysis and dechlorination. The protein AtzB performs the dehydrochlorination reaction that converts the hydroxyatrazine to the *N*-isopropyl cyanuric acid. AtzC catalyzes the cyanuric acid produced in the above step. Finally, the end product remnants are carbon dioxide and ammonia. It had been strongly suggested that genes encoding these degradative enzymes are expressed by the extra chromosome known as a plasmid. For example, the enzymes degrading the 2–4D are absolutely encoded by the bacterial plasmid. Several genes are retained on the main bacterial chromosome for the expression of degradative enzymes (Bhatt et al. 2021d).

8.5 Pesticides Used in the Agriculture

Organochlorine pesticides are widely used in the agricultural sector due to their rapid action and easy to use. These include hexachlorocyclohexane, 111, DDT (dichloro-diphenyl-trichloroethane), methoxychlor, dieldrin, chlordane, toxaphene, mirex, kepone, lindane, etc.

8.5.1 Steps in Biodegradation Mechanism

Degradation mechanism is broadly categorized into mainly three parts: adsorption, penetration, and enzyme catalysis reaction. Adsorption is a physical phenomenon that is only on the surface accumulation of substrate also known as physisorption. Penetration includes the entry of the target chemical compound into the desired cell. Finally, at last, the enzyme catalysis reaction involves the rapid degradation into the simpler non-toxic product. The bio-stimulation-based degradation process can be accelerated via the supplementation of appropriate and additional stimulating agents, e.g., surfactant and electron donor (water, hydrogen, sulfide). The transformation of the organic compound into the non-toxic or less toxic inorganic compound under the action of soil biota via microbial activity is partially or completely degraded. Through this process, it can degrade the various toxic chemical compounds from the environment. Co-metabolism also known as the co-substrate inoculation, the exogenous addition of suitable biomass or supplementation, rapidly breaks down the toxic chemical pesticides. Synergistic effect consortium of microbial biomass has increased efficiency rather than the single microbial degradation (Sulbhi et al. 2021; Bhandari et al. 2021; Bhatt et al. 2021a, b, c). Synergistic or additive effects can degrade the toxic chemical compound rapidly comparable to single microbial degradation of pesticides. For example, the insecticide parathion is degraded synergistically with the combined effect of *P. aeruginosa* and *P. stutzeri*. Genetically, modified organisms are used to increase the rate of degradation. RDT is the possible way to manipulate the appropriate gene and subsequently degradative process. Bioventing is a process of bio-stimulation by which stimulating agents are added into the soil to increase the microbial catabolic activity (Van Eerd et al. 2003; Nadeau et al. 1994).

8.5.2 Factors Affecting Pesticide Degradation

There are various parameters that influence the efficient degradation. Providing appropriate condition has increased microbial activity to degrade the pesticides (Table 8.3).

8.5.3 Chemistry of Pesticide Degradation

The following are some of the reasons with respect to chemistry for the degradation of pesticides (Table 8.4).

- Linear or aliphatic hydrocarbon compounds are easy to degrade rather than aromatic compounds.
- Ring restricts the ease of its biodegradation.
- High molecular weight, larger chain compounds are difficult to be degraded.
- Branching decreases the efficiency of degradation.
- Molecular position of functional substituent group has large effect on degradation.

Table 8.3 Factors influencing the microbial degradation

Factors	Examples
Types of microbial species	*Pseudomonas* strain specific to atrazine
Effect of pesticide chemistry	Functional groups, molecular weight, chemical nature
Environmental factor	pH, salinity, humidity, nutrition, oxygen availability, substrate concentration

Table 8.4 Microorganism specific to the chemical pollutant

Microorganism	Toxic chemical compound
Pseudomonas sp.	Linear (aliphatic) and aromatic hydrocarbons
Corynebacterium sp.	Halogenated hydrocarbons
Bacillus sp.	High molecular wt. hydrocarbons
Candida sp.	PCBs
Fusarium sp.	Propanil
Nocardia sp.	Naphthalene

Table 8.5 Microorganism with specialized catabolic gene

Microorganism	Catabolic gene
E. coli	pepA
Bacillus cereus	phn
SMSP-1	opdB
Nocardia sp.	adpB
Pseudomonas monteilii	hocA
A. radiobacter	opdA

8.5.4 Successful Biodegradation Process of Chemical Compound

- Microorganism catabolic activity on the target substrate.
- The toxic target compound must be bioavailable.
- Soil conditions must favor microbial growth and activity.
- The cost of remediation should be less expensive.

8.6 Microbial Enzyme System

Intracellular and extracellular enzymes play an important and key role in the degradation of xenobiotics. Various enzymes have been reported to degrade the organochlorine such as the dioxygenases (Nadeau et al. 1994), cytochrome P450 (the great variety of reactions catalyzed by P450s was well recognized, ligninases) (Pang et al. 2020), dehydrogenases (Bourquin 1977), esterases, and glycosidases. These enzymes have been documented for the degradation and detoxification of xenobiotics. Carbamate degradation (Parekh et al. 1995) has been reported by a wide genus of bacterial species (Table 8.5).

8.7 Enzyme System for Organophosphate Degradation

Organophosphate compounds are present everywhere in the surroundings and used in high quantity in the agricultural sector. It can also be used as a weapon tool to threaten other nation as well. So, there is a need to completely remove these toxic compounds or convert them to simpler non-toxic. The phosphate ester of alcohol and the phosphoric acid combines and resulting in the formation of organophosphate. A wide range of pesticides contains this organophosphate compound, and that can also affect our nervous system severely. OP can cause chronic or acute symptoms of diseases depending upon its concentration and exposure to the organs. Organophosphate can be removed from the environment by the use of chemical decontamination, physical adsorbent, and finally an efficient microbial system. In human neuromuscular junction or synapse, an enzyme acetyl-choline-esterase found which break acetyl choline (Ach) and in similar way organo-phosphate compound bind irreversibly with the acetyl choline. Organophosphate chemical compounds can be degraded through the nucleophilic attack of the phosphorous center of compound. Some of the common enzymes are peroxidases, laccase, esterase, and oxidoreductase. Pesticide degradation characteristics are analyzed by following techniques GC–MS, GC–MS/MS, and LC–MS/MS for two filed trials, high-resolution MS-based methods, liquid chromatography (LC) with ultraviolet, and diode array (Mishra et al. 2020; Zhang et al. 2020b; Feng et al. 2020; Lin et al. 2020; Zhan et al. 2020; Ye et al. 2019; Huang et al. 2019, 2020).

8.8 Cyclic Wave Voltammetry

It is the combination of the square wave potential and staircase potential applied to a static electrode. It is a form of linear sweep voltammetry. In a simple word, it is the combination of a square wave superimposed onto a staircase potential. This SWV involves sweeping the potential linearly with time at a rate between 10 and 100 V/s. The advantageous properties of this technique are fast scan rate, speed analysis, sensitivity, low detection limits, and reduced analysis time and reject background signal (Khati et al. 2017a, b; Van Eerd et al. 2003).

8.8.1 Amperometry

It determines diffusion current; potential fixed electrode and reducible and non-reducible agents can be determined. Amperometry is based on the principle of polarography with the exception that the voltage is maintained constant during titration (Castro et al. 1985; Parte et al. 2017; Pizzul et al. 2009). Here the electroreducible and electro-non-reducible analytes can be analyzed (Table 8.6).

Table 8.6 Examples of analytes with their biosensing elements and transducers

Analytes	Biosensing elements	Transducers	Samples
2,4-D	Acetyl cholinesterase	Amperometric	Soil
Diuron, paraquat	Cyanobacterial	Bioluminescence	Soil
Atrazine	Antibody	Amperometric	Orange juice
2,4-D	Antibody	Impedance	Soil
Isoproturon	Antibody	Potentiometric	Drinking water

8.8.2 *Methodology*

- The potential of the instrument is fixed.
- Take sample solution and add supporting electrode.
- The titrant is added and current flow is measured in mA.
- It can determine concentration up to 0.0001 molars of analyte in the target sample.

The other voltammetry techniques for the pesticide's detection are differential pulse voltammetry, electrochemical impedance spectra (EIS), differential pulse polarography (DPP), anodic stripping voltammetry, Kaloussek's switcher, and oscillopolarography.

8.9 Conclusion

As the population natality rate is exponentially increasing, to meet their demands agricultural sector has to take care and deal to meet their demands. The agricultural sector relies on technological advancement to decrease the productivity loss and enhance productivity to prevent a food crisis. Thus, fulfilling these above requirements, modern methods of agro-farming using chemical compounds are taking place. These modern methods that practice to use pesticides besides harmful effects on the ecosystem can also cause short- and long-term effects of diseases and disorder, with acute and probably chronic symptoms. The only way to cope up with the harmful impacts of modern chemical compounds is by monitoring their use and analyzing their negative effects on the environment. Government and the agricultural sector should take care of the use of modern chemical compounds and also their constant monitoring.

References

Agrawal A, Pandey RS, Sharma B (2010) Water pollution with special reference to pesticide contamination in India. J Water Resour Protect 2:432–448

Barbosa P, Fernanda P (2019) Int J Electrochem Sci 14:3418–3433

Bhandari G, Bhatt P (2020) Concepts and application of plant microbe interaction in remediation of heavy metals. In: Sharma A (ed) Microbes and signaling biomolecules against plant stress. Rhizosphere biology. Springer, Singapore. https://doi.org/10.1007/978-981-15-7094-0_4

Bhandari G, Bagheri AR, Bhatt P, Bilal M (2021) Occurrence, potential ecological risks, and degradation of endocrine disrupter, nonylphenol from the aqueous environment. Chemosphere 275:130013. https://doi.org/10.1016/j.chemosphere.2021.130013

Bhatt P (2018) *In silico* tools to study the bioremediation in microorganisms. In: Pathak V, Navneet (eds) Handbook of research on microbial tools for environmental waste management. IGI Global, Hershey, PA, pp 389–395. https://doi.org/10.4018/978-1-5225-3540-9.ch018

Bhatt P, Barh A (2018) Bioinformatic tools to study the soil microorganisms: an *in silico* approach for sustainable agriculture. In: Choudhary D, Kumar M, Prasad R, Kumar V (eds) In silico approach for sustainable agriculture. Springer, Singapore. https://doi.org/10.1007/978-981-13-0347-0_10

Bhatt K, Bhatt P (2020) Rhizosphere biology: alternate tactics for enhancing sustainable agriculture. In: Phytomicrobiome interactions and sustainable agriculture. John Wiley & Sons, New York, NY. ISBN: 9781119644620

Bhatt P, Nailwal TK (2018) Chapter: 11Crop improvement through microbial technology: a Step towards sustainable agriculture. In: Prasad R, Gill SS, Tuteja N (eds) Crop improvement through microbial biotechnology. Elsevier, Amsterdam, pp 245–253. https://doi.org/10.1016/B978-0-444-63987-5.00011-6

Bhatt P, Bisht TS, Pathak VM, Barh A, Chandra D (2015a) Optimization of amylase production from the fungal isolates of Himalayan region Uttarakhand. Ecol Environ Conserv 21 (3):1517–1521

Bhatt P, Negi G, Gangola S, Khati P, Srivastava A, Sharma A (2015b) Optimization of sulfosulfuron biodegradation through response surface methodology using indigenous bacterial strain isolated from contaminated agriculture field. Int J Curr Microbiol App Sci 4(8):105–112

Bhatt P, Negi G, Gangola S, Khati P, Kumar G, Srivastava A, Sharma A (2016a) Differential expression and characterization of cypermethrin degrading potential proteins in *Bacillus thuringiensis* strain, SG4. 3 Biotech 6:225

Bhatt P, Sharma A, Gangola S, Khati P, Kumar G, Srivastava A (2016b) Novel pathway of cypermethrin biodegradation in *a Bacillus* sp. strain SG2 isolated from cypermethrin-contaminated agriculture field. 3 Biotech 6:45

Bhatt P, Huang Y, Zhan H, Chen S (2019a) Insight into microbial applications for the biodegradation of pyrethroid insecticides. Front Microbiol 10:1778

Bhatt P, Pal K, Bhandari G, Barh A (2019b) Modeling of methyl halide biodegradation on bacteria and its effect on other environmental systems. Pestic Biochem Physiol 158:88–100

Bhatt P, Gangola S, Chaudhary P, Khati P, Kumar G, Sharma A, Srivastava A (2019c) Pesticide induced up-regulation of esterase and aldehyde dehydrogenase in Indigenous *Bacillus* spp. Bioremed J 23(1):42–52

Bhatt P, Joshi D, Kumar N, Kumar N (2019d) Recent trends to study the functional analysis of mycorrhizosphere. In: Varma A, Choudhary D (eds) Mycorrhizosphere and pedogenesis. Springer, Singapore. https://doi.org/10.1007/978-981-13-6480-8_11

Bhatt P, Pathak VM, Joshi S, Bisht TS, Singh K, Chandra D (2019e) Chapter 12: Major metabolites after degradation of xenobiotics and enzymes involved in these pathways. In: Smart bioremediation technologies: microbial enzymes, pp 205–215. https://doi.org/10.1016/B978-0-12-818307-6.00012-3

Bhatt P, Rene ER, Kumar AJ, Kumar AJ, Zhang W, Chen S (2020a) Binding interaction of allethrin with esterase: bioremediation potential and mechanism. Bioresour Technol 315:13845. https://doi.org/10.1016/j.biortech.2020.123845

Bhatt P, Verma A, Verma S, Anwar MS, Prasher P, Mudila H, Chen S (2020b) Understanding phytomicrobiome: a potential reservoir for better crop management. Sustainability 12:5446

Bhatt P, Huang Y, Rene ER, Kumar AJ, Chen S (2020c) Mechanism of allethrin biodegradation by a newly isolated *Sphingomonas trueperi* strain CW3 from wastewater sludge. Bioresour Technol 305:123074

Bhatt P, Zhang W, Lin Z, Pang S, Huang Y, Chen S (2020d) Biodegradation of allethrin by a novel fungus *Fusarium proliferatum* strain CF2, Isolated from Contaminated. Soils. Microorganisms 8:593

Bhatt P, Huang Y, Zhang W, Sharma A, Chen S (2020e) Enhanced cypermethrin degradation kinetics and metabolic pathway in *Bacillus* thuringiensis strain, SG4. Microorganisms 8:223

Bhatt P, Bhatt K, Huang Y, Ziqiu L, Chen S (2020f) Esterase is a powerful tool for the biodegradation of pyrethroid insecticides. Chemosphere 244:125507

Bhatt P, Zhou X, Huang Y, Zhang W, Chen S (2021a) Characterization of the role of esterases in the biodegradation of organophosphate, carbamate and pyrethroid group pesticides. J Hazard Mater 411:125026

Bhatt P, Joshi T, Bhatt K, Zhang W, Huang Y, Chen S (2021b) Binding interaction of glyphosate oxidoreductase and C-P lyase: molecular docking and molecular dynamics simulation studies. J Hazard Mater 409:124927

Bhatt P, Bhatt K, Sharma A, Zhang W, Mishra S, Chen S (2021c) Biotechnological basis of microbial consortia for the removal of pesticides from the environment. Crit Rev Biotechnol 41:317. https://doi.org/10.1080/07388551.2020.1853032

Bhatt P, Sethi K, Gangola S, Bhandari G, Verma A, Adnan M, Singh Y, Chaube S (2021d) Modeling and simulation of atrazine biodegradation in bacteria and its effect in other living systems. J Biomol Struct Dyn. https://doi.org/10.1080/07391102.2020.1846623

Bhatt P, Gangola S, Bhandari G, Zhang W, Maithani D, Mishra S, Chen S (2021e) New insights into the degradation of synthetic pollutants in contaminated environments. Chemosphere 268:128827. https://doi.org/10.1016/j.chemosphere.2020.128827

Bhatt P, Sharma A, Rene ER, Kumar AJ, Zhang W, Chen S (2021f) Bioremediation mechanism, kinetics of fipronil degradation using *Bacillus* sp. FA3 and resource recovery potential from contaminated environments. J Water Process Eng 39:101712

Bourquin AW (1977) Degradation of malathion by salt-marsh microorganisms. Appl Environ Microbiol 33:356–362

Castro CE, Wade RS, Balser NO (1985) Bio-dehalogenation: reactions of cytochrome P-450 with polyhalomethanes. Biochemistry 24:204–210

Fan X, Ye T, Li Q, Bhatt P, Zhang L, Chen S (2020) Potential of a quorum quenching bacteria isolate Ochrobactrum intermedium D-2 against soft rot pathogen *Pectobacterium carotovorum* subsp. Carotovora. Front Microbiol 11:898

Feng Y, Huang Y, Zhan H, Bhatt P, Chen S (2020) An overview of strobilurin fungicide degradation: current status and future perspective. Front Microbiol 11:389

Gangola S, Bhatt P, Chaudhary P, Khati P, Kumar N, Sharma A (2018a) Bioremediation of industrial waste using microbial metabolic diversity. In: Pankaj, Sharma A (eds) Microbial biotechnology in environmental monitoring and cleanup. IGI Global, Hershey, PA, pp 1–27. https://doi.org/10.4018/978-1-5225-3126-5.ch001

Gangola S, Sharma A, Bhatt P, Khati P, Chaudhary P (2018b) Presence of esterase and laccase in *Bacillus subtilis* facilitates biodegradation and detoxification of cypermethrin. Sci Rep 8:12755. https://doi.org/10.1038/s41598-018-31082-5

Gupta S, Bhatt P, Chaturvedi P (2018) Determination and quantification of asiaticoside in endophytic fungus from Centella asiatica (L.) Urban. World J Microbiol Biotechnol 34:111

Huang Y, Zhan H, Bhatt P, Chen S (2019) Paraquat degradation from contaminated environments: current achievements and perspectives. Front Microbiol 10:1754

Huang Y, Lin Z, Zhang W, Pang S, Bhatt P, Rene ER, Kumar AJ, Chen S (2020) New insights into the microbial degradation of D-cyphenothrin in contaminated water/soil environments. Microorganisms 8:473

Huang Y, Zhang W, Pang S, Chen J, Bhatt P, Mishra S, Chen S (2021) Insights into the microbial degradation and catalytic mechanism of chlorpyrifos. Environ Res 194:110660

Hussain S, Siddique T, Saleem M, Arshad M, Kha-lid A (2009) Impact of pesticides on soil microbial diversity, enzymes, and biochemical reactions. Adv Agron 102:159–200
Khati P, Sharma A, Gangola S, Kumar R, Bhatt P, Kumar G (2017a) Impact of agriusable nanocompounds on soil microbial activity: an indicator of soil health. Soil, Air, Water 45 (5):1600458
Khati P, Parul, Gangola S, Bhatt P, Sharma A (2017b) Nanochitosan induced growth of Zea Mays with soil health maintenance. 3 Biotech 7(1):81
Khati P, Gangola S, Bhatt P, Kumar R, Sharma A (2018a) Application of nanocompounds for sustainable agriculture system. In: Pankaj, Sharma A (eds) Microbial biotechnology in environmental monitoring and cleanup. IGI Global, Hershey, PA, pp 194–211. https://doi.org/10.4018/978-1-5225-3126-5.ch012
Khati P, Parul, Bhatt P, Nisha Kumar R, Sharma A (2018b) Effect of nanozeolite and plant growth promoting rhizobacteria on maize. 3 Biotech 8:141
Kumar G, Arya K, Verma A, Pankaj, Khati P, Gangola S, Kumar R, Sharma A, Singh H (2017) Bioremediation of petrol engine oil polluted soil using microbial consortium and wheat crop. J Pure Appl Microbiol 11(3):1583–1588
Lin Z, Zhang W, Pang S, Huang Y, Mishra S, Bhatt P, Chen S (2020) Current approaches to and future perspectives on methomyl degradation in contaminated soil/water environments. Molecules 25:738
Mishra S, Zhang W, Lin Z, Pang S, Huang Y, Bhatt P, Chen S (2020) Carbofuran toxicity and its microbial degradation in contaminated environments. Chemosphere 259:127429
Nadeau LJ, Fu-Min M, Breen A (1994) Sayler GS Aerobic degradation of (1,1,1 trichloro-2,2-bis (4-chlorophenyl) ethane) DDT by Alcaligenes eutrophus A5. Appl Environ Microbiol 60:51–55
Pang S, Lin Z, Zhang W, Mishra S, Bhatt P, Chen S (2020) Insights into the microbial degradation and biochemical mechanisms of neonicotinoids. Front Microbiol 11:868
Parekh NR, Hartmann A, Charnay MP, Fournier JC (1995) Diversity of carbofuran-degrading soil bacteria and detection of plasmid-encoded sequences homologous to the mcd gene. FEMS Microbiol Ecol 17:149–160
Parte GS et al (2017) Microbial degradation of pesticide: a review. Afr J Microbiol Res 11 (24):992–1001
Pizzul L, Castillo MP, Stenström J (2009) Degradation of glyphosate and other pesticides by ligninolytic enzymes. Biodegradation 20:751–759
Sharma A, Bhatt P (2016) Chapter 3: Bioremediation: a microbial technology for improvising wildlife. In: Wildlife management concept analysis and conservation, pp 29–40
Sharma A, Pankaj, Khati P, Gangola S, Kumar G (2016) Chapter 6: Microbial degradation of pesticides for environmental cleanup. In: Bioremediation of industrial pollutants. Write and Print Publication, New Delhi, pp 178–205
Singh K, Gera R, Sharma R, Maithani D, Chandra D, Bhat AM, Kumar R, Bhatt P (2021) Mechanism and application of *Sesbania* root nodulating bacteria: an alternative for chemical fertilizers and sustainable development. Arch Microbiol 203:1259. https://doi.org/10.1007/s00203-020-02137-x
Sulbhi V, Bhatt P, Verma A, Mudila H, Prasher P, Rene ER (2021) Microbial technologies for heavy metal remediation: effect of process conditions and current practices. Clean Techn Environ Policy. https://doi.org/10.1007/s10098-021-02029-8
Torres RD (2003) El papel de los microorganismo en la biodegradación de compuestos tóxicos. Ecosistemas 2:1–5
Van Eerd LL, Hoagland RE, Zablotowicz RM, Hall JC (2003) Pesticide metabolism in plants and microorganisms. Weed Sci 51:472–495
Vischetti C, Monaci E, Cardinali A, Casucci C, Perucci P (2008) The effect of initial concentration, co-application and repeated applications on pesticide degradation in a biobed mixture. Chemosphere 72:1739–1743

Ye T, Zhou T, Fan X, Bhatt P, Zhang L, Chen S (2019) *Acinetobacter lactucae* strain QL-1, a novel quorum quenching candidate against bacterial pathogen *Xanthomonas campestris* pv. Campestris. Front Microbiol 10:2867

Zhan H, Huang Y, Lin Z, Bhatt P, Chen S (2020) New insights into the microbial degradation and catalytic mechanism of synthetic pyrethroids. Environ Res 182:109138

Zhang W, Pang S, Lin Z, Mishra S, Bhatt P, Chen S (2020a) Biotransformation of perfluoroalkyl acid precursors from various environmental system: advances and perspectives. Environ Pollut 272:115908

Zhang W, Lin Z, Pang S, Bhatt P, Chen S (2020b) Insights into the biodegradation of lindane (γ-Hexachlorocyclohexane) using a microbial system. Front Microbiol 11:522

Chapter 9
Biofortification of Plants by Using Microbes

Ankur Adhikari, Kamal Pandey, Vinita Pant, Tara Singh Bisht, and Himanshu Punetha

Abstract In the twenty-first century, a key obstacle is meeting the hunger needs of rising population every day. Thus, the efficiency of cultivated land and food fortification needs to be improved. The sole objective of the latest agronomic practices, particularly with regard to food crops, is to maximize grain size and weight, even at the price of nutritional value, aiding in few cases to scarcity in micronutrients (termed as hidden hunger). Life-threatening health situations and ailments such as birth defects, heart disease, cancer, nerve disorders, and several others can be caused by hidden hunger. Micronutrient deficiencies are caused by the lessen amount of micronutrients in food crops and are one of the major threats affecting more than two million people worldwide. Plant growth-promoting microbes (PGPM) are categorized in two major groups: plant growth-promoting rhizobacteria (PGPR) and plant growth-promoting fungi (PGPF), which are feasible ways to solve this issue. In staple foods, the use of biofortified nutrients and mineral content can be enhanced. These microbes help to increase micronutrient uptake (e.g., phosphorous, iron, selenium, and zinc). Plant-microbe relationships and our expanding understanding of these interactions can help to biofortify crops with micronutrients in a sustainable and environmentally friendly manner. Biofortification of food crops is indeed espoused as a novel method not only to raise micronutrient concentration in edible food crops but also to boost crop productivity on scarce fertile soils. PGPR are soil bacteria that are responsible of colonizing the rhizosphere and strengthening the development of plants through a

A. Adhikari (✉) · H. Punetha
Department of Biochemistry, College of Basic Sciences and Humanities, Gobind Ballabh Pant University of Agriculture and Technology, Pantnagar, Uttarakhand, India

K. Pandey
Department of Biotechnology, Kumaun University, Bhimtal, India

V. Pant
Indian Council of Agricultural Research-Directorate of Cold Water Fisheries Research, Bhimtal, Uttarakhand, India

T. S. Bisht
Patanjali Research Institute, Haridwar, Uttarakhand, India

P. Bhatt et al. (eds.), *Microbial Technology for Sustainable Environment*,
https://doi.org/10.1007/978-981-16-3840-4_9

wide range of different mechanisms, such as organic matter mineralization, biological control of soil-borne pathogens, N_2 biological fixation, and root growth development. The nutrients are mobilized by microorganisms through multiple methods, such as chelation, acidification, exchange reactions, and organic acid release. In addition, several pieces of information reveal that plant inoculation at both physiological and molecular levels with PGPR will have significant effects on plants.

Keywords Biofortification · Plant growth-promoting microorganisms · Micronutrients · Siderophores · Rhizosphere

9.1 Introduction

A lot of pressure has been put on agriculture to support the continually growing human population, by combating not only the hunger but also the "hidden hunger." Given the increased food supply-demand and abiotic stresses related to climate change, plant's growth, productivity, and nutritional value are getting affected to a great extent (Roriz et al. 2020). Till now, the concern was to increase the overall yield to mitigate the food insecurity issues which was significantly resolved by green revolution. Another upcoming global issue which needs to be addressed is of malnutrition due to insufficient dietary intake of micronutrients which is regarded as hidden hunger. Various symptoms manifested by humans, resulting from micronutrient-deficiency, are stunted growth, respiratory infections, anemia, impaired mental development, and increased risk of mortality during childbirth (Singh et al. 2020). It is predominant in poor developing countries because their major staple food is deprived of some essential micronutrients and they cannot afford dietary diversification, so the focus has now been shifted to producing food that is nutritionally rich in a sustainable way. The State of Food Security and Nutrition in the World has penned in its latest edition that around 690 million people went hungry in 2019, up by ten million from 2018, and by ~60 million in the last 5 years, according to a WHO 2020 press release. A warning has been put forth by the heads of five principal agencies, namely, FAO, UNICEF, IFAD, WFP, and WHO, "five years after the global commitment to end hunger, food insecurity and all forms of malnutrition, this goal is still off-track by 2030." It is also reported that COVID-19 pandemic will intensify this issue. There is a dire need to respond in a sustainable manner to meet the current societal obstacles of ensuring the production of foods with an elevated nutritional content associated with a low environmental diet.

Biofortification has been deployed as a common agricultural practice that will curb hidden hunger, fulling all the constraints mentioned above. Current strategies of biofortification like organic and chemical fertilizer application, elite germplasm breeding, and metabolic engineering come with various drawbacks. Breeding and CRISPR/Cas9 technologies, for example, can be exploited to improve the micronutrient value of crops; however, they are time-consuming and costly and encounter multiple challenges with genetically modified organisms (GMOs). In addition, the utilization of microbes for biofortification has been advocated as a "greener" substitute for sustainable agriculture due to the environmental hazards from the unnecessarily high usages of chemical fertilizers. Given their nutrient mobilization potential, microbes-based biofortification is the most inexpensive, quickest, and sustainable way to increase the bioavailability of micro- and macronutrients in our diets while simultaneously restoring soil health (Singh et al. 2011).

9.1.1 Plants' Hidden Hunger

There are 17 essential elements, required by plants in order to complete their life cycles, categorized into two groups: macronutrient, required in large quantities, and micronutrient, required in trace amounts. Macronutrient consists of nine elements N, P, K, C, O, H, Ca, S, and Mg, whereas micronutrient consists of eight elements, viz., Fe, Mo, B, Zn, Mn, Ni, Cl, and Cu (Blevins 2009). The micronutrient is an integral part of plant's metabolic pathways as they facilitate catalytic reactions, function as protein cofactors, and/or stabilize structural domains. Various studies have shown that even a deficiency of any of the soil's micronutrients can restrict the ideal plant yield and contribute to low food products, amidst the adequate presence of all other nutrients. Plants suffering from hidden hunger become vulnerable to various biotic and abiotic stresses (Ahmad and Prasad 2011). It is predicted that the issue of hidden hunger is projected to be amplified by climate change (Ku et al. 2019). Climate change, typically rising temperatures, changing precipitation patterns, varying levels of soil water, and elevated atmospheric concentration (CO_2) have long been established to have intense repercussions for the nutritional content of crop products (Nakandalage and Seneweera 2018).

9.1.2 Biofortification

Biofortification is the strategy of strengthening the nutritional efficiency of food crops through agronomical practices, traditional plant breeding, or modern biotechnology, as defined by the WHO (WHO 2016). Plant biofortification is pertinent, as micronutrients are inadequate in the three staple crops, rice, maize, and wheat, which make a significant contribution almost half of the calories devoured by humans. Conventional and genetic breeding are the prime approaches; the objective is to maximize the bioavailable micronutrient content in principal food crops such as maize, rice, wheat, pearl millet, etc. (Velu et al. 2014; Prasanna et al. 2016).

The insight of the complicated plant-microbe rhizospheric relationships has fueled studies into the role and application of PGPB in the crop biofortification. The role of PGPB in plant growth and defense was first stated by Kloepper (Kloepper et al. 1980), and since then, a lot of studies has been conducted. They are now popularly being considered as an efficient and environmental-friendly alternatives to chemical pesticides and fertilizers (Olivares et al. 2015). Exploiting this plant-microbe interaction can further benefit in achieving biological substrate enrichment, and this is imperative as the micronutrients are inadequate in three staple crops rice, maize, and wheat, which help in providing almost half of the calories ingested by humans. Biofortification can be achieved through various approaches:

1. *Gene manipulation*
2. *Transgenic approaches*
3. Organic fertilizer
4. *Chemical fertilizer*
5. PGPB mediated

In 1999 came the first biofortified staple crop, "golden rice," the genetically modified rice that additionally contains beta-carotene. One of the most important systematic and symbolic breeding programs, the Harvest Plus program, was initiated in 2004 to mitigate micronutrient malnutrition in Africa and Asia. The scheme centered on enriching rice, beans, cassava, maize, sweet potatoes, and pearl millet, focusing on three main nutrients, Fe, Zn, and vitamin A (Pfeiffer and McClafferty 2007). Carvalho and Vasconcelos (2013) and Garg et al. (2018) stated that the efficiency of any food fortification programs depends heavily on the acquiescence of farmers and the public and also on political assistance to determine their cost/ benefit. This lack of public acceptance is a sole reason why golden rice is not yet commercialized, in spite of its clearly identified benefits.

9.1.3 Plant Growth-Promoting Bacteria (PGPB)

PGPB are those bacteria that intensify the growth of plant and also protect it from various diseases and abiotic stresses by inducing number of compounds such as phytohormones (cytokines, gibberellic, ethylene, and indole-3-acetic acid), atmospheric nitrogen fixation, siderophores, organic acid, and phosphate solubilization and by producing antibiotics which help to decrease the detrimental effect of pathogens (Ji et al. 2019; Glick 2012; Felestrino et al. 2017). These ecofriendly and beneficial bacteria inhabit near the plant roots area which is known as rhizosphere. Plant release numerous nutrients (sugar, organic acid, amino acids, vitamin, flavonoid, and lipid), which is known as root exudates, into the rhizosphere that leads to relocation of soil bacteria toward these nutrients (as these root exudates attract the bacteria), and therefore, plethora of microorganism are found inhabited in rhizosphere (Schillaci et al. 2019). Proteobacteria (*Pseudomonas*, *Acinetobacter*, *Serratia*, *Pantoea*, *Psychrobacter*, *Enterobacter*, and *Rahnella*) and Firmicutes (*Bacillus* sp.) are the two phylum to which PGPB belongs (Ramakrishna et al. 2019). They encourage plant growth by two mechanism, i.e., direct and indirect. Direct mechanism includes N_2 fixation, siderophore production, phosphate solubilization, phytohormone production, and 1-Aminocyclopropane-1-carboxylate (ACC) deaminase, whereas in indirect one, PGPB create hindrance in the growth of pathogen (Ahemad and Kibret 2014).

N_2 plays important role in the photosynthesis of plant as it is the vital component of chlorophyll pigment (Wagner 2011). Even though abundant amount of N_2 is present in the atmosphere, plants utilize only reduced form of it. Biological nitrogen fixation (BNF) is a process in which prokaryotic microorganisms convert the atmospheric N_2 into plant usable reduced form, i.e., ammonia (NH_3) (O'Hara 1998; Cheng 2008). Bacteria that carried out nitrogen fixation are known as diazotrophs, and they carry a nitrogenase enzyme that acts as catalyst, hence fasten the transformation of N_2 into ammonia (Santi et al. 2013). Genes which control the symbiotic N_2 fixation in rhizobia are *nod*, *nif*, and *fix* genes (Bano and Iqbal 2016). N_2-fixing bacteria are classified as symbiotic N_2-fixing bacteria such as rhizobia in leguminous plant and *Frankia* in non-leguminous plant and non-symbiotic N_2-fixing bacteria (*Cyanobacteria*, *Azospirillum*, *Azotobacter*, etc.) that are free living and endophytes (Ahemad and Kibret 2014). Legume crop supply food and energy to rhizobia and in lieu of that rhizobia provide ammonia to crop (Lindström and Mousavi 2020), and this ammonia is used by plant for the synthesis of biomolecules such as nucleic acid and proteins that is required for the proper growth and development of plant (Soumare et al. 2020; Babalola et al. 2017). Siderophores are iron (Fe)-chelating compound having low molecular weight produced by soil organism. Its function is to encourage plant growth and hamper the phytopathogens by arresting Fe from the environment (Maheshwari et al. 2019). Phosphate-solubilizing bacteria (*Pseudomonas*, *Rhizobium*, *Bacillus*, etc.), another direct mechanism, convert the insoluble form of phosphate into plant utilizable form, i.e., H_2PO^{4-} and $HPO_4{}^{2-}$ via process of chelation, exchange reaction, acidification, as well as producing gluconic acid (Rodríguez et al. 2006). Phytohormones (auxin, cytokinins, gibberellins, and inhibiting ethylene production) are produced by bacteria and liberated in the rhizosphere by rhizobacteria or in plant tissues by endophytes. These hormones protect plant from biotic as well as abiotic stresses. Apart from that, it also helps in the growth, development, and physiological activity of plant (Esitken et al. 2010; Shilev 2020). High concentration of ethylene can inhibit the plant growth and can even kill the plant. ACC deaminase enzyme degrades the precursor of plant ethylene, i.e., ACC into ammonia and α-ketobutyrate. As a result, concentration of ethylene lowers (Glick 2014; Hao et al. 2007) that helps in encouraging the growth of plant in adverse conditions.

9.2 Microbe-Mediated Biofortification

In order to improve yield and soil fertility, biofortification of crops via the effects of PGPM can be seen as a successful concomitant indicator which increased micronutrient concentrations in the food crop may occur along with transgenic cultivars. In addition to strengthening soil fertility and crop yield, PGPM has been observed to biofortify the micronutrient content in food crops (Rana et al. 2012). In addition to insoluble phosphorus solubilization, through N_2 fixation, phytohormone synthesis, reduction of ethylene concentration, synthesis, and induction of systemic resistance to antibiotics and antifungal metabolites, plant growth will be enhanced by PGPM as well (Singh and Prasad 2014). Thus, biofortification of crops utilizing PGPMs can be seen as a potential additional way to enhance micronutrient concentrations throughout the cultivation of wheat, in addition to improvement of soil fertility and yield alongside breeding varieties (Singh et al. 2017). All the more as of late, a comprehension of the perplexing interaction among microbial and plant networks in the rhizosphere has fueled the exploitation of soil microorganisms for the crop biofortification. Bacteria and mycorrhizal fungi have the significant contribution in the assimilation of the micronutrients in plants which is reported in the literature (Wu et al. 2015; Berruti et al. 2016). When the connection between beneficial bacteria and plants is utilized, the rhizosphere, a crucial layer among plant roots and the soil, can make a substantial contribution to agriculture. In particular, this chapter will be focusing on biofortification of crop through mycorrhizal and bacterial mobilization of the rhizosphere-based micronutrients (Prakash and Verma 2016; Jha and Subramanian 2016). The biofortification of major crops by exploiting these PGPR has been the focus of plant-microbe interactions. To ensure that the human body works properly, micronutrients such as Se, Zn, and Fe are vitally important, and these constitute a major part of PGPR-mediated biofortification (Priyadharsini and Muthukumar 2016; Kumar et al. 2017; Raghavendra et al. 2016; Zahedi 2016; Dominguez-Nuñez et al. 2016; Dotaniya et al. 2016).

Colonized plants of arbuscular mycorrhizal fungi (AMF) can efficiently procure nutrients from a greater volume of soil, well outside the nutrition depletion region to which the roots of the plant unable to reach. AMF then stimulates the plant to improve and enhance its efficiency. As a result, the host plant supplies AMF along the sugar necessary to finish its life cycle. The host plant, in exchange, supplies AMF with the carbohydrates in order to accomplish its life cycle (Zhang et al. 2015). In process of extracting pivotal nutrients from the soil, such as Fe, K, Cu, P, N, and Zn, the relationship of AMF with roots provides plants with exposure to absolutely vital nutrients; this contributes to the solubilization, mobilization, and uptake of such principal nutrients needed for development (Clark and Zeto 2000; Chen et al. 2020). Incorporation of AMF under field condition, in the soil of *Cicer arietinum* plants, enhanced the biomass, yield, and dietary component of the legume and biofortification of zinc and iron (Pellegrino and Bedini 2014). The AMF also accelerates the relationships accompanied by rhizobacteria, along with valuable rhizobacteria that can directly induce AMF spore germination and root colonization. Hyphal exudates of AMF greatly impact microbial communities and therefore can accelerate beneficial bacterial-plant interactions and therefore strengthen the health and plant growth (Qin et al. 2016; Cruz and Ishii 2012; Schillaci et al. 2019). Co-inoculation of AMF and advantageous bacteria like *Pseudomonas* and selenobacteria vastly enhances wheat production and phosphorous, iron, manganese, copper, potassium, zinc, and selenium mineral nutrient concentrations in wheat grains (Mäder et al. 2011; Durán et al. 2013). In the soybean/maize intercropping system, twofold supplementation of arbuscular mycorrhizal fungi and rhizobium substantially accelerated the pace of soybean N_2 fixation and strengthened the transport of N_2 from soybean to maize, contributing to improved yields for both crops (Meng et al. 2015). An even more significant benefit of arbuscular mycorrhizal fungi is the discharge of large molecular weight glycoprotein that stabilizes soil aggregation, aggrandizes a healthier soil and plant-soil network, and improves phytoextraction of aluminum, arsenic, cadmium, mercury, and lead (Wu et al. 2015; He and Nara 2007). AMF's primary function in nutrient supply and crop production is the main determinant in the biofortification of crops utilizing soil-based essential mineral components. Via symbiotic relationship with AMF, crops benefit immensely from plant-microbial interactions, and this could assist in reducing the utilization of chemical fertilizers and raising nutrient availability. As a consequence, AMF could perform a major contribution in the biofortification of micronutrient-containing crops, while also contributing to better nutrition and sustainable crops that can tackle global malnutrition. Agriculture integrating AMF as well as other valuable microbes is essential, which include increased biofortification of Fe, Zn, provitamin-A, carotenoids, aminoacids, and proteins in the crops that are fiscally valuable include rice, wheat, sorghum, cassava, potatoes, maize, etc. (Mayer et al. 2008; Khush et al. 2012).

9.2.1 General Mechanisms

In attempt to maximize plant growth and yield, multiple PGPM pathways had also been reported. Selenium is categorized as a metalloid and is available as elemental Se(0), selenide(2^-), selenite(4^+), and selenate(6^+) in different valences in the environment, and plant roots do have tendency to process Se as compounds of selenate, selenite, or oganoselenium, such as selenocysteine and selenomethionine (White et al. 2004, 2007). Selenate is perhaps the most frequently used source of Se in plants. The ability of PGPR to respond as Se-biofortification agents has been analyzed on the basis of research carried out over the past two decades. To improve selenium accumulation and volatilization, *Brassica juncea* could well be assisted by PGPR (De Souza et al. 2002). In axenic plant experiments, rhizobacterial strains classified as BJ15 and BJ2 had been employed to evaluate the influence of rhizobacteria in accumulation and volatilization of Se. While these two strains had been applied in the rhizosphere of axenic *Brassica juncea* plants, the intensity of Se volatilization from selenite (4^+) was significantly four times significantly higher than those of the axenic control plants, and when comparison to axenic controls, plants managed to accumulate more Se in their tissues (De Souza et al. 2002). Isolates either from *Bacillus* and *Klebsiella* (strains E5 and E1), *Bacillus* (strain E8.1), or *Acinetobacter* (strain E6.2) genera impeded in vitro propagation of *Gaeumannomyces graminis* variety tritici mycelia by ~100%, 50%, and 30%, respectively, increasing the efficiency of PGPR for biofortification agents which not only improve the accumulation of selenium in the plant but also act as feasible crop protection agents (Durán et al. 2014). PGPM-synthesized compounds are siderophores (low molecular weight compounds (<10 kDa)) which are eventually used by crops escorting the Fe molecule to the plant cells. Through optimizing Fe ingestion, and hindering plant pathogen through competition, microbial siderophores can boost plant growth eventually resulting in Fe biofortification of plants (Srivastava et al. 2013). A broad spectrum of bacterial species, viz., *Azotobacter*, *Azospirillum*, *Burkholderia*, *Bacillus*, *Enterobacter*, *Arthrobacter*, *Rhodospirillum*, *Pseudomonas*, *Serratia*, and *Rhizobium*, and fungal species, viz., *Rhizopus*, *Syncephalastrum*, *Aspergillus*, and *Penicillium*, produce siderophores (Leong and Neilands 1982; Das et al. 2007; Durán et al. 2016; Srivastava et al. 2013). Masalha et al. (2000) experimented that the sterile soil and non-sterile soil contribute significant affects in the uptake of iron in *Helianthus annuus* L. (Sunflower). The concentration of iron in aseptic soil was ~248 ± 29 μg g^{-1}, but in non-sterile soil (~1748 ± 48 μg g^{-1}), the uptake was significantly high, showing that the iron transport from the soil to the plant was enhanced by microbes. In a biocontrol capacity, siderophores can also be engaged, with the exception of the rhizospheric soil, absorbing Fe and making it inaccessible to any other pathogen (Siddiqui 2005). By triggering Fe-deficient reaction in the plant, microbes provoke hormonal compounds similar to plant hormones that strengthen Fe increase. By inducing transcriptional upregulation of FIT1 (Fe deficiency-induced transcription factor 1), *Bacillus subtilis* GB03 has the capability to exacerbate Fe procurement in *Arabidopsis*. The FIT1 is required for the acquisition of iron at the plant for the commencement of ferric reductase FRO2 and IRT1 Fe transporters (Zhang et al. 2009). In the mere existence of *Acinetobacter* isolates, *Cicer arietinum* L (chickpea) has the capability for iron biofortification, which greatly increased the mineral content in PGPR inoculated *Cicer arietinum* L. comparison to un-inoculated controls, but post-harvesting and cooking can impair grain micronutrient concentrations by limiting from 5% to 30% and in some cases achieving ~21% (Sathya et al. 2016; Gopalakrishnan et al. 2016). Therefore, we can conclude that exploiting PGPM-producing siderophores is a suitable strategy than other traditional methods like chemical fertilizers to significantly raise Fe content in plants and grains. Staple grain processing, for example, wheat and rice, reduces the grain micronutrient, particularly in case of wheat Zn concentration (Kutman et al. 2011; Zhang et al. 2010). Like other micronutrients, Zn is also one of the most essential micronutrients which is involved in the various mechanisms. Microorganisms which are beneficial for the plants support various mechanisms for the zinc solubilization by chelation, soil pH decrement, or ameliorating root growth and absorptive area. The presence of "antinutrient" phytic acid reduces the uptake of heavy metals because it can form insoluble compounds to essential nutrients like Fe^{3+} and Zn^{2+}, and hinders Zn grain concentration (Urbano et al. 2000; Srivastava 2016). The consortium of two strains of *Azospirillum lipoferum*, two strains of the *Pseudomonas* sp., and one strain of the *Agrobacterium* sp. have the potential to increase Zn concentration in rice (Tariq et al. 2007). Zn-solubilizing bacteria are said to enhance the concentration of Zn in soybean and wheat shoots and roots. The consortium of *Bacillus* sp. SH-10 and *Bacillus cereus* SH-17 provoked the zinc translocation in basmati by ~31 mg/kg as compared to control (~18 mg/kg) (He et al. 2010; Shakeel et al. 2015). Microbes generate chelating compounds that, when bonded with Zn, form complexes. Furthermore, they discharge chelated Zn just at root surface and inevitably improve the accessibility of Zn to result to biofortification of Zn in plants.

9.2.2 Gene Upregulation

To facilitate growth and development, various biochemical and molecular mechanisms are utilized by microbes. Plant species live in soil, participating in a vast array of situations with soil microorganisms. Those very interaction may involve an advantage, vulnerability, or null impact on plant development and nutrient absorption, and this impact also relies on soil properties, the accessibility of nutrients for plants and microorganisms in specific. The vital aspects of the microbial communities are AMF and PGPR. Different nutrient acquisition pathways can also be provided by AMF, which, in particularly, are significant for plant growth if nutrient accessibility is limited. The monobasic H_2PO^{4-} is one of the plants' most commonly acquired phosphorus components (Marschner et al. 2011). Pi transport, for example, is limited, particularly in alkaline soils, and its absorption quickly escorts to the formation of depletion zones around the roots, even farther limiting the take-up of phosphorus (Schachtman et al. 1998). The acquisition of Pi in plants is maintained by plasma membrane members of the Pi/H^+ symporter belonging to the phosphate transporter family 1 (Pht1), which would be primarily exhibited in the roots, particularly in the root hair, rhizodermis, and in the outer cortical cells (Liu et al. 1998; Kobae et al. 2010; Daram et al. 1998; Schünmann et al. 2004; Mudge et al. 2002; Karthikeyan et al. 2002; Chiou et al. 2001). Soils may accommodate huge quantities of phosphorous, but they are not very accessible for need on plants, both inorganic and organic P being compounds that are extremely insoluble. Various bacteria of the *Azospirillum*, *Bacillus*, *Azotobacter*, *Pseudomonas*, *Burkholderia*, *Rhizobium*, and *Serratia* genera are being identified as microorganisms that solubilize P. Pi migration between plant cells and tissues and even some Pi remobilization from senescent to novel onset organs were engaged (Lambers et al. 2008). In several species of plants, homologous Pht1 genes were identified, which include *Arabidopsis thaliana* (Misson et al. 2004), tomatoes (Daram et al. 1998), *Zea mays* L. (Nagy et al. 2006), and *Triticum aestivum* (Liu et al. 2013). With regard to the Pi uptake root pathways, there is still no evidence that PGPR directly affects Pi. Interestingly, a complicated exchange of signals among host and fungal plants is triggering the emergence of root/AMF symbiosis which also causes cell reprogramming. In fact, AMF-colonized plants can take advantage of an added pathway to acquire P, which occurs in non-identical types of cells depending on multiple molecular setup (i.e., transporters) and accesses phosphorous in distinct soil areas. In colonized cortical cells, Pi absorption is accomplished mostly by expression of Pi transporters that will either be explicitly triggered by AMF symbiosis or strenuously stimulated all through symbiosis, and basal activation is also present in non-mycorrhizal roots (Chen et al. 2007; Nagy et al. 2005).

Utilizing mechanisms that are distinct among both monocots and dicots, plants are able to receive Fe (Kobayashi and Nishizawa 2012). The acquisition of Fe is entirely focused on a plasma membrane level mechanism that really incorporates Fe^{3+} to Fe^{2+} reduction and the accumulation of Fe^{2+} through the electrochemical transmembrane gradient assured by the activity of plasma membrane H^+-ATPase in dicots, categorized as plant strategy I (Marschner and Römheld 1994). *Strategy I* plants significantly boost the proton transfer in the rhizosphere in the particular instance of Fe deficit, triggering the Fe concentration to raise in the roots close vicinity (Colombo et al. 2014). Additionally, the *strategy I* forms significantly raise the Fe^{3+} reduction activity performed by the Fe-regulated transporting protein-like (IRT) of FRO (ferrochelate-reductase oxidase) and the transport of Fe^{2+} across the membranes (Connolly et al. 2003). Furthermore, grasses, also referred to as Strategy II plants, have a fierce chelation affinity for Fe^{3+} and are capable of taking Fe from the biosynthesis and exudation of phytosiderophores (PSs) (Schaaf et al. 2004). In the rhizosphere, PSs are discharged via the mugineic acid family phytosiderophores1 (TOM1) transporter (Nozoye et al. 2011), while the Fe^{3+}-PS aggregates are therefore transported via specific transporters, the YS1 (yellow stripe1) and YSL (YS1-like) transporters, into root cells.

In the soil, Zn is usually found in Fe-Mg minerals such as sulfide in crystalline form (ZnS). The Zn is slightly mobile in plants, and mobility may vary between species. In the rhizosphere, Zn^{2+} is drawn by the plant roots from the soil with an expenditure of energy after solubilization. The absorption of Zn^{2+} takes place in plasmalemma via the production of intermediates of transiently dissociable ion carriers (Bowen 1986; Gupta et al. 2016). In compartments of the apoplast and symplast, Zn absorbed by the roots is racked up in three separate fractions: the interchangeable one, accumulated to its negatively charged pectin matrix including its cell wall in the apoplast; the labile one, which is associated to the cytoplasm nutrient; and the non-labile one, in the vacuole-deposited nutrient, not translocated in plant (Wang et al. 2014). Two physiological schemes are engaged in Zn absorption based on the ligand secreted by roots: (Abbaspour et al. 2014) entail efflux of reducing agents, H^+ ions, and organic acids, which enhance Zn bioavailability for root epidermal cell uptake, and (Acuña et al. 2013) require phytosiderophore efflux forming stable complexes with Zn^{2+}. Proteins that pass Zn across membranes and are thus suitable for nutrient homeostasis are clustered in plants into three families required to carry metal cations (Abbaspour et al. 2014) Zrt/Irt-like protein (ZIP) or Zn-Fe permease, proteins which promote cytosol Zn entry and likely arbitrate the absorption of Zn accessible in the soil (White and Broadley 2011; Olsen and Palmgren 2014), Heavy metal ATPases (HMAs) or P-type ATPase held accountable for the stacking of xylem, regulation of plastid efflux Zn, and vacuolar scavenging (Hussain et al. 2004; Hanikenne et al. 2008; Kim et al. 2009; Nagy et al. 2005). At all levels of phylogenetics, together with bacteria, fungi, plants, and mammals, ZIP proteins are expressed (Eide 2006). The specific mechanism that controls the

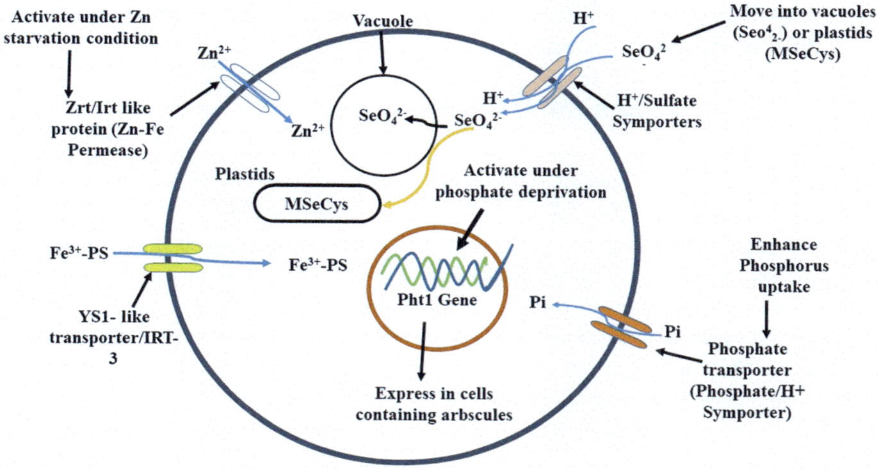

Fig. 9.1 General mechanisms for the uptake of micronutrients from the soil with different transporters involved (*PS* phytosiderophores, *MSeCys* methyl selenocysteine)

transporter's response to the reduction or rise of Zn in the soil solution still remains obscure.

Se absorption, transfer, and allocation varied from the species to the physiological conditions (soil and salinity pH), the activities of membrane carriers, and plant translocation mechanisms (Renkema et al. 2012; Li et al. 2008; Zhao et al. 2005). Transporters available in root cell membrane initiate the absorption of Se in plants. Selenite ($SeO^{2-}{}_3$) is observed to be transferred through the transport mechanism of phosphate (Li et al. 2008). In contrast, selenate ($SeO^{2-}{}_4$) is transported and channeled through sulfate (Zhang et al. 2003). SULTR1;2 and SULTR1 transporters were identified to convey ($SeO^{2-}{}_4$) inside the plant in *Arabidopsis thaliana* sulfate transporters (El Kassis et al. 2007). Sulfur deprivation improved in the *Triticum aestivum* Se absorption (Bowen 1986). In younger leaf tissues, the concentration of Se was observed to be higher compared to the older ones throughout seedling growth (Cappa et al. 2014; Harris et al. 2014). Selenium is often accrued in its vacuoles within the plant cells and thus can be effluxed via sulfate transporters available in the tonoplast (Gigolashvili and Kopriva 2014; Mazej et al. 2008). The very initial step in selenium accumulation is to convert inorganic selenium to $SeO^{2-}{}_3$. It requires two enzymes, APS (ATP sulfurylase) and APR (APS reductase), to perform sequentially. APS catalyzed ATP hydrolysis to produce adenosine phosphoselenate, which has been reduced by APR to $SeO^{2-}{}_3$. $SeO^{2-}{}_3$ is then transformed by enzymatic sulfite reductase to selenide (Se^{2-}) (Sors et al. 2005). Through coupling with *O*-acetylserine (OAS) in the presence of the enzyme cysteine synthase, selenide is then converted to Se-Cys. Se-Cys can then be processed to elemental Se in the availability of the enzyme Se-Cys lyase, or can be methylated to methyl-Se-Cys by selenocysteine methyltransferase, or can be switched to selenomethionine by a series of enzymes, based on plant species and environmental conditions (Fig. 9.1).

9.3 Common Nutritional Deficiency and Corresponding Biofortification Strategies

9.3.1 Biofortification for Iron Deficiency

A huge amount of iron present in the earth's crust is available in the form of Fe^{3+} which cannot be accessed by plants. Thus, despite the abundance of iron in soil, its uptake by plants is hampered due to very low bioavailability. Iron plays important role in human health as it affects multiple metabolic functions, including oxidative metabolism, oxygen transportation, DNA synthesis, hemoglobin and myoglobin synthesis, etc. (Abbaspour et al. 2014). Low iron also affects the agricultural yield due to lower chlorophyll production, which ultimately leads to necrosis of leaves and retards the growth of plants (Zhang et al. 2015). Variety of microbes produce siderophores that help in alleviation of iron deficiency (Patel et al. 2018). The siderophores form siderophore-Fe^{3+} complex that help in increasing the bioavailability of free Fe^{3+} present in the soil (Jin et al. 2006). Plant growth-promoting actinobacteria when inoculated with chickpea seed were found to increase the iron content by 10–38% in the processed seeds postharvest. Real-time PCR results supported the hypothesis for siderophore production being responsible for the increased iron levels in the seeds. Genes responsible for siderophore production were found to be upregulated by 1.4–25 times as compared to the corresponding control (Sathya et al. 2016). Similarly, multiple plant growth-promoting bacteria isolated from organic soils not only enhanced the crop yield but also produced biofortified chickpea and pigeon pea grains with higher contents of micronutrients including iron, zinc, copper, and calcium (Gopalakrishnan et al. 2016). *Providencia* sp. bacteria isolated from the rhizosphere of wheat plant also exhibited plant growth characteristics in field trials. Furthermore, it was also found to increase the protein content by 18.6% and iron, manganese, and copper by 105.3%, 36.7%, and 150%, respectively (Rana et al. 2012). Later, *Providencia* sp. when used either individually or in combination with three different strains of cyanobacteria showed synergistic benefits with promising results in the crop yield. Furthermore, a better crop yield and micronutrient levels were observed when the bacterial isolate was inoculated in wheat crops as compared to rice plants (Rana et al. 2015). Combination of microbial endophytes with foliar spraying of micronutrients enhanced plant height, area of leaf, plant biomass, and iron and zinc concentrations 10.14% and 37.93%, respectively, as compared to control without foliar spraying of micronutrients (Yaseen et al. 2018). *Arthrobacter sulfonivorans*, a siderophore-producing endophytic bacterial treatment to wheat, enhanced the iron content by overexpression of TaZIP3 gene (Singh et al. 2018). Furthermore, combination treatment of native growth-promoting bacteria alone and in combination with arbuscular mycorrhizae provokes significant rise in the levels of iron in wheat plants as compared to the controls (Yadav et al. 2020). Thus, plant growth-promoting bacteria, arbuscular mycorrhiza, and endophytes alone or in combination could provide future perspectives to deal with low bioavailability of iron.

9.3.2 Biofortification for Zinc Deficiency

The bioavailability of zinc being very low in the soil affects its concentration in the crops and also affects the human health. Zinc deficiency in the plants directly impacts the growth rate and yield of the crops. Zinc-solubilizing bacteria have been essential in alleviating zinc deficiency by enhancing the assimilation of zinc in the seeds and simultaneously maintaining the productivity. Zinc-solubilizing *Bacillus* strains exhibited both enhancement of zinc and also increased the crop yield in soybean and wheat plants (Khande et al. 2017). Similar results with increased crop productivity and zinc assimilation were observed when for the first time *Bacillus aryabhattai*-related bacterial isolates were studied in soybean and wheat (Ramesh et al. 2014). In another study conducted in wheat plants, *Bacillus subtilis* enhanced zinc amount in grains by 24% relative to untreated control (Moreno-Lora et al. 2019). Biofortification of maize with enhanced levels of zinc was also studied using *Bacillus* strains, which produced ammonia, hydrogen cyanide, siderophores, exopolysaccharides, and cellulase (Goteti et al. 2013; Mumtaz et al. 2017). Zinc-solubilizing bacteria isolated from the rhizosphere region of rice plants revealed similar conclusions regarding their plant growth-promoting potential and zinc solubilization. The isolates were identified as *Pseudomonas aeruginosa*, *Ralstonia pickettii*, *Burkholderia cepacia*, and *Klebsiella pneumoniae* (Gontia-Mishra et al. 2017). Bacterial isolates were screened from rhizosphere of wheat plants, and out of the five selected strains, *Enterobacter cloacae* and *Pantoea agglomerans* had the highest amount of zinc when the wheat plants were harvested after a month, while after a 3-month harvesting, *Pseudomonas fragi* exhibited the highest amount of zinc in the grains of the plants, estimated using atomic absorption spectroscopy (Kamran et al. 2017). *Burkholderia* and *Acinetobacter* species isolated from zinc-deficient soil were tested for their plant growth promotion and zinc solubilization. Zinc levels were estimated with the straw and grain samples of rice and saw an increase in the zinc uptake by 52.5% (Vaid et al. 2014). Five endophytic bacteria inoculated in rice plants grown using hydroponics were studied for their plant growth and zinc biofortification. The endophytes were able to colonize the plant roots in 72 h and showed increase in zinc concentration by 44.4% and 51.1% in shoots and by 73.6% and 83.4% in roots treated with Sphingomonas and Enterobacter sp., respectively. However, when grown in soil, the endophytes Sphingomonas and Enterobacter sp. increased zinc concentrations by 20.3% and 21.9%, respectively (Wang et al. 2014). In another study with endophytes, two bacterial isolates (*Bacillus subtilis* and *Arthrobacter* sp.) were used for the treatment of wheat genotypes that had either low or high zinc accumulation rates in soils sufficient and deficient in zinc. The treatment with endophytes doubled the amount of zinc present in the grains as compared to the untreated controls regardless of the zinc-accumulating genotypes (Singh et al. 2017). Biofortification of zinc can also be done by easily priming the plant seeds with zinc and *Pseudomonas* sp. in combination. Moreover, soil or foliar application of zinc resulted in the highest amount of grain yield and also increased the bioavailability of zinc (Rehman et al. 2018). Thus, application of zinc-solubilizing bacteria enhances and promotes the growth of plants, and additionally, these bacteria could help in zinc biofortification and as zinc translocators for feasible agricultural practice.

9.3.3 Biofortification for Selenium Deficiency

Selenium is an important micronutrient necessary for the vital functions of the humans and mammals. Golubkina et al. studied the impact of selenium biofortification and arbuscular mycorrhiza fungi on the yield, quality and antioxidant properties of onion variety. They showed that the inoculation of both selenium and mycorrhiza enhanced the overall quality, yield, and selenium content of onions (Golubkina et al. 2019). In a recent research, volatile organic compounds (VOCs) extracted from *Bacillus amyloliquefaciens* were found to be very beneficial in enhancing the selenium and iron uptake by plants inoculated with VOCs. They were also shown to enhance the growth and photosynthesis of Arabidopsis plants and upregulate the expression of genes responsible for iron uptake, sulfur, and amino acid transport. Surprisingly, when plants with sulfur transport mutants were inoculated with VOCs, no significant change in the selenium absorption was observed. Thus, upregulated expression of sulfate transporter genes was essential in increasing the uptake of selenium by plants (Wang et al. 2017). Selenobacteria isolated from the rhizosphere also serve as potential tool for enhancing the plant growth, yield, and selenium biofortification. Multiple selenium-tolerant bacteria (*Stenotrophomonas*, *Bacillus*, *Enterobacter*, and *Pseudomonas*) have been found to be beneficial in wheat crops. Acuña et al. demonstrated that inoculation of these salt-tolerant bacteria under greenhouse conditions increased the translocation of selenium in the leaves of the wheat plants (Acuña et al. 2013). Another study also reported salt-tolerant bacteria from *Bacillus*, *Paenibacillus*, *Klebsiella*, and *Acinetobacter* genera which helped in selenium biofortification as well as the biocontrol of *Gaeumannomyces graminis*, a soil-borne pathogen in wheat plants (Durán et al. 2014). Similarly, *Bacillus pichinotyi* isolated from the Roohi Nala drain increased the selenium concentration in wheat kernels (167%) and stems (252%) as compared to the un-inoculated plants. The inoculated plants also showed a significant enhancement in the acid phosphatase activity which might be responsible for the enhanced growth of wheat crops (Yasin et al. 2015). Furthermore, combinatorial utilization of arbuscular fungi (*Funneliformis mosseae* and *Glomus versiforme*) and selenium fertilizer synergistically enhanced the bioavailable selenium in the soil and its accumulation in the rice grains (Chen et al. 2020). In another combination study, inoculation of selenobacteria (*Acinetobacter* or *Bacillus* sp.) and arbuscular fungi (*Rhizophagus intraradices*) enhanced the growth of lettuce under drought conditions and also promoted the uptake of selenium as compared with the un-inoculated controls (Durán et al. 2016). Recently, Trivedi et al. demonstrated the utilization of endophytic selenobacteria MGT9 in soybean for biofortification of selenium. Moreover,

Table 9.1 Microorganisms employed to combat common nutritional deficiencies

S. No.	Nutrient deficiency	Crop	Microbes utilized	Contribution to biofortification	References
1.	Iron	Chickpea	*Enterobacter ludwigii* SRI-229	Increased iron content up to 18% and plant growth-promoting traits	Gopalakrishnan et al. (2016)
		Pigeon pea	*Enterobacter ludwigii* SRI-229	Increased iron content up to 12% and plant growth-promoting traits	Gopalakrishnan et al. (2016)
		Maize	*Arthrobacter globiformis*	Increased iron and phosphate uptake and protein and chlorophyll contents	Sharma et al. (2016)
			Pseudomonas chlororaphis, *Pseudomonas* spp.	Increase in germination percentage and plant growth	Sharma and Johri (2003)
			Pseudomonas aeruginosa	Increased iron transportation	Sah et al. (2017)
		Wheat	*Arthrobacter globiformis*	Increased iron and phosphate uptake and protein and chlorophyll contents	Sharma et al. (2016)
			Arthrobacter sulfonivorans	Upregulation of TaZIP3 and TaZIP7 genes resulted in higher iron and zinc translocation	Singh et al. (2020)
			Bacillus pichinotyi	Increase in Fe content and increase in acid phosphatase activity, which contributed to the enhanced growth	Yasin et al. (2015)
			Providencia sp.	Increased iron concentration	Rana et al. (2015)
			Pseudomonas fluorescens, *Pseudomonas putida*	Enhanced iron concentration, enhanced seed germination, root length and shoot length	Sayyed et al. (2005)
2.	Zinc	Rice	*Acinetobacter* sp., *Burkholderia* sp.	Enhanced zinc content as well as shoot and root length and weight	Vaid et al. (2014)

(continued)

Table 9.1 (continued)

S. No.	Nutrient deficiency	Crop	Microbes utilized	Contribution to biofortification	References
			Burkholderia cepacia, *Klebsiella pneumonia*, *Pseudomonas aeruginosa*, *Ralstonia pickettii*	Zinc solubilization, ACC utilization, EPS production, Ammonia production, production of lytic enzymes	Gontia-Mishra et al. (2017)
		Wheat	*Aeromonas* sp., *Arthrobacter* sp., *Trabulsiella* sp.	Increased in zinc and iron percentage	Shaikh and Saraf (2017)
			Arthrobacter sp., *Bacillus subtilis*	Zinc solubilization enhanced the translocation and enrichment of zinc	Singh et al. (2018)
			Providencia sp.	Zinc solubilization; increase zinc concentration	Rana et al. (2015)
			Pseudomonas sp.	Increase grain yield and zinc uptake	Rehman et al. (2018)
		Soybean	*Bacillus amyloliquefaciens*, *Bacillus firmus*	Increase zinc concentration	Sharma et al. (2011)
			Bacillus anthracis, *Bacillus cereus*	Increased yield as well as zinc assimilation in soybean	Khande et al. (2017)
			Bacillus aryabhattai	Improved growth, yield and zinc concentration	Ramesh et al. (2014)
		Maize	*Bacillus aryabhattai*, *Bacillus subtilis*, *Bacillus aryabhattai*	Promoted growth and zinc uptake	Mumtaz et al. (2017)
		Sugarcane	*Enterobacter cloacae*, *Pantoea agglomerans*, *Pantoea dispersa*, *Pseudomonas fragi*	Increased zinc content and growth promotion	Kamran et al. (2017)
			Gluconacetobacter	Solubilization of insoluble zinc compounds	Saravanan et al. (2007)
		Chickpea	*Rhizophagus irregularis*, *Funneliformis mosseae*	Increased zinc and iron concentration, fungal root colonization, as well as plant biomass and yield	Pellegrino and Bedini (2014)
3.	Selenium	Wheat		Increased selenium concentration,	Durán et al. (2014)

(continued)

Table 9.1 (continued)

S. No.	Nutrient deficiency	Crop	Microbes utilized	Contribution to biofortification	References
			Acinetobacter sp., *Bacillus* sp., *Klebsiella* sp.	auxin and siderophore production, phytate mineralization, and tricalcium phosphate solubilization	
			Bacillus sp., *Enterobacter* sp., *Pseudomonas*, *Stenotrophomonas*	Increased selenium content	Acuña et al. (2013)
			Bacillus pichinotyi	Elevated selenium and increase in acid phosphatase activity resulting in plant growth	Yasin et al. (2015)
		Lettuce	*Acinetobacter* sp., *Bacillus* sp., *Rhizophagus intraradices*	Enhanced the growth of lettuce and uptake of selenium	Durán et al. (2016)
		Rice	*Funneliformis mosseae*, *Glomus versiforme*	Enhanced the bioavailable selenium	Chen et al. (2020)

selenobacteria also enhanced the growth of soybean under water stressed conditions (Table 9.1) (Trivedi et al. 2020).

9.4 Role of Biofortified food in human health

Although micronutrients are required in trace amounts, their role in the human health is very prominent (Imtiaz et al. 2010). The depletion of micronutrient from diet causes deficiency of micronutrients or hidden hunger in humans, which has dramatically affected over two billion people globally. Surprisingly, more than 60% of the global population are iron, 30% zinc and 15% selenium deficient. In South African continent children below 9 years are at the risk of having iron, zinc deficiency of 11% and 45.3% respectively (Siwela et al. 2020). Biofortification allows to increase the concentration of selectively essential nutrients in the food that can be beneficial to minimize micronutrient deficiency and health problems associated with. Consumption of biofortified food has found to be beneficial for positive restructuring of the gut microbiota. Reed et al. demonstrated that those humans which consumed wheat biofortified with zinc in their diet had enhanced microbial β diversity, with a naturally associated increase short chain fatty acid producing lactic acid bacteria (Reed et al. 2018). Deficiency of a particular micronutrient has shown to develop multiple health-related concerns. In the initial stage of pregnancy, the low levels of selenium in the blood of mother might result in low birth weight of a newly born child (Pieczyńska and Grajeta 2015). Foods enriched in micronutrients have also been useful in fighting against viral infections. Zinc and selenium supplements against SARS-CoV-2-infected people with reduced selenium or zinc in their blood could serve as an alternative natural remedy against the virus (Zhang and Liu 2020). In this context, the development of biofortified food to enhance the availability of micronutrients is very essential for the improvement of human health.

9.5 Conclusion and Future Outlook

Numerous researchers to date have illustrated the potential risks to food security. PGPMs interact with plants and promote plant development activities and strengthen the plant's potential to absorb micronutrients from the soil. The development of PGPM all across agriculture is a substantial benefit of cost-effective and being environmentally harmless. In order to tackle the challenge of micronutrient deficiency, the expansion of crops with high concentrations of micronutrients is enormously and promptly deemed necessary. In the future, biofortification tactics to address the problem of hidden hunger may be used to formulate microorganisms with multiple beneficial characteristics. PGPMs communicate with plants and promote plant developmental activities and strengthen the plant's capacity to absorb micronutrients from the soil. In the different edible segments of crop plants, Zn solubilization and siderophore secretion microbes significantly raise the concentration of Zn and Fe and offer an efficient approach to fortify micronutrients and produce food prosperous in micronutrients. In addition to genetic modification, other effective techniques have been used to biofortify crops, such as traditional breeding strategies. The implementation of inoculants from such microorganisms lessens dependence on expensive strategies to biofortification, i.e., agronomic and genetic approaches. In this perspective, the utilization of PGPR as bio-inoculants could be a very successful technique, given that the both plant growth (root and shoot biomass enhancement, more branched root system) and nutrient bioavailability could be enhanced. Even after such empowering findings, there are still many elements to be investigated so that all interactions among plants and PGPRs can be better understood, with a view to improving the effectiveness of nutrients. The rise of innovative agricultural fertilization practices focused on biotechnological methodologies for rhizosphere engineering and management could allow the molecular mechanisms underlying the implications of microbes on plants to be clarified and vice versa, as well as the physiological and ecological effects that are mediated.

References

Abbaspour N, Hurrell R, Kelishadi R (2014) Review on iron and its importance for human health. J Res Med Sci 19(2):164

Acuña JJ, Jorquera MA, Barra PJ, Crowley DE, de la Luz Mora M (2013) Selenobacteria selected from the rhizosphere as a potential tool for Se biofortification of wheat crops. Biol Fertil Soils 49 (2):175–185

Ahemad M, Kibret M (2014) Mechanisms and applications of plant growth promoting rhizobacteria: current perspective. J King Saud Univ Sci 26(1):1–20

Ahmad P, Prasad MN (eds) (2011) Abiotic stress responses in plants: metabolism, productivity and sustainability, vol 16. Springer Science & Business Media, New York, NY

Babalola OO, Olanrewaju OS, Dias T, Ajilogba CF, Kutu FR, Cruz C (2017) Biological nitrogen fixation: the role of underutilized leguminous plants. In: Microorganisms for green revolution. Springer, Singapore, pp 431–443

Bano SA, Iqbal SM (2016) Biological nitrogen fixation to improve plant growth and productivity. Int J Agric Innovat Res 4:596–599

Berruti A, Lumini E, Balestrini R, Bianciotto V (2016) Arbuscular mycorrhizal fungi as natural biofertilizers: let's benefit from past successes. Front Microbiol 6:1559

Blevins DG (2009) More hidden hunger: special nutrient needs of plants based on their structure and function. In: The Proceedings of the International Plant Nutrition Colloquium

Bowen JE (1986) Kinetics of zinc uptake by two rice cultivars. Plant Soil 94(1):99–107

Cappa JJ, Cappa PJ, El Mehdawi AF, McAleer JM, Simmons MP, Pilon-Smits EA (2014) Characterization of selenium and sulfur accumulation across the genus Stanleya (Brassicaceae): a field survey and common-garden experiment. Am J Bot 101(5):830–839

Carvalho SM, Vasconcelos MW (2013) Producing more with less: strategies and novel technologies for plant-based food biofortification. Food Res Int 54(1):961–971

Chen A, Hu J, Sun S, Xu G (2007) Conservation and divergence of both phosphate-and mycorrhiza-regulated physiological responses and expression patterns of phosphate transporters in solanaceous species. New Phytol 4:817–831

Chen X, Zhang Z, Gu M, Li H, Shohag MJ, Shen F, Wang X, Wei Y (2020) Combined use of arbuscular mycorrhizal fungus and selenium fertilizer shapes microbial community structure and enhances organic selenium accumulation in rice grain. Sci Total Environ 748:141166

Cheng Q (2008) Perspectives in biological nitrogen fixation research. J Integr Plant Biol 50 (7):786–798

Chiou TJ, Liu H, Harrison MJ (2001) The spatial expression patterns of a phosphate transporter (MtPT1) from *Medicago truncatula* indicate a role in phosphate transport at the root/soil interface. Plant J 25(3):281–293

Clark RÁ, Zeto SK (2000) Mineral acquisition by arbuscular mycorrhizal plants. J Plant Nutr 23 (7):867–902

Colombo C, Palumbo G, He JZ, Pinton R, Cesco S (2014) Review on iron availability in soil: interaction of Fe minerals, plants, and microbes. J Soils Sediments 14(3):538–548

Connolly EL, Campbell NH, Grotz N, Prichard CL, Guerinot ML (2003) Overexpression of the FRO2 ferric chelate reductase confers tolerance to growth on low iron and uncovers posttranscriptional control. Plant Physiol 133(3):1102–1110

Cruz AF, Ishii T (2012) Arbuscular mycorrhizal fungal spores host bacteria that affect nutrient biodynamics and biocontrol of soil-borne plant pathogens. Biology Open 1(1):52–57

Daram P, Brunner S, Persson BL, Amrhein N, Bucher M (1998) Functional analysis and cell-specific expression of a phosphate transporter from tomato. Planta 206(2):225–233

Das A, Prasad R, Srivastava A, Giang PH, Bhatnagar K, Varma A (2007) Fungal siderophores: structure, functions and regulation. In: Microbial siderophores. Springer, Berlin, pp 1–42

De Souza MP, Pickering IJ, Walla M, Terry N (2002) Selenium assimilation and volatilization from selenocyanate-treated Indian mustard and muskgrass. Plant Physiol 128(2):625–633

Dominguez-Nuñez JA, Benito B, Berrocal-Lobo M, Albanesi A (2016) Mycorrhizal fungi: role in the solubilization of potassium. In: Potassium solubilizing microorganisms for sustainable agriculture. Springer, New Delhi, pp 77–98

Dotaniya ML, Meena VD, Basak BB, Meena RS (2016) Potassium uptake by crops as well as microorganisms. In: Potassium solubilizing microorganisms for sustainable agriculture. Springer, New Delhi, pp 267–280

Durán P, Acuña JJ, Jorquera MA, Azcón R, Borie F, Cornejo P, Mora ML (2013) Enhanced selenium content in wheat grain by co-inoculation of selenobacteria and arbuscular mycorrhizal fungi: a preliminary study as a potential Se biofortification strategy. J Cereal Sci 57(3):275–280

Durán P, Acuña JJ, Jorquera MA, Azcón R, Paredes C, Rengel Z, de la Luz Mora M (2014) Endophytic bacteria from selenium-supplemented wheat plants could be useful for plant-growth promotion, biofortification and *Gaeumannomyces graminis* biocontrol in wheat production. Biol Fertil Soils 50(6):983–990

Durán P, Acuña JJ, Armada E, López-Castillo OM, Cornejo P, Mora ML, Azcón R (2016) Inoculation with selenobacteria and arbuscular mycorrhizal fungi to enhance selenium content in lettuce plants and improve tolerance against drought stress. J Soil Sci Plant Nutr 16 (1):211–225

Eide DJ (2006) Zinc transporters and the cellular trafficking of zinc. Biochimica et Biophysica Acta (BBA)-Molecular. Cell Res 1763(7):711–722

El Kassis E, Cathala N, Rouached H, Fourcroy P, Berthomieu P, Terry N, Davidian JC (2007) Characterization of a selenate-resistant Arabidopsis mutant. Root growth as a potential target for selenate toxicity. Plant Physiol 143(3):1231–1241

Esitken A, Yildiz HE, Ercisli S, Donmez MF, Turan M, Gunes A (2010) Effects of plant growth promoting bacteria (PGPB) on yield, growth and nutrient contents of organically grown strawberry. Sci Hortic 124(1):62–66

Felestrino ÉB, Santiago IF, Freitas LD, Rosa LH, Ribeiro SP, Moreira LM (2017) Plant growth promoting bacteria associated with *Langsdorffia hypogaea*-rhizosphere-host biological interface: a neglected model of bacterial prospection. Front Microbiol 8:172

Garg M, Sharma N, Sharma S, Kapoor P, Kumar A, Chunduri V, Arora P (2018) Biofortified crops generated by breeding, agronomy, and transgenic approaches are improving lives of millions of people around the world. Front Nutr 5:12

Gigolashvili T, Kopriva S (2014) Transporters in plant sulfur metabolism. Front Plant Sci 5:442

Glick BR (2012) Plant growth-promoting bacteria: mechanisms and applications. Scientifica 2012:963401

Glick BR (2014) Bacteria with ACC deaminase can promote plant growth and help to feed the world. Microbiol Res 169(1):30–39

Golubkina N, Zamana S, Seredin T, Poluboyarinov P, Sokolov S, Baranova H, Krivenkov L, Pietrantonio L, Caruso G (2019) Effect of selenium biofortification and beneficial microorganism inoculation on yield, quality and antioxidant properties of shallot bulbs. Plants. 8(4):102

Gontia-Mishra I, Sapre S, Tiwari S (2017) Zinc solubilizing bacteria from the rhizosphere of rice as prospective modulator of zinc biofortification in rice. Rhizosphere 3:185–190

Gopalakrishnan S, Vadlamudi S, Samineni S, Kumar CS (2016) Plant growth-promotion and biofortification of chickpea and pigeon pea through inoculation of biocontrol potential bacteria, isolated from organic soils. Springerplus 5(1):1–1

Goteti PK, Emmanuel LD, Desai S, Shaik MH (2013) Prospective zinc solubilizing bacteria for enhanced nutrient uptake and growth promotion in maize (*Zea mays* L.). Int J Microbiol 2013:869697

Gupta N, Ram H, Kumar B (2016) Mechanism of Zinc absorption in plants: uptake, transport, translocation and accumulation. Rev Environ Sci Biotechnol 15(1):89–109

Hanikenne M, Talke IN, Haydon MJ, Lanz C, Nolte A, Motte P, Kroymann J, Weigel D, Krämer U (2008) Evolution of metal hyperaccumulation required cis-regulatory changes and triplication of HMA4. Nature 453(7193):391–395

Hao Y, Charles TC, Glick BR (2007) ACC deaminase from plant growth-promoting bacteria affects crown gall development. Can J Microbiol 53(12):1291–1299

Harris J, Schneberg KA, Pilon-Smits EA (2014) Sulfur–selenium–molybdenum interactions distinguish selenium hyperaccumulator *Stanleya pinnata* from non-hyperaccumulator *Brassica juncea* (Brassicaceae). Planta 239(2):479–491

He X, Nara K (2007) Element biofortification: can mycorrhizas potentially offer a more effective and sustainable pathway to curb human malnutrition? Trends Plant Sci 12(8):331–333

He CQ, Tan GE, Liang X, Du W, Chen YL, Zhi GY, Zhu Y (2010) Effect of Zn-tolerant bacterial strains on growth and Zn accumulation in *Orychophragmus violaceus*. Appl Soil Ecol 44(1):1–5

Hussain D, Haydon MJ, Wang Y, Wong E, Sherson SM, Young J, Camakaris J, Harper JF, Cobbett CS (2004) P-type ATPase heavy metal transporters with roles in essential zinc homeostasis in *Arabidopsis*. Plant Cell 16(5):1327–1339

Imtiaz M, Rashid A, Khan P, Memon MY, Aslam M (2010) The role of micronutrients in crop production and human health. Pak J Bot 42(4):2565–2578

Jha Y, Subramanian RB (2016) Regulation of plant physiology and antioxidant enzymes for alleviating salinity stress by potassium-mobilizing bacteria. In: Potassium solubilizing microorganisms for sustainable agriculture. Springer, New Delhi, pp 149–162

Ji SH, Kim JS, Lee CH, Seo HS, Chun SC, Oh J, Choi EH, Park G (2019) Enhancement of vitality and activity of a plant growth-promoting bacteria (PGPB) by atmospheric pressure non-thermal plasma. Sci Rep 9(1):1–6

Jin CW, He YF, Tang CX, Wu P, Zheng SJ (2006) Mechanisms of microbially enhanced Fe acquisition in red clover (*Trifolium pratense* L.). Plant Cell Environ 29(5):888–897

Kamran S, Shahid I, Baig DN, Rizwan M, Malik KA, Mehnaz S (2017) Contribution of zinc solubilizing bacteria in growth promotion and zinc content of wheat. Front Microbiol 8:2593

Karthikeyan AS, Varadarajan DK, Mukatira UT, D'Urzo MP, Damsz B, Raghothama KG (2002) Regulated expression of Arabidopsis phosphate transporters. Plant Physiol 130(1):221–233

Khande R, Sharma SK, Ramesh A, Sharma MP (2017) Zinc solubilizing Bacillus strains that modulate growth, yield and zinc biofortification of soybean and wheat. Rhizosphere 4:126–138

Khush GS, Lee S, Cho JI, Jeon JS (2012) Biofortification of crops for reducing malnutrition. Plant Biotechnol Rep 6(3):195–202

Kim YY, Choi H, Segami S, Cho HT, Martinoia E, Maeshima M, Lee Y (2009) AtHMA1 contributes to the detoxification of excess Zn (II) in Arabidopsis. Plant J 58(5):737–753

Kloepper JW, Schroth MN, Miller TD (1980) Effects of rhizosphere colonization by plant growth-promoting rhizobacteria on potato plant development and yield. Phytopathology 70 (11):1078–1082

Kobae Y, Tamura Y, Takai S, Banba M, Hata S (2010) Localized expression of arbuscular mycorrhiza-inducible ammonium transporters in soybean. Plant Cell Physiol 51(9):1411–1415

Kobayashi T, Nishizawa NK (2012) Iron uptake, translocation, and regulation in higher plants. Annu Rev Plant Biol 63:131–152

Ku YS, Rehman HM, Lam HM (2019) Possible roles of rhizospheric and endophytic microbes to provide a safe and affordable means of crop biofortification. Agronomy 9(11):764

Kumar A, Maurya BR, Raghuwanshi R, Meena VS, Islam MT (2017) Co-inoculation with Enterobacter and Rhizobacteria on yield and nutrient uptake by wheat (*Triticum aestivum* L.) in the alluvial soil under indo-gangetic plain of India. J Plant Growth Regul 36(3):608–617

Kutman UB, Yildiz B, Cakmak I (2011) Improved nitrogen status enhances zinc and iron concentrations both in the whole grain and the endosperm fraction of wheat. J Cereal Sci 53 (1):118–125

Lambers H, Raven JA, Shaver GR, Smith SE (2008) Plant nutrient-acquisition strategies change with soil age. Trends Ecol Evol 23(2):95–103

Leong SA, Neilands JB (1982) Siderophore production by phytopathogenic microbial species. Arch Biochem Biophys 218(2):351–359

Li HF, McGrath SP, Zhao FJ (2008) Selenium uptake, translocation and speciation in wheat supplied with selenate or selenite. New Phytol 178(1):92–102

Lindström K, Mousavi SA (2020) Effectiveness of nitrogen fixation in rhizobia. Microb Biotechnol 13(5):1314–1335
Liu C, Muchhal US, Uthappa M, Kononowicz AK, Raghothama KG (1998) Tomato phosphate transporter genes are differentially regulated in plant tissues by phosphorus. Plant Physiol 116 (1):91–99
Liu X, Zhao X, Zhang L, Lu W, Li X, Xiao K (2013) TaPht1; 4, a high-affinity phosphate transporter gene in wheat (*Triticum aestivum*), plays an important role in plant phosphate acquisition under phosphorus deprivation. Funct Plant Biol 40(4):329–341
Mäder P, Kaiser F, Adholeya A, Singh R, Uppal HS, Sharma AK, Srivastava R, Sahai V, Aragno M, Wiemken A, Johri BN (2011) Inoculation of root microorganisms for sustainable wheat–rice and wheat–black gram rotations in India. Soil Biol Biochem 43(3):609–619
Maheshwari R, Bhutani N, Suneja P (2019) Screening and characterization of siderophore producing endophytic bacteria from *Cicer arietinum* and *Pisum sativum* plants. J Appl Biol Biotechnol 7:7–14
Marschner H, Römheld V (1994) Strategies of plants for acquisition of iron. Plant Soil 165 (2):261–274
Marschner P, Crowley D, Rengel Z (2011) Rhizosphere interactions between microorganisms and plants govern iron and phosphorus acquisition along the root axis–model and research methods. Soil Biol Biochem 43(5):883–894
Masalha J, Kosegarten H, Elmaci Ö, Mengel K (2000) The central role of microbial activity for iron acquisition in maize and sunflower. Biol Fertil Soils 30(5):433–439
Mayer JE, Pfeiffer WH, Beyer P (2008) Biofortified crops to alleviate micronutrient malnutrition. Curr Opin Plant Biol 11(2):166–170
Mazej D, Osvald J, Stibilj V (2008) Selenium species in leaves of chicory, dandelion, lamb's lettuce and parsley. Food Chem 107(1):75–83
Meng L, Zhang A, Wang F, Han X, Wang D, Li S (2015) Arbuscular mycorrhizal fungi and rhizobium facilitate nitrogen uptake and transfer in soybean/maize intercropping system. Front Plant Sci 6:339
Misson J, Thibaud MC, Bechtold N, Raghothama K, Nussaume L (2004) Transcriptional regulation and functional properties of Arabidopsis Pht1; 4, a high affinity transporter contributing greatly to phosphate uptake in phosphate deprived plants. Plant Mol Biol 55(5):727–741
Moreno-Lora A, Recena R, Delgado A (2019) *Bacillus subtilis* QST713 and cellulose amendment enhance phosphorus uptake while improving zinc biofortification in wheat. Appl Soil Ecol 142:81–89
Mudge SR, Rae AL, Diatloff E, Smith FW (2002) Expression analysis suggests novel roles for members of the Pht1 family of phosphate transporters in Arabidopsis. Plant J 31(3):341–353
Mumtaz MZ, Ahmad M, Jamil M, Hussain T (2017) Zinc solubilizing *Bacillus* spp. potential candidates for biofortification in maize. Microbiol Res 202:51–60
Nagy R, Karandashov V, Chague V, Kalinkevich K, Tamasloukht MB, Xu G, Jakobsen I, Levy AA, Amrhein N, Bucher M (2005) The characterization of novel mycorrhiza-specific phosphate transporters from *Lycopersicon esculentum* and *Solanum tuberosum* uncovers functional redundancy in symbiotic phosphate transport in solanaceous species. Plant J 42(2):236–250
Nagy R, Vasconcelos MJ, Zhao S, McElver J, Bruce W, Amrhein N, Raghothama KG, Bucher M (2006) Differential regulation of five Pht1 phosphate transporters from maize (Zea mays L.). Plant Biol 8(2):186–197
Nakandalage N, Seneweera S (2018) Micronutrients use efficiency of crop-plants under changing climate. In: Plant micronutrient use efficiency. Academic Press, New York, NY, pp 209–224
Nozoye T, Nagasaka S, Kobayashi T, Takahashi M, Sato Y, Sato Y, Uozumi N, Nakanishi H, Nishizawa NK (2011) Phytosiderophore efflux transporters are crucial for iron acquisition in graminaceous plants. J Biol Chem 286(7):5446–5454
O'Hara GW (1998) The role of nitrogen fixation in crop production. J Crop Prod 1(2):115–138

Olivares FL, Aguiar NO, Rosa RC, Canellas LP (2015) Substrate biofortification in combination with foliar sprays of plant growth promoting bacteria and humic substances boosts production of organic tomatoes. Sci Hortic 183:100–108

Olsen LI, Palmgren MG (2014) Many rivers to cross: the journey of zinc from soil to seed. Front Plant Sci 5:30

Patel P, Trivedi G, Saraf M (2018) Iron biofortification in mungbean using siderophore producing plant growth promoting bacteria. Environ Sustain 1(4):357–365

Pellegrino E, Bedini S (2014) Enhancing ecosystem services in sustainable agriculture: biofertilization and biofortification of chickpea (*Cicer arietinum* L.) by arbuscular mycorrhizal fungi. Soil Biol Biochem 68:429–439

Pfeiffer WH, McClafferty B (2007) HarvestPlus: breeding crops for better nutrition. Crop Sci 47: S88

Pieczyńska J, Grajeta H (2015) The role of selenium in human conception and pregnancy. J Trace Elem Med Biol 29:31–38

Prakash S, Verma JP (2016) Global perspective of potash for fertilizer production. In: Potassium solubilizing microorganisms for sustainable agriculture. Springer, New Delhi, pp 327–331

Prasanna R, Nain L, Rana A, Shivay YS (2016) Biofortification with microorganisms: present status and future challenges. In: Biofortification of food crops. Springer, New Delhi, pp 249–262

Priyadharsini P, Muthukumar T (2016) Interactions between arbuscular mycorrhizal fungi and potassium-solubilizing microorganisms on agricultural productivity. In: Potassium solubilizing microorganisms for sustainable agriculture. Springer, New Delhi, pp 111–125

Qin H, Brookes PC, Xu J (2016) Arbuscular mycorrhizal fungal hyphae alter soil bacterial community and enhance polychlorinated biphenyls dissipation. Front Microbiol 7:939

Raghavendra MP, Nayaka SC, Nuthan BR (2016) Role of rhizosphere microflora in potassium solubilization. In: Potassium solubilizing microorganisms for sustainable agriculture. Springer, New Delhi, pp 43–59

Ramakrishna W, Yadav R, Li K (2019) Plant growth promoting bacteria in agriculture: two sides of a coin. Appl Soil Ecol 138:10–18

Ramesh A, Sharma SK, Sharma MP, Yadav N, Joshi OP (2014) Inoculation of zinc solubilizing *Bacillus aryabhattai* strains for improved growth, mobilization and biofortification of zinc in soybean and wheat cultivated in Vertisols of central India. Appl Soil Ecol 73:87–96

Rana A, Joshi M, Prasanna R, Shivay YS, Nain L (2012) Biofortification of wheat through inoculation of plant growth promoting rhizobacteria and cyanobacteria. Eur J Soil Biol 50:118–126

Rana A, Kabi SR, Verma S, Adak A, Pal M, Shivay YS, Prasanna R, Nain L (2015) Prospecting plant growth promoting bacteria and cyanobacteria as options for enrichment of macro-and micronutrients in grains in rice–wheat cropping sequence. Cogent Food Agric 1(1):1037379

Reed S, Knez M, Uzan A, Stangoulis JC, Glahn RP, Koren O, Tako E (2018) Alterations in the gut (*Gallus gallus*) microbiota following the consumption of zinc biofortified wheat (*Triticum aestivum*)-based diet. J Agric Food Chem 66(25):6291–6299

Rehman A, Farooq M, Naveed M, Nawaz A, Shahzad B (2018) Seed priming of Zn with endophytic bacteria improves the productivity and grain biofortification of bread wheat. Eur J Agron 94:98–107

Renkema H, Koopmans A, Kersbergen L, Kikkert J, Hale B, Berkelaar E (2012) The effect of transpiration on selenium uptake and mobility in durum wheat and spring canola. Plant Soil 354 (1):239–250

Rodríguez H, Fraga R, Gonzalez T, Bashan Y (2006) Genetics of phosphate solubilization and its potential applications for improving plant growth-promoting bacteria. Plant Soil 287(1):15–21

Roriz M, Carvalho SM, Castro PM, Vasconcelos MW (2020) Legume biofortification and the role of plant growth-promoting bacteria in a sustainable agricultural era. Agronomy 10(3):435

Sah S, Singh N, Singh R (2017) Iron acquisition in maize (*Zea mays* L.) using Pseudomonas siderophore. 3 Biotech 7(2):1–7

Santi C, Bogusz D, Franche C (2013) Biological nitrogen fixation in non-legume plants. Ann Bot 111(5):743–767

Saravanan VS, Madhaiyan M, Thangaraju M (2007) Solubilization of zinc compounds by the diazotrophic, plant growth promoting bacterium *Gluconacetobacter diazotrophicus*. Chemosphere 66(9):1794–1798

Sathya A, Vijayabharathi R, Srinivas V, Gopalakrishnan S (2016) Plant growth-promoting actinobacteria on chickpea seed mineral density: an upcoming complementary tool for sustainable biofortification strategy. 3 Biotech 6(2):1–6

Sayyed RZ, Badgujar MD, Sonawane HM, Mhaske MM, Chincholkar SB (2005) Production of microbial iron chelators (siderophores) by fluorescent pseudomonads. Indian J Biotechnol 4 (4):484

Schaaf G, Ludewig U, Erenoglu BE, Mori S, Kitahara T, Von Wirén N (2004) ZmYS1 functions as a proton-coupled symporter for phytosiderophore-and nicotianamine-chelated metals. J Biol Chem 279(10):9091–9096

Schachtman DP, Reid RJ, Ayling SM (1998) Phosphorus uptake by plants: from soil to cell. Plant Physiol 116(2):447–453

Schillaci M, Gupta S, Walker R, Roessner U (2019) The role of plant growth-promoting bacteria in the growth of cereals under abiotic stresses. In: Root biology - growth, physiology, and functions. IntechOpen, London, pp 1–21

Schünmann PH, Richardson AE, Smith FW, Delhaize E (2004) Characterization of promoter expression patterns derived from the Pht1 phosphate transporter genes of barley (*Hordeum vulgare* L.). J Exp Bot 55(398):855–865

Shaikh S, Saraf M (2017) Biofortification of *Triticum aestivum* through the inoculation of zinc solubilizing plant growth promoting rhizobacteria in field experiment. Biocatal Agric Biotechnol 9:120–126

Shakeel M, Rais A, Hassan MN, Hafeez FY (2015) Root associated Bacillus sp. improves growth, yield and zinc translocation for basmati rice (*Oryza sativa*) varieties. Front Microbiol 6:1286

Sharma A, Johri BN (2003) Growth promoting influence of siderophore-producing Pseudomonas strains GRP3A and PRS9 in maize (*Zea mays* L.) under iron limiting conditions. Microbiol Res 158(3):243–248

Sharma SK, Ramesh A, Joshi OP (2011) Characterization of zinc-solubilizing Bacillus isolates and their potential to influence zinc assimilation in soybean seeds. J Microbiol Biotechnol 22:352

Sharma M, Mishra V, Rau N, Sharma RS (2016) Increased iron-stress resilience of maize through inoculation of siderophore-producing *Arthrobacter globiformis* from mine. J Basic Microbiol 56(7):719–735

Shilev S (2020) Plant-Growth-Promoting Bacteria Mitigating Soil Salinity Stress in Plants. Appl Sci 10(20):7326

Siddiqui ZA (2005) PGPR: prospective biocontrol agents of plant pathogens. In: PGPR: biocontrol and biofertilization. Springer, Dordrecht, pp 111–142

Singh AV, Prasad B (2014) Enhancement of plant growth, nodulation and seed yield through plant growth promoting rhizobacteria in lentil (*Lens culinaris* Medik cv. VL125). Int J Curr Microbiol App Sci 3(6):614–622

Singh JS, Pandey VC, Singh DP (2011) Efficient soil microorganisms: a new dimension for sustainable agriculture and environmental development. Agric Ecosyst Environ 140 (3–4):339–353

Singh D, Rajawat MV, Kaushik R, Prasanna R, Saxena AK (2017) Beneficial role of endophytes in biofortification of Zn in wheat genotypes varying in nutrient use efficiency grown in soils sufficient and deficient in Zn. Plant Soil 416(1):107–116

Singh D, Geat N, Rajawat MV, Mahajan MM, Prasanna R, Singh S, Kaushik R, Singh RN, Kumar K, Saxena AK (2018) Deciphering the mechanisms of endophyte-mediated biofortification of Fe and Zn in wheat. J Plant Growth Regul 37(1):174–182

Singh D, Prasanna R, Sharma V, Rajawat MV, Nishanth S, Saxena AK (2020) Prospecting plant-microbe interactions for enhancing nutrient availability and grain biofortification. In: Wheat and barley grain biofortification. Woodhead Publishing, New Delhi, pp 203–228

Siwela M, Pillay K, Govender L, Lottering S, Mudau FN, Modi AT, Mabhaudhi T (2020) Biofortified crops for combating hidden hunger in South Africa: availability, acceptability, micronutrient retention and bioavailability. Foods 9(6):815

Sors TG, Ellis DR, Na GN, Lahner B, Lee S, Leustek T, Pickering IJ, Salt DE (2005) Analysis of sulfur and selenium assimilation in Astragalus plants with varying capacities to accumulate selenium. Plant J 42(6):785–797

Soumare A, Diedhiou AG, Thuita M, Hafidi M, Ouhdouch Y, Gopalakrishnan S, Kouisni L (2020) Exploiting biological nitrogen fixation: a route towards a sustainable agriculture. Plants 9 (8):1011

Srivastava RP (2016) Antinutrients restraining biofortification. In: Biofortification of food crops. Springer, New Delhi, pp 333–348

Srivastava MP, Tiwari R, Sharma N (2013) Effect of different cultural variables on siderophores produced by *Trichoderma* spp. Int J Adv Res 1:1–6

Tariq M, Hameed S, Malik KA, Hafeez FY (2007) Plant root associated bacteria for zinc mobilization in rice. Pak J Bot 39(1):245

Trivedi G, Patel P, Saraf M (2020) Synergistic effect of endophytic selenobacteria on biofortification and growth of Glycine max under drought stress. S Afr J Bot 134:27–35

Urbano G, Lopez-Jurado M, Aranda P, Vidal-Valverde C, Tenorio E, Porres J (2000) The role of phytic acid in legumes: antinutrient or beneficial function? J Physiol Biochem 56(3):283–294

Vaid SK, Kumar B, Sharma A, Shukla AK, Srivastava PC (2014) Effect of Zn solubilizing bacteria on growth promotion and Zn nutrition of rice. J Soil Sci Plant Nutr 14(4):889–910

Velu G, Ortiz-Monasterio I, Cakmak I, Hao Y, Singh RÁ (2014) Biofortification strategies to increase grain zinc and iron concentrations in wheat. J Cereal Sci 59(3):365–372

Wagner SC (2011) Biological nitrogen fixation. Nat Educ Knowl 3:15

Wang Y, Yang X, Zhang X, Dong L, Zhang J, Wei Y, Feng Y, Lu L (2014) Improved plant growth and Zn accumulation in grains of rice (*Oryza sativa* L.) by inoculation of endophytic microbes isolated from a Zn Hyperaccumulator, Sedum alfredii H. J Agric Food Chem 62(8):1783–1791

Wang J, Zhou C, Xiao X, Xie Y, Zhu L, Ma Z (2017) Enhanced iron and selenium uptake in plants by volatile emissions of *Bacillus amyloliquefaciens* (BF06). Appl Sci 7(1):85

White PJ, Broadley MR (2011) Physiological limits to zinc biofortification of edible crops. Front Plant Sci 2:80

White PJ, Bowen HC, Parmaguru P, Fritz M, Spracklen WP, Spiby RE, Meacham MC, Mead A, Harriman M, Trueman LJ, Smith BM (2004) Interactions between selenium and sulphur nutrition in *Arabidopsis thaliana*. J Exp Bot 55(404):1927–1937

White PJ, Broadley MR, Bowen HC, Johnson SE (2007) Selenium and its relationship with sulfur. In: Sulfur in plants an ecological perspective. Springer, Dordrecht, pp 225–252

WHO (2016) Biofortification of staple crops. WHO, Geneva

Wu Z, Bañuelos GS, Lin ZQ, Liu Y, Yuan L, Yin X, Li M (2015) Biofortification and phytoremediation of selenium in China. Front Plant Sci 6:136

Yadav R, Ror P, Rathore P, Ramakrishna W (2020) Bacteria from native soil in combination with arbuscular mycorrhizal fungi augment wheat yield and biofortification. Plant Physiol Biochem 150:222–233

Yaseen M, Abbas T, Aziz MZ, Wakeel A, Yasmeen H, Ahmed W, Ullah A, Naveed M (2018) Microbial assisted foliar feeding of micronutrients enhance growth, yield and biofortification of wheat. Int J Agric Biol 20:353–360

Yasin M, El-Mehdawi AF, Anwar A, Pilon-Smits EA, Faisal M (2015) Microbial-enhanced selenium and iron biofortification of wheat (*Triticum aestivum* L.)-applications in phytoremediation and biofortification. Int J Phytoremed 17(4):341–347

Zahedi H (2016) Growth-promoting effect of potassium-solubilizing microorganisms on some crop species. In: Potassium solubilizing microorganisms for sustainable agriculture. Springer, New Delhi, pp 31–42

Zhang L, Liu Y (2020) Potential interventions for novel coronavirus in China: a systematic review. J Med Virol 92(5):479–490

Zhang Y, Pan G, Chen J, Hu Q (2003) Uptake and transport of selenite and selenate by soybean seedlings of two genotypes. Plant Soil 253(2):437–443

Zhang H, Sun Y, Xie X, Kim MS, Dowd SE, Paré PW (2009) A soil bacterium regulates plant acquisition of iron via deficiency-inducible mechanisms. Plant J 58(4):568–577

Zhang Y, Song Q, Yan J, Tang J, Zhao R, Zhang Y, He Z, Zou C, Ortiz-Monasterio I (2010) Mineral element concentrations in grains of Chinese wheat cultivars. Euphytica 174(3):303–313

Zhang ZZ, Lou YG, Deng DJ, Rahman MM, Wu QS (2015) Effects of common mycorrhizal network on plant carbohydrates and soil properties in trifoliate orange–white clover association. PLoS One 10(11):e0142371

Zhao C, Ren J, Xue C, Lin E (2005) Study on the relationship between soil selenium and plant selenium uptake. Plant Soil 277(1):197–206

Chapter 10
Microbial Biopesticides: Development and Application

H. T. Mandakini and Dimuthu S. Manamgoda

Abstract Application of bio-based pesticides has been a growing trend in organic agriculture globally. The use of natural or genetically modified microorganisms as biopesticides can be considered as an effective and sustainable approach in disease control. The discovery and the development of a microbial biopesticide is a process of two major phases consisting experimental component and commercialization of the product. The experimental component begins with the field collection of the potential microbial samples and isolation of potential microbes and to evaluate their bio-control efficiency. Once a potential candidate microorganism is selected for the production of biopesticides, the candidate should be accurately identified and characterized. Genetic modifications can be done to improve the efficiency of the organism. The commercialization of the product includes mass production, formulation, and field testing and safety evaluation. Finally, a biopesticide can be registered and introduced to the market upon the completion of safety evaluation and regulatory approval. This chapter summarizes the discovery and development of biopesticides with special reference to nematophagous fungi and their applications.

Keywords Entomopathogens · Formulation · Mass production · Pesticides

10.1 Introduction

Chemical pesticides had been used to fight effectively against pests and diseases for a long period of time. However, long-term wide use of chemical-based pesticides has resulted to several adverse effects such as pesticide contamination of soil and water leaving residues on crop produce, development of insecticide resistance, and biomagnifications of insecticides. As a result, regulatory measures are currently

H. T. Mandakini · D. S. Manamgoda (✉)
Department of Botany, Faculty of Applied Sciences, University of Sri Jayewardenepura, Nugegoda, Sri Lanka
e-mail: dsmanamgoda@sjp.ac.lk

P. Bhatt et al. (eds.), *Microbial Technology for Sustainable Environment*,
https://doi.org/10.1007/978-981-16-3840-4_10

established globally to restrict the export of chemical-based pesticides. Thus, there is a tremendous demand for biopesticides around the world (Leng et al. 2011).

Biopesticides can be defined as the types of pesticides derived from natural materials as bacteria, plants, animals, and certain minerals (US Environmental Protection Agency, Regulating Pesticides 2008). During the last three decades, development in the fields of molecular biology, genetics, protein engineering, and genome sequencing has improved the production procedures of biopesticides. As a result, biopesticides are gradually replacing toxic chemical pesticides in the market (Leng et al. 2011). The numbers of newly developed and registered biopesticides are increasing at a rate of 4% annually. Similarly, the annual market share of biopesticides is rising approximately 30% (Cheng et al. 2010). Biopesticides can be classified as botanical pesticides, microbial pesticides, and zooid pesticides. Among them, microbial pesticides can be considered as the first developed formulations and widely used biopesticides (Leng et al. 2011).

Many studies are available focusing on screening the pesticide activity of microbes. However, comparatively a few of them are leading toward successful commercialization of a microbial biopesticide. The discovery, development, and commercialization of a microbial biopesticide is a long-term process which should be carefully handled and managed. This chapter summarizes the discovery, development, and application of microbial pesticides in commercial scale with special reference to nematophagous fungi that have the potential to be used as biopesticides.

10.2 Discovery and Development of Microbial-Based Pesticides

The development of microbial-based biopesticide is a complex, time, and money-consuming multistep process. It can be divided into two major phases in which Phase 1 consists of extensive experimental and research works, where Phase 2 is mainly considered the commercialization of microbial-based pesticides (Fig. 10.1). Phase 1 contains six major steps which are screening for suitable sampling site and sampling, isolation of potential microbial candidates, evaluating bio-control efficacy of isolates, identification and characterization of isolates, strain improvement, and selecting the microbial candidate of interest.

After selecting a suitable microbial candidate for biopesticide production, the next steps would involve the production of biopesticides on a commercial scale. The first phase (Phase 1) can be carried out inside a laboratory, whereas the second phase (Phase 2) will be carried out in an industrial-scale facility.

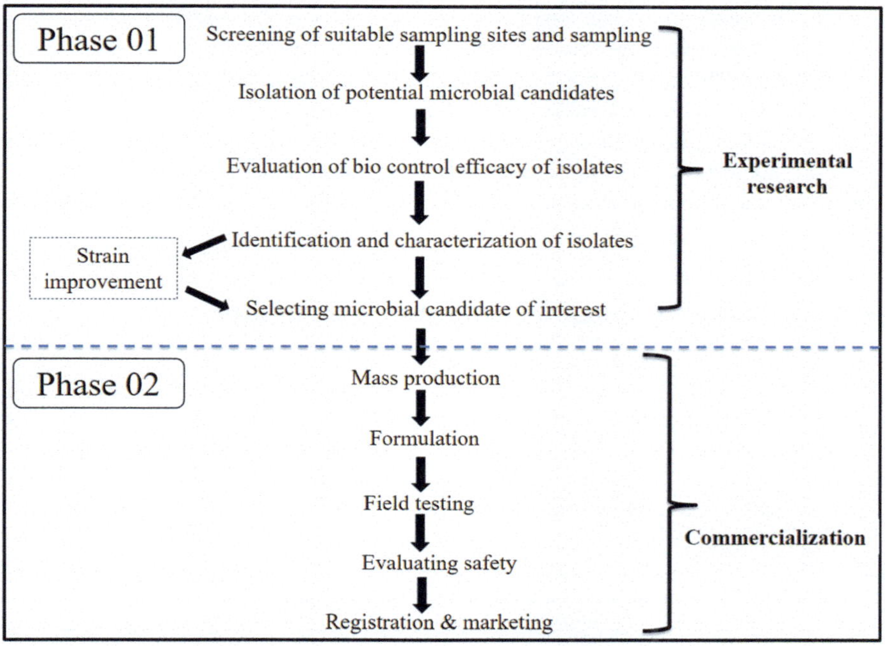

Fig. 10.1 Steps involved in developing microbial-based bio-control agent

10.3 Screening of Suitable Sampling Sites and Sampling

The first step in developing a microbial-based biopesticide is the selection of a suitable sampling site to isolate potential microbial candidates. It is important to identify the biology and the nature of the pest problem in a particular area to which a microbial biopesticide is expected to be developed. Then, the information of pest biology including the most susceptible stages of the host, host range, nature of the symptoms, and the most susceptible period of the year must be explored (Ravensberg 2011).

The background information of the pest problem narrows down the blind isolation of microorganisms and increases the probability of obtaining useful candidates. After elaborately defining the host, pathogen, and epidemiology stages of the disease, the next step is to collect potential microbial candidates (usually bacteria or fungi) from proper sampling sites. The potential microbial candidate may suppress the pest by producing secondary metabolites or it can cause serious diseases on the pest. Keeping this in mind, healthy plants from the field with the disease, disease suppressive soil, and also the dead insects in the field are considered as good sampling sites and materials for potential candidates (Glare and Moran-Diez 2016). Besides those, areas with fewer disease developments or no disease development at all even with the presence of susceptible hosts are also considered to be potential grounds to isolate prospective biological control agents (Schisler and Slininger 1997). Once the sampling site is selected, sampling must be carried out in a manner where it covers a wide area in the selected sampling site or sampling materials to avoid biased collection and missing any opportunities (Montesinos 2003; Köhl et al. 2011). Collected soil samples are considered as a source for free-living strains of beneficial microorganisms while plant parts are sources for beneficial endophytes (Glare and Moran-Diez 2016).

10.4 Isolation of Potential Microbial Candidates

After a successful sampling, the next step is to isolate microbial candidates on natural or synthetic microbiological media or to a biological system like cell cultures and trap organisms (Montesinos 2003). To isolate all the types of potential microorganisms representing each microbial taxon, isolation needs to be carried out on different media that facilitate the growth of all of them.

As applying microbial isolation techniques can affect and favor some characteristics of microorganisms, isolation techniques have to be designed carefully. It is recommended to isolate them under suitable environmental conditions, from correct plant parts without the use of selective enrichment media (Schisler and Slininger 1997). In some cases, the use of highly selective media and selective incubation conditions can promote certain types of microbial growth over others and will result in losing the chance to isolate many potential microbial candidates (Köhl et al. 2011). In some cases, generalized media are used for initial isolation. For an example, potato dextrose agar (PDA) can be used to isolate entomopathogenic fungi like *Beauveria* and *Metarhizium* (Tuininga et al. 2014). However, microbial candidates with beneficial traits that are most important in the later stages of the biopesticide development process can be preselected by carefully selecting appropriate isolation techniques. For instance, microorganisms with the ability to grow on a commercially inexpensive medium can be preselected by using an isolating medium more similar to a commercially feasible medium which uses at later stages the biopesticide production process (Schisler and Slininger 1997).

10.5 Evaluating Bio-control Efficacy

Followed by the isolation of microbial candidates, the next step is to experimentally evaluate their efficacy as bio-controlling agent, against pathogens under both in vitro and in vivo conditions. This is one of the critical steps of biopesticide development as the potent strains can be separated from a large number of microbial candidates (Glare and Moran-Diez 2016). In this step, a large number of microbial isolates are screened for potential bio-control agents by subjecting to bio-control bioassays. Bio-controlling ability of microorganisms can be achieved through diverse mechanisms as antibiosis, parasitism, competition, and induction of plant systemic resistance (Montesinos 2003; Köhl et al. 2019). It is required to know the mode of action of potential microbial candidates as it is important in achieving optimal disease controlling and also to address potential risks for humans and the environment (Köhl et al. 2019).

According to Harvey and Spurr, two types of bioassays are designed to evaluate bio-control activity: primary in vitro bioassays and secondary in vivo bioassays (Spurr 1985). In the process of development of new bio-control, isolates are first screened for bio-control activity based on their potential to inhibit pathogen of interest in in vitro through antagonistic effect. In vitro bioassays are comparatively fast, resource-efficient, and easy to perform, and results can be easy. In many cases, if the isolates showed no in vitro efficacy, those species would not be promoted for in vivo testing (Besset-Manzoni et al. 2019). Some of the in vitro bioassays are in vitro petri plate antagonism tests which are performed on agar, such as dual culture plate test, volatile and nonvolatile compound tests (Glare and Moran-Diez 2016), or spore germination and mycelia growth-inhibitory test performed in liquid media (Besset-Manzoni et al. 2019).

However, the literature suggests that results of in vitro bio-control efficacy tests are not always correlated with results of in vivo bio-control experiments. One of the main reasons behind this is that in vitro assays only evaluate the bio-control potential mediated mainly through antibiosis mode of action in an artificial environment. The other mechanisms like competition, hyperparasitism, and induced systemic resistance are not able to be detected by that means (Schisler and Slininger 1997; Köhl et al. 2019). Also, another drawback of in vitro bio-control efficacy test is the inability of mimicking the natural environment where the proposed bio-control is truly expected to act. Generally, nutrient concentration, composition, and the porous nature of the media result in comparatively higher production and higher diffusion rate of antibiotic compounds which are responsible for bio-controlling activity (Köhl et al. 2019).

In vivo bioassays, also known as secondary bioassays, are considered to be more realistic as they test the bio-control activity of candidates in the presence of a host plant, grown on field soil or an environment that is close to the field environment. They are designed to closely mimic the field environment. Literature reveals many diversely designed in vivo bioassays. In the study of Schisler and Slininger, antagonists of Fusarium dry rot were selected on wounded potato tubers incubated at 15 °C. Aqueous soil pastes which were prepared by enrichment of potato tuber periderm and inoculated with *Gibberella pulicaris* were applied onto wounded tubers (Schisler and Slininger 1997). Verhaar et al. used a bioassay to select potential bio-control of cucumber powdery mildew on detached cucumber leaves that had previously been inoculated with *Sphaerotheca fuliginea.* They tested the *Verticillium lecanii* isolates with high antagonistic potential at different humidity (Verhaar et al. 1998). Another successful in vivo bioassay performed by Clarkson et al. is the use of onion seedlings to select potential bio-control of white rot caused by sclerotium-forming *Sclerotium cepivorum.* Onion seedling was potted in a soil mixture that has been previously inoculated with conditioned sclerotia of

S. cepivorum and potential bio-control fungi isolated either as wheat bran cultures or spore suspension (Clarkson et al. 2002). After performing a successful bioassay, the next immediate step is to exclude the candidates that lose bio-control efficacy after repeated laboratory cultivation. Therefore, candidates who pass bio-control efficacy testing are repeatedly sub-cultured and subjected to bioassays testing of bio-control efficacy (Köhl et al. 2011). However, it is known that both the knowledge on the mode of action of bio-control agent and the activity related gene/s together can improve high-throughput screening approaches. Some of the rapid screening methods are use of phenotypic or genetic markers, DNA arrays, or specific secondary metabolites (Glare and Moran-Diez 2016; Köhl et al. 2011).

10.6 Identification and Characterization

Accurate identification of the isolates is extremely important before commercialization. Isolated candidates should be characterized and identified up to species level by macroscopic and microscopic morphologies, physiology, carbon utilization and growth rates, biochemical characteristics, secondary metabolites profiling, and molecular methods. Taxonomic literature, morphological keys, guides, and illustrations can be used to identify species using morphological techniques. Patterns of carbon utilization and relative growth rates on different media have been used to characterize fungi and bacteria. Biochemical characterization is commonly used in bacterial identification, and there are kits that are developed for easy biochemical characterization of bacterial strains. Fatty acid methyl ester analysis (FAME) and secondary metabolite profiles on high-performance TLC (HPTLC) and high-performance liquid chromatography (HPLC) have been used for both bacterial and fungal identification. DNA extraction, PCR, and sequencing can be considered as a gold strand of microorganism identification. Once sequences are generated, they can be identified through the BLAST comparison with the data on public databases. Phylogenetic analysis is always preferred in species identification since true phylogeny represents the correct taxonomy. Other than that, the polyphasic approach of combining all the methods mentioned above has been also practiced in fungal and bacterial identification.

Once the species are identified, they are screened in numerous databases for safety information to assess potential risks of allergy, toxicity, and pathogenicity to humans, animals, and crops, and also to find out the candidates' bio-control application is patent protected or not (Köhl et al. 2011). The candidates associated with potential risks and already patent protected are then discarded from further screening.

10.7 Strain Improvement

If a microbial strain is suitable for a commercial level production, it would lead to the possibility of improving strains' performance. Due to the main concern on the economic setting in developing a biopesticide, it is always challenging to obtain the biologically perfect strain. The most promising candidate is the one that shows the best compromise between bio-control efficacy and economic factors (Ravensberg 2011). Hence, improvement of strains to achieve positive production aspects such as the quick speed of bio-control action, high mass production, and environmental tolerance is beneficial.

Until the last two decades, there were no many studies involved in the strain improvement of the pathogenic fungi due to a lack of available information on the genes that control either pathogenesis or specificity (St Leger and Screen 2001). With the development of molecular biological techniques, this situation changed drastically. These techniques helped to elucidate pathogenic process in important bio-control agents; pathogenicity related genes could be cloned and express just as microorganisms are induced by physical or chemical stimuli to change its saprobic habit to pathogenic or parasitic phase. Some of these pathogenic genes encode hydrolytic enzymes and toxins which can act on the target pathogens. This would allow them to develop transgenic microorganisms with improved pathogenic qualities.

10.8 Selecting Microbial Candidate of Interest

In the final selecting process, many criteria are taken under consideration other than the efficacy and consistency of results in repeated bioassays. Some of these additional criteria are positive growth kinetics in the field, favorable toxicological profiles of microorganisms, the requirement of low effective doses for bioactivity, the specificity of action to a target pest or range of pests, tolerance to market available pesticides, and estimated cost-effectiveness of technologies for mass production (Köhl et al. 2011). These additional limiting conditions further narrow down a large number of candidate isolates and result in only a fewer isolate. Literature reveals that only less than 1% of candidate isolates satisfy all those criteria and make it a successful product (Montesinos 2003).

10.9 Commercialization of Microbial-Based Biopesticides

In Phase 2 of the development of biopesticides, mass scale production, storage, and perseveration of selected bio-control agents and also the formulation are considered as major initial steps in the commercialization process (Powell and Jutsum 1993). In the selection of production process, it is required to pay attention to the nature of the organisms which are expected to be produced and expected active final product as the required conditions can differ from one another (Glare and Moran-Diez 2016). Generally, there are two types of production processes: liquid phase fermentation or solid phase fermentation. Usually, liquid phase fermentation is widely used for bacteria and yeast production, while some other fungi are fermented in solid state.

10.10 Mass Production

Production of inoculum is considered as the major step in making a biopesticide commercially feasible. The goal of mass production is to achieve the highest yield of stable, quality biomass of bio-control agent under the lowest possible cost. The end product of the production process can be either microbial cells or biological structures as conidia, microsclerotia, or specific secondary metabolites like toxins which are expected to be used as bioactive agents (Glare and Moran-Diez 2016).

The production process should be designed as a reliable, cost-effective one with a higher production rate. The ideal production process for commercial level production should be obtained by experimenting with changing parameters. Key aspects to be considered here are the type of fermentation, growth media which is intended to be used, and the choice of the suitable equipment and conditions; for production, downstream processing, and storage of the end product (Ravensberg 2011).

The type of fermentation and the growth medium greatly depends on the nature of the expected bio-control agent to be produced. Normally, liquid phase fermentation is chosen for the production of bacteria and yeast, while solid-phase fermentation is preferred for the production of filamentous fungi (Köhl et al. 2011). However, in many cases, solid-state fermentation uses the inoculum produced from prior submerged fermentation (Glare and Moran-Diez 2016). *Bacillus thuringiensis* is a commercially available, successful biopesticidal agent and industrially produced in a larger scale via liquid-state fermentation. However, Devi et al. show the semi-solid and solid-state fermentation of *B. thuringiensis* in a small scale in developing countries for home usage (Devi et al. 2005). Liquid-state fermentation is considered as the most suitable choice for large-scale biomass production for industrial purposes due to easy manipulation of growth parameter gradients such as temperature, pH, and dissolved oxygen, wherein the solid-state fermentation faces difficulty in controlling due to deprived level of water availability. However, the production of fungal spores by aerial hyphae in small-scale solid-state fermentation is considered as the best method (Glare and Moran-Diez 2016). At the end of the fermentation process, produced inocula are recovered and purified by downstream processing, and the end product is called "technical product" which may be composed of pure inoculum or inocula with the medium remnant, secreted metabolites, and compounds added during downstream processing (Ravensberg 2011).

However, before scaling up the mass production, preliminary experiments on mass production are performed on a small laboratory scale using a set of chosen microbial candidates from the end of Phase 1. During these preliminary experiments, inexpensive growth media and fermentation conditions (temperature, pH, dissolved oxygen concentration, etc.) are evaluated and optimized. Also, the viability and the bioactivity of the inocula after several downstream processes are determined. This allows producers to have a rough estimation of production costs. The selected strains from research and experiments are further screened, and strains producing low inconsistent yield which are greatly dependent on specific growth conditions are excluded (Köhl et al. 2011). Köhl et al. suggest performing the step of evaluating bio-control efficacy and pilot mass production step parallel to each other, as it can reduce estimated costs for each step considerably (Köhl et al. 2011). During a parallel performance of both steps, candidates that are unable to produce sufficient biomass can be discarded avoiding further efficacy testing.

10.11 Formulation of a Product

Once the biopesticide inoculum is produced, maintaining the stability during storage, distribution, and at the target site is a prerequisite. The formulation allows stabilization of the technical product along with persistence, safety, high efficacy, and user-friendliness (Ravensberg 2011; Kala et al. 2020). Hence, it is considered a vital step in developing high quality, viable biopesticide. During formulation, the active ingredient (produced inoculum) is mixed with inert ingredients such as carriers, surface active ingredients (stickers, spreaders), and other additives (stabilizers, coloring agents, etc.) (Kala et al. 2020).

There are different types of formulations of biopesticides available in the market. The choice of a formulation is mainly determined by the type of active ingredients and the environment where it is applied (Glare and Moran-Diez 2016). Physiochemical properties and biological activity of active ingredients, method of intended delivery, and safety in usage, cost-effectiveness, and market preferences are further taken into account when selecting the type. Once all the factors have been determined, the final formulation type is selected with the most appropriate inert ingredients where formulated biopesticides have at least 2 years of shelf life under dynamic climatic conditions (Knowles 2008).

With respect to its physical state, biopesticide formulation can be categorized as liquid formulation and dry formulations which are further divided according to their properties. Each and every type of formulation can have its own challenges and benefits. Generally, dry formulations are manufactured by adding wetting agents, dispersants, and binding agents (Knowles 2008). They are available in the form of direct application such as dusts (DP), powder for seed dressing (DS), granules (GR), micro granules (MG) or in the form of concentrates such as water dispersible granules (WG) and wettable powders (WP), which require dilution in water prior to use (Kala et al. 2020).

Dust formulation: Dusts are made up by mixing the active ingredient with fine, dry, inert powder carriers like talc, clay, with anticaking agents, and ultraviolet protectant as additives. Percentage of the active ingredient is comparatively low and 10% or less by weight. The particle size of dust can be in the range of 50–100μ. They can be delivered to the target site either directly or mechanically. However, due to possible inhalation exposure risk to handlers, usage is limited (Kala et al. 2020).

Granular formulation (GR): Except for large size and heavyweight, granular formulations are also the same as the dust formulation but made of coarse particles of carriers. Granule particle size can range from 100 to 1000μm. Minerals like silica, kaolin, and ground plant residues as corn cobs and walnut shells can serve as carriers, while the active ingredient (5–20% by weight) either coated to the outer surface or absorb into them. Once the active ingredients adhere to the granules, they are coated in a layer of inert material which facilitates the slow release of the active ingredients (Kala et al. 2020; Singh and Arora 2016). After granules are applied, some require the presence of water to release the active ingredient, while others do as they gradually decompose (Knowles 2008).

Wettable powder (WP): Dry, finely ground solid mineral carriers, wetting, and dispersing agents blended with active ingredients and are generally applied as a spray after being diluted in water. It can contain 5–95% of active ingredients, generally more than 50%. Application is convenient using spraying equipment with agitation. Therefore, such application is known to possess significant residual activity and pose no harm for treated surfaces.

Water dispersible granules (WG): This is an advanced form of a wettable powder that is compressed into dust free granules. Once dissolved in water, granules are broken apart into a fine powder and form dust-free uniform suspension. Compared to other dust formulations, WG poses a low handler exposure risk and is convenient in usage.

Liquid formulations can be based on water, oil, polymer, or a combination where water-based formulations need some additives such as surfactants, coloring agents, and antifreezing materials (Kala et al. 2020). They are formulated in the forms of emulsions: suspension concentrates (SC), oil dispersions (OD), suspo-emulsions (SE), capsule suspension (CS), and ultra-low volume formulations (Kala et al. 2020; Knowles 2008).

Emulsion (EC or E): Emulsion comprises droplets of a liquid active ingredient (0.1–10μm in size) which is oil-soluble dispersed in an immiscible liquid, usually a petroleum-based solvent. Emulsions are multipurpose formulations and highly adaptable as an application of ECs can be done from small portable sprayers to aircraft sprayers. Emulsions have comparatively lower shelf-life (Knowles 2008).

Suspension concentrate (SC): SC is a solid active ingredient dispersed in a liquid, usually water. The solid active ingredient should be insoluble and is ground finely before it is being suspended in water. Other inert ingredients are as follows: wetting agents, dispersants, viscosity modifiers (thickeners), preservatives, and antifreeze.

They often require agitation before use and leave a residue on the treated surfaces. SCs have low formulation cost and are popular since they are easy to use and comparatively safe (Knowles 2008).

Capsule suspension: A capsule suspension (CS) formulation is a combination of an active ingredient encapsulated in a polymer shell suspended in a liquid medium with a dispersant and wetting agent. These are needed to be diluted before use. The residual stability of these is enhanced by using living cells for the encapsulation. CSs are an advanced type of formulation which is highly beneficial, but the production is complex and expensive (Knowles 2008).

Ultra-low volume liquids: These are usually used as it is without dilution and comprises high concentrations of the active ingredient (mostly 100%). ULVs are easy to handle, and the residues on treated surfaces are almost invisible as these are applied at very low rates in fine droplets (Knowles 2008).

At this stage of biopesticide development, candidates with sufficient biomass production that show disease control ability repeatedly under control conditions in bioassays are selected. They are further screened in the formulation step. Shelf-life studies of the formulated product are performed for differently produced and formulated inocula with sufficient viability. Formulated inocula are stored under various expected storage temperatures such as room temperature and in freezer and refrigerators for at least 1 year time duration with regular testing of the viability of the product (Ali et al. 2013). The stress tolerance of the product is also evaluated by applying experimental stress conditions. After that, diversely produced, formulated, and stored products are subjected to bioassays under laboratory conditions to compare each of their actions and to compare with the action of fresh inoculum which is produced at Phase 1 to select the most promising formulated products (Köhl et al. 2011).

10.12 Field Testing

After the production and formulation, the end product is needed to successfully deliver to the target pest under field conditions favorable for disease development. There are various application and delivery techniques based on the type of formulation of the product including direct delivery, seed dressing delivery, delivery by dilution in water, etc. (Mishra et al. 2019). In some cases, the pathogen is artificially inoculated to the field, and growth conditions are controlled to facilitate the disease development. At this stage, several treatments such as different microbial candidates, formulations, and concentrations and applying times are tested and compared in field conditions (Ali et al. 2013).

The success of field trials depends on the intensity of disease during the experiments which frequently vary in accordance with the season of the year and location of the field. The efficacy of bio-control products is varying with environmental conditions. Hence, evaluating the efficacy of a developed product under multiple locations and different seasons is required to predict the consistency of the product in terms of disease control (Köhl et al. 2011).

10.13 Evaluating Safety

Assessment of biosafety of microbial pesticide is a compulsory aspect in the biopesticide development process to make sure there is no negative effect on non-target organisms and also on the environment. Global regulatory authorities of biopesticide production have posed strict procedures for toxicological assessment of produced biopesticides. Hence, thorough toxicological assessment at strain level is essential. Possible toxicity of microbial candidates can be identified at earlier stages of screening microbial candidates by database mining and then discard toxicologically suspected microbial candidates from further screening. If the microbial candidate is a well-studied one with the absence of clinical or veterinary history, it can be excluded from the development process without further thorough toxicological studies (Köhl et al. 2011).

After data mining for potential risks of microbial isolates, primary toxicological studies can be carried out with non-human model organisms such as *Caenorhabditis elegans* in order to get an idea of toxicological profiles of selected microbial candidates (Zachow et al. 2009).

Furthermore, the detailed experiments on human toxicology and ecotoxicology data of developed biopesticides are carried out as requested by some regulatory bodies. These toxicological data can be derived from numerous acute, subchronic, and chronic studies. The toxicology studies of biopesticide on mammals are a prerequisite to safeguard the handlers and consumers. One of the well-known toxicological tests is the oral acute test in which the objective is to find out the median lethal dose (LD50) and to minimize the lethality. When considering bacterial bio-control agents, LD50 is considered to be more than 10^{11} of colony-forming units per 1 kg of animal weight (Montesinos 2003).

10.14 Registration and Marketing

Many of the global regulatory authorities concern about the production of biopesticide and information about the composition, performance, and safety. The major aims of regulation of biopesticides are to safeguard consumers and the environment and also to characterize products to ensure a continuous supply of consistent quality products by manufacturers.

Legislation background on regulating the use and development differs from one country to another. The US Environmental Protection Agency (EPA) and European Union (EU) are considered as the largest regulatory bodies worldwide and show a contrasting difference in the biopesticide regulation process. International Organization for Biological Control (IOBC), European and Mediterranean Plant Protection Organization (EPPO) Economic, and Organization for Co-operative Development (OECD) are few global regulatory agencies developed to give some flexibility for the biopesticide regulation process (Arora et al. 2016).

Once the biopesticide is developed, registration is done in the relevant regional or national regulatory body. However, Glare and Moran-Diez (2016) state that the lack of specifically developed guidelines for microbial-based biopesticides by most regulatory bodies results in unfitting registration procedures that are mainly derived from synthetic chemical-based pesticides (Glare and Moran-Diez 2016). Hence, registration is known to be a very complex, time- and money-consuming, cumbersome process that significantly differs from country to country. Major characteristics of the registration processes of EPA and EU are briefly described below.

The registration process of EPA: The Biopesticide and Pollution Prevention Division (BPPD) that comes under EPA is responsible for all regulatory activities of biopesticides. The process of registration starts with a pre-submission consultation meeting, in which the applicant is informed about guidelines and types of study required and the labeling. A dossier is then compiled with required data and formally submitted to BPPD for further evaluation. Upon the formal submission, the content will initially screen for completeness followed by preliminary technical screening to determine further adequacy and completeness of data. If sufficient data is present with no other issues, registration is granted after reviewing. Generally, the full process takes up to 12–18 months (Arora et al. 2016).

The registration process of EU: In the EU, registration process is a comparatively complex and long procedure which is composed of two stages where both the active ingredient and product undergo separate evaluation. In contrast to EPA, the EU does not identify biopesticide as a separate regulatory category but under the plant protection product.

Registration of active ingredients, which is the first stage of EU registration, is composed of three subsequent phases: Rapporteur Member State phase (RMS phase), risk assessment phase, and risk management phase.

In the RMS phase, a compiled dossier containing all the information on active substance and at least one representative PPP is submitted to designate RMS for registration. If the application is admissible, then RMS proceed with the evaluation and produce a draft assessment report which is then submitted to other responsible European Committee (EC) and European Food Safety Authority (EFSA) for risk assessment on the active compound. After the peer-reviewing by EFSA, they released a conclusive scientific risk assessment report. Using the risk assessment report, EC presents a review report to the Standing Committee for Food Chain and Animal health (SCFCAH), and voting is carried out for approval or non-approval of the active substance. This is called the risk management phase. Depending on the weight and complexity of the dossier, the process will take up to 2.5–3.5 years.

The second stage or the registration of PPP is carried out by the relevant MS after the applicant specifies the MS. Once the multiple dossiers for the product, every active ingredient is submitted; the relevant MS evaluates and grants the authorization if it is acceptable. Then the application for mutual recognition within one of the three geographical zones (North, Central, and South) with the same use under similar conditions can be done by the applicant company (Frederiks and Wesseler 2019).

Although registration processes of both EPA and EU show contrasting differences including required data for dossier preparation, Glare and Moran-Diez (2016) present generalized data requirements for dossier compiling as follows.

Data requirements for an active substance are as follows:

Identity and purity, physical and chemical or biological properties, information on usage and production process, analytical methods for identification of active substance(s), human and domestic animals' health effects, residue chemistry, fate and behavior in the applied environment, and possible non-target effects.

Data requirements for the formulated product are as follows:

Type of formulation and composition details, physical and chemical properties, application techniques, labeling requirements, and packaging details; analytical methods used, efficacy data, and toxicology and exposure data; residues, fate, and the behavior of the product in applied environment; and possible effects on non-target organisms (Glare and Moran-Diez 2016).

10.15 Development and Application of Nematophagous Fungi as Biopesticides

Plant parasitic nematodes cause significant damage to vegetables and other agricultural crops throughout the world. It was estimated that plant parasitic nematodes could cause much more damage annually compared to other insect pests. The estimated annual crop loss due to nematodes is around 12.3% (157 billion dollars) worldwide (Singh and Kumar 2015). Based on the data generated by the All India Coordinated Research Project (AICRP) on nematodes, plant-parasitic nematodes may cause 21.3% crop losses amounting to Rs. 102,039.79 million (1.58 billion USD) annually in India (Kumar et al. 2020).

During the last decades, concerns about using nematicidal chemicals were raised, and as a result, biological control agents to control nematodes have been discussed. Nematophagous fungi as natural enemies of the nematodes offer a propitious eco-friendly approach in the control of plant parasitic nematodes.

10.16 Nematophagous Fungi as Biopesticides

Nematophagous fungi naturally can be found on soil. These are carnivorous fungi that are specialized to trap and digest nematodes. Thus, nematophagous fungi can be used to control the soil nematodes attacking crops and other plants. The activity of nematodes controlled by several ways as described further:

1. Fungi produce specialized structures (adhesive or mechanical traps) that can trap nematodes.
2. Fungi can produce mycotoxins that immobilize nematodes before invasion.
3. Endoparasitic and egg parasitic fungi can attack all life stages of nematodes.

A nematode trapping fungus *Arthrobotrys oligospora* was first described in 1852; however, by the time, its important ability in control nematode diseases was not discovered.

Nematode trapping fungi are taxonomically heterogeneous; they belong into class Ascomycota, Basidomycota, and Zygomycota. Around 350 different species are recorded around the world. Some of the examples are *Arthrobotrys*, *Cystopage*, *Dactylellina*, *Dactylella*, *Drechslerella*, *Hohenbuehelia*, *Hyphoderma*, *Monacrosporium*, *Nematoctonus*, *Orbilia*, *Stylopage*, *Tridentaria*, *Triposporina*, and *Zoophagus*.

Nematodes trapping fungi infect their hosts through a sequence of events (Nordbring-Hertz et al. 2001). First, the nematodes were isolated and trapped by the fungi. According to the literature, fungi produce traps constitutively. However, the trap formation is initiated as a response to the prey (Nordbring-Hertz 1977). The recognition of the presence of the nematodes and increment of producing trapping devices depend on a molecular mechanism. Lectins is reported to be involved in the recognition process. However, recent studies based on *A. oligospora* suggested that these fungi may be capable of compensating even for the absence of the lectin by expressing other proteins with similar function(s) to the lectin (Jiang et al. 2017).

The trapping devices of these fungi are derived from the vegetative mycelium. Trapping adheres nematode or traps the nematode with non-adhesive mycelia so that the nematode cannot move. Types of trapping devices can be described as adhesive networks, adhesive knobs, constricting rings, non-constricting rings, adhesive branches, undifferentiated or unmodified adhesive hyphae, stephanocysts, spiny balls, and acanthocytes (Fig. 10.2). The type of trapping device varies with the type of fungi. There is a correlation of the morphology of the trapping device with the phylogeny of nematode trapping fungi (Jiang et al. 2017).

Just like other pathogenic fungi, after recognition and adhesion, nematode trapping fungi enter the host by enzymatic degradation and by adding mechanical pressure on the nematode body. Extracellular hydrolytic enzyme produced by the nematode itself is a key virulence factor in nematode trapping fungi. After the penetration, fungi obtain nutrients decomposing the nematodes. Proteases are involved mainly in the penetration of the fungi into the nematodes. The first study on nematophagous proteases was done based on *Arthrobotrys oligospora* by Tunlid et al. (1991). Later, Lopez-Llorca (1990) was able to purify and identify the first nematophagous serine protease from the nematode egg parasite *Verticillium suchiasporium*. After that, many different serine proteases were detected, characterized, and cloned.

Several nematophagous fungi are reported to produce nematotoxins that immobilize or kill nematodes. For example, *Pleurotus* species such as *Pleurotus cystidiosus*, *P. cornucopiae*, *Pleurotus djamor*, *P. levis*, *P. populinus*,

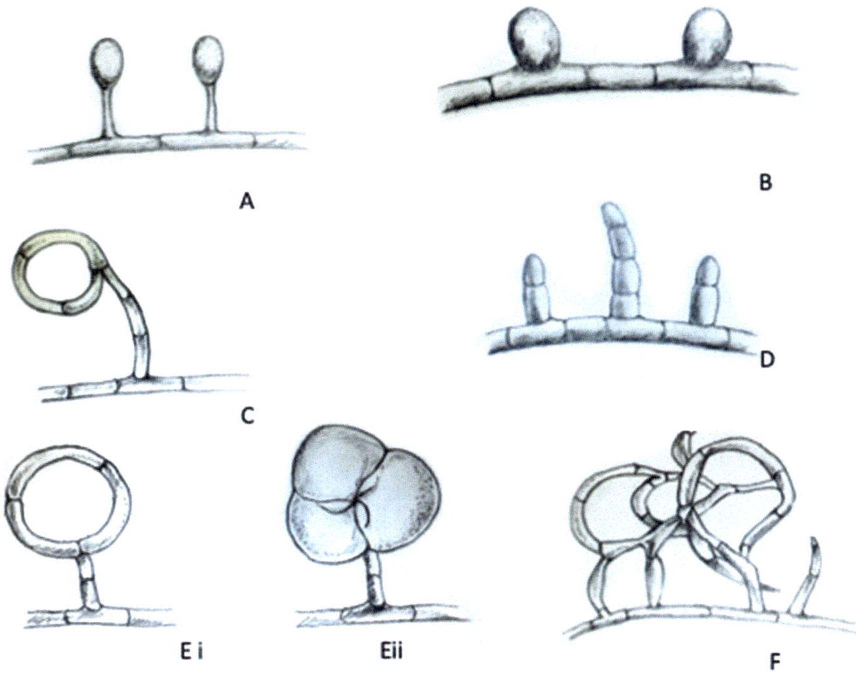

Fig. 10.2 Trapping devices of nematophagous fungi. (**a**) Stalked adhesive knobs, (**b**) sessile adhesive knobs, (**c**) non-constricting rings, (**d**) adhesive columns Ei and Eii: constricting rings, (**f**) adhesive nets. (Source: Redrawn based on Jiang et al. 2017)

P. tuberregium, and *P. subareolatus* are reported to produce toxins that can kill or immobilize nematodes (Lopez-Llorca et al. 2006; de Freitas Soares et al. 2018). Some fungi in the order Agaricales (*Megacollybia platyphylla*, *Cyathus striatus*, *Kuehneromyces mutabilis*) and order Polyporales (*Daedalea quercina*, *Fomitopsis pinicola*, *Gymnopilus junonius*) are reported to produce toxins that would paralyze the nematodes (Balaeş and Tănase 2016). Other than the one mentioned above, there are numerous other fungi that are capable of producing toxins against nematodes as a defense mechanism to prevent the consumption of fungal colony by the nematodes. Li et al. (2007) reviewed 179 fungal producing chemical compounds with nematicidal action.

Some nematophagous fungi prey on eggs. For example, *Myrothecium verrucaria* strain X-16 has been demonstrated to prey on eggs, second stage juveniles, and also adult nematodes of *Meloidogyne hapla* (Dong et al. 2015).

10.17 Productions of Biopesticides Using Nematophagous Fungi

The first attempt to use nematophagous fungi as a bio-control agent was carried out in 1930s (Linford and Oliveira 1937). According to Soares (2006) biological control of nematodes has many benefits over chemical control. Some of these advantages are easy application, less or no harm to the environment, not leaving any residue on the crop, not favoring the emergence of resistant nematode verities, not adversely affecting soil microbiota, and the potential ability to convert conducive soil into a suppressive one.

When considering the development of biopesticides, Phase 1 begins with the isolation of potential nematophagous fungi. Fungi can be isolated from soil, nematode body, and eggs (Aminuzzaman et al. 2013). The in vitro predatory activity of the fungi can be evaluated by estimating the reduction of nematode population on assay plates after a given incubation time (Aminuzzaman et al. 2013). Also, assays are developed to detect the activity of extracellular hydrolytic enzymes of the fungi (Yang et al. 2007).

After selecting a suitable candidate with a great predatory activity, the next steps in Phase 1 will continue. As well as the other fungi, nematophagous fungi can be identified using morphological and molecular methods. Morphological structures of these groups of fungi are mostly identified using the keys prepared by Cooke and Godfrey (1964) and Philip (2002). Sequencing the fungal barcode (nuclear ribosomal internal transcribed spacers and 5.8S), ITS region is commonly practiced. DNA sequences are initially identified using NCBI blast comparison, and later phylogenetic analysis using only ITS or combining the sequences data of protein coding gene regions has been carried out (Falbo et al. 2013).

Phase 2 of the biopesticide development starts with mass production and formulation. The most favorable temperature for the growth of nematophagous fungi is 25 °C (Velvis and Kamp 1996; Castro et al. 2000). Nematophagous fungi can be mass produced on solid or liquid media. The amount of biomass produced is generally higher in the liquid media. The formulation step can be carried out by several means. One of the common formulation techniques used in the literature is listed below.

10.18 Capsule Formulation Using Sodium Alginate

This technique would encapsulate the conidia of nematophagous fungi. Fungal strains will be multiplied on solid or liquid media until conidia are produced. The conidia are quantified on sterile water. In order to capsule formation, clay (5% sodium bentonite), sodium alginate (1%), and streptomycin sulfate (1%) were used according to the methodology mentioned by (Carneiro and Gomes 1997). Conidia were then mixed with a homogenized mixture of above-mentioned three ingredients.

The formulation concentration depends on the added concentration of conidia there for concentration should be adjusted accordingly.

After that, a $CaCl_2$ solution (0.25 M) is added to induce the formation of solid aggregates or pellets. These solid aggregates were then submitted to a drying process in a laminar flow chamber for about 6 h. According to the literature, process will yield approximately 3.4 g pellets for each 100 mL of suspension (Carneiro and Gomes 1997). These formulated aggregates can be stored at room temperature, away from light, for about 12 months. The viability of the conidia may vary according to the strain of the fungi.

10.19 Formulation Using Rice and Sorghum Grain

Formulation of fungi on shell rice or sorghum grains is regularly practiced. The technique can be used for different nematophagous fungi such as *Paecilomyces*, *Pochonia*, *Arthrobotrys*, *Monacrosporium*, *Dactylella*, and *Dactylaria*. Grains should be soaked overnight and then cleaned and sterilized in 121 °C for 15 min at 15 PSI. If there is excess water remain after sterilization, it can be drained, and inoculation of the fungi can be carried out once the grains are cooled.

10.20 Formulation Using Sugarcane Bagasse and Rice Bran

Sugarcane bagasse which is a byproduct of sugar cane production can be enriched with rice bran to be used as formulation material. A sterilized mixture of sugarcane to rice bran 2:1 ratio is generally used. After cooled to room temperature, fungi can be inoculated.

10.21 Cocktail Formulation

Since there are different parasitism/predation forms of nematophagous fungi, researchers came up with an idea of formulating of some nematophagous fungi with different capturing strategies (dos Santos and Ferraz 2000). dos Santos and Ferraz (2000) named this mixture as "nematophagous fungi cocktail." Instead of developing formulation from one strain of fungi, nematophagous fungi are separately multiplied and then mixed in equal proportions. The first such "nematophagous fungi cocktail" was comprised of the species of *P. lilacinus*, *A. musiformis*, *A. oligospora*, *D. leptospora*, and *Monacrosporium robustum*, and it was used against root-knot nematodes, *Meloidogyne* spp. (dos Santos and Ferraz 2000).

Another cocktail containing fungi *M. eudermatum*, *D. leptospora*, *A. musiformis*, *A. conoides*, and *P. lilacinus* was tested against phytopathogenic nematodes of *T. semipenetrans* and *P. jaehni* on citrus plants (Martinelli 2008; Martinelli et al. 2012). A significant reduction in the nematode populations was observed compared to the use of chemicals against these nematodes on citrus.

10.22 Application of Nematophagous Fungi Formulations

Previous studies indicate that before application of formulations under field conditions, it is extremely important to evaluate some characteristics on soil such as organic matter content, relative humidity, temperature, and the presence of nematodes in the area (Barbosa et al. 2011). Time of application is also critical, because microorganisms would not survive in dry soil. Application during the midday under direct sunlight may lead to death of nematophagous fungi prior to its action (dos Santos and Ferraz 2000). Several studies using cocktail of nematophagous fungi on crops of chrysanthemum, lettuce, peppers, and okra were conducted by Soares (2006), and those experiments presented very impressive results in controlling nematodes. However, the efficacy of nematophagous fungi is inversely proportional to the longevity of the crop so that a long duration crop requires more than one application of the fungal formulation (Martinelli et al. 2015).

10.23 Commercialization of Nematophagous Fungi Formulations

Although there are a large number of fungi with nematicidal activity, only several fungi can be actually used in the commercial level production. Screening for suitable fungal strains and strain improvement strategies to develop high virulent strains has been researched by academic scientists (Casas-Flores and Herrera-Estrella 2007). The development of these biopesticides and commercialization have been tried by enraptures; however, still only a few products are successfully introduced to the market. The main challenges faced by these newly introduced biopesticides are the low productivity and the inability to establish under a wide range of soil conditions (Moosavi et al. 2011). However, suppressive soil demonstrates that these bio-control agents could generate durable and robust nematode control even though genetic factors that contribute to this development are not fully understood. Genomic studies on nematophagous fungi and biology of the infection procedure into the phytopathogenic nematodes would provide more evidence about these pathogenic factors.

Several other intrinsic, ecological, and environmental factors may affect the efficacy of the nematophagous fungi when they are applied on a larger scale. Even though some nematophagous fungi show excellent activity under a small scale, after the applied a larger commercial scale under field conditions, it does not provide expected results. Therefore, it is recommended to start research with few potentially effective strains without restricting into one because only a highly virulent nematophagous fungus with highly competitive saprobic ability may provide the actual acceptable control level. The major environmental factors that affect the ability of the nematophagous fungi are soil temperature, pH, and the nutritional state of the soil. Also the biotic factors such as the host plant and soil microflora may affect the activity of the fungi under the field conditions. A better overall understanding of all of these factors would help to develop a successful bioproduct. Considering the commercialization of the product, favorable traits of the fungi are the ability to mass produce, persistent and stable formulations, the easy applicability, and the high productivity. The new scientific and technological developments would provide easy screening and development procedures for nematophagous fungi. On the other hand, farmers should also be informed about the environmental benefits and economic benefits of organic farming for switching into bio-based nematicidal products (Moosavi and Askary 2015).

10.24 Conclusion

Microbe-based biopesticides are eco-friendly and less hazardous to the environment. Therefore, these formulations are considered as an integral part of sustainable agricultural practices. However, the development pathway is challenging due to high research and production costs and time- consuming legislative requirements. However, considering the rate of rising market value and global demand and future of microbe-based biopesticides will be promising alternatives to hazardous chemicals. Thus, with the development of technology, there is a greater potential in the research areas focusing on isolation, characterization, and development of suitable bio-based pesticides.

References

Ali SRA, Kamarudin NH, Mslim R, Wahid MB (2013) Microbial approach in pests control. In: Wahid MB, May CY, Weng CK (eds) Further advances in oil palm research (2000-2010), vol 1. Malaysian Palm Oil Board, Bandar Baru Bangi, pp 407–456

Aminuzzaman FM, Xie HY, Duan WJ, Sun BD, Liu XZ (2013) Isolation of nematophagous fungi from eggs and females of Meloidogyne spp. and evaluation of their biological control potential. Biocontrol Sci Tech 23(2):170–182

Arora NK, Verma M, Prakash J, Mishra J (2016) Regulation of biopesticides: global concerns and policies. In: Arora NK, Mehnaz S, Balestrini R (eds) Bioformulations: for sustainable agriculture. Springer, New Delhi, pp 283–299

Balaeş T, Tănase C (2016) Basidiomycetes as potential biocontrol agents against nematodes. Rom Biotechnol Lett 21(1):11185

Barbosa BFF, Barbosa JPS, Santos JM, Barbosa JC, Negri MA, Soares PLM (2011) Eficacia do controle de nematoides em cana de acucar com produtos organicos e biologicos. In: Congresso Brasileiro de Fitossanidade, *Jaboticabal*. FUNEP, Jaboticabal, pp 64–67

Besset-Manzoni Y, Joly P, Brutel A, Gerin F, Soudiere O, Langin T, Prigent-Combaret C (2019) Does in vitro selection of biocontrol agents guarantee success in planta? A study case of wheat protection against Fusarium seedling blight by soil bacteria. PLoS One 14(12):e0225655

Carneiro RMDG, Gomes CB (1997) Encapsulacao do fungo *Paecilomyces lilacinus* em matrizes de alginato-argila e avaliacao da viabilidade dos conídios em duas temperaturas. Nematologia Brasileira 21:85–91

Casas-Flores S, Herrera-Estrella A (2007) Antagonism of plant parasitic nematodes by fungi. In: Kubicek CP, Druzhinina IS (eds) Environmental and microbial relationships, The mycota, vol 4. Springer, Berlin, pp 147–158

Castro JMC, de Lima RD, Ferraz S, Neves JCL (2000) Capacidade de predacao de Arthrobotrys musiformis a fitonematoides. Summa Phytopathol 26:58–62

Cheng XL, Liu CJ, Yao JW (2010) The current status, development trend and strategy of the bio-pesticide industry in China. Hubei Agric Sci 49:2287–2290

Clarkson J, Payne T, Mead A, Whipps J (2002) Selection of fungal biological control agents of Sclerotium cepivorum for control of white rot by sclerotial degradation in a UK soil. Plant Pathol 51(6):735–745

Cooke RC, Godfrey BES (1964) A key to the nematode-destroying fungi. Trans Br Mycol Soc 47 (1):61–74

de Freitas Soares FE, Sufiate BL, de Queiroz JH (2018) Nematophagous fungi: far beyond the endoparasite, predator and ovicidal groups. Agric Nat Resour 52(1):1–8

Devi PV, Ravinder T, Jaidev C (2005) Cost-effective production of Bacillus thuringiensis by solid state fermentation. J Invertebr Pathol 88(2):163–168

Dong H, Zhou XG, Wang J, Xu Y, Lu P (2015) Myrothecium verrucaria strain X-16, a novel parasitic fungus to Meloidogyne hapla. Biol Control 83:7–12

dos Santos MA, Ferraz S (2000) Detection, isolation and maintenance of some endoparasitic nematophagous fungi and in vitro evaluation of their pathogenicity. Nematologia Brasileira 24(2):183–191

Falbo MK, Soccol VT, Sandini IE, Vicente VA, Robl D, Soccol CR (2013) Isolation and characterization of the nematophagous fungus Arthrobotrys conoides. Parasitol Res 112 (1):177–185

Frederiks C, Wesseler JH (2019) A comparison of the EU and US regulatory frameworks for the active substance registration of microbial biological control agents. Pest Manag Sci 75 (1):87–103

Glare TR, Moran-Diez ME (2016) What are microbial-based biopesticides. In: Glare TR, Moran-Diez ME (eds) Microbial-based biopesticides: methods and protocols. Humana Press, New Jersey, pp 1–9

Jiang X, Xiang M, Liu X (2017) Nematode-trapping fungi. In: Heitman J, Howlett BJ, Crous PW, Stukenbrock EH, James TY, Gow NAR (eds) The fungal kingdom. Wiley, Hoboken, pp 963–974

Kala S, Sogan N, Agarwal A, Naik SN, Patanjali PK, Kumar J (2020) Biopesticides: formulations and delivery techniques. In: Egbuna C, Sawicka B (eds) Natural remedies for pest, disease and weed control, natural remedies for pest, disease and weed control. Academic Press, Amsterdam, pp 209–220

Knowles A (2008) Recent developments of safer formulations of agrochemicals. Environmentalist 28(1):35–44

Köhl J, Postma J, Nicot P, Ruocco M, Blum B (2011) Stepwise screening of microorganisms for commercial use in biological control of plant-pathogenic fungi and bacteria. Biol Control 57 (1):1–12

Köhl J, Kolnaar R, Ravensberg W (2019) Mode of action of microbial biological control agents against plant diseases: relevance beyond efficacy. Front Plant Sci 10:845

Kumar V, Khan MR, Walia RK (2020) Crop loss estimations due to plant-parasitic nematodes in major crops in India. Natl Acad Sci Lett 43(5):409–412
Leng P, Zhang Z, Pan G, Zhao M (2011) Applications and development trends in biopesticides. Afr J Biotechnol 10(86):19864–19873
Li G, Zhang K, Xu J, Dong J, Liu Y (2007) Nematicidal substances from fungi. Recent Pat Biotechnol 1(3):212–233
Linford MB, Oliveira JM (1937) The feeding of hollow-spear nematodes on other nematodes. Science 85(2203):295–297
Lopez-Llorca LV (1990) Purification and properties of extracellular proteases produced by the nematophagous fungus Verticillium suchlasporium. Can J Microbiol 36(8):530–537
Lopez-Llorca LV, Jansson HB, Vicente JGM, Salinas J (2006) Nematophagous fungi as root endophytes. In: Schulz BJE, Boyle CJC, Sieber TN (eds) Microbial root endophytes, Soil biology, vol 9. Springer, Berlin, pp 191–206
Martinelli PRP (2008) Studies on biological control of the citrus nematode in the State of Sao Paulo. Dissertation, College of Agricultural Sciences and Veterinary, State University of Sao Paulo, Brazil
Martinelli PRP, Santos JM, Barbosa JC (2012) Efficacy of formulations containing five nematophagous fungi for the management of Pratylenchus jaehni in Citrus. Nematologia Brasileira 36:1–8
Martinelli PRP, Soares PLM, dos Santos JM, da Silveira AJ (2015) Nematophagous fungi: formulation, mass production and application technology. In: Askary TH, Martinelli PRP (eds) Biocontrol agents of phytonematodes. CABI, Wallingford, p 175
Mishra A, Arshi A, Mishra SP, Bala M (2019) Microbe-based biopesticide formulation: a tool for crop protection and sustainable agriculture development. In: Arora P (ed) Microbial technology for the welfare of society, microorganisms for sustainability, vol 17. Springer, Singapore, pp 125–145
Montesinos E (2003) Development, registration and commercialization of microbial pesticides for plant protection. Int Microbiol 6(4):245–252
Moosavi MR, Askary TH (2015) Nematophagous fungi-commercialization. In: Askary TH, Martinelli PRP (eds) Biocontrol agents of phytonematodes. CABI, Wallingford, pp 187–202
Moosavi MR, Zare R, Zamanizadeh HR, Fatemy S (2011) Pathogenicity of *Verticillium epiphytum* isolates against *Meloidogyne javanica*. Int J Pest Manag 57:291–297
Nordbring-Hertz B (1977) Nematode-induced morphogenesis in the predacious fungus Arthrobotrys oligospora. Nematologica 23(4):443–451
Nordbring-Hertz B, Jansson HB, Tunlid A (2001) Nematophagous fungi. In: Encylopedia of life sciences. Published on 2001 May 30. https://doi.org/10.1038/npg.els.0004293
Philip J (2002) Nematophagous fungi: guide by Philip Jacobs, BRIC-Version online. www.biological-research.com. Accessed Jun 2008
Powell KA, Jutsum AR (1993) Technical and commercial aspects of biocontrol products. Pestic Sci 37(4):315–321
Ravensberg W (2011) The development of microbial pest control products for control of arthropods: a critical evaluation and a roadmap to success. Dissertation, Wageningen University
Schisler D, Slininger P (1997) Microbial selection strategies that enhance the likelihood of developing commercial biological control products. J Ind Microbiol Biotechnol 19(3):172–179
Singh R, Arora NK (2016) Bacterial formulations and delivery systems against pests in sustainable agro-food production. Food Sci 1:1–11
Singh R, Kumar U (2015) Assessment of nematode distribution and yield losses in vegetable crops of Western Uttar Pradesh in India. Int J Sci Res 4(5):2812–2816
Soares PLM (2006) Estudos do controle biologico de fitonematoides com fungos nematofagos. Tese de Doutorado, Faculdade de Ciencias Agrárias e Veterinarias. Universidade Estadual Paulista (FCAV-UNESP), Jaboticabal, 217 pp
Spurr HW (1985) Bioassays: critical to biocontrol of plant disease. J Agric Entomol 2(1):117–122

St Leger RJ, Screen S (2001) Prospects for strain improvement of fungal pathogens of insects and weeds. In: Butt TM, Jackson CW, Magan N (eds) Fungi as biocontrol agents. CABI, Wallingford, pp 219–237

Tuininga AR, Miller JL, Morath SU, Daniels TJ, Falco RC, Marchese M, Sahabi S, Rosa D, Stafford KC (2014) Isolation of entomopathogenic fungi from soils and Ixodes scapularis (Acari: Ixodidae) ticks: prevalence and methods. J Med Entomol 46(3):557–565

Tunlid A, Johansson T, Nordbring-Hertz B (1991) Surface polymers of the nematode-trapping fungus Arthrobotrys oligospora. Microbiology 137(6):1231–1240

US Environmental Protection Agency, Regulating Pesticides (2008) What are biopesticides? http://www.epa.gov/pesticides/biopesticides/whatarebiopesticides.html. Accessed 25 Jan 2021

Velvis H, Kamp P (1996) Suppression of potato cyst nematode root penetration by the endoparasitic nematophagous fungus Hirsutella rhossiliensis. Eur J Plant Pathol 102:115–122

Verhaar MA, Hijwegen T, Zadoks JC (1998) Selection of Verticillium lecanii isolates with high potential for biocontrol of cucumber powdery mildew by means of components analysis at different humidity regimes. Biocontrol Sci Tech 8(4):465–477

Yang J, Tian B, Liang L, Zhang KQ (2007) Extracellular enzymes and the pathogenesis of nematophagous fungi. Appl Microbiol Biotechnol 75(1):21–31

Zachow C, Pirker H, Westendorf C, Tilcher R, Berg G (2009) The Caenorhabditis elegans assay: a tool to evaluate the pathogenic potential of bacterial biocontrol agents. Eur J Plant Pathol 125 (3):367–376

Chapter 11
Microbial Consortia and Their Application for the Development of a Sustainable Environment

Sneha Trivedi, Naresh Butani, Helina Patel, Manoj Nath, and Deepesh Bhatt

Abstract Abundant diversity of microbial communities existing in nature follows the rule of coexistence demonstrating interspecies and interkingdom interactions for their proliferation. These communities exhibit highly complex and interwoven web of metabolic communications. They also perform division of labor managing their metabolic load for the survival of different co-inhabitants of that area. These microbial consortia are involved in various interactions such as mutualism and syntropy, commensalism, and antagonistic behavior as a part of their struggle for existence. Due to their unique properties of division of labor, diversified metabolic capacities, and ability of co-survival, microbial consortia can be exploited for approaches applied for maintaining sustainability of the environment. They can be utilized in agricultural practices that can reduce the use of chemical fertilizers and pesticides. Microbial mixed cultures also have potential for waste management and bioremediation. Additionally, they can prove to be an attractive alternative as a source of biofuel and bioenergy generation.

Keywords Arbuscular mycorrhizal fungi (AMF) · Microbial consortia · Microbial interactions · Plant growth-promoting rhizobacteria (PGPR) · Soil ecosystem

S. Trivedi · N. Butani · H. Patel
Department of Microbiology, Shree Ramkrishna Institute of Computer Education and Applied Sciences, Sarvajanik University, Surat, Gujarat, India

M. Nath
ICAR-Directorate of Mushroom Research, Solan, Himachal Pradesh, India

D. Bhatt (✉)
Department of Biotechnology, Shree Ramkrishna Institute of Computer Education and Applied Sciences, Sarvajanik University, Surat, Gujarat, India

P. Bhatt et al. (eds.), *Microbial Technology for Sustainable Environment*,
https://doi.org/10.1007/978-981-16-3840-4_11

11.1 Introduction

Due to enormous human activities which have caused massive impact on environment, its sustainability has become a matter of great concern. Increased use of chemical fertilizers and pesticides has created questions regarding sustainability of soil ecosystems. Increasing efforts are being made to develop such processes which can lead to sustainable agriculture practices. Along with that, there is an increasing awareness about solving pollution issues through sustainable approach. Additionally, bioproduction sector also has their own problems regarding the sustainability of processes, which can cause economic load on the industries. Microbial consortia can prove to be an attractive alternative to solve these issues.

Microorganisms are the most diverse forms of living beings that thrive in any ecosystem ranging from extreme environments to water and soil. In all these ecosystems, different microbial communities flourish simultaneously creating complex web of biochemical interactions that function as a key to ecological architecture. Enormous diversified species in large number forms complex interactive webs of metabolic activities in natural ecosystems such as soil, gastrointestinal system, etc. (Shahab et al. 2020). Hence, it leads to a fact that "Mother Nature" has designed the microbial existence as a large consortium.

Microbial consortia refers to the coexistence or co-cultivation of two or more species in terms of growth, metabolism, and division of labor, creating conditions for their own composition (Jia et al. 2016). Even though, in recent times, microbial consortia have attracted the attention of bioprocess engineers by offering certain advantages, it was also used in the ancient eras, from wineries in Africa to biogas generation from undefined microbial mixtures used for heating bath in Assyria (Peng et al. 2016). These multiple microbes together possess the capacity to perform difficult metabolic tasks and provide a miraculous outcome. It is advantageous over single species-operated bioprocesses in certain aspects. Firstly, combining suitable species can fulfil complex tasks without much difficulty. Secondly, co-adaptability of multiple strains leads to co-stability and can resists environmental fluctuations. This is because of their internal beneficial interactions (Nuti and Geovannetti 2015). Additionally, metabolic requirements of different species can be fulfilled by the fellow microbial members in the consortia. This occurs when a metabolite produced by one member of consortia is utilized as limiting nutrient for other species, thereby manifesting complete mineralization of the components along with favoring their coexistence. Also, through gene regulation via quorum sensing, these microbial consortia demonstrate multiple unexplored biochemical mechanisms that contribute in their functionality (Odoh et al. 2020).

There is an enormous array of mechanisms that occurs continuously among the members of the consortia which ranges from highly beneficial or synergistic to extreme damaging or antagonistic. These interactions occur as a part of processes microbes develop while thriving for their existence in the natural environment, while dealing with space limitations, nutrient depletion, and other natural or co-inhabitant generated stressful environment. Hence, these interactive mechanisms can be studied and exploited as a tool for the search of newer metabolites, novel enzymes, and other components. Furthermore, study of these interactions can provide an insight into gene regulation which assists in altering the production yields of certain components at the industrial level.

Microbial consortia can be categorized broadly in two types: (1) natural communities and (2) man-made co-cultures (Canon et al. 2020). The former types are present in the natural ecosystem and are known to establish a spontaneous association where the dynamics are governed by the laws of natural selection. They are substantially complex, and the species flourishes according to the release of metabolites by one member which can be subsequently utilized as substrate by other members of the community. Natural communities are employed in number of processes such as methane production in activated sludge process, fermentation such as cheese or kimchi production, as well as waste management (Canon et al. 2020). The second type of microbial consortia includes artificially mixed cultures as well as synthetic co-cultures that are developed by human conciliation. They are customized mixtures framed according to the process requirements. Hence, it can be deciphered that these mixtures of microorganisms hold enormous applications in variety of fields which include agriculture (Sekar et al. 2016), bioproduction (Canon et al. 2020), bioremediation, and even in waste management (Al-Dhabi et al. 2019). In addition, the potentiality of these microbial consortia had recently opened up the gates for a newer aspect of synthetic biology that deals with designing and analyzing synthetic microbial consortia, having applicability in the field of bioprocessing. They are also added as a part of ingredients which are supplied as plant bio-stimulants (Woo and Pepe 2018).

Traditional bioprocesses are focused on exploiting the potentiality of monoculture for fermentation; however, when it comes to complex biotransformations, the mentioned approach faces some of the drawbacks, namely, extensive metabolic burden, cellular space limitation, and toxic intermediates (Shahab et al. 2020). This has diverted microbial engineers toward exploring the possibilities of utilizing microbial consortia to get the job done.

11.2 Microbial Interactions

Cellular interactions among microbial species are ubiquitous in nature and are prime sculptors of ecosystem dynamics (Kong et al. 2018). Interaction between members of microbial consortia, either natural or artificially designed, is a pivotal factor that plays a decisive role in the success of any bioprocess. Furthermore, emergence and sustainability of metabolic interactions depend on factors such as nutrient availability, diffusion constraints, and microbial community spatial structure and structure. In their habitat, even though microbes tend to demonstrate individualism, the consortia respond to environmental stress as a unique organism (Odoh et al. 2020). Positive interactions are mutualistic and improve the fitness among co-cultures and favor product formation (Canon et al. 2020). When microbial species flourish in their natural habitat, they may have to overcome through negative interactions such as competition, parasitism, etc. However, when polycultures are applied purposefully, the designing is done to selectively enhance beneficial interactions, excluding the members involved in damaging mechanisms (Fig. 11.1).

11.2.1 Mutualism

Established as one of the most common interactions found in biosphere, mutualism refers to the interaction in which different species are dependent on each other for fulfilling their nutritional requirements or need for space (Kouzuma et al. 2015). Also termed as symbiotic relationship, such interactions are widespread in nature. The word "symbiose" was coined by De Bary in 1879, for dissimilar organisms that are living together, and his definition covered both mutualists and parasites (Hirsch 2004).

In agriculture, microbial consortia use symbiotic interactions which can contribute for enhancing soil fertility. In soil ecosystem, where bacteria accounts for 95% of the microbial community (Odoh et al. 2020), multiple microbial species colonizes around the rhizosphere either as free living, epiphytes, or endophytes. They establish close interactions with root system which may prove beneficial to plants and microbes both. The signature example of such interactions can be understood in terms of rhizobium-root legume relationship, where rhizobium is involved in nitrogen fixation benefiting plants and in return uses inhabiting space and biochemical metabolites released by plant roots as carbon and energy source. Additionally, there are members of plant growth-promoting rhizobacteria (PGPR) community that is involved in such co-relations (Khan et al. 2020). Members of genera *Bacillus*, *Paenibacillus*, *Burkholderia*, *Azotobacter*, *Rhizobium*, and *Pseudomonas* (Ahmad et al. 2018) inhabit plant rhizosphere where they indulge in diversified mechanism which benefits the plants by solubilization of minerals, production of phytohormones, releasing antagonistic substances that inhibit the growth of pathogenic organisms and many other. In return, they reap the benefits in the form of exudates released by plant roots as their nutrient and energy source (Menéndez and Paço 2020), have termed beneficial bacteria of rhizosphere as plant probiotic bacteria (PPB), and have classified them into two subgroups: (1) rhizobial bacterial endophytes (RBEs) and (2) non-rhizobial bacterial endophytes (NRBEs). Their studies suggested that those symbiotic interactions between plants and microbes are highly complex, and microbial consortia in the form of PPBs have ensured good results in terms of plant benefits. They have also stated improvement in symbiotic effectiveness between RBEs and NRBEs in the form of consortia along with mycorrhiza increased crop yields and reduced the usage of chemical fertilizers.

Another popular mutualism is demonstrated by mycorrhizal fungal species. They are named as arbuscular mycorrhizal fungi (AMF) and belong to subphyla Glomeromycotina. Under symbiotic association with plants, they acquire lipids and carbohydrates from the plants through special structures such as arbuscules

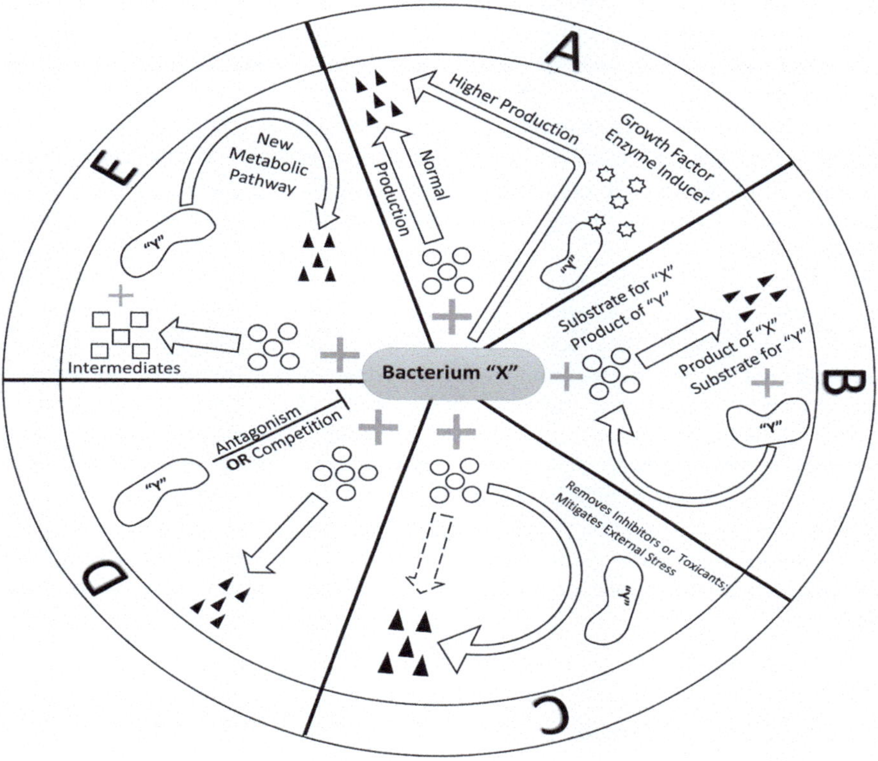

Bacterium "X"	"Y"			
The Bacterium present in consortium	Another bacterium from the consortium	Substrate (For Figure part A,C,D,E)	Intermediate	Product (For Figure part A,C,D,E)

Fig. 11.1 Interactive mechanisms observed among the members of microbial consortia: The figure deciphers the release of growth factors from one of the members can help in better utilization of the substrate by other member (Section A). Other possibility is of synergistic relationship where both the species are interdependent for their substrate requirements as shown in section B of the figure. Section C explains the favors extended from one member through removal of inhibitors of the reaction, thereby making substrate utilization possible for co-member. Antagonism is also a major interaction observed in microbial communities in which competition for resources can inhibit other organism or else this can divert the organism to produce novel metabolites for the purpose of their survival (Section D). Also, biotransformations can occur as a result of interactions in which the intermediate produced by one microbe can act as starting metabolite for fellow member as shown in Section E

and vesicles. As a return gift, these fungal species are involved in rapid uptake of plant nutrients mainly phosphorous, nitrogen, and potassium, thereby increasing crop yields. Mixture of such species provides better outputs compared to single mycorrhizal species (Paredes et al. 2020).

In addition to the above examples, one more popularized example of symbionts is lichens. They can be considered as tripartite or quadripartite symbionts which comprise of photosynthetic organisms as cyanobacteria or algae, fungus as another member, and a nitrogen-fixing root symbiont which also includes mycorrhiza (Hirsch 2004).

Kouzuma et al. (2015) have reported findings on specific type of mutualism known as "syntrophy," which is referred to as the interspecies interactions in which tropic benefits are procured by both the partners. They have studied the establishment of interspecies communication in syntrophic microbial consortium related with methanogenesis and stated that the rate-limiting step of methanogenesis is the mutualistic interactions between secondary fermentative bacteria and hydrogenotrophic methanogens in which intercellular exchange of reducing equivalents occurs. In this metabolic event, H_2 is generated by propionate-oxidizing bacteria as a part of energy generation through oxidation of propionic acid. Hydrogenotrophic methanogens trap this hydrogen for methane production thereby reducing the product concentration, favoring the reaction to move in forward direction.

11.2.2 Commensalism

Commensalism is an interaction between microbial communities in which one member termed as commensal can have substances released or captured by the host due to special proximity (Bogitsh et al. 2018). In this relationship, host remains unaffected and releases the exudates as a part of its normal metabolic activities. Microbial communities in the biofilms may be governed by such interactions. Use of high-end technologies such as fluorescent in situ hybridization (FISH) and microelectrodes applications has demonstrated the clustering of nitrate-oxidizing bacteria around ammonia-oxidizing bacteria in nitrite-oxidizing zones in the waste water treatment where nitrifying biofilm is involved (Christensen et al. 2002). Hirsch (2004) has suggested that commensalism has resulted in increased bacterial survival in rhizosphere as well as bulk soil.

Kong et al. (2018) demonstrated ecosystem dynamics of commensalism using synthetically designed two strain consortia. They created one strain CmA from *L. lactis* which had full nisin pathway and constitutively expressed tetracycline resistance. It was co-cultured with CmB which contained nisin inducible tetracycline resistance power. Their studies determined that when two strains were grown as mono-cultures in the presence of tetracycline, CmA, because of its constitutive expression of tetR gene, was able to grow; in contrast, CmB could not grow by itself as its resistance was not self-dependent. However, when both the strains were combined, CmB grew well, suggesting that the presence of CmA benefited CmB for its growth. Studies of detailed mechanism conferred that tetracycline resistance was the mechanism of commensalism for CmB which was achieved using nisin released by CmA as a part of its normal metabolism.

Hirsch (2004) has suggested that microbial interactions such as mutualism, commensalism, and parasitism should be considered in continuum. Mutualists like rhizobium species initially flourish in the soil or plant roots as commensals and can form biofilms on inanimate surfaces. This may be for the survival of the bacterium until the suitable host is found.

11.2.3 Antagonism

Antagonism is the mechanism of causing harm to other organisms primarily for the purpose of self-survival or as a product of normal metabolism. It has a substantial contribution for framing the bacterial community by inhibiting one community while opening the doors for other bacterial niches (Long et al. 2013). Competitiveness between different species in the consortia usually occurs due to nutrient depletion and space requirements which result in the synthesis of secondary metabolites, synthesis of enzymes that can combat the situation or activation, and expression of silent genes present in the genome of the species. Under such situations, microbial residents of the same niche excrete such substances that can cause cell damage or oxidative stress in the surroundings. The target species may generate substances that can combat those stressful situations as a part of microbial defense mechanism (Zhang et al. 2018). Hence forth, these mechanisms can provide enormous potential in search of newer secondary metabolites or study of gene expression and regulation or improvement of product yields. It can also be favorable for combating plant pathogens in agricultural practices.

Maiyappan et al. (2010) have studied the effect of microbial consortia Omega on the growth promotion of black gram and found its antagonistic effect on pathogenic fungi like *M. phaseolina*, *S. rolfsii*, *F. oxysporum*, and *R. solani*. Extracellular antagonistic substances secreted by bacteria in combination with plant exudates may be responsible for keeping pathogens in check. Long et al. (2013) conferred the existence of antagonistic interactions between self-sustaining microbial communities in the microbial mats of hypersaline lake and suggested that antagonism which is driven by chemicals is a major contributor in designing of bacterial communities. They supported their hypothesis by co-culturing assay and determined that 57% of bacterial isolates expressed damaging behavior toward one or more species, while 5% of the isolates showed antagonism toward 80% of the isolates.

11.3 Applications of Microbial Consortia

There are number of important areas where unique potential of microbial consortia is exploited for various aspects. Majorly microbial consortia are used for sustainable agricultural practices (Al-Dhabi et al. 2019; Paredes et al. 2020; Maiyappan et al. 2010), bioproduction processes (Canon et al. 2020; Sabra et al. 2010; Ghosh et al. 2016), and bioremediation strategies (Paerl and Pinckney 1996). Furthermore, they are now being explored as a tool to unravel the secrets of coexistence in microbial ecology as well as synthetic biology (Haruta and Yamamoto 2018).

11.3.1 Sustainable Agriculture

In the wave of urbanization, there has been a drastic increase in the usage of chemical fertilizers and chemical pesticides that has raised serious environmental concerns. Also, many of these components have entered the food chain causing several health issues in the people consuming such food. Overall, such harmful chemicals have entered the water bodies and deteriorated soil quality thereby impacting the biodiversity of that particular habitat. Microbial consortia can prove a remedy to such problem providing an eco-friendly alternative (Sekar et al. 2016; Paredes et al. 2020).

Plant growth depends on the availability of various nutrients such as nitrogen, phosphorus, and potassium available in solubilized form as well as growth-promoting hormones, namely, indole acetic acid, gibberellic acid, and others. One of the prime characteristics of microbial consortia is to act synergistically among each other and to secret the components that support plant growth promotion. Plant growth-promoting microorganisms are heterogenous flora encompassing diverse species including bacteria, actinomycetes, fungi, etc., demonstrating cross kingdom interactions. Additionally, those pool of microbes provide immunity to plants by keeping phytopathogenic organisms at bay through various antagonistic mechanisms.

Maiyappan et al. (2010) have isolated, identified, and evaluated the plant growth-promoting efficiency of nine stains of microorganisms that were the members of *Bacillus* sp., *Streptomyces* sp., *Azotobacter* sp., and *Fruateria* sp., developed a consortia Omega, and studied its effect of plant growth of black gram (*Vigna Mungo* L.). Pot culture studies demonstrated significant increase in shoot length of the plant compared to chemical fertilizer-added plants and control plants. Also, the root volume as well as total dry mass was higher in the consortium-fed plants compared to chemical fertilizers.

The potentiality of *Streptomyces* sp. associated with tomato plant roots and the effect of microbial consortia developed from composting the waste were studied by Al-Dhabi et al. (2019). They stated that use of such approach increased the tomato weight up to 15%. The presence of phytohormone such as indole acetic acid and siderophores was also analyzed.

Sekar et al. (2016), Woo and Pepe (2018), and Vishwakarma et al. (2020) have summarized outcomes of various studies carried out in the field of analyzing effects of PGPR and microbial consortia on plant growth and agricultural practices. They have stated that co-inoculation of rhizobia with other microbial species and PGPR proved to be favorable for nodulation as well as below-ground and above-ground development of plant species. Furthermore, such consortia were able to reduce the occurrence of as well as severity of certain plant diseases. Inoculation of AMF (arbuscular mycorrhizal fungi) along with PGPR increased plant yields both in nursery and field studies. They have also suggested that application of microbial cocultures can provide a solution to alleviate plant abiotic stress and enable more growth and productivity of plants under stressful environment such as high salinity, dryness, and rising temperature. This approach can improve the deteriorated soil quality thereby increasing crop production that can combat food shortage in the future.

However, the influence of management practices in agriculture decides the direction of the crop productivity as plants are major selectors of the specific taxa in their phyllosphere. In addition to microbial inoculants, practices such as organic farming, crop rotation, intercropping, automation, and technology implementation are key co-factors that determine the sustainability of agriculture.

11.3.2 Waste Treatment and Pollution Control

A major challenge the world is facing in the present time is rotten fruits of industrialization, in the form of pollutants that has accumulated in soil, water, and air. This is causing serious threat to the sustainability of environment along with risking human health. The processes used currently for reducing toxic compounds involves burning, recycling, land-filling, and pyrolysis. However, such processes are releasing even more toxic and non-degradable compounds in the environment alongside creating difficulty in execution for environmental decontamination (Ahmad et al. 2018).

Microbial communities as weapons of green technology are more efficient in metabolizing chemically complex and toxic metabolites compared to single species. The explanation of this potential of co-cultures lies in the enormous interactions occurring among the multiple species in which metabolite released by one species can be utilized by other co-inhabitant of the community. Hence, this capability of microbial mixed cultures can be utilized in bioremediation of complex polluted sites such as dye polluted water, heavy metal polluted sites, oil spilled sites, and many others (Ahmad et al. 2018; Ghosh et al. 2016). Furthermore, artificial designing of ecological niche is a newer concept emerging in this area where resource specific niche can be sculptured, thereby achieving desirable biotransformations of the target substrates (Shahab et al. 2020).

11.3.2.1 Dyes

Dyes are present in industrial effluents discharged from textile industries. Dyes comprise of recalcitrant compounds, mainly azo dyes that are hard to degrade due to their strong bonding capacity, causing serious concerns. There are certain bacterial species that produces enzymes such as azo-reductase that can act upon azo dyes. Some species produce enzymes that can degrade the intermediate components formed during azo dye reduction reaction. There are number of reports and studies proving potential of different microbial species including fungi, algae, and bacteria for dye degradation and other such xenobiotics excreted in the effluents of paper pulp industries (Ahmad et al. 2018).

11.3.2.2 Organic Domestic Wastes

Domestic waste generated in households also covers a major section of the waste generation. The usual procedure applied for such waste disposal including dumping on landfills which can contaminate the soil of that site in a long run. Household-generated waste is a rich source of organic nutrients, and if treated with microorganisms, they can be transformed into biofertilizers and source of biofuels or animal feed. Microbial communities have physiological diversity that can be manipulated and designed for the customized treatment of such organic wastes which can reduce the treatment time and provide with desired output in waste management.

Sarkar et al. (2011) have developed eleven different consortia which were analyzed for their capacity to produce suitable cocktail of hydrolytic enzymes, namely, amylase, lipase, protease, and cellulase, that can breakdown those components present in kitchen wastes. In their studies, they found effective reduction of kitchen waste in reduced time span, that too without the generation of foul smell.

11.3.3 Bioenergy Generation

Biofuels and bioenergy resources such as bioethanol and biogas are equally important options for sustainable energy generation alongside the renewable energy sources like wind energy, hydro energy, and solar energy. These biological processes depend on the biochemical activities performed by diverse species of microorganisms. As lignocellulosic wastes are major substrate used for biofuel production, monoculture of microbes cannot degrade them efficiently. A single bioreactor equipped with correct combination of microbes can convert complex feedstocks to simpler sugars that can be utilized for the production of bioenergy. For example, when pair of *Bacillus* and *Clostridium* sp. were co-cultured in a bioreactor on lignocellulosic wastes, *Bacillus* sp. produced hydrolytic enzymes that acted upon complex polymeric substrate and released simple sugars which were further utilized by *Clostridium* sp. for efficient generation of hydrogen (Ghosh et al. 2016).

Methanogenic microbes demonstrate a type of mutualistic relationship termed as syntropy, where reducing equivalents such as formate and hydrogen are transferred between syntrophic partners. Microbial species involved in such communication have developed mechanisms at molecular level for the establishment of interspecies partnership. There are evidences that suggest that these transfers can occur in the form of electric current which can be exploited for bioenergy generation (Kouzuma et al. 2015).

11.4 Future Aspects

Deciphering the mysterious potentials of microbial consortia has opened new avenues for its applications in the areas of applied microbiology. Tailor-made designing of microbial co-cultures has enabled the bioprocesses to achieve custom-made desirability; hence continuous attempts are carried out for upgradation of such systems for achieving increased product yields. Engineering of microbial consortia has lot a greater number of future prospects in environmental restoration, in improvement of soil and water quality, as well as in biosystem analysis and industrial biotechnology. Synthetic microbial consortia can be applied in unraveling the intricate mechanisms involved in gene circuits and quorum sensing. Microbial consortia engineering and in silico approach hold huge potential that can reframe ecological niche and can provide novel biomolecules for better environmental sustainability and ecosystem management.

References

Ahmad M, Pataczek L, Hilger TH, Zahir ZA, Hussain A, Rasche F, Solberg SØ (2018) Perspectives of microbial inoculation for sustainable development and environmental management. Front Microbiol 9:2992

Al-Dhabi NA, Esmail GA, Mohammed Ghilan AK, Valan Arasu M (2019) Composting of vegetable waste using microbial consortium and biocontrol efficacy of Streptomyces Sp. Al-Dhabi 30 isolated from the Saudi Arabian environment for sustainable agriculture. Sustainability 11(23):6845

Bogitsh BJ, Carter CE, Oeltmann TN (2018) Human parasitology. Academic, Amsterdam

Canon F, Nidelet T, Guédon E, Thierry A, Gagnaire V (2020) Understanding the mechanisms of positive microbial interactions that benefit lactic acid bacteria co-cultures. Front Microbiol 11:2088

Christensen BB, Haagensen JA, Heydorn A, Molin S (2002) Metabolic commensalism and competition in a two-species microbial consortium. Appl Environ Microbiol 68(5):2495–2502

Ghosh S, Chowdhury R, Bhattacharya P (2016) Mixed consortia in bioprocesses: role of microbial interactions. Appl Microbiol Biotechnol 100(10):4283–4295

Haruta S, Yamamoto K (2018) Model microbial consortia as tools for understanding complex microbial communities. Curr Genomics 19(8):723–733

Hirsch AM (2004) Plant-microbe symbioses: a continuum from commensalism to parasitism. Symbiosis 37(1–3):345–363

Jia X, Liu C, Song H, Ding M, Du J, Ma Q, Yuan Y (2016) Design, analysis and application of synthetic microbial consortia. Synth Syst Biotechnol 1(2):109–117

Khan RA, Singh AV, Upadhyay VK, Prasad B (2020) Development of potential microbial consortia and their assessment on wheat (*Triticum aestivum*) seed germination. Environ Ecol 38(1):6–16

Kong W, Meldgin DR, Collins JJ, Lu T (2018) Designing microbial consortia with defined social interactions. Nat Chem Biol 14(8):821–829

Kouzuma A, Kato S, Watanabe K (2015) Microbial interspecies interactions: recent findings in syntrophic consortia. Front Microbiol 6:477

Long RA, Eveillard D, Franco SL, Reeves E, Pinckney JL (2013) Antagonistic interactions between heterotrophic bacteria as a potential regulator of community structure of hypersaline microbial mats. FEMS Microbiol Ecol 83(1):74–81

Maiyappan S, Amalraj ELD, Santhosh A, Peter AJ (2010) Isolation, evaluation and formulation of selected microbial consortia for sustainable agriculture. J Biofertil Biopestic 2(109):2

Menéndez E, Paço A (2020) Is the application of plant probiotic bacterial consortia always beneficial for plants? Exploring synergies between rhizobial and non-rhizobial bacteria and their effects on agro-economically valuable crops. Life 10(3):24

Nuti M, Geovannetti G (2015) Borderline products between biofertilizers/bio-effectors and plant protectants: the role of microbial consortia. J Agric Sci Technol 5:305–315

Odoh CK, Sam K, Zabbey N, Eze CN, Nwankwegu AS, Laku C, Dumpe BB (2020) Microbial consortium as biofertilizers for crops growing under the extreme habitats. In: Plant microbiomes for sustainable agriculture. Springer, Cham, pp 381–424

Paerl HW, Pinckney JL (1996) A mini-review of microbial consortia: their roles in aquatic production and biogeochemical cycling. Microb Ecol 31(3):225–247

Paredes AA, Valdes G, Nuti M (2020) Ecosystem functions of microbial consortia in sustainable agriculture. Agronomy 10:1902

Peng XN, Gilmore SP, O'Malley MA (2016) Microbial communities for bioprocessing: lessons learned from nature. Curr Opin Chem Eng 14:103–109

Sabra W, Dietz D, Tjahjasari D, Zeng AP (2010) Biosystems analysis and engineering of microbial consortia for industrial biotechnology. Eng Life Sci 10(5):407–421

Sarkar P, Meghvanshi M, Singh R (2011) Microbial consortium: a new approach in effective degradation of organic kitchen wastes. Int J Environ Sci Dev 2(3):170

Sekar J, Raj R, Prabavathy VR (2016) Microbial consortial products for sustainable agriculture: commercialization and regulatory issues in India. In: Agriculturally important microorganisms. Springer, Singapore, pp 107–132

Shahab RL, Brethauer S, Luterbacher JS, Studer MH (2020) Engineering of ecological niches to create stable artificial consortia for complex biotransformations. Curr Opin Biotechnol 62:129–136

Vishwakarma K, Kumar N, Shandilya C, Mohapatra S, Bhayana S, Varma A (2020) Revisiting plant–microbe interactions and microbial consortia application for enhancing sustainable agriculture: a review. Front Microbiol 11:3195

Woo SL, Pepe O (2018) Microbial consortia: promising probiotics as plant biostimulants for sustainable agriculture. Front Plant Sci 9:1801

Zhang S, Merino N, Okamoto A, Gedalanga P (2018) Interkingdom microbial consortia mechanisms to guide biotechnological applications. Microb Biotechnol 11(5):833–847

Chapter 12
Microbial Engineering and Applications for the Development of Value-Added Products

Ashutosh Paliwal, Abhishek Verma, Ashwini Kumar Nigam, Jalaj Kumar Gour, Manoj Kumar Singh, and Rohit Kumar

Abstract Downstream is a very affluent process for fermentation. It usually involves complicated equipment and processes to obtain desired chemicals or materials from intra- and/or extracellular spaces of microorganisms. Recently, it becomes possible to simplify the microbial cell separation processes by morphologically engineering the shapes of small microorganisms. Biologically engineered entities have enabled discoveries in the past decade and a half, spanning from novel routes for the syntheses of drugs and value-added products to carbon capture. The precise cellular reprogramming has extended to the production of nanomaterials owing to their ever-growing demand. Additionally, nutraceuticals are important natural bioactive compounds that confer health-promoting and medical benefits to humans. Globally, growing demands for value-added nutraceuticals for prevention and treatment of human diseases have rendered nutraceuticals a multi-billion dollar market. However, supply limitations and extraction difficulties from natural sources such as plants, animals, or fungi restrict the large-scale use of nutraceuticals. Metabolic engineering via microbial production platforms has been advanced as an eco-friendly alternative approach for production of value-added nutraceuticals from simple carbon sources. Microbial platforms like the most widely used

---

A. Paliwal · A. Verma
Department of Biotechnology, Kumaun University, Sir J. C. Bose Technical Campus, Bhimtal, Uttarakhand, India

A. K. Nigam
Department of Zoology, Udai Pratap College, Varanasi, Uttar Pradesh, India

J. K. Gour
Department of Biochemistry, Faculty of Science, University of Allahabad, Prayagraj, Uttar Pradesh, India

M. K. Singh (✉)
National Centre for Disease Control, New Delhi, India

R. Kumar
Cell and Tumor Biology Group, Advanced Centre for Treatment Research & Education in Cancer, Tata Memorial Centre, Navi Mumbai, Maharashtra, India

P. Bhatt et al. (eds.), *Microbial Technology for Sustainable Environment*,
https://doi.org/10.1007/978-981-16-3840-4_12

Escherichia coli and *Saccharomyces cerevisiae* have been engineered as versatile cell factories for production of diverse and complex value-added chemicals such as phytochemicals, prebiotics, polysaccharides, and poly amino acids. This chapter highlights the recent progresses in biological production of value-added nutraceuticals via metabolic engineering approaches.

Keywords Microbial engineering · Microbes · Nutraceuticals · Phytochemicals

12.1 Introduction

Microbial engineering involves the use of biological, chemical, and engineering aspect of biotechnology that results in manipulations and development of microbes to get the desired products in different fields (Peralta-Yahya et al. 2012). The technology which is employed in microbiological systems and their derivatives to transform products used in daily need are highly beneficial for humankind (Okafor 2007). India is a country where more than half of the population is engaged in agricultural practices. India is also considered the second-largest producer of agricultural products worldwide (Gulati and Juneja 2018). Statics data reported that India produces approximately 81.285 million metric tonnes of fruits and 162.187 million tonnes of vegetables, respectively, in year 2013 (Negi 2014). Most of the production is consumed fresh; however, a larger quantity which accounts for approximately 25–40% gets rotten due to unavailability of proper postharvest facilities. This wastage causes a huge loss to crop yield and also exhibits great impact on economy. Henceforth, reducing the postharvest wastages requires utmost consideration for making a chain among consumption and supply. Microbial biotechnology has been used in handling food since ancient times such as in making bread or beverage. Though various metabolites are produced by microbes due to the introduction of modern biotechnology, microbial molecular structures possess strong potential that could be used in food industry particularly in fermentation of foods, enzymes, ingredients of food, testing of food, and postharvest administration of agricultural yields. However, microbial biotechnology in food processing division represents collection and advancement of microbes by ideas of refining regulated production, effectiveness, as well as the quality, safety, and consistency of bioprocessed foodstuffs. Microbes have a pivotal role in the production of fermented food. Microbial cultures could be genetically modified by traditional and molecular tactics.

A deliberate breakdown practice which is induced by microorganisms for the transformation of carbohydrates to alcohols or organic acids is generally referred to as fermentation (Battcock and Azam-Ali 1998). Fermentation is worldwide functional in the conservation of various raw agricultural products including vegetables, fruits, tubers and cereals, fish, milk, meat, etc. Some microorganisms accompanying the fermented foodstuffs such as *Lactobacillus* sp. are probiotic used as live microbial food complements or food constituents that help in improvement of the metabolic process of gastrointestinal tract's flora. Therefore, microbes are believed to be advantageous in maximum fermentations. Nowadays, nanotechnology is considered as innovative field of science which deals with the synthesis and use of material with nanoscale size in numerous aspect of life. However, the number of microbes is naturally proficient in generating nanoparticles either intracellularly or extracellularly while confronted with various metal salts. Accessibility of many biotechnological tools including synthetic biology, genetic, and protein engineering increases usage of microbial systems to up-skill synthesis of nanoparticles.

12.2 Nanoparticles Synthesis by Microbial Engineering

The idea of the natural synthesis of nanoparticles was first started in the 1960s and recently has seen an evolution in the last one and a half decade. Biological nanoparticle synthesis represents an extensive range of biological methods for generating nanoparticles through biotic hosts which are not restricted to bacteria, yeast, fungi, algae, and plants. Efficient synthesis of nanoparticle does require a compatible host that comprises molecular machinery to convert the raw material into the nanoparticle and can efficiently accommodate the end product, i.e. synthesized nanoparticle. Majorly few cellular proteins that have a protective role may interfere in the cellular metabolism and hinder the uptake of metallic ions and their conversion into nanoparticles with precise size and morphology. Therefore, the manipulation of such host and proteins should be achievable. The ultimate benefit of the biological synthesis of nanoparticles is their synthesis at ambient temperature and pressure, and no involvement of chemicals are required in the synthesis which could perhaps be hazardous.

Biosynthesis of nanoparticle through microbes occurs in two ways that further comprise two sub-modes, i.e. (a) intracellular in non-template or template mode and (b) extracellular in culture or membrane adherent mode. During intracellular synthesis, the cell culture is incubated with the metal salt solution where metal ions pass across the cell membrane and synthesis takes place inside the cells. Subsequently, the cells are lysed and nanoparticles are purified. On the other side, extracellular synthesis as the name suggests involves the synthesis of nanoparticle on the cell membrane or in the culture broth. Therefore, extracellularly synthesized nanoparticles are easier to retrieve and require lesser downstream processing phases. The very first report on biosynthesis of nanoparticle using genetically engineered bacteria was documented in 2006–2007 (Sambhy et al. 2006; Vigneshwaran et al. 2007). Later, Kang et al. (2008) genetically modified *E. coli* strain JM109 to express phytochelatin synthase of *Schizosaccharomyces pombe* along with improved g-glutamyl cysteine synthetase (GSHI) to synthesize cadmium sulfide (CdS) nanocrystals. The GSHI is responsible for catalysing the glutathione synthesis which is also a precursor of metal-binding peptide phytochelatin that in turn assists as capping agent for CdS nanocrystals. Phytochelatin synthesis of *S. pombe* is the best characterized natural defence mechanism towards cadmium toxicity. Further development in the above-mentioned approach was achieved in another strain, *E. coli* R189, where uniform CdS quantum dot (QDs) nanocrystals of 3–4 nm size were synthesized (Kang et al. 2008). Mi et al. (2011) expressed the transgene encoding CdS-binding histidine-rich peptide (CDS7) reported to bind with CdS (Peelle et al. 2005) to induce the formation of CdS QDs. Noble nanoparticles from silver and gold, alkali-earth (Cs, Sr), magnetic (Fe, Co, Ni, Mn), semiconducting (Cd, Se, Zn, Te) metals, as well as rare earth fluorides (Pr, Gd) were successfully synthesized using genetically engineered *E. coli*-expressing recombinant metallothionein from *Pseudomonas putida* and phytochelatin from *Arabidopsis thaliana* (Park et al. 2010; Ashraf et al. 2021). Some of the extremophiles including Antarctic bacteria have also been exploited to synthesize the fluorescent nanoparticle due to their natural resistant to cadmium and tellurides (Plaza et al. 2016). Several strains of *E. coli* have a CusCFBA silver/copper system that promotes the synthesis of silver nanoparticle in periplasmic space (Lok et al. 2008). Shi et al. (2007) also used a similar type of strain for the synthesis of silver nanoparticle in periplasmic space using anaerobic conditions. The procedure generates reduced metal nanoparticle using oxidized metal ions as electron acceptors with the assistance of cytochrome-c (Shi et al. 2007; Suresh et al. 2010).

12.3 Microbial Enzymes

Enzymes can be defined as the biotic catalyst which is involved in various biosynthetic reactions and metabolic processes (Li et al. 2021; Kumar et al. 2021). Microbes serve as a major source of enzymes. Microbes can replicate easily and rapidly and could be genetically engineered as per the desired requirement of the product (Bhandari et al. 2021; Verma et al. 2021). Microbial enzymes are relatively more active and stable compared to that of isolated from plants or animal sources (Gopinath et al. 2013; Anbu et al. 2017; Bhatt et al. 2021). Various extremophilic bacterial and fungal strains have been isolated from unfavourable pH and temperature as well as high salt and heavy metal conditions for the synthesis of different useful enzymes comprising properties of higher yield (Gopinat et al. 2003; Gopinath et al. 2005).

Microbial enzymes can be isolated from different microorganisms including thermophilic (requires a higher temperature for growth), acidophilic (optimally active in acidic pH range), and alkalophilic (activates at higher pH range) bacteria. The synthesis of these microbial enzymes can be carried out at extreme conditions that can decrease the possibility of contamination during large-scale production (Banat et al. 1992; Cadet et al. 2016). Revolution in enzymes acquired from microorganisms creates a great opportunity for the enhancement of low liveliness consuming improvements that could be applied for biotransformation of poultry waste into beneficial harvests. Enzymatic events may be accommodating to recycle waste rich in protein unconfined by the poultry industry, besides these lines protecting environment by declining waste (Atuanya and Aigbirior 2002). There are some enzymes that have various roles in industrial applications (Table 12.1).

Table 12.1 Various enzymes and their role in industries

Sr. no.	Enzyme	Role(s)
1.	Protease	Breaks proteins into their simple form
2.	Keratinase	Decomposes keratin found in hairs, nails, etc.
3.	Amylase	Breaks starch into sugars
4.	Xylanase	Converts polysaccharides into xylose
		Catalytic breakdown of hemicellulose
5.	Ligninase	Degrades lignin
6.	Cellulase	Breakdown of cellulose
7.	Lipase	Hydrolysis of fats, triglycerides
8.	Pectinase	Breakdown of pectin

12.4 Nutraceuticals

Nutraceutical, a fusion of nutrition and pharmaceuticals, is defined as 'a material which possesses the nutritional value of a diet and delivers pharmaceutical or health assistances such as preclusion and disease management' (DeFelice 1995). Further revision quoted nutraceuticals as 'a product isolated or purified from foods that are generally sold in medicinal forms not usually associated with food' (Pandey et al. 2010). Nutraceuticals have been obtained from various sources ranging from microbes (e.g. poly amino acids), plants (e.g. phytochemicals, vitamins), and animals (polysaccharides) as well as marine sources (glucosamine and chitosan) (Rasmussen and Morrissey 2007; Lordan et al. 2011; Wang et al. 2016). Nutraceuticals are potentially helpful in health up-gradation and disease prevention especially in avoiding age-related disorders such as depression, oxidative damages, inflammation, diabetes, gastrointestinal diseases, and even cancer (Jain and Ramawat 2013). The increasing demands and benefits of microbial supplements having health benefits have significantly stimulated advancement in the market of nutraceuticals. The global nutraceutical market has rapidly grown, and in 2014, it was valued at $171.8 billion. The market is expected to reach $722.49 billion in the next 6–7 years with a compound annual progress rate (CAGR) of 8.3% over the forecast period (NMSS&TA 2020). Though, the growing market of nutraceutical could barely be contented through the efficiency of straight nutraceutical industries. Direct extraction approaches are restricted with accessibility and price of raw ingredients, quality check of goods, and less content and pureness of nutraceuticals. While synthesis by chemicals could be another method, it is unsuitable to generate adequate quantity and quality of biochemicals and certainly not feasible for composite biochemicals (De Luca et al. 2012). To overcome the issue, metabolic engineering of microbes is considered a promising methodology that has recently attained prodigious improvement towards production of value-added nutraceuticals. We have further discussed the recent advances of microbial-based metabolic engineering and their role in nutraceutical production including phytochemicals, prebiotics, polysaccharides, as well as poly amino acids.

12.5 Phytochemicals

Phytochemicals are the broad spectrum of secondary bioactive metabolites obtained from different parts of the plants including stem, leaf, fruits, beans, and grains. Phytochemicals are often involved in plant defence mechanism against adverse biotic and abiotic conditions or may exert health-promoting or disease-resistant properties (Jain and Ramawat 2013). Some of the major types of phytochemicals are discussed further.

12.5.1 Alkaloids

Alkaloids are amino acid-derived nitrogenous complexes with various beneficial properties including antimalarial to anticancer effects (Marienhagen and Bott 2013). Due to long biosynthetic pathways and the complex structure, alkaloid production was limited to the plants for past few years. The most commonly used alkaloids are (a) monoterpene indole alkaloids (MIAs) derived from tryptophan and glucosinolates and (b) benzylisoquinoline alkaloids (BIAs) derived from tyrosine.

Due to the recent advancement and knowledge of the BIA biosynthetic pathway, the synthesis is now carried out in various microorganisms like *E. coli* and *S. cerevisiae* (Nakagawa et al. 2011, 2014; Fossati et al. 2014). The (*S*)-reticuline biologically synthesized from simple carbon sources is an intermediate of BIA pathway (Nakagawa et al. 2011). Apart from biosynthesis of (*S*)-reticuline in *E. coli*, *S. cerevisiae* also facilitate to synthesizes of (*R*, *S*)-reticuline that in turn engineered to produce salutaridine from (*R*)-reticuline and scoulerine, tetrahydroberberine, and tetrahydrocolumbamine from (*S*)-reticuline (Hawkins and Smolke 2008).

Metabolic engineering of MIA alkaloids in microbes is inadequate and not much diverse like that of BIA alkaloids. Strictosidine, a de novo MIA alkaloid, has been successfully produced in yeast by the deletion of three genes and the introduction of 21 new genes in yeast genome (Brown et al. 2015). Yeast has also been bioengineered by introducing eight genes of plants into its genome for the production of tryptophan-derived indolylglucosinolate (IG) (Mikkelsen et al. 2012). Tryptophan-derived IG is a sulphur-rich, amino acid-derived natural composites of glucosinolates. For metabolic engineering in microbes and large-scale alkaloid production, genes isolated from plant platforms are most stable and show promising potential for biosynthesis of plant-derived complexes (Brown et al. 2015; Mora-Pale et al. 2013).

12.5.2 Terpenoids

It serves as the largest class of phytonutrients with various beneficial properties like anti-infectious, anti-inflammatory, and anticancer properties (Jain and Ramawat 2013; Mora-Pale et al. 2013). It is generally present in cereals, soy plants, and green foods. Terpenoids are dimethylallyl pyrophosphate (DMAPP) or isopentenyl pyrophosphate (IPP) derived broad carbon skeleton compounds, like monoterpenes (e.g. menthol), diterpenes (e.g. paclitaxel), triterpenes (e.g. steroid saponins, oleanane, ursane), tetraterpenes (e.g. carotenoids), sesquiterpenes (e.g. artemisinin), and polyterpenes (Marienhagen and Bott 2013). Terpenoids production in microbes illustrates the success and advancement of metabolic engineering for the synthesis of terpenoid drugs. Most common terpenoids that are used in pharmaceuticals are (a) artemisinic acid which is a precursor of antimalarial drug known as artemisinin and (b) taxadiene which is an intermediate of anticancer drug known as paclitaxel (Besumbes et al. 2004; van Herpen et al. 2010).

In nutraceutical industries, carotenoids (a tetraterpene) including astaxanthin, α-carotene and β-carotene, and lycopene act as feed supplements and natural food colourants (Marienhagen and Bott 2013). For a long time, combinatorial carotenoid biosynthesis has been done in heterologous non-carotenogenic hosts including *E. coli* and *S. cerevisiae* due to large-scale production of carotenoid so the metabolically engineered efforts are generally focused on it only. Strain improvement through gene knockout technique generally increases the production of lycopene to a large extent in *E.coli* (Lin et al. 2014). Metabolic engineered *E. coli* have high supply of ATP and NADPH which help in the production of upgraded β-carotene up to 2.1 g/L β-carotene and increase the harvest up to 60 mg/g DCW (Zhao et al. 2013). A high amount of astaxanthin has been produced through harvest of 1.4 mg/g DCW when the biosynthetic genes of xanthophylls are chromosomally integrated with a plasmid-free *E. coli* (Lemuth et al. 2011). Lycopene *E. coli* has generally used for the production of carotenoids because it not only is involved in the production of novel carotenoid like 4-ketozeinoxanthin but also produces some rare carotenoids like decaprenoxanthin, sarcinaxanthin, and sarprenoxanthin (Netzer et al. 2010; Maoka et al. 2014).

12.6 Prebiotics

Prebiotics are nonviable components in food that encourage the growth or activity of useful microorganisms in the gastrointestinal (Pineiro et al. 2008). Prebiotics are polysaccharide with 3–10 monomeric units of sugar which will not further dissociate in the body; hence, it is nondigestible. Prebiotics show a beneficial effect on the metabolic activity and diversity of the gut microbiota, and this leads to major effect on the immune system of host. Prebiotics can also be used in treatment of diverse inflammation-induced diseases by improving the gut microbiota using probiotics such as *Bifidobacteria* or *Lactobacillus* sp. (Lin et al. 2014). A general example of prebiotics is soluble dietary fibres such as inulin, fructooligosaccharides (FOS), and lactose-based galactooligosaccharides (GOS). Inulin and fructooligosaccharides are produced by probiotic *Lactobacillus gasseri* strain (Anwar et al. 2010), whereas galactooligosaccharides (GOS) are short-chain and lactose-derived galactose polymer synthesized by *Kluyveromyces lactis* (Rodriguez-Colinas et al. 2011). Conversion of lactose to GOS is done when the codon-optimized β-galactosidase expresses from hyper-thermophile *Sulfolobus solfataricus* in *Lactococcus lactis*. In terms of infants and toddler, human milk has been accepted as a best nutritive substance due to the presence of most abundant oligosaccharide present in it, i.e. 2′-fucosyl lactose (2′-FL). 2′-FL can also be produced from lactose and glycerol with the help of *E. coli* by overexpressing the fucosyl transferase or by increasing the availability of GDP-L-fructose for the high yield (Lee et al. 2012; Baumgartner et al. 2013).

12.7 Polysaccharides

Polysaccharides are sugar polymers composed of the large number of small monomeric sugar units with highly versatile structure. Polysaccharides are produced by most of the microorganism, e.g. bacteria, fungi, and yeast, or may be extracted from plant and animal tissues. Due to their health beneficial properties, microbial polysaccharides including bacterial polysaccharides and fungal polysaccharides are referred to as the best source for nutraceuticals. Commercialization production of bacterial polysaccharides like gellan, dextran, xanthan, and alginate can be carried out through microbial engineering and refinement, respectively (Giavasis 2013). For dairy product usage, exopolysaccharides are produced through the metabolic engineering of *Streptococcus* and *Lactococcus* species (Jolly et al. 2002). Other than bacterial and fungal polysaccharide shows various extensive properties like immunostimulatory, antitumor, antimicrobial, antioxidant, hypocholesterolaemic, and hypoglycaemic benefits (Giavasis 2014).

Due to these properties, fungal polysaccharides show great potential in pharmaceutical and nutraceutical applications (Giavasis 2014). Scleroglucan excreted by mycelia of the fungus *Sclerotiumrol fsii* is a potent antiviral and antitumor glucan-based polysaccharide and the yield of the polysaccharide can easily be increased by the adding of L-lysine and uridine monophosphate (UMP) (Giavasis 2014). Hyaluronic acid (HA), chondroitin, and heparosan are animal-based polysaccharides which have been produced by microbial host instead of extracting it from the animal tissues. *E. coli*, *L. lactis*, and *Streptomyces albulus* are the major microorganisms used for the production of hyaluronic acid (Yu and Stephanopoulos 2008; Sheng et al. 2015; Yoshimura et al. 2015). Some therapeutically essential polysaccharides like heparosan and chondroitin could be synthesized from engineered *E. coli* at a relatively high titter (He et al. 2015; Zhang et al. 2012).

12.8 Poly Amino Acids

Poly amino acids comprising one or two amino acids are produced by microorganism through a ribosome-independent enzymatic reaction that differentiates them from polypeptides which are generally synthesized by translation. There are three poly amino acid found naturally, viz. poly-γ-glutamic acid (γ-PGA), multi-L-arginyl-poly (L-aspartic acid), and poly-ε-L-lysine (ε-PL). γ-PGA is a biodegradable polymer that is soluble in aqueous solutions and therefore used as drug carriers or hydrogels (Khalil et al. 2017). Few genetically engineered *Bacillus* species produced γ-PGA in high quantity ranging from 31.7 to 107.7 g/L, when it feeds on L-glutamic acid (Hsueh et al. 2017). An equimolar amount of arginine and aspartic acid is used to produce cyanophycin by cyanobacteria or some chemotrophic bacteria like *Acinetobacter calcoaceticus*, etc. Cyanophycin is used as a dipeptide precursor for therapeutic and nutritional applications (Watzer and Forchhammer 2018). Cyanophycin can be produced in *E. coli* by overexpressing the cyanophycin synthetase *cphA* gene isolated from *Synechocystis* sp. PCC 6803 with productivity of 120 mg/g CDW (Tseng et al. 2012). *Streptomyces* species like *Streptomyces albulus* are exclusively involved in the production of ε-PL and could reach up 35.14 g/L when glucose and glycerol are used as carbon source (Chen et al. 2012; Dodd et al. 2018). Poly-ε-L-lysine or ε-PL is a homo poly amino acid that is produced by the polymerization of lysine via ε-PL synthetase (PLS). ε-PL has been approved as food preservatives or dietary agents due to having antibacterial anticancer activities in developed countries like the United States and Japan (El-Sersy et al. 2012).

12.9 Conclusion

Nowadays, due to green manufacturing and sustainable development, microbial production of different substances is widely used, but this development is limited because of its high cost. For developing countries, it creates an opportunity for using microbes and their derivatives from small places like household and village-level production to large-scale industrial productions. These microbial processes need more exploration to be exploited with full intensity with their beneficiary effects. The last decades have recorded extraordinary advancements in production of nutraceuticals by metabolic engineering of microorganisms. Nutraceuticals own countless application including strengthening of immunity of human beings. Further studies are recommended for exploration of different microbial explorations in which microbes are directly involved in enhancing the productivity of processed food or food products. The use of metabolically engineered microbes opens a promising door not only in laboratory-based production but also for the industry-based production of intricate natural compounds from simple carbon sources. The emerging role of synthetic biology will promote the progression of this field in upcoming years, and hopefully, this will deliver requisite tools for tuneable synthesis and optimization of nutraceutical synthesis in biological hosts.

References

Ashraf A, Ramamurthy R, Sayavedra SM, Bhatt P, Gangola S, Noor T, Desmarais M, Rabbani A, Rene ER (2021) Integrating photobioreactor with conventional activated sludge treatment for nitrogen removal from sidestream digestate: Current challenges and opportunities. J Environ Chem Eng:106171. https://doi.org/10.1016/j.jece.2021.106171

Anbu P, Gopinath SCB, Chaulagain BP, Lakshmipriya T (2017) Microbial enzymes and their applications in industries and medicine. Biomed Res Int 2016:2195808

Anwar MA, Kralj S, Pique AV, Leemhuis H, van der Maarel M, Dijkhuizen L (2010) Inulin and levan synthesis by probiotic *Lactobacillus gasseri* strains: characterization of three novel fructansucrase enzymes and their fructan products. Microbio (Reading) 156:1264–1274

Atuanya E, Aigbirior M (2002) Mesophilic biomethanation and treatment of poultry waste-water using pilot scale UASB reactor. Environ Monit Assess 77:139–147

Banat IM, Nigam P, Marchant R (1992) Isolation of thermotolerant, fermentative yeasts growing at 52 degrees C and producing ethanol at 45 degrees C and 50 degrees C. World J Microbiol Biotechnol 8:259–263

Battcock M, Azam-Ali S (1998) Fermented fruits and vegetables. A global perspective. Food and Agriculture Organization (FAO), Rome

Baumgartner F, Seitz L, Sprenger GA, Albermann C (2013) Construction of *Escherichia coli* strains with chromosomally integrated expression cassettes for the synthesis of 2′-fucosyllactose. Microb Cell Fact 12:40

Besumbes O, Sauret-Gueto S, Phillips MA, Imperial S, Rodriguez-Concepcion M, Boronat A (2004) Metabolic engineering of isoprenoid biosynthesis in *Arabidopsis* for the production of taxadiene, the first committed precursor of Taxol. Biotechnol Bioeng 88:168–175

Bhandari G, Bhatt P, Gangola S, Srivastava A, Sharma A (2021) Degradation mechanism and kinetics of carbendazim using Achromobacter sp. strain GB61. Biorem J:1–17. https://doi.org/10.1080/10889868.2021.1911921

Bhatt P, Verma A, Gangola S, Bhandari G, Chen S (2021) Microbial glycoconjugates in organic pollutant bioremediation: recent advances and applications. Microb Cell Factories 20(1):1–8. https://doi.org/10.1186/s12934-021-01556-9

Brown S, Clastre M, Courdavault V, O'Connor SE (2015) *De novo* production of the plant-derived alkaloid strictosidine in yeast. Proc Natl Acad Sci U S A 112:3205–3210

Cadet M, Gounel S, Stines-Chaumeil C, Brilland X, Rouhana J, Louerat F, Mano N (2016) An enzymatic glucose/O2 biofuel cell operating in human blood. Biosens Bioelectron 83:60–67

Chen XS, Ren XD, Dong N, Li S, Li F, Zhao FL, Tang L, Zhang JH, Mao ZG (2012) Culture medium containing glucose and glycerol as a mixed carbon source improves epsilon-poly-L-lysine production by *Streptomyces* sp. M-Z18. Bioprocess Biosyst Eng 35:469–475

De Luca V, Salim V, Atsumi SM, Yu F (2012) Mining the biodiversity of plants: a revolution in the making. Science 336:1658–1661

DeFelice SL (1995) The nutraceutical revolution: its impact on food industry R&D. Trends Food Sci Technol 6:59–61

Dodd A, Swanevelder D, Zhou N, Brady D, Hallsworth JE, Rumbold K (2018) *Streptomyces albulus* yields epsilon-poly-L-lysine and other products from salt-contaminated glycerol waste. J Ind Microbiol Biotechnol 45:1083–1090

El-Sersy NA, Abdelwahab AE, Abouelkhiir SS, Abou-Zeid DM, Sabry SA (2012) Antibacterial and anticancer activity of epsilon-poly-L-lysine (epsilon-PL) produced by a marine *Bacillus subtilis* sp. J Basic Microbiol 52:513–522

Fossati E, Ekins A, Narcross L, Zhu Y, Falgueyret JP, Beaudoin GA, Facchini PJ, Martin VJ (2014) Reconstitution of a 10-gene pathway for synthesis of the plant alkaloid dihydrosanguinarine in *Saccharomyces cerevisiae*. Nat Commun 5:3283

Giavasis I (2013) 16—Production of microbial polysaccharides for use in food. In: McNeil B, Archer D, Giavasis I, Harvey L (eds) Microbial production of food ingredients, enzymes and nutraceuticals. Woodhead Publishing, Sawston, pp 413–468

Giavasis I (2014) Bioactive fungal polysaccharides as potential functional ingredients in food and nutraceuticals. Curr Opin Biotechnol 26:162–173

Gopinat SCB, Hilda A, Lakshmi Priya T, Annadurai G, Anbu P (2003) Purification of lipase from *Geotrichum candidum* : conditions optimized for enzyme production using Box–Behnken design. World J. Microbiol Biotechnol 19:681–689

Gopinath SCB, Hilda A, Anbu P (2005) Extracellular enzymatic activity profiles in fungi isolated from oil-rich environments. Mycoscience 46:119–126

Gopinath SC, Anbu P, Lakshmipriya T, Hilda A (2013) Strategies to characterize fungal lipases for applications in medicine and dairy industry. Biomed Res Int 2013:154549

Gulati A, Juneja R (2018) Innovations and revolutions in Indian agriculture: a review. J Agri Sci Tech B 8:473–482

Hawkins KM, Smolke CD (2008) Production of benzylisoquinoline alkaloids in *Saccharomyces cerevisiae*. Nat Chem Biol 4:564–573

He W, Fu L, Li G, Andrew Jones J, Linhardt RJ, Koffas M (2015) Production of chondroitin in metabolically engineered *E. coli*. Metab Eng 27:92–100

Hsueh YH, Huang KY, Kunene SC, Lee TY (2017) Poly-gamma-glutamic acid synthesis, gene regulation, phylogenetic relationships, and role in fermentation. Int J Mol Sci 18(12):2644

Jain N, Ramawat KG (2013) Nutraceuticals and antioxidants in prevention of diseases. In: Ramawat KG, Mérillon J-M (eds) Natural products: phytochemistry, botany and metabolism of alkaloids, phenolics and terpenes. Springer, Berlin, pp 2559–2580

Jolly L, Vincent SJ, Duboc P, Neeser JR (2002) Exploiting expolysaccharides from lactic acid bacteria. Antonie Van Leeuwenhoek 82:367–374

Kang SH, Bozhilov KN, Myung NV, Mulchandani A, Chen W (2008) Microbial synthesis of CdS nanocrystals in genetically engineered *E. coli*. Angew Chem Int Ed 47:5186–5189

Khalil IR, Burns AT, Radecka I, Kowalczuk M, Khalaf T, Adamus G, Johnston B, Khechara MP (2017) Bacterial-derived polymer poly-y-glutamic acid (y-PGA)-based micro/nanoparticles as a delivery system for antimicrobials and other biomedical applications. Int J Mol Sci 18(2):313

Kumar G, Lal S, Bhatt P, Ram RA, Bhattacherjee AK, Abhay RDS (2021) Mechanisms and kinetics for the degradation of paclobutrazol and biocontrol action of a novel Pseudomonas putida strain T7. Pestic Biochem Physiol 175:104846. https://doi.org/10.1016/j.pestbp.2021.104846

Lee WH, Pathanibul P, Quarterman J, Jo JH, Han NS, Miller MJ, Jin YS, Seo JH (2012) Whole cell biosynthesis of a functional oligosaccharide, 2′-fucosyllactose, using engineered *Escherichia coli*. Microb Cell Fact 11:48

Lemuth K, Steuer K, Albermann C (2011) Engineering of a plasmid-free Escherichia coli strain for improved in vivo biosynthesis of astaxanthin. Microb Cell Fact 10:29

Li J, Zhang W, Lin Z, Huang Y, Bhatt P, Chen S (2021) Emerging strategies for the bioremediation of the phenylurea herbicide diuron. Front Microbiol 12:686509. https://doi.org/10.3389/fmicb.2021.686509

Lin Y, Jain R, Yan Y (2014) Microbial production of antioxidant food ingredients via metabolic engineering. Curr Opin Biotechnol 26:71–78

Lok CN, Ho CM, Chen R, Tam PK, Chiu JF, Che CM (2008) Proteomic identification of the Cus system as a major determinant of constitutive Escherichia coli silver resistance of chromosomal origin. J Proteome Res 7:2351–2356

Lordan S, Ross RP, Stanton C (2011) Marine bioactives as functional food ingredients: potential to reduce the incidence of chronic diseases. Mar Drugs 9:1056–1100

Maoka T, Takemura M, Tokuda H, Suzuki N, Misawa N (2014) 4-Ketozeinoxanthin, a novel carotenoid produced in *Escherichia coli* through metabolic engineering using carotenogenic genes of bacterium and liverwort. Tetrahedron Lett 55:6708–6710

Marienhagen J, Bott M (2013) Metabolic engineering of microorganisms for the synthesis of plant natural products. J Biotechnol 163:166–178

Mi C, Wang Y, Zhang J, Huang H, Xu L, Wang S, Fang X, Fang J, Mao C, Xu S (2011) Biosynthesis and characterization of CdS quantum dots in genetically engineered *Escherichia coli*. J Biotechnol 153:125–132

Mikkelsen MD, Buron LD, Salomonsen B, Olsen CE, Hansen BG, Mortensen UH, Halkier BA (2012) Microbial production of indolylglucosinolate through engineering of a multi-gene pathway in a versatile yeast expression platform. Metab Eng 14:104–111

Mora-Pale M, Sanchez-Rodriguez SP, Linhardt RJ, Dordick JS, Koffas MA (2013) Metabolic engineering and in vitro biosynthesis of phytochemicals and non-natural analogues. Plant Sci 210:10–24

Nakagawa A, Minami H, Kim JS, Koyanagi T, Katayama T, Sato F, Kumagai H (2011) A bacterial platform for fermentative production of plant alkaloids. Nat Commun 2:326

Nakagawa A, Matsuzaki C, Matsumura E, Koyanagi T, Katayama T, Yamamoto K, Sato F, Kumagai H, Minami H (2014) (R,S)-tetrahydropapaveroline production by stepwise fermentation using engineered *Escherichia coli*. Sci Rep 4:6695

Negi S (2014) Supply chain efficiency: an insight from fruits and vegetables sector in India. J Oper Supp Chain Manag 7:154

Netzer R, Stafsnes MH, Andreassen T, Goksoyr A, Bruheim P, Brautaset T (2010) Biosynthetic pathway for gamma-cyclic sarcinaxanthin in *Micrococcus luteus*: heterologous expression and evidence for diverse and multiple catalytic functions of C(50) carotenoid cyclases. J Bacteriol 192:5688–5699

NMSS&TA (2020) Nutraceutical market size, share & trends analysis report by product (Dietary supplements, functional foods, functional beverages), by region, and segment forecasts, 2020–2027. https://www.businesswire.com/news/home/20200520005477/en/722-Billion-Nutraceutical-Market-Size-and-Share-Breakdownby-Product-and-Region%2D%2D-ResearchAndMarkets.com

Okafor N (2007) Modern industrial microbiology and biotechnology. Science Publishers, Enfield

Pandey M, Verma RK, Saraf SA (2010) Nutraceuticals: new era of medicine and health. Asian J Pharm Clin Res 3:11–15

Park TJ, Lee SY, Heo NS, Seo TS (2010) *In vivo* synthesis of diverse metal nanoparticles by recombinant *Escherichia coli*. Angew Chem Int Ed 49:7019–7024

Peelle BR, Krauland EM, Wittrup KD, Belcher AM (2005) Probing the interface between biomolecules and inorganic materials using yeast surface display and genetic engineering. Acta Biomater 1:145–154

Peralta-Yahya PP, Zhang F, del Cardayre SB, Keasling JD (2012) Microbial engineering for the production of advanced biofuels. Nature 488:320–328

Pineiro M, Asp NG, Reid G, Macfarlane S, Morelli L, Brunser O, Tuohy K (2008) FAO technical meeting on prebiotics. J Clin Gastroenterol 42(Suppl 3 Pt 2):S156–S159

Plaza DO, Gallardo C, Straub YD, Bravo D, Pérez-Donoso JM (2016) Biological synthesis of fluorescent nanoparticles by cadmium and tellurite resistant Antarctic bacteria: exploring novel natural nanofactories. Microb Cell Fact 15:76

Rasmussen RS, Morrissey MT (2007) Marine biotechnology for production of food ingredients. Adv Food Nutr Res 52:237–292

Rodriguez-Colinas B, de Abreu MA, Fernandez-Arrojo L, de Beer R, Poveda A, Jimenez-Barbero J, Haltrich D, Ballesteros Olmo AO, Fernandez-Lobato M, Plou FJ (2011) Production of galacto-oligosaccharides by the beta-galactosidase from *Kluyveromyces lactis* : comparative analysis of permeabilized cells versus soluble enzyme. J Agric Food Chem 59:10477–10484

Sambhy V, MacBride MM, Peterson BR, Sen A (2006) Silver bromide nanoparticle/polymer composites: dual action tunable antimicrobial materials. J Am Chem Soc 128:9798–9808

Sheng J, Ling P, Wang F (2015) Constructing a recombinant hyaluronic acid biosynthesis operon and producing food-grade hyaluronic acid in *Lactococcus lactis*. J Ind Microbiol Biotechnol 42:197–206

Shi L, Squier TC, Zachara JM, Fredrickson JK (2007) Respiration of metal (hydr)oxides by *Shewanella* and *Geobacter*: a key role for multihaem c-type cytochromes. Mol Microbiol 65:12–20

Suresh AK, Pelletier DA, Wang W, Moon JW, Gu B, Mortensen NP, Allison DP, Joy DC, Phelps TJ, Doktycz MJ (2010) Silver nanocrystallites: biofabrication using *Shewanella oneidensis*, and an evaluation of their comparative toxicity on gram-negative and gram-positive bacteria. Environ Sci Technol 44:5210–5215

Tseng WC, Fang TY, Cho CY, Chen PS, Tsai CS (2012) Assessments of growth conditions on the production of cyanophycin by recombinant *Escherichia coli* strains expressing cyanophycin synthetase gene. Biotechnol Prog 28:358–363

van Herpen TW, Cankar K, Nogueira M, Bosch D, Bouwmeester HJ, Beekwilder J (2010) *Nicotiana benthamiana* as a production platform for artemisinin precursors. PLoS One 5: e14222

Verma S, Bhatt P, Verma A, Mudila H, Prasher P, Rene ER (2021) Microbial technologies for heavy metal remediation: effect of process conditions and current practices. Clean Techn Environ Policy:1–23. https://doi.org/10.1007/s10098-021-02029-8

Vigneshwaran N, Kathe AA, Varadarajan PV, Nachane RP, Balasubramanya RH (2007) Silver-protein (core-shell) nanoparticle production using spent mushroom substrate. Langmuir 23:7113–7117

Wang J, Guleria S, Koffas MA, Yan Y (2016) Microbial production of value-added nutraceuticals. Curr Opin Biotechnol 37:97–104

Watzer B, Forchhammer K (2018) Cyanophycin: a nitrogen-rich reserve polymer. IntechOpen, London

Yoshimura T, Shibata N, Hamano Y, Yamanaka K (2015) Heterologous production of hyaluronic acid in an epsilon-poly-L-lysine producer, *Streptomyces albulus*. Appl Environ Microbiol 81:3631–3640

Yu H, Stephanopoulos G (2008) Metabolic engineering of *Escherichia coli* for biosynthesis of hyaluronic acid. Metab Eng 10:24–32

Zhang C, Liu L, Teng L, Chen J, Liu J, Li J, Du G, Chen J (2012) Metabolic engineering of *Escherichia coli* BL21 for biosynthesis of heparosan, a bioengineered heparin precursor. Metab Eng 14:521–527

Zhao J, Li Q, Sun T, Zhu X, Xu H, Tang J, Zhang X, Ma Y (2013) Engineering central metabolic modules of *Escherichia coli* for improving beta-carotene production. Metab Eng 17:42–50

Chapter 13
Plant Growth-Promoting Rhizobacteria and Their Application in Sustainable Crop Production

Parul Chaudhary, Heena Parveen, Saurabh Gangola, Govind Kumar, Pankaj Bhatt, and Anuj Chaudhary

Abstract The world population is increasing at an alarming rate, and to feed such a huge population from an exhausted arable land with depleted water resources would be a great concern and challenge in near future for agriculture scientists. Present agricultural practices are dependent on seed quality, irrigation facilities, and chemical fertilizers that affect the fertility of soil significantly. To overcome from this problem, some alternative methods must be employed. The use of several rhizospheric bacteria and fungi as bioinoculant with plant growth-promoting ability is under practice by some farmers, forming the base of upcoming green revolution. Plant growth-promoting microbes (PGPM) are usually present around the root zone, rhizosphere, and inside the plants and found responsible for enhanced plant growth by way of providing essential nutrition and hormones and controlling pathogens by producing antifungal and antimicrobial compounds. Soil is the most important component of a terrestrial ecosystem and essentially required to establish, support, and enhance plant growth by providing nutrition and water. The cross signaling between the plants and rhizobacteria for mutual benefits is possible because of the involvement of different types of root exudates, genes, and several known and

P. Chaudhary (✉)
Department of Animal Biotechnology, NDRI, Karnal, Haryana, India

H. Parveen
Department of Dairy Microbiology, NDRI, Karnal, Haryana, India

S. Gangola
Department of School of Agriculture, Graphic Era Hill University, Bhimtal, Uttarakhand, India

G. Kumar
Department of Crop Production, Central Institute for Subtropical Horticulture, Lucknow, India

P. Bhatt
State Key Laboratory for Conservation and Utilization of Subtropical Agro-bioresources, Guangdong Laboratory for Lingnan Modern Agriculture Integrative Microbiology Research Centre, South China Agricultural University, Guangzhou, China

A. Chaudhary
School of Agriculture and Environmental Sciences, Shobhit University, Gangoh, Uttar Pradesh, India

P. Bhatt et al. (eds.), *Microbial Technology for Sustainable Environment*,
https://doi.org/10.1007/978-981-16-3840-4_13

unknown factors which make a network for a better functioning of plant and bacteria. Conservation of biogeochemical cycles consisting of carbon, nitrogen, phosphorus, and sulfur in relation to recycling of essential nutrients in an ecosystem is only possible by the involvement of numerous soil microbes. Biological activity of a soil or soil health must be known to get good agriculture produce. In this chapter, we have undertaken various aspects of rhizospheric microbes including their involvement in higher food production and maintenance of soil and plant health.

Keywords Biofertilizers · Soil health · Agriculture · Soil fertility

13.1 Introduction

Agriculture, the backbone of Indian economy, provides sufficient food and feed to satisfy the need of burgeoning human population (Bhatt et al. 2019a, b). Application of crop protectants, fertilizers, and pesticides to enhance crop production is found to be a boon for the agriculture system, but indiscriminate use of pesticides not only contaminates agriculture produce but also affects soil health and quality of ground water which ultimately affects human health (Pankaj et al. 2015a, b, 2016a, b). Hence, it is important to improve soil health by increasing the appliance of beneficial microorganisms which are involved in growth and productivity of plants by means of recycling of essential nutrients via biogeochemical cycles (Huang et al. 2021; Singh et al. 2021; Zhang et al. 2021). Sustainable agriculture fascinates the increasing demand of biological based natural fertilizers substitute to agrochemicals. Plant growth-promoting bacteria (PGPB) can be used as a green technology which will diminish the application of chemical fertilizers and improve soil health by involving in cycling of biological elements wherein the complex biological is converted into simpler ones (Di Benedetto et al. 2017). Rhizomicrobiome plays an important part in improving plant health by influencing the growth of their eukaryotic hosts and providing essential nutrients to the soil. It is highly influenced through root exudates of the host plant (Backer et al. 2018). Rhizomicrobiome is composed of diversified microbial community and highly influenced by plant type, soil type, and environmental conditions. Microbial associations around the roots in soil are complex and facilitate the plants to get the nutrients from soil through different mechanism. PGPM are the soil bacteria which make nutrients available, fix atmospheric nitrogen and are commonly used as biofertilizers which enhanced the productivity of different crops (Santoyo et al. 2016). Rhizospheric microbes such as bacteria, cyanobacteria, actinomycetes, and arbuscular mycorrhiza fungi directly or indirectly allied with plants or soil (Kumar et al. 2016). PGPR that inhabits in the rhizosphere belongs to *Azotobacter*, *Azospirillum*, *Bacillus*, *Pseudomonas*, *Paenibacillus*, *Pantoea*, *Serratia*, *Streptomyces*, etc. These PGPRs are generally used as bioinoculants for the biocontrol, biostimulation, and biofertilization (Kaushal and Wani 2016). The use of biofertilizers not only assures food security but in addition helps in maintenance of the soil microbial diversity.

Soil is the most suitable, dynamic, and complex surface environment which provides substrate for growth of plants, animal, and microorganisms. Microorganisms are metabolically more diverse, and soil offers a variety of microhabitat; microbial diversity in soil is much higher than other environments. Soil quality indicators such as microbial population, soil enzyme activities, soil pH, and micro and macronutrients play a major role in sustenance of soil health status and quality. Soil physicochemical properties exhibit seasonal variation and alter the structure and composition of bacterial population (Lopez-Monejar et al. 2015). Among soil physicochemical properties, pH has been known to influence the bacterial communities via availability of organic and inorganic nutrients. Soil enzymes are derived from plant root exudates, animal remains, and soil microorganisms of which soil microbes have the major involvement (Thomson et al. 2015). They play a vital function in catalyzing reactions necessary for maintaining microbial life in soil, stabilize soil structure, disintegrate organic matter, and are involved in nutrient cycling. Enzyme activities act as biomarker for detecting changes in soil quality and measure of microbial diversity and community structure. Thus, represent the functioning of the entire microbial community (Yang et al. 2017). This chapter reports the prospective use of PGPM for enhancing the crop productivity and maintaining the soil heath in a sustainable manner.

13.2 Rhizosphere and Plant Growth-Promoting Microbes

Rhizospheric term was given by Hiltner to explain the narrow zone of soil adjacent to roots where microbial populations are stimulated by root activities and manage plant growth (York et al. 2016). Rhizosphere is a hot spot for microbial action contributed generally by native bacteria and fungi. PGPM can be separated into two major groups: plant growth-promoting bacteria (PGPR) and plant growth-promoting fungi (PGPF). PGPRs were initially identified by Klopper and Schorth (1978) to illustrate bacteria which colonize the plant roots and enhance plant growth. Bacteria are the most dominant microorganism in the soil. There are two groups of PGPR on the basis of interaction of microbes with plants. Endophytes (symbiotic) colonize internal plant tissues. Endophytes sense the chemical signal and reach at the root surface and produce counter signal that makes the plant root vulnerable for microbial penetration into the root and improve the growth of plants (Defez et al. 2017). Epiphytes, in which a microbe lives on the surface of plants, are also known as non-symbiotic. Besides bacterial species, fungal species such as *Trichoderma*, *Penicillium*, *Aspergillus*, and *Alternaria* are used as PGPR.

13.3 PGPR and Their Mode of Action

PGPRs are the group of soil bacteria which to be found around the root surface of plants and support plant health using different mechanism. They affect plant health parameters by producing various growth-promoting hormones within rhizosphere. PGPR directly facilitates the plant growth by improving uptake of macro and micro nutrients like nitrogen, phosphorus, and potassium as well as by modulating phytohormone like IAA, cytokinin, and gibberellins, vitamins, and enzymes and solubilization of different minerals like phosphate, iron, and potassium which are directly involved in increasing root length (Liu et al. 2017). PGPR protects plant health indirectly from plant pathogens by producing different bacterial/fungal metabolites like HCN, phenazines, and tensin and lytic enzymes and boosts plant growth and productivity (Bhardwaj et al. 2014). Interaction of PGPRs with plant root induces resistance against pathogens (fungal, bacteria, and viruses) by promoting ethylene and jasmonate signaling. Different bacteria like *Pseudomonas, Bacillus*, *Azospirillum*, *Burkholderia*, *Enterobacter*, *Arthrobacter*, *Serratia*, *Burkholderia*, and *Klebsiella* are used as bioinoculants to improve the yield of diverse agricultural crops (Ghevariya and Desai 2014). Population density of rhizospheric bacteria depends on the pH, water potential of soil, and plant root exudates. These microorganisms are also helpful in synthesis and degradation of organic matter. Calvo et al. (2019) reported that *Bacillus* helps in regulation of ammonium and nitrate uptake genes in *Arabidopsis thaliana* which increased the overall growth of plant. Number of bacterial strains has the capability to induce plant growth and enhance yield in different cropping systems (Nieto-Jacobo et al. 2017; Chaudhary et al. 2021a). *Pseudomonas taiwanensis* and *Pantoea agglomerans* are used as soil inoculants in form of biofertilizers in agriculture and enhanced the plant growth by solubilization of different nutrients (Chaudhary and Sharma 2019). Khati et al. (2019b) reported that *Bacillus* spp. (PS2 and PS10) is also involved in IAA and siderophore production which improved the plant health parameters. Shen et al. (2015) observed that application of bioinoculant enhances total bacterial population in soil and helps in plant growth promotion via way of posing beneficial interactions among bioinoculants, plants, and microbes. Application of PGPRs enhanced the shoot and root length of wheat plants by releasing metabolites and mineralization of nutrients which are simply accessible to plants (Sheirdil et al. 2019). de-Lima et al. (2019) reported that inoculation with *Bacillus subtilis* increased leaf area and water use efficiency in maize and common bean plants.

13.4 Effect of PGPRs on Plant Health Parameters

PGPRs hold promise opportunity for sustainable agriculture. Biofertilizers are artificially preserved cultures isolated from the soil used as soil inoculants which improve the soil fertility and enhance the productivity. They colonize the central part of plants and enhanced plant growth by escalating the accessibility of nutrients to plants and also considered an integrated nutrient management system (Malusa et al. 2012). PGPR has been proven as safe and efficient methods to enhance the crop yields (Vejan et al. 2016). Agriculture can take advantage of symbiotic relationship of microbes with plants which helps in maintenance of the particular microbial flora in the plant rhizosphere (Asei et al. 2015).

Bacillus megaterium helped in growth promotion of tea plants and disease reduction by producing siderophore (Chakraborty et al. 2006). Use of *Pseudomonas*, *Micrococcus*, *Staphylococcus*, and *Bacillus* increases the growth and productivity of *Triticum aestivum (*Ali et al. 2009*)*. These microbes showed the production of indole acetic acid. Mishra et al. (2010) found that *Pseudomonas fluorescens* and *Bacillus subtilis* as bioinoculant (ammonia producers) increased the biomass of *Geranium* (a medicinal plant). Genetically, modified strains of *Pseudomonas* are reported to improve plant development, yield, and disease resistance in agricultural field (Vessy 2013). Inoculation of seeds using *Burkholderia cepacia*, *Azotobacter chroococcum*, and *Bacillus subtilis* improved maize and wheat productivity (Zhao et al. 2014). Application of *Pseudomonas putida*, *Bacillus megaterium*, and *Mesorhizobium* showed positive effect on *Cicer arietinum* and significantly enhanced plant biomass and seed production (Fernández and Alexander 2017). Bioinoculation of *Bacillus* sp. in mung bean and rice enhanced plant growth and NPK content in soil which supported plant growth (Pahari et al. 2017). Rozier et al. (2019) found that bioinoculation of *Azospirillum lipoferum* CRT1 enhanced the seed germination, root surface area, and photosynthetic pigment in maize plants.

Khati et al. (2017, 2018) reported that *Bacillus* spp. improved the maize health parameters. Bioinoculation of *Sinorhizobium meliloti* and *Pseudomonas fluorescens* enhances the seed weight and morphological and phytochemical parameters of Fenugreek plants (Sharghi et al. 2019). Rahman et al. (2018) observed that *Bacillus amyloliquefaciens* and *Paraburkholderia fungorum* increased the strawberry fruit growth, yield, and antioxidant contents. Andrade et al. (2019) reported that *Azospirillum brasilense* with *Burkholderia cepacia* and *Enterobacter cloacae* improved the growth of strawberry crop.

13.5 Role of PGPR as Biocontrol Agent

The predominant factor in reducing crop yields is disease in plants. To protect plants from phytopathogens, farmers rely on pesticides which improve the plant growth and productivity. The over usage of these chemicals in soil has established resistance in phytopathogens and poses many environmental issues. Biological control using PGPRs is an alternative way to control phytopathogens. The use of PGPR as a biological agent enhanced plant growth and improved health without causing any toxic effect. There are various mechanism by which plant growth-promoting bacteria (PGPB) damage the effect of phytopathogens. Production of antibiotics like tensin, pyrrolnitrin, and cyclic lipopeptides produced by *Pseudomonas* inhibits plant pathogens. Ahmad et al. (2017) found that *Bacillus subtilis* inhibit the growth of fungal pathogens like *Rhizoctonia solani*, *Botrytis cinerea*, and *Fusarium oxysporum*. Radhakrishan et al. (2017) reported that *Bacillus* sp. stimulates plant growth and prevents pathogen infection. Mokrani et al. (2019) observed that combination of *Pseudomonas pseudo* and *Pseudomonas cichorii* hinders the growth of phytopathogenic fungi. Liu et al. (2018) reported that application of *Bacillus amyloliquefaciens* with bio-organic additives suppressed the tomato disease. Zhang et al. (2019) observed that consortium of *Bacillus cereus*, *Bacillus subtilis*, and *Serratia* acts as environmentally friendly biocontrol agent on sweet pepper. These three bacteria reduced the prevalence of phytopthora blight and improved the quality of fruit. Hashem et al. (2019) found that *Bacillus subtilis* secrete antibiotics and hydrolytic enzymes and have potential biocontrol activity against pathogens.

Passera et al. (2017) observed that *Paenibacillus pasadenensis* have the capability to prevent plant diseases. Wu et al. (2016) reported that *Pseudomonas saponiphila* increased the disease control of plant pathogens. Sharma et al. (2019) reported that halotolerant species, *Klebsiella* species MBE02, regulated the genes which involved in pathogen defense mechanism and jasmonic acid/ethylene signal pathways. These bacteria have growth stimulating effect on peanut and controlled the infection caused by *Aspergillus* and several other fungal phytopathogens.

13.6 Role of PGPR Under Abiotic Stress Conditions

Stresses have significantly reduced the fertility of soil and adversely influence the microbial community and therefore affected the plant growth and agriculture production. Plant-microbe interactions play significant roles in the maintenance of plant and soil productivity under stress conditions (Vimal et al. 2019). Under the influence of stress conditions, plants become vulnerable to osmotic stress and ionic toxicity and start overproducing reactive oxygen species (ROS) which damages the proteins, lipids, and other cellular components (Kyei-Boahen et al. 2017). To cope up stress, use of rhizobacteria as bioinoculants represents powerful techniques to alleviate various environmental stresses like salinity stress, drought, weed infestation, and nutrient deficiency. Heydarian et al. (2016) reported that PGPB produce ACC deaminase in high salt condition and improved the *Camelina sativa* plant growth and seed production. Bharti et al. (2016) observed that *Dietzia natronolimnaea* protected the wheat plants from salt stress by upregulation of the ABA signaling. Cura et al. (2017) found that *Azospirillum brasilense* and *Herbaspirillum seropedicae* under drought condition improved the chlorophyll and biomass production in corn. Numan et al. (2018) reported that bacteria elicit plants to produce growth hormones under salt stress condition. Application of *Bradyrhizobium* in cowpea plant minimizes the harmful effects of salt stress by stimulation of enzymatic and non-enzymatic antioxidant enzymes (Santos et al. 2018). Ansari et al. (2019) reported that *Bacillus pumilus* tolerated 250 mM salt concentration and enhanced the plant growth and yield of wheat plant. They observed decline in stress markers like catalase, superoxidase dismutase, proline, and malondialdehyde and reduced salt stress in wheat. Bidgoli et al. (2019) reported that application of PGPR (*Pseudomonas fluorescens*) improved the yield of medicinal plant *Rosmarinus officinalis* under salinity condition @ 10 g/L NaCl. *Curtobacterium albidum* inoculation in soil alleviates the salinity stress and improves the paddy yield (Vimal et al. 2017). He et al. (2019) observed that *Pseudomonas putida* Rs-198 increased the shoot/root length of pepper in slightly saline soil.

13.7 Plant Growth-Promoting Fungi

Different studies showed the beneficial effects of rhizospheric fungi on plant quality and quantity and their positive role with the ecological environment (Murali et al. 2012). Endophytic fungi are colonizing the interior of plant parts. Plant-microbe interactions resulted in adaptation, plant growth promotion, and production of different secondary metabolites and bioactive compounds with potential application in agriculture (Yadav et al. 2018). Among the PGPF, *Penicillium*, *Aspergillus*, *Penicillium*, and *Chaetomium* play a valuable position in promoting plant growth and disease repression (Kumar et al. 2021). PGPF is involved in production of plant hormones, organic matter decomposition, and protection of plants from stress conditions and helps in solubilization of different minerals (Sangamesh et al. 2018). Ripa et al. (2019) reported that *Trichoderma aureoviride*, *T. harzianum*, *Aspergillus flavus*, *A. tenuissima*, *Alternaria alternate*, *Fusarium equiseti*, and *F. proliferatum* produce IAA, ACCD, and phosphate solubilization, tolerate the abiotic stress, and showed resistance against the antibiotics.

It was observed that *Alternaria* sp. A13 enhanced the root growth of *Salvia miltiorrhiza* and promotes secondary metabolite accumulation (Zhou et al. 2018). *Penicillium* enhances the plant growth by production of phytohormones, solubilization of phosphate, synthesis of amino acids, and secretion of antioxidants enzymes and is involved in secondary metabolite production which helps in coping up stress situation (Chaudhary et al. 2018). Shah et al. (2019) observed that *Fusarium* sp. and *Hypoxylon* sp. increased the shoot/root length and chlorophyll in *Rhynchostylis retusa* plants. Wang et al. (2018) reported that *Aspergillus niger* CS-1 promoted the plant growth of wheat. Mycorrhizal fungi *Rhizophagus intraradices* benefits host plant by improving translocation of mineral nutrients (Chitarra et al. 2016). Tyagi et al. (2017) found that *Rhizophagus intraradices* and *Piriformospora indica* tolerate drought stress and enhanced the plant biomass and chlorophyll content in finger millet.

13.8 Bioformulation

In several fields, bioformulation for the promotion of plant growth continues to encourage research and development. The key goals of the bioformulation industry are increased soil fertility, promotion of plant growth, and control of phytopathogens, which contribute to the creation of an eco-friendly climate. *Bacillus* and *Pseudomonas* sp. are reported as best PGPRs to enhance the plant growth and productivity in different cropping systems (Chaudhary et al. 2021c, d; Agri et al. 2021). Bioformulation provides an environmentally sustainable approach to enhance crop productivity by using microbial bioinoculants and does not affect microbial activity in soil. PGPRs can be used to enhance the plant productivity under field condition. It may be an efficient approach to replace the chemical fertilizers with bioformulation to develop a sustainable agriculture system in India (Aslani et al. 2014).

13.9 Soil Health

Soil is the basis in terms of food, water security, climate alleviation, and protection of biodiversity (McBratney et al. 2014). Soil health has usually judged in terms of production and provides overall picture of soil functionality and affects agricultural production significantly (Haney et al. 2018). Depending on biological and chemical constituent of the soil that affects the ecosystem and environmental quality, soil health is characterized as the net product of ongoing conservation and degradation processes (Bünemann et al. 2018). Healthy soil supports and maintains diverse microbial community which helps in maintenance of beneficial association of different microbes with plant roots, controls plant diseases, and recycles plant nutrients. Soil health is a measure index of soil texture, nutrient cycling, soil protein, soil respiration, surface hardness, water stable aggregates, and availability of water capacity (Khati et al. 2019a; Mann et al. 2019).

Soil quality is the ability of a soil to function, within a natural/managed ecosystem limit to enhance productivity of plants and retain water quality and human health (Karlen et al. 1997). Number of aspects influence soil fertility, but biological designators are found as vitally consequential because soil microorganisms directly manipulate processes of soil ecosystem. Hence, protection of microbial diversity and biomass in the soil is one of the most important challenges for sustainable resource use as higher level of microbial diversity leads to more nutrient turnover (Torsvik and Øvreas 2002). Soil health refers to physical, chemical, and biological texture of the soil which is vital to get protracted sustainable agricultural productivity with least environmental impact (Moebius-Clune et al. 2016).

13.10 Physical Indicators

Physical indicators such as bulk density, porosity, texture, and aggregate stability of the soil are involved in erosion and water holding facility (Schoenholtz et al. 2000). Soil texture is determined by the arrangement of major soil particles like silt, sand, and clay and can be affected by cropping system(s) (Dexter 2004).

13.11 Chemical Indicators

Different chemicals are considered to provide nutrients to the soil and take part in determining healthy status of the soil. Organic matter and ion-exchange capacity of soil help in supplying nutrients like calcium and magnesium, and electrical conductivity indicates the number of ions in soil. The soil pH affects the solubility of minerals, activity of microorganisms, and nutrient status of the soil which affects soil health (Kelly et al. 2009). Chemical indicators are helpful in maintaining nutrient cycles and organic matter.

13.12 Microorganisms as Biological Indicator of Soil Health

Microorganisms have the capability to provide an integrated evaluation of soil health that cannot be acquired from higher organisms or by physical/chemical steps. Microbes respond rapidly to changes and thus adapt quickly to environmental situations and can be used to analyze the soil health. Population changes and microbial activities are an indication of changes in soil quality. Microorganisms respond quickly to environmental stress, because of their close relations. Alteration in microbial activity in soil can precede an early indication of soil perfection and degradation (Pankhurst et al. 1995).

13.13 Role of Microbial Enzymes in Maintenance of Soil Health

Activities of soil enzymes are directly related to microbial population to express metabolic requirements and available nutrients in a soil system (Nannipieri et al. 2002). These enzymes take part in the disintegration of unprocessed material in the soil. Enzymes catalyze many critical reactions needed for soil microorganism life processes and soil structure stabilization, organic waste disintegration, and nutrient cycling processes (Garcia et al. 2002) (Table 13.1). Application of various approaches to assess functional diversity in the soil through estimating enzyme

Table 13.1 Soil enzymes used as indicator of soil health

Enzymes involved	Function	Microbes involved	References
Amylase	Helps in breakdown of starch into simpler form (glucose and maltose and involved in carbon recycling)	Bacteria: *Bacillus subtilis*, *Pseudomonas* sp., *Bacillus amyloliquefaciens*, *Bacillus licheniformis*	Thomas et al. (1971)
		Fungi: *Aspergillus niger*	
Arylsulfatase	Helps in breakdown of complex sulfate esters into phenols/sulfur and involved in sulfur cycling	Bacteria: *Pseudomonas* sp., *Klebsiella* sp., *Raoultella* sp.	McGill and Cole (1981)
		Fungi: *Trichoderma* sp., *Eupenicillium* sp.	
β-glucosidase	Helps in cellobiose hydrolysis and involved in C-cycling	Bacteria: *Flavobacterium johnsoniae*, *Lactobacillus plantarum*, *Dyella koreensis*	Eivazi and Tabatabai (1988)
		Fungi: *Penicillium purpurogenum*, *Trichoderma*	
Cellulase	Helps in cellulose degradation and involved in C-cycling	*Bacteria: Bacillus subtilis*, *Acinetobacter junii*, *Cellulomonas biazotea*, *Pseudomonas cellulose*	Deng and Tabatabai (1994)
		Fungi: Aspergillus niger, *A. oryzae*, *Trichoderma reesei*, *Phanerochaete chrysosporium*, *Agaricus arvensis*	
Chitinase	Helps in chitin hydrolysis	Bacteria: *Bacillus*, *Streptomyces plicatus*, *S. halstedii*	Deshpande (1986)
		Fungi: *Aspergillus* sp., *Trichoderma* sp.	
Dehydrogenase	Helps in electron transport system, soil respiration, and C-cycling	Bacteria: *Pseudomonas entomophila*	Trevors (1984)
		Fungi:	
Fluorescein diacetate	Helps in hydrolysis of lipase, protease, and esterase in soil (overall microbial activities)	Bacteria: *Pseudomonas denitrificans*	Schnurer and Rosswall (1982)
		Fungi: *Fusarium culmorum*	
Invertase	Helps in sucrose hydrolysis, involved in C-cycling	Bacteria:	Shi et al. (2008)
		Fungi: *Aspergillus niger*, *Saccharomyces cerevisiae*, *Candida utilis*	
Phosphatase	Helps in conversion of insoluble phosphate to soluble form and involved in phosphorus cycling	Bacteria: *Bacillus*, *Pseudomonas*	Eivazi and Tabatabai (1977)
		Fungi: *Aspergillus*, *Penicillium*	
Protease	Helps in protein hydrolysis (amino acids) and involved in N-cycling	Bacteria: *Bacillus subtilis*	Ladd and Jackson (1982)
		Fungi: *Aspergillus niger*	
Urease	Helps in conversion of urea into NH_3 and CO_2 and involved in N-cycling	Bacteria: *Bacillus*	Rotini (1935)
		Fungi: *Trichosporon cutaneum*	

activities supports our perceptive of the relations among accessibility of resources, microbial community function, and ecosystem processes (Kumari et al. 2021).

Fluorescein diacetate is a nonspecific enzyme assay used for the estimation of esterase, lipase, and proteases which converts it into fluorescein. Overall, it provides information about the active microbial activity (bacteria, fungi, and living protist) in soil (Schnurer and Rosswall 1982). Dehydrogenase is also called DH or DHase. It is not extracellular but an integral part of intact cells. It provides information about the biological and microbial population activity in the soil which helps maintain soil fertility and health (Burns 1982). Phosphatases catalyze the ester and anhydride hydrolysis and release free phosphate in the soil. Two main types of phosphatases are found in the soil. Phosphomonoesterases are low molecular weight compounds and have monoester bond. Acid and alkaline phosphatases play important roles in plant nutrition. They are involved in mineralization of organic phosphorus to inorganic phosphorus in soil. Phosphodiesterases are less studied enzymes. They are occupied in degradation of phospholipids and nucleic acid and provide greater part of fresh organic P in the soil. Both enzymes are important to release free phosphate from phosphate diester (Benjamin and Philip 2005). β-glucosidase is the most important enzyme and plays key role in C-cycle in catalyzing the hydrolysis of a variety of β-glucosides present in plant waste (Eivazi and Tabatabai 1988). This enzyme is included in the group of disaccharide-hydrolyzing glucosidases. Also included among the glucosidases is alpha-glucosidase, which catalyzes the hydrolysis of alpha-D-glucopyranoside. α-Galactosidase and β-galactosidase (also referred to as lactase) are other glucosidases. In soil, β-glucosidase is more prevalent as compared to alpha-glucosidase and β-galactosidases. This enzyme catalyzes cellulose degradation to glucose, an essential life source of C energy for microorganisms in soil. Kumari et al. (2020) observed twofold increase in soil enzyme activities when inoculated with *Bacillus* spp. as compared to control which improves the soil quality and helped in plant growth of maize. Kukreti et al. (2020) found that application of PGPRs improved soil health by increasing the soil physicochemical analysis and soil FDA, dehydrogenase, and alkaline phosphatase activities. Chaudhary et al. (2021b) also observed that application of *Bacillus* sp. improved the soil health by raising the useful bacteria community of maize rhizospheric soil. Parul (2019) observed that application of *Pseudomonas*, *Pantoea agglomerans*, and *Bacillus* spp. improved the soil microbial population and enhanced maize growth and productivity. The indigenous microbes producing the soil enzymes play an important role in sustainable development and increased crop production (Goel et al. 2020; Kumar et al. 2020; Suyal et al. 2019a, b).

13.14 Conclusion

Application of chemical fertilizers and pesticides enhanced plant growth and protects plants from diseases but affects soil microflora as well as humans. PGPR opens a new door for the farmers to enhance productivity and protect plants from biotic and abiotic conditions. They play a vital function in nutrient cycling within the soil and improve the soil quality and soil health without any harmful effect on microbial population. Maintenance of the microbial diversity leads to sustainable development as soil ecosystem forms complex interrelations among the various entities of nature. For this reason, there is a significant need for research to clearly define the useful and required bacterial traits for different environmental conditions and plants in order to be able to select optimal bacterial strains for bioformulation. However, to provide a better perceptive of the biological efficacy of increased yields in the crop system, field experiments using bioformulations are required.

References

Agri U, Chaudhary P, Sharma A (2021) In vitro compatibility evaluation of agriusable nanochitosan on beneficial plant growth-promoting rhizobacteria and maize plant. Natl Acad Sci Lett. https://doi.org/10.1007/s40009-021-01047-w

Ahmad Z, Wu J, Chen L (2017) Isolated *Bacillus subtilis* strain 330-2 and its antagonistic genes identified by the removing PCR. Sci Rep 7:1777

Ali B, Sabri AN, Ljung K, Hasnainl S (2009) Auxin production by plant associated bacteria: impact on endogenous IAA content and growth of *Triticum aestivum L.* Lett Appl Microbiol 48:542–547

Andrade FM, Pereira TA, Souza TP, Guimaraes PHS, Martins AD, Schwan RF, Pasqual M, Doria J (2019) Beneficial effects of inoculation of growth-promoting bacteria in strawberry. Microbiol Res 223–225:120–128

Ansari FA, Ahmad I, Pichtel J (2019) Growth stimulation and alleviation of salinity stress to wheat by the biofilm forming *Bacillus pumilus* strain FAB10. Appl Soil Ecol 143:45–54

Asei R, Ewusi-Mensah N, Abaidoo RC (2015) Response of soybean (*Glycine max* L.) to rhizobia inoculation and molybdenum application in the Northern savannah zones of Ghana. J Plant Sci 3:64–70

Aslani F, Bagheri S, Julkapli NM, Juraimi AS, Hashemi FSG, Baghdadi A (2014) Effects of engineered nanomaterials on plants growth: an overview. Sci World J 2014:641759

Backer R, Rokem JS, Ilangumaran G, Lamont J, Praslickova D, Ricci E, Subramanian S, Smith DL (2018) Plant growth-promoting rhizobacteria: context, mechanisms of action, and roadmap to commercialization of biostimulants for sustainable agriculture. Front Plant Sci 9:1473

Benjamin LT, Philip MH (2005) Phosphatase activity in temperate pasture soils: potential regulation of labile organic phosphorus turnover by phosphodiesterase activity. Science Total Environ 344:27–36

Bhardwaj D, Ansari MW, Sahoo RK, Tuteja N (2014) Biofertilizers function as key player in sustainable agriculture by improving soil fertility, plant tolerance and crop productivity. Microb Cell Factories 13:66

Bharti N, Pandey S, Barnawal D (2016) Plant growth promoting rhizobacteria *Dietzia natronolimnaea* modulates the expression of stress responsive genes providing protection of wheat from salinity stress. Sci Rep 6:34768

Bhatt P, Pal K, Bhandari G, Barh A (2019a) Modeling of methyl halide biodegradation on bacteria and its effect on other environmental systems. Pest Biochem Physiol 158:88–100

Bhatt P, Gangola S, Chaudhary P, Khati P, Kumar G, Sharma A, Srivastava A (2019b) Pesticide induced up-regulation of esterase and aldehyde dehydrogenase in indigenous *Bacillus* spp. Biorem J 23(1):42–52

Bidgoli D, Azarnezhad N, Akhbari M, Ghorbani M (2019) Salinity stress and PGPR effects on essential oil changes in *Rosmarinus officinalis* L. Agric Food Secur 8:2

Bünemann EK, Bongiorno G, Bai Z, Creamer RE, de Deyn G, de Goede RG, Fleskens L, Geissen V, Kuyper TW, Mäder P, Pulleman MM, Sukkel W, van Groenigen JW, Brussaard L (2018) Soil quality—a critical review. Soil Biol Biochem 120:105–125

Burns RG (1982) Enzyme activity in soil: location and a possible role in microbial ecology. Soil Biol Biochem 14:423–427

Calvo P, Zebelo S, McNear D, Kloepper J, Fadamiro H (2019) Plant growth-promoting rhizobacteria induce changes in *Arabidopsis thaliana* gene expression of nitrate and ammonium uptake genes. J Plant Interact 14:224–231

Chakraborty U, Chakraborty B, Basnet M (2006) Plant growth promotion and induction of resistance in *Camellia sinensis* by *Bacillus megaterium.* J Basic Microbiol 46:186–195

Chaudhary P, Sharma A (2019) Response of nanogypsum on the performance of plant growth promotory bacteria recovered from nanocompound infested agriculture field. Environ Ecol 37:363–372

Chaudhary S, Shankar A, Singh A, Prasad V (2018) Usefulness of *Penicillium* in enhancing plants resistance to abiotic stresses: an overview. In: Gupta V (ed) New and future developments in microbial biotechnology and bioengineering. Elsevier, Amsterdam, pp 277–284

Chaudhary P, Khati P, Chaudhary A, Gangola S, Kumar R, Sharma A (2021a) Bioinoculation using indigenous *Bacillus* spp. improves growth and yield of *Zea mays* under the influence of nanozeolite. 3 Biotech 11:11

Chaudhary P, Sharma A, Chaudhary A, Sharma P, Gangola S, Maithani D (2021b) Illumina based high throughput analysis of microbial diversity of rhizospheric soil of maize infested with nanocompounds and *Bacillus* sp. Appl Soil Ecol 159:103836

Chaudhary P, Khati P, Gangola S, Kumar A, Kumar R, Sharma A (2021c) Impact of nanochitosan and Bacillus spp. on health, productivity and defence response in Zea mays under field condition. 3 Biotech 11:237. https://doi.org/10.1007/s13205-021-02790-z

Chaudhary P, Khati P, Chaudhary A, Maithani D, Kumar G, Sharma A (2021d) Cultivable and metagenomic approach to study the combined impact of nanogypsum and Pseudomonas taiwanensis on maize plant health and its rhizospheric microbiome. PLoS One 16(4): e0250574. https://doi.org/10.1371/journal.pone.0250574

Chitarra W, Pagliarani C, Maserti B, Lumini E, Siciliano I, Cascone P (2016) Insights on the impact of arbuscular mycorrhizal symbiosis on tomato tolerance to water stress. Plant Physiol 171:1009–1023

Curá JA, Franz DR, Filosofía JE, Balestrasse KB, Burgueño LE (2017) Inoculation with *Azospirillum* sp and *Herbaspirillum* sp bacteria increases the tolerance of maize to drought stress. Microorganisms 5:41

de Lima BC, Moro AL, Santos ACP, Bonifacio A, Araujo ASF, de Araujo FF (2019) *Bacillus subtilis* ameliorates water stress tolerance in maize and common bean. J Plant Interact 14:432–439

Defez R, Andreozzi A, Dickinson M (2017) Improved drought-stress response in alfalfa plants nodulated by an IAA over-producing *Rhizobium* strain. Front Microbiol 8:2466

Deng SP, Tabatabai MA (1994) Cellulose activity in soils. Soil Biol Biochem 26:1347–1354

Deshpande HV (1986) Enzymatic degradation of chitin and its biological applications. J Sci Ind Res 45:273–281

Dexter AR (2004) Soil physical quality: Part I. Theory, effects of soil texture, density, and organic matter, and effects on root growth. Geoderma 120:201–214

Di Benedetto NA, Corbo MR, Campaniello D, Cataldi MP, Bevilacqua A, Sinigaglia M, Flagella Z (2017) The role of plant growth promoting bacteria in improving nitrogen use efficiency for sustainable crop production: a focus on wheat. AIMS Microbiol 3:413–434

Eivazi F, Tabatabai MA (1977) Phosphatases in soils. Soil Biol Biochem 9:167–172

Eivazi F, Tabatabai MA (1988) Glucosidases and galactosidases in soils. Soil Biol Biochem 20:601–606

Fernández MP, Alexander V (2017) Enhanced plant performance in *Cicer arietinum* L. due to the addition of a combination of plant growth-promoting bacteria. Agriculture 7:40

Garcia C, Hernandez T, Roldan A, Martin A (2002) Effect of plant cover decline on chemical and microbiological parameters under Mediterranean climate. Soil Biol Biochem 34:635–642

Ghevariya KK, Desai PB (2014) Rhizobacteria of sugarcane: *in vitro* screening for their plant growth promoting potentials. Res J Rec Sci 3:52–58

Goel R, Debbarma P, Kumar P, Suyal DC, Kumar S, Mahapatra BS (2020) Assessment of soil chemical quality, soil microbial population and plant growth parameters under organic and conventional rice–wheat cropping system. Agric Res 10:193–204. https://doi.org/10.1007/s40003-020-00499-8

Haney RL, Haney EB, Smith DR, Harmel RD, White MJ (2018) The soil health tool—theory and initial broad-scale application. Appl Soil Ecol 125:162–168

Hashem A, Tabassum B, Fathi Abd Allah E (2019) *Bacillus subtilis*: a plant-growth promoting rhizobacterium that also impacts biotic stress. Saudi J Biol Sci 26:1291–1297

He Y, Wu Z, Wang W, Ye BC, Zhang F, Liu X (2019) Different responses of *Capsicum annuum* L. root and shoot to salt stress with *Pseudomonas putida* Rs-198 inoculation. J Plant Growth Regul 38:799–811

Heydarian Z, Yu M, Gruber M, Glick BR, Zhou R, Hegedus DD (2016) Inoculation of soil with plant growth promoting bacteria producing 1-aminocyclopropane-1-carboxylate deaminase or expression of the corresponding acdS gene in transgenic plants increases salinity tolerance in *Camelina sativa*. Front Microbiol 7:966

Huang Y, Zhang W, Pang S, Chen J, Bhatt P, Mishra S, Chen S (2021) Insights into the microbial degradation and catalytic mechanism of chlorpyrifos. Environ Res 192:110660

Karlen DL, Mausbach MJ, Doran JW, Cline RG, Harris RF, Schuman GE (1997) Soil quality: a concept, definition, and framework for evaluation (a guest editorial). Soil Sci Soc Am J 61:4–10

Kaushal M, Wani SP (2016) Plant growth-promoting rhizobacteria: drought stress alleviators to ameliorate crop production in drylands. Ann Microbiol 66:35–42

Kelly B, Allan C, Wilson BP (2009) Soil indicators and their use by farmers in the Billabong Catchment, southern New South Wales. Aust J Soil Res 47:234–242

Khati P, Chaudhary P, Gangola S, Bhatt P, Sharma A (2017) Nanochitosan supports growth of *Zea mays* and also maintains soil health following growth. 3 Biotech 7:81

Khati P, Parul, Bhatt P, Nisha KR, Sharma A (2018) Effect of nanozeolite and plant growth promoting rhizobacteria on maize. 3 Biotech 8:141

Khati P, Sharma A, Gangola S, Kumar R, Bhatt P, Kumar G (2019a) Impact of some agriusable nanocompounds on soil microbial activity: an indicator of soil health. Clean Soil Air Water 45:1600458

Khati P, Chaudhary P, Gangola S, Sharma P (2019b) Influence of nanozeolite on plant growth promotory bacterial isolates recovered from nanocompound infested agriculture field. Environ Ecol 37:521–527

Kloepper JW, Schroth MN (1978) Proceedings of the 4th international conference on plant pathogenic bacteria, Gibert-Clorey, pp 879–882

Kukreti B, Sharma A, Chaudhary P, Agri U, Maithani D (2020) Influence of nanosilicon dioxide along with bioinoculants on *Zea mays* and its rhizospheric soil. 3 Biotech 10:345

Kumar A, Patel JS, Bahadur I, Meena VS (2016) The molecular mechanisms of KSMs for enhancement of crop production under organic farming. In: Meena VS, Maurya BR, Verma JP, Meena RS (eds) Potassium solubilizing microorganisms for sustainable agriculture. Springer, New Delhi, pp 61–75

Kumar S, Suyal DC, Yadav A, Shouche Y, Goel R (2020) Psychrophilic *Pseudomonas helmanticensis* proteome under simulated cold stress. Cell Stress Chaperones 25 (6):1025–1032. https://doi.org/10.1007/s12192-020-01139-4

Kumar A, Sharma A, Chaudhary P, Gangola S (2021) Chlorpyrifos degradation using binary fungal strains isolated from industrial waste soil. Biologia. https://doi.org/10.1007/s11756-021-00816-8

Kumari S, Sharma A, Chaudhary P, Khati P (2020) Management of plant vigor and soil health using two agriusable nanocompounds and plant growth promotory rhizobacteria in Fenugreek. 3 Biotech 10:461

Kumari H, Khati P, Gangola S, Chaudhary P, Sharma A (2021) Performance of plant growth promotory rhizobacteria on maize and soil characteristics under the influence of TiO2 nanoparticles. Pantnagar J Res 19:28–39

Kyei-Boahen S, Savala CEN, Chikoye D, Abaidoo R (2017) Growth and yield responses of cowpea to inoculation and phosphorus fertilization in different environments. Front Plant Sci 8:646

Ladd JN, Jackson RB (1982) In: Stevenson FJ (ed) Nitrogen in agricultural soils. American Society of Agronomy, Madison, pp 173–228

Liu D, Yang Q, Ge K, Hu Z, Guozhen Q, Binghai D, Liu K, Ding Y (2017) Promotion of iron nutrition and growth on peanut by *Paenibacillus illinoisensis* and *Bacillus* sp. strains in calcareous soil. Braz J Microbiol 48:656–670

Liu H, Xiong W, Zhang R, Hang X, Wang D, Li R (2018) Continuous application of different organic additives can suppress tomato disease by inducing the healthy rhizospheric microbiota through alterations to the bulk soil microflora. Plant Soil 423:229–240

Lopez-Monejar R, Voriskov AJ, Vetrovský T, Baldrian P (2015) The bacterial community inhabiting temperate deciduous forests is vertically stratified and undergoes seasonal dynamics. Soil Biol Biochem 87:43–50

Malusa E, Sas-Paszt L, Ciesielska J (2012) Technologies for beneficial micro-organisms inoculation used as biofertilizers. Sci World J 2012:491206

Mann C, Lynch D, Fillmore S, Mills A (2019) Relationships between field management, soil health, and microbial community composition. Appl Soil Ecol 144:12–21

McBratney A, Field DJ, Koch A (2014) The dimensions of soil security. Geoderma 213:203

McGill WB, Cole CV (1981) Comparative aspects of cycling of organic C, N, S and P through soil organic matter. Geoderma 26:267–268

Mishra RK, Prakash O, Alam M, Dikshit A (2010) Influence of plant growth promoting rhizobacteria (PGPR) on the productivity of *Pelargonium graveolens*. Rec Res Sci Technol 2 (5):53–57

Moebius-Clune BN, Moebius-Clune DJ, Gugino BK, Idowu OJ, Schindelbeck RR, Ristow AJ, van Es HM, Thies JE, Shayler HA, McBride MB, Kurtz KSM, Wolfe DW, Abawi GS (2016) Comprehensive assessment of soil health: the Cornell framework manual, 3.1 edn. Cornell University, Ithaca

Mokrani S, Belabid L, Nabti E (2019) Antifungal activity of some non-fluorescent *Pseudomonas* bacteria isolated from *Phaseolus vulgaris* L rhizosphere. Eur J Biotechnol Biosci 7:38–44

Murali M, Amruthesh KN, Sudisha J, Niranjana SR, Shetty HS (2012) Screening for plant growth promoting fungi and their ability for growth promotion and induction of resistance in pearl millet against downy mildew disease. J Phytology 4:30–36

Nannipieri P, Kandeler E, Ruggiero P (2002) Enzyme activities and microbiological and biochemical processes in soil. In: Burns RG, Dick R (eds) Enzymes in the environment: activity, ecology, and applications. Marcel Dekker, New York, pp 1–34

Nieto-Jacobo MF, Steyaert JM, Salazar-Badillo FB, Nguyen DV, Rostás M, Braithwaite M, De Souza JT, Jimenez-Bremont JF, Ohkura M, Stweart A (2017) Environmental growth conditions of *Trichoderma* spp affects indole acetic adic derivates, volatile organic compounds, and plant growth promotion. Front Plant Sci 8:102

Numan M, Bashir S, Khan Y, Mumtaz R, Shinwari ZK, Khan AL, Khan A, Harrasi AAL (2018) Plant growth promoting bacteria as an alternative strategy for salt tolerance in plants: a review. Microbiol Res 209:21–32

Pahari A, Pradhan A, Maity S, Mishra BB (2017) Carrier based formulation of plant growth promoting *bacillus* species and their effect on different crop plants. Int J Curr Microbiol Appl Sci 6:379–385

Pankaj, Bisht TS, Pathak VM, Barh A, Chandra D (2015a) Optimization of amylase production from the fungal isolates of Himalayan region Uttarakhand. Ecol Environ Conserv 21 (3):1517–1521

Pankaj, Negi G, Gangola S, Khati P, Srivastava A, Sharma A (2015b) Optimization of sulfosulfuron biodegradation through response surface methodology using indigenous bacterial strain isolated from contaminated agriculture field. Int J Curr Microbiol Appl Sci 4(8):105–112

Pankaj, Negi G, Gangola S, Khati P, Kumar G, Srivastava A, Sharma A (2016a) Differential expression and characterization of cypermethrin degrading potential proteins in *Bacillus thuringiensis* strain, SG4. 3 Biotech 6:225

Pankaj, Sharma A, Gangola S, Khati P, Kumar G, Srivastava A (2016b) Novel pathway of cypermethrin biodegradation in a *Bacillus* sp. strain SG2 isolated from cypermethrin-contaminated agriculture field. 3 Biotech 6:45

Pankhurst CE, Hawke BG, McDonald HJ, Kirkby CA, Buckerfield JC, Michelsen P, O'Brien KA, Gupta VVSR, Doube BM (1995) Evaluation of soil biological properties as potential bioindicators of soil health. Austr J Exp Agric 35:1015–1028

Parul (2019) Application of conventional and metagenomic approaches to analyse the effect of nanocompounds and indigenous bioinoculants on the health of soil and *Zea mays*. PhD thesis. G.B.P.U.A & T, Pantnagar

Passera A, Venturini G, Battelli G, Casati P, Penaca F, Quaglino F, Bianco PA (2017) Competition assays revealed *Paenibacillus pasadenensis* strain r16 as a novel antifungal agent. Microbiol Res 198:16–26

Radhakrishnan R, Hashem A, Abd Allah EF (2017) *Bacillus*: a biological tool for crop improvement through bio-molecular changes in adverse environments. Front Physiol 8:667

Rahman M, Sabir AA, Mukta JA (2018) Plant probiotic bacteria Bacillus and Paraburkholderia improve growth, yield and content of antioxidants in strawberry fruit. Sci Rep 8:2504

Ripa FA, Cao W, Tong S, Sun J (2019) Assessment of plant growth promoting and abiotic stress tolerance properties of wheat endophytic fungi. Biomed Res Int 12:6105865

Rotini OT (1935) La trasformazioneen zimaticadell'ureanelterreno. Ann Labor Ric Ferm Spallanrani 3:143–154

Rozier C, Gerin F, Czarnes S, Legendre L (2019) Biopriming of maize germination by the plant growth-promoting rhizobacterium *Azospirillum lipoferum* CRT1. J Plant Physiol 237:111–119

Sangamesh MB, Jambagi S, Vasanthakumari MM (2018) Thermotolerance of fungal endophytes isolated from plants adapted to the Thar Desert, India. Symbiosis 75:135–147

Santos ADA, Silveira JAGD, Bonifacio A, Rodrigues AC, Figueiredo MD, Santos VB (2018) Antioxidant response of cowpea co-inoculated with plant growth-promoting bacteria under salt stress. Braz J Microbiol 49:513–521

Santoyo S, Moreno-Hagelsieb M, Orozco-Mosqueda DC, Glick BR (2016) Plant growth-promoting bacterial endophytes. Microbiol Res 183:92–99

Schnurer J, Rosswall T (1982) Fluorescein diacetate hydrolysis as a measure of total microbial activity in soil and litter. Appl Environ Microbiol 43:1256–1261

Schoenholtz SH, VamMiegroet H, Burger JA (2000) A review of chemical and physical properties as indicator of forest soil quality: challenge and opportunities. For Ecol Manag 138:335–356

Shah S, Shrestha R, Maharjan S, Selosse M-A, Pant B (2019) Isolation and characterization of plant growth-promoting endophytic fungi from the roots of *Dendrobium moniliforme*. Plan Theory 8:5

Sharghi A, Badi HN, Bolandnazar S, Mehrafarin A, Reza M, Sarikhani MZ (2019) Morphophysiological and phytochemical responses of fenugreek to plant growth promoting rhizobacteria (PGPR) under different soil water levels. Folia Horticult 30:215–228

Sharma S, Chen C, Navathe S (2019) A halotolerant growth promoting rhizobacteria triggers induced systemic resistance in plants and defends against fungal infection. Sci Rep 9:4054

Sheirdil RA, Hayat R, Zhang X, Abbasi NA, Ali S, Ahmed M, Khattak JZK, Ahmad S (2019) Exploring potential soil bacteria for sustainable wheat (*Triticum aestivum* L.) production. Sustainability 11:3361

Shen Z, Ruan Y, Chao X, Zhang J, Li R, Shen Q (2015) Rhizosphere microbial community manipulated by 2 years of consecutive biofertilizer application associated with banana *Fusarium* wilt disease suppression. Biol Fertil Soils 51:553–562

Shi JZ, Lu Y, Xu ZG, Fu SL (2008) Enzyme activities of urban soils under different land use in the Shenzhen city, China. Plant Soil Environment 54(8):341–346

Singh K, Gera R, Sharma R, Maithani D, Chandra D, Bhat AM, Kumar R, Bhatt P (2021) Mechanism and application of *Sesbania* root nodulating bacteria: an alternative for chemical fertilizers and sustainable development. Arch Microbiol 203:1259–1270. https://doi.org/10.1007/s00203-020-02137-x

Suyal DC, Joshi D, Kumar S, Soni R, Goel R (2019a) Differential protein profiling of soil diazotroph *Rhodococcus qingshengii* S10107 towards low-temperature and nitrogen deficiency. Sci Rep 9(20378):1–9. https://doi.org/10.1038/s41598-019-56592-8

Suyal DC, Kumar S, Joshi D, Yadav A, Shouche Y, Goel R (2019b) Comparative overview of red kidney bean (*Phaseolus valgaris*) rhizospheric bacterial diversity in perspective of altitudinal variations. Biologia 74(10):1405–1413. https://doi.org/10.2478/s11756-019-00292-1

Thomas JA, Rao GV, Brothers C, Spradlin J, Li LH (1971) Subsite mapping of enzymes. Correlation of product patterns with Michaelis parameters and substrate-induced strain. J Biol Chem 246:5621–5635

Thomson BC, Tisserant E, Plassart P, Uroz S, Griffiths RI, Hannula SE, Buee M, Mougel C, Ranjard L, Van Veen JA, Martin F, Bailey MJ, Lemanceau P (2015) Soil conditions and land use intensification effects on soil microbial communities across a range of European field sites. Soil Biol Biochem 88:403–413

Torsvik V, Øvreas L (2002) Microbial diversity and function in soil: from genes to ecosystems. Curr Opin Microbiol 5:240–245

Trevors JT (1984) Effect of substrate concentration, inorganic nitrogen, O_2 concentration, temperature, pH on dehydrogenase activity in soil. Plant Soil 77:285

Tyagi J, Varma A, Pudake RN (2017) Evaluation of comparative effects of arbuscular Mycorrhiza (*Rhizophagus intraradices*) and endophyte (*Piriformospora indica*) association with finger millet (*Eleusine coracana*) under drought stress. Eur J Soil Biol 81:1–10

Vejan P, Abdullah R, Khadiran T, Ismail S, Boyce AN (2016) Role of plant growth promoting rhizobacteria in agricultural sustainability—a review. Molecules 21:1–17

Vessy JK (2013) Plant growth promoting rhizobacteria as biofertilizers. Plant Soil 255(2):571–586

Vimal R, Singh JS, Arora NK, Singh S (2017) Soil-plant-microbe interactions in stressed agriculture management: a review. Pedosphere 27:177–192

Vimal SR, Patel VK, Singh JS (2019) Plant growth promoting *Curtobacterium albidum* strain SRV4: an agriculturally important microbe to alleviate salinity stress in paddy plants. Ecol Indic 105:553–562

Wang X, Wang C, Sui J (2018) Isolation and characterization of phosphofungi, and screening of their plant growth-promoting activities. AMB Expr 8:63

Wu L, Shang H, Wang Q, Gu H, Liu G, Yang S (2016) Isolation and characterization of antagonistic endophytes from *Dendrobium candidum* Wall ex Lindl., and the biofertilizing potential of a novel *Pseudomonas saponiphila* strain. Appl Soil Ecol 105:101–108

Yadav AN, Kumar V, Dhaliwal HS, Prasad R, Saxena AK (2018) Microbiome in crops: diversity, distribution, and potential role in crop improvement. In: Crop improvement through microbial biotechnology. Elsevier, Amsterdam, pp 305–332

Yang S, Xu Z, Wang R, Zhang Y, Yao F, Zhang Y (2017) Variations in soil microbial community composition and enzymatic activities in response to increased N deposition and precipitation in Inner Mongolian grassland. Appl Soil Ecol 119:275–285

York LM, Carminati A, Mooney SJ, Ritz K, Bennett MJ (2016) The holistic rhizosphere: integrating zones, processes, and semantics in the soil influenced by roots. J Exp Bot 67:3629–3643

Zhang LN, Wang DC, Hu Q, Dai XQ, Xie YS, Li Q, Liu HM, Guo JH (2019) Consortium of plant growth-promoting rhizobacteria strains suppresses sweet pepper disease by altering the rhizosphere microbiota. Front Microbiol 10:1668

Zhang W, Pang S, Lin Z, Mishra S, Bhatt P, Chen S (2021) Biotransformation of perfluoroalkyl acid precursors from various environmental system: advances and perspectives. Environ Pollut 272:115908

Zhao LJ, Peralta-Videa JR, Rico CM, Hernandez-Viezcas JA, Sun YP, Niu GH (2014) CeO_2 and ZnO nanoparticles change the nutritional qualities of cucumber (*Cucumis sativus*). J Agric Food Chem 62:2752–2759

Zhou LS, Tang K, Guo SX (2018) The plant growth-promoting fungus (PGPF) Alternaria sp. A13 markedly enhances Salvia miltiorrhiza root growth and active ingredient accumulation under greenhouse and field conditions. Int J Mol Sci 19:270

Chapter 14
Reinstating Microbial Diversity in Degraded Ecosystems for Enhancing Their Functioning and Sustainability

Sachini Wayanthimali Meepegamage, Ambalangodage Thilini Dhanushka Rathnathilake, Mahesh Premarathna, and Gamini Seneviratne

Abstract Biodiversity is the variety of life on earth, starting from genes, individual species, and communities to the whole ecosystems. The biodiversity and complex interaction networks of its components play a crucial role in regulating processes in different ecosystems' functioning and sustainability. Global biodiversity has been declining rapidly due to human impacts like land-use change, urbanization, environmental pollutions, and also the resultant climate change, leading to losing ecosystem functioning and sustainability. This chapter discusses the biodiversity, the causes of its degradation, and ways to reverse it. Interestingly, application of advanced microbial formulations to the soil has been shown to be capable of reinstating the lost biodiversity in agroecosystems. One such formulation is biofilm microbial ameliorators [BMAs, e.g., biofilm biofertilizers (BFBFs)]. Once applied to the soil, they break the dormancy of microbial seed bank formed to circumvent the stress of agricultural practices, thus re-establishing the biodiversity to a considerable extent for improved ecosystem functioning and sustainability. The same mechanism has been shown to be instrumental in environmental bioremediation. Fascinatingly, the potential of BMAs even in reinstating the biodiversity of disease-proven human gut microbiota has been reported for improved human health. It is also important to note that the past social impacts on decreasing biodiversity have now boomeranged to humans' existence. The application of microbial biotechnologies like developed BMAs could mitigate such devastating events in the future.

Keywords Biodiversity · Biofilms · Health · Sustainability

S. W. Meepegamage · A. T. D. Rathnathilake · M. Premarathna · G. Seneviratne (✉)
Microbial Biotechnology Unit, National Institute of Fundamental Studies, Kandy, Sri Lanka
e-mail: sachini.me@nifs.ac.lk; thilini.ra@nifs.ac.lk; mahesh.pr@nifs.ac.lk; gamini.se@nifs.ac.lk

P. Bhatt et al. (eds.), *Microbial Technology for Sustainable Environment*,
https://doi.org/10.1007/978-981-16-3840-4_14

14.1 Introduction

Biodiversity is the variety and variability of life on earth at all its levels, starting from genes, individual species, and communities to whole ecosystems. Genetic diversity is the variety of genes in a given species. Individuals within a species contain their genetic compositions and lead to make different populations in a species. The variety of species within a habitat is known as species diversity. According to similar characteristics shared by species, they group into genera, families, order, etc. About 99% of animal species are invertebrates such as insects, worms, snails, crabs, corals, and sea stars. Most of the invertebrates are insects which are responsible for the pollination of plants and recycling of nutrients. Mammals represent only less than 1% of the animal species. As a whole, mainly plants and animals constitute the visible ecosystem diversity, which is the variety of individuals in a particular area or a place. However, the functional ecosystem is an outcome of both visible and invisible difference and its interactions with the physical environment.

Soil ecosystem consists of its loose outermost layer of soil and the living and non-living component in it. The living component includes root of the plant and soil fauna and flora. The soil fauna is divided into three categories: microfauna, mesofauna, and macrofauna (Kulasooriya et al. 2017). The microfauna includes nematodes, protozoa, ciliates, and so on, and mesofauna consists of ants, insects, and microarthropods. Earthworms, snails, spiders, and mammals belong to macrofauna. Generally, the soil fauna is involved mainly in manipulating carbon fluxes in the ecosystem (Lavelle 1996). Microflora includes microscopic alga, archaeobacteria, eubacteria, actinobacteria, and fungi. The soil microflora continuously interacts with the plant root system and involves in the process like organic molecular decomposition, nitrogen fixation, phosphorous solubilization, etc.

Primarily, all the functions of ecosystems are influenced by diverse microbial communities in the soil (Wittebolle et al. 2009). Human impacts on ecosystems like agricultural and environmental activities tend to decrease soil microbial diversity and abundance, leading to depleting ecosystem processes. Therefore, restoration of the degraded soil microbiome is vital for the sustainability of managed ecosystems like agroecosystems. The conventional method of replenishing the lost soil microbial diversity is the inoculation of individual microbes as mono or mixed cultures, known as biofertilization. Except for legume-rhizobium symbiosis, this has not been that attractive to non-legumes due to low efficacy. However, developed microbial biofilms have proven the potential to be used as a novel biotechnological means to reinstate the lost microbial diversity in agroecosystems and possibly in natural ecosystems (Herath et al. 2017). Therefore, this chapter discusses the effect of the developed biofilms on re-establishing the lost microbial diversity in agriculture and the environment and also the health implications of the biofilm use.

14.2 Ecosystem Networks

Interactions in ecosystems lead to developing biological networks which are grouped into molecular, cellular, and population levels. Networks in population level include ecological networks, epidemiological networks, and food-web networks (He and Petoukhov 2011). Networks in the cellular level consist of neuronal networks and immunological networks. Molecular-level networks include gene regulatory networks, metabolic networks, and protein interaction networks. Gene regulatory systems include interactions among DNA, RNA, protein, and other molecules. For example, each mRNA molecule is responsible for the production of a specific protein or a set of proteins, and the mRNA molecules and proteins interact with each other, and the outcome of it decides whether they are diffused around the cell or bound to the cell membrane. In the protein interaction networks, proteins with the same function interact through protein-to-protein interactions to connect physically and to make larger macromolecular assemblies. These protein complexes link to each other via the next level protein-to-protein interactions to form interaction networks, which are involved in the cellular process. Metabolic systems consist of chemical reactions of metabolism and regulatory interactions which guide these chemical reactions. In immunological networks, plant immune signaling network is stimulated due to pathogen attack, and the stimulus regulates the defense to the attack (Sato et al. 2010). Ecological systems consist of biotic interactions between organisms in ecosystems. For example, these biotic interactions can modify stressor effects, which may be transferred to a distant group of organisms, thus generating new stressor interactions (Bruder et al. 2019).

Ecological networks can also be seen in microbial community interactions. The interactions among different taxa within a soil microbial community are linked directly or indirectly through intermediate species (Lupatini et al. 2014). These inter-taxa connections allow understanding of soil microbial community structure. In soil bacterial communities, most of the soil bacterial genera are essential for network connectance. However, only a few genera like proteobacteria and actinobacteria play a crucial role as connectors. Moreover, each ecosystem consists of a different set of keystone genera, and the other genera only show a general distribution. Species interactions are significant to soil processes, especially in complex ecosystems.

Biodiversity and their interaction networks play a crucial role in regulating processes in different ecosystems' functioning and sustainability. In soil ecosystems, diverse soil organisms are involved in ecosystem functions and sustainability. Invertebrates contribute to the formation of soil structure which governs hydrologic processes and gas exchange. Microbes and plant root interaction control soil carbon sequestration, detoxification, nutrient cycling, organic matter decomposition, and suppression of pests, parasites, and diseases. The interaction also helps produce sources of medicines (Bunning and Jiménez 2003). Among soil organisms, microbial diversity plays an essential role in ecosystem functioning. Most of the soil-borne diseases are caused by pathogenic fungi which are suppressed by the high beneficial microbial biomass, thus generating competition for nutrients.

Moreover, pathogenic species are suppressed by specific antagonists (Brussaard et al. 2007). Rhizobia, mycorrhizae, actinobacteria, diazotrophic bacteria, other rhizosphere microorganisms and ants are responsible for symbiotic relationships with plants and their roots. Different types of litter feeding invertebrates (detritivores), fungi, bacteria, actinobacteria, and some other microorganisms are involved in organic matter decomposition (Bunning and Jiménez 2003). Moreover, soil microbes are involved even in determining plant species diversity in natural ecosystems like forests (Mangan et al. 2010). In grasslands, high plant species richness contributes to increased plant productivity and soil nitrogen utilization, thus lowering nitrogen leaching loss. In deep-sea ecosystems, too, biodiversity is the key for marine ecosystem functioning and sustainability. Therefore, all the above-stated evidence supports the fact that biodiversity is the impetus for functioning and sustainability of any ecosystem.

14.3 Degradation of Biodiversity

Biodiversity in different ecosystems is declining rapidly due to habitat fragmentation, land-use change, deforestation, climate change, urbanization, agricultural practices, and environmental pollutions, leading to losing ecosystem functioning and the sustainability through disrupting all types of the networks. Habitat fragmentation is a process in which a significant habitat is broken into small patches of areas which are then not connected as the original habitat. It reduces habitat area and species connectivity. This can happen through natural or anthropogenic impacts in any ecosystem. Degradation of forest ecosystem biodiversity is mainly driven by forest fire, acid rains, and human activities such as deforestation. Acid rains occur due to climatic change and oil exploration (Nduka et al. 2008; Wei et al. 2017), altering soil faunal community and vertical distribution of soil faunal groups. To avoid these, acid stress nematodes and other soil faunas move deep into the soil, which adversely affects them, thus eventually altering the decomposition of organic matter and greenhouse gas emission (Wei et al. 2017). Land-use change refers to the transformation of land to farmlands, grazing areas, human habitation, and urban centers. These alternations are interconnected with loss of biodiversity and degradation of land. It has been found that the diversity of indigenes plants and animals and plant cover are lost due to loss of native vegetation (Maitima et al. 2009). Land-use change due to agriculture intensification causes the species loss from soil food webs and reduction of average body mass of soil fauna. Further, alteration of the diversity of a single group in a food web can adversely effect on the diversity of another group (Tsiafouli et al. 2014). Eventually, this may lead to an increase in the pressure of pathogens and pests in agriculture (Plaas et al. 2019).

Growing demand for agricultural produces leads to more and more transformation of land cover to agroecosystems. In turn, the expansion of croplands with intensified agriculture tends to decrease soil fertility and moisture while increasing soil erosion (Maitima et al. 2009). Here, the degradation of microbial diversity in agroecosystems is mainly due to tillage and excessive applications of chemical fertilizers and other agrochemicals. These chemical inputs are dispersed through abiotic processes such as diffusion or transport by water or biotic processes such as meta-ecosystem effects, thus changing diversity and composition of natural communities in agroecosystems (Seneviratne and Kulasooriya 2013). Consequently, degradation of agro-biodiversity influences adversely on ecosystem functioning and sustainability.

14.4 Reinstating the Lost Biodiversity

There is a renewed interest in searching alternatives to reduce the use of chemical inputs in agriculture and to reintroduce the lost soil microbiome. Among other countries, Sri Lankan government has also taken actions to reduce the use of chemical inputs. However, such actions should be acceptable to farmers in terms of the yields of the crop and income of the farmer. The non-synthetic, organic fertilizers named green manure, animal dung, and compost manure which are prepared on a domestic or industrial level are the traditional alternatives for chemical fertilizers. Due to bulkiness of those alternatives, they are not much attractive to farmers who have been using less bulky granulated or liquid forms of chemical inputs. Transportation, storage, and field application of organic fertilizers would be time consuming and need additional labor and cost. The organic fertilizers take time to decompose and release nutrients to the specific crop plants, and the rate of nutrient release is not enough for the demand of short-term crops. To address this, some research studies have been conducted in certain countries to increase the decomposition of the organic fertilizers using soil microbes named efficient microbial inoculants (EMIs) (Kulasooriya et al. 2017). It has been found out that imported EMIs do not give the predicted outcomes in Sri Lanka (Kulasooriya et al. 2017), and minimal studies are in progress in this vital field. Optionally, a desirable substitute has been the establishment and introduction of microbial biofertilizers which are living microbial formulations that live in close contact with crop plants and increase nutrient use efficiency and provide other benefits to the ecosystem. One such formulation is biofilm microbial ameliorators [BMAs, e.g., biofilm biofertilizers (BFBFs)], a novel concept introduced by the National Institute of Fundamental Studies (NIFS) in Sri Lanka.

14.5 Developed BMAs as Agents of Reinstating Biodiversity

It is a known fact that soil microbes play a crucial role in the ecosystem, even though they are microscopic. They are not free living, swimming (planktonic) single cells in nature. Patel et al. (2014) explained that microorganisms tend to live in communities due to numerous benefits they can acquire in them. They stick together and form film-like structures, named microbial biofilms. Different microbial genera are capable of developing biofilms through effective communication. There are three types of biofilms that can be identified in the soil, viz., bacterial biofilms, fungal biofilms, and fungal-bacterial biofilms (FBBs) (Seneviratne et al. 2008). Nitrogen-fixing bacteria which are attached to fungal surface surrounded by extracellular polymeric substances (EPSs) are generally called FBBs. Seneviratne and Jayasinghearachchi (2003) for the first time reported the in vitro development of FBBs by colonizing fungal mycelia using bradyrhizobial and azorhizobial strains.

The EPS that covers the FBBs is less permeable to gases, and hence it generates a microaerobic state near N_2-fixing bacteria (Seneviratne et al. 2017). That enables them to fix N_2 that is moderately dissolved in the EPS. Nutrients generated via N_2 fixation in the N_2 fixers such as ammonium (NH_4^+), amino acids, and some proteins may be transferred to neighboring fungal mycelium to fulfill the nutritional requirement. In return, carbon sources may be provided by the fungal mycelium as the energy source to the N_2 fixers. In this way, fungi and bacteria in the FBBs develop a metabolic relationship in the symbiosis. Moreover, it is also reported that the FBBs secrete hormones such as IAA, many organic acids, and glycoproteins to the vicinity (Bandara et al. 2006). Up on decomposition, glycoproteins gradually release their nutrients in to the soil, and it can be taken up by plants and microbes. Therefore, the FBBs play an essential role in growth and nutrition of the plant. Also, they are more effective than monocultures in their activity and function (Bandara et al. 2006). The FBBs which attach to plant root systems establish root-biofilm associations, and that benefits both the plant and microbes in the biofilm (Seneviratne et al. 2009). This association supplies sufficient amount of nitrogen to the soil-plant system and increases soil nutrient uptake, which results in significant plant growth. It has been found that some biomolecules of the FBBs exudates are responsible for breaking the dormancy of soil microbial seed bank which was developed under stress in conventional agricultural practices (Seneviratne and Kulasooriya 2013). This helps to reinstate the depleted microbial diversity in degraded agroecosystems and consequently to improve ecosystem functioning and sustainability (Fig. 14.1).

It has also been shown that the inoculation of FBBs maintains a higher cell density of rhizobia on the legume root system than the inoculation of monocultures of rhizobia (Seneviratne and Jayasinghearachchi 2005). On the roots of non-legumes, nodule-like structures which are known as "pseudo-nodules" are formed by FBBs, and they fix N_2 biologically. In addition, these microbial communities associate with the root system to defend the plant from adverse environmental conditions and pathogenic infections (Seneviratne et al. 2010).

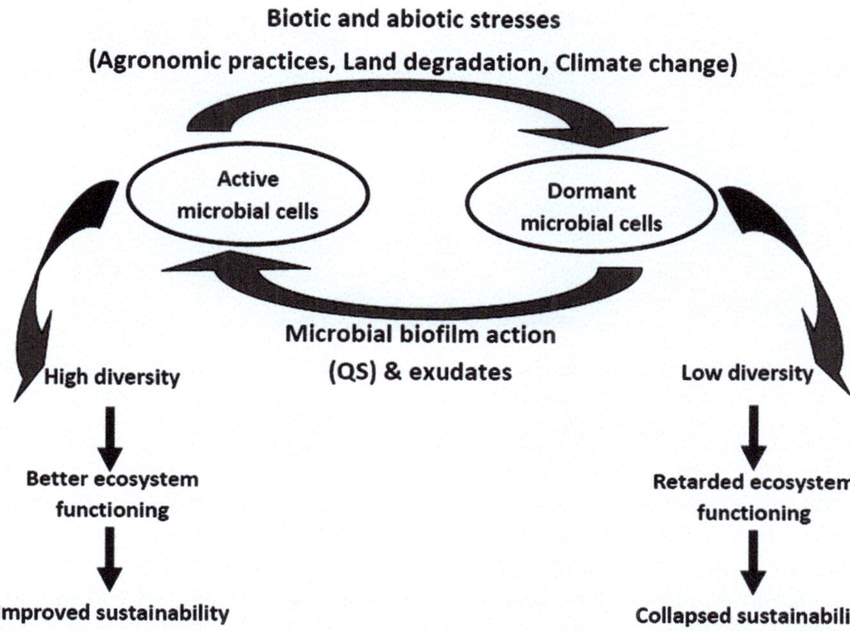

Fig. 14.1 Biofilm microbial ameliorators (BMAs) for improving sustainability in agroecosystems through increased microbial diversity and ecosystem functioning. (Reproduced from Seneviratne et al. 2013)

14.6 Developed BMAs for Sustainable Agriculture

The BFBFs are essential in solving many issues that directly affect the agroecosystems' and other terrestrial ecosystems' sustainability. They provide a variety of biochemicals and physiological benefits to the growth of plant under stress conditions and enhance soil quality, thus leading to a reduction of chemical fertilizer such as NPK use even up to 50% of the recommended doses in different crops (Buddhika et al. 2016). Moreover, reduction of the chemical fertilizers does not lead to drop crop yields; instead, it increases the crop yields up to ca. 30% on average. Also, BFBFs contribute to various health, economic, and environmental benefits in agroecosystems, particularly in regaining of the lost biodiversity. Therefore, BFBFs lead to restoration of degraded croplands rather than acting just as a biofertilizer. Hence, their application is essential for sustaining the productivity of croplands, even in the midst of different nutrient sources, chemical or organic.

Generally, the endophytic microbial colonization is considered to be essential in environmental stress tolerance of plants and their growth and yield increase. The endophytes perform this by inducing stress genes and also by producing biomolecules like reactive oxygen species scavengers (Lata et al. 2018). Application of BFBFs to rice plants under soil moisture stress condition illustrated improved

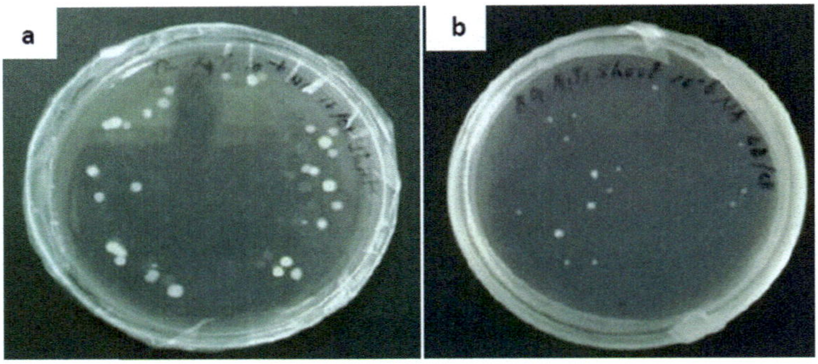

Fig. 14.2 Rice leaf endophytic bacterial colonies that were isolated on combined carbon medium. (**a**) Rice plants treated with biofilm biofertilizer (BFBF) and (**b**) those treated with chemical fertilizers alone. Large colonies in plate (**a**) are due to initial, large colony-forming units, which reflect the bacterial cell aggregation and endophytic biofilm formation with the BFBF application

seedling and root growth and also a reduction of leaf drying and rolling (Weerarathne and Seneviratne 2013). Leaf transpiration during dry periods was significantly reduced by BFBFs in a multi-location experiment of tea cultivation (De Silva et al. 2014). These results are attributable to enhanced colonization of endophytic microbes with the BFBF applications, and also a possible endophytic biofilm formation (Fig. 14.2).

Improving drought-resistant crops through selective breeding and novel molecular techniques is a common practice at present (Seneviratne et al. 2017). However, more natural biological techniques such as the use of BMAs, which have promising effects, have gained little attention thus far.

14.7 Developed BMAs for Environmental Bioremediation

The land and resources of the earth are rapidly losing its overall quality due to the carelessness and negligence in using them, and that has resulted in contaminated sites. It is considered that a contaminated site is harmful to human health. For a considerable period of time, scientists have been seeking novel remedies other than conventional techniques such as landfill, application of chemical, high-temperature incineration, etc. (Vidali 2001). Among all, bioremediation is a promising alternative since traditional methods have some drawbacks. The term bioremediation can be defined as the degradation of hazardous or contaminated pollutants to non-hazardous substances by using biological agents (Sasikumar and Papinazath 2003). Bioremediation process can be followed either in contaminated sites in situ or by bringing out contaminants from the sites, i.e., ex situ, and then applying fauna and flora (Sharma 2012). Some plants are capable of absorbing heavy metals or toxic materials, the process of which is known as phytoremediation. However, microbial techniques are much more effective than phytoremediation, because the production of energy by microbes during the formation of nonhazardous substances is utilized for their metabolic activities, and also the substances are used as nutrient sources (Tang et al. 2007). Some naturally occurring beneficial bacteria and fungi in the soil ecosystem are occupied in the bioremediation process. During the transformation of the contaminants to substances which are non-hazardous, the microbes attack the pollutants enzymatically (Vidali 2001). However, these microbes can withstand within a limited range of the chemical contaminant concentrations, because their utilization varies depending upon the type of contaminants (Prescott et al. 2002).

Moreover, there must be favorable environmental conditions for microbial growth and activity and practical bioremediation (Kumar et al. 2011). Considerable attention has been made on microbial biofilms and their ability of bioremediation (Decho 2000). Therefore, developed BMAs have been tested in environmental applications (Seneviratne et al. 2017), and this research field is yet in an early stage.

14.8 Perspectives

As explained above, nature is made out of continuously interacting living and non-living objects. Balance of those interactions determines the stability and hence sustainability of ecosystems. However, functional strength is predominantly maintained by diverse microbial communities and their signal-mediated interactions with plants. Thus, they show a mechanistic relationship between plant diversity and ecosystem functioning (Seneviratne and Premarathna 2020). Being the focal point of the biosphere, microbes govern human, animal, plant, and microbial health (Seneviratne 2020). Therefore, they control diversity and abundance of the organisms, which lead to the functioning and sustainability of the biosphere, including soil which is the main home and source of diverse microbes. Human impacts on the world after green and industrial revolutions altered/decreased the natural diversity and abundance of microbes, which has led to an imbalance of the other living beings too, again through their decreased diversity or abundance in the biosphere.

Further, increased greenhouse gas emissions, CO_2, in particular, in the post-green and industrial revolutions have led to global warming and climate changes. Interestingly, as a remedial measure to mitigate the climate effects, BFBFs have shown the potential of sequestering a considerable amount of C in the soil in large-scale rice ecosystems. Other environmental benefits like locking toxic compounds and heavy metals in the soil organic matter and reduced N_2O emissions (Seneviratne et al. 2017) also contribute immensely to lower health cost and mitigate climate change, respectively.

Good health, as the complete physical, mental, and social well-being, is a dividend that one can obtain in his/her lifespan. However, the health has been widely vulnerable to diseases such as deadly cancer in the present context. Recent studies have proven that human gut microbiome is directly related to the conditions and illnesses encountered by us. This has been attributed partly to the intake of antimicrobials like antibiotics, which degrade the native gut microbiota. Sometimes, people use probiotics after intake of antibiotics. They are mixtures of beneficial gut bacteria, but not in the biofilm mode, which may not be that effective in regaining the native gut bacterial diversity, as explained above. Thus, in future medicines, developed microbial communities in biofilm mode and their exudates, known as biofilm medicines, should be adopted for treating ailments (Seneviratne and Premarathna 2020). Then, they would help increase native biodiversity of gut bacteria, thus resulting in enhanced functioning of human body ecosystem for a healthy life.

Scientists have found that a microbiome assembling via food web takes place according to a specific pattern (Premarathna 2019). Generally, the environment gifts a subset of their microbiome to plants. Thus, the herbivores, including humans, get a subset of the microbiome from the plants that are associated with the soil-animal-microbial network interactions in the environment. Therefore, reinstating the degraded networks using the BFBFs in the soil may lead to an enriched human microbiome consisting of least number of pathogens and plenty of beneficial microbes. Ultimately, it may result in improved human health (Premarathna 2019).

It is also important to note here that the past human impacts on decreasing biodiversity have now boomeranged to the humans' existence. As explained above, the application of microbial biotechnologies like developed BMAs could mitigate such devastating events in the future.

14.9 Conclusion

BMAs like BFBF are capable enough to reinstate the degraded ecosystems for enhancing their function and sustainability. It holds promise for increasing food security, mitigating climate change, and improving human health. Thus, the application of BMAs is a holistic, important approach to make the globe a safer place to continue life. Doing it is an urgent requirement because options like Mars colonization in an abrupt event like a climate calamity on the earth are not near. Therefore, microbial interventions like BMAs should be researched rapidly and adopted for the sustainability of the earth system.

References

Bandara WMMS, Seneviratne G, Kulasooriya SA (2006) Interactions among endophytic bacteria and fungi effects and potentials. J Biosci 31:645–650. https://doi.org/10.1007/BF02708417

Bruder A, Frainer A, Rota T, Primicerio R (2019) The importance of ecological networks in multiple-stressor research and management. Front Environ Sci 7:59. https://doi.org/10.3389/fenvs.2019.00059

Brussaard L, De Ruiter PC, Brown GG (2007) Soil biodiversity for agricultural sustainability. Agric Ecosyst Environ 121:233–244

Buddhika UVA, Seneviratne G, Ekanayake MHGSE, Senanayake DMN, Igalavithane AD, Weeraratne N, Jayasekara APDA, Weerakoon WL, Indrajith A, Gunaratne HMAC, Kumara

RKGK, De Silva MSDL, Kennedy IR (2016) Biofilmed biofertilizers: application in agroecosystems. In: Gupta VK, Thangdurai D, Sharma GD (eds) Microbial bioresources. CAB International, Wallingford, pp 96–106

Bunning S, Jiménez JJ (2003) March indicators and assessment of soil biodiversity/soil ecosystem functioning for farmers and governments. In: Proceedings of OECD expert meeting on soil erosion and soil biodiversity indicators. OECD, Rome, pp 1–16

De Silva MSDL, Jayasekera APDA, Seneviratne G, Abeysekera UP, Premathunga EWTP, Wijesekera SN (2014) Soil fertility improvement through biofilmed biofertilizers: potential for field applications in tea cultivations. Sri Lanka J Tea Sci 79:46–61

Decho AW (2000) Microbial biofilms in intertidal systems: an overview. Cont Shelf Res 20:1257–1273

He M, Petoukhov S (2011) Mathematics of bioinformatics: theory, methods and applications. John Wiley & Sons, New Jersey

Herath L, Seneviratne G, Jayasinghe JAWW, Senanayake DMN (2017) Microbial biofilms and mitigation of loss of agro-biodiversity in degraded soils. J Natl Sci Found Sri Lanka 45:329–335

Kulasooriya SA, Seneviratne G, Ekanayake EMHGS (2017) Soil microbial diversity and its utilization in agriculture in Sri Lanka. In: Patra JK, Vishnuprasad CN, Das G (eds) Microbial biotechnology. Springer Nature, Singapore, pp 203–224

Kumar A, Bisht BS, Joshi VD, Dhewa T (2011) Review on bioremediation of polluted environment: a management tool. Int J Environ Sci 1:1079–1083

Lata R, Chowdhury S, Gond SK, White JF Jr (2018) Induction of abiotic stress tolerance in plants by endophytic microbes. Lett Appl Microbiol 66:268–276

Lavelle P (1996) Diversity of soil fauna and ecosystem function. Biol Int 33:3–16

Lupatini M, Suleiman AK, Jacques RJ, Antoniolli ZI, de Siqueira FA, Kuramae EE, Roesch LF (2014) Network topology reveals high connectance levels and few key microbial genera within soils. Front Environ Sci 2:10. https://doi.org/10.3389/fenvs.2014.00010

Maitima JM, Mugatha SM, Reid RS, Gachimbi LN, Majule A, Lyaruu H, Pomery D, Mathai S, Mugisha S (2009) The linkages between land use change, land degradation and biodiversity across East Africa. Afr J Environ Sci Technol 3:310–325

Mangan SA, Schnitzer SA, Herre EA, Mack KM, Valencia MC, Sanchez EI, Bever JD (2010) Negative plant–soil feedback predicts tree-species relative abundance in a tropical forest. Nature 466:752–755

Nduka JKC, Orisakwe OE, Ezenweke LO, Ezenwa TE, ChendoMN ENG (2008) Acid rain phenomenon in Niger delta region of Nigeria: economic, biodiversity, and public health concern. Sci World J 8:811–818

Patel I, Patel V, Thakkar A, Kothari V (2014) Microbial biofilms: microbes in social mode. Int J Agric Food Res 3:34–49

Plaas E, Meyer-Wolfarth F, Banse M, Bengtsson J, Bergmann H, Faber J, Potthoff M, Runge T, Schrader S, Taylor A (2019) Towards valuation of biodiversity in agricultural soils: a case for earthworms. Ecol Econ 15:291–300

Premarathna M (2019) Modified human microbiome towards better health. J Brief Ideas. https://doi.org/10.5281/zenodo.3445543

Prescott LM, Harley JP, Klein DA (2002) Microbiology, 5th edn. McGraw Hill, New York

Sasikumar SC, Papinazath T (2003). Environmental management-bioremediation of polluted environment. In: Martin J, Bunch V, Suresh M, Kumaran TV (eds) Proceedings of the third international conference on environment and health, Chennai, India. Department of Geography, University of Madras and Faculty of Environmental Studies, York University, pp 465–469

Sato M, Tsuda K, Wang L, Coller J, Watanabe Y, Glazebrook J, Katagiri F (2010) Network modeling reveals prevalent negative regulatory relationships between signaling sectors in Arabidopsis immune signaling. PLoS Pathog 6(7):e1001011. https://doi.org/10.1371/journal.ppat.1001011

Seneviratne G (2020) Microbes govern the world: their adverse effects like current COVID-19 outbreak can only be addressed again by microbial interventions. ResearchGate. https://doi.org/10.13140/RG.2.2.14494.08006
Seneviratne G, Jayasinghearachchi HS (2003) Mycelial colonization by bradyrhizobia and azorhizobia. J Biosci 28:243–247
Seneviratne G, Jayasinghearachchi HS (2005) A rhizobial biofilm with nitrogenase activity alters nutrient availability in a soil. Soil Biol Biochem 37:1975–1197
Seneviratne G, Kulasooriya SA (2013) Reinstating soil microbial diversity in agroecosystems: the need of the hour for sustainability and health. Agric Ecosyst Environ 164:181–182
Seneviratne G, Premarathna M (2020) Biofilm medicines: next-generation drugs. ResearchGate. https://doi.org/10.13140/RG.2.2.24088.14081
Seneviratne G, Zavahir JS, Bandara WMMS, Weerasekara MLMAW (2008) Fungal-bacterial biofilms: their development for novel biotechnological applications. World J Microbiol Biotechnol 24:739–743
Seneviratne G, Thilakaratne RMMS, Jayasekara KACN, Padmathilake KRE, De Silva MSDL (2009) Developing beneficial microbial biofilms on roots of non-legumes: a novel biofertilizer technique. In: Khan MS, Zaidi A, Mussarat J (eds) Microbial strategies for crop improvement. Springer-Verlag, Heidelberg, pp 51–61
Seneviratne G, Weerasekara MLMAW, Seneviratne KACN, Zavahir JS, Kecskés ML, Kennedy IR (2010) Importance of biofilm formation in plant growth promoting rhizobacterial action. In: Maheshwari DK (ed) Plant growth and health promoting bacteria. Springer, New Delhi, pp 81–95
Seneviratne G, Weeraratne N, Buddhika UVA (2013) Diversity of plant root associated microbes: its regulation by introduced biofilms. In: Arora NK (ed) Plant microbe symbiosis—fundamentals and advances. Springer, New Delhi, pp 351–372
Seneviratne G, Wijepala PC, Chandrasiri KPNK (2017) Developed biofilm-based microbial ameliorators for remediating degraded agroecosystems and the environment. In: Ahmad I, Husain FM (eds) Biofilms in plant and soil health. John Wiley & Sons Ltd., Chichester, pp 327–335
Sharma S (2012) Bioremediation: features, strategies and applications. Asian J Pharm Life Sci 2:202–213
Tang CY, Criddle QS, Fu CS, Leckie JO (2007) Effect of flux (transmembrane pressure) and membranes properties on fouling and rejection of reverse osmosis and nanofiltration membranes treating perfluoro-octane-sulfonate containing waste water. Environ Sci Technol 41:2008–2014
Tsiafouli MA, Thébault E, Sgardelis SP, De Ruiter PC, Van Der Putten WH, Birkhofer K, Hemerik L, De Vries FT, Bardgett RD, Brady MV, Bjornlund L (2014) Intensive agriculture reduces soil biodiversity across Europe. Glob Chang Biol 21:973–985
Vidali M (2001) Bioremediation: an overview. Pure Appl Chem 73:1163–1172
Weerarathne N, Seneviratne G (2013) Biofilmed biofertilizers induce drought tolerance of rice. In: Interdrought–IV conference, Perth, Australia
Wei H, Liu W, Zhang J, Qin Z (2017) Effects of simulated acid rain on soil fauna community composition and their ecological niches. Environ Pollut 220:460–468
Wittebolle L, Marzorati M, Clement L, Balloi A, Daffonchio D, Heylen K, De Vos P, Verstraete W, Boon N (2009) Initial community evenness favours functionality under selective stress. Nature 458:623–626

Chapter 15
Recent Advancements and Mechanism of Microbial Enzymes in Sustainable Agriculture

Pankaj Bhatt, Saurabh Gangola, Charu Joshi, Parul Chaudhary, Govind Kumar, Geeta Bhandari, Saurabh Kumar, Samiksha Joshi, Avikal Kumar, Narendra Singh Bhandari, and Samarth Tewari

Abstract To make the environment sustainable, using microbial technology is an important aspect. Indigenous microbial strains have been investigated for their application in bioremediation and sustainable development. The microbial enzymes are found effective for the bioremediation of the xenobiotic compounds from the environment. The enzymes can be produced extracellularly and intracellularly by microbial cells and catalyze the degradation of the toxic chemicals. Catalysis of the enzymes is performed via the respective amino acids located onto the binding site. The enzymes belonging to the esterase, laccase, dehydrogenase, oxygenase etc. are

P. Bhatt (✉)
State Key Laboratory for Conservation and Utilization of Subtropical Agro-bioresources, Guangdong Laboratory for Lingnan Modern Agriculture Integrative Microbiology Research Centre, South China Agricultural University, Guangzhou, China

S. Gangola · A. Kumar · N. S. Bhandari · S. Tewari
School of Agriculture, Graphic Era Hill University, Bhimtal, Uttarakhand, India

C. Joshi
Department of Biotechnology, Kumaun University Nainital, Bhimtal, Uttarakhand, India

P. Chaudhary
Department of Animal Biotechnology, NDRI, Karnal, Haryana, India

G. Kumar
Division of Crop Production, ICAR-Central Institute for Subtropical Horticulture, Lucknow, Uttar Pradesh, India

G. Bhandari
Department of Biotechnology, Sardar Bhagwan Singh University, Dehradun, Uttarakhand, India

S. Kumar
ICAR-Research Complex for Eastern Region, Patna, India

S. Joshi
Department of Bioscience, Shri Ram Group of Colleges, Muzaffarnagar, India

P. Bhatt et al. (eds.), *Microbial Technology for Sustainable Environment*,
https://doi.org/10.1007/978-981-16-3840-4_15

found effective for the bioremediation and sustainable development in agricultural fields. Here in this book chapter, we have discussed the microbial enzymes and their application for the sustainable development.

Keywords Microbial technology · Agriculture · Sustainable development · Enzymes

15.1 Introduction

To fulfill the global food and feed demands, the agriculture is the main component to nourish the global society (Bhatt et al. 2021a, b). The various crops, vegetables, and fruits have been used as a sole source of nutrition by humans globally (Bhatt et al. 2020a; Goel et al. 2020; Kumar et al. 2019). The microorganism is the key component of the agricultural system (Bhatt et al. 2020b; Suyal et al. 2019a). Microbes are found effective for the transfer of the macro and micro element from the soil to the plant rhizospheric region (Bhatt and Barh 2018; Bhatt and Bhatt 2021; Verma et al. 2021). Plant-microbe interaction makes the beneficial effects into the ecosystem (Ye et al. 2019; Rawat et al. 2019; Kukreti et al. 2020). The earlier research focuses on the beneficial effects and depth of molecular mechanism involved into the plant-microbe interaction for the sustainable agricultural development (Kumar et al. 2019; Jin et al. 2019; Suyal et al. 2019b). Microbial cells are able to produce the various extracellular and intracellular enzymes participated into contrasting biochemical mechanism (Bhatt et al. 2020c, d, e, f). This biochemical mechanism includes the bioremediation of toxic chemicals, complex reactions into the plant cells, and biodeterioration (Bhatt et al. 2020b, g, h, i, 2021a, b; Lin et al. 2020; Bhandari et al. 2021).

Microbes act as the cellular factory for the production of the enzymes (Bhatt et al. 2020a, 2021b). Microbial enzymes catalyze the reactions on the basis of the preferred substrates. The complex polymer lignin has been degraded by the microbial cells using the enzymes such as lignin peroxidase and laccase. Bacterial and fungal strains have been found effective for the degradation of the wood in ecological niches. Previous researchers investigated the fungi as the potential lignin degrader. Fungal strains belonging to the Basidiomycota, Aphyllophorales, and Ascomycota are considered as the potential wood degrading fungi (Bhatt et al. 2020c; Anasonye et al. 2015; Zhang et al. 2020a).

Bioremediation using microbial enzymes is the most promising approach for the sustainable agricultural development (Bhatt et al. 2019a, b; Pankaj et al. 2016a). The organophosphate, carbamate, organochlorine, pyrethroids, and neonicotinoids group of the pesticides are degraded by using the microbial enzymes such as hydrolases, dehydrogenases, laccases, monooxygenases, and dioxygenases (Mishra et al. 2020; Pankaj et al. 2016b). Immobilized enzymes have been documented for their more potential approach as compared to the individual strains (Bhatt et al. 2019c, 2020i; Pankaj et al. 2015).

15.2 Microbial Enzymes Used in Bioremediation Process

15.2.1 Microbial Oxidoreductases

Microorganisms (bacteria and fungi) use various oxidoreductase enzymes and reduce the toxicity of the various organic contaminants by oxidation coupling reactions. During the reaction, electrons are transferred from the reduced substrate (donor) to another substrate (acceptor) by cleaving the bonds and generating energy. So, in that way the toxic compounds are converted into the less toxic compounds (Mishra et al. 2021). Oxidoreductase enzymes have a wide range of activity for degradation of different organic substrate such as azo dyes, pesticides, lignin, and polyhydrocarbons (Karigar and Rao 2011; Park et al. 2006; Bhatt et al. 2015).

15.2.2 Microbial Oxygenases

Oxygenase enzymes belong to oxidoreductase family (Huang et al. 2021). It oxidizes the reduced substrate by adding oxygen from the molecular oxygen in the presence of co-substrate (FAD/NADH/NADPH). Generally, the oxygenases are subdivided into two groups on the basis of number of molecules of oxygen used for the oxidation of substrate. If one molecule of oxygen is added in the substrate, it is called monooxygenases, while when the two molecules of the oxygen are added in the substrate, it is called dioxygenases. These are the two enzymes which play key role in the metabolism of the various organic substances by increasing the substrate reactivity and solubility in water and help to break down the cyclic ring structure of the chemical compounds by adding oxygen molecule. Most of the enzymes used in the bioremediation process are mono-oxygenases or di-oxygenases (Arora et al. 2009; Singh et al. 2021).

15.2.3 Microbial Laccases

Laccase (*p*-diphenol:dioxygen oxidoreductase) is a multicopper protein generally found in plants, fungi, and bacteria. Laccase has a wide range of activity that catalyzes the oxidation of reduced substrate with concomitant oxidation reaction coupled to four electron reduction of molecular oxygen to water (Gianfreda et al. 1999; Mai et al. 2000; Zhang et al. 2020b). Laccase has multiple form of isoenzymes encoded by different genes (Mishra et al. 2020; Giardina et al. 1995). Many microbial cells are reported to produce extracellular or intracellular laccase enzyme. *Bacillus subtilis* produces laccase which help to degrade the pesticides present in agricultural fields (Gangola et al. 2018). Laccase helps to metabolize the organic contaminant and provides the nutrients for the microorganisms (Kim et al. 2002).

Among all the oxidoreductase enzymes, laccase has the biotechnological and bioremediation applications (Karigar and Rao 2011; Gianfreda et al. 1999).

15.2.4 Microbial Peroxidase

Microbial peroxidases are highly oxidizing agents that oxidize lignin, lignocellulose, and other phenolic substrate into substances devoid of hydrogen atom. Multiple isoenzymes of the fungal associated peroxidases are found such as lignin peroxidase and manganese peroxidase (Bansal and Kanwar 2013; Cocco et al. 2017). Degradation of lignocellulose is a very important step for the carbon recycle in land ecosystem. Fungi generally basidiomycetes are known to degrade lignin and help in the carbon recycling process. White and brown rot basidiomycetes digest the lignocellulose as a sole source of carbon. Manganese peroxidase is an extracellular enzyme, generally found in the basidiomycetes, which also helps in digestion of lignin and other phenolic compounds. For initiating the activity of manganese peroxidase, Mn^{2+} acts as substrate and helps to oxidize phenolic compounds (Lin et al. 2020; Zhang et al. 2020a; Chowdhary et al. 2019; Feng et al. 2020).

15.2.5 Hydrolases

Hydrolytic enzymes reduce the toxicity of xenobiotic compounds by breaking the linkage in between. Hydrolase enzymes are also known to degrade oil spill and agricultural pesticides degradation (Karigar and Rao 2011). The key feature of the enzyme is its broad range of activity, availability, and high tolerability. The enzymes such as lipase, DNases, proteases, xylanases, and pullulanase are the hydrolytic enzyme generally used in food industry and biomedical sciences. The hemicellulase, cellulase, and glycosidase are the hydrolytic enzymes and actively participate in biomass degradation (Peixoto et al. 2011; Thakur et al. 2019).

The microbial systems are found effective for the development of the sustainable agriculture by using the enzymes. Application of enzymes in agricultural fields accelerates the agricultural wastes and helps in resource recovery from the environment. Here we are enlightening the impact of the microbial enzymes for their resource recovery potential. Accumulation of different kinds of wastes has become a severe environmental and public concern. It is urgent to develop innovative approaches for the removal/degradation of disposal. Microbial enzymes are extremely effective to catalyze the biochemical reaction. Microbial enzymes have achieved importance for their extensive utilization in food, agriculture, and pharmaceuticals. These enzymes are capable of degrading complex compounds into simpler ones through the process of bioremediation. The researchers focus on the resource recovery potential of the microbial enzymes. These enzymes could be used directly for the xenobiotics compound degradation and cleaning of soil and water environments Zhan et al. 2020; Ye et al. 2019; Huang et al. 2019; Danso et al. 2018; Fan et al. 2020).

15.3 Importance and Mechanism of Microbial Enzymes in Maintenance of Soil Health

Soil enzymes secreted by microbes are natural mediators which catalyze many processes like organic matter decomposition, soil humus formation, and release of different minerals involved in different cycles and help in the maintenance of soil health and plant growth (Wallenstein et al. 2012; Khati et al. 2019). Thus, it is crucial to find out the enzymatic activities of soil enzymes and their mechanism to illustrate metabolic prospective of soil fertility (Astner et al. 2020). Activity of enzymes is an insightful marker; any changes in soil environment can affect their activity in agricultural farming (Chaudhary et al. 2021). There are different bacterial and fungal enzymes which involve in the breakdown of complex form of mineral nutrients into simpler ones which are easily taken up by plants and enhance their growth (Kukreti et al. 2020; Khati et al. 2017). Kwiatkowski et al. (2020) reported that dehydrogenase, urease, and protease activity improves the fertility of soil and biological properties of soil and involves in nutrient cycling (Kwiatkowski et al. 2020).

Cellulose is a largely abundant polysaccharide which is hydrolyzed by cellulase enzyme into D-glucose. It consists of three enzymes such as endo-1,4-β-glucanase which attacks at random on cellulose chain, exo-1,4-β-glucanase which removes glucose/cellobiose from cellulose chain, and β-D-glucosidase which hydrolyses cellobiose into glucose. These enzymes are found in various bacteria and fungi such as *Bacillus subtilis*, *Aspergillus niger*, *Phanerochaete chrysosporium*, *Clostridium thermocellum*, and *Poria placenta* (Deng and Tabatabai 1994). Amylase enzyme is involved in the breakdown of starch into glucose and maltose which also play an important role in carbon nutrient cycling. This enzyme is secreted by various bacterial and fungal species in soil such as *Pseudomonas*, *Bacillus* spp., and *Aspergillus niger*. Arylesterase enzyme helps in degradation of complex sulfate esters into sulfur which takes part in sulfur cycle. *Klebsiella* spp., *Raoultella* spp., and *Trichoderma* sp. release this enzyme in soil and helps for maintenance of soil fertility and plant development.

Phosphorus is usually entrapped in the complex soil system and becomes unavailable for plant uptake. To enhance the phosphorus bioavailability, *phosphatase* enzyme is involved. The conversion of organic P to inorganic P by microbial associated *phosphatase* activity leads to P cycling. *Phosphatase* activity can be measured as acid and alkaline *phosphatase* at pH range 4–6 and 9–11, respectively. After the action of such enzymes, the available P (inorganic P) had good correlation with nutrient transport and plant vigor. Phosphatase enzyme is involved in the conversion of insoluble form of phosphate into soluble form and helps in phosphorus cycling (Khati et al. 2019). Phosphomonoesters and phosphodiesters are involved in

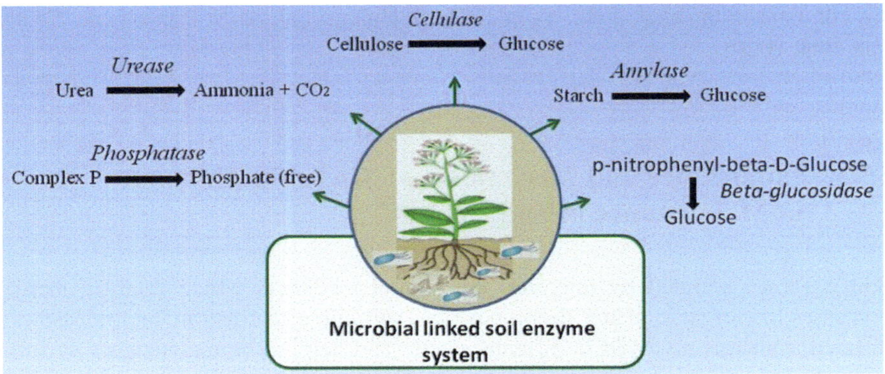

Fig. 15.1 Mechanism of different soil enzymes in nutrient cycling

the breakdown of sugar phosphates and phospholipids and release free phosphate from phosphate diester (Turner and Haygarth 2005). Alkaline and acid phosphatase are good indicators of soil microbial activity and correlated to organic content of soil. These enzymes are secreted by *Anabaena oryzae*, *Bacillus*, *Pseudomonas*, and *Penicillium* (Pankaj et al. 2016a; Khati et al. 2019, 2017). Various factors are involved for phosphatase activity: (1) if pH increases the activity of alkaline, P increases, while acid phosphatase activity reduces significantly and vice versa; (2) water content of the soil is better linked with alkaline phosphatase activity as compared to the acid phosphatase activity; and (3) salinity stress is better correlated with alkaline phosphatase activity in all seasons.

Urease enzyme takes part in the conversion of urea into ammonia and CO_2 with rise in pH and helps in nitrogen cycle which is produced by fungi, bacteria, and yeast such as *Pseudomonas* and *Trichosporon cutaneum.* These enzymes are used as biological indicators because they are influenced by soil factors like organic matter content, heavy metals, and nanocompounds and due to the cropping history (Fig. 15.1). B-glucosidase enzyme is the most common enzyme in soil which catalyzes the hydrolysis of glucosides present in soil. This enzyme activity was observed in various plant species and microbes such as *Flavobacterium*, *Trichoderma* spp., and *Lactobacillus plantarum* (Kwiatkowski et al. 2020).

For the health and sustainable soil system, the soil enzymatic activity plays a vital role in nutrient cycling. To establish system for organic matter decomposition to channelizing available nutrient to the plant system, the soil enzyme system plays a key role. Different enzymes are available with the soil system that showed overall health of the soil system. Dehydrogenase (DHA) enzyme is one of them that plays a significant role to indicate the oxidative power of the soil system associated with the existed microbiome (Zhang et al. 2020a). Dehydrogenase is a subclass of *oxidoreductase* enzymes that catalyze the oxidation and reduction process. DHA activity is directly linked with the soil organic matter (SOM) that is available for microbial activity.

The enzymatic activity of the microbes in soil can be measured by fluorescein diacetate (FDA) hydrolysis assay. In FDA analysis, membrane-bound nonspecific kind of enzymes like esterases, proteases, and lipases hydrolyzes the 3,6-diacetyle fluorescein (FDA). After hydrolysis, a yellow green color is produced that can be quantified by using spectrophotometer at 490 nm (Gilan et al. 2004). The total enzymatic activity (TEA) including oxygen utilization, microbial biomass, amount of ATP, etc. can be directly correlated with the FDA hydrolysis (Dzionek et al. 2020). Soil protease enzyme is one of the major enzymatic actions involved in N-cycle due to availability of huge protein content in the soil system. The protein components of organic nitrogen in the soil system are usually hydrolyzed by protease enzyme action. In nitrogen cycle, degradation of protein is an essential component. Protease enzyme is mainly identified in microorganisms that are responsible for conversion of protein to polypeptide to amino acids (de Morais et al. 2018).

15.4 Circular Economy for SPs and Sustainable Development

Synthetic pollutants (SPs) have a detrimental effect on the environment and long-term economic aspects. The existing unsustainable "linear" material and energy flow model are also prominent contributing factors to these issues (Blomsma and Brennan 2017). However, in recent years, the sustainable development strategies have been designed for the reduction of the SPs from the environment that favors economic security without simultaneously degrading the environment. The concept of the circular economy has become a promising alternative model for sustainable development by reducing the SPs and promoting ecological design. The circular economy aims at minimizing waste production, enhancing product life, optimizing reuse, and utilizing energy sources such as biomass (Saldarriaga-Hernandez et al. 2020).

The methods used for bioremediation of the SPs can be based on the cyclical flow of the resources. The bacteria, fungi, and algae are used in a circular perspective for the bioremediation of the SPs from soil and water systems. Microbial strains can use the SPs as a source of nutrition for their growth and development and convert the toxic metabolites into the environmentally accepted form. The metabolic end products can be exploited for the production of useful metabolites and as a resource for recovery for value-added products. Aerobic and anaerobic digestion of SPs generates humus and digestate which could be applied to soil as a fertilizer. During the anaerobic digestion process, biogas that is rich in methane and carbon dioxide is produced, which could be used as a fuel for combustion in transport or energy production (Soobhany 2019). Various volatile fatty acids are also formed as intermediates during the anaerobic degradation process and can find various industrial applications (Singh et al. 2012). Bio-electrochemical systems (BESs), integrating microbial–electro-chemical removal mechanisms, have been intensively investigated for the organic compound removal in wastewater and simultaneous generation of electricity and biofuel. Different microbial strains can also synthesize biopolymers by utilizing various synthetic pollutants (Pagliano et al. 2017). New emerging strategies are also allowed production of value-added products such as enzymes (Steiner and Schwab 2012), single-cell oil (de Oliveria Finco et al. 2017), and building block chemicals (FitzPatrick et al. 2010) from biodegradation end products. The methods used for the conversion of the SPs from the environment when aligned in a sustainable model can generate more fruitful results in the long-term future. However, still, it is a big challenge to convert the SPs into the environmentally accepted form and use them for the production of the value-added products.

15.5 Conclusion

Many xenobiotic compounds are present in our surrounding environment which are released from the different source such as textile industry releases many azo dye and excess application of pesticides in agricultural field. The physical and chemical techniques are available to remove these xenobiotic compounds from the environment. However, these techniques are very costly and expensive. Using the microbial system for the degradation of these xenobiotic compounds is ecofriendly and cost-effective technique. Application of indigenous microorganism transforms the toxic xenobiotic compounds into less toxic or nontoxic and environmental accepted form. These microorganisms produce different class of extracellular and intracellular enzymes which help to metabolize the environmental pollutant and use them as a source of energy. Generally, the enzymes such as lipase, esterase, laccase, dehydrogenase, and oxygenase play a key role in bioremediation processes. In a similar way, microbes are also playing key role in the agricultural fields. They produce different plant growth-promoting hormones, solubilize the insoluble compounds, and make them available for the plants and enhance the plant productivity and quality. Therefore, the use of microbial enzymes is important for the development of sustainable environment. In addition to that, this microbial technique is efficient, cost-effective, and ecofriendly.

References

Anasonye F, Winquist E, Räsänen M, Kontro J, Björklöf K, Vasilyeva G, Jørgensen KS, Steffen KT, Tuomela M (2015) Bioremediation of TNT contaminated soil with fungi under laboratory and pilot scale conditions. Int Biodeterior Biodegrad 105:7–12. https://doi.org/10.1016/j.ibiod.2015.08.003

Arora PK, Kumar M, Chauhan A, Raghava GP, Jain RK (2009) OxDBase: a database of oxygenases involved in biodegradation. BMC Res Notes 2:67. https://doi.org/10.1186/1756-0500-2-67

Astner AF, Hayes DG, Pingali SV, O'Neill HM, Littrell KC, Evans BR, Urban VS (2020) Effects of soil particles and convective transport on dispersion and aggregation of nanoplastics via small-angle neutron scattering (SANS) and ultra SANS (USANS). PLoS One 15:1–14. https://doi.org/10.1371/journal.pone.0235893

Bansal N, Kanwar SS (2013) Peroxidase(s) in environment protection. Sci World J 2013:714639. https://doi.org/10.1155/2013/714639

Bhandari G, Bagheri AR, Bhatt P, Bilal M (2021) Occurrence, potential ecological risks, and degradation of endocrine disrupter, nonylphenol, from the aqueous environment. Chemosphere 275:130013. https://doi.org/10.1016/j.chemosphere.2021.130013

Bhatt P, Barh A (2018) Bioinformatic tools to study the soil microorganisms: an in silico approach for sustainable agriculture. In: In silico approach for sustainable agriculture. Springer, Singapore, pp 169–182. https://doi.org/10.1007/978-981-13-0347-0_10

Bhatt K, Bhatt P (2021) Rhizospheric biology. In: Phytomicrobiome interactions and sustainable agriculture. Wiley, New York, pp 164–186. https://doi.org/10.1002/9781119644798.ch9

Bhatt PG, Negi G, Gangola S, Khati P, Srivastava A, Sharma A (2015) Optimization of sulfosulfuron biodegradation through response surface methodology using indigenous bacterial strain isolated from contaminated agriculture field. http://www.ijcmas.com

Bhatt P, Pal K, Bhandari G, Barh A (2019a) Modelling of the methyl halide biodegradation in bacteria and its effect on environmental systems. Pestic Biochem Physiol 158:88–100. https://doi.org/10.1016/j.pestbp.2019.04.015

Bhatt P, Gangola S, Chaudhary P, Khati P, Kumar G, Sharma A, Srivastava A (2019b) Pesticide induced up-regulation of esterase and aldehyde dehydrogenase in indigenous *Bacillus* spp. Bioremediat J 23:42–52. https://doi.org/10.1080/10889868.2019.1569586

Bhatt P, Gangola S, Chaudhary P, Khati P, Kumar G, Sharma A, Srivastava A (2019c) Pesticide induced up-regulation of esterase and aldehyde dehydrogenase in indigenous *Bacillus* spp. Biorem J 23:42–52. https://doi.org/10.1080/10889868.2019.1569586

Bhatt P, Huang Y, Zhang W, Sharma A, Chen S (2020a) Enhanced cypermethrin degradation kinetics and metabolic pathway in Bacillus thuringiensis strain SG4. Microorganisms 8:1–15. https://doi.org/10.3390/microorganisms8020223

Bhatt P, Sethi K, Gangola S, Bhandari G, Verma A, Adnan M, Singh Y, Chaube S (2020b) Modeling and simulation of atrazine biodegradation in bacteria and its effect in other living systems. J Biomol Struct Dyn 38:1–11. https://doi.org/10.1080/07391102.2020.1846623

Bhatt P, Bhatt K, Huang Y, Lin Z, Chen S (2020c) Esterase is a powerful tool for the biodegradation of pyrethroid insecticides. Chemosphere 244:125507. https://doi.org/10.1016/j.chemosphere.2019.125507

Bhatt P, Verma A, Verma S, Anwar MS, Prasher P, Mudila H, Chen S (2020d) Understanding phytomicrobiome: a potential reservoir for better crop management. Sustainability 12:1–20. https://doi.org/10.3390/su12135446

Bhatt P, Huang Y, Rene ER, Kumar AJ, Chen S (2020e) Mechanism of allethrin biodegradation by a newly isolated Sphingomonas trueperi strain CW3 from wastewater sludge. Bioresour Technol 305:123074. https://doi.org/10.1016/j.biortech.2020.123074

Bhatt P, Zhang W, Lin Z, Pang S, Huang Y, Chen S (2020f) Biodegradation of allethrin by a novel fungus fusarium proliferatum strain cf2, isolated from contaminated soils. Microorganisms 8:593. https://doi.org/10.3390/microorganisms8040593

Bhatt P, Rene ER, Kumar AJ, Zhang W, Chen S (2020g) Binding interaction of allethrin with esterase: bioremediation potential and mechanism. Bioresour Technol 315:123845. https://doi.org/10.1016/j.biortech.2020.123845

Bhatt P, Gangola S, Bhandari G, Zhang W, Maithani D, Mishra S, Chen S (2020h) New insights into the degradation of synthetic pollutants in contaminated environments. Chemosphere 268:128827. https://doi.org/10.1016/j.chemosphere.2020.128827

Bhatt P, Sharma A, Rene ER, Kumar AJ, Zhang W, Chen S (2020i) Bioremediation of fipronil using Bacillus sp. FA3: mechanism, kinetics and resource recovery potential from contaminated environments. J Water Process Eng 39:101712. https://doi.org/10.1016/j.jwpe.2020.101712

Bhatt P, Zhou X, Huang Y, Zhang W, Chen S (2021a) Characterization of the role of esterases in the biodegradation of organophosphate, carbamate, and pyrethroid pesticides. J Hazard Mater 411:125026. https://doi.org/10.1016/j.jhazmat.2020.125026

Bhatt P, Joshi T, Bhatt K, Zhang W, Huang Y, Chen S (2021b) Binding interaction of glyphosate with glyphosate oxidoreductase and C–P lyase: molecular docking and molecular dynamics simulation studies. J Hazard Mater 409:124927. https://doi.org/10.1016/j.jhazmat.2020.124927

Blomsma F, Brennan G (2017) The emergence of circular economy: a new framing around prolonging resource productivity. J Ind Ecol 21:603–614. https://doi.org/10.1111/jiec.12603

Chaudhary P, Khati P, Chaudhary A, Gangola S, Kumar R, Sharma A (2021) Bioinoculation using indigenous Bacillus spp. improves growth and yield of Zea mays under the influence of nanozeolite, 3. Biotech 11:1–12. https://doi.org/10.1007/s13205-020-02561-2

Chowdhary P, Shukla G, Raj G, Ferreira LFR, Bharagava RN (2019) Microbial manganese peroxidase: a ligninolytic enzyme and its ample opportunities in research. SN Appl Sci 1:45. https://doi.org/10.1007/s42452-018-0046-3

Cocco G, Cocco A, Sollai F, Sanjust E, Zucca P (2017) Bioinspired versus enzymatic oxidation of some homologous thionine dyes in the presence of immobilized metalloporphyrin catalysts and ligninolytic enzymes. Int J Mol Sci 18:2553. https://doi.org/10.3390/ijms18122553

Danso D, Schmeisser C, Chow J, Zimmermann W, Wei R, Leggewie C, Li X, Hazen T, Streit WR (2018) New insights into the function and global distribution of polyethylene terephthalate (PET)-degrading bacteria and enzymes in marine and terrestrial metagenomes. Appl Environ Microbiol 84:e02773-17. https://doi.org/10.1128/AEM.02773-17

Deng SP, Tabatabai MA (1994) Cellulase activity of soils. Soil Biol Biochem 26:1347–1354. https://doi.org/10.1016/0038-0717(94)90216-X

Dzionek A, Wojcieszyńska D, Adamczyk-Habrajska M, Guzik U (2020) Enhanced degradation of naproxen by immobilization of Bacillus thuringiensis B1(2015b) on loofah sponge. Molecules 25:872. https://doi.org/10.3390/molecules25040872

Fan X, Ye T, Li Q, Bhatt P, Zhang L, Chen S (2020) Potential of a quorum quenching bacteria isolate ochrobactrum intermedium D-2 against soft rot pathogen Pectobacterium carotovorum subsp. carotovorum. Front Microbiol 11:1–14. https://doi.org/10.3389/fmicb.2020.00898

Feng Y, Huang Y, Zhan H, Bhatt P, Chen S (2020) An overview of strobilurin fungicide degradation: current status and future perspective. Front Microbiol 11:1–11. https://doi.org/10.3389/fmicb.2020.00389

FitzPatrick M, Champagne P, Cunningham MF, Whitney RA (2010) A biorefinery processing perspective: treatment of lignocellulosic materials for the production of value-added products. Bioresour Technol 101:8915–8922. https://doi.org/10.1016/j.biortech.2010.06.125

Gangola S, Sharma A, Bhatt P, Khati P, Chaudhary P (2018) Presence of esterase and laccase in Bacillus subtilis facilitates biodegradation and detoxification of cypermethrin. Sci Rep 8:1–11. https://doi.org/10.1038/s41598-018-31082-5

Gianfreda L, Xu F, Bollag JM (1999) Laccases: a useful group of oxidoreductive enzymes. Biorem J 3:1–26. https://doi.org/10.1080/10889869991219163

Giardina P, Cannio R, Martirani L, Marzullo L, Palmieri G, Sannia G (1995) Cloning and sequencing of a laccase gene from the lignin-degrading basidiomycete Pleurotus ostreatus. Appl Environ Microbiol 61:2408–2413. https://doi.org/10.1128/aem.61.6.2408-2413.1995

Gilan I, Hadar Y, Sivan A (2004) Colonization, biofilm formation and biodegradation of polyethylene by a strain of Rhodococcus ruber. Appl Microbiol Biotechnol 65:97–104. https://doi.org/10.1007/s00253-004-1584-8

Goel R, Debbarma P, Kumari P, Suyal DC, Kumar S, Mahapatra BS (2020) Assessment of soil chemical quality, soil microbial population and plant growth parameters under organic and conventional Rice–wheat cropping system. Agric Res 10(2):1–12. https://doi.org/10.1007/s40003-020-00499-8

Huang Y, Zhan H, Bhatt P, Chen S (2019) Paraquat degradation from contaminated environments: current achievements and perspectives. Front Microbiol 10:1754. https://doi.org/10.3389/fmicb.2019.01754

Huang Y, Zhang W, Pang S, Chen J, Bhatt P, Mishra S, Chen S (2021) Insights into the microbial degradation and catalytic mechanisms of chlorpyrifos. Environ Res 194:110660. https://doi.org/10.1016/j.envres.2020.110660

Jin J, Wang M, Lu W, Zhang L, Jiang Q, Jin Y, Lu K, Sun S, Cao Q, Wang Y, Xiao M (2019) Effect of plants and their root exudate on bacterial activities during rhizobacterium–plant remediation of phenol from water. Environ Int 127:114–124. https://doi.org/10.1016/j.envint.2019.03.015

Karigar CS, Rao SS (2011) Role of microbial enzymes in the bioremediation of pollutants: a review. Enzyme Res 2011:805187. https://doi.org/10.4061/2011/805187

Khati P, Chaudhary P, Gangola S, Bhatt P, Sharma A (2017) Nanochitosan supports growth of Zea mays and also maintains soil health following growth. 3 Biotech 7:81. https://doi.org/10.1007/s13205-017-0668-y

Khati P, Sharma A, Chaudhary P, Singh AK, Gangola S, Kumar R (2019) High-throughput sequencing approach to access the impact of nanozeolite treatment on species richness and evenness of soil metagenome. Biocatal Agric Biotechnol 20:101249. https://doi.org/10.1016/j.bcab.2019.101249

Kim JS, Park JW, Lee SE, Kim JE (2002) Formation of bound residues of 8-hydroxybentazon by oxidoreductive catalysts in soil. J Agric Food Chem 50:3507–3511. https://doi.org/10.1021/jf011504z

Kukreti B, Sharma A, Chaudhary P, Agri U, Maithani D (2020) Influence of nanosilicon dioxide along with bioinoculants on Zea mays and its rhizospheric soil. 3 Biotech 10:1–11. https://doi.org/10.1007/s13205-020-02329-8

Kumar S, Suyal DC, Yadav A, Shouche Y, Goel R (2019) Microbial diversity and soil physiochemical characteristic of higher altitude. PLoS One 14:e0213844. https://doi.org/10.1371/journal.pone.0213844

Kwiatkowski CA, Harasim E, Feledyn-Szewczyk B, Antonkiewicz J (2020) Enzymatic activity of loess soil in organic and conventional farming systems. Agriculture 10:135. https://doi.org/10.3390/agriculture10040135

Lin Z, Zhang W, Pang S, Huang Y, Mishra S, Bhatt P, Chen S (2020) Current approaches to and future perspectives on methomyl degradation in contaminated soil/water environments. Molecules 25:1–16. https://doi.org/10.3390/molecules25030738

Mai C, Schormann W, Milstein O, Huttermann A (2000) Enhanced stability of laccase in the presence of phenolic compounds. Appl Microbiol Biotechnol 54:510–514. https://doi.org/10.1007/s002530000452

Mishra S, Zhang W, Lin Z, Pang S, Huang Y, Bhatt P, Chen S (2020) Carbofuran toxicity and its microbial degradation in contaminated environments. Chemosphere 259:127419. https://doi.org/10.1016/j.chemosphere.2020.127419

Mishra S, Lin Z, Pang S, Zhang W, Bhatt P, Chen S (2021) Recent advanced technologies for the characterization of xenobiotic-degrading microorganisms and microbial communities. Front Bioeng Biotechnol 9:632059. https://doi.org/10.3389/fbioe.2021.632059

de Morais CR, Travençolo BAN, Carvalho SM, Beletti ME, Vieira Santos VS, Campos CF, de Campos Júnior EO, Pereira BB, Carvalho Naves MP, de Rezende AAA, Spanó MA, Vieira CU, Bonetti AM (2018) Ecotoxicological effects of the insecticide fipronil in Brazilian native stingless bees Melipona scutellaris (Apidae: Meliponini). Chemosphere 206:632–642. https://doi.org/10.1016/j.chemosphere.2018.04.153

de Oliveria Finco AM, Mamani LDG, de Carvalho JC, de Melo Pereira GV, Thomaz-Soccol V, Soccol CR (2017) Technological trends and market perspectives for production of microbial oils rich in omega-3. Crit Rev Biotechnol 37:656–671. https://doi.org/10.1080/07388551.2016.1213221

Pagliano G, Ventorino V, Panico A, Pepe O (2017) Integrated systems for biopolymers and bioenergy production from organic waste and by-products: a review of microbial processes. Biotechnol Biofuels 10:113. https://doi.org/10.1186/s13068-017-0802-4

Pankaj T, Singh Bisht M, Pathak A, Barh A, Sharma GB (2015) Article-68 optimization of amylase production from the fungal isolates of Himalayan region Uttarakhand, 2015

Pankaj, Negi G, Gangola S, Khati P, Kumar G, Srivastava A, Sharma A (2016a) Differential expression and characterization of cypermethrin-degrading potential proteins in Bacillus thuringiensis strain, SG4. 3 Biotech 6:225. https://doi.org/10.1007/s13205-016-0541-4

Pankaj, Sharma A, Gangola S, Khati P, Kumar G, Srivastava A (2016b) Novel pathway of cypermethrin biodegradation in a Bacillus sp. strain SG2 isolated from cypermethrin-contaminated agriculture field. 3 Biotech 6:45. https://doi.org/10.1007/s13205-016-0372-3

Park JW, Park BK, Kim JE (2006) Remediation of soil contaminated with 2,4-dichlorophenol by treatment of minced shepherd's purse roots. Arch Environ Contam Toxicol 50:191–195. https://doi.org/10.1007/s00244-004-0119-8

Peixoto RS, Vermelho AB, Rosado AS (2011) Petroleum-degrading enzymes: bioremediation and new prospects. Enzyme Res. 2011:475193. https://doi.org/10.4061/2011/475193

Rawat N, Sharma M, Suyal DC, Singh DK, Joshi D, Singh P, Goel R (2019) Psychrotolerant bio-inoculants and their co-inoculation to improve Cicer arietinum growth and soil nutrient status for sustainable mountain agriculture. J Soil Sci Plant Nutr 19:639–647. https://doi.org/10.1007/s42729-019-00064-5

Saldarriaga-Hernandez S, Hernandez-Vargas G, Iqbal HMN, Barceló D, Parra-Saldívar R (2020) Bioremediation potential of Sargassum sp. biomass to tackle pollution in coastal ecosystems: circular economy approach. Sci Total Environ 715:136978. https://doi.org/10.1016/j.scitotenv.2020.136978

Singh B, Kaur J, Singh K (2012) Microbial remediation of explosive waste. Crit Rev Microbiol 38:152–167. https://doi.org/10.3109/1040841X.2011.640979

Singh K, Gera R, Sharma R, Maithani D, Chandra D, Amin M (2021) Mechanism and application of Sesbania root - nodulating bacteria: an alternative for chemical fertilizers and sustainable development. Arch Microbiol 203:1259–1270. https://doi.org/10.1007/s00203-020-02137-x

Soobhany N (2019) Insight into the recovery of nutrients from organic solid waste through biochemical conversion processes for fertilizer production: a review. J Clean Prod 241:118413. https://doi.org/10.1016/j.jclepro.2019.118413

Steiner K, Schwab H (2012) Recent advances in rational approaches for enzyme engineering. Comput Struct Biotechnol J 2:e201209010. https://doi.org/10.5936/csbj.201209010

Suyal DC, Joshi D, Kumar S, Soni R, Goel R (2019a) Differential protein profiling of soil diazotroph Rhodococcus qingshengii S10107 towards low-temperature and nitrogen deficiency. Sci Rep 9:1–9. https://doi.org/10.1038/s41598-019-56592-8

Suyal DC, Kumar S, Joshi D, Yadav A, Shouche Y, Goel R (2019b) Comparative overview of red kidney bean (Phaseolus vulgaris) rhizospheric bacterial diversity in perspective of altitudinal variations. Biologia (Bratisl) 74:1405–1413. https://doi.org/10.2478/s11756-019-00292-1

Thakur M, Medintz IL, Walper SA (2019) Enzymatic bioremediation of organophosphate compounds—progress and remaining challenges. Front Bioeng Biotechnol 7:289. https://doi.org/10.3389/fbioe.2019.00289

Turner BL, Haygarth PM (2005) Phosphatase activity in temperate pasture soils: potential regulation of labile organic phosphorus turnover by phosphodiesterase activity. Sci Total Environ 344:27–36. https://doi.org/10.1016/j.scitotenv.2005.02.003

Verma S, Bhatt P, Verma A, Mudila H, Prasher P, Rene ER (2021) Microbial technologies for heavy metal remediation: effect of process conditions and current practices. Clean Techn Environ Policy 1–23. https://doi.org/10.1007/s10098-021-02029-8

Wallenstein MD, Haddix ML, Lee DD, Conant RT, Paul EA (2012) A litter-slurry technique elucidates the key role of enzyme production and microbial dynamics in temperature sensitivity of organic matter decomposition. Soil Biol Biochem 47:18–26. https://doi.org/10.1016/j.soilbio.2011.12.009

Ye T, Zhou T, Fan X, Bhatt P, Zhang L, Chen S (2019) Acinetobacter lactucae strain QL-1, a novel quorum quenching candidate against bacterial pathogen Xanthomonas campestris pv. campestris. Front Microbiol 10:1–11. https://doi.org/10.3389/fmicb.2019.02867

Zhan H, Huang Y, Lin Z, Bhatt P, Chen S (2020) New insights into the microbial degradation and catalytic mechanism of synthetic pyrethroids. Environ Res 182:109138. https://doi.org/10.1016/j.envres.2020.109138

Zhang W, Lin Z, Pang S, Bhatt P, Chen S (2020a) Insights into the biodegradation of lindane (γ-hexachlorocyclohexane) using a microbial system. Front Microbiol 11:1–12. https://doi.org/10.3389/fmicb.2020.00522

Zhang W, Pang S, Lin Z, Mishra S, Bhatt P, Chen S (2020b) Biotransformation of perfluoroalkyl acid precursors from various environmental systems: advances and perspectives. Environ Pollut 272:115908. https://doi.org/10.1016/j.envpol.2020.115908

Chapter 16
Application of Microbial Technology for Waste Removal

Ravi Ranjan Kumar, Chitra Bhattacharya, and Nutan Prakash Vishwakarma

Abstract The continuous rise of waste in the environment becomes a global burden as it decreases the natural balance of waste recycle. It has further accelerated due to quality and amount of waste added in the environment in the last century. Increasing human population, introduction of xenobiotic compounds, overexploitation of natural resources and alarming increased waste generation rate are major threats to environmental safety. Several waste management practices have been implemented to decrease the harmful impacts of waste. Microorganisms are inhabitants of nature that play a major role in biodegradation, bioremediation, nutrient cycling and detoxification to maintain a sustainable environment. Microbial technology utilizes a wide range of selective microorganisms in specific condition for removal of waste from the environment. The utilization of microbes is only limited to culture-dependent method, and the majority of undiscovered microbes has also been explored using culture-independent techniques. Technological advancement has increased the exploration of microbial diversity for their utilization in solid and liquid waste management. Traditional and advanced techniques such as composting, anaerobic digestion and bioremediation techniques have been implemented in solid waste management. Waste from wastewater has been successfully removed using fixed-film processes, activated sludge, biosorption technology and microbial electrochemical technology. Notorious chemicals such as synthetic dyes and oil spillage have been also removed from wastewater using microbial technology. Microbial technology has been magnificently implemented around the world for removal of waste from the environment. This chapter represents traditional and advanced microbial technology in both solid and liquid waste treatments.

Keywords Microbes · Microbial technology · Waste disposal · Aerobic processes · Anaerobic processes Solid and liquid waste management

---

R. R. Kumar · N. P. Vishwakarma (✉)
Department of Biotechnology, Atmiya University, Rajkot, Gujarat, India

C. Bhattacharya
Department of Microbiology, Atmiya University, Rajkot, Gujarat, India
e-mail: chitra.bhattacharya@atmiyauni.ac.in

P. Bhatt et al. (eds.), *Microbial Technology for Sustainable Environment*,
https://doi.org/10.1007/978-981-16-3840-4_16

16.1 Introduction

Waste management and environmental sustainability are major global concerns of the present society. In the past, there was a balance between the biosphere's natural cycle and mankind activity. The amount and quality of waste were not a burden, hence recycled by natural phenomenon. Human activity always generated waste products either as a by-product of their activity or generation of product which is assumed to be not useful for their life. Despite the generation of waste throughout their life, nature has treated the waste using their own treatment process such as waste degradation, dispersion and dilution. But, the activity of modern society creates disturbance in the equilibrium that continuously accelerates the burden of waste. There are three primary causes that play a synergistic effect on the continuous increase of waste: exponential growth of human population, extensive utilization and diffusion of toxic metals into the environment and tremendous utilization and dispersion of notorious chemicals such as xenobiotic compounds (Gandolla and Aragno 1992). The amounts of different wastes rise in such a way that they create intense damage to the environment and possibility of return to the environment is also reduced. Rapid consumption rate of raw materials leads to an upsurge of huge quantities of physical, chemical and radioactive wastes in the environment, which damage the biosphere rapidly. The accumulation of waste is the core factor for change of environmental consequences and loss of materials (Kumar et al. 2017). There is an urgent interference needed to tackle this critical situation.

Waste is difficult to address, as it becomes useful for someone and not for others. It was also defined as "any material or product which is useless to the producer is considered as waste". According to Dijkema, peoples want to dispose the generated materials even if it requires them to pay for their disposal (Dijkema et al. 2000). Brunner and Rechberger mentioned that despite the fact that waste is always a major concern for the environment and mankind, most human activities generate waste (Brunner and Rechberger 2015). Waste can be classified on the basis of some common characteristics such as physical states, reusable and recyclable potentials, physical properties, source of production, biodegradable potentials and degree of environmental impact (Demirbas 2011). It can be further classified on the basis of source as domestic waste, industrial waste, commercial waste, agriculture waste, mining waste and construction waste. Waste is also classified on the basis of environmental impact as hazardous waste and non-hazardous waste.

Waste has been categorized into various categories in different countries and their treatment is also based on its quality. The reduction and disposal of wastes have become a global concern that needs to be managed. All the waste treatment strategy is based on the return of waste mass to the environment. Disposal of waste into the environment can be done by either concentration of waste at a target site or dispersal of the waste substance into the biosphere. Dispersal is a good strategy but only environmentally acceptable substance can be dispersed. Therefore, waste treatment is aimed to either produce dispersible derivatives or concentrate the harmful ingredients. According to United Nations Environment Programme, waste management programmes should focus on the 3R concept which is expanded as "Reduce, Reuse and Recycle". The proper waste management planning and control is necessary for the prevention of the harmful impacts of waste on the environment (Ghiani et al. 2014).

16.2 Waste Management Practices

There are three categories of waste management practices: physical, chemical and biological. Physical methods comprise of incineration, compacting and sorting for solid waste management, whereas sediment dredging, artificial aeration, mechanical algae removal and water diversion are methods of waste water treatment (Wang et al. 2016). These traditional methods require a lot of time, material, effort and financial sustenance. These methods are able to remove large-size particles but not all pollutants from land or waterbodies competently.

The second technique is chemical method that includes gaseous emissions, gasification and pyrolysis for solid waste treatment. Liquid waste is chemically treated by flocculation, precipitation, chemical oxidation and chemical algae removal (Ma et al. 2015). Chemical methods require the support of a lot of chemical agents that include oxidizing agents, precipitants and coagulant salts. These methods are considered as emergency treatment, as they are not much efficient for the treatment of organic matter. Chemical methods have several other disadvantages such as consumption of high energy, foul odour and generation of toxic methane gas which are hazardous for the ecosystem.

The third approach is biological process that includes anaerobic digestion, composting, vermicomposting, biofiltration, microbial technology, biological-ecological methods, plant purification technology, combinatorial biotechnology, artificial wetland technology and biofilm technology (Xu et al. 2017). Biological methods are environmental-friendly that improve the natural process of waste removal and self-purification process of the polluted water ecosystem (Ravikumar et al. 2017). Technical supports are needed to overcome minor difficulties and cost of waste removal process in the future. Among various biological methods, microbial technology positions its own advantages and is considered as a highly efficient method.

16.3 Microbial Technology

Microorganisms have long since been explored in the food and beverage industry, pharmaceuticals, healthcare products, agriculture and industrial applications and for environment protection. A large number of microbes still need to be discovered and explored for their ecological application. Microbes play an important role in environmental protection and sustainability. Proper selection and effective utilization of microbes are crucial steps for the development of microbial technology. Microbial technology has been utilized for nutrient cycling, biodegradation, bioremediation, environmental detoxification, production of biocatalysts, bio-detergent, biomass fuel production, bioaccumulation, bioleaching, biomonitoring and so on. It has several advantages such as highly competent degradation ability, free from secondary pollution, reduced energy consumption, enduring viability, ease in technical operation and low capital investment.

Microbial technology has been magnificently implemented in several countries for environmental safety and waste management. Microorganisms especially bacteria, fungi, yeasts and their products were effectively used to remove different quality of waste and returned back into the atmosphere. Microbial technology was also used for the production of energy and biofuel from waste; biotreatment of pulp, oil and textiles; as well as production of valuable end products by fermentation process. They were also used for treatment of sewage water using aerobic and anaerobic microbes sequentially. Engineered microbes were recently used as additive for treatment of heavily polluted tidal river in China which shows improvement in treatment up to 70% (Sheng et al. 2012). Advanced microbial technology uses microalgae-bacteria consortium for treatment of nitrogen and phosphorus contamination from surface water (Liu et al. 2017). These photosynthetic bacteria were easily grown, more viable, more environment friendly, economical and highly effective in bioremediation process (Idi et al. 2015). The salient features of application of microbial technology seem to be highly effective in waste removal, and therefore, the present chapter emphasizes primarily on effective microorganisms and recent methods of microbial technology for removal of waste substance from terrestrial and wastewater.

16.4 Effective Microorganism

Microorganisms are cosmopolitan, present dominantly in soil where they a play vital role for sustainability of the environment. They are also residing in fresh- and marine water, plant, animals as well as air. These microbes are highly diverse, considered to be the largest unexplored reservoir. They perform numerous functions in the biosphere. Effective microorganisms play several roles in the biosphere such as recycling of waste from soil and water, plant growth-promoting activity, inhibition of soil-borne pathogens, enhancement of anti-oxidation capability in soil, nutrient recycling and so on (Shalaby 2011). They have functional diversity; some are known, while most need to be explored. Attention has been paid for the exploration of desired microorganisms from unique and less explored habitats. Selection of effective microorganisms is a fundamental part of microbial technology. It is usually followed by strain improvement using mutagenesis or genetic engineering to make them more capable for desired purposes.

The microbial activity can be further accelerated by application of microbial accelerating agents. Microbial accelerating agents are formulation of microbial growth promoters such as trace elements, amino acids, humic acid enzymes and vitamins in proportionate amount. These agents are always harmless to soil environment. These nutrients can accelerate microbial growth and stimulate them for biological reactions. They promote the proliferation and activity of indigenous microorganisms while inhibiting the anaerobic decomposition of pollutants (Wu and Xie 2012).

16.5 Microbial Technology for Waste Removal

Microbial technology is a scientific technique which utilizes a wide range of microorganisms in meticulous condition without distressing the ecosystem. An eclectic variety of microorganisms have been effectively used for waste removal practices. Microbial technology is fairly used for waste management practices in very effective way. It is eco-friendly, cost-effective and a better substitute compared to expensive physico-chemical remediation processes. The efficient methods adopted for waste removal process using microorganisms are composting, bioremediation, biodegradation, bioaccumulation, biotransformation, anaerobic digestion and wastewater treatment.

16.6 Microbial Technology for Solid Waste Treatment

16.6.1 Composting

Composting is an aerobic microbial decomposition process in which organic matter hydrolyzes into stable residue (Wei et al. 2017). This process is generally used through metabolic activity of microbial consortium to produce safe and stabilized form of organic compost for various agricultural practices. During the process temperature increases spontaneously; it would help to eradicate the pathogenic organisms; hence, finally generated compost becomes safe for usage (Rastogi et al. 2020). Several composting divisions exist at various locations in India such as Mumbai, Bengaluru, Indore, Vadodara, Kanpur and Delhi. These units have 150–-300 tonnes/day as installation capacity (Sharholy et al. 2008).

16.6.1.1 Framework of Composting Process

The following phases are included in compositing process:

1. **Mesophilic phase:** In this phase mesophilic fungi and bacteria degrade the complex compounds such as carbohydrate and amino acids into simple one by rapidly elevating the temperature.

2. **Thermophilic phase:** This is the second phase of composting in which thermophiles degrade the organic matters (lignin, cellulose, fats and hemicellulose). During this phase, thermophilic microbes would utilize the organic carbon content for their feedstock and metabolic activities.
3. **Cooling phase:** In the last phase microbial activity is diminished and temperature has also decreased. During this stage mesophilic microorganisms degrade the residual substances such as cellulose, hemicellulose, sugars and humus (Albrecht et al. 2010).

16.6.1.2 Factors Affecting the Composting Process

Some physiochemical factors such as nutrient balance, C/N ratio, pH, particle size, moisture, temperature, porosity, oxygen, moisture content and nutrient availability play a foremost role in various phases to determine the development of microbial populations during composting (Leow et al. 2018).

16.6.1.2.1 C/N Ratio

As a nutritional parameter, the optimum balance of carbon-nitrogen ratio is essential for the formulation of compost. Ideally, C/N ratio essentially range between 25 and 35, declaring that microbial requisite is around 30 parts carbon to 1 part nitrogen (Kutsanedzie 2015). However, some authors state that C/N value at the range of 20–50 also gives a worthy result (Petric et al. 2015). During the composting process, carbon is transformed to carbon dioxide during organic degradation. Due to higher C/N ratio, there is deficiency in nutrient level to microbiota and composting speed decreases. Lower C/N gives rise to increased nitrogen content compared to degradable carbon and that results in loss of ammonia in soil through leaching or volatilization (Zhang and Sun 2016). Generally, declined C/N ratio indicates higher waste degradation over the mineralization process (Wang et al. 2016).

16.6.1.2.2 Temperature

Composting is an exothermic technique, which includes organic matter degradation by bio-oxidative microbial activity. In this process a huge magnitude of energy is produced. But microorganisms utilize only 40–50% of the generated energy for the synthesis of ATP; the rest of the energy is lost in the form of heat. The huge quantity of heat is the basic reason for increased temperature during the process and it can reach up to 70–90 °C. This process was named as "microbial suicide" by Finstein (Waszkielis et al. 2013). This high temperature inhibits the microbial growth and slows down the biodegradation process. The organic matter decomposition rate is dependent on the temperature of the raw material. In the higher temperature (above 70 °C), only few thermophilic bacteria can survive and show metabolic activity. So, for the destruction of pathogens, weed seeds and fly larvae, maximum temperature would be maintained at least 3–4 days (Garg and Tothill 2009). Increased microbial diversity will be necessary to obtain higher biodegradation rate and the required temperature should be in the range of 30–45 °C (Finstein et al. 1983). The set point for feedback temperature control ranged between 30 and 45 °C during the process of composting.

16.6.1.2.3 Aeration

For the aerobic decomposition of raw material, sufficient oxygen supply is required. Anaerobic condition could be developed in the deficiency of oxygen in the environment that will create pungent odour of methane (CH_4) gas. Oxygen would be refilled in the waste materials by using perforated pipes (Garg and Tothill 2009).

16.6.1.2.4 Moisture Content

Water is the most essential part of composting cycle and microbial activity. The maintenance of optimal moisture content is dependent on particle size, physical state of the initial material and the type of composting system. Usually, 60% of moisture content is considered as suitable for starting material. Because different materials have different water-holding capacities, an exact generalization of moisture levels cannot be made. If the moisture content is too low, early dehydration occurs which arrests the microbial activity, giving physically stable but biologically unstable compost. In modern composting systems, water can be added during the process. In newly designed plant system having a capacity to remove a large amount of water, evaporative cooling system and high rate of heat generation, this dried material required the addition of water for sustainability of microbial activity. This will be only possible in conjunction with mechanical turning. The biological activity of stabilized end material can be prevented by lowering (about 30%) the moisture content of composting process (Diaz and Savage 2007).

16.6.1.2.5 Time

The period of active composting: dairy waste can be converted into compost around 1–14 weeks; this is followed by 3–4 weeks of curing period (Garg and Tothill 2009).

16.6.1.2.6 pH

The estimated pH range for the decomposition of organic matter is 5.5–8.0 (Zhang and Sun 2016). If the pH is more than 8.0, lime or bleaching powder is used at the collection/storage points. If the pH decreases less than 5.5, then microbial nitrification and volatilization occur, which results in the production of enormous quantity of acids and CO_2 (Wang et al. 2016). Volatilization of ammonia leads to disappearance of protein mineralization and slowdown of nitrogen during consequent stages of composting (Guo et al. 2012). Alkalization of compost is associated with preeminent pH (more than 9). It may obstruct the survivability of pH-sensitive microbes that have plentiful contribution in compost sanitation (Hachicha et al. 2009). Herbocel or sanitreat process is applied to control flies and foul odour (Raza and Ahmad 2016).

16.6.1.3 Composting Methods

There are numerous composting methods that suit the goal of the researchers and the nature of waste materials to be composted can be adopted for their purpose of utilization. Some of the composting methods are enumerated below.

1. *Vessel Composting*

   This method depends on the variety of forced aeration and mechanical turning techniques for the enhancement of the composting process. A modified version of in-vessel composting refers to confining the materials within an enclosed area such as a container, building or vessel (Gonawala and Jardosh 2018). This method is too costly and labour-intensive.
2. *Windrow Composting*

   In this composting process, raw materials are employed in a long narrow piles or windrows that curved frequently. Aeration is utilized for the mixing of materials into the setup. In the windrow composting process, system arrangement should be started from 3 ft. height for manures (dense materials) and 12 ft. height for fluffy materials such as leaves. It is a rapid process due to the materials' heat holding capacity and expensiveness in nature (Ayilara et al. 2020).
3. *Vermicomposting*

   Vermicomposting is a fundamentally accomplished process for digesting organic matter by combined action of earthworms and microorganisms to transform the organic waste material into a compost for soil amendment. Earthworms are insatiable consumers that biodegrade matter as vermi-castings or excreta. Earthworms are key players in vermicomposting but they also stimulate microbial activity through fragmentation of organic matter and aeration, and they increase the surface area for microbes. During composting, microbial colonization begins that leads to succession of microbial community composition (Rogayan et al. 2010). Vermicompost provides growth-promoting hormones and nutrients to plants and also plays a responsive role in improving the soil structure by increasing nutrient capacities and moisture content of the soil. With the usages of vermicompost, fruits, flowers, vegetables and other plant products are grown in better quality (Arumugam et al. 2017).

16.6.1.4 Advantages and Disadvantages of Composting

Despite the advantages of compost, some disadvantages also occur (Beffa 2002).

Advantages

- It is cost-effective and very sophisticated.
- Humus and plant litter nutrients can be recycled into the soil.
- It increased the microbial diversity into the soil.
- Compost can be used as biofertilizer as well as biocontrol agent, as they compete with phytopathogens.
- It would be beneficial for biodegradation of toxic compounds and pollutants.

Disadvantages

- Pungent odour nuisances.
- Propagation and dispersal of potential pathogens or allergenic microorganisms.
- If heavy metal content is too high in compost, it could affect the soil.
- This process could be used only for sewage sludge that originate from non-industrial sources.

16.6.2 Aerobic Digestion

In the aerobic biological treatment system, organic matter in waste material is digested under humid and warm condition present. Activated sludge, trickling filters and oxidation ponds are the foremost categories of aerobic systems. Aerobic digestion system works by employing the basic biochemical reaction applied for all the microorganisms (McKinney 1957).

$$\begin{aligned}\text{Food} + \text{Microorganisms} + \text{Oxygen} \rightarrow\ & \text{Increased Microorganisms}\\ & + \text{Carbon dioxide} + \text{Water}\\ & + \text{Waste products } (+\text{Energy})\end{aligned}$$

All the digestion system operates on the fundamentals of biochemical ethics, but some vary from each other by oxygen transfer method. Compressed air is utilized in activated sludge for mixing of oxygen source. Generally, in trickling filter microorganisms are attached to the stones and attained their oxygen from the diffused air. Details of aerobic digestion treatment process have been included in wastewater treatment process in this chapter.

16.6.3 *Anaerobic Digestion*

Anaerobic absorption is a fermentative decomposition technique which converts the organic waste into composites that can be utilized as biofertilizers and soil conditioners (Rastogi et al. 2020; D'Annibale et al. 2006). Anaerobic absorption is defined as a sequence of organic developments, where microbes can catalyze organic waste material in anaerobic condition. Stages included in anaerobic digestion are acidogenesis, hydrolysis, methanogenesis and acetogenesis.

16.6.3.1 Hydrolysis

Hydrolysis is the preliminary juncture of anaerobic absorption, in which insoluble organic substances such as polymers of amino acids, carbohydrates and fats are converted into soluble matter such as long chain of fatty acids, amino acid and sugar molecules (carbohydrates). Hydrolase enzymes are supported in the hydrolysis reactions. The microbes which synthesize these enzymes are termed as hydrolytic microorganisms such as *Bacteroides*, *Clostridium*, *Bacillus*, *Proteus vulgaris*, *vibrio* and *Micrococcus* and *Staphylococcus* bacteria (Amani et al. 2010). Some hydrolase enzymes are utilized in the form of peptidases, esterase and glycosidases. During anaerobic digestion process, hydrolysis is considered as a rate-limiting stage where insoluble complex organic substance is slowly depolymerized.

16.6.3.2 Acidogenesis

The second stage of anaerobic digestion process involves acidogenesis, where acid-gas-producing bacteria can hydrolyze the soluble molecules to alcohols, lactate, volatile fatty acids and carbon dioxide. Numerous fermentation pathways are involved in this process. Different bacterial genera are responsible to carry out acidogenesis. Some bacterial species that perform fermentation under anaerobic digestion are as follows: for the production of alcohol, *Saccharomyces* is used; for lactate fermentation *Lactobacillus* and *Streptococcus* were utilized; *Butyribacterium* species are employed for butyrate fermentation; and *Clostridium* is used for propionate fermentation. The group of bacteria such as *Clostridium*, *Sporomusa*, *Acetobacterium* and *Eubacterium* is utilized for the production of acetate. These acetate-forming fermentative bacteria are distinctive species for acidogenesis (Amani et al. 2010).

16.6.3.3 Acetogenesis

Acetogenesis is a process involved in the third stage of anaerobic digestion method. Acetogenic bacteria called acetogens can transform the alcohols and volatile acids into acetate in the form of by-product hydrogen and carbon dioxide. Acetogenic bacteria such as *Syntrophobacter wolinii* and *Smithella propionica* are acetate-producing microbes that abrade the butyrate and propionate, respectively. On the other hand, some bacterial species such as *Pelotomaculum thermopropionicum* and *Syntrophobacter fumaroxidans* convert volatile fatty acid into formic acid with the liberation of CO_2 and H_2. *Clostridium aceticum* has the ability to produce acetate from H_2 and CO_2 (Amani et al. 2010).

16.6.3.4 Methanogenesis

Methanogenesis is the ultimate step of anaerobic digestion. In this phase methanogens play a significant role in the consumption of acetogenesis products. It can convert the acetate and molecular hydrogen to methane gas and terminate the activity of acetate-forming bacteria. Few pathways are involved for the production of methane from acetate:

- Acetoclastic methanogenic pathway: In this pathway bacterial species *Methanothrix concilii* and *Methanosaeta soehngenii* are able to transform acetic acid to carbon dioxide and methane.
- Hydrogenotrophic methanogenic pathway: Hydrogen and carbon dioxide are used for the production of methane by employing *Methanobacterium bryantii*, *Methanobrevibacter arboriphilus* and *Methanobacterium thermoautotrophicum.*

Some another bacterial species such as *Methanococcus voltae*, *Methanobacterium formicicum* and *Methanobrevibacter smithii* are applied for the production of methane from methanoate, hydrogen and carbon dioxide (Amani et al. 2010).

16.6.3.5 Factors That Affect the Anaerobic Digestion Process

Temperature: Temperature is the most significant cause for the continued existence and microbial growth in anaerobic digestion. There are two ranges of temperature for the growth of microorganism such as mesophilic (30–40 °C) and thermophilic (50–60 °C) bacteria that optimize the digestion of organic matter. In mesophilic environment (35 °C), the organic part of municipal solid waste can easily be assimilated in the atmospheric condition. Maximum microbial growth rate, microbial activity and production of methane are accomplished by anaerobic digestion under thermophilic conditions (Mata-Alvarez 2002).

Moisture: Moisture content is necessary for controlling the cell turgidity, transport of nutrients, enzymes and hydrolysis of complex organic matters (Khalid et al. 2011). The methanogenic activity increased in high-solid sludge at 90–96% moisture content of mesophilic anaerobic digestion process (Jiunn-Jyi et al. 1997).

Retention Time: In mesophilic condition, 8–40 days of retention time is necessary for anaerobic digestion. According to the report of Fdez, thermophilic anaerobic digestion process of solid organic matter is also accomplished in the range of 8–40 days (Fdez-Güelfo et al. 2011).

pH: The ideal pH range for anaerobic digestion process is 6.8–7.2. Hydrolysis and acidogenesis process enhanced at pH 5.5–6.5, whereas methanogenesis requires pH 7.0 for optimum activity (Ward et al. 2008).

Carbon and Nitrogen Content**:** Carbon is the energy source for the growth of microbial community, whereas nitrogen assists in the enhancement of microbial growth. The requirement of carbon source by microorganisms is 25–30 times more greater than that of nitrogen, chiefly at a ratio of 20–30:1 (Ward et al. 2008). Elsewhere, for methanization process, the required elements C/N/P/S (carbon/nitrogen/phosphorus/sulphur) have a nutrient ratio of 600:15:5:3 for microbial growth (Khalid et al. 2011).

16.6.4 Bioremediation

Bioremediation is distinct as a technique involved in the degradation of organic matter primarily by microorganisms that can utilize for the transformation of less toxic substances. In this process, naturally occurring microorganisms such as bacteria and fungi and chlorophyll-containing plants can detoxify hazardous substances which are harmful for human health and the environment. The microorganisms are isolated from the contaminated site by the physiological factors that influence the optimization of bioremediation method. Some of the environment factors such as type of soil, temperature, pH, aeration and nutrients are incorporated for the availability of contaminants to the microbial inhabitants. Based on the utilization of microbes, it can subcategorize microorganisms into various groups:

Aerobic bacteria: Microorganisms which survive in the existence of oxygen and utilize the contaminant carbon as the energy source for their growth. Some of the aerobic bacterial species such as *Alcaligenes* sp., *Mycobacterium* sp., *Rhodococcus* sp. and *Pseudomonas* species have the abilities to degrade under aerobic condition. Aerobic bacterial species can also degrade the chemical constituents such as hydrocarbons and pesticides, both poly-aromatic compounds and alkanes.

Anaerobic bacteria: For the degradation of polychlorinated biphenyls (PCBs), anaerobic bacteria can be employed for the bioremediation process of river sediments. Dechlorination process is involved in the bioremediation by anaerobic bacteria. In this process, solvent trichloroethylene (TCE) and chloroform are degraded in the absence of oxygen molecule.

Ligninolytic fungi: Fungi such as *Phanerochaete chrysosporium* are involved in the process of degradation of the extremely miscellaneous range of toxic environmental pollutants. They utilize some easily available substrates including straw, sawdust or corn cobs for their growth and survival.

Methylotrophs: Methylotrophs are a group of aerobic bacteria that utilizes methane and converts it to carbon as energy source for their augmentation and survival. In this pathway the initial enzyme, i.e. methane monooxygenase, is used as a catalyst to produce chemical compounds such as chlorinated aliphatic trichloroethylene and 1,2-dichloroethane.

16.6.4.1 Bioremediation Strategies

Various methods are involved in the process of bioremediation and are employed depending on the degree of saturation and aeration of an area as follows.

16.6.4.1.1 Culture-Dependent Approach

In the process of culture-dependent method, microorganisms are cultured into the nutrient growth media and further processed maintaining an objective. This technique involves microbes for the degradation of poisonous composites and manufacture of by-products (Gupta et al. 2019). There are two categories of bioremediation process.

In Situ Remediation

In this technique, generally site-specific microflora plays a significant role in the bioremediation process of the polluted site, which involves without any excavation and transport of contaminants. This method is eco-friendly and cost-effective (Singh 2014). Generally, three methods are used for in situ remediation: bioventing, biosparging and bioaugmentation. These methods are favourable for less polluted sites (Gupta et al. 2019).

1. *Biosparging*

   Petroleum hydrocarbons containing benzene, toluene, ethylbenzene, and xylene isomers (BTEX) are the major components of hazardous fuel that can be accidentally spilled into the environment with the release of gasoline products from the leakage of pipelines and fuel storage tanks. This is one of the primary sources of groundwater impurity. Intrinsic bioremediation process is considered as a potential method for the cleanup and removal for the sites of petroleum-hydrocarbon-contaminated areas (Chen et al. 2006). In the process of biosparging, indigenous microorganisms are used to degrade the organic substances in the saturated zone. Oxygen and nutrient are introduced into the saturated zone for the enhancement of microbial activity. Biosparging technique is utilized for the reduction of petroleum components which are dissolved in groundwater and within the capillary fringe. Biosparging differs from air sparging techniques in the sense that in the air sparging process contaminants are removed through volatilization method, while biosparging stimulates biodegradation of ingredients. If any volatile constituent is present in the form of contaminants, biosparging is frequently combined with soil vapour extraction and other remedial technologies. Whenever combinational process of biosparging and extraction of vapour occurs, the vapour extraction system can produce negative pressure in the unsaturated zone of groundwater through the extraction wells, which could also control the vapour plume migration.

2. *Bioventing*

   This is one of the important in situ remediation processes in which microorganisms can be utilized for the degradation of organic matter which is usually adsorbed with soil in the unsaturated zone. Bioventing system essentially has similar components as present in the soil venting system. It consists of a series of air induction (influent), blower and venting (effluent) wells. Bioventing usually occurs at specific site where air is injected into the soil or extracted from the soil. It can also be accomplished with excavated soils. In the sparging method, air is utilized to provide oxygen for the biodegradation and it can transfer the volatile contaminants from groundwater to the vadose zone in the soil. In the bioventing process, withdrawn air (off-gas) needs treatment (Hoeppel et al. 1991). The off-gas is usually treated by activated carbon, adsorption against thermal pressure and catalysis in a biofilter. During carbon adsorption the air treatment contains a very high cost of venting project (Hoeppel et al. 1991); therefore, as an alternate option, the vented air is used as a biofilter that can pass through unsaturated and uncontaminated soil (Miller and Bartha 1989). With the usage of bioventing process, the cost remediation process can be minimized and the need for off-gas treatment is also eliminated (Lee and Swindoll 1993).

3. *Bioaugmentation*

   Bioaugmentation is an in situ method for the removal of undesired hazardous compounds through the genetically modified microorganisms (hydrocarbon-degrading microbes) from soil and groundwater (Mrozik and Piotrowska-Seget 2010). The foundation of this approach is the augmentation of pollutant-degrading microorganisms for degradation of complex pollutants (Omokhagbor Adams et al. 2020). In this process microorganisms containing the specific metabolic activity are augmented to enhance the rate of waste degradation at the contaminated site. Contamination of chlorinated ethenes, such as tetrachloroethylene and trichloroethylene in the groundwater or soil, can be completely eradicated using in situ microbes. These contaminants are converted to non-toxic substances such as chloride and ethylene (Singh 2014).

4. *Biostimulation*

   Biostimulation is a cost-effective, eco-friendly and extremely efficient bioremediation process. It refers to the addition of micronutrients such as phosphorus, nitrogen, oxygen and electron donors in harshly polluted area for stimulation and to improve the efficient degradation ability of inhabitant microorganisms. Biostimulation technique is applied to accelerate the rate of decontamination and degradation of the toxic contaminants and hazardous compounds. The rate of biostimulation is influenced by various environmental conditions such as moisture content, pH and temperature which predominate the ecological physiology (Abdulsalam and Omale 2009). Biostimulation technique plays an important role in the degradation of petroleum products, hydrocarbons and their derivatives (Kaouther Zaafouri 2014). In the petroleum-contaminated areas, the microbial population and their activity are very low. Poor metabolic activity of microbial population can be remediated through the process of biostimulation (Tyagi et al. 2011).

Ex Situ Remediation

In ex situ remediation, the process is carried out in any other place from the contaminated site (Singh 2014). By using this method, it gives favourable result in decreasing the total load of toxic xenobiotic compounds from the waste (Kumari et al. 2014). Ex situ bioremediation technologies can be facilitated by different methods such as biopiling and landfarming and bioreactors.

1. *Biopiling*

   Biopiling-mediated remediation process is employed for the piling of excavated polluted soil above the ground and nutrient modification, followed by aeration. It enhances the remediation process by increasing the microbial activities. Ex situ technique has been always considered for their cost efficiency. Biopiling can control the nutrient condition, aeration rate and temperature during the process (Whelan et al. 2015). Biopile can also be useful for the limited volatilization of low molecular weight pollutants. It can be effectively utilized for the remediation of pollutants of psychrophilic environments, i.e. cold regions (Dias et al. 2015). In the biopile heating system, augmentation of microbial activity is combined with contaminant availability which enhances the rate of biodegradation. Moreover, heated air is inserted into biopile design to deliver the air and heat in tandem, to facilitate enhanced bioremediation (Aislabie et al. 2006). In the processing of contaminated soil, sieving and aeration are provided by biopile (Delille et al. 2008). However, biopile systems have disadvantages over ex situ bioremediation process that include robust engineering, lack of power supply, maintenance cost, and land farming, especially at remote areas. Further, excessive heating of air can lead to drying of soil undergoing bioremediation, which results in the inhibition of microbial activities, reduction of biodegradation process and promotion of volatilization (Sanscartier et al. 2009).
2. *Landfarming*

   Landfarming is useful mainly for the control of pollutions from the soil by pesticides. In the landfarming process, a bilayer of fresh soil, clay or concrete are prepared from the fresh soil and combination of clay and concrete. The fresh soil is placed at the bottom part and the concrete layer is always placed in the superior layer. Then, natural microbial deprivation is permitted. In this process air, nutrient and humidity are necessary to exhibit the protocol by maintaining the 7.0 pH through liming (Singh 2014).

3. *Bioreactors*

   In the ex situ treatment, slurry bioreactors are used for the removal of stained soil and water pumped up which contaminated the column. Bioremediation by bioreactors involves the removal of contaminated solid materials such as soil, residues and sludge. In this reactor system the rate of biodegradation is higher than in situ remediation, because the contained environment is more convenient, predictable and manageable. The disadvantages of bioreactor is pretreatment of contaminated soil by excavation or vacuum extraction method (Mary 2011).

16.6.4.1.2 Culture-Independent Approach

In this approach, sequencing technique is being utilized for the analysis of genomes of all nucleotides such as DNA, RNA and protein. Various processes are involved in this culture-independent method like processing of whole DNA sequence from specific environmental condition known as metagenomics, whole-genome sequencing, whole-transcriptome sequencing (isolation of RNA from a single and pure culture), metatranscriptomics (RNA isolated from a sample of ecological unit) and so on (Rathour et al. 2017). In nature microorganisms are omnipresent; therefore, to explore their microbial diversity, "next-generation sequencing" would be in assistance to overcome the concealed potential of microbiome functionally, taxonomically and morphologically (Gupta et al. 2019).

16.6.4.1.3 Bioremediation of Rubber Waste

The use of rubber is increasing day by day in vehicles (Holst et al. 1998). Tire is composed of high-grade black carbon and synthetic polymers for maintaining the strength of the rubber tire (Larsen et al. 2006). About 12% of rubber is considered as solid waste. With regard to its physical composition, rubber can neither be degraded easily nor recycled (Conesa et al. 2004). A major environmental problem arises due to the burning of rubber; it produces a large number of toxic chemical components such as zinc oxides and carbon monoxide (Zabaniotou and Stavropoulos 2003). The toxic component from rubber has been preliminarily eliminated by using *Recinicium* fungi. In the second step, rubber can be devulcanized by sulphur-oxidizing or sulphur-reducing bacteria like *Thiobacillus ferrooxidans* and *Pyrococcus furiosus*. Devulcanized rubber can be recycled (Stevenson et al. 2008). The best way of waste management is to control the combustion of rubber and the liberated heat is also utilized for the generation of energy (Conesa et al. 2004).

16.7 Roles of Microorganisms in Wastewater Treatment

Application of microbial technology is now being also focused on the removal of heavy metals, toxic substance, dissolved inorganic dyes and nutrients. Microorganisms in sewage treatment are manifested through their metabolic activities, degradation potential as well as detoxification capacity. Microorganisms regulate a series of chemical reactions, and metabolize the pollutants by utilizing organic matter as a nutrient source present in wastewater. Microbial technology also utilizes degradation of organic matter into inorganic substance and detoxification. On the basis of metabolic capacity of microbes under anaerobic or aerobic conditions, microbial process is divided into aerobic and anaerobic biological treatment.

Wastewater treatment process is associated with the following group of microorganisms:

1. ***Bacteria:*** Bacteria play a pulsating role in the treatment of wastewater. They hold the principal responsibility for the removal of organic compounds. Organic matters are utilized by them to get energy and they use this energy for their growth. Most common bacteria present in wastewater are *Achromobacter* sp., *Arthrobacter* sp., *Acinetobacter* sp., *Alcaligenes* sp., *Bacillus* sp., *Chromobacterium* sp., *Cronobacter* sp., *Citramonas* sp., *Enterobacter* sp., *Escherichia coli*, *Flavobacterium* sp., *Klebsiella* sp., *Kosakonia oryzae*, *Leclercia* sp., *Pseudomonas* sp. and *Serratia* sp. (Silva-Bedoya et al. 2016).
2. ***Filamentous bacteria:*** They are normally associated with biomass present in activated sludge. Their presence is noteworthy for the formation of floc. Their population size varies with nutrient conditions such as the amount of nutrients, DO, sludge age, pH and temperature of wastewater system. Common filamentous bacteria found in activated sludge are *Alcanivorax* sp., *Beggiatoa* sp., *Microthrix* sp., *Sphaerotilus natans* and *Thiothrix* sp. (Paillard et al. 2005).
3. ***Algae:*** They are photosynthetic organisms that have a substantial role in nutrient (nitrogen and phosphorous) removal and accumulation of some xenobiotic compounds, toxic substances (both organic and inorganic) and heavy metals. Some common algae present in wastewater are *Chlamydomonas* sp., *Euglena* sp., *Limnothrix*, *Lyngbya*, *Microcystis*, *Oscillatoria*, *Phormidium autumnale* and *Synechocystis* (Martins et al. 2010).
4. ***Protozoa:*** Protozoans are unicellular eukaryotic organisms that digest and eliminate free-swimming bacteria and other suspended particles from wastewater. Requirement of oxygen varies among different species of protozoan for their survival. Common ciliated protozoans found in wastewater are *Aspidisca* sp., *Carchesium* sp., *Chilodonella* sp., *Opercularia coarctata*, *Trachelophyllum* sp. and *Vorticella* sp. (Amaral et al. 2004).
5. **Fungi:** They are multicellular eukaryotes having ability to hydrolyze organic matter even at low pH. They secrete hydrolytic enzymes to degrade substrate and adsorb suspended solids through mycelia to accomplish their nutrient requirements during wastewater treatment. Common fungi in waste removal process are

Aspergillus, *Absidia*, *Fusarium*, *Sphaerotilus*, *Penicillium* and so on (Akpor et al. 2013).

16.8 Microbial Technology for Wastewater Treatment

The strategy for treatment of wastewater begins in the early twentieth century. Technology application for wastewater treatment has been focused on the quality of waste present in sewage. The major pollutants present in wastewater are biodegradable and volatile organic compounds, nutrients (nitrogen and phosphorus), suspended solids, toxic metals, recalcitrant xenobiotics as well as microbial pathogens. Microbial technologies are environment friendly and sensible choice for hazardous waste from wastewater. Waste present in water is usually treated primarily by physical treatment methods to remove the physical pollutants. Microbes were usually applied as a secondary treatment to remove organic matter present in waste. The choice of technique is based on the quality of waste, aerobic or anaerobic method of degradation and several other parameters. The British Royal Commission declares their goal for waste water treatment to decrease BOD of wastewater up to 20 mg/L and the final yield of effluent and suspended solids should also decrease to 30 mg/L. BOD (biochemical oxygen demand) is defined as "the amount of oxygen required by microbes present in waste water for the oxidation of organic nutrients". It measures the strength of the organic waste present in wastewater; the more organic matter present, the higher the BOD value. We included some common and advanced technology used for wastewater treatment.

16.8.1 Fixed-Film Processes

Fixed-film microbiological processes are based on attachment of high concentration of microorganisms on a solid support material such as plastic, gravels, sand or stone particles. Microbial growth is influenced on the substratum by factors such as geometric configuration of particles and the flow rate of wastewater. There are several advantages of fixed-film processes such as low specific growth rate of microorganisms, suitability for small size reactor and lower operation cost. The disadvantage associated with this process is overgrowth of microorganisms on the solid substratum and biofouling which affects heat exchange process unpleasantly. Two major categories of fixed-film reactors, trickling filters and rotating biological contactors, are described here.

1. *Trickling Filters*

   The treatment of wastewater by trickling filters is conducted in a rectangular or circular reservoir filled with filter medium with a depth of approximately 100–250 cm. Large surface area available in filter medium is suitable for growth of microorganisms and also provides adequate void space for diffusion of air.

Usually crushed stones, granite, hard coal, ceramic material, plastic and treated wood are used in filter media. The size of support matrix is based on microbial attachment and void space. Smaller sized matrix has higher surface area for microbial attachment, but void space is less. Wastewater is sprayed upon the surface of the bed with sprinklers that allow uniform hydraulic load. Wastewater containing organic matter is dripped above the filter media that provides nutrients for the growth of microorganisms on filter surface. During the flow, the wastewater undergoes aerobic decomposition by microorganisms. An underdrain system is also connected that collects the treated wastewater as well as sloughed off microbial biomass. A final clarifier separates the microbial biomass from the treated wastewater. Depending on the amount of treated water, the trickling filters may be subdivided into flushing and percolating filters.

Microbes in trickling filter: Biofilm formation occurring at filter surface is called zoogleal film which consists of algae, bacteria, protozoa and fungi. Biofilm formation on the surface of filter media is similar to naturally occurring aquatic environments. Common bacteria found in zoogleal film are *Achromobacter*, *Alcaligenes*, *Flavobacterium*, filamentous bacteria, nitrifying bacteria (*Nitrosomonas* and *Nitrobacter*), *Pseudomonas* and *Zooglea*. Bacteria adsorb and anchor on the substratum using polymer-containing matrix, glycocalyx. After acclimatizing to the substratum, the filter surface colonized by bacteria is also further occupied by successional life forms (Rani et al. 2019). Common fungi found in zoogleal film are *Aspergillus*, *Geotrichum*, *Fusarium*, yeasts, *Mucor* and *Penicillium*. The growth of fungal hyphae supports the transfer of oxygen at the depths of the biofilm. Several types of algae also flourish in biofilm such as *Anacystis*, *Chlorella*, *Euglena*, *Phormidium* and *Ulothrix*. Algae produce oxygen and some of them also fix nitrogen. The protozoa occurring in biofilms are amoeba, arcella, ciliates (*Colpidium*, *Vorticella*) and flagellates (Bodo, Monas). Microbial biofilm degrades the organic matter present in wastewater and continues to grow (Arezoo et al. 2017). The increase in thickness of biofilm acts as a limiting condition for diffusion of oxygen in the deeper layers, hence creating an anaerobic environment. It also reduces supply of organic substrates into the deeper layer. These biofilms are later sloughed off and further a new biofilm forms.

Advantages and Disadvantages: Trickling filters are reliable, are easy to operate and have little maintenance costs. Both domestic and toxic industrial effluents can be treated. Its disadvantage is associated with the clogging of filter under higher organic load. It results into excessive growth of slime bacteria that cause restriction in air circulation, low oxygen availability and hence foul odour.

2. *Rotating Biological Contactors (RBCs)*

This is a fixed-film bioreactor in which the disk surface is used for adsorption of microbes. A thin biofilm of 1–4 mm thickness is accountable to decrease BOD level. RBC comprises a horizontal shaft with a series of microbial immobilized disks straddling on it. Disks are around 40% submerged and rotate slowly in the wastewater. The rotation enhances oxygen transfer required for microbial activity. It also improves contact between immobilized biomass and wastewater that reduces BOD at a faster rate. There is successional growth of microorganism on the rotating disk. Biofilms on RBCs encompass diverse microbial community such as filamentous and eubacteria, metazoan, protozoa and filamentous algae such as *Oscillatoria* (Aguilera et al. 2007). RBCs mostly remove organic materials and also oxidized ammonia. Several advantages associated with RBCs are low operation cost, short residence time, low maintenance costs and release of dewatered sludge.

16.8.2 Activated Sludge

This is the most extensively used suspended-growth wastewater treatment process. It utilizes microbial culture to degrade organic matter under aerobic condition. The treatment consists of oxidation of organic matter to NH_4, CO_2 and H_2O with the formation of cellular biomass. Activated sludge forms floc for separation of solid in settling tank during the aeration of wastewater. Activated sludge process is constructed with two major objectives:

1. By using aerobic oxidation process, soluble organic matter is converted into new biomass of biodegradable organic matter
2. Activated sludge technique is applied for the separation of biomass from treated effluent

Activated sludge system consists of aeration and sedimentation tank. Aerobic oxidation of organic matter using microorganism is carried out in aeration tank. Wastewater treated after preliminary method is pass in the tank and assorted with sludge or return activated sludge (RAS) to form mixed suspended solids. Aeration is provided by mechanical method. Air is utilized by microorganisms to develop a biological floc. During return activated sludge, a huge amount of the microbial biomass is recycled, maintaining an enormous number of intial microbes for effective oxidation of organic matters in a reasonably short time. The total content activated sludge present in aeration tank is called mixed liquor, whereas the total amount of microorganisms, mineral and organic suspended solids in the assorted with liquor is called mixed liquor suspended solids (MLSS). MLSS possess both organic and inorganic portions. The organic portion comprises live or dead microorganisms, lacking microbes in the organic matter, and cellular debris is denoted by MLVSS.

The MLSS is shifted from the aeration tank to the sedimentation tank for the separation of sludge from the treated effluent. Sedimentation tank is utilized for the sedimentation of microbial sludge formed in aeration tank during oxidation of organic substance. The amounts of regimented solids can be varied by waste activated sludge (WAS) and returned activated sludge (RAS). A rational settled activated sludge is returned back to the aeration tank for the treatment of incoming raw wastewater (RAS). The remnants or excess sludge is removed to balance the ratio of food for the microorganisms (F/M). The food to microorganism ratio is significant to maintain the balance between organic loads to biomass generation in activated sludge system conveyed in terms of kilogram BOD per kilogram of MLSS per day. Mean cell resident time (MCRT) or solid retention time (SRT) is an important component of activated sludge that measures the contact time of microorganism with substrate. Activated sludge process is controlled by several parameters including food to microorganism ratio, sludge volume index, mean cell residence time, sludge age, dissolved oxygen and biochemical oxidation demand (Johnston et al. 2019).

The solid part of flocs absorbs impurities present in wastewater, while microorganisms oxidize the absorbed substances. Activated sludge has a loose and porous structure. These microbial cells occur as flocs or agglomerates whose density is abundant for sedimentation in the sedimentation tank. Sedimentation is monitored by using "secondary clarification" for the separation of microorganism and solids from treated wastewater.

Microbes in Activated Sludge: The activated sludge flocs contain numerous microorganisms but mostly bacteria, organic substance and inorganic particles. The size of floc varies between <1 mm and >1000 mm. Microelectrodes are used to measure microbial activity and estimate the concentration of oxygen, ammonia, nitrate, redox potential, pH or sulphide profiles within flocs. The major bacterial genera present in flocs are *Achromobacter*, *Acinetobacter*, *Alcaligenes*, *Bacillus*, *Brevibacterium*, *Corynebacterium*, *Comamonas*, *Flavobacterium*, *Pseudomonas* and *Zooglea*. Fungal growth is usually not favoured; however, some genera such as *Alternaria*, *Cladosporium*, *Cephalosporium*, *Geotrichum* and *Penicillium* are present in activated sludge (Yang et al. 2020). Several conditions such as oxygen deficiency and overloading of aeration tanks do not favour proper floc formation; this phase is called as active-sludge swelling. There are two discrete types of sludge swelling: fibrous and non-fibrous swelling. The growth of *Sphaerotilus*, *Beggiatoa* or *Thiothrix* is observed during fibrous swelling and the amount of mucous secretion is increased during non-fibrous swelling.

16.8.3 Biosorption Technology

Microorganism-based sorbent materials have been recently implicated for the retrieval of heavy metals from wastewater. Biosorption is a specific property of some microbes that concentrate the organic or inorganic substances from liquid solutions. Though biosorption techniques are principally applied to sequester heavy metals, metalloids, radionuclides and rare earth elements, recently it has been implicated for the removal of organic dyes (Kaushik and Malik 2009). Microbial EPS (extracellular polymeric substance) and cell wall play an important role in absorption and act as an alternative of synthetic adsorptive substances.

EPS is a high molecular weight natural polymer secreted by several microorganisms, and their chemical composition also varies with microbial genera. EPS is utilized by microbe primarily to protect themselves from metal toxicity. Diverse structure, chemical stability, metal binding property, selectivity and high reactivity of microbial EPS advocate them as great competitor. Factors for biosorption include the type of metal and its ionic form, metal binding site and external environmental of microbes. Biomass of algae, bacteria, cyanobacteria, filamentous fungi and yeast was commonly used for biosorption.

These microbes could be isolated from their natural habitat or obtained from waste by-product of fermentation industries to make the process economic. The surface area of bacteria and yeast has been increased for enhanced absorption using genetic engineering methods. Water inhabitants as algae and cyanobacteria were reported to absorb a variety of toxic metals in natural water and wastewater. They show various degrees of binding and specificity with Ni, Pb, Cd, Zn, Co, Cu, As, Mn, Mg and Zn. *Anabaena doliolum* Ind1, having surface group's carboxyl, hydroxyl, carbonyl, sulphate and amide groups, has great ability to bind with Cd (II) (Goswami et al. 2015). Similarly, *Providencia vermicola* strain SJ2 and *Paenibacillus peoriae* strain TS7 have shown specificity to bind with Pb (II) (Arumugam et al. 2017). *Streptomyces rimosus* and *Rhodococcus opacus* also show their ability to bind with Al (III) (Cayllahua and Torem 2010). *Rhizobium radiobacter* strain VBCK1062 which is commonly found in contaminated soil is a highly specific strain that binds with (V).

Metal ions were extraordinarily fixed by bacterial cell wall. Two mechanisms are conveyed in metal binding capacity of microbial EPS and cell wall:

1. Ion exchange performed by high quantity of negatively charged functional group in microbial EPS.
2. Complex formation with the charged group present in EPS or cell wall.

A thick layer of peptidoglycan found in gram-positive bacterial cell wall and lipopolysaccharide in gram-negative bacteria plays an important role in binding with metal ions (Flemming and Wingender 2001). Gram-positive bacteria and actinobacteria are comparatively more capable of adsorbing metal on their cell wall when exposed to selective metals. Copper was efficiently removed by desulphurization bacteria from wastewater. White-rot fungi and yeasts were also used to absorb toxic substances such as chromium, lead and other constituents of wastewater in China. Microbial technology for adsorption was also integrated with activated sludge treatment.

Advantage: Biosorption process offers several advantages such as low operating cost, high efficiency for even low metal concentrations, minimization of chemical uses, free from nutritional requirements, free from disposal of organic or inorganic sludge and avoidance from metal toxicity issues.

16.8.4 Microbial Electrochemical Technology

Microbial electrochemical technology (MET) is an emerging technology that amalgamates microbiology with electrochemistry (Schröder et al. 2015). In MET method, electroactive bacteria are capable of using a solid electrode as electron acceptor or electron donor (Rabaey et al. 2006). This electrode acts as an alternative to traditionally used nitrate/oxygen as electron acceptor or hydrogen/organic matter as electron donor (Karanasios et al. 2010). Depending on the quality of groundwater or pollutant present in them, MET system can be operated as microbial electrolysis cell (MEC) or as microbial fuel cell (MFC) (Schröder et al. 2015). MFC device differs from MEC in the sense that in MFC, energy can be extracted, while in MEC energy is supplied to allow or enhance bio-electrochemical process. MFC differs substantially from the conformist fuel cells. For conversion of fuel cell microorganisms act as biocatalysts for the cathodic and anodic substrate to catalyze the electrochemical reactions. Direct electron transfer in MFC was initially demonstrated in *Saccharomyces cerevisiae*, where it was grown in enriched medium, for the separation the platinum cathode and anode electrodes; porous cylinder is utilized in the field.

Microbial fuel cells emerged as a new bioremediation technology that is primarily used to recover toxic metal or mobilize pollutants present in wastewater. Several groups of bacteria, yeast, algae and fungi were found to remediate the heavy metal ions (Pous et al. 2017). Microbial electrochemical technologies are effectively applied for in situ or ex situ treatment of groundwater contamination. However, ex situ MET is widely applicable. During ex situ treatment, wastewater or groundwater has to be pumped to the other location, where intensive treatment is applied for reckless removal of pollutants. Different MET patterns have been utilized for the treatment of different groundwater pollutants such as aromatic hydrocarbons, chlorinated hydrocarbons (Aulenta et al. 2007) and metals (Huang et al. 2013).

16.8.5 Wastewater Treatment Using Oleaginous Microorganisms

Oleaginous microorganisms are well-known sources to produce microbial biofuels of comparable fatty acid configuration present in higher plants and animal oils. However, a major limitation of synthesis of biofuel from oleaginous microorganisms is the high cost of raw material (Azócar et al. 2010). Therefore, nutrient-rich wastewater was employed as cheaper substrates for oleaginous microorganisms. It was not only feasible economically but also important for environment-friendly biodiesel production (Huang et al. 2013). Different oleaginous microorganisms have been studied for simultaneous biofuel production along with simultaneous wastewater treatment. Certain groups of microbes such as bacteria, yeast microalgae and fungi are acting as single cell oils (SCOs) or oleaginous microorganisms (Arous et al. 2019).

Yeast: Oleaginous yeast *Debaryomyces etchellsii* is able to accumulate a substantial quantity of lipids from agroindustrial wastewaters such as wastewater obtained from olive mill, expired soft drinks and confectionary industries. Wastewater obtained from milk candy was effectively used by *Rhodosporidium toruloides* to produce sufficient quantity of reserve lipids (Zhou et al. 2013). Strain *Rhodotorula glutinis* TISTR 5159 was found to convert corn starch wastewater into lipids under semicontinuous fermentation conditions with higher efficiency of 65% reduction of COD value. Oleaginous yeasts belonging to genus *Trichosporon* also showed their high ability to remove pollutants, removing more than 55% COD from industrial wastewaters. Wastewater obtained from bioethanol industry was efficiently utilized by oleaginous yeast *Rhodosporidium toruloides* Y2 to decrease 72% BOD level. Other group of oleaginous yeast such as *Cryptococcus* sp. (Fernandes et al. 2014), *Lipomyces starkeyi*, *Rhodosporidium toruloides* and *Yarrowia lipolytica* (Louhasakul et al. 2016) was found to be highly efficient for utilization of wastewater.

Algae: Oleaginous microalgae show significant advantages over yeasts such as natural habitat to grow in wastewater and the requirement of low nutrition owing to their autotrophic characteristics (Cai et al. 2013). Some oleaginous microalgae such as *Neochloris*, *Arthrospira*, *Botryococcus*, *Chlorella* and *Scenedesmus* are able to remove pollutants from wastewater along with the production of microbial lipids (Perez-Garcia et al. 2011).

Fungi: An oleaginous fungus *Aspergillus oryzae* is able to produce biodiesel using starch-rich industrial wastewater (Muniraj et al. 2013).

Bacteria: An oleaginous bacterial strain of *Alcanivorax* was able to solve environmental pollution by removal of spilled petroleum. Their population significantly increased in oil spilled water by utilizing them as nutrients. Researchers also observed that soil-borne bacteria *Geobacter* are able to electroplate uranium, rendering it in insoluble form; therefore, it cannot dissolve and contaminate groundwater. There will be a wide application of *Geobacter* at the uranium contamination sites such as nuclear plants and mines in order to limit the catastrophic spillages.

16.8.6 Removal of Synthetic Dye Using Microbial Technology

Synthetic dyes with various structural diversities are frequently used in textile, paper, leather tanning, agricultural research and photochemical research. They cause extensive environmental pollution, deteriorate the quality of wastewater, increase toxicity and are also hazardous for health. Several wastewater treatment methods have been adopted to handle artificial dyes. The solicitation of microbial technology for degradation of synthetic dye is highly effective, economic and environment friendly. Different wastewater treatment methods using microbes have been adopted for the removal of synthetic dye. Rhodamine B and methyl violet have been removed during activated sludge process using cattle dung derived from microorganisms (Kanekar and Sarnaik 1991). Certain microbial culture was used for degradation of Acid Orange 7 dye in biofilm. For the assimilation of immobilization and decomposition of azo dye, multistage rotating biological reactor is used (Ogawa and Yatome 1990).

Removal of synthetic dye tartrazine and azo dye from wastewater was done by some bacterial species under anaerobic conditions. Both individual strains as well as mixed cultures were used under aerobic and anaerobic conditions for removal of synthetic dye. The combined aerobic and anaerobic sequential method has been effectively engaging on the disintegration of anthraquinone and monochlotriazine dyes (Panswad and Luangdilok 2000). More than 100 fungal and some bacterial laccase producers such as *Bacillus subtilis*, *Coprinus cinereus*, *Melanocarpus albomyces*, *Pycnoporus cinnabarinus*, *Thielavia arenaria* and so on have been reported that efficiently decolourize dye from wastewater (Mishra et al. 2019). Some dye decolourizer and degrader group of microorganisms have been mentioned below.

Fungi: White-rot fungi have attracted researchers in recent years due to their potent biodegradation capacity of highly stable natural molecule such as cellulose, lignin and hemicellulose. They are able to synthesize the wide range of extracellular enzymes such as laccase, phenol oxidase, lignin peroxidase, manganese-dependent peroxidase and manganese-independent peroxidase. They have been extensively investigated for decolourization of wastewater (Young and Yu 1997). A white-rot fungus, *Phanerochaete chrysosporium*, has been employed for the decolouration of Azure B, Congo red, Orange II and Tropaeolin O dye under aerobic conditions. This fungus is able to remove these dyes from wastewater between 6 and 9 days (Cripps et al. 1990). *Trametes versicolor* were able to decompose azo, anthraquinone and indigo-based dyes. Highly sensitive azo dyes have been efficiently removed from the wastewater by *Aspergillus foetidus*. Another dye Remazol Brilliant Blue R has been successfully decolourized by fungus *Pycnoporus cinnabarinus* using packed-bed bioreactor (Schliephake and Lonergan 1996). Anthraquinone dyes *and* triarylmethane and indigoid dye present in environment were efficiently degraded by *Trametes hirsute* (Abadulla et al. 2000). One of the efficient fungi *Kurthia* sp. is capable to decolourize the synthetic dyes such as brilliant green, pararosaniline, crystal violet, magenta and malachite green (Sani and Banerjee 1999).

Bacteria: *Bacillus subtilis* is able to degrade *p*-aminoazobenzene to produce *p*-phenylenediamine and aniline compound under anoxic condition (Zissi and Lyberatos 1996). Under fixed-film reactor conditions, *Pseudomonas mendocina* are used for the decolourization of methyl violet in the textile industrial effluent wastewater. *Pseudomonas luteola* and *Klebsiella pneumoniae* bacteria were used to decolourize reactive azo dyes from wastewater.

Algae: Reactive azo dyes have been also degraded by algae such as *Chlorella vulgaris*, *Chlorella pyrenoidosa* and *Oscillatoria tenuis*. An enzyme "azo reductase" found in some algae is responsible for the breakdown of azo linkage, hence converting azo dyes into aromatic amines.

16.9 Conclusions and Future Prospects

Hazardous waste is a major threat for environmental safety that needs to be eradicated in various ways. Microbial technology is the amalgamation of microbial process with technology in which selective microbes are used by scientific methods. Microbial technology has been extensively used for the removal of waste from terrestrial area and wastewater. Removal of waste using microbes is not only cost-effective, efficient but also environment friendly. A major section of the waste is organic matter which can be properly handled by suitable microorganisms using the phenomena of reduction, reutilization and recycling of waste. Besides the organic matter, several microorganisms have been explored that are able to remove oil, synthetic dyes, toxic heavy metals and xenobiotic compounds. Modern techniques such as biosorption and microbial electrochemical technologies were implemented for toxic metal recovery. Exponential increase of hazardous waste still creates pressure to further explore newer microbial technology or extend the existing techniques with underexplored microbes to solve the global problem. Therefore, genetically engineered microbes and high-throughput screening of specific microorganism have been promoted for microbial technology as more convenient and effective for waste removal. Careful monitoring of multiple strain-based inoculums could also reduce waste in rapid rate. Generation of energy using microbes during waste management is an economic way that needs to be implemented in the future. The potentials of hidden domains of microbes have great prospect that can be further explored for the treatment of waste.

References

Abadulla E, Tzanov T, Costa S, Robra KH, Cavaco-Paulo A, Gubitz GM (2000) Decolorization and detoxification of textile dyes with a laccase from Trametes hirsuta. Appl Environ Microbiol 66 (8):3357–3362

Abdulsalam S, Omale AB (2009) Comparison of biostimulation and bioaugmentation techniques for the remediation of used motor oil contaminated soil. Brazilian Arch Biol Technol 52 (3):747–754

Aguilera A, Souza-Egipsy V, Gómez F, Amils R (2007) Development and structure of eukaryotic biofilms in an extreme acidic environment, Río Tinto (SW, Spain). Microb Ecol 53(2):294–305

Aislabie J, Saul DJ, Foght JM (2006) Bioremediation of hydrocarbon-contaminated polar soils. Extremophiles 10(3):171–179

Akpor OB, Adelani-Akande TA, Aderiye BI (2013) The effect of temperature on nutrient removal from wastewater by selected fungal species. Int J Curr Microbiol 2(9):328–340

Albrecht R, Périssol C, Ruaudel F, Le Petit J, Terrom G (2010) Functional changes in culturable microbial communities during a co-composting process: carbon source utilization and co-metabolism. Waste Manage 30(5):764–770. https://doi.org/10.1016/j.wasman.2009.12.008

Amani T, Nosrati M, Sreekrishnan TR (2010) Anaerobic digestion from the viewpoint of microbiological, chemical, and operational aspects - a review. Environ Rev 18(1):255–278

Amaral AL, Da Motta M, Pons MN, Vivier H, Roche N, Mota M et al (2004) Survey of Protozoa and Metazoa populations in wastewater treatment plants by image analysis and discriminant analysis. Environmetrics 15(4):381–390

Arezoo D, Mohammed Maikudi U, Kang Tzin L, Rita Devi V, Nasser S, Pravin V et al (2017) Microbial aspects in wastewater treatment – a technical review. Environ Pollut Prot 2(2):75–84

Arous F, Jaouani A, Mechichi T (2019) Oleaginous microorganisms for simultaneous biodiesel production and wastewater treatment: a review. In: Microbial wastewater treatment. Elsevier, Amsterdam, pp 153–174. https://doi.org/10.1016/B978-0-12-816809-7.00008-7

Arumugam K, Renganathan S, Renganathan K, Sharma NK, Babalola OO (2017) Enhancing the post consumer waste management through vermicomposting along with bioinoculumn. Int J Eng Trends Technol 44(4):179–182

Aulenta F, Catervi A, Majone M, Panero S, Reale P, Rossetti S (2007) Electron transfer from a solid-state electrode assisted by methyl viologen sustains efficient microbial reductive dechlorination of TCE. Environ Sci Technol 41(7):2554–2559

Ayilara MS, Olanrewaju OS, Babalola OO, Odeyemi O (2020) Waste management through composting: challenges and potentials. Sustainability 12(11):1–23

Azócar L, Ciudad G, Heipieper HJ, Navia R (2010) Biotechnological processes for biodiesel production using alternative oils. Appl Microbiol Biotechnol 88(3):621–636

Beffa T (2002) The composting biotechnology: a microbial aerobic slid substrate fermentation complex process. HappyswissCom [Internet], pp 1–37. http://www.happyswiss.com/compag/pdf/THECOMPOSTING BIOTECHNOLOGY COMPLETE.pdf

Brunner PH, Rechberger H (2015) Waste to energy - key element for sustainable waste management. Waste Manag 37(April 2018):3–12

Cai T, Park SY, Li Y (2013) Nutrient recovery from wastewater streams by microalgae: status and prospects. Renew Sustain Energy Rev 19:360–369. https://doi.org/10.1016/j.rser.2012.11.030

Cayllahua JEB, Torem ML (2010) Biosorption of aluminum ions onto Rhodococcus opacus from wastewaters. Chem Eng J 161(1):1–8. http://www.sciencedirect.com/science/article/pii/S1385894710002445

Chen KF, Kao CM, Chen TY, Weng CH, Tsai CT (2006) Intrinsic bioremediation of MTBE-contaminated groundwater at a petroleum-hydrocarbon spill site. Environ Geol 50(3):439–445

Conesa JA, Martín-Gullón I, Font R, Jauhiainen J (2004) Complete study of the pyrolysis and gasification of scrap tires in a pilot plant reactor. Environ Sci Technol 38(11):3189–3194

Cripps C, Bumpus JA, Aust SD (1990) Biodegradation of azo and heterocyclic dyes by Phanerochaete chrysosporium. Appl Environ Microbiol 56(4):1114–1118

D'Annibale A, Rosetto F, Leonardi V, Federici F, Petruccioli M (2006) Role of autochthonous filamentous fungi in bioremediation of a soil historically contaminated with aromatic hydrocarbons. Appl Environ Microbiol 72(1):28–36

Delille D, Duval A, Pelletier E (2008) Highly efficient pilot biopiles for on-site fertilization treatment of diesel oil-contaminated sub-Antarctic soil. Cold Reg Sci Technol 54(1):7–18

Demirbas A (2011) Waste management, waste resource facilities and waste conversion processes. Energy Convers Manag [Internet] 52(2):1280–1287. https://doi.org/10.1016/j.enconman.2010.09.025

Dias RL, Ruberto L, Calabró A, Lo BA, Del Panno MT, Mac Cormack WP (2015) Hydrocarbon removal and bacterial community structure in on-site biostimulated biopile systems designed for bioremediation of diesel-contaminated Antarctic soil. Polar Biol 38(5):677–687

Diaz LF, Savage GM (2007) Chapter 4 factors that affect the process. Waste Manage Ser 8:49–65

Dijkema GPJ, Reuter MA, Verhoef EV (2000) A new paradigm for waste management. Waste Manag 20(8):633–638

Fdez-Güelfo LA, Álvarez-Gallego C, Sales Márquez D, Romero García LI (2011) Dry-thermophilic anaerobic digestion of simulated organic fraction of municipal solid waste: process modeling. Bioresour Technol 102(2):606–611

Fernandes L, Lucas MS, Maldonado MI, Oller I, Sampaio A (2014) Treatment of pulp mill wastewater by Cryptococcus podzolicus and solar photo-Fenton: a case study. Chem Eng J 245:158–165. https://doi.org/10.1016/j.cej.2014.02.043

Finstein MS, Miller FC, Strom PF (1983) Composting ecosystem management for waste treatment. Rev Med Suisse Romande 103(6):347–353

Flemming H-C, Wingender J (2001) Relevance of microbial extracellular polymeric substances (EPSs) - Part II: technical aspects. Water Sci Technol 43(6):9–16. https://doi.org/10.2166/wst.2001.0328

Gandolla M, Aragno M (1992) The importance of microbiology in waste management. Experientia 48:362–366

Garg A, Tothill IE (2009) A review of solid waste composting process-the UK perspective. J Dyn Soil Dyn Plant 3(1):57–63. http://www.ecochem.com/t_compost_faq2.html

Ghiani G, Laganà D, Manni E, Musmanno R, Vigo D (2014) Operations research in solid waste management: A survey of strategic and tactical issues. Comput Oper Res [Internet] 44:22–32. https://doi.org/10.1016/j.cor.2013.10.006

Gonawala SS, Jardosh H (2018) Organic waste in composting: a brief review. Int J Curr Eng Technol 8(01):36–38

Goswami S, Syiem MB, Pakshirajan K (2015) Cadmium removal by Anabaena doliolum Ind1 isolated from a coal mining area in Meghalaya, India: associated structural and physiological alterations. Environ Eng Res 20(1):41–50

Guo X, Gu J, Gao H, Qin Q, Chen Z, Shao L et al (2012) Effects of Cu on metabolisms and enzyme activities of microbial communities in the process of composting. Bioresour Technol 108:140–148. https://doi.org/10.1016/j.biortech.2011.12.087

Gupta J, Tyagi B, Rathour R (2019) Microbial diversity in ecosystem sustainability and biotechnological applications. Springer, New York

Hachicha S, Sellami F, Cegarra J, Hachicha R, Drira N, Medhioub K et al (2009) Biological activity during co-composting of sludge issued from the OMW evaporation ponds with poultry manure-Physico-chemical characterization of the processed organic matter. J Hazard Mater 162 (1):402–409

Hoeppel RE, Hinchee RE, Arthur MF (1991) Bioventing soils contaminated with petroleum hydrocarbons. J Ind Microbiol 8(3):141–146

Holst O, Stenberg B, Christiansson M (1998) Biotechnological possibilities for waste tyre-rubber treatment. Biodegradation 9(3–4):301–310

Huang B, Feng H, Ding Y, Zheng X, Wang M, Li N et al (2013) Microbial metabolism and activity in terms of nitrate removal in bioelectrochemical systems. Electrochim Acta 113:29–36

Idi A, Md Nor MH, Abdul Wahab MF, Ibrahim Z (2015) Photosynthetic bacteria: an eco-friendly and cheap tool for bioremediation. Rev Environ Sci Biotechnol 14(2):271–285

Jiunn-Jyi L, Yu-You L, Noike T (1997) Influences of pH and moisture content on the methane production in high-solids sludge digestion. Water Res 31(6):1518–1524

Johnston J, LaPara T, Behrens S (2019) Composition and dynamics of the activated sludge microbiome during seasonal nitrification failure. Sci Rep 9(1):1–15

Kanekar P, Sarnaik S (1991) An activated sludge process to reduce the pollution load of a dye-industry waste. Environ Pollut 70(1):27–33

Kaouther Zaafouri AA (2014) Feasibility of a bioremediation process using biostimulation with inorganic nutrient NPK for hydrocarbon contaminated soil in Tunisia. J Bioremediat Biodegrad 5(4):224

Karanasios KA, Vasiliadou IA, Pavlou S, Vayenas DV (2010) Hydrogenotrophic denitrification of potable water: a review. J Hazard Mater 180(1–3):20–37. https://doi.org/10.1016/j.jhazmat.2010.04.090

Kaushik P, Malik A (2009) Fungal dye decolourization: recent advances and future potential. Environ Int 35(1):127–141. http://www.sciencedirect.com/science/article/pii/S0160412008001037

Khalid A, Arshad M, Anjum M, Mahmood T, Dawson L (2011) The anaerobic digestion of solid organic waste. Waste Manage 31(8):1737–1744. https://doi.org/10.1016/j.wasman.2011.03.021
Kumar S, Smith SR, Fowler G, Velis C, Kumar SJ, Arya S et al (2017) Challenges and opportunities associated with waste management in India. R Soc Open Sci 4(3):160764
Kumari D, Li M, Pan X, Xin-Yi Q (2014) Effect of bacterial treatment on Cr(VI) remediation from soil and subsequent plantation of Pisum sativum. Ecol Eng 73:404–408. http://www.sciencedirect.com/science/article/pii/S0925857414005047
Kutsanedzie F (2015) Maturity and safety of compost processed in HV and TW composting systems. Int J Sci Technol Soc 3(4):232
Larsen MB, Schultz L, Glarborg P, Skaarup-Jensen L, Dam-Johansen K, Frandsen F et al (2006) Devolatilization characteristics of large particles of tyre rubber under combustion conditions. Fuel 85(10–11):1335–1345
Lee MD, Swindoll CM (1993) Bioventing for in situ remediation. Hydrol Sci J 38(4):273–282
Leow CW, Van Fan Y, Chua LS, Muhamad II, Klemes JJ, Lee CT (2018) A review on application of microorganisms for organic waste management. Chem Eng Trans 63:85–90
Liu J, Wu Y, Wu C, Muylaert K, Vyverman W, Yu HQ et al (2017) Advanced nutrient removal from surface water by a consortium of attached microalgae and bacteria: a review. Bioresour Technol 241:1127–1137. https://doi.org/10.1016/j.biortech.2017.06.054
Louhasakul Y, Cheirsilp B, Prasertsan P (2016) Valorization of palm oil mill effluent into lipid and cell-bound lipase by marine yeast Yarrowia lipolytica and their application in biodiesel production. Waste Biomass Valor 7(3):417–426
Ma X, Wang Y, Feng S, Wang S (2015) Comparison of four flocculants for removing algae in Dianchi Lake. Environ Earth Sci 74(5):3795–3804
Martins J, Peixe L, Vasconcelos V (2010) Cyanobacteria and bacteria co-occurrence in a wastewater treatment plant: absence of allelopathic effects. Water Sci Technol 62(8):1954–1962
Mary KV (2011) Bioremediation - an overview. J Ind Pollut Control 27(2):161–168
Mata-Alvarez J (2002) The biomethanization of the organic fraction of municipal solid waste. Water 21(October):59–61
McKinney RE (1957) Activity of microorganisms in organic waste disposal. II. Aerobic processes. Appl Microbiol 5(3):167–174
Miller RM, Bartha R (1989) Evidence from liposome encapsulation for transport-limited microbial metabolism of solid alkanes. Appl Environ Microbiol 55(2):269–274
Mishra A, Kumar S, Bhatnagar A (2019) Potential of fungal laccase in decolorization of synthetic dyes. In: Microbial wastewater treatment. Elsevier, Amsterdam, pp 127–151. https://doi.org/10.1016/B978-0-12-816809-7.00007-5
Mrozik A, Piotrowska-Seget Z (2010) Bioaugmentation as a strategy for cleaning up of soils contaminated with aromatic compounds. Microbiol Res 165(5):363–375. https://doi.org/10.1016/j.micres.2009.08.001
Muniraj IK, Xiao L, Hu Z, Zhan X, Shi J (2013) Microbial lipid production from potato processing wastewater using oleaginous filamentous fungi Aspergillus oryzae. Water Res 47(10):3477–3483. https://doi.org/10.1016/j.watres.2013.03.046
Ogawa T, Yatome C (1990) Biodégradation of azo dyes in multistage rotating biological contactor immobilized by assimilating bacteria. Bull Environ Contam Toxicol 44(4):561–566
Omokhagbor Adams G, Tawari Fufeyin P, Eruke Okoro S, Ehinomen I (2020) Bioremediation, biostimulation and Bioaugmention: a review. Int J Environ Bioremediat Biodegrad 3(1):28–39
Paillard D, Dubois V, Thiebaut R, Nathier F, Hoogland E, Caumette P et al (2005) Occurrence of Listeria spp. in effluents of French urban wastewater treatment plants. Appl Environ Microbiol 71(11):7562–7566
Panswad T, Luangdilok W (2000) Decolorization of reactive dyes with different molecular structures under different environmental conditions. Water Res 34(17):4177–4184
Perez-Garcia O, Escalante FME, de-Bashan LE, Bashan Y (2011) Heterotrophic cultures of microalgae: metabolism and potential products. Water Res 45(1):11–36

Petric I, Avdihodžić E, Ibrić N (2015) Numerical simulation of composting process for mixture of organic fraction of municipal solid waste and poultry manure. Ecol Eng 75:242–249

Pous N, Puig S, Balaguer MD, Colprim J (2017) Effect of hydraulic retention time and substrate availability in denitrifying bioelectrochemical systems. Environ Sci Water Res Technol 3 (5):922–929

Rabaey K, Van De Sompel K, Maignien L, Boon N, Aelterman P, Clauwaert P et al (2006) Microbial fuel cells for sulfide removal. Environ Sci Technol 40(17):5218–5224

Rani N, Sangwan P, Joshi M, Sagar A, Bala K (2019) Microbes: a key player in industrial wastewater treatment. In: Microbial wastewater treatment. Elsevier, Amsterdam, pp 83–102. https://doi.org/10.1016/B978-0-12-816809-7.00005-1

Rastogi M, Nandal M, Khosla B (2020) Microbes as vital additives for solid waste composting. Heliyon 6(2):e03343. https://doi.org/10.1016/j.heliyon.2020.e03343

Rathour R, Gupta J, Kumar M, Hiloidhari M, Mehrotra AK, Thakur IS (2017) Metagenomic sequencing of microbial communities from brackish water of Pangong lake of the Northwest Indian Himalayas. Genome Announc 5(40):40–41

Ravikumar S, Baylon MG, Park SJ, Choi J (2017) Engineered microbial biosensors based on bacterial two-component systems as synthetic biotechnology platforms in bioremediation and biorefinery. Microb Cell Factories 16(1):1–10

Raza S, Ahmad J (2016) Composting process: a review. Int J Biol Res 4(2):102

Rogayan JD V, Tomboc EHF, Paje AV, Lim KLP, Ararro JAR, Ocampo JG, et al. Vermiculture and vermicomposting 2010;6(October 2010):1–23

Sani RK, Banerjee UC (1999) Decolorization of triphenylmethane dyes and textile and dye-stuff effluent by Kurthia sp. Enzym Microb Technol 24(7):433–437

Sanscartier D, Zeeb B, Koch I, Reimer K (2009) Bioremediation of diesel-contaminated soil by heated and humidified biopile system in cold climates. Cold Reg Sci Technol 55(1):167–173. https://doi.org/10.1016/j.coldregions.2008.07.004

Schliephake K, Lonergan GT (1996) Laccase variation during dye decolonisation in A 200 L packed-bed bioreactor. Biotechnol Lett 18(8):881–886

Schröder U, Harnisch F, Angenent LT (2015) Microbial electrochemistry and technology: terminology and classification. Energy Environ Sci 8(2):513–519

Shalaby EA (2011) Prospects of effective microorganisms technology in wastes treatment in Egypt. Asian Pac J Trop Biomed 1(3):243–248

Sharholy M, Ahmad K, Mahmood G, Trivedi RC (2008) Municipal solid waste management in Indian cities - a review. Waste Manag 28(2):459–467

Sheng Y, Chen F, Sheng G, Fu J (2012) Water quality remediation in a heavily polluted tidal river in Guangzhou, South China. Aquat Ecosyst Health Manag 15(2):219–226

Silva-Bedoya LM, Sánchez-Pinzón MS, Cadavid-Restrepo GE, Moreno-Herrera CX (2016) Bacterial community analysis of an industrial wastewater treatment plant in Colombia with screening for lipid-degrading microorganisms. Microbiol Res 192:313–325. https://doi.org/10.1016/j.micres.2016.08.006

Singh SPTG (2014) Application of bioremediation on solid waste management: a review. J Bioremediation Biodegrad 5(6):248

Stevenson K, Stallwood B, Hart AG (2008) Tire rubber recycling and bioremediation: a review. Biorem J 12(1):1–11

Tyagi M, da Fonseca MMR, de Carvalho CCCR (2011) Bioaugmentation and biostimulation strategies to improve the effectiveness of bioremediation processes. Biodegradation 22 (2):231–241

Wang Y, Zhang W, Zhao Y, Peng H, Shi Y (2016) Modelling water quality and quantity with the influence of inter-basin water diversion projects and cascade reservoirs in the middle-lower Hanjiang River. J Hydrol [Internet] 541:1348–1362. https://doi.org/10.1016/j.jhydrol.2016.08.039

Ward AJ, Hobbs PJ, Holliman PJ, Jones DL (2008) Optimisation of the anaerobic digestion of agricultural resources. Bioresour Technol 99(17):7928–7940

Waszkielis KM, Wronowski R, Chlebus W, Białobrzewski I, Dach J, Pilarski K et al (2013) The effect of temperature, composition and phase of the composting process on the thermal conductivity of the substrate. Ecol Eng 61:354–357. https://doi.org/10.1016/j.ecoleng.2013.09.024

Wei Y, Li J, Shi D, Liu G, Zhao Y, Shimaoka T (2017) Environmental challenges impeding the composting of biodegradable municipal solid waste: a critical review. Resour Conserv Recyclying 122:51–65. https://doi.org/10.1016/j.resconrec.2017.01.024

Whelan MJ, Coulon F, Hince G, Rayner J, McWatters R, Spedding T et al (2015) Fate and transport of petroleum hydrocarbons in engineered biopiles in polar regions. Chemosphere 131:232–240. https://doi.org/10.1016/j.chemosphere.2014.10.088

Wu X, Xie YB (2012) The field application of microbial technology used in bioremediation of urban polluted river. Adv Mater Res 518–523:2906–2911

Xu L, Yu H, Yu J, Luo H, Li Z (2017) Study on ecological remediation method of purification of rural polluted slow-flow water. 2017;3(Bep):275–282

Yang Y, Wang L, Xiang F, Zhao L, Qiao Z (2020) Activated sludge microbial community and treatment performance of wastewater treatment plants in industrial and municipal zones. Int J Environ Res Public Health 17(2):436

Young L, Yu J (1997) Ligninase-catalysed decolorization of synthetic dyes. Water Res 31 (5):1187–1193

Zabaniotou AA, Stavropoulos G (2003) Pyrolysis of used automobile tires and residual char utilization. J Anal Appl Pyrolysis 70(2):711–722

Zhang L, Sun X (2016) Influence of bulking agents on physical, chemical, and microbiological properties during the two-stage composting of green waste. Waste Manage 48:115–126. https://doi.org/10.1016/j.wasman.2015.11.032

Zhou W, Wang W, Li Y, Zhang Y (2013) Lipid production by Rhodosporidium toruloides Y2 in bioethanol wastewater and evaluation of biomass energetic yield. Bioresour Technol 127:435–440. https://doi.org/10.1016/j.biortech.2012.09.067

Zissi U, Lyberatos G (1996) Azo-dye biodegradation under anoxic conditions. Water Sci Technol 34(5–6):495–500. https://doi.org/10.1016/0273-1223(96)00684-1

Chapter 17
Metagenomics: Insights into Microbial Removal of the Contaminants

Dipti Singh, Shruti Bhasin, Anshi Mehra, Manali Singh, Neha Suyal, Nasib Singh, Ravindra Soni, and Deep Chandra Suyal

Abstract Metagenomics has changed the microbial world completely. It has provided new insights to analyze microbial genes and metabolites. In metagenomics, the influence of genomics is applied to the entire communities, by avoiding the requirement of their isolation and selection. It requires several interconnected approaches and methods to get the maximum information. It offers an outstanding way to characterize the microbes, their genes, proteins, and metabolic pathways, which can be explored in the bioremediation of various contaminants. Recently, this technique is being explored to identify the unique microbial groups in an ecosystem which are later utilized in the development of microbial consortia for biodegradation. With the emergence of new sequencing techniques, the field has completely revolutionized. Moreover, new bioinformatic and statistical tools will always be in demand to analyze the huge metagenomic data and transformed it into meaningful results.

D. Singh
Department of Biochemistry, Govind Ballabh Pant University of Agriculture and Technology, Pantnagar, India

S. Bhasin
Department of Biotechnology, Banasthali Vidyapith, Tonk, Rajasthan, India

A. Mehra · M. Singh
Department of Biotechnology, Invertis Institute of Engineering and Technology (IIET), Invertis University, Bareilly, India

N. Suyal
Government Nursing College, Haldwani Nainital, Uttarakhand, India

N. Singh · D. C. Suyal (✉)
Department of Microbiology, Akal College of Basic Sciences, Eternal University, Sirmaur, Himachal Pradesh, India

R. Soni
Department of Agricultural Microbiology, College of Agriculture, Indira Gandhi Krishi Vishwa Vidyalaya, Raipur, Chhattisgarh, India

P. Bhatt et al. (eds.), *Microbial Technology for Sustainable Environment*,
https://doi.org/10.1007/978-981-16-3840-4_17

Keywords Bioremediation · Shotgun metagenomics · DNA sequencing · Bioinformatics · Gene prediction

17.1 Introduction

The term metagenome, introduced by Handelsman et al. (1998), describes the genome that is directly extracted from an environmental sample. It corresponds to all the cells present in that particular sample. Its isolation and characterization constitute the field of metagenomics that is being explored for energy production, human health, and food security (Joshi et al. 2017; Goel et al. 2018). Traditional microbiology relies upon the cultivation of the microbes, while metagenomics requires genomic sequences in order to characterize molecular taxonomy. It is a well-known fact that the majority of the microbial population is still uncultured; therefore, traditional microbiology fails to explore them properly (Goel et al. 2017; Yadav et al. 2019). The alternative technology is microbial community analysis, called metagenomics (Suyal et al. 2014; Soni et al. 2016, 2017). It involves the direct extraction of the DNA from environmental samples, their cloning, and screening for diversity or the functions. Nowadays, the function-driven approach becomes popular for searching of newer metabolites. The combination of the metagenomic data with diversity analysis, meta-transcriptomics, meta-proteomics, and environmental parameters has opened a new way for integrated ecosystem studies.

Metagenomics has overcome the issues related to the genomic diversity and uncultivability of several microbial groups, the major obstructions to proceed in environmental and clinical microbiology. The combination of metagenomics and environmental parameters has also been used to screen stress-tolerant microorganisms. Moreover, metagenomics has also shown its potential for bioremediation, antibiotic discovery, biodegradation, etc.

17.2 Structural and Functional Metagenomics

Metagenomics depend on the characteristics of the native microbial communities and their adjacent environmental conditions. It has two ideas: structural and functional metagenomics. Structural metagenomics deals with the diversity analysis of the microbial communities by using phylogenetic markers. It focuses on the microbial community structure and unculturable microbial groups. While functional metagenomics analyzes the gene functions, enzymes, and proteins of the microbiome, it targets the role of microbial communities in the ecology and biogeochemical cycles. It employs nucleic acid and protein databases for interpreting the functional attributes. It is considered low throughput and time-consuming. Besides these, metagenomics employs two different strategies: shotgun metagenomics and targeted metagenomics. Shotgun approach analyzes the complete metagenome with the help of restriction digestion and/or sequencing methods. Contrary to it, targeted metagenomics involves the analysis of a specific gene(s) and/or enzyme(s) in the metagenome with the help of PCR techniques (Suyal et al. 2015a, b). Both strategies have their own limitations—targeted metagenomics is dependent on the PCR primers, whereas shotgun metagenomics is dependent on the strength of sequencing technologies (Yadav et al. 2019).

Metagenomics starts with DNA extraction from the environmental sample and represents the blueprint of the whole community. The contaminants must be removed from the DNA as they can hinder the sequencing and other analytical applications. The main challenge of the metagenomics is to extract maximum information from the huge DNA libraries. New bioinformatics and biotechnological technologies will make metagenomics easier.

17.3 Steps of Metagenomics

The major steps of metagenomics are:

- Metagenome extraction directly from the environmental sample.
- PCR amplification of the marker genes (marker-based metagenomics) or restriction digestion of the metagenome (shotgun metagenomics).
- Cloning of the amplicons or DNA fragments in the molecular vectors.
- Transformation of the recombinant vectors into a suitable host strain.
- Preparation of the metagenomic library.
- Sequencing of the clones and diversity analysis (structural metagenomics).
- Screening of the clones for important metabolites, enzymes, proteins, antibiotics, etc. (functional metagenomics).

The combined use of meta-transcriptomics, meta-proteomics, metabolomics, and other meta-approaches in association with metagenomics could be promising to identify novel genes and metabolic pathways for bioremediation of the contaminants.

17.3.1 Designing of the Metagenomic Experiments

To get accurate and high-quality reliable data, the researchers need to focus on the proper experimental design (Fig. 17.1). It must employ updated, cost-effective, replicative, and accurate technologies. More specifically, during experiment designing, one must consider the replicates of samples, budget for sequencing, paramount protocols for DNA extraction, and updated sequencing platforms (Cooke et al. 2017).

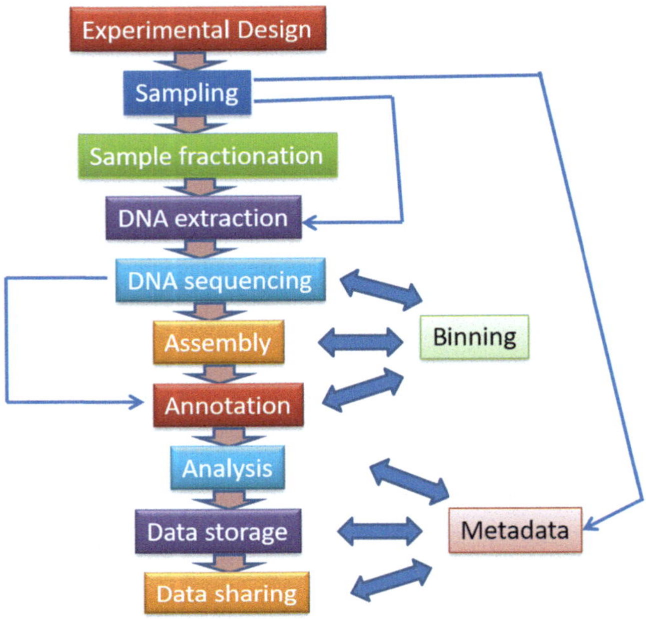

Fig. 17.1 The major steps in metagenomic studies

17.3.2 Sampling

Sampling is an important step in the metagenomics. Any environmental sample can be used as the sample to isolate the DNA, viz., plant tissues, animal tissues, water, soils, and air. The contaminated samples need to be sampled properly. The quality of the sample defines the DNA quality and thus is a crucial step. Moreover, sampling should be done in such a way that it represents whole population of the ecosystem and must be reproducible. The details should be mentioned on the samples properly, viz., date, time, season, conditions, etc. The statistical methods and experimental designs should be defined before starting the experiments.

17.3.3 Sample Fractionation

It is a procedure to treat the cells for lysis and metagenomic DNA extraction. It is a preparatory step and depends upon the type of the cell wall and cell membranes. This process can be done by using physical methods, chemical methods, and enzymatic methods. The suitable nuclease degradation reagents should be used in the fractionation buffer to avoid DNA lysis.

17.3.4 DNA Extraction

It is the most crucial step in metagenomics. For getting the good quality and yield of DNA, it is important to select proper DNA extraction protocols. The DNA may present in the free form or in the bound form (within the cells/tissues). Therefore, selection of extraction protocols is crucial to get the maximum DNA. Two strategies can be used for DNA extraction: direct and indirect. In direct strategy, cell lysis and DNA extraction occur in situ (within a sample), while during indirect strategy, separation of sample is done from non-cellular material before lysis, and thus, it is ex situ. By the indirect method, the yield of DNA product comes very low, approximately 100 times; however, it covers higher microbial diversity.

17.3.5 Preparation of the Inserts and Gene Cloning

The extracted DNA can be PCR amplified (PCR cloning) or restricted digested (shotgun cloning) as per the requirements and objectives. The DNA inserts thus formed are ligated in the suitable molecular vector. These recombinant vectors are transformed in the host cells especially in the bioengineered *E. coli* cells. In that way, one can prepare a gene library that can be screened or sequenced further.

17.3.6 DNA Sequencing

Two different approaches can be employed for sequencing DNA molecules, which are shotgun sequencing (restriction digestion-based sequencing) and amplicon sequencing (PCR-based sequencing). Amplicon sequencing utilizes PCR techniques and is frequently used for the microbial identification and characterization. It employs phylogenetic markers, viz., 16SrRNA gene, for microbial diversity and community structure analysis (Kumar et al. 2019). On the other hand, shotgun sequencing explores the ability of restriction enzymes to cut the DNA into smaller fragments. These fragments are then sequenced individually and later assembled again to get the whole sequence. Advancement in the next-generation sequencing technologies has tried to answer the above-raised questions, therefore identifying the microbial world with new dimensions.

17.3.7 Quality Control

It typically involves recognition and filters out those sequencing which are contaminating reads and low in quality. In metagenomic data, sequences of high eukaryotic species are typically measured as contaminations, which should be identified and filtered out before further analysis to avoid incorrect results and interferences. For quality control analysis, various tools are available, viz., FastX-Toolkit, PRINSEQ, NGS QC Toolkit, FastQC, etc.

17.3.8 Marker Gene Analysis

Marker gene analysis is the best way for the characterization of the taxonomic diversity of the microbial communities. This will help in making out the richness and profusion about microbe's population in a particular community (Suyal et al. 2019a, b). Taxonomic diversity can be analyzed by examining the marker genes, grouping sequences into a distinct taxonomic class called binning, and assembling sequences into separate genomes. It helps in the comparison of gene families, phylogenetic resemblance, and taxonomic annotation of the homologs. The ribosomal genes that are present in a single copy in the microbial genomes are preferred as a marker gene. Two methods can be used to annotate metagenome with the help of the marker genes: MetaPhyler (Liu et al. 2011) and AMPHORA (Wu and Scott 2012). MetaPhyler is based on sequence similarity between the sequences and the marker genes. The alignment and comparative analysis between the metagenomic sequences and available databases provides fruitful information. Moreover, sequence attributes are analyzed to get an idea about the taxonomy. Further, AMPHORA is based on phylogenetic information, which takes much time to calculate, but it provides higher accuracy. Further, it employs hidden Markov models (HMMs) for homolog recognition and taxonomy.

17.3.9 Assembly

In this process, small DNA fragments that are generated from the sequencer are reconstructed. It is used to obtain full-length protein-coding sequences (CDS) or full genomes of the uncultivable microorganisms. It provides longer genomic contigs. Assembly can be done by using two approaches: de novo assembly or comparative assembly. In the first approach the reads data are directly used to reconstruct the genome, while in the latter case, the reference genomes are needed. Sometimes, variations may be observed in the metagenome due to polymorphism, insertion, and/or deletion. It results in the fragmented assembly and error-full results.

17.3.10 Gene Prediction

Gene prediction is an important step in metagenomics. It utilizes the following three approaches.

(a) De Novo Gene Prediction

It employs the different properties of the genes, viz., GC content, length, codon usage, etc. to predict the genes. It can recognize the diverged genes in the metagenome.

(b) Protein Family Classification

This method identifies the protein-making ability of the gene. The sequences are converted into protein-coding frames and compared with the existing database. From this method, novel proteins can be identified by comparing with the reference database. Moreover, it helps in identifying the protein homologs which are diverged during the evolution.

(c) Gene Fragment Recruitment

This gene prediction method is based on a reading of algorithm mapping that quickly evaluates whether a genomic fragment is almost similar to a sequence of database. But this can be concluded after analysis of different homologs of a known gene. Various tools are available nowadays that help in gene prediction, viz., MetaGene, GeneMark, MetaGeneAnnotator, Gliommer MG, etc.

17.3.11 Annotation

Annotation step represents the analysis of gene locations, functions, homologs, metabolic pathways, and sequence-specific biological information. It involves the prediction of structure and functions of the metagenome based on the available information and databases. Several metagenome annotation tools are available online, viz., RAST (Aziz et al. 2008) or IMG (Markowitz et al. 2009). This process involves two sub-steps: structural annotation and functional annotation. Structural annotation identifies the gene of interest and its coding sequences (CDS). It is followed by functional annotation in which the functions of the respected genes are predicted. WebMGA, IMG/M, MG-RAST, STAMP, CoMet, RAMMCAP, and CAMERA are some online tools that can be explored for functional annotation.

17.3.12 Binning

In this process, metagenomic sequences from the population are arranged in the groups and assigned to the particular individual. Such groups represent the temporary metagenome of the whole population. This step is important as it has potential to identify the novel non-homologous genes and proteins. It may be achieved by using FAMeS (Mavromatis et al. 2007).

17.4 Meta-science and Bioremediation

The metagenome sequencing is not enough to provide the answer "who is doing what," because a lot of genes have unknown functions. The meta-transcriptome tells only about the expression of a gene in a particular sample at a certain instant and only in certain circumstances by executing the total mRNA. Now it is advanced in metagenomics to carry out meta-transcriptomic analysis. This meta-approach can deliver the functional and expression profile of the microbial communities. It can use two approaches, viz., first, de novo assembly of the transcripts and, second, using a reference transcriptome. The first approach is restricted by the capability of software programs to collect the transcripts, while the second approach is restricted by the availability of the reference transcriptome. Similarly, meta-proteomics offers an additional way to characterize the native microbiome along with metagenome. In it, environmental proteins are extracted, analyzed, and compared for identifying the functional attributes of the population. Metabolomics is another technique that complements metagenomics. It involves the study of all the metabolites that are identified and quantified which are released by the organism into the instantaneous environment. Due to various environmental stresses, microbial cells released various kinds of primary and secondary metabolites. The metabolomics strategy is based on the analysis of the practical values of these low-molecular-weight metabolites. Through the analysis of metabolome, it is directly told about the conditions of an environment or of any changes in homeostasis. Any change in the signature metabolites represents the deviation in the metabolic pathways and can be linked to the environmental factors. It involves metabolite detection, extraction, analysis, and characterization. However, one of its main limitations is complexity.

17.5 Role of Metagenomics in Bioremediation

As urbanization has increased around the globe by human activities, the environment gets polluted to a great extent from various sources. Further, industrialization has also disturbed environmental balance and has resulted in severe health issues that are enhanced by large-scale environmental pollutions. Bioremediation offers an eco-friendly way to remediate environmental contaminants ex situ or in situ (Bhatt et al. 2020a, b, c). Bioremediation is usually based on the identification of which microbial communities can degrade pollutants efficiently. It can be used for functional screening of the potential microorganisms.

Ex situ and in situ experiments were conducted to compare hydrocarbon-degrading efficiency of Arctic soil metagenome with others (Bell et al. 2011).

Table 17.1 Exploration of metagenomics in the isolation of the biodegradation-related enzymes

Enzymes	DNA source	Screening method	References
Carboxylesterase	Seawater	Function-based	Zhang et al. (2017)
β-Glucosidase	Soil	Function-based	Matsuzawa and Yaoi (2017)
Lipases, EstATII Esterase	Red Sea Atlantis II brine pool	Function-based	Yasmine et al. (2013)
Proteases Serine protease	Tannery sludge	Function-based	Devi et al. (2016)
Dioxygenase	Soil	Function-based	dos Santos et al. (2015)
Oxygenases	Artificially polluted soil	Function-based	Nagayama et al. (2015)
Amylases Amylopullulanase	Insect gut	Function-based	Lee et al. (2016)
Chitinase	Soil	Function-based	Thimoteo et al. (2017)
Phytase	Peat soil	Sequence-based	Tan et al. (2016)

Further, metagenomic analyses showed that mixed culture is more efficient than monocultures (Debbarma et al. 2017). Through this analysis, it could make recognition of genes that are concerned in the degradation of specific pollutants possible. Sequencing is also very selective with this method because the sequencing of rRNA and housekeeping genes is restricted (Bell et al. 2015). In most cases, bioremediation aims to identify novel genes, proteins, and metabolic pathways that could remove the contaminants (Table 17.1). In many bioremediation analyses, various oxidizing, reducing, and definite catabolic genes are the topics of concern. In that situation, it is enviable to merely amplify and sequence these targeted genes. Various studies revealed the role of metagenomics in bioremediation under a variety of environmental conditions like soil and marine environments. Xenobiotics are human-made chemical compounds that get accumulated in the environment due to their lower degradation rate (Bhatt et al. 2021a, b). Microorganisms respond to those chemicals through different mechanisms: (1) Due to the random mutation, toxic xenobiotic is formed. (2) These mutations could increase the microbial capability to degrade a xenobiotic. (3) Through horizontal gene transfer, it attains new genes encoding catabolic enzymes. Some species of microbes are known to neutralize a huge amount of xenobiotics, like halogenated, polyaromatic, and polyester compounds (Jeffries et al. 2018). Iwai et al. (2010) have analyzed novel dioxygenase genes in the contaminated soils. Similarly, Bell et al. (2011) have analyzed alkane monooxygenases from the Arctic soil and revealed that monoammonium phosphate affected the microbial population to a greater extent. Recently, Jaswal et al. (2019) have identified novel *Burkholderia* and other proteobacterial phyla that were able to degrade environmental contaminants from uraniferous soils. In another study,

Kachienga et al. (2018) have analyzed the microbial diversity in African petroleum-contaminated water samples. These analyses have given enhanced reorganization in different fields like climate change, agriculture, and bioremediation. In a characteristic hydrocarbon-polluted site, a metagenomics molecular strategy has been performed to describe the bacterial population. It has revealed the dominance of proteobacteria and *Clostridia* on those sites (Costeira et al. 2019). From industrial wastewater, various harmful chemicals has been observed like dibutyl phthalate, benzeneacetamide, resorcinol, benzene-1,2,4-triol, and benzoic acid. Since these chemicals are very harmful and caused genotoxicity, it is not safe to discharge them openly into the river or aquatic systems (Yadav et al. 2019). Such harmful compounds can be managed by using a metagenomics mechanism. The help of a group of diverse microbes that act as active biomass contributed a chief role in this mechanism (Jadeja et al. 2019). Additionally, the metagenomic approach has also been used for isolation of the potential genes toward wastewater treatment. Cabral et al. (2019) studied metagenomic methods to understand the degradation process through the genes which are responsible for the biotransformation of antibiotic resistance and metals in the oil mangrove microbiome.

In bioremediation-related studies, different samples and the various approaches are essential to attain a rich metagenomic library. However, the collection of samples and preparations for metagenome extraction differ on the type of environments, viz., water, soil, and air. Sampling from the contaminated water samples involves the proper filtration and separation. The excessive use of agrochemicals, industrial waste, and sewage contaminates the water and poses a serious threat to humankind. These contaminants can be removed by using bioremediation techniques, especially microorganisms. In this perspective, water metagenome plays a very important role to identify the novel genes and/or enzymes. Hamner et al. (2019) have performed a metagenomic study on the polluted water to analyze the microbial diversity and search for novel genes.

The soil system perhaps has the maximum quantity of microbial diversity in the environment. Soil microbial communities have been explored for agricultural and environmental purposes especially in the degradation of xenobiotics. In recent years, this approach has given wide knowledge of the contaminated soil metagenomes. For examining the contaminated soils, samples should be collected from different layers and transported immediately in the sterile bags and under cold conditions. It prevents any damage to the samples. The soil DNA should be extracted by using appropriate methods in good quality and quantity. The collected metagenome then further be analyzed by using any metagenomic approach either to analyze the microbial diversity or to identify the bioremediation-associated genes/enzymes. Metagenomics has also been used to analyze the microbiome of contaminated air. Moreover, the role of air microbial diversity on climatic variations has also been assessed (Schloss et al. 2016).

17.6 Clinical Waste Management and Metagenomics

Any type of waste material generated during the work of healthcare personnel at the hospitals and laboratories comes under clinical waste. It may include both biodegradable and non-biodegradable materials, viz., culture plates, plasticwares, glasswares, gloves, needles, swabs, bandages, tissues, etc. These waste materials may be infectious and hazardous. Therefore, it is very important to manage these wastes properly to avoid any environment as well as health risks. In view of the limitations and environmental concerns associated with traditional methods, bioremediation offers an eco-friendly and effective way for clinical waste management (Wen et al. 2021). In this perspective, metagenomics has proven itself a crucial technique to identify the potential microorganisms and their enzymes (Anitha and Jayraaj 2012). Mwaikono et al. (2015) have analyzed the microbial diversity in the biomedical waste to isolate the bacteria having biodegradation potential. Further, Marathe et al. (2018) have revealed the novel antibiotic resistance genes in the biomedical-waste-contaminated rivers by using functional metagenomics.

17.7 Future Perspective

Metagenomics has the immense possibility in the field of bioremediation. Integrated use of metagenomics with pure-culture techniques will be more effective for the screening of novel microbes, genes, enzymes, and proteins toward environmental cleaning. Exploration of microorganisms for bioremediation requires the understanding of their diversity, community structure, and mutual interactions. Further, prediction of degraded intermediates, metabolites, and their pathways is the major concern in bioremediation. This can only be achieved with the help of metagenomic technologies. Further, processing and analysis of the huge metagenomic data are among the major challenges in the field. New and updated bioinformatic softwares, tools, and databases are inevitable for the success of meta-technologies.

References

Anitha J, Jayraaj IA (2012) Isolation and identification of bacteria in biomedical wastes (BMW). Int J Pharm Pharm Sci 4:386–388

Aziz RK, Bartels D, Best AA, DeJongh M, Disz T, Edwards RA, Formsma K, Gerdes S, Glass EM, Kubal M, Meyer F, Olsen GJ, Olson R, Osterman AL, Overbeek RA, McNeil LK, Paarmann D, Paczian T, Parrello B, Pusch GD, Reich C, Stevens R, Vassieva O, Vonstein V, Wilke A, Zagnitko O (2008) The RAST server: rapid annotations using subsystems technology. BMC Genomics 9:75

Bell TH, Yergeau E, Martineau C et al (2011) Identification of nitrogen-incorporating bacteria in petroleum-contaminated Arctic soils by using [(15)N] DNA-based stable isotope probing and pyrosequencing. Appl Environ Microb 77:4163–4171

Bell TH, Greer CW, Yergeau E (2015) Metagenomics potential for bioremediation. In: Nelson KE (ed) Encyclopedia of metagenomics. Springer, Boston, pp 4–6

Bhatt P, Sharma A, Rene ER, Kumar AJ, Zhang W, Chen S (2020a) Bioremediation mechanism, kinetics of fipronil degradation using Bacillus sp. FA3 and resource recovery potential from contaminated environments. J Water Process Eng 39:101712

Bhatt P, Gangola S, Bhandari G, Zhang W, Maithani D, Mishra S, Chen S (2020b) New insights into the degradation of synthetic pollutants in contaminated environments. Chemosphere 268:128827

Bhatt P, Huang Y, Rene ER, Kumar AJ, Chen S (2020c) Mechanism of allethrin biodegradation by a newly isolated Sphingomonas trueperi strain CW3 from wastewater sludge. Bioresour Technol 305:123074

Bhatt P, Joshi T, Bhatt K, Zhang W, Huang Y, Chen S (2021a) Binding interaction of glyphosate oxidoreductase and C-P lyase: molecular docking and molecular dynamics simulation studies. J Hazard Mater 409:124927

Bhatt P, Zhou X, Huang Y, Zhang W, Chen S (2021b) Characterization of the role of esterases in the biodegradation of organophosphate, carbamate and pyrethroid group pesticides. J Hazard Mater. 411:125026

Cabral L, Noronha MF, de Sousa STP, Lacerda-Junior GV, Richter L, Fostier AH et al (2019) The metagenomic landscape of xenobiotics biodegradation in mangrove sediments. Ecotoxicol Environ Saf 179:232–240

Cooke SJ, Birnie-Gauvin K, Lennox RJ, Taylor JJ, Rytwinski T, Rummer JL, Franklin CE, Bennett JR, Haddaway NR (2017) How experimental biology and ecology can support evidence-based decision-making in conservation: avoiding pitfalls and enabling application. Conserv Physiol 5 (1):cox043

Costeira R, Doherty R, Allen CCR, Larkin MJ, Kulakov LA (2019) Analysis of viral and bacterial communities in groundwater associated with contaminated land. Sci Total Environ 656:1413–1426

Debbarma P, Raghuwanshi S, Singh J, Suyal DC, MGH Z, Goel R (2017) Comparative in situ biodegradation studies of polyhydroxybutyrate film composites. 3 Biotech 7(178):1–9

Devi SG, Fathima AA, Sanitha M et al (2016) Expression and characterization of alkaline protease from the metagenomic library of tannery activated sludge. J Biosci Bioeng 122:694–700

dos Santos DF, Istvan P, Noronha EF et al (2015) New dioxygenase from metagenomic library from Brazilian soil: insight into antibiotic resistance and bioremediation. Biotechnol Lett 37:1809–1817

Goel R, Suyal DC, Narayan DB, Soni R (2017) Soil metagenomics: a tool for sustainable agriculture. In: Kalia V, Shouche Y, Purohit H, Rahi P (eds) Mining of microbial wealth and metagenomics. Springer, Singapore, pp 217–225

Goel R, Kumar V, Suyal DC, Narayan DB, Soni R (2018) Toward the unculturable microbes for sustainable agricultural production. In: Meena VS (ed) Role of rhizospheric microbes in soil. Springer, Singapore, pp 107–123

Hamner S, Brown BL, Hasan NA, Franklin MJ, Doyle J, Eggers MJ, Colwell RR, Ford TE (2019) Metagenomic profiling of microbial pathogens in the Little Bighorn River, Montana. Int J Environ Res Public Health 16:1097

Handelsman J, Rondon MR, Brady SF, Clardy J, Goodman RM (1998) Molecular biological access to the chemistry of unknown soil microbes: a new frontier for natural products. Chem Biol 5 (10):245–249

Iwai S, Chai B, Sul WJ, Cole JR, Hashsham SA, Tiedje JM (2010) Gene-targeted-metagenomics reveals extensive diversity of aromatic dioxygenase genes in the environment. ISME J 4 (2):279–285

Jadeja NB, Purohit HJ, Kapley A (2019) Decoding microbial community intelligence through metagenomics for efficient wastewater treatment. Funct Integr Genomics 19:839–851

Jaswal R, Pathak A, Edwards B III, Lewis R III, Seaman JC, Stothard P et al (2019) Metagenomics-guided survey, isolation, and characterization of uranium resistant microbiota from the Savannah River site, USA. Genes (Basel) 10(5):E325

Jeffries TC, Rayu S, Nielsen UF, Lai K, Ljaz A, Nazaries L, Singh BK (2018) Metagenomic functional potential predicts degradation rates of a model organophosphorus xenobiotic in pesticide contaminated soils. Front Microbiol 9:147

Joshi D, Kumar S, Suyal DC, Goel R (2017) The microbiome of the Himalayan ecosystem. In: Kalia V, Shouche Y, Purohit H, Rahi P (eds) Mining of microbial wealth and metagenomics. Springer, Singapore, pp 101–116

Kachienga L, Jitendra K, Momba M (2018) Metagenomic profiling for assessing microbial diversity and microbial adaptation to degradation of hydrocarbons in two South African petroleum-contaminated water aquifers. Sci Rep 8(1):7564

Kumar S, Suyal DC, Yadav A, Shouche Y, Goel R (2019) Microbial diversity and soil physiochemical characteristic of higher altitude. PLoS One 14(3):e0213844

Lee YS, Seo SH, Yoon SH et al (2016) Identification of a novel alkaline amylopullulanase from a gut metagenome of *Hermetia illucens*. Int J Biol Macromol 82:514–521

Liu B, Gibbons T, Ghodsi M, Treangen T, Pop M (2011) Accurate and fast estimation of taxonomic profiles from metagenomic shot gun sequences. BMC Genomics 12:S4

Marathe NP, Janzon A, Kotsakis SD, Flach CF et al (2018) Functional metagenomics reveals a novel carbapenem-hydrolyzing mobile beta-lactamase from Indian river sediments contaminated with antibiotic production waste. Environ Int 122:279–286

Markowitz VM, Mavromatis K, Ivanova NN, Chen IM, Chu K, Kyrpides NC (2009) IMG ER: a system for microbial genome annotation expert review and curation. Bioinformatics 25 (17):2271–2278

Matsuzawa T, Yaoi K (2017) Screening, identification, and characterization of a novel saccharide-stimulated β-glycosidase from a soil metagenomic library. Appl Microbiol Biot 101:633–646

Mavromatis K, Ivanova N, Barry K, Shapiro H, Goltsman E, McHardy AC et al (2007) Use of simulated data sets to evaluate the fidelity of metagenomic processing methods. Nat Methods 4:495–500

Mwaikono KS, Maina S, Sebastian A, Kapur V, Gwakisa P (2015) 16S rRNA amplicons survey revealed unprecedented bacterial community in solid biomedical wastes. Am J Microbiol Res 3 (4):135–143

Nagayama H, Sugawara T, Endo R et al (2015) Isolation of oxygenase genes for indigo-forming activity from an artificially polluted soil metagenome by functional screening using *Pseudomonas putida* strains as hosts. Appl Microbiol Biotechnol 99(10):4453–4470

Schloss PD, Girard RA, Martin T, Edwards J, Thrash JC (2016) Status of the archaeal and bacterial census: an update. MBio 7(3):e00201–e00216

Soni R, Suyal DC, Sai S, Goel R (2016) Exploration of *nif*H gene through soil metagenomes of the western Indian Himalayas. 3 Biotech 6(1):1–4

Soni R, Kumar V, Suyal DC, Jain L, Goel R (2017) Metagenomics of plant rhizosphere microbiome. In: Singh R, Kothari R, Koringa P, Singh S (eds) Understanding host-microbiome interactions - an omics approach. Springer, Singapore, pp 193–205

Suyal DC, Agrawal K, Goel R (2014) Identification of soil microbial communities: approaches and methodologies. In: Singh DP, Singh HB (eds) Trends in soil ecology. Studium Press LLC, Houston, pp 93–114

Suyal DC, Yadav A, Shouche Y, Goel R (2015a) Bacterial diversity and community structure of Western Indian Himalayan red kidney bean (*Phaseolus vulgaris* L.) rhizosphere as revealed by 16S rRNA gene sequences. Biologia 70(3):305–313

Suyal DC, Yadav A, Shouche Y, Goel R (2015b) Diversified diazotrophs associated with the rhizosphere of Western Indian Himalayan native red kidney beans (*Phaseolus vulgaris* L.). 3 Biotech 5(4):433–441

Suyal DC, Joshi D, Debbarma P, Soni R, Dash B, Goel R (2019a) Soil metagenomics: unculturable microbial diversity and its function. In: Varma A, Choudhary D (eds) Mycorrhizosphere and Pedogenesis. Springer, Singapore, pp 355–362

Suyal DC, Kumar S, Joshi D, Yadav A, Shouche Y, Goel R (2019b) Comparative overview of red kidney bean (*Phaseolus valgaris*) rhizospheric bacterial diversity in perspective of altitudinal variations. Biologia 74(10):1405–1413

Tan H, Wu X, Xie L et al (2016) A novel phytase derived from an acidic peat-soil microbiome showing high stability under acidic plus pepsin conditions. J Mol Microbiol Biotechnol 26:291–301

Thimoteo SS, Glogauer A, Faoro H et al (2017) A broad pH range and processive chitinase from a metagenomic library. Braz J Med Biol Res 50:e5658

Wen P, Huang Y, Qiu Z, Li Q (2021) Microbial response during treatment of different types of landfill leachate in a semi-aerobic aged refuse biofilter. Chemosphere 262:127822

Wu M, Scott AJ (2012) Phylogenomic analysis of bacterial and archaeal sequences with AMPHORA2. Bioinformatics 28(7):1033–1034

Yadav BS, Yadav AK, Singh S, Singh NK, Mani A (2019) Methods in metagenomics and environmental biotechnology. In: Gothandam K, Ranjan S, Dasgupta N, Lichtfouse E (eds) Nanoscience and biotechnology for environmental applications, environmental chemistry for a sustainable world, vol 22. Springer, Cham, pp 85–113

Yasmine MM, Mohamed AG, Ahmed S, Amged O, El-Dorry H, Rania S (2013) Isolation and characterization of a heavy metal-resistant, thermophilic esterase from a Red Sea Brine Pool. Sci Rep 3:3358

Zhang Y, Hao J, Zhang YQ, Chen XL, Xie BB et al (2017) Identification and characterization of a novel salt-tolerant esterase from the deep-sea sediment of the South China Sea. Front Microbiol 8:441

Chapter 18
Methods of Strain Improvement for Crop Improvement

Jyoti Rawat and Veena Pande

Abstract Biofertilizers are suitable substitutes to chemical-based fertilizers and pesticides, which cause serious environmental problems around the world. Therefore, there is an important requirement to implement ecological regulation using native microorganisms. These beneficial microorganisms are inexpensive, consistent, and more effective than synthetic fertilizers in terms of plant protection against pathogens. These beneficial microorganisms protect plants against pathogens and enhance nutrient availability. Hence, to achieve this goal, better quality strains are needed. Crop improvement relies on the modulation of genes and genomic regions that underlie crucial characteristics, either directly or indirectly. Recombinant biotechnology intends to benefit in reducing the use of synthetic fertilizers; for this function genetically improved microbes could be used. By using recombinant DNA technology, genes of microbes are improved via several genetic modifications depending on the recognition and selection of the desirable characteristics or genes of interest. The current investigation is focused on different strategies used to improve beneficial strain for crop productivity.

Keywords Agriculture · Crop improvement · Gene information · Molecular approaches

18.1 Introduction

Agriculture relies heavily on the use of chemical or synthetic fertilizers and insecticides to achieve higher yields. Issues such as environmental pollution, health threats, disruption of the natural cycle of ecological inputs, and the destruction of biological ecosystems that otherwise support agricultural production are correlated with this reliance. There is a growing use of biological resources to replace chemical

J. Rawat · V. Pande (✉)
Department of Biotechnology, Sir J. C. Bose Technical Campus Bhimtal, Kumaun University, Nainital, Uttarakhand, India

P. Bhatt et al. (eds.), *Microbial Technology for Sustainable Environment*,
https://doi.org/10.1007/978-981-16-3840-4_18

fertilizers and pesticides. Agricultural development and pest and disease management must therefore be carried out with fewer harmful inputs at shorter periods. In this sense, plant growth-promoting rhizobacteria (PGPR) are potential resources to bring significant benefits to agriculture. Studies have shown that PGPR have great potential to improve crop growth and yield. Cereals are the primary source of food for human nutrition and constitute more than two-fifths of the world population's staple diet. Environmental and genetic factors influence crop productivity (Radhakrishnan et al. 2017). The usage of beneficial microbes alone or as microbial consortia to selected plants with multifunctional properties is a good method to stimulate strength and crop productivity (Ahmad et al. 2018; Maron et al. 2018). Investigation on isolation and characterization of advantageous microbes to plants has been extensively cited, but some of them have been commercialized. It has been demonstrated that many commercial bioinoculants were not effective in the agriculture field but they were effective in laboratory or greenhouse experiments (Vassilev et al. 2015; Arora and Mishra 2016; Sulbhi et al. 2021; Bhandari et al. 2021) due to their low stability and quality of formulation (Bhatt et al. 2021a, b, c). Newly, selective use of beneficial microbiome plants and their combinations to combat biotic and abiotic stress is gaining traction and becoming a stimulating research frontier (Malusá et al. 2016; Bashan et al. 2016; Baez-Rogelio et al. 2017; Stamenkovic et al. 2018). Biofertilizers are formulated from nitrogen-fixing rhizobacteria naturally present in the legume nodules or microbes that are responsible for plant growth promotion. However, these bio-formulations would not be proficient enough for supplying nitrogen to non-leguminous plants. In that circumstances, the practice of genetic engineering is of particular significance, for developing efficient management systems is needed. Consequently, the non-leguminous plants could be cultivated with symbiotic rhizobia root nodules without applying external nitrogen fertilizers (Santi et al. 2013). Foreign genes used to transform microorganisms could be integrated into the genome of the host. For that, the regulatory region of the gene is altered at the promoter or the terminator sites to augment the inserted gene function in the host. The addition of a particular gene that can confer biological control capacity could improve the biological control capacity of microbes that lack these genes (Dash et al. 2016). Various rhizobacteria possess biological control activity that simultaneously produces chitinases. However, few rhizobacteria such as *P. putida* and *R. meliloti* are root colonizer but devoid of chitinase synthesis (Bagwan et al. 2010). Hence, the chitinase gene assimilation into their genome made them competent in the defense of the plants against fungi (Huang et al. 2001). Biofertilizers, when formulated using molecular tools, can enhance cellular pathways for phytohormone production, such as cytokinin, auxin, etc. as well as help in plant growth and development (Fuentes-Ramirez and Caballero-Mellado 2005). Most breeding approaches for biotic and abiotic stress resistance are based on the insertion of a single resistant gene into plants, and therefore crop resistance only lasts for a short duration (Kottapalli et al. 2010; Bhatt et al. 2020a, b, 2021d, e, f). Therefore, the development of multi-stress resistant genotypes is now demonstrated by combining multiple genes from different sources in a single plant (Bhatt et al. 2019a, 2020c, d, e, f). The process of manipulating and improving microbial strains to enhance their metabolic capabilities is called strain improvement.

Protoplast fusion is an important tool in the selection of strains to provide genetic recombination and develop hybrid strains in filamentous fungi (Steiner et al. 2019). It is used to produce interspecific or even intergeneric hybrids. It has become an important tool for genetic manipulation, as it breaks down obstacles to the genetic exchange performed by conventional mating systems. This technique has great potential for genetic analysis and strain improvement. The stress tolerance capacity in crops has been explained in many studies using the pyramid of various resistance genes (Suresh and Malathi 2013; Muthurajan and Balasubramanian 2010). Abiotic stress also affects the growth and yield of the crop (Pancaldi and Trindade 2020) and can even disturb the survival of plants (Rana et al. 2019; Verma et al. 2008). Salinity is one of the major problems for crop productivity, and maximum crops are subtle to salt during their lifespan and particularly at the seedling period (Bai et al. 2018). Certain varieties of crops that are salt resistant express salt-sensitive genes to tolerate excess salts, and the quantitative trait locus (QTL) linked to these genes can be mapped by microsatellite markers for the selection of salt-tolerant lines (Ruengphayak et al. 2015; Llorens et al. 2020). Numerous drought-tolerant genes have been well discovered and modified in various crops to develop drought resistance (Yu and Yang 2016; Waqas et al. 2020). In many plants, cold resistance genes (OsRAN1 and QTL) have also been acknowledged which is further used in developing cold tolerance in plant varieties using molecular marker enhancement tools (Thitisaksakul et al. 2015; Tiwari et al. 2016; Bimpong et al. 2016). Plant tissue cultures (PTC) also have an important part in modern biotechnology. They are widely used in studies of plant development processes (Sandhu et al. 2019), genetic function (Rai et al. 2018), micropropagation (Zhang et al. 2014), and generation of transgenic plants with specific industrial and agronomic characteristics (Shinada et al. 2014). In this chapter, various techniques involved in beneficial microbes/strain improvement, for the production of biotic and abiotic stress resistivity in different plant varieties, are described. Molecular biological applications for crop improvement like genetic engineering (GE)/recombinant DNA technology (RDT) to adopt better traits of agronomic importance are too elaborated (Almeida et al. 2016; Firn et al. 1994; Bhatt et al. 2015a, b, 2016a, b, 2019b, c; Deng et al. 2010; Kumar 2011; Loyola-Vargas and Ochoa-Alejo 2018).

18.2 Crop Improvement by Genetic Engineering

For many decades, gene transfer among distinct species of plants has played a fundamental role in crop improvement. By transforming genes, many useful traits, such as insect, stress, and disease resistance, have been shifted to many varieties of non-cultivated plant crops (Akhtar et al. 2014; Amin et al. 2014; Dar et al. 2014; Tariq et al. 2014). Recombinant DNA methods and many other methods are used for the transformation of genetic information. Genetic engineering is a technique that has made possible the transfer of genes between different genera or species using recombinant DNA. This method is an exceptional selection method of expanding the genetic base as compared to conventional breeding. Additionally, because it avoids the skidding problem associated with conventional farming, it is more efficient and takes less time (Khan et al. 2015a). Until now, many genetically modified crops have been developed and commercialized, resulting in higher production efficiency, a greater focus on the market, and better conservation of the environment. These crops include longer postharvest storage tomatoes, insect-resistant cotton and corn, virus-resistant potatoes, herbicide-resistant soybeans, and canola, and many others (Puspito et al. 2015). To improve crops through genetic engineering, an efficient processing system is needed. Different approaches are used to transform different cultures such as recombinant DNA technology, which is used to manipulate genes of microbes via various genetic modifications (Tabashnik et al. 2011). Also, many *Pseudomonas* spp. chelate Fe ions by producing siderophores, thus increasing Fe uptake in plants. *S. meliloti* (RMBPC-2), a genetically modified strain, was made by introducing the genes that drive the plant nitrogenase enzyme to the bacteria (Boccia and Sarnacchiaro 2015). *T. harzianum* is a very effective colonizer that is widely present in soil and also can parasitize pathogenic fungi. In fact, many extracellular enzymes like chitinases, proteases, and glucanases are synthesized by *Trichoderma* which are enhanced by adding chitinase genes. Many extracellular enzymes such as glucanases, chitinases, and proteases synthesized by *Trichoderma* have been enhanced by the addition of chitinase genes (Tabashnik et al. 2011; Boccia and Sarnacchiaro 2015; Awais et al. 2010). Therefore, these genetically altered strains could effectively act as biofertilizers and improve crop yield and quality.

18.2.1 Genetically Modified Microbes

GM microbes provide better access to nutrients for crops and therefore increase plant development. The most important beneficial microbes that are used as biofertilizers are nitrogen-fixing bacteria, such as *Rhizobium* and *Azospirillum*. *Rhizobium* and *Sinorhizobium* are the symbiotic bacteria that form root nodules in legumes and fix nitrogen. It has been reported that these bacteria can stay in soil alive for a long time and in certain cases even without a defined host (Ngwako 2008). These microbes have been widely used as bioinoculants to enhance the growth and yield of crops. However, the improvement in yield is variable and the success of the inoculants appears to depend on the competition with the native strains which are generally the least effective (Qaim 2009). Mycorrhizal fungi signify the group of fungi that form a symbiotic association with plants. An investigation has been carried out to identify if transgenic *Rhizobium* strains enhance nodulation or interfere with a symbiotic association in plants. It was noticed that the strain GM *S. meliloti* does not interfere with the formation of mycorrhizae but improves nodulation. GM sweet clover increased colonization units of arbuscular mycorrhiza and increased the nutrient acquisition capability of mycorrhizal plants (Papagianni 2004; Van Loon 2007).

Azospirillum is recognized for its capability of plant growth promotion by augmenting nitrogen uptake, through phytohormone production (Gonzalez et al. 2015). *Sinorhizobium meliloti* has been genetically modified to promote nodulation in alfalfa roots. This genetic modification includes modification of the expression of nifA gene which is responsible for the management of all other nitrogen fixation (nif) genes (Bakshi 2003). It is assumed that nifA regulates the gene expression other than nif cluster that aids in nodule development (Beyer et al. 2002). In the rhizosphere region of *Pisum sativum*, GM *Rhizobium leguminosarum* strains, labeled with HgCb resistance genes (mer genes) and lacZ genes, were inoculated. In order to observe its impact on crop productivity, *Alcaligenes faecalis*, a non-nodule-forming bacterium, has been genetically engineered and introduced into rice fields in China. By introducing a constitutively expressed nifA regulatory gene, *A. faecalis* was genetically modified and nitrogen fixation got increased as compared to treated fields (Gray and Smith 2004; Huang et al. 2021).

18.3 Intraspecific and Interspecific Gene Transfer

In the nineteenth century, plant breeding began with discoveries about how plant traits are inherited. Plant breeding could be achieved by selecting plants with interesting attributes and manipulation in cross-fertilization. An improved variety with the desired characteristics is formed when a cultivated variety is backcrossed with a wild variety (Goodman et al. 1987; Khan et al. 2015b). In the twentieth century, plant breeders used interspecies hybridization to transfer genes from a non-cultivated plant species to other convertible crop species. For example, *Avena sativa* (oats) and *Beta vulgaris* (sugar beet) were processed and resulted in increased yields of 25–30% and resistance to sugar beet nematodes, respectively (Sharma and Gill 1983). In the 1940s, methods for transferring DNA directly from one organism to another organism were developed as DNA was established as a chemical base of genetic inheritance. Genes can be obtained from plant, animal, bacterial, and viral sources and injected into crops. Tissue specificity, timing, and expression level of genes are under control and they can be modified by gene modification into a fresh host. These methods provide the basis of diversity and permit the regulation of expression of genes (Qamar et al. 2015). In recent times, the expansion of molecular methods has generated different options for the assortment and genetic improvement of livestock (Godrat et al. 2005).

18.4 Genetic Modification Through Somatic Hybridization

18.4.1 Protoplast Fusion

Somatic hybridization is the best technique aimed at the production of interspecific and intergeneric hybrids for plant breeding and crop improvement. In this technique fusion of protoplasts from two different genomes followed by the selection of the desired somatic hybrid cells is carried out for regeneration of hybrid plants (Evans and Bravo 1988). Therefore, it is accepted as an effective approach to generate hybrids by joining two different protoplasts from different plant species or varieties, and hybrids produced via this method are called somatic hybrids. Protoplast fusion is a commonly used method for introducing a group of biosynthetic genes or entire chromosomes into a recipient cell for subsequent genetic manipulation or directed evolutionary approaches. It facilitates the transmission of mitochondrial genomes among taxonomically associated species (Vincelli 2016). This is one of the important or widely studied approaches as a technique to improve fungal strains (Assefa 2018; Nagoshi et al. 2018). In physiology, genetic study and genetic manipulation fungal protoplast are important tools that can be successfully carried by fusing protoplasts into filamentous fungi that lack sexual reproductive ability (Kage et al. 2016; Sharifzadeh et al. 2018). It is admitted as one of the recombinant DNA technologies that provide the tools to increase gene dosage and gene expression from strong promoters, remove unwanted genes from the fungal genome, manipulate the metabolic pathways, and develop fungal strains for the production of heterologous proteins. Several reports have confirmed the isolation and regeneration of protoplasts in different fungi. Protoplast fusion is found to be good for improvement of *Trichoderma* spp. and development of hybrid strains in other filamentous fungi (Atique et al. 2018; Mwobobia et al. 2020). The isolation, fusion, and regeneration of protoplasts were carried out in the genus *Trichoderma* primarily to improve its cellulolytic activity (Federico et al. 2019; Pandeya et al. 2018) and chitinase production (Bowman and Zilberman 2013). However, partial attempts have been done to improve *Trichoderma* species and increase enzyme production (Pandeya et al. 2018; Waddington et al. 2010). Ogawa and his team (Ogawa et al. 1989) revealed an increased cellulase production in *Trichoderma reesei* through interspecific protoplast fusion, while Prabavathy et al. reported an increase in chitinase and biological control activity in *Trichoderma harzianum* through protoplast autofusion; nevertheless, little research has been done on the application of chitinase in the degradation of shellfish waste applying this method (Prabavathy et al. 2006).

18.4.2 Agrobacterium*-Mediated Gene Transfer*

Agrobacterium tumefaciens is a phytopathogenic bacterium capable of transferring part of its genetic material to other plant species through a simple process called transformation. The genes are encoded in a region of the Ti plasmid called T-DNA. This causes the growth of a tumor termed "crown gall" disease in plants (Gordon and Christie 2015). This bacterium is altered in the laboratory and transfers the gene of interest to plants without causing disease symptoms. The *Agrobacterium* system is quite attractive due to the easy protocol that is associated with minimal cost in terms of equipment and also the resulting transgenic plants have single-copy insertion (Gordon and Christie 2015; Hansen and Wright 1999). With this method, genes for resistance to insects and diseases were transferred. Using recombinant DNA technology, many plant and bacterial genes encoding enzymes have been engineered to make crop plants tolerant of broad-spectrum herbicides and safer for the environment. Because this bacterial gene is designed in such a way that its enzyme is insensitive to the herbicide and then transfers it to the plant, it can also be done by having plants express genes that detoxify the herbicide. The genes obtained from *Bacillus thuringiensis* have been modified and transferred to plants that act as insecticides (Shahid et al. 2016).

18.4.3 Non-Agrobacterium-Based Gene Transfer

Four decades before it was identified, some members of the Rhizobiaceae family also can transfer the gene to the host. *Ensifer adhaerens*, *Ochrobactrum haywardense*, and *Rhizobium etli* are some of the *Agrobacterium*-related species that have been used in gene transfer but have the disadvantage of a limited host range (Mullins et al. 2006).

18.4.4 Viral-Mediated Gene Transfer

Viruses carry complex arrangements and life cycles; many are pathogenic but act as very efficient vehicles in gene delivery (Patel and Misra 2011). RNA and DNA viruses that infect plants can be used as a vector to transfer genes to the target. The gene to be transferred is integrated into the viral genome, and at this instant, the virus acts as a vector to transfer the gene. The virus with the transferred gene infects the target cell and results in a successful transformation. The main disadvantage is the high number of copies per cell, and virus-mediated gene transfer can only produce transient transfer and not stable transformation—that means they cannot be transferred to the offspring. Some of the viral vectors used are a retrovirus, an adenovirus (Chailertvanitkul and Pouton 2010), adeno-associated virus, herpes virus, smallpox virus, human moss virus (HFV), and lentivirus (Patel and Misra 2011; Fiandaca and Federoff 2014).

18.5 Mutagenesis and Crop Improvement

18.5.1 Site-Directed Mutagenesis

In a study, chemical mutagenesis was used to attain fungicide benomyl-resistant strains of *Trichoderma harzianum* (Ahmad and Baker 1987). Remarkably, the mutant strains were better colonizers of the rhizosphere than wild-type strains. The mutation technique will undoubtedly contribute to the upgradation of biological control agents. Genetic engineering proposes stimulating possibilities for the genetic manipulation of fungi both to improve biological control strains and to understand how biological control works. Transformation of filamentous fungi was first reported in the laboratories of Tatum (Mishra and Tatum 1973) and Case (Case et al. 1979). Since then, molecular techniques have become more accessible for use by possible biological control fungi (Fincham 1989; Bhatt et al. 2019d; Sharma and Bhatt 2016; Sharma et al. 2016; Bhatt and Nailwal 2018). There is no doubt that the expansion and use of molecular practices will persist to advance rapidly (Khati et al. 2018a; Gangola et al. 2018a; Bhatt 2018; Bhatt and Barh 2018; Bhatt et al. 2019e; Bhandari and Bhatt 2020; Bhatt and Bhatt 2021).

18.6 Bioinformatics Tools in Crop Improvement

Bioinformatics resources, in addition to various web databases, provide extensive information on genomic data that is widely needed for research purposes. Improving crops using bioinformatics tools is more promising these days (Singh et al. 2021; Zhang et al. 2020a, b; Mishra et al. 2020; Feng et al. 2020; Lin et al. 2020; Zhan et al. 2020; Ye et al. 2019; Huang et al. 2019, 2020). Over time, technology has improved to a surprising level, bioinformatics provides crucial information about crop genomic data, and this technology explores the sequence of many genes. This could help us to sequence the economically important crop and the more beneficial traits. Whole-genome comparisons are accelerating the pace of competent research (Fan et al. 2020; Pang et al. 2020; Gangola et al. 2018b; Gupta et al. 2018; Khati et al. 2017a, 2018b; Kumar et al. 2017). Projects of genome sequencing of economically important crops have been accomplished and are seen as the access to new research. Database of specific data sets in a compiled form with enriched annotations helps to study gene families with greater precision. Genomic comparisons of different crops help pinpoint the conserved regions between crops, providing common adaptation strategies for plants (Nagoshi et al. 2018). After completing the sequencing of the cultures, the data generated was used to create modeled proteomic data that helped to understand the content of certain gene families. Major events, such as gene duplication, as well as other abnormalities, are manipulated using bioinformatics tools (Khati et al. 2017a, b, 2018b; Kumar et al. 2017). Additionally, access to critical data to improve crop traits is positively simplified at a great end by using advances in technology and data acquisition sites. Therefore, efficient use of genetic data supports sustainable crop improvement. Different techniques, such as high-throughput sequencing, generate a stack of crop data. Omics research works on the prediction of candidate genes and, therefore, on the predicted functions (Mochida and Shinozaki 2010; Lockhart and Winzeler 2000). Data on transcriptomics and metabolomics have also elucidated the regulatory networks that are crucial against plant stressors. As a result, several crops were protected from biotic and abiotic stressors and yield was restored.

18.7 Plant Tissue Culture in Crop Improvement

Advancements in tissue culture methods have very important part in breeding various crops. These in vitro tissue culture techniques offer cloning, screening, micropropagation, micrografting, organogenesis, etc. to assist plant breeders in several ways. In tissue culture practices, the phenomenon of totipotency capacity of the plants (explants) is exploited to introduce variance in genetic organization of plants (Brown and Thorpe 1995). Explants or plants are treated with appropriate treatments such as thermotherapy to eradicate viruses and diseases and allowed to divide to forms a colorless undifferentiated mass of cells (callus) (Jain 2001). The epigenetic alterations induced during tissue culture processes are known as somaclonal variations. Together with molecular and biotechnological interventions, several techniques have been developed to transfer necessary genetic traits that are commercially favored. Clonal multiplication ornamental crop industries operate massively and thus greatly increase cultivars. Plant traits are thus evaluated against different plants in plant breeding (Tazeb 2017). Several genetically modified plants have been established during the last 20 years utilizing technological advancements in genetic engineering (Bawa and Anilakumar 2013). These plants have been developed such that often use either a transforming vector or other techniques that require chemical and enzyme action coupled to favor transformation such as use of liposomes, biolistic particle gun, microinjection, and electroporation techniques (Bhalla 2006). Transformation vector such as *Agrobacterium tumefaciens* induces tumors with its Ti plasmid and subsequently transfers T-DNA (transfer DNA) into host plant parts (typically leaves). The DNA segment of interest was inserted into the T-DNA (transfer DNA), eliminating the nonessential part (the portion of the plasmid that is not required for the act of transfer) (Gheysen et al. 1998; Gelvin 2003). Transformation success modifies the cells, followed by cell harvesting and finally regenerated in vitro into complete plantlets. However, assembly of necessary and advantageous crop traits for any crop enhancement program is certainly most critical and is usually performed by genetic transformation or hybridization program. Single genes are favored for transfer by most molecular and genetic methods. Hybridization is preferred for the successful transfer of more genes in a single reaction time. Tissue culture techniques facilitate the hybridization process when the embryo is aborted and therefore does not favor plant establishment. Tissue culture embryo rescue has been used successfully to overcome the problem of embryo abortion or the inability of seeds to develop (Tazeb 2017).

There are so many important advantages of plant tissue culture over crops. A wide variety of cultures have been recovered by IVF using pistil pollination and self-pollination and cross-pollination of ovules. A wide series of plants have been recovered by IVF using pistil pollination and self-pollination and cross-pollination of ovules, such as tobacco, corn, clover, poppy, canola, cabbage, cotton, etc. Another type used to give value to cultures is embryo culture and orchids, roses, and bananas are formed by embryo culture. Several other varieties are also successfully formed, such as stress, drought, and heat-tolerant varieties. In vitro propagation by meristem, cell organ and tissue culture, organogenesis, and somatic embryogenesis are presented. These techniques certainly may make breeding programs simpler and overcome some important economic and agronomic factors that might have never occurred with conventional plant breeding and improvement methods (Wang et al. 2016; De Filippis 2013). The method of plant tissue culture plays a dominant role in the second green revolution in which plant biotechnology is considered desirable crops. The yield and quality of the crops are greatly increased through the extensive use of this technology. Increasing nutrition and food safety are the basic points to consider before implementing tissue culture techniques.

18.8 Immobilization of Microbes to Improve Soil Health and Crop Yield

The use of beneficial microbes as bioinoculant increases their number in soil, which in turn increases the availability of nutrients to the plants. Yet, complications in technical handling are often observed with fungal cells when employed as bioinoculants for practical purposes, since satisfactory results are observed during in vitro conditions, but not typically realized in natural agricultural systems (Jain et al. 2010). A number of factors attribute to poor survivability and colonizing ability in rhizosphere, such as competition with native microbiota and abiotic stresses. Encapsulation of the cells in biodegradable capsules can be useful to overcome such hindrances.

18.8.1 Encapsulation of Bacterial Cells

Cell encapsulation facilitates sustainability and stability of biological functions; hence, enhanced cellular activities are realized (Juarez-Jimenez et al. 2012). Besides stability, encapsulation also aids in protecting the cells against all contrary ecological factors and facilitates slow release of cells into the soil in a controlled method, thus improving the efficiency of microbial fertilizers or biofertilizers (Vassilev and Vassileva 2003).

18.9 Conclusion

The world population is growing rapidly. Thus, in the next few years, it will be the biggest challenge to feed a huge population. Global warming, restricted environmental conditions, and biotic factors limit crop yields. The main challenge for researchers working on different crops is to increase agricultural productivity to counter the demand for foodstuff supply to a rapidly expanding global population. Therefore, crop improvement is the main element of agricultural progress, and there are still a lot of zones left to work on in the field of crop improvement. Applications of RDT or genetic engineering to crop improvement are well suited to deciphering the problem of world hunger and depriving sustainable intensification.

References

Ahmad JS, Baker R (1987) Rhizosphere competence of *Trichoderma harzianum*. Phytopathology 77:182–189

Ahmad M, Pataczek L, Hilger TH, Zahir ZA, Hussain A et al (2018) Perspectives of microbial inoculation for sustainable development and environmental management. Front Microbiol 9:2992

Akhtar S, Shahid AA, Rao AQ, Bajwa KS, Muzaffar A et al (2014) Genetic effects of Calotropis procera CpTIP1 gene on fiber quality in cotton (*Gossypium hirsutum*). Adv Life Sci 1 (4):223–230

Almeida DM, Almadanim MC, Lourenço T, Abreu IA, Saibo NJ et al (2016) Screening for abiotic stress tolerance in rice: salt, cold, and drought. Methods Mol Biol 1398:155–182

Amin H, Arain BA, Amin F, Surhio MA (2014) Analysis of growth response and tolerance index of Glycine max (L.) Merr. under hexavalent chromium stress. Adv Life Sci 1(4):231–241

Arora NK, Mishra J (2016) Prospecting the roles of metabolites and additives in future bioformulations for sustainable agriculture. Appl Soil Ecol 107:405–407

Assefa F (2018) Status of fall Armyworm (*Spodoptera frugiperda*), biology and control measures on maize crop in Ethiopia: a review. Int J Entomol Res 6:75–85

Atique R, Farooq M, Rashid A, Nadeem F, Stuerz S et al (2018) Boron nutrition of rice in different production systems: a review. Agron Sustain Dev 38:25

Awais MA, Pervez A, Yaqub A, Sarwar R, Alam F et al (2010) Current status of biotechnology in health. Am Eurasian J Agric Environ Sci 7:210–220

Baez-Rogelio A, Morales-Garcia YE, Quintero-Hernandez V, Muños-Rojas J (2017) Next generation of microbial inoculants for agriculture and bioremediation. Microb Biotechnol 10:19–21

Bagwan JD, Patil SJ, Mane AS, Kadam VV, Vichare S (2010) Genetically modified crops: food of the future. Int J Adv Biotechnol Res 1(1):21–30

Bai Y, Kissoudis C, Yan Z, Visser RGF, van der Linden G (2018) Plant behaviour under combined stress: tomato responses to combined salinity and pathogen stress. Plant J 93:781–793

Bakshi A (2003) Potential adverse health effects of genetically modified crops. J Toxicol Environ Health 7:211–226

Bashan Y, de-Bashan LE, Prabhu SR (2016) Superior polymeric formulations and emerging innovative products of bacterial inoculants for sustainable agriculture and the environment. In: Singh HB, Sarma BK, Keswani C (eds) Agriculturally important microorganisms-commercialization and regulatory requirements in Asia. Springer, Berlin, pp 15–46

Bawa AS, Anilakumar KR (2013) Genetically modified foods: safety, risks and public concerns—a review. J Food Sci Technol 50(6):1035–1046

Beyer P, Al-Babili S, Ye X, Lucca P, Schaub P, Welsch R, Potrykus I (2002) Golden rice: introducing the b-carotene biosynthesis pathway into rice endosperm by genetic engineering to defeat vitamin a deficiency. J Nutr 132:506–510

Bhalla PL (2006) Genetic engineering of wheat, current challenges and opportunities. Trends Biotechnol 24:305–311

Bhandari G, Bhatt P (2020) Concepts and application of plant microbe interaction in remediation of heavy metals. In: Sharma A (ed) Microbes and signaling biomolecules against plant stress. Rhizosphere biology. Springer, Singapore. https://doi.org/10.1007/978-981-15-7094-0_4

Bhandari G, Bagheri AR, Bhatt P, Bilal M (2021) Occurrence, potential ecological risks, and degradation of endocrine disrupter, nonylphenol from the aqueous environment. Chemosphere. https://doi.org/10.1016/j.chemosphere.2021.130013

Bhatt P (2018) *Insilico* tools to study the bioremediation in microorganisms. In: Pathak V, Navneet (eds) Handbook of research on microbial tools for environmental waste management. IGI Global, Hershey, pp 389–395. https://doi.org/10.4018/978-1-5225-3540-9.ch018

Bhatt P, Barh A (2018) Bioinformatic tools to study the soil microorganisms: an *in silico* approach for sustainable agriculture. In: Choudhary D, Kumar M, Prasad R, Kumar V (eds) In silico approach for sustainable agriculture. Springer, Singapore. https://doi.org/10.1007/978-981-13-0347-0_10

Bhatt K, Bhatt P (2021) Rhizospheric biology: alternate tactics for enhancing sustainable agriculture. In: Phytomicrobiome interactions and sustainable agriculture, pp 164–186. https://doi.org/10.1002/9781119644798.ch9

Bhatt P, Nailwal TK (2018) Chapter 11. Crop improvement through microbial biotechnology: a step towards sustainable agriculture. In: Prasad R, Singh Gill S, Tuteja N (eds) Crop improvement through microbial biotechnology. Elsevier, Amsterdam, pp 245–253. https://doi.org/10.1016/B978-0-444-63987-5.00011-6

Bhatt P, Bisht TS, Pathak VM, Barh A, Chandra D (2015a) Optimization of amylase production from the fungal isolates of Himalayan region Uttarakhand. Ecol Environ Conserv 21 (3):1517–1521

Bhatt P, Negi G, Gangola S, Khati P, Srivastava A, Sharma A (2015b) Optimization of sulfosulfuron biodegradation through response surface methodology using indigenous bacterial strain isolated from contaminated agriculture field. Int J Curr Microbiol App Sci 4(8):105–112

Bhatt P, Negi G, Gangola S, Khati P, Kumar G, Srivastava A, Sharma A (2016a) Differential expression and characterization of cypermethrin degrading potential proteins in *Bacillus thuringiensis* strain, SG4. 3 Biotech 6:225

Bhatt P, Sharma A, Gangola S, Khati P, Kumar G, Srivastava A (2016b) Novel pathway of cypermethrin biodegradation in *a Bacillus* sp. strain SG2 isolated from cypermethrin-contaminated agriculture field. 3 Biotech 6:45

Bhatt P, Huang Y, Zhan H, Chen S (2019a) Insight into microbial applications for the biodegradation of Pyrethroid insecticides. Front Microbiol 10:1778

Bhatt P, Pal K, Bhandari G, Barh A (2019b) Modeling of methyl halide biodegradation on bacteria and its effect on other environmental systems. Pestic Biochem Physiol 158:88–100

Bhatt P, Gangola S, Chaudhary P, Khati P, Kumar G, Sharma A, Srivastava A (2019c) Pesticide induced up-regulation of esterase and aldehyde dehydrogenase in indigenous *Bacillus* spp. Biorem J 23(1):42–52

Bhatt P, Joshi D, Kumar N, Kumar N (2019d) Recent trends to study the functional analysis of mycorrhizosphere. In: Varma A, Choudhary D (eds) Mycorrhizosphere and pedogenesis. Springer, Singapore. https://doi.org/10.1007/978-981-13-6480-8_11

Bhatt P, Pathak VM, Joshi S, Bisht TS, Singh K, Chandra D (2019e) Major metabolites after degradation of xenobiotics and enzymes involved in these pathways. In: Smart bioremediation technologies: microbial enzymes, chap 12, pp 205–215. https://doi.org/10.1016/B978-0-12-818307-6.00012-3

Bhatt P, Rene ER, Kumar AJ, Kumar AJ, Zhang W, Chen S (2020a) Binding interaction of allethrin with esterase: bioremediation potential and mechanism. Bioresour Technol 315:13845. https://doi.org/10.1016/j.biortech.2020.123845

Bhatt P, Verma A, Verma S, Anwar MS, Prasher P, Mudila H, Chen S (2020b) Understanding phytomicrobiome: a potential reservoir for better crop management. Sustainability 12:5446

Bhatt P, Huang Y, Rene ER, Kumar AJ, Chen S (2020c) Mechanism of allethrin biodegradation by a newly isolated *Sphingomonas trueperi* strain CW3 from wastewater sludge. Bioresour Technol 305:123074

Bhatt P, Zhang W, Lin Z, Pang S, Huang Y, Chen S (2020d) Biodegradation of allethrin by a novel fungus *Fusarium proliferatum* strain CF2, isolated from contaminated. Soils Microorg 8:593

Bhatt P, Huang Y, Zhang W, Sharma A, Chen S (2020e) Enhanced cypermethrin degradation kinetics and metabolic pathway in *Bacillus* thuringiensis strain, SG4. Microorganisms 8:223

Bhatt P, Bhatt K, Huang Y, Ziqiu L, Chen S (2020f) Esterase is a powerful tool for the biodegradation of pyrethroid insecticides. Chemosphere 244:125507

Bhatt P, Zhou X, Huang Y, Zhang W, Chen S (2021a) Characterization of the role of esterases in the biodegradation of organophosphate, carbamate and pyrethroid group pesticides. J Hazard Mater 411:125026

Bhatt P, Joshi T, Bhatt K, Zhang W, Huang Y, Chen S (2021b) Binding interaction of glyphosate oxidoreductase and C-P lyase: molecular docking and molecular dynamics simulation studies. J Hazard Mater 409:124927

Bhatt P, Bhatt K, Sharma A, Zhang W, Mishra S, Chen S (2021c) Biotechnological basis of microbial consortia for the removal of pesticides from the environment. Crit Rev Biotechnol 41:317–338. https://doi.org/10.1080/07388551.2020.1853032

Bhatt P, Sethi K, Gangola S, Bhandari G, Verma A, Adnan M, Singh Y, Chaube S (2021d) Modeling and simulation of atrazine biodegradation in bacteria and its effect in other living systems. J Biomol Struct Dyn. https://doi.org/10.1080/07391102.2020.1846623

Bhatt P, Gangola S, Bhandari G, Zhang W, Maithani D, Mishra S, Chen S (2021e) New insights into the degradation of synthetic pollutants in contaminated environments. Chemosphere. https://doi.org/10.1016/j.chemosphere.2020.128827

Bhatt P, Sharma A, Rene ER, Kumar AJ, Zhang W, Chen S (2021f) Bioremediation mechanism, kinetics of fipronil degradation using *Bacillus* sp. FA3 and resource recovery potential from contaminated environments. J Water Process Eng 39:101712

Bimpong IK, Manneh B, Sock M, Diaw F, Amoah NKA (2016) Improving salt tolerance of lowland rice cultivar 'Rassi' through marker-aided backcross breeding in West Africa. Plant Sci 242:288–299

Boccia F, Sarnacchiaro PA (2015) Genetically modified foods and consumer perspective. Recent Pat Food Nutr Agric 7:28–34

Bowman MS, Zilberman D (2013) Economic factors affecting diversified farming systems. Ecol Soc 18:33

Brown DW, Thorpe TA (1995) Crop improvement through tissue culture. World J Microbiol Biotechnol 11:409–415

Case ME, Schweizer M, Kushner SR, Giles NH (1979) Efficient transformation of Neurospora crassa by utilizing hybrid plasmid DNA. Proc Natl Acad Sci U S A 76:5259–5263

Chailertvanitkul VA, Pouton CW (2010) Adenovirus: a blueprint for non-viral gene delivery. Curr Opin Biotechnol 21(5):627–632

Dar AI, Saleem F, Ahmad M, Tariq M, Khan A et al (2014) Characterization and efficiency assessment of PGPR for enhancement of rice (*Oryza sativa* L.) yield. Adv Life Sci 2(1):38–45

Dash A, Kundu D, Das M, Bose D, Adak S et al (2016) Food biotechnology: a step towards improving nutritional quality of food for Asian countries. Recent Pat Biotechnol 10:43–57

De Filippis L (2013) Bioinformatic tools in crop improvement. In: Crop Improvement. Springer, New York, pp 49–122
Deng Y, Johnson DR, Guan X, Ang CY, Ai J, Perkins EJ (2010) In vitro gene regulatory networks predict in vivo function of liver. BMC Syst Biol 4:153
Evans DA, Bravo JE (1988) Agricultural applications of protoplast fusion. In: Marby TI (ed) Plant biotechnology. Austin, pp 51–91
Fan X, Ye T, Li Q, Bhatt P, Zhang L, Chen S (2020) Potential of a quorum quenching bacteria isolate Ochrobactrum intermedium D-2 against soft rot pathogen *Pectobacterium carotovorum* subsp. Carotovora. Front Microbiol 11:898
Federico AC, Darci AG, Patrick JT (2019) Empirical investigation of mutation rate for herbicide resistance. Weed Sci 67:361–368
Feng Y, Huang Y, Zhan H, Bhatt P, Chen S (2020) An overview of strobilurin fungicide degradation: current status and future perspective. Front Microbiol 11:389
Fiandaca MS, Federoff HJ (2014) Using viral-mediated gene delivery to model Parkinson's disease: do nonhuman primate investigations expand our understanding? Exp Neurol 256:117–125
Fincham JRS (1989) Transformation in fungi. Microbiol Rev 53:148–170
Firn RD, Sharma N, Digby J (1994) Physiology, growth and development of plants and cells in culture-the way ahead. In: Lumsden PJ, Nicholas JR, Davies WJ (eds) Physiology, growth and development of plants in culture. Springer, Dordrecht, pp 409–421
Fuentes-Ramirez LE, Caballero-Mellado J (2005) Bacterial biofertilizers. In: PGPR: biocontrol and biofertilization. Springer, Dordrecht, pp 143–172
Gangola S, Bhatt P, Chaudhary P, Khati P, Kumar N, Sharma A (2018a) Bioremediation of industrial waste using microbial metabolic diversity. In: Pankaj, Sharma A (eds) Microbial biotechnology in environmental monitoring and cleanup. IGI Global, Hershey, pp 1–27. https://doi.org/10.4018/978-1-5225-3126-5.ch001
Gangola S, Sharma A, Bhatt P, Khati P, Chaudhary P (2018b) Presence of esterase and laccase in *Bacillus subtilis* facilitates biodegradation and detoxification of cypermethrin. Sci Rep 8:12755. https://doi.org/10.1038/s41598-018-31082-5
Gelvin SB (2003) Agrobacterium-mediated plant transformation: the biology behind the "gene-jockeying" tool. Microbiol Mol Biol Rev 67:16–37
Gheysen G, Angenon G, van Montagu M (1998) Agrobacterium-mediated plant transformation: a scientifically intriguing story with significant applications. In: Lindsey K (ed) Transgenic plant research. Harwood Academic, Amsterdam, pp 1–33
Godrat R, Alireza K, Ardeshir NJ, Saeid S (2005) Evaluation of genetic variability in a breeder flock of native chicken based on randomly amplified polymorphic DNA markers. Iran J Biotechnol 3:231–234
Gonzalez AJ, Larraburu EE, Llorente BE (2015) *Azospirillum brasilense* increased salt tolerance of jojoba during in vitro rooting. Ind Crop Prod 76:41–48
Goodman RM, Hauptli H, Crossway A, Knauf VC (1987) Gene transfer in crop improvement. Science 236(4797):48–54
Gordon JE, Christie PJ (2015) The agrobacterium Ti plasmids. In: Plasmids: biology and impact in biotechnology and discovery, pp 295–313. https://doi.org/10.1128/9781555818982.ch17
Gray EJ, Smith DL (2004) International and extracellular PGPR: commonalities and distinction in the plant bacterium signalling processes. Soil Biol Biochem 37:395–412
Gupta S, Bhatt P, Chaturvedi P (2018) Determination and quantification of asiaticoside in endophytic fungus from Centella asiatica (L.) urban. World J Microbiol Biotechnol 34:111
Hansen G, Wright M (1999) Recent advances in the transformation of plants. Trends Plant Sci 4:226–231
Huang J, Rozelle S, Pray C, Wang Q (2001) Plant biotechnology in China. Science 295:674–677
Huang Y, Zhan H, Bhatt P, Chen S (2019) Paraquat degradation from contaminated environments: current achievements and perspectives. Front Microbiol 10:1754

Huang Y, Lin Z, Zhang W, Pang S, Bhatt P, Rene ER, Kumar AJ, Chen S (2020) New insights into the microbial degradation of D-cyphenothrin in contaminated water/soil environments. Microorganisms. 8:473

Huang Y, Zhang W, Pang S, Chen J, Bhatt P, Mishra S, Chen S (2021) Insights into the microbial degradation and catalytic mechanism of chlorpyrifos. Environ Res 194:110660

Jain SM (2001) Tissue culture-derived variation in crop improvement. Euphytica 118:153–166

Jain R, Saxena J, Sharma V (2010) The evaluation of free and encapsulated *Aspergillus awamori* for phosphate solubilization in fermentation and soil–plant system. Appl Soil Ecol 46(1):90–94

Juarez-Jimenez B, Reboleiro-Rivas P, Gonzalez-Lopez J, Pesciaroli C, Barghini P et al (2012) Immobilization of *Delftia tsuruhatensis* in macro-porous cellulose and biodegradation of phenolic compounds in repeated batch process. J Biotechnol 157:148–153

Kage U, Kumar A, Dhokane D, Karre S, Kushalappa AC (2016) Functional molecular markers for crop improvement. Crit Rev Biotechnol 36:917–930

Khan MAU, Shahid AA, Rao AQ, Shahid N, Latif A et al (2015a) Defense strategies of cotton against whitefly transmitted CLCuV and Begomoviruses. Adv Life Sci 2(2):58–66

Khan MAU, Shahid AA, Rao AQ, Shahid N, Latif A, ud Din S, Husnain T (2015b) Defense strategies of cotton against whitefly transmitted CLCuV and Begomoviruses. Adv Life Sci 2 (2):58–66

Khati P, Sharma A, Gangola S, Kumar R, Bhatt P, Kumar G (2017a) Impact of agriusable nanocompounds on soil microbial activity: an indicator of soil health. Soil Air Water 45 (5):1600458

Khati P, Parul, Gangola S, Bhatt P, Sharma A (2017b) Nanochitosan induced growth of Zea Mays with soil health maintenance. 3 Biotech 7(1):81

Khati P, Gangola S, Bhatt P, Kumar R, Sharma A (2018a) Application of nanocompounds for sustainable agriculture system. In: Pankaj, Sharma A (eds) Microbial biotechnology in environmental monitoring and cleanup. IGI Global, Hershey, pp 194–211. https://doi.org/10.4018/978-1-5225-3126-5.ch012

Khati P, Parul, Bhatt P, Nisha Kumar R, Sharma A (2018b) Effect of nanozeolite and plant growth promoting rhizobacteria on maize. 3 Biotech 8:141

Kottapalli KR, Lakshmi NM, Jena KK (2010) Effective strategy for pyramiding three bacterial blight resistance genes into fine grain rice cultivar, Samba Mahsuri, using sequence tagged site markers. Biotechnol Lett 32:989–996

Kumar N (2011) Redd MP in vitro plant propagation: a review. J For Sci 27:61–72

Kumar G, Arya K, Verma A, Pankaj, Khati P, Gangola S, Kumar R, Sharma A, Singh H (2017) Bioremediation of petrol engine oil polluted soil using Microbial Consortium and wheat crop. J Pure Appl Microbiol 11(3):1583–1588

Lin Z, Zhang W, Pang S, Huang Y, Mishra S, Bhatt P, Chen S (2020) Current approaches to and future perspectives on methomyl degradation in contaminated soil/water environments. Molecules 25:738

Llorens E, González-Hernández AI, Scalschi L, Fernández-Crespo E, Camañes G et al (2020) Chapter 1-Priming mediated stress and cross-stress tolerance in plants: concepts and opportunities. In: Priming-mediated stress and cross-stress tolerance in crop plants. Academic, Cambridge, pp 1–20

Lockhart DJ, Winzeler EA (2000) Genomics, gene expression and DNA arrays. Nat 45:827–836

Loyola-Vargas VM, Ochoa-Alejo N (2018) An introduction to plant tissue culture: advances and perspectives. In: Loyola-Vargas VM, Ochoa-Alejo N (eds) Plant cell culture protocols. Springer, New York, pp 3–13

Malusá E, Pinzari F, Canfora L (2016) Efficacy of biofertilizers: challenges to improve crop production. In: Singh D, Singh H, Prabha R (eds) Microbial inoculants in sustainable agricultural productivity, vol 2, Berlin, Springer, pp 17–40

Maron PA, Sarr A, Kaisermann A, Leveque J, Mathieu O, Guigue J et al (2018) High microbial diversity promotes soil ecosystem functioning. Appl Environ Microbiol 84:e002738–e002717

Mishra NC, Tatum EL (1973) Non-mendelian inheritance of DNA-induced inositol independence in Neurospora. Proc Natl Acad Sci U S A 70:3875–3879
Mishra S, Zhang W, Lin Z, Pang S, Huang Y, Bhatt P, Chen S (2020) Carbofuran toxicity and its microbial degradation in contaminated environments. Chemosphere 259:127429
Mochida K, Shinozaki K (2010) Genomics and bioinformatics resources for crop improvement. Plant Cell Physiol 51(4):497–523
Mullins E, Milbourne D, Petti C, Doyle-Prestwich BM, Meade C (2006) Potato in the age of biotechnology. Trends Plant Sci 11:254–260
Muthurajan R, Balasubramanian P (2010) Pyramiding genes for enhancing tolerance to abiotic and biotic stresses. In: Jain S, Brar D (eds) Molecular techniques in crop improvement. Springer, Dordrecht
Mwobobia EG, Sichangi AW, Thiong'o KB (2020) Characterization of wheat production using earth-based observations: a case study of Meru County, Kenya. Model Earth Syst Environ 6:13–25
Nagoshi RN, Goergen G, Tounou KA, Agboka K, Koffi D, Meagher RL (2018) Analysis of strain distribution, migratory potential, and invasion history of fall armyworm populations in northern Sub-Saharan Africa. Sci Rep 8:3710
Ngwako S (2008) Mapping quantitative trait loci using marker regression and interval mapping methods. Pak J Biol Sci 11:553–558
Ogawa K, Ohara H, Koide T, Toyama N (1989) Interspecific hybridization of T. reesei by protoplast fusion. J Ferment Bioeng 67:207–209
Pancaldi F, Trindade LM (2020) Marginal lands to grow novel bio-based crops: a plant breeding perspective. Front Plant Sci 11:227
Pandeya D, López-Arredondo DL, Janga MR, Campbell LM, Estrella HP et al (2018) Selective fertilization with phosphite allows unhindered growth of cotton plants expressing the ptxD gene while suppressing weeds. Proc Natl Acad Sci U S A 115:201804862-E6955
Pang S, Lin Z, Zhang W, Mishra S, Bhatt P, Chen S (2020) Insights into the microbial degradation and biochemical mechanisms of neonicotinoids. Front Microbiol 11:868
Papagianni M (2004) Fungal morphology and metabolite production in submerged mycelia processes. Biotechnol Adv 22:189–259
Patel DH, Misra A (2011) Gene delivery using viral vectors. In: Challenges in delivery of therapeutic genomics and proteomics, vol 207. Elsevier, Amsterdam, p 270
Prabavathy V, Mathivanan N, Sagadevan E, Murugesan K, Lalithakumari D (2006) Self-fusion of protoplasts enhances chitinase production and biocontrol activity in *Trichoderma harzianum*. Bioresour Technol 97:2330–2334
Puspito AN, Rao AQ, Hafeez MN, Iqbal MS, Bajwa KS et al (2015) Transformation and evaluation of Cry1Ac+ Cry2A and GTGene in Gossypium hirsutum L. Front Plant Sci 6:943
Qaim M (2009) The economics of genetically modified crops. Ann Rev Resour Econ 1:665–693
Qamar Z, Aaliya K, Nasir IA, Farooq AM, Tabassum B, Qurban A et al (2015) An overview of genetic transformation of glyphosate resistant gene in Zea mays. Nat Sci 13(3):80–90
Radhakrishnan R, Hashem A, Abd Allah EF (2017) A biological tool for crop improvement through bio-molecular changes in adverse environments. Front Physiol 8:667
Rai N, Bellundagi A, Kumar PKC, Kalasapura TR, Rani S, Sinha N, Krishna H, Jain N, Singh GP, Singh PK et al (2018) Marker-assisted backcross breeding for improvement of drought tolerance in bread wheat (*Triticum aestivum* L. em Thell). Plant Breed 137:514–526
Rana M, Sood A, Hussain W, Kaldate R, Sharma TR et al (2019) Chapter 6: Gene pyramiding and multiple character breeding. In: Singh M (ed) Lentils. Academic, Cambridge, pp 83–124
Ruengphayak S, Chaichumpoo E, Phromphan S, Kamolsukyunyong W, Sukhaket W et al (2015) Pseudo-backcrossing design for rapidly pyramiding multiple traits into a preferential rice variety. Rice 8:7
Sandhu N, Dixit S, Swamy BPM, Raman A, Kumar S et al (2019) Marker assisted breeding to develop multiple stress tolerant varieties for flood and drought prone areas. Rice. 12:8

Santi C, Bogusz D, Franche C (2013) Biological nitrogen fixation in non-legume plants. Ann Bot 111(5):743–767

Shahid AA, Bano S, Khalid S, Samiullah TR, Bajwa KS, Ali MA (2016) Biosafety assessment of transgenic Bt cotton on model animals. Adv Life Sci 3(3):97–108

Sharifzadeh MS, Abdollahzadeh G, Damalas CA, Rezaei R (2018) Farmers' criteria for pesticide selection and use in the Pest control process. Agriculture 8:24

Sharma A, Bhatt P (2016) Bioremediation: a microbial technology for improvising wildlife. In: Wildlife management concept analysis and conservation, chap 3. Daya Publisher, New Delhi, pp 29–40

Sharma HC, Gill BS (1983) Current status of wide hybridization in wheat. Euphytica 32(1):17–31

Sharma A, Pankaj, Khati P, Gangola S, Kumar G (2016) Microbial degradation of pesticides for environmental cleanup. In: Bioremediation of industrial pollutants, chap 6. Write and Print Publication, New Delhi, pp 178–205

Shinada H, Iwata N, Sato T, Fujino K (2014) QTL pyramiding for improving of cold tolerance at fertilization stage in rice. Breed Sci 63:483–488

Singh K, Gera R, Sharma R, Maithani D, Chandra D, Bhat AM, Kumar R, Bhatt P (2021) Mechanism and application of *Sesbania* root nodulating bacteria: an alternative for chemical fertilizers and sustainable development. Arch Microbiol. https://doi.org/10.1007/s00203-020-02137-x

Stamenkovic S, Beskoski V, Karabegovic I, Lazic M, Nikolic N (2018) Microbial fertilizers: a comprehensive review on current findings and future perspectives. Spanish J Agric Res 16: e09R01. https://doi.org/10.5424/sjar/2018161-2012117

Steiner B, Michel S, Maccaferri M, Lemmens M, Tuberosa R et al (2019) Exploring and exploiting the genetic variation of Fusarium head blight resistance for genomic-assisted breeding in the elite durum wheat gene pool. Theor Appl Genet 132:969–988

Sulbhi V, Bhatt P, Verma A, Mudila H, Prasher P, Rene ER (2021) Microbial technologies for heavy metal remediation: effect of process conditions and current practices. Clean Technol Environ Policy. https://doi.org/10.1007/s10098-021-02029-8

Suresh S, Malathi D (2013) Gene pyramiding for biotic stress tolerance in crop plants. Wkly Sci Res J 1:1–14

Tabashnik BE, Huang F, Ghimire BR, Leonard BD, Siegfried et al (2011) Efficacy of genetically modified BT toxins against insects with different genetic mechanisms of resistance. Nat Biotechnol 29:1128–1131

Tariq M, Ali Q, Khan A, Khan GA, Rashid B et al (2014) Yield potential study of *Capsicum annuum* L. under the application of PGPR. Adv Life Sci 1(4):202–207

Tazeb A (2017) Plant tissue culture technique as a novel tool in plant breeding: a review. Am-Euras J Agric Environ Sci 17(2):111–118

Thitisaksakul M, Tananuwong K, Shoemaker CF, Chun A, Tanadul OUM et al (2015) Effects of timing and severity of salinity stress on rice (Oryza sativa L.) yield, grain composition, and starch functionality. J Agric Food Chem 63:2296–2304

Tiwari S, Sl K, Kumar V, Singh B, Rao AR et al (2016) Mapping QTLs for salt tolerance in rice (Oryza sativa L.) by bulked segregant analysis of recombinant inbred lines using 50K SNP chip. PLoS One 11:e0153610

Van Loon LC (2007) Plant response to plant growth-promoting rhizobacteria. Eur J Plant Pathol 119:243–254

Vassilev N, Vassileva M (2003) Biotechnological solubilization of mineral phosphates on media containing agro-industrial wastes. Appl Microbiol Biotechnol 61:435–440

Vassilev N, Vassileva M, Lopez D, Martos V, Reyes A et al (2015) Unexploited potential of some biotechnological techniques for biofertilizer production and formulation. Appl Microbiol Biotechnol 99:4983–4996

Verma N, Bansal MC, Kumar V (2008) Protoplast fusion technology and its biotechnological applications. Chem Eng Trans 14:113–120

Vincelli P (2016) Genetic engineering and sustainable crop disease management: opportunities for case-by-case decision-making. Sustainability 8:495

Waddington SR, Li X, Dixon J, Hyman G, de Vicente MC (2010) Getting the focus right: production constraints for six major food crops in Asian and African farming systems. Food Secur 2:27–48

Wang GP, Yu XD, Sun YW, Jones HD, Xia LQ (2016) Generation of marker- and/or backbone-free transgenic wheat plants via Agrobacterium-mediated transformation. Front Plant Sci 7:1324

Waqas M, Shahid L, Shoukat K, Aslam U, Azeem F et al (2020) Chapter 1-Role of DNA-binding with one finger (Dof) transcription factors for abiotic stress tolerance in plants. In: Transcription factors for abiotic stress tolerance in plants. Academic, Cambridge, pp 1–14

Ye T, Zhou T, Fan X, Bhatt P, Zhang L, Chen S (2019) *Acinetobacter lactucae* strain QL-1, a novel quorum quenching candidate against bacterial pathogen *Xanthomonas campestris* pv. Campestris. Front Microbiol 10:2867

Yu Y, Yang Z (2016) The isolation and characterization of a putative dehydrin gene in Triticum aestivum L. Biochem Syst Ecol 66:173–180

Zhan H, Huang Y, Lin Z, Bhatt P, Chen S (2020) New insights into the microbial degradation and catalytic mechanism of synthetic pyrethroids. Environ Res 182:109138

Zhang H, Zhang J, Wei P, Zhang B, Gou F et al (2014) The CRISPR/Cas9 system produces specific and homozygous targeted gene editing in rice in one generation. Plant Biotechnol J 12:797–807

Zhang W, Pang S, Lin Z, Mishra S, Bhatt P, Chen S (2020a) Biotransformation of perfluoroalkyl acid precursors from various environmental system: advances and perspectives. Environ Pollut 272:115908

Zhang W, Lin Z, Pang S, Bhatt P, Chen S (2020b) Insights into the biodegradation of lindane (γ-hexachlorocyclohexane) using a microbial system. Front Microbiol 11:522

Chapter 19
Microbial Technologies in Pest and Disease Management of Tea (*Camellia sinensis* (L.) O. Kuntze)

Ganga Devi Sinniah and Padmini Dharmalatha Senanayake

Abstract Tea, the natural beverage prepared from *Camellia sinensis* (L). O. Kuntze, is the most consumed drink next to water. Commercially grown tea plants are attacked by pest and diseases leading to considerable crop loss. Synthetic pesticides have been the preferred method of pest and disease control in tea cultivation. The emerging consumer demand on pesticide-free tea, environmental and human impacts felt due to continuous pesticide use, forced us to look for alternatives. Microbial biopesticides have been identified as one of the alternatives in the integrated pest management strategies. This chapter discusses about the microbe-based techniques used in pest and disease management of tea and highlights the opportunities for further development.

Keywords Biopesticides · Entomopathogens · Microbial metabolites · Nanoparticles

19.1 Introduction to Tea Pest and Diseases

Tea (*Camellia sinensis* (L.) O. Kuntze) is a non-alcoholic natural beverage prepared from infusion of tea leaves. Due to its economic value, tea is cultivated as a commercial plantation crop mainly in tropical and subtropical areas of Asia, Africa and America. The world tea production in 2018 accounts to six million tonnes to which China, India, Kenya, Sri Lanka, Vietnam, Argentina and Indonesia alone contributed to about 90% in production (Anonymous 2019). Tea is grown as monoculture in large contiguous area under specific climatic conditions of temperature of 10–30°C, minimum annual rainfall of 1250 mm, acidic soils of 4.5–5.5 pH,

G. D. Sinniah (✉)
Plant Pathology Division, Tea Research Institute of Sri Lanka, Talawakelle, Sri Lanka

P. D. Senanayake
Entomology and Nematology Division, Tea Research Institute of Sri Lanka, Talawakelle, Sri Lanka

P. Bhatt et al. (eds.), *Microbial Technology for Sustainable Environment*,
https://doi.org/10.1007/978-981-16-3840-4_19

0.5°–10° slopes and elevations up to 2000 m above the sea level. Tea is a perennial plant. Under commercial cultivation, it is subjected to continuous agronomic practices. Climatic and anthropogenic factors make tea plant host for several disease-causing organisms and insect pests. It has been reported that over 350 fungal diseases, 1 algal disease, 1034 species of arthropods and 82 species of nematodes are associated with tea plants worldwide (Chen and Chen 1989).

19.2 Important Diseases of Tea

Fungi, bacteria, algae and viruses are known to be associated with diseases in tea. Among them fungal pathogens are considered as the most important causes of tea diseases. Blister blight (*Exobasidium vexans* Massee) (Fig. 19.1a), brown blight/anthracnose (*Colletotrichum camelliae* and other *Colletotrichum* spp.) and grey blight (*Pestalotiopsis* spp.) (Fig. 19.1b) are important fungal diseases which attack foliage of a tea plant. Bacterial blight caused by *Pseudomonas syringae* and an algal disease caused by *Cephaleuros parasiticus* Karsten and *C. mycoidea* Karsten are also important in causing considerable crop loss in tea (Fig. 19.1c). Stems of a tea plant are attacked by different types of cankers. Fungi that belong to family Botryosphaeriaceae (*Macrophoma theicola* Petch, *Lasiodiplodia theobromae* (Pat.) Griffon & Maubl.) (Fig. 19.1d), *Poria hypobrunnea* Petch, *Phomopsis theae* Petch and members of *Fusarium solani* species complex are the important canker-causing fungi in tea. Root rots caused by *Poria hypolateritia* Berk. ex Cooke (red root rot, Fig. 19.1e, f), *Rosellinia arcuata* Petch (black root rot), *Phellinus noxius* (Corner) G. Cunn. (brown root rot), *Ustulina deusta* (Hoffm.) Maire (charcoal root rot) and *Armillaria mellea* (Vahl) P. Kumm. are the important diseases which affect the root system of tea plants.

19.3 Important Pests of Tea

Several insect and mite pests cause damage to foliage, stem and root systems of tea plants. These tea pests can be classified as perennial pests, seasonal pest and occasional pests depending on their occurrence in tea (Fig. 19.2). Shot-hole borer (SHB, *Euwallacea fornicatus* Eichhoff) and live wood termites (*Glyptotermes dilatatus* Bugnion & Popoff, *Postelectrotermes militaris* Desneux, *Neotermes greeni* Desneus) are adapted to tea-growing environments and considered as perennial pests. The seasonal pests infest tea plants only during specific environmental conditions favourable for them. Leaf-eating caterpillars such as Tea tortrix (*Homona coffearia* Nietner), looper caterpillar (*Buzura suppressaria* Guenée), tea mosquito bug (*Helopeltis theivora* Waterhouse), fringed nettle grubs (*Macroplectra nararia* Moore), red slug (*Eterusia aedea cingala* Moore) and different mite species (*Oligonychus coffeae* Nietner, *Acaphylla theae* Watt, *Calacarus carinatus* Green,

Fig. 19.1 Common diseases of tea. (**a**) Blister blight (*Exobasidium vexans*). (**b**) Grey blight (*Pestalotiopsis theae*). (**c**) Red rust (*Cephaleuros parasiticus*). (**d**) Canker on main stem (*Macrophoma theicola*). (**e**) Death of bush due to root infection. (**f**) Red root rot (*Poria hypolateritia*)

Fig. 19.2 Common pests of tea. (**a**) Shot-hole borer (*Euwallacea fornicates*). (**b**) Tea tortrix (*Homona coffearia*. (**c**) White grubs (*Holotrichia disparilis*). (**d**) Fringed nettle grubs (*Macroplectra nararia*). (**e**) Live wood termites (*Postelectrotermes militaris*)

Brevipalpus californicus Banks, *Hemitarsonemus latus* Banks) are important seasonal pests in tea. Occasional pests such as red borer (*Zeuzera coffeae* Nietner), grubs (*Holotrichia disparalis* Arrow and *Microtrichia costata* Wlk.), etc. only cause damage under specific environmental conditions or anthropogenic influences.

19.4 Microbes in Integrated Pest Management of Tea

Integrated pest management (IPM) in tea focuses on sustainable pest and disease management by combining available management tactics such as cultural practices, use of resistance plant materials, chemical application and biological control methods. Though application of synthetic pesticides has been a promising method of pest and disease management in tea, the detrimental effects of chemicals in soil, water, humans, microbes and all other living and non-living component in the environment have been felt. Thus, reduction of chemicals is promoted in the tea plantations and in agriculture sector as a whole. Biological control methods captured the attention of scientists and growers due their eco-friendly nature, specificity, biodegradability and negligible effect on host plants, environment and humans. Broadly, biological control is managing a target pest or pathogen using its natural antagonists or enemies by suppressing its growth, infection or reproduction. Microbial inoculants, parasites and predators are generally used as biological control agents. Several virulent bacterial, fungal and viral microbial inoculants (microbial pesticides) are used in crop protection purposes. This chapter mainly focuses on microbial inoculants and other technological application of microbes against diseases and insect pests of tea.

In the past decades researches and commercialization of microbial biopesticides have been given more importance globally. Cernava et al. (2019) showed that phyllosphere bacterial diversity of tea has been altered with the use of chemical application. They elucidated non-target effects by showing distinct microbial fingerprints between plants that were treated with synthetic chemical pesticides and plants treated with a *Piriformospora indica* spore solution. Increasing consumer awareness on chemical residues and technical advancements in analytical methods for detecting very low levels of chemical residues lead many countries to establish maximum residue limit (MRL) set for pesticides as food safety standard. Tea being a major export-oriented crop in major producing countries is challenged by strict regulations on MRLs in the global export market. Therefore, several tea-producing countries adopt microbial pesticides as one of the potential non-chemical alternatives to minimize or replace the use of synthetic chemicals in a sustainable manner to ensure food, environmental and health safety.

Microbial biopesticides employ different modes of actions to suppress target organisms. They can directly antagonize organisms by hyperparasitism and predation. Hyperparasitic biocontrol agents directly kill the target organism or its propagules (Köhl et al. 2019). Hypoviruses and obligate bacterial and fungal organisms are the examples of hyperparasites. Several fungal and bacterial biocontrol agents produce antibiotics, which are antimicrobial substances that can suppress or kill other microorganisms and their action is dose dependent. *Pseudomonas fluorescens*, *Bacillus subtilis* and *Trichoderma* spp. are known antibiotic-producing microbial biocontrol agents (Lo 1998). Many microorganisms release lytic enzymes targeting polymers such as chitin, proteins, cellulose, hemicellulose and DNA of the target organisms (Pal and Gardener 2006). Microbial biocontrol agents also compete for physical space and nutrients with target pathogenic organism, thereby minimizing the damage caused to the host plants. Non-pathogenic microorganisms known to trigger induced systemic resistance (ISR) in plants (Yang et al. 2008a). Induction of host resistance mechanisms and plant growth promotion are reported when plants are colonized by plant growth-promoting rhizobacteria (PGPR). Direct parasitism and production of toxin and hydrolytic enzymes are mainly employed by entomopathogens against insect hosts (Sandhu et al. 2012).

19.5 Microbial Biopesticides Used in Disease Management in Tea

19.5.1 Biopesticides of Fungal Origin

Trichoderma based products are the most common biopesticides used in the management of diseases of tea. *Trichoderma harzianum*, *T. atroviride* and *T. asperellum* are the commonly used microbial biopesticides. *Trichoderma* has shown antagonistic ability against leaf, stem and root pathogens which infect tea plants. Foliar application of *T. harzianum* formulations has been used to control bird's-eye spot caused by *Cercospora theae* (Gnanamangai and Ponmurugan 2012). *Trichoderma viride*, *T. asperellum* and *T. harzianum* have been used in India for managing *Fusarium* dieback (Kumhar and Babu 2015; Kumhar et al. 2015, 2020; Sarmah et al. 2017). *Trichoderma* spp. have also been successfully tested against grey blight caused by *P. theae* (Sarkar et al. 2015) and brown blight caused by *Glomerella cingulata* (Kuberan et al. 2012).

Trichoderma atroviride (Anita and Ponmurugan 2011) and *T. harzianum* + *Gliocladium virens* (*T. virens*) (Ponmurugan et al. 2002) are reported to control Phomopsis canker in India. Both wood dressing with a paste of talc formulation and soil application of *T. harzianum* + *G. virens* were effective in protecting from collar canker (Ponmurugan et al. 2002). *Hypoxylon serpens*, a wood rot-causing fungus in tea, was successfully controlled by *T. viride* (Nepolean et al. 2014). Balasuriya and Pradeepa (2005) reported *T. harzianum* to suppress root rot pathogen *P. hypolateritia* (red root rot) and horse hair blight (*Marasmius equicrinis*). Recently *T. asperellum* isolated from tea soil in Sri Lanka showed high efficacy against *P. hypolateritia*, *R. arcuata* (black root rot) and *P. noxius* (brown root rot) (Fig. 19.3). Biocontrol of brown root and charcoal stump rot with *Trichoderma* was effective in India (Hazarika et al. 2000).

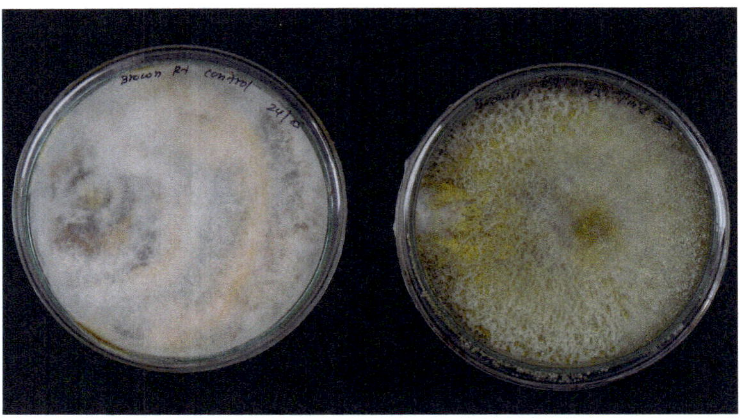

Fig. 19.3 Biocontrol activity of *Trichoderma asperellum* on brown root rot pathogen *Phellinus noxius*

In addition to *Trichoderma*-based biocide, Phong et al. (2014) reported that *Chaetomium cupreum* strain significantly inhibited grey blight. *Gliocladium virens* was found to be better than *T. harzianum* in managing Phomopsis canker (Ponmurugan and Baby 2007). Baby et al. (2004) showed *G. virens* was effective in controlling red root disease in new planting area and protected the young plants. *Aspergillus niger* has been evaluated against black rot disease caused by *Corticium theae* and algal red rust disease under field conditions (Premkumar et al. 2009).

19.5.2 Biopesticides of Bacterial Origin

In Northeast India, *B. subtilis* and two strains of actinomycetes (MM AC/02 and 05) were found to give satisfactory control of blister blight disease. The disease protection achieved with *B. subtilis* was 40% (Sarmah et al. 2020), while actinomycetes strains gave above 50% protection when compared to 84% protection achieved with standard copper oxychloride (COC) treatment (Premkumar et al. 2009). *Bacillus subtilis* have been used in India for the potential reduction (74–87%) of *Fusarium* dieback (Kumhar et al. 2020; Kumhar and Babu 2019) and algal red rust under field conditions (Sarmah et al. 2020). *Bacillus subtilis* has also been tested against *P. theae* under in vitro conditions. Hong et al. (Hong et al. 2005) reported that *B. subtilis* TL2 isolated from C. sinensis cv. Tie-guanyin in China showed 83.6%, 83.3%, 90.3% and 86.5% inhibitory effect against *Phyllosticta gemmiphliae*, *P. theae*, *Gloeosporium theae-sinensis* and *Neocapnodium theae*, respectively.

The bacterium *Micrococcus luteus* showed antagonism to *G. cingulate*, a causal agent of brown blight (Chakraborty et al. 1998). *Ochrobactrum anthropi* (1.9×10^8 CFU/ml) was found to be effective against blister blight (Sowndhararajan et al. 2013a, b) and brown root rot disease (Chakraborty et al. 2009). Rhizosphere inhabitant *Serratia marcescens* strain ETR17 was effective in suppressing root rot disease caused by *Rhizoctonia solani* in tea plants. Broad-spectrum antifungal activity, production of lytic enzymes such as HCN, IAA, siderophore and antibiotics production by *S. marcescens* inhibited growth of *R. solani* (Purkayastha et al. 2018). Application of bioformulations of *P. fluorescens* and *B. subtilis* at the rate of 100 g/ planting pit is also equally effective to *Trichoderma* in controlling primary root diseases of tea (Premkumar and Baby 2005).

Plant growth-promoting rhizobacterial strain-mediated induced systemic resistance has been reported against brown root rot and charcoal stump rot (Mishra et al. 2014). Islam et al. (2018) reported *Bacillus*, *Pseudomonas* and *Streptomyces* strains are capable of 10–14% reduction of black rot disease severity.

19.6 Microbial Biopesticides in Insect Pest Management in Tea

19.6.1 Entomopathogenic Viruses

Nucleopolyhedroviruses (NPVs) and granuloviruses (GVs) (baculoviruses) are the most commonly used viral biopesticides in tea. Nucleopolyhedroviruses mainly occur on lepidopteran, dipteran and hymenopteran hosts, while GVs are more specific to lepidopteran hosts.

Viruses isolated from natural infection of insect pests such as *B. suppressaria* NPV (BusuNPV), *Ectropis obliqua* NPV (EcobNPV), *Euproctis pesudoconspersa* NPV (EupeNPV), *Andraca bipunctata* GV (AnbiGV) and *Adoxophyes orana* GV (AdorGV) have been used as large-scale biocontrol agents in tea in China (Roy and Muraleedharan 2014). Among the above viruses, EcobNPV and EupeNPV formulations have been accepted by the Ministry of Agriculture of China as commercial insecticides against tea geometrid *E. obliqua* and tea caterpillar *E. pseudoconspersa*, respectively (Ye et al. 2014; Sun 2015). EupeNPV has shown 80% mortality in *E. pseudoconspersa* (Sun et al. 1988, 1996). A biopesticide based on *A. orana* GV (AdorGV) and *H. magnanima* GV (HomaGV) has been registered as the commercial biopesticide, Hamaki-Tenteki, in Japan (Takahashi et al. 2015).

Application of BusuNPV has resulted in more than 90% mortality of the first and second generations of *B. suppressaria* in China (Peng et al. 1998). Outbreak of NPV disease in *B. suppressaria* has also been reported in India (Hazarika and Puzari 2001). In Japan, NPVs, GVs and entomopoxviruses (EPVs) have been tested against *Adoxophyes honmai* and *H. magnanima* (Kodomari 1987; Ishii et al. 2002; Nakai 2009). An entomopoxvirus isolated from *A. honmai* (AHEV) has shown broad host range including *H. magnanima* and *A. insularis* (Takatsuka et al. 2010). Nucleopolyhedroviruses have also been isolated and characterized from *A. bipunctata* (Hazarika et al. 1995), *Arctornis submarginata* (Mukhopadhyay and De 2009), *Hyposidra talaca* NPV (HytaNPV) (Mukhopadhyay et al. 2011) and *H. infixaria* NPV (HyinNPV) (Antony et al. 2011) in India. Natural populations of entomopathogenic viruses are prevalent in *H. coffearia* in Sri Lanka. These reports show widespread occurrence of viruses, and so far, 82 species of viruses have been reported from insects associated with tea plant (Ye et al. 2014). Antony (2014) has shown that *H. infixaria* virus can be passed on to the next generation, and thus it may be useful for the long-term pest control.

19.6.2 Entomopathogenic Fungi

Entomopathogenic fungi commonly used in tea are *Beauveria bassiana* (Bals.) Vuill., *Metarhizium anisopliae* (Metschn.) Sorokin, *Verticillium lecanii* (*Lecanicillium lecanii*), *Paecilomyces* spp., *Hirsutella thompsonii*, *Cladosporium*

Table 19.1 Entomopathogenic fungi reported against tea pests

Biocontrol agents	Target pests	References
Aegertia weberri	Black spiny whitefly, *Aleurocanthus spiniferus* Quaint.	Chen et al. (1997)
Aspergillus sp.	Aphids	Barua (1983)
	Tea mosquito bug, *H. theivora*	Bordoloi et al. (2012)
Beauveria bassiana	Brown weevil, *M. aurolineatus*	Wu and Sun (1994)
	Tea mosquito bug	Kumhar et al. (2020)
	SHB *E. fornicatus*	Selvasundaram and Muraleedharan (2000), Senanayake and Kulathunga (2015)
	Tea looper, *B. suppressaria*	Ghatak and Reza (2007)
	Tea termites *Microtermes obesi* *Microcerotermes* spp.	Singha et al. (2011), Roy et al. (2020)
Beauveria brongniartii	*H. picea*	Yaginuma et al. (2006)
Metarhizium anisopliae	Red spider mite *Oligonychus coffeae*	Kumhar et al. (2020)
	Termites	Singha et al. (2011), Roy et al. (2020)
	Tea thrips	Shanmugrapriyan and Mathew (2011), Shanmugrapriyan et al. (2010)
Entomophthora sp.	Scale insects	Barua (1983)
Hirsutella thompsonii	Red spider mite	Muraleedharan (2001), Babu (2010)
	Tea aphid *Toxoptera aurantii*	Wahab (2004)
	Scale insect *Coccus viridis*	Hazarika and Puzari (2001), Barua (1983)
Paecilomyces fumosoroseus	Red spider mite	Muraleedharan (2001)
P. carneus	Mole cricket *Gryllotalpa africana*	Hazarika et al. (1994)
P. cinnamomeus	Whitefly *Aleurocanthus camelliae*	Saito et al. (2012)
P. tenuipes	Flushworm *Cydia leucostoma*, *Laspeyresia leucostoma*	Debnath (1986)
P. lilacinus	Bunch caterpillar *Andraca bipunctata*	Debnath (1998)
	Red spider mite	Debnath (1997)
	Tea termite *Odontotermes* spp.	Debnath (1997)
Verticillium lecanii	Tea thrips	Babu et al. (2008)
	Red spider mite	Muraleedharan (2001), Mamun et al. (2014)

sp., *A. niger*, *A. flavus*, *Cephalosporium* sp., *Aschersonia aleyrodis* Webb., *Aegerita webber* Fawcett, etc. (Table 19.1). *Beauveria bassiana* has been used to control a few pests of tea successfully in the global tea industry (Table 19.1). *Beauveria*

Fig. 19.4 *Beauveria brongniartii* infection on a white grub larva (*Holotrichia disparilis*)

bassiana (strain 871) gave more than 95% mortality of the brown weevil (*Myllocerus aurolineatus*) (Wu and Sun 1994) and 80% control of the weevil in the field (Wu et al. 1995) in China. Ghatak and Reza (2007) observed that the bioefficacy of *B. bassiana* against tea looper, *B. suppressaria*, was comparable to that of the synthetic insecticides. *Beauveria amorpha* (Hiromori et al. 2004) and *B. brongniartii* (Yaginuma et al. 2006) have been reported as alternatives to chemical insecticides for *Heptophylla picea* larvae in a tea plantation of Japan. Effective strains of *B. brongniartii* have also been isolated from *H. disparilis* (white grub) from Sri Lanka (Fig. 19.4).

Metarhizium anisopliae was found to be effective in controlling red spider mite (Kumhar et al. 2020), tea termite (*Microtermes obesi*) (Singha et al. 2011) and tea thrips (Shanmugrapriyan and Mathew 2011; Shanmugrapriyan et al. 2010). Application of *M. anisopliae* (2×10^7 cfu/g) at 10% in September to October during initial infestation of termite for organic gardens has been recommended by Tocklai Tea Research Institute. Commercial formulation of *M. anisopliae* var. *anisopliae* (BIO 1020) has been evaluated against white grubs and found to be effective in controlling white grubs in tea in Sri Lanka (Vitarana et al. 1997).

Paecilomyces fumosoroseus from south Indian tea fields was found effective against red spider mite, *O. coffeae*, and is now being marketed as a commercial formulation (Muraleedharan 2001). The potentiality of utilizing *V. lecanii* has also been tested against red spider mite (Muraleedharan 2001) and tea thrips (Babu et al. 2008). Natural infection of *Entomophthora* sp. and *Verticillium* sp. has been reported in *O. coffeae* in Sri Lanka (Vitarana 2000).

19.6.3 Entomopathogenic Bacteria

Bacillus thuringiensis (Bt) is the mostly used bacterial biopesticide used solely or in combination with other biopesticides and chemicals in tea. *Bacillus thuringiensis* has been tested against *B. suppressaria*, *D. baibarana*, *E. pseudoconspersa* and *E. obliqua* in China and *H. magnanima*, *Adoxophyes* sp., *Agromyza theae* and *Gracillaria theivora* in India (Ye et al. 2014; Hazarika et al. 1995). Two kinds of *B. thuringiensis* preparations, TOAROW-CTR (spore-dead *B. thuringiensis*) and BACILEXR (living spore crystal mixture), are registered for use in tea plantations in Japan (Kodomari 1987).

Bacillus subtilis formulation at 5% is recommended for tea mosquito bug in India. The *Enterobacter* sp. has been identified as an effective biocontrol agent of leaf rollers (De et al. 2008). Roobakkumar et al. (2011) showed efficacy of *P. fluorescens* against *O. coffeae* and also reported that *P. fluorescens* produced bacterial chitinases which had been reported to be effective in controlling mites. *Serratia marcescens* was reported to be pathogenic to two species of scale insects, *P. tubulorum* and *C. ficus* (Wang et al. 2010).

19.7 Bioconsortium and Compatibility of Biopesticides with Chemicals

Different species of microbial biopesticides are used in combination (bioconsortium) or as a cocktail mixture with synthetic pesticides and/or botanicals in order to enhance bioefficacy and maintain their efficacy under different environmental conditions. For example, a consortium of *Pseudomonas*, *Trichoderma* and VAM fungi *Glomus fasciculatum* was used, and the level of disease control achieved against *U. zonata* was higher than the application of individual biocontrol agents (Hazarika and Phookan 2003). The highest reduction of Macrophoma canker size was reported with an integrated treatment of Companion (carbendazim 12% + mancozeb 63% WP) + *B. amyloliquefaciens* and COC + *B. amyloliquefaciens* than application of Companion and COC alone (Jeyaraman and Robert 2018). Application of two rounds of 5% of *M. anisopliae* + *B. bassiana* spore suspension (1:1) and *P. fumosoroseus* 2 kg/ha + jaggery 2 kg/ha at 7-day interval showed promising control of red spider mite in mild to moderately infested tea fields in North India (Roy et al. 2018). Disease and pest control can be improved when biocontrol agents are combined with synthetic chemicals at a dose which is not harmful to the biological agents. These treatment combinations may have the potential to develop new strategies for integrated pest management by lowering the chances of resistance development and reduce the chemical dose compared to conventional treatment with single pesticides (Ons et al. 2020).

19.8 Microbe-Synthesized Nanoparticles in Tea Disease Management

The use of nanotechnology in agriculture is getting attention of both researchers and farmers following its successful application in other fields such as medicine, pharmacology, automobile, etc. Nanoparticles are materials that range from 10 to 100 nm in size and they can be produced with unique physical, chemical and biological properties (Yang et al. 2008b). Silver, gold, copper, zinc oxide and titanium oxide nanoparticles have shown to possess an excellent biocompatibility and low toxicity, and they have antibacterial/antifungal/antiviral properties (Ali et al. 2020). Nanoparticles themselves can act as plant protectant directly or they can be used as carrier materials for existing or new pesticides (Worrall et al. 2018). The use of microorganisms in the synthesis of nanoparticles is considered as an eco-friendly biological approach (Ali et al. 2020).

The nanoparticles synthesized from antagonistic microbes have been tested in controlling tea diseases under both laboratory and field levels. Silver and gold nanoparticles synthesized by *T. atroviride* showed the maximum growth inhibition of 75.7% and 80.3%, respectively, in *P. theae* on the seventh day of incubation (Ponmurugan 2017). The shelf-life of both nanoparticles synthesized from *T. atroviride* was 3 months. Inhibition of *P. theae* by silver nanoparticle was about 75%. The maximum disease control was observed at 2 ppm dosage when compared with untreated control.

Trichoderma atroviride and *S. sannanensis* were evaluated for copper and silver nanoparticle synthesis against *Cercospora theae* isolates (Gnanamangai et al. 2017) causing bird's-eye spot disease. Initially, the freshly prepared extracellular silver nanoparticles showed high disease control (59.42–79.76%), but the stability of antagonistic property in stored nanoparticles was significantly high in copper nanoparticles (58.71–73.81%). Ponmurugan et al. (2016) tested field effectiveness of copper nanoparticles biosynthesized extracellularly by using *S. griseus* through soil application in controlling *P. hypolateritia*. The maximum disease control was observed at 2.5 ppm dosage with 52.7% disease reduction when compared to copper fungicide application which resulted in 45.3% disease reduction. In addition, maximum leaf yield, improvement in soil macronutrients and difference in population dynamics of microbes were noted in plants treated with nano-copper in comparison to synthetic copper and carbendazim fungicide-treated plants. These studies show the potential of using green-synthesized nanoparticles in pest and disease management of tea. However, toxic potentials of synthesized nanoparticles, if any, to plants and beneficial microbes, human and environment are to be evaluated before large-scale adoption of this technique.

19.9 Microbial Metabolites in Tea Pest and Disease Management

As discussed above, microbial pesticides produce an array of secondary metabolites including enzymes, hormones and toxins to exploit the pathogenic microbes and insect pests. Antimicrobial metabolites from fungi and bacteria involve in antibiosis on phytopathogens and also trigger the plant's defence systems (Ongena et al. 2007). Lipases, chitinases and protease enzymes that digest the insect's body shells help entomopathogens to penetrate, and other secondary metabolites such as cyclic depsipeptides, peptides, amino acid derivatives, polyketides, peptide hybrids and terpenoids cause paralysis and disrupt the host's physiological processes enabling the microbes to colonize the hosts (Litwin et al. 2020).

Therefore, researchers have targeted microbe-based metabolites alone as pesticides in plant protection. Microbial metabolites tested against tea pathogens include volatile compounds of *Trichoderma* isolate T4, which inhibited the growth of *G. cingulate* Gc3 by 50.0%. The nonvolatile compound of *Trichoderma* cell-free culture at 10% concentration showed 74.28% growth inhibition of the same isolate Gc3 (Kuberan et al. 2012). *Trichoderma atroviride* has shown to produce volatile and nonvolatile secondary metabolites to suppress the Phomopsis canker-causing pathogen (Anita and Ponmurugan 2011). Cell-free culture filtrate of non-pathogenic *C. gloeosporioides* CgloTINO1 isolated from the tea garden of Assam in India had a strong antagonistic effect on the tea pathogens *P. theae* and *C. camelliae*. The chitinase and protease enzymes released by *C. gloeosporioides* played a role in the process of antagonism (Rabha et al. 2014).

19.10 Endophytic Organisms in Tea Pest and Disease Management

Endophytes are microorganisms (fungi, bacteria) associated with plants which do not cause any damage to the host plant (Zhang et al. 2019). Endophytes have attracted attention of researchers as they promote plant growth and protect the host plants from herbivores/insect pest and diseases. Several fungi and bacteria reported as biocontrol agents have shown endophytic colonization in tea and other plants. *Beauveria bassiana*, *M. anisopliae*, *Cladosporium* spp., *Acremonium* spp., *V. lecanii*, *Bacillus* sp. and *Streptomyces* sp. have been isolated as endophytes (Xie et al. 2020).

Utilization of endophytes in plant protection as indirect biocontrol mechanisms has been considered as a novel approach. Entophytic colonization of *B. bassiana* and *L. lecanii* on cotton leaves were found to reduce feeding and reproduction of *Aphis gossypii* (Gurulingappa et al. 2010). Endophytic colonization of *B. bassiana* in banana plants significantly reduced the survival of banana weevil larvae and protected them from damage (Akello et al. 2008). Similarly, *M. anisopliae* was successfully established as endophyte in tea by Kaushik and Dutta (2016). They recorded less infestation of aphid (6.3%), tea mosquito bug (13.8%) and red spider mite (12.5%) in foliar application and less infestation of termites and carpenter worm in soil application of *M. anisopliae*. This reveals the influence of mode of endophytic fungal application in targeting pests. It is evident in other crops that *B. bassiana* applied via seed treatment in tomato and cotton induced protection from *Pythium myriotylum* and *R. solani*, pathogens that cause damping off of seedlings and root rot of older plants (Ownley et al. 2008).

Shan et al. (2018) reported 13 genera of endophytic actinomycetes including *Streptomyces*, *Actinomadura*, *Kribbella*, *Nocardia*, *Kytococcus*, *Leifsonia*, *Microbacterium*, *Micromonospora*, *Mobilicoccus*, *Mycobacterium*, *Nocardiopsis*, *Piscicoccus* and *Pseudonocardia* belonging to 11 families in 15 tea cultivars collected from Fujian province in China. These endophytic actinomycetes have shown a high potential for producing antimicrobial metabolites, and their growth-promoting ability was proven by indoleacetic acid (IAA) production and 1-aminocyclopropane-1-carboxylic acid (ACC) deaminase activities. Groups of *Herbaspirillum* spp., *Methylobacteria* spp. and *Bacillus* spp. are reported as the most important endophytic bacteria for tea cultivars of Zijuan and Yunkang-10 in China (Yan et al. 2018). Forty three strains of endophytic fungi belonging to 14 genera were isolated and morphologically identified from the roots, stems and leaves of tea (You 2008). The endophytic community, their abundance and their activity depend on tea cultivars, age of the plants and environmental conditions (Yan et al. 2018). Use of biocontrol agents and entomopathogens as endophytes could be a novel alternative for management of insect pests and plant pathogens in tea.

19.11 Conclusions and Future Prospects

Potential use of rhizospheric, phyllospheric and entomopathogenic microorganisms for managing pests and disease-causing pathogens is increasing in global tea industry. *Trichoderma* sp., *B. subtilis* and *Pseudomonas* spp. are the common microbial pesticides used in disease management. Entomopathogenic fungi *B. bassiana* and *M. anisopliae*, entomopathogenic bacterium *B. thuringiensis* and NVPs are the common microbial inoculants in insect pest control. In addition to direct use of microbial biopesticides, microbial metabolites, microbe-synthesized nanoparticles and endophytic colonization have been tried in pest and disease management in tea.

Most of the techniques discussed above are so far restricted to experimental stage and field studies. Hence, adoption and commercial application of microbiological techniques including the use of microbial biopesticides in global tea industry are limited. Commercial application of microbial biopesticides is relatively high in China and Japan followed by India. Commercial-scale utilization of microbial pesticides in other major tea-growing countries such as Sri Lanka, Indonesia, Malaysia, Kenya, etc. is scarce.

Recent studies and reviews on rhizosphere, phyllosphere and endophytic microbiome (Pandey and Palni 2002; Bhattacharyya and Sarmah 2018; Tanti et al. 2016; Wu et al. 2020; Borah and Thakur 2020; Xu et al. 2020) unravel information on microbial dynamics and interaction of microorganism with tea plants. The new information has opened new horizons for research and shows the potential use of microbes in most beneficial ways in plant protection.

Though microbial pesticides have been considered as potential alternatives in crop protection, variability in their efficacy, environmental influence on efficacy, limited availability, attitude of growers towards microbial biopesticides and government policies and registration constrain have hampered the wider use of microbial biopesticides in global tea industry. Advancement in molecular genetics and biotechnology has allowed improving the efficacy of wild strains of biocontrol agents. Mutation, protoplast fusion, genetic modification and transformation allow genetic manipulation of microbial biopesticides. Protoplast fusion enhanced carboxymethylcellulase activity in *Trichoderma reesei* (Prabhavathi et al. 2006). Genetically modified *P. fluorescens* F113Rif (pCUGP) increased 2,4-diacetylphloroglucinol (Phl) production for biocontrol efficacy against *Polymyxa betae* on sugar beet (Resca et al. 2001). Genetic modifications change the properties of these biological agents to enhance production of toxins and secondary metabolites, survive under stress conditions, grow under varied environmental condition and enhance their efficiency and spectrum of biocontrol activity.

Lengthy, complicated data requirement and the high cost dissuade registration and commercialization of microbial biopesticidal products. Laws and policies regulating biopesticidal development and use vary from country to country (Arora et al. 2016). Simpler and harmonized regulatory policies would encourage commercialization of new products. Further, as discussed by Verma et al. (2020), microbial metabolites could be used as primary molecules for the synthesis of plant protective chemicals opening up new vistas for entrepreneurs and industrialists. Ready availability at affordable price and convincing large-scale farmer field trials promoting the use of microbial biopesticide could encourage the practical use of these products. Implementation of other IPM practices is also equally necessary to overcome the limitations and problems associated with microbial technologies and ensure sustainable crop protection in tea.

Acknowledgements The authors sincerely thank Ms. R.D.S.M. Gamlath and the staff of Plant Pathology and Entomology and Nematology Divisions, Tea Research Institute of Sri Lanka, for their support and field photographs.

References

Akello J, Dubois T, Coyne D, Kyamanywa S (2008) Endophytic *Beauveria bassiana* in banana (*Musa* spp.) reduces banana weevil (*Cosmopolites sordidus*) fitness and damage. Crop Prot 27:1437–1441

Ali MA, Ahmed T, Wu W, Hossain A, Hafeez R, Islam Masum MM, Wang Y, An Q, Sun G, Li B (2020) Advancements in plant and microbe-based synthesis of metallic nanoparticles and their antimicrobial activity against plant pathogens. Nanomaterials 10(6). https://doi.org/10.3390/nano10061146

Anita S, Ponmurugan P (2011) *In vitro* evaluation of *Trichoderma atroviride* against *Phomopsis theae* a causal agent of collar canker disease in tea plants. Int J Agric Res 6:620–631

Anonymous (2019) International Tea Committee. Annual Bulletin of Statistics, September 2019

Antony B (2014) Detection of nucleopolyhedroviruses in the eggs and caterpillars of tea looper caterpillar *Hyposidra infixaria* (Walk.) (Lepidoptera: Geometridae) as evidence of transovarial transmission. Arch Phytopathol Plant Protect 47(12):1426–1430. https://doi.org/10.1080/03235408.2013.845469

Antony B, Sinu AP, Das S (2011) New record of nucleorpolyhedrosis viruses in tea looper caterpillar in India. J Invertebr Pathol 108:63–67

Arora NK, Verma M, Prakash J, Mishra J (2016) Regulation of biopesticides: global concerns and policies. In: Arora N, Mehnaz S, Balestrini R (eds) Bioformulations: for sustainable agriculture. Springer, New Delhi. https://doi.org/10.1007/978-81-322-2779-3_16

Babu A (2010) Pest management in tea: the South Indian scenario. Bull UPASI Tea Res Found 55:23–30

Babu A, James SP, Subramaniam MSR, Shanmughapriyan R, Achuthan R, Mathew S (2008) Enhanced efficacy of *Verticillium lecania* against tea thrips by the addition of jaggery. Newslett UPASI Tea Res Found 18(2):3

Baby UI, Premkumar R, Ajay D, Sanjay R, Sasidhar R (2004) Chemical and biological control of red root disease. Newslett UPASI Tea Res Foundations 14(2):1–2

Balasuriya A, NHL P (2005) Towards biological control of tea diseases. In: Proceedings of 211th meeting of the experiments and extension forum, 29 July 2005. Tea Research Institute, Sri Lanka

Barua GCS (1983) Fungi in biocontrol of tea pest and disease in North-East India. Two Bud 30:5–7

Bhattacharyya PN, Sarmah SR (2018) The role of microbes in tea cultivation. In: Sharma VS, Gunasekare MTK (eds) Global tea science: current status and future needs. Burleigh Dodds Science Publishing, Cambridge. https://doi.org/10.19103/AS.2017.0036.24

Borah A, Thakur D (2020) Phylogenetic and functional characterization of culturable endophytic Actinobacteria associated with *Camellia* spp. for growth promotion in commercial tea cultivars. Front Microbiol 11:318. https://doi.org/10.3389/fmicb.2020.00318

Bordoloi M, Madhab M, Dutta P, Borah T, Nair SC, Phukan I, Debnath S, Borthakur BK (2012) Potential of entomopathogenic fungi for the management of *Helopeltis theivora* (Waterhouse). Two Bud 59(1):24–26

Cernava T, Chen X, Krug L, Li H, Yang M, Berg G (2019) The tea leaf microbiome shows specific responses to chemical pesticides and biocontrol applications. Sci Total Environ 667:33–40

Chakraborty U, Das G, Das SK, Chakraborty BN (1998) Evaluation of *Micrococcus luteus as* a potential biocontrol agent of brown blight disease of tea caused by *Glomerella cingulata*. J Plant Crops 26:50–57

Chakraborty U, Chakraborty BN, Basnet M, Chakraborty AP (2009) Evaluation of *Ochrobactrum anthropi* TRS-2 and its talc based formulation for enhancement of growth of tea plants and management of brown root rot disease. J Appl Microbiol 107(2):625–634. https://doi.org/10.1111/j.1365-2672.2009.04242.x

Chen Z, Chen X (1989) An analysis of the world tea pest fauna. J Tea Sci 9(1):13–22

Chen XF, Sun JD, Wu GY, Jin JZ, Zeng MS (1997) Integrated management technology of citrus black spiny whitefly (*Aleurocanthus spiniferus* Quaintance). J Tea Sci 17:15–20

De D, Mukhopadhyay A, Chakraborty R (2008) A novel bacterial pathogen (*Enterobacter* sp.) isolated from the leaf roller, *Caloptilia theivora* of tea of Darjeeling foothills. World J Microbiol Biotechnol 24:2727–2729

Debnath S (1986) A new record of entomogenous fungus *Paecilomyces tenuipes* (Peck) Samson on various psychid pests of tea in Darjeeling. Two Bud 33(1–2):46–47

Debnath S (1997) Studies on pathogenicity of *Paecilomyces tenuipes* (Peck) Samson to *Lesperysia leucostoma* a flush worm pest of tea. Two Bud 44(1):25–26

Debnath S (1998) Occurrence of native entomogenous fungus *Paecilomyces lilacinus* (Thom) Samson on eggs and larvae of bunch caterpillar, *Andraca bipunctata*. Two Bud 45(2):24–25

Ghatak SS, Reza W (2007) Bio-efficacy of *Beauveria bassiana* against tea looper caterpillar, *Buzura suppressaria* Guen. (Lepidopera:Geometridae). Acta Entomol Sin 50(9):962–966

Gnanamangai BM, Ponmurugan P (2012) Evaluation of various fungicides and microbial based biocontrol agents against bird's eye spot disease of tea plants. Crop Prot 32:111–118

Gnanamangai BM, Ponmurugan P, Jeeva SE, Manjukarunambika K, Elango V, Hemalatha K, Kakati JP, Mohanraj R, Prathap S (2017) Biosynthesised silver and copper nanoformulation as foliar spray to control bird's eye spot disease in tea plantations. IET Nanobiotechnol 11 (8):917–928. https://doi.org/10.1049/iet-nbt.2017.0023

Gurulingappa P, Sword GA, Murdoch G, McGee PA (2010) Colonization of crop plants by fungal entomopathogens and their effects on two insect pests when in planta. Biol Control 55:34–41

Hazarika DK, Phookan A (2003) Combination of *Glomus fasciculatum* with *Pseudomonas fluorescens* and *Trichoderma harzianum*: effect on biocontrol potential and growth promotion in tea seedling. In: Sixth international workshop on PGPR, IISR, Calicut, pp 289–294

Hazarika LK, Puzari KC (2001) Microbials in tea pest management. In: Ignacimuthu S, Sen A (eds) Microbials in insect pest management. New Delhi, Oxford, p p174

Hazarika LK, Puzari KC, Barua G (1994) Fungi as microbial insecticides on mole cricket. Two Bud 41(2):22–25

Hazarika LK, Borthakur M, Sing K (1995) A new pathogen of tea bunch caterpillar. Two Bud 42 (1):40–41

Hazarika OK, Phookan AK, Saikia GK, Borthakur BK, Sarma D (2000) Management of charcoal stump rot of tea with biocontrol agents. J Plant Crops 28:149–153

Hiromori H, Yaginuma D, Kajino K, Hatsukade M (2004) The effects of temperature on insecticidal activity of *Beauveria amorpha* to Hepthophylla picea. Appl Entomol Zool 39:389–392

Hong Y, Xin W, Lai Y et al (2005) Isolation of endophytic antifungal and pesticide degrading bacteria from tea plant. J Tea Sci 25(3):183–188. [Abstract]

Ishii T, Takatsuka J, Nakai M (2002) A comparative study of the growth characteristics and competitive abilities of a nucleopolyhedrovirus and an entomopoxvirus in larvae of the smaller tea tortrix, *Adoxophyes honmai* (Lepidoptera: Tortricidae). Biol Control 23:96–105

Islam M, Ahmad I, Ali M (2018) Biocontrol studies on rhizospheric microorganisms against black rot disease of tea caused by *Corticium theae* Bernard. Bangladesh J Bot 47(4):985–991. https://doi.org/10.3329/bjb.v47i4.47399

Jeyaraman M, Robert PSA (2018) Bio efficacy of indigenous biological agents and selected fungicides against branch canker disease of (*Macrophoma theicola*) tea under field level. BMC Plant Biol 18:222. https://doi.org/10.1186/s12870-018-1445-8

Kaushik H, Dutta P (2016) Establishment of *Metarhizium anisopliae,* an entomopathogen as endophyte for biological control in tea. Res Crop 17:375–387. https://doi.org/10.5958/2348-7542.2016.00063.2

Kodomari S (1987) Control of leaf rollers with granulosis viruses in tea field. 1. Effectiveness of application of GV mixture of oriental tea tortrix and smaller tea tortrix. Bull Shizuoka Tea Exp Stn 13:39–48

Köhl J, Kolnaar R, Ravensberg WJ (2019) Mode of action of microbial biological control agents against plant diseases: relevance beyond efficacy. Front Plant Sci 10:845. https://doi.org/10.3389/fpls.2019.00845

Kuberan T, Vidhyapallavi RS, Balamurugan A, Nepolean P, Jayanthi R, Premkumar R (2012) Isolation and biocontrol potential of phylloplane *Trichoderma* against *Glomerella cingulata* in tea. J Agric Sci Technol 8(3):1039–1050

Kumhar KC, Babu A (2015) Effect of indigenous *Trichoderma* strain on tea crop. Two Bud 62 (2):3–6

Kumhar KC, Babu A (2019) Biocontrol potency of *Trichoderma* isolates against tea (*Camellia* sp.) pathogens and their susceptibility towards fungicides. Int J Chem Stud 7:4192–4195

Kumhar KC, Babu A, Bordoloi M, Banerjee P, Dey T (2015) Biological and chemical control of *Fusarium solani*, causing dieback disease of tea *Camellia sinensis* (L): an in vitro study. Int J Curr Microbial Appl Sci 4(8):955–963

Kumhar KC, Babu A, Arulmarianathan JP et al (2020) Role of beneficial fungi in managing diseases and insect pests of tea plantation. Egypt J Biol Pest Control 30:78. https://doi.org/10.1186/s41938-020-00270-9

Litwin A, Nowak M, Różalska S (2020) Entomopathogenic fungi: unconventional applications. Rev Environ Sci Biotechnol 19:23–42. https://doi.org/10.1007/s11157-020-09525-1

Lo CT (1998) General mechanisms of action of microbial biocontrol agents. Plant Pathol Bull 7:155–166

Mamun MSA, Hoque M, Ahmed M (2014) *In vitro* and *in vivo* screening of some entomopathogens against red spider mite, *Oligonychus coffeae* Nietner (Acarina: Tetranychidae) in tea. Tea J Bangladesh 43:34–44

Mishra AK, Morang P, Deka M, Nishanth Kumar S, Dileep Kumar BS (2014) Plant growth-promoting rhizobacterial strain-mediated induced systemic resistance in tea (*Camellia sinensis* (L.) O. Kuntze) through defense-related enzymes against brown root rot and charcoal stump rot. Appl Biochem Biotechnol 174(2):506–521. https://doi.org/10.1007/s12010-014-1090-0

Mukhopadhyay A, De D (2009) Pathogenicity of a baculovirus isolated from *Arctornis submarginata* (Walker) (Lepidoptera:Lymantriidae), a potential pest of tea growing in the Darjeeling foothills of India. J Invertebr Pathol 100:57–60

Mukhopadhyay A, Khewa S, De D (2011) Characteristics and virulence of NPV isolated from *Hyposidra talaca*. Int J Trop Insect Sci 31(1–2):13–19

Muraleedharan N (2001) Evaluation of three entomopathogenic fungi for the control of red spider mites. http://www.o-cha.net/english/conference2/pdf/2001/files/PROC/II-082.pdf

Nakai M (2009) Biological control of tortricidae in tea fields in Japan using insect viruses and parasitoids. Virol Sin 24(4):323–332

Nepolean P, Balamurugan A, Jayanthi R, Mareeswaran J, Premkumar R (2014) Bioefficacy of certain chemical and biofungicides against *Hypoxylon* spp. causing wood rot disease in tea. J Plant Crops 42(3):341–347

Ongena M, Jourdan E, Adam A, Paquot M, Brans A, Joris B, Arpigny JL, Thonart P (2007) Surfactin and fengycin lipopeptides of *Bacillus subtilis* as elicitors of induced systemic resistance in plants. Environ Microbiol 9:1084–1090

Ons L, Bylemans D, Thevissen K, Cammue BPA (2020) Combining biocontrol agents with chemical fungicides for integrated plant fungal disease control. Microorganisms 8(12). https://doi.org/10.3390/microorganisms8121930

Ownley BH, Griffin MR, Klingeman WE, Gwinn KD, Moulton JK, Pereira RM (2008) *Beauveria bassiana*: endophytic colonization and plant disease control. J Invertebr Pathol 98(3):267–270

Pal KK, Gardener MB (2006) Biological control of plant pathogens. Plant Health Instructor. https://doi.org/10.1094/PHI-A-2006-1117-02

Pandey A, Palni LMS (2002) Tea rhizosphere: characteristics, features, microbial diversity and applications. Int J Tea Sci 1(4):10–24

Peng HY, Zeng YT, Chen XW, Ge YH, Jin F, Xie TE (1998) Efficacy analysis of viral insecticide of *Buzura suppressaria* nuclear polyhedrosis virus (BsSNPV). Chin J Appl Environ Biol 4:258–262

Phong NH, Wattanachai P, Kasem S (2014) Antifungal substances from *Chaetomium cupreum* against *Pestalotia* spp. causing gray blight disease of tea. In: Proceeding of the third international conference on integration of science and technology, 27–28 November 2014, Champasak Grand Hotel Pakse, Laos, PDR, p 38

Ponmurugan P (2017) Biosynthesis of silver and gold nanoparticles using *Trichoderma atroviride* for the biological control of Phomopsis canker disease in tea plants. IET Nanobiotechnol 11 (3):261–267. https://doi.org/10.1049/iet-nbt.2016.0029

Ponmurugan P, Baby UI (2007) Evaluation of fungicides and biocontrol agents against Phomopsis canker of tea under field condition. Australas Plant Pathol 36:68–72

Ponmurugan P, Baby UI, Premkumar R, Radhakrishnan B (2002) Integrated control of Phomopsis canker disease of tea. In: Proceedings of the 15th plantation crops symposium Placrosym XV, Mysore, 10–13 December, pp 599–602

Ponmurugan P, Manjukarunambika K, Elango V, Gnanamangai BM (2016) Antifungal activity of biosynthesized copper nanoparticles evaluated against red root-rot disease in tea plants. J Exp Nanosci 11(13):1019–1031. https://doi.org/10.1080/17458080.2016.1184766

Prabhavathi VR, Mathivanan N, Sagadevan E, Murugesan K, Lalithakumari D (2006) Intra-strain protoplast fusion enhances carboxymethyl cellulase activity in *Trichoderma reesei*. Enzym Microb Technol 38:719–723

Premkumar R, Baby UI (2005) Recommendations on the control of root and stem diseases of tea. In: Handbook of tea culture section. UPASI Tea Research Institute, India, pp 15–16

Premkumar R, Ajay D, Muraleedharan N (2009) Biological control of tea diseases-a review. In: Premkumar R, Ajay D, Muraleedharan N (eds) Role of biocontrol agents for disease management in sustainable agriculture. Research India Publications, New Delhi, pp 223–230

Purkayastha GD, Mangar P, Saha A, Saha D (2018) Evaluation of the biocontrol efficacy of a *Serratia marcescens* strain indigenous to tea rhizosphere for the management of root rot disease in tea. PLoS One 13(2):e0191761. https://doi.org/10.1371/journal.pone.0191761

Rabha AJ, Naglot A, Sharma GD et al (2014) In vitro evaluation of antagonism of endophytic *Colletotrichum gloeosporioides* against potent fungal pathogens of *Camellia sinensis*. Indian J Microbiol 54:302–309. https://doi.org/10.1007/s12088-014-0458-8

Resca R, Basaglia M, Poggiolini S, Vian P, Bardin S, Walsh UF et al (2001) An integrated approach for the evaluation of biological control of the complex *Polymyxa betae*/beet necrotic yellow vein virus, by means of seed inoculants. Plant Soil 12:215–226. https://doi.org/10.1023/A:1010310825363

Roobakkumar A, Babu A, Kumar DV, Rahman VJ, Sarkar S (2011) *Pseudomonas fluorescens* as an efficient entomopathogen against *Oligonychus coffeae* Nietner (Acari: Tetranychidae) infesting tea. J Entomol Nematol 3(7):73–77

Roy S, Muraleedharan N (2014) Microbial management of arthropod pests of tea: current state and prospects. Appl Microbiol Biotechnol 98(12):5375–5386. https://doi.org/10.1007/s00253-014-5749-9

Roy S, Rahman A, Sarma M, Babu A, Deka B (2018) Special bulletin-integrated management of major tea pests of Northeast India. Tocklai Tea Research Association, Tocklai

Roy S, Prasad AK, Neave S et al (2020) Nonchemical based integrated management package for live-wood eating termites in tea plantations of north-east India. Int J Trop Insect Sci 40:435–440. https://doi.org/10.1007/s42690-019-00095-6

Saito T, Takatsuka J, Shimazu M (2012) Characterization of *Paecilomyces cinnamomeus* from the camellia whitefly, *Aleurocanthus camelliae* (Hemiptera: Aleyrodidae), infesting tea in Japan. J Invertebr Pathol 110(1):14–23

Sandhu SS, Sharma AK, Beniwal V, Goel G, Batra P, Kumar A, Jaglan S, Sharma AK, Malhotra S (2012) Myco-biocontrol of insect pests: factors involved, mechanism, and regulation. J Pathog. https://doi.org/10.1155/2012/126819

Sarkar A, Das RC, Rashid MH, Ali M, Islam MS, Asaeda T, Wang Q (2015) Biocontrol potentiality of isolated *Trichoderma* spp. against *Pestalozzia theae* Saw. in tea. Acta Phytopathol Entomol Hung 50(2):179–186

Sarmah SR, Baruah PK, Das SC (2017) Practical utilization of botanical extracts and microbial in controlling dieback disease of tea [*Camellia sinensis* (L) O. Kuntze] caused by *Fusarium solani* (Mart.) Sacc. J Tea Sci Res 7(2):11–19. https://doi.org/10.5376/jtsr.2017.07.0002

Sarmah SR, Bhattacharyya PN, Barooah K (2020) Microbial biocides- viable alternatives to chemicals for tea disease management. J Biol Control 34(2):144–152. https://doi.org/10.18311/jbc/2020/22689

Selvasundaram R, Muraleedharan N (2000) Occurrence of the entomogenous fungus *Beauveria bassiana* on the short hole borer of tea. J Plant Crop 28(3):229–230

Senanayake PD, Kulathunga SD (2015) Potential entomopathogenic fungus for shot-hole borer control in tea cultivation. In: NSF research summit conference proceedings; empowered by research and innovation. BMICH, Colombo, p 123

Shan W, Zhou Y, Liu H, Yu X (2018) Endophytic Actinomycetes from tea plants (*Camellia sinensis*): isolation, abundance, antimicrobial, and plant-growth-promoting activities. BioMed Res Int. https://doi.org/10.1155/2018/1470305

Shanmugrapriyan R, Mathew S (2011) Bioefficacy of *Metarhizium anisopliae* and Derrimax against tea thrips. Newslett UPASI Tea Res Foundation 20(1):6

Shanmugrapriyan R, Mathew S, Babu A (2010) Bioefficacy of *Metarhizium anisopliae* on tea thrips. Newslett UPASI Tea Res Foundation 21:2

Singha D, Singha B, Dutta B (2011) Potential of *Metarhizium anisopliae* and *Beauveria bassiana* in the control of tea termite *Microtermes obesi* Holmgren in vitro and under field conditions. J Pest Sci 84(1):69–75

Sowndhararajan K, Marimuthu S, Manian S (2013a) Integrated control of blister blight disease in tea using the biocontrol agent *Ochrobactrum anthropi* strain BMO-111 with chemical fungicides. J Appl Microbiol 114(5):1491–1499

Sowndhararajan K, Marimuthu S, Manian S (2013b) Biocontrol potential of phylloplane bacterium *Ochrobactrum anthropi* BMO -111 against blister blight disease of tea. J Appl Microbiol 114:209–218

Sun X (2015) History and current status of development and use of viral insecticides in China. Viruses 7(1):306–319. https://doi.org/10.3390/v7010306

Sun JD, Xu DY, Zhang YM (1988) *Euproctis pseudoconspersa* nuclear polyhedrosis virus and its application in tea plantation. Fujian J Agric Sci 3:1–9

Sun JD, Wu GY, Lin AX, ZengMS XDY, Wang QS (1996) Efficiency of/*Euproctis pseudoconspersa* nuclear polyhedrosis virus (EpNPV) preparation. Entomol J East China 5:55–59

Takahashi M, Nakai M, Saito Y, Sato Y, Ishijima C, Kunimi Y (2015) Field efficacy and transmission of fast- and slow-killing nucleopolyhedroviruses that are infectious to *Adoxophyes honmai* (Lepidoptera: Tortricidae). Viruses 7(3):1271–1283. https://doi.org/10.3390/v7031271

Takatsuka J, Okuno S, Ishii T, Nakai M, Kunimi Y (2010) Fitness-related traits of entomopoxviruses isolated from *Adoxophyes honmai* (Lepidoptera: Tortricidae) at three localities in Japan. J Invertebr Pathol 105(2):121–131

Tanti A, Bhattacharyya P, Dutta P, Sarmah S, Madhab M, Saikia D, Kachari A, Berceroyjyrwa R (2016) Diversity of phylloplane microflora in certain tea cultivars of Assam, North-East India. Eur J Biol Res 6:287–292

Verma C, Jandaik S, Gupta BK, Kashyap N, Suryaprakash VS, Kashyap S, Kerketta A (2020) Microbial metabolites in plant disease management: review on biological approach. Int J Chem Stud 8(4):2570–2581. https://doi.org/10.22271/chemi.2020.v8.i4ad.10026

Vitarana SI (2000) 75 years of research in entomology, acarology and nematology. In: Modder WWD (ed) Twentieth century tea research in Sri Lanka. Tea Research Institute of Sri Lanka, Talawakelle, pp 111–160

Vitarana SI, Nagahaulla SM, Abeysinghe LS, Vitana SB (1997) Environment friendly pest control agents: use of bio 1020 to control white grub damage on tea. Sri Lanka J Tea Sci 65(1–2):34–45

Wahab S (2004) Tea pests and their management with bio-pesticides. Int J Tea Sci 3:21–30

Wang GH, Xia SM, Han BY (2010) Survey on resources and appraisal of dominant species of natural enemies in tea plantations in Guizhou province. J Anhui Agric Univ 37:772–780

Worrall EA, Hamid A, Mody KT, Mitter N, Pappu HR (2018) Nanotechnology for plant disease management. Agronomy 8:285. https://doi.org/10.3390/agronomy8120285

Wu G, Sun JD (1994) A study and application of *Beauveria bassiana* (Bals.) Vuill. in the control of the tea brown weevil. China Tea 16(1):30–31

Wu GY, De SJ, Shen ZM, Aug LA, Yean XD, Shen WQ (1995) Application of *Beauveria bassiana* strain 871 in controlling tea weevil (*Myllocerus aurolineatus*). J Fujian Acad Agric Sci 10 (2):39–43

Wu Z, Su Q, Cui Y et al (2020) Temporal and spatial pattern of endophytic fungi diversity of *Camellia sinensis* (cv. Shu Cha Zao). BMC Microbiol 20:270. https://doi.org/10.1186/s12866-020-01941-1

Xie H, Feng X, Wang M, Wang Y, Kumar Awasthi M, Xu P (2020) Implications of endophytic microbiota in *Camellia sinensis*: a review on current understanding and future insights. Bioengineered 11(1):1001–1015. https://doi.org/10.1080/21655979.2020.1816788

Xu P, Stirling E, Li W et al (2020) Trade-off between stochastic and deterministic processes shifts from soil to leaf microbiome of tea plant. Res Square. https://doi.org/10.21203/rs.3.rs-28737/v1

Yaginuma D, Hiromori H, Hatsukade M (2006) Virulence of the entomopathogenic fungus *Beauveria brongniartii* to several life stages of the yellowish elongate chafer *Heptophylla picea* Motschulsky (Coleoptera: Scarabaeidae). Appl Entomol Zool 41(2):287–293

Yan X, Wang Z, Mei Y, Wang L, Wang X, Xu Q, Peng S, Zhou Y, Wei C (2018) Isolation, diversity, and growth-promoting activities of endophytic bacteria from tea cultivars of Zijuan and Yunkang-10. Front Microbiol 9:1848. https://doi.org/10.3389/fmicb.2018.01848

Yang J, Kloepper JW, Ryu CM (2008a) Rhizosphere bacteria help plants tolerate abiotic stress. Trends Plant Sci 14:1–4. https://doi.org/10.1016/j.tplants.2008.10.004

Yang W, Peters JI, Williams RO (2008b) Inhaled nanoparticles-a current review. Int J Pharm 356:239–247

Ye GY, Xiao Q, Chen M et al (2014) Tea: biological control of insect and mite pests in China. Biol Control 68:73–81

You J (2008) Dynamic distributes of endophytic fungi from *Camellia sinensis*. Guihaia 28 (1):82–85

Zhang Y, Yu X, Zhang W, Lang D, Zhang X, Cui G et al (2019) Interactions between endophytes and plants: beneficial effect of endophytes to ameliorate biotic and abiotic stresses in plants. J Plant Biol 62:1–13. https://doi.org/10.1007/s12374-018-0274-5

Chapter 20
Field Application of the Microbial Technology and Its Importance in Sustainable Development

Saloni Kunwar, Shristi Bhatt, Deepa Pandey, and Neha Pandey

Abstract Microorganisms are ubiquitous in nature and are a rich source of primary as well as secondary metabolites. The uniqueness of microorganisms and their unpredictable nature attracts them for more and more exploration for the welfare of the humans and society. Products formed by microbes are natural and have the ability to reduce problems like high cost of synthetic chemicals, environmental pollution, hazards to human health, etc., and are helpful in sustaining the environment by applying different microbial technologies and sustainability goals. These indigenous microorganisms are involved in biotechnological field applications such as sustainable agriculture (biofertilizers and PGPR), food technology, chemical technology, recombinant technology, and sustainable environment (wastewater treatment, micro- and nanoparticle synthesis, oil remediation, and radioactive treatment). Apart from this, various strains of microbes are also being modified genetically for defending many environmental sustainability aspects. This review focuses upon the applications of microbial technologies for sustainable development of environment that meets the needs of current generations without compromising the ability of future generation to meet their own needs. Microorganisms like *Micrococcus*, *Pseudomonas*, *Chromobacterium*, *Bacillus*, and many others play a major role in the development of a sustainable environment.

Keywords Microorganisms · Sustainable development · Microbial technologies · Environment

---

S. Kunwar · S. Bhatt · N. Pandey (✉)
Department of Biotechnology and Life Sciences, Graphic Era Deemed to be University, Dehradun, Uttarakhand, India
e-mail: neha.pandey@geu.ac.in

D. Pandey
Department of Zoology, Government PG College, Ranikhet, Uttarakhand, India

P. Bhatt et al. (eds.), *Microbial Technology for Sustainable Environment*,
https://doi.org/10.1007/978-981-16-3840-4_20

20.1 Introduction

Microorganisms are small creatures and consist of protozoa, fungi, viruses, microalgae, and bacteria. These microorganisms live in water, animal intestines, food, soil, and different environments (Mosttafiz and Rahman 2012). Microbes can survive extreme environmental challenges. There are many reasons why microorganisms are important, especially because what they produce is valuable to us (Liu 2020). These substances may be very immense substances (such as nucleic acids, proteins, carbohydrate polymers, and cells), or they may be smaller molecules. They are usually divided into metabolites (mainly) essential for vegetative growth and non-essential metabolites (minor). There are many kinds of microorganisms. They play an important contribution to develop sustainable environment and also play a vital role in a series of green processes and cleaner technologies (from biogeochemical cycles to various industrial productions) (Kuhad 2012).

The advanced era provides various means of technology in various fields, and the emergence of the microbes establishes themselves as an advantage for this century. Though the microbes came forefront during the 1660s, their uses in association with technologies marked the beginning of the microbial technology dimension. A microbial technology is defined as a technology that uses microbial system or the organisms or its derivatives for the manufacturing of a product or its modification for any specific purpose. The current situation of the biosphere drives the attention of using any newer technology for its sustainability, and since microbes are a part of the ecosystem, their application for the sustainable development has surrey advantage and acts as a feasible remedy for the global eradication of problems in agriculture and environment. Various microbial technologies including GMOs (genetically modified organisms), plant growth-promoting rhizobacteria (PGPR), biopesticides, and biofertilizer have the potential to resolve many environmental and agricultural issues such as in bioremediation of nutrients in soil, promoting healthy plant growth and maintaining good health of the soil. The deterioration of the biosphere calls for an urgent need to replenish it in a feasible manner.

20.2 Microbial Biotechnology and Its Applications

The application of Microbial Biotechnology in science aims at treatment of materials by microorganisms to produce useful products or processes. The technologies along with their applications are mentioned below.

20.2.1 Agriculture Technology

The widespread use of microorganisms in sustainable agriculture is due to the genetic dependence of plants on the beneficial functions provided by symbiotic inhabitants (Noble and Ruaysoongnern 2010). Plant microbe symbiosis is carried out through the analysis of their ecological effects, which is the best research method for fixing nitrogen (N_2) (Franche et al. 2009). Industrial microbiology has revolutionized agriculture through genetic engineering and related disciplines. The commonly used bacteria, *Bacillus thuringiensis* (insecticidal bacteria) and *Agrobacterium tumefaciens* usually produce corona choloma in dicots. The tumor-forming gene of *Agrobacterium tumefaciens* exists in tumor-inducing plasmid (Ti plasmid). These genes instruct plants to form opines (nutrient factors that bacteria cannot produce by themselves). Ti vector has extremely important value for introducing foreign genes into dicot plants to produce transgenic plants. However, Ti plasmids cannot successfully transfer genes into monocots. For example, they bypassed a problem and developed particle accelerator, which injects metal particles encapsulated with DNA into the host cells or plant cell, thereby avoiding this problem. Along with this, different techniques for microbial applications in agriculture sector are as follows.

20.2.1.1 Biofertilizers

Biofertilizers consist of microorganisms that facilitate the growth of the plant and is also required to meet the growing demands for food by protecting the crops from pathogens with the use of naturally derived fertilizers in soil (Youssef and Eissa 2014). Organic farming is one of the practices that allow microbes to maintain the biodiversity of the soil. Biofertilizers aid in nutrient uptake such as nitrogen and phosphorus and recycle the nutrients back to the soil via various mechanisms including nitrogen uptake, phosphate solubilization, mineralization, production of antibiotics, or degradation of compounds (Gopal et al. 2013). A metagenomic study revealed a core microbiome transfer therapy which provides resistance to crops from the diseases. The technique involves complete transfer of microorganisms by mixing disease-inhibiting soil with the diseased favorable soil. For example, in an experiment by Mendes and colleagues, soils suppressive to *Rhizoctonia solani* were mixed with diseased conducive soil in the ratio of 9:1 which proved to be effective against sugar beet infection. Other successful examples of this treatment include infection of common scab of potato and tobacco black root rot infection (Rosenzweig et al. 2012).

20.2.1.2 Plant Growth-Promoting Rhizobacteria (PGPR)

They are always in a symbiotic relationship with plant- and root-related bacteria. PGPR improve the utilization of nutrients by dissolving unusable forms of nutrients and by producing siderophores, thereby contributing to the transportation of nutrients. There are some examples how PGPR work as a microbial technology:

1. *PGPR as Disease-Suppressive Agents*. To enhance the disease suppression ability of the soil, PGPR secrete metabolites that aid in the protection of plants from various diseases. *Bacillus subtilis* GBO3 produces salicylic acid and jasmonic acid for defense pathway (Ryu et al. 2004). PGPR with *B. amyloliquefaciens* 937b are effective for tomato mottle virus (Murphy et al. 2003). *B. megaterium* IISRBP x17 from black pepper acts against *Phytophthora capsici (*Aravind et al. 2009*)*. *Bacillus subtilis* N11 together with composts controls *Fusarium* infection on banana roots (Zhang et al. 2011). *B. subtilis* (UFLA285) shows resistance against *R. solani* and induces leaf and root growth of cotton plants (Medeiros et al. 2011). *Paenibacillus polymyxa* SQR-21 controls the *Fusarium* wilt in watermelon (Ling et al. 2011). PGPR plays a vital role in restoring plants from blight virus of tomato,cucumber mosaic virus and pepper mottle virus and banana bunchy top virus (Harish et al. 2009). *Glomus mosseae* is found to be used against *Fusarium oxysporum*, the pathogen of basil plant root rot disease (Toussaint et al. 2008). *Psuedomonas fluorescens* and arbuscular mycorrhiza (fungi) reduced root rot disease and also assist in increasing the productivity of *Phaseolus vulgaris* (common bean) (Neeraj 2011).
2. PGPR as Plant Growth inducers

   PGPR showed much effective results when applied alone but was effective much more than usual when used in combination with arbuscular mycorrhiza (fungi) like *Glomus intraradices*; it leads to better nutrient absorption and improves physiological processes in lettuce under stress conditions. *Trifolium alexandrinum* is inoculated with *Rhizobium trifolii*; it increased nodulation under saline stressed condition (Antoun and Prevost 2005). Paul and Nair found that *Pseudomonas fluorescens* (MSP-393) have the ability to overcome the influence of soil with the production of osmolytes and salt-stress-induced proteins. *P. putida* can withstand high alkaline and saline condition by uptake of K^+, Mg^+, and Ca^{2+} and lowering the uptake of Na^+ which gained effectiveness in cotton production (Yao et al. 2010). Arbuscular mycorrhiza (fungi) along with nitrogen-fixing bacteria was effective for legumes in drought conditions, and also inoculation of rice crops with AM increased antioxidant and photosynthetic efficiency (Ruiz-Sanchez et al. 2010). *Pseudomonads* sp. improved the photosynthetic pigments and antioxidants in basil plants under drought condition. They increase the enzymes catalase, glutathione peroxidase, and ascorbate peroxidase activity and chlorophyll content in leaves under drought conditions (Heidari and Golpayegani 2012).

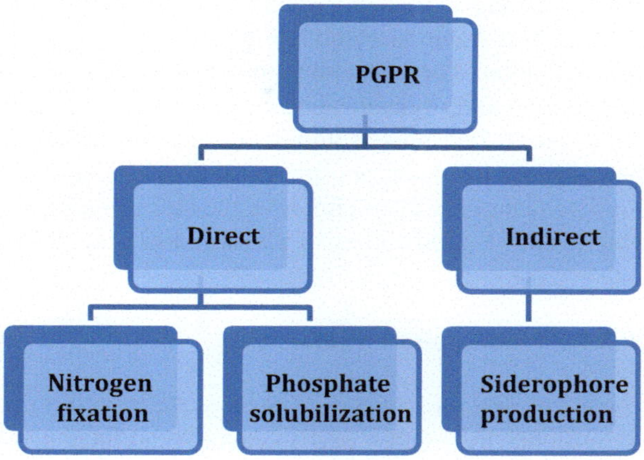

Fig. 20.1 Mechanism of PGPR

20.2.1.2.1 Mechanisms of Plant Growth PGPR

- Direct mechanism (by promoting nutrient acquisition such as nitrogen, phosphorus) or secretion of plant hormone levels.
- Indirect mechanism (by interfering with plant pathogen and promoting plant growth and development). The flowchart of PGPR mechanism is shown in Fig. 20.1.

20.2.1.3 Nitrogen Fixation

Plants cannot utilize atmospheric nitrogen directly, so it needs to be converted by a process called nitrogen fixation, where nitrogenase enzyme is used to produce bacteria such as *Rhizobium* and *Cyanobacteria*. *Rhizobium* forms a symbiotic association with leguminous plant roots by the formation of nodules which act as colonizing sites of rhizobia (Giordano and Hirsch 2004). In non-leguminous plants, rhizobacteria called as diazotrophs form a non-mandatory interaction with the host (Glick 2012).

20.2.1.4 Phosphate Solubilization

Phosphorus occurs in insoluble form in soil which is not assimilated by plants. So insoluble phosphorus needs to be converted into soluble form to be utilized by plants. Soluble phosphorus as monobasic (HPO_4) and dibasic (H_2PO_4) is utilized by plants. Phosphate is provided to soil in fertilizers that is converted into insoluble form in soil, causing phosphorus deficiency in soil; only 0.1% of phosphorus is available for plant usage. Bacteria such as *Flavobacterium*, *Bacillus*, *Azotobacter*, *Erwinia*, *Microbacterium*, *Enterobacter*, *Beijerinckia*, *Pseudomonas*, *Burkholderia*,

Rhizobium, and *Serratia* are phosphate-solubilizing bacteria (Bhattacharyya and Jha 2012). These bacteria solubilize phosphorus by two mechanisms: solubilization (hydrolyzation of organic and inorganic insoluble phosphorus compounds to soluble phosphorus) and mineralization (conversion of organic phosphorus into inorganic phosphorus). Phosphate-solubilizing bacteria (PSB) secrete phosphatase enzyme that converts insoluble phosphorus into soluble phosphorus by dissolving it. PSB promote plant growth and improve the utilization of other trace elements in the soil which yields a good variety of plants (Zaidi et al. 2009).

20.2.1.5 Siderophore Production

Iron is important for all organisms. Iron (Fe^{3+}) in the environment is present in form of insoluble hydroxides and oxyhydroxides that makes it difficult for plants to uptake them from soil. Bacteria secrete water-soluble siderophores (low-molecular mass iron chelators) which complex with iron and form siderophore-Fe complex. This complex reduces Fe^{3+} to Fe^{2+} and it is introduced into cell through channels in membrane and ultimately siderophore gets destroyed or used up once again (Rajkumar et al. 2010). Bacterial siderophores also reduce heavy metals such as aluminum, cadmium, copper, gallium, indium, lead, zinc, uranium, and neptunium (Neubauer et al. 2000).

20.2.2 Food Technology

Microorganisms have two different roles. First, they play a prime role in fermentation (in this case, genetically modified organisms are not allowed). Second, they become absorbed in the industries to produce food ingredients. Genetic engineering is used to modify the yeast and improve its performance in the fermentation process. Yeasts are optimized to changes in temperature, pH, and high yields on a wide range of products. Amylases are acquired from fungus *Aspergillus niger* (Adejuwon et al. 2015) or bacteria *Bacillus subtilis*; for example, they were used to replace the chemical additives for processing wheat flour, improving dough preparation for baking foods to be possessed (Bueno et al. 2016). The protein extracted from SCP is used as supplements in whole foods, replacing valuable traditional sources and solving the protein deficiency problem. In animal and human food, the single cell protein is used as a source of protein (Sadiku et al. 2019).

Emerging need of the increase in production of food allowed the use of agricultural farming techniques with chemical-based fertilizers, but this did not prove itself to be beneficial for the environment as it offered itself a large number of drawbacks with its extensive and prolonged use. Here the microbial technology stepped forward and in comparison to conventional methods is a much safer practice; it showed a 10–20% increase in economically important crop production. Despite its great advantage, limits itself to the need of the strain of the organism, the selection, and the application of the technology required for a better understanding of the relationship between the inoculants and the microbiome. The understanding of their relationship is important for the use of new technologies involving microfluidics-based technologies like 'microbiome on a chip'; this foregrounds multitrophic plant-microbiome interactions with the expression of environmental parameters and the host response against the treatments (Stanley and van der Heijden 2017). It can induce long shelf-lives and also enhance the effectiveness of the microbial product. One way of accomplishing this goal is by modifying the plant at an early stage (seedling stage) by incorporating the desired bacteria (Mitter et al. 2017). This technique is more reliable and has great advantages including the expression of desired and favorable traits and protection of the plants from another microbiota. The use of the synthetic microbial communities promotes early flowering, enhances nutrient acquisition by plants, and induces resistance to plants (Gopal et al. 2013), but the maintenance of the microbe in plant is achieved by genetic engineering and plant breeding techniques. The plant interacts with the microbes in soil by the release of components which are specific in nature and allows interaction with only the required organism, but the incorporation of the desired organism makes it easy for plants to interact and come up with expected outcomes.

20.2.3 Chemical Technology

Chemicals like organic acids via activity of microbes are very bright. Most organic acids are natural products present in important metabolic pathways or intermediates of microbial metabolism (Sauer et al. 2008). For example, global annual industrial-scale production of citric acid is demanded by the market as a food additive is through glucose fermentation using *Aspergillus niger* to form cane molasses, corn starch, or beet molasses (Wang et al. 2016). In addition to this, the lactic acid fermentation process has recently received more attention due to the increasing demand for new biomaterials, such as biocompatible polylactics and biodegradable products (Gao et al. 2011). The production of butanol and acetone, which was effectively carried out by the genus *Clostridium*, was one of the industrial fermentation processes of global importance, but this production led to chemical synthesis. Likewise, the inability of a chemical compound to compete with a petrochemical has affected the microbial composition of glycerol feedstocks.

20.2.4 Recombinant Technology

The main microbial hosts for the production of recombinant proteins are *Hansenula polymorpha*, *Bacillus subtilis*, *Aspergillus niger*, *S. cerevisiae*, *E. coli*, and *Pichia pastoris*. The recombinant microorganisms have provide the methods for the host to produce glycosylated recombinant proteins with high bacterial content, such as mammalian and insect cell cultures, as well as transgenic animals and plants. Plant breeding along with genetic engineering has become a common practice for the development of in-demand product, and its establishment has contributed to eliminate hunger and poverty on a global scale. The introduction of the GM crops was first developed in the mid-1990s in the USA. Today this technique is adopted globally to meet the increasing demand of the population and a way to move forward toward attaining sustainable production of crops. The GM crops are regulated by three agencies: EPA (Environmental Protection Agency), FDA (Food and Drug Administration), and USDA (US Department of Agriculture). The development of GM crops decreased the use of chemically originate herbicides and pesticides. Besides, the newly modified crop was herbicide resistant naturally. The administration of a soil bacterium called *Bacillus thuringiensis* (Bt) allowed modification in the plants to make them insect resistant; such plants are popularly known as Bt crops. Globally today many forms of Bt plants are available in the market; the advantage of such engineering is that it produced plants of superior quality with the extension of its shelf-life. Some of the crops considered under GMO crops are rice, tomato, cotton, soybean, maize, etc. The crops to be modified are targeted with the bacterium containing the desired gene; primarily the gene is introduced in the crop in vitro, and once the crop attains the growth, it is subjected to the field and allowed to grow in natural conditions. Such crops are usually herbicide tolerant and insect resistant; the commonly used bacteria are *Agrobacterium tumefaciens* and *Bacillus thuringiensis*.

20.2.5 Environmental Health and Microbial Technology

Environmental microbiology is the study of the composition and physiology of the microbial community in the environment. The various microorganisms have been recorded during the solid waste composting process, including autotrophic or heterotrophic aerobic bacteria, fecal coliform bacteria, thermophiles, yeast, actinomycetes, and other fungi (Tiquia et al. 2002). Treatment of waste is based on enzymatic processes and is inexpensive; however, these enzymes are biodegradable, so further research is needed for microbial enzymes that are thermally stable or resistant to significant changes in pH (Hasan et al. 2006).

20.2.6 Wastewater Treatment

Water is essential to sustain the life-forms on earth. Water accounts for 71% of the surface of earth; about 96.5% of water found in the oceans is unfit for the sustainability of life-forms. About 0.3% water is available as suitable for human consumption, which is obtained from rivers and lakes, and only 0.61% of water is available in groundwater. According to a study, 321 billion gallons of water is consumed per day by human beings and 77 billion gallons of water is taken up from groundwater.

Wastewater contains organic and inorganic pollutants which are lethal to the biota system; therefore, prior to discharge the wastewater requires treatment. Modern approach used to treat wastewater is microbial fuel cell (MFC) (Rabaey et al. 2010). MFC has several advantages such as:

1. It restores electrical energy and valuable products.
2. It generates nontoxic effluents.
3. It is easy to monitor and control and provides easy monitoring of the system.

MFC uses microbes at the electrode (either cathode or anode) to catalyze oxido-reduction and generate electric current. The reducing microbes reduce the electrode and donate electrons to the negative electrode (anode). These microbes reduce substrate from the wastewater required for the reaction by microbes to produce energy and provide a healthy biosphere (Clark and Pazdernik 2016).

20.2.7 Oil Remediation

With the increase in the world's population, the demand for the resources also increases; therefore, there is a subsequently increased need for petroleum and petroleum-based products, but accidents related to petroleum and petroleum-based products are also hazardous to all life-forms including marine lives and humans and plants (Strong and Burgess 2008). There have been numerous incidents pertaining to petroleum products as the 1971 incident in Pennsylvania where a gasoline pipeline was damaged and it released 100,000 gallons of the gasoline into the nearby water supply.

The first application of the use of this technique was the Exxon Valdez oil spill which proved to be effective to clean up the petroleum-contaminated ground system; this gained the interest in the use of bioremediation to clean up the environment in a safer way. Petroleum is a natural resource obtained beneath the earth layer which is an admixture of nitrogen, sulfur, and oxygen. It constitutes a variety of compounds such as aromatic compounds, some metals (iron, copper, nickel, vanadium), alkanes, and cycloalkanes as shown in Table 20.1.

The permeation of these compounds into the groundwater or surface water can impose serious health hazards to the biosphere, and the uptake of these compounds by the living organisms can affect their health. Gasoline is considered as a cancer-causing agent to humans by the International Agency for Research on Cancer (IARC) (2000) and tends to cause irritation in the eyes and mucous membrane.

Table 20.1 Composition of compounds present in petroleum by weight

Hydrocarbon	Amount present (%)
Alkanes	30
Naphthalenes	49
Aromatics	15
Asphaltics	6

Table 20.2 Illustrates various petroleum compounds and their effects with the concentration

Petroleum compounds	Effect on humans	Concentration (parts per million)
Benzene	Leukemia	<1
Toluene	Memory loss and coordination impairment Palpations	200–500 500–1000
Gasoline	Cancer	>2000
Cyclohexane	Polyneuropathy	<1000
Other hexanes	Narcosis	1000

Table 20.2 shows various compounds of petroleum and their harmful effects on humans.

20.2.7.1 Remediation of the Petroleum Products

The conventional method of the remediation of petroleum products includes dispersion, sorption, volatilization, recovery, dilution, and abiotic modification of hydrocarbon removal. These require high capital investment and large machineries and also dispose the residues into the environment (Matsumiya and Kubo 2007). On the other hand, bioremediation generates ecofriendly residues and does not require much capital; bioremediation techniques of petroleum products make the affected site free of every contaminant.

20.2.7.2 Petroleum Hydrocarbon Degradation Mechanism

The initial step of the degradation in petroleum hydrocarbons is the oxygenation process in which the organic pollutants are attacked intracellularly and the oxygen is transferred within the cell by the enzymes like oxygenase and peroxidases, followed by the conversion of the pollutants into various intermediates through a number of degradation pathways such as TCA (tricarboxylic cycle). This method is mediated by the enzymes and hence is considered as an enzymatic-dependent mechanism. Below are some of the examples that show the enzymes produced by various microorganisms:

1. *Methylococcus* produces soluble methane and particulate methane that act on C1 to C8 of cycloalkanes, alkenes, and alkanes (McDonald et al. 2006).
2. *Burkholderia* species produces alkanes that act on C5–C16 of fatty acids, alkanes, cycloalkanes, and alkyl benzenes. *Rhodococcus* and *Mycobacterium* produce hydroxylases that act on C5–C16 of alkyl benzenes, fatty acids, cycloalkanes, and alkanes (Jan et al. 2003).

Table 20.3 Microorganisms producing biosurfactants

Biosurfactant	Microorganism	References
Lipomannan	*Candida tropicalis*	Ilori et al. (2005)
Glycolipid	*Aeromonas* sp.	Youssef et al. (2007)
Surfactin	*Bacillus subtilis*	Daverey and Pakshirajan (2009)
Sophorolipids	*Candida bombicola*	Kumar et al. (2008)
Rhamnolipids	*Pseudomonas aeruginosa*	Das and Chandran (2011)

3. *Acinetobacter* produces dioxygenases that act on C10–C30 of alkanes (Maeng et al. 1996).
4. Species of *Caulobacter*, *Mycobacterium*, and *Acinetobacter* produce bacterial oxygenase P450 that acts upon C5 to C16 of cycloalkanes and alkanes (Van Beilen et al. 2006).

Other methods include the production of biosurfactants. Biosurfactants are agents produced by some microorganisms that form micelles by reducing the surface area given in Table 20.3. In this method the cell surface is encapsulated with microdroplets that absorb the petroleum products and degrade them. It was observed that 90% of hydrocarbons was degraded within a time period of 6 weeks in vitro (Cameotra and Singh 2008); with further alteration it was shown that most of the hydrocarbons is degraded with the application of crude biosurfactant (Muthuswamy et al. 2008).

The recent advancement offers application of plasmid DNA incorporation for the biodegradation of hydrocarbons. Plasmid DNA is a mobile DNA that can be transferred by the processes of conjugation and transformation and has the ability to express the quality of the incorporated DNA and change the phenotypic expression of the host. This has been successfully exhibited on *Pseudomonas* sp. by incorporating plasmid DNA for the nutrition of various compounds such as octane, naphthalene, toluene, camphor, salicylate, and xylene. Further incorporation of plasmid DNA has been able to degrade the recalcitrant compounds of petroleum. The plasmid DNA enhances the potential of the recipient after getting incorporated inside the recipient (Okoh 2006).

20.2.8 Radioactive Waste

Radioactive compounds are widely used in every field like industry, research and development, and medical sectors as an innovative technique but their disposal is very heedful. Accumulation of radioactive waste poses a threat to all life-forms. To remediate the pollutant in soil and water, bioremediation has become a successful method. Microbial-associated techniques aim to reduce radionuclides significantly; the genetically engineered (GE) microorganisms affect the properties of

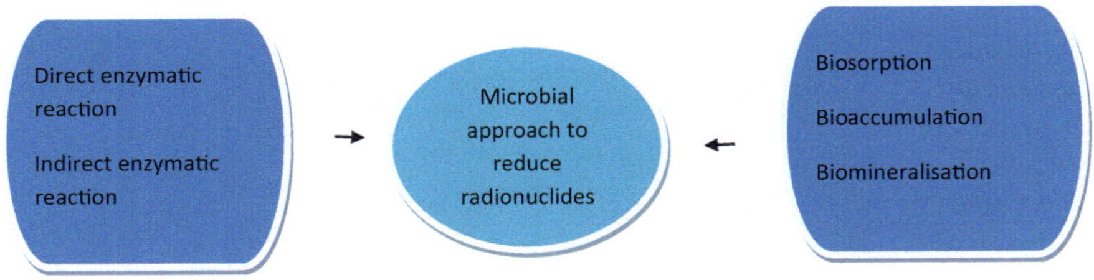

Fig. 20.2 Various microbial techniques for bioremediation of radionuclides

radionuclides and thereby reduce their concentration. The following are the common radionuclides that are released by industrial and biomedical wastes:

Cobalt-60
Uranium-238
Thorium-232
Radium-226
Radon-222
Plutonium-239
Technetium-99

Among this technetium is the most commonly used isotope in medical imaging and has a shelf-life of 6 h, but other isotopes like Uranium-238 have a half-life of 1600 years (Kurnaz et al. 2007). The radioactive wastes not only affect the health of life-forms but also deteriorate the nutrient level in soil if remained for a longer period.

A common way by which soil gets contaminated with radioisotopes is by dumping of radionuclides into the environment along with other wastes. The physiochemical ways are practiced for a long time but failed to be resolved; therefore, bioremediation came forth to act as an alternative way to remediate the soil from radionuclides. In view to this, Fig. 20.2 gives details of different methods for bioremediation of radionuclides. Some microorganisms act directly on the polluted site and convert the radionuclides into soluble product by either oxidation or reduction, allowing their fast removal from soil. Bacteria like *Rhodanobacter* sp. and *Desulfuromusa ferrireducens* act directly on the pollutant (Green et al. 2012).

20.3 Importance of Microbial Technology in Sustainable Development

The microorganisms play important roles in the environment. In recent studies, every possible industrial process, the use of chemicals, the increased use of nonrenewable energy, and the uncontrolled production of waste products pose a huge threat to environmental sustainability. Now, the world has a greater responsibility to adopt cleaner production, green technologies, and sustainable measures in order to protect the earth's ecology for future generations (Kuhad 2012).

Microorganisms play an important role in the urban ecosystem. There are various resident microorganisms in the urban ecosystem, which play an important role like waste management, soil management, industrial productivity, bioremediation, health and disease and marine pollution management, etc. (King 2014).

- By using wastes such as municipal waste, agricultural waste, and sewage sludge, microorganisms can also be used to produce bioenergy (Appels et al. 2011). Microorganisms such as *Penicillium*, *Aspergillus*, *Trichoderma*, and *Clostridium* are highly effective (Elshahed 2010).
- Solid waste management is the biological conversion of its organisms into useful products such as biofuels and biogas. Using microbial *Thermoactinomyces*, *Pseudomonas*, *Actinobifida*, *Microbispora*, and *Bacillus* to compost solid waste as an economically feasible method to convert its organic compounds into useful products. Compost is used as fertilizer for the production of crops, thereby increasing their product activity and developing a sustainable environment.
- Microorganisms such as *Rhizobium* are used as biological synthetic materials and increase the productivity of industrial agriculture.
- The algae *Nostoc* and *Azolla* are used as economical viable sources of bioremediation, which can then be used to produce biofuels.
- GM *E. coli* produces large amounts of insulin.
- Microorganisms show their role in infant health. *Bifidobacterium* and *Lactobacillus* are microbes which are important for the regulation of human immune system (Romano-Keeler and Weitkamp 2015).
- Microorganisms are also used to generate clean electricity. For example, *Geobacter sulfurreducens* and *Shewanella oneidensis* are used to generate usable electricity (Lal 2013).
- Microorganisms such as *Micrococcus*, *Pseudomonas*, *Chromobacterium*, *Bacillus*, *Arthrobacter*, *Candida*, and *Burkholderia* degrade hydrocarbons and crude oil through a method called intrinsic repair without any artificial enhancement (Kumar and Gopal 2015).
- Microorganisms like *Marinobacter*, *Pseudomonas*, *Bacillus*, *E. coli*, and *Streptomyces* help in the remediation of heavy metals (arsenic, mercury, and lead) from waterbodies and control the pollution in marine waters.

20.4 Conclusion

Microbial technology has unique applications in majority of areas including its effectiveness in achieving a sustainable environment. It has not limited its application only in healthcare but has evolved in other sectors too. The present scenario regarding the issues on agricultural and environmental sustainability has increased abruptly with the advancement of urbanization. Meeting the challenges of modernization has greatly deteriorated the conditions of the biosphere. In agriculture, microbial technology has contributed to the production of improved crop varieties. It produces crops with an admixture of different types of crops, resulting in improved version of the crops containing the dominant and admirable traits of the different crops. It has also contributed in increasing nutritional content of the product by manipulating the genes of the product through genetic engineering. In environment, microbial technology has contributed very much in eliminating the pollutants from the contaminated sites which includes the removal of oil spills from massive oceans without generation of any additional toxic residues and also in significantly removing heavy metals from unfertile land.

Resources are also replenished more conveniently by microbial technology than conventional methods since it used microbes that naturally conserve resources or generate such product. Although microbial technology has some drawbacks due to which it has not come forward more distinctively, with the pace in the biotechnology, it may become the only possible, reliable, and ultimate method to resolve all the issues relating to sustainability in agriculture and environment and accomplish the sustainability goals of the UN to meet the demands of the developing urbanization.

References

Adejuwon AO, Oluduro AO, Agboola FK, Olutiola PO, Burkhardt BA, Segal SJ (2015) Expression of a-amylase by Aspergillus niger: effect of nitrogen source of growth medium. Adv Biosci Bioeng 3:12–19

Antoun H, Prevost D (2005) Ecology of plant growth promoting rhizobacteria. In: Siddiqui ZA (ed) PGPR: biocontrol and biofertilization. Springer, Netherlands, pp 1–38

Appels L, Lauwers J, Degrve J, Helsen L, Lievens B, Willems K, Van Impe J, Dewil R (2011) Anaerobic digestion in global bio-energy production: potential and research challenges. Renew Sust Energ Rev 15:4295–4301

Aravind R, Kumar A, Eapen SJ, Ramana KV (2009) Endophytic bacterial flora in root and stem tissues of black pepper (Piper nigrum L.) genotype: isolation, identification and evaluation against Phytophthora capsici. Lett Appl Microbiol 48:58–64

Bhattacharyya PN, Jha DK (2012) Plant growth-promoting rhizobacteria (PGPR): emergence in agriculture. World J Microbiol Biotechnol 4:1327–1350

Bueno MM, Thys RCS, Rodrigues RC (2016) Microbial enzymes as substitutes of chemical additives in baking wheat flour-part II: combined effects of nine enzymes on dough rheology. Food Bioprocess Technol 9:1598–1611

Cameotra SS, Singh P (2008) Bioremediation of oil sludge using crude biosurfactants. Int Biodeterior Biodegrad 62:274–280

Clark DP, Pazdernik NJ (2016) Environmental biotechnology: biotechnology (second edition) applying the genetic revolution. Elsevier, Amsterdam, pp 393–418

Das N, Chandran P (2011) Microbial degradation of petroleum hydrocarbon contaminants: an overview. Biotechnol Res Int 2011:941810

Daverey A, Pakshirajan K (2009) Production of sophorolipids by the yeast Candida bombicola using simple and low cost fermentative media. Food Res Int 42:499–504

Elshahed MS (2010) Microbiological aspects of biofuel production: current status and future directions. J Adv Res 1:103–111

Franche C, Lindstrom K, Elmerich C (2009) Nitrogen-fixing bacteria associated with leguminous and nonleguminous plants. Plant Soil 321:35–59

Gao C, Ma C, Xu P (2011) Biotechnological routes based on lactic acid production from biomass. Biotechnol Adv 29:930–939

Giordano W, Hirsch AM (2004) The expression of MaEXP1, a Melilotus alba expansin gene, is upregulated during the sweet clover- Sinorhizobium meliloti interaction. Mol Plant-Microbe Interact 6:613–622

Glick BR (2012) Plant growth-promoting bacteria: mechanisms and applications. Hindawi Publishing Corporation, Scientifica, Cairo

Gopal M, Gupta A, Thomas GV (2013) Bespoke microbiome therapy to manage plant diseases. Front Microbiol 5:15

Green SJ, Prakash O, Jasrotia P, Overholt WA, Cardenas E, Hubbard D et al (2012) Denitrifying bacteria from the genus Rhodanobacter dominate bacterial communities in the highly contaminated subsurface of a nuclear legacy waste site. Appl Environ Microbiol 78:1039–1047

Harish S, Kavino M, Kumar N, Balasubramanian P, Samiyappan R (2009) Induction of defense-related proteins by mixtures of plant growth promoting endophytic bacteria against Banana bunchy top virus. Biol Control 51:16–25

Hasan F, Shah AA, Hameed A (2006) Industrial applications of microbial lipases. Enzym Microb Technol 39:235–251

Heidari M, Golpayegani A (2012) Effects of water stress and inoculation with plant growth promoting rhizobacteria (PGPR) on antioxidant status and photosynthetic pigments in basil (Ocimum basilicum L.). J Saudi Soc Agric Sci 11:57–61

Ilori MO, Amobi CJ, Odocha AC (2005) Factors affecting biosurfactant production by oil degrading Aeromonas spp isolated from a tropical environment. Chemosphere 61:985–992

International Agency for Research on Cancer (IARC) (2000) Monographs on the valuation of the carcinogenic risk of some industrial chemicals to humans. IARC, Lyon

Jan B, Beilen V, Neuenschwunder M, Suits THM, Roth C, Balada SB, Witholt B (2003) Rubredoxins involved in alkane degradation. J Bacteriol 184:1722–1732

King GM (2014) Urban microbiomes and urban ecology: how do microbes in the built environment affect human sustainability in cities? J Microbiol 52(9):721–728

Kuhad R (2012) Microbes and their role in sustainable development. Indian J Microbiol 52:309–313

Kumar BL, Gopal DVRS (2015) Effective role of indigenous microorganisms for sustainable environment. 3 Biotech 5(6):867–876

Kumar M, Leon V, De Sisto MA, Ilzins OA, Luis L (2008) Biosurfactant production and hydrocarbon degradation by halotolerant and thermotolerant Pseudomonas sp. World J Microbiol Biotechnol 24:1047–1057

Kurnaz A, Kucukomeroglu B, Keser R, Okumusoglu NT, Korkmaz F, Karahan G, Cevik U (2007) Determination of radioactivity levels and hazards of soil and sediment samples in Firtina Valley (Rize, Turkey). Appl Radiat Isot 65:1281–1289

Lal D (2013) Microbes to generate electricity. Indian J Microbiol 53:120e122

Ling N, Huang Q, Guo S, Shen Q (2011) Paenibacillus polymyxa SQR-21 systemically affects root exudates of watermelon to decrease the conidial germination of Fusarium oxysporum f.sp. niveum. Plant Soil 341:485–493

Liu X (2020) Microbial technology for the sustainable development of energy and environment. Biotechnol Rep 27:486

Maeng JHO, Sakai Y, Tani Y, Kato N (1996) Isolation and characterization of a novel oxygenase that catalyzes the first step of n-alkane oxidation in Acinetobacter sp. strain M-1. J Bacteriol 178:3695–3700

Matsumiya Y, Kubo M (2007) Bioprocess handbook, 1st edn. NTS Publishing, Tokyo

McDonald IR, Miguez CB, Rogge G, Bourque D, Wendlandt KD, Groleau D, Murrell JC (2006) Diversity of soluble methane monooxygenase-containing methanotrophs isolated from polluted environments. FEMS Microbiol Lett 255:225–232

Medeiros FHV, Souza RM, Medeiros FCL, Zhang H, Wheeler T, Payton P, Ferro HM, Paré PW (2011) Transcriptional profiling in cotton associated with Bacillus subtilis (UFLA285) induced biotic-stress tolerance. Plant Soil 347:327–337

Mitter B, Pfaffenbichler N, Flavell R, Compant S, Antoniellli L, Petric A et al (2017) A new approach to modify plant microbiomes and traits by introducing beneficial bacteria at flowering into progeny seeds. Front Microbiol 8:11

Mosttafiz S, Rahman M (2012) Biotechnology: role of microbes in sustainable agriculture and environmental health. Internet J Microbiol 10:1937

Murphy JF, Reddy MS, Ryu CM, Kloepper JW, Li R (2003) Rhizobacteria mediated growth promotion of tomato leads to protection against cucumber mosaic virus. Phytopathology 93:1301–1307

Muthuswamy K, Gopalakrishnan S, Ravi TK, Sivachidambaram P (2008) Biosurfactants: properties, commercial production and application. Curr Sci 94:736–747

Neeraj KS (2011) Organic amendments to soil inoculated arbuscular mycorrhizal fungi and Pseudomonas fluorescens treatments reduce the development of root-rot disease and enhance the yield of Phaseolus vulgaris L. Eur J Soil Biol 47:288–295

Neubauer U, Furrer G, Kayser A, Schulin R (2000) Siderophores, NTA, and citrate: potential soil amendments to enhance heavy metal mobility in phytoremediation. Int J Phytoremediat 2 (4):353–368

Noble AD, Ruaysoongnern S (2010) The nature of sustainable agriculture. In: Soil microbiology and sustainable crop production. Springer, Dordrecht, pp 1–25

Okoh AI (2006) Biodegradation alternative in the cleanup of petroleum hydrocarbon pollutants. Biotechnol Mol Biol Rev 1:38–50

Rabaey K, Angenent L, Schroder U, Keller J (2010) Bioelectrochemical systems: from extracellular electron transfer to biotechnological application. IWA Publishing, London

Rajkumar M, Ae N, Prasad MNV, Freitas H (2010) Potential of siderophore-producing bacteria for improving heavy metal phytoextraction. Trends Biotechnol 28:142–149

Romano-Keeler J, Weitkamp J-H (2015) Maternal influences on fetal microbial colonization and immune development. Pediatr Res 77:189–195

Rosenzweig N, Tiedje JM, Quensen JF, Meng QX, Hao JJJ (2012) Microbial communities associated with potato common scab-suppressive soil determined by pyrosequencing analyses. Plant Dis 96:718–725

Ruiz-Sanchez M, Aroca R, Munoz Y, Polon R, Ruiz-Lozano JM (2010) The arbuscular mycorrhizal symbiosis enhances the photosynthetic efficiency and the antioxidative response of rice plants subjected to drought stress. J Plant Physiol 167:862–869

Ryu CM, Farag MA, Hu CH, Reddy MS, Kloepper JW, Pare PW (2004) Bacterial volatiles induce systemic resistance in Arabidopsis. Plant Physiol 134:1017–1026

Sadiku MNO, Ashaolu TJ, Musa SM (2019) Food microbiology. Int J Trend Sci Res Dev 3:837–838

Sauer M, Porro D, Mattanovich D, Branduardi P (2008) Microbial production of organic acids: expanding the markets. Trends Biotechnol 26:100–108

Stanley CE, van der Heijden MGA (2017) Microbiome-on-a-chip: new frontiers in plant-microbiota research. Trends Microbiol 25:610–613

Strong PJ, Burgess JE (2008) Treatment methods for wine-related ad distillery wastewaters: a review. Biorem J 12:70–87

Tiquia SM, Wan HC, Tam NFY (2002) Microbial population dynamics and enzyme activities during composting. Compost Sci Util 10:150–161

Toussaint JP, Kraml M, Nell M, Smith SE, Smith FA, Steinkellner S, Schmiderer H, Novak V (2008) Effect of Glomus mosseae on concentrations of rosmarinic and caffeic acids and essential oil compounds in basil inoculated with Fusarium oxysporum f. sp. basilica. Plant Pathol 57:1109–1116

Van Beilen JB, Funhoff EG, Van Loon A, Just A, Kaysser L, Bouza M, Holtackers R, Röthlisberger M, Li Z, Witholt B (2006) Cytochrome P450 alkane hydroxylases of the

CYP153 family are common in alkane-degrading eubacteria lacking integral membrane alkane hydroxylases. Appl Environ Microbiol 72:59–65

Wang L, Cao Z, Hou L, Yin L, Wang D, Gao Q et al (2016) The opposite roles of agdA and glaA on citric acid production in Aspergillus niger. Appl Microbiol Biotechnol 100:5791–5803

Yao L, Wu Z, Zheng Y, Kaleem I, Li C (2010) Growth promotion and protection against salt stress by Pseudomonas putida Rs-198 on cotton. Eur J Soil Biol 46:49–54

Youssef MMA, Eissa MFM (2014) Biofertilizers and their role in management of plant parasitic nematodes. A review. E3 J Biotechnol Pharm Res 5:1–6

Youssef N, Simpson DR, Duncan KE, McInerney MJ, Folmsbee M, Fincher T, Knapp RM (2007) In situ biosurfactant production by Bacillus strains injected into a limestone petroleum reservoir. Appl Environ Microbiol 73:1239–1247

Zaidi A, Khan MS, Ahemad M, Oves M (2009) Plant growth promotion by phosphate solubilizing bacteria. Acta Microbiol Immunol Hung 56(3):263–284

Zhang N, Kai W, He X, Li S, Zhang Z, Shen B, Yang X, Zhang R, Huang Q, Shen Q (2011) A new bioorganic fertilizer can effectively control banana wilt by strong colonization with Bacillus subtilis N11. Plant Soil 344:87–97

Chapter 21
Solubilization of Micronutrients Using Indigenous Microorganisms

A. D. Sarangi N. P. Athukorala

Abstract Out of the 17 elements essential for plant growth and reproduction, 8 are micronutrients. The soil supplies relatively large amounts of nitrogen, phosphorus, potassium, calcium, magnesium, and sulfur as macronutrients and relatively small amounts of iron, manganese, boron, molybdenum, copper, zinc, chlorine, and cobalt, as micronutrients. Both deficiency and excess of micronutrients negatively impact the growth and productivity of plants and therefore should be supplied in sufficient amounts in appropriate ratios. A number of biotic and abiotic factors and their relationships affect the appropriate balance of macro- and micronutrient pool in the soil. The focus on addressing micronutrient deficiencies in soil in relation to agriculture has not been adequate in comparison to that for macronutrients. However, with the recent recognition on their impact on crop productivity and the efficiency of NPK uptake by plants, much attention was drawn to regulating micronutrient content in soil with chemical supplements. Environmental concerns encountered with the use of chemical supplements have directed the world into eco-friendly and sustainable approaches in addressing issues in many fields including agriculture. The use of naturally inhabiting microorganisms, "indigenous microorganisms," has been one of such eco-friendly approaches in agriculture. This review discusses the approaches that have been researched and used with indigenous microorganisms having micronutrient solubilization ability to regulate micronutrient availability in soil and the potential of developing them to optimize the crop productivity while maintaining a sustainable environment.

Keywords Micronutrients · Solubilization · Bioavailability · Indigenous microorganisms

A. D. S. N. P. Athukorala (✉)
Department of Botany, Faculty of Science, University of Peradeniya, Peradeniya, Sri Lanka
e-mail: sarangiathu@pdn.ac.lk

P. Bhatt et al. (eds.), *Microbial Technology for Sustainable Environment*,
https://doi.org/10.1007/978-981-16-3840-4_21

21.1 Introduction

The nutrients which are required by organisms in comparatively small quantities are called micronutrients. They are required in plant tissues at concentrations of less than 100μ g^{-1} dry weight (Welch and Shuman 1995). The attention to micronutrients has increased in the recent past with the understanding of their important role in disease resistance in plants and stress resistance in roots especially in crop plants (Welch and Shuman 1995; Graham and Webb 1991; Miller et al. 1991; Van Campen 1991; Nielsen 1992). Micronutrients are as equally important as macronutrients for plant growth, yield, and quality (Maurya et al. 2018; Yadav et al. 2018). There are eight micronutrients which have currently been recognized to be essential for higher plants, namely, boron (B), iron (Fe), copper (Cu), zinc (Zn), manganese (Mn), chlorine (Cl), molybdenum (Mo), and nickel (Ni) (Welch and Shuman 1995). However, an exact number of the micronutrients vital for higher plants cannot be strictly proposed since intense molecular and physiological studies are required for each and every nutrient before such conclusions and for elimination of controversies. One such example is Si where there are arguments whether it is to be categorized as an essential or a beneficial element (Maathuis 2013; Mengel et al. 2001; Barker and Pilbeam 2015). Micronutrients play a role in primary as well as in secondary metabolism, energy metabolism, cell defense, signal transduction, hormone perception, and gene regulation (Maathuis 2013; Barker and Pilbeam 2015; Maathuis and Diatloff 2013; Vatansever et al. 2017). They also enhance the chemical composition and the quality of plants including crops and are known to act as catalyst in various organic reactions in plants (Karthick et al. 2018). Since these play an important role in plant growth and development, the deficiencies result in several physiological disorders and diseases in plants that reduce the yield and quality of plant produce (Sharma 2006). It should be taken into account that a number of physiological events in relation to plant metabolism are directly or somewhat related to the mineral elements. Deficiencies of mineral elements or toxicities intensely affect the life cycle of plants, and their availability to plants especially crops should be closely looked at in order to address the food safety and food security issues that would upswing in the near future. With the United Nations sustainability goals to be met by 2030 with the already identified and not yet identified ecological, environmental, and health problems associated with chemical fertilization, eco-friendly alternatives for soil augmentation are being investigated from which mineral-solubilizing microorganism is one. Deficiencies in micronutrients in soil will ultimately be related to human and animal health risks by crops not having the required amount of micronutrients present in crop produce. This chapter will be focused on different mechanisms and methods by which soil indigenous microflora increases the micronutrient availability in soil, enhancing the soil fertility and crop production and thereby human health through a sustainable approach. When considering a sustainable environment, the role of microbes in other processes such as bioremediation and bioleaching in addition to biofortification and biofertilization should also be discussed. Therefore, the chapter will also discuss situations where indigenous microbes have been used in the above processes in relation to eight micronutrients.

21.2 Micronutrient Deficiencies

Deficiency of micronutrients in soil is a global issue with slight variations in particular to different micronutrients (Monreal et al. 2015; Voortman and Bindraban 2015; Dimkpa and Bindraban 2016). In addition, their low crop use efficiency (i.e., low crop response per unit of micronutrient, relative to no micronutrient application), typically $<10\%$ in comparison to 20% and 80% for N, P, and K (Baligar et al. 2001), has also contributed to their deficiency in global agro-ecosystems. Many arable lands are affected by deficiencies of more than one micronutrient (Monreal et al. 2015; Voortman and Bindraban 2015; Oliver and Gregory 2015). This issue has further been made complicated with extensive extraction by high yielded crops supplemented with NPK fertilizers in addition to inadequate micronutrient fertilization. Yield increments reported with the addition of micronutrient supplements in different crops both with and without NPK fertilizers (Dimkpa and Bindraban 2016; Katyal and Ponamperuma 1974; Kanwar and Youngdahl 1985; Rietra et al. 2015) suggest the importance of micronutrients in crop productivity and for effectiveness of NPK fertilization (Dimkpa and Bindraban 2016). An amount ranging from 0.01 to 4.9 kg ha^{-1} of micronutrients has been estimated to be collectively removed from the soil annually by different crops (Rietra et al. 2015) and the type of micronutrient (Mallarino et al. 2011; Marschner 2012). A normal growth of a plant requires an amount of each micronutrient ranging between 0.1 and 100 mg kg^{-1} with mean levels of 90 mg kg^{-1} present in DTPA-extractible form (Samourgiannidis and Matsi 2013; Sobral et al. 2013). Therefore, soils do not fulfill the micronutrient requirement of a normal healthy plant. These deficiencies in soil result in low crop productivity and nutritional quality which will ultimately affect human health (Marschner 2012; Alloway 2009; Itelima et al. 2018; Dhaliwal et al. 2019; Kaura et al. 2020).

Since human nutrition is directly or indirectly based on plants, any micronutrient deficiency in food crops could cause micronutrient deficiency in humans, referred to as "hidden hunger" (Oliver and Gregory 2015; Kaura et al. 2020; White and Broadley 2009; Joy et al. 2015; Riaz et al. 2020). For example, Zn deficiency in soil has been shown to cause Zn-deficient symptoms in humans, such as stunting and child death (Monreal et al. 2015; Cakmak 2008).

21.3 Micronutrients in Plants

The ability of a plant to obtain sufficient levels of vital minerals is also a function of specific characteristics in the plasma membranes of root cells such as the presence of relevant transport proteins and related acquisition mechanisms of ions (Vatansever et al. 2017; Dimkpa and Bindraban 2016; Kochian 1991). Modification of the rhizosphere by plant roots can also affect nutrient availability, through the release of protons, chelators, phytosiderophores, and/or chemical reductants and also by elaboration of extensive root systems (Dimkpa and Bindraban 2016; White et al. 2013; Keuskamp et al. 2015). Micronutrient availability for plants is not entirely dependent on the amount of mineral present in the soil matrix but also depends on the molar fraction existing in soil solution and on the variation of the ions of a particular mineral (Vatansever et al. 2017; Lindsay 1991). Speciation and solubility of a particular mineral are influenced by abiotic factors such as redox state, pH, and temperature as well as by biotic factors such as phenolic compounds and organic acids which are metabolically generated or released through degradation of soil organic matter by indigenous microorganisms. In the soil, some micronutrients react with compounds such as phosphates and carbonates, to form chemical precipitates or interact with clay particles and other mineral complexes, making them unavailable (Dimkpa and Bindraban 2016; Marschner 2012; Allen 2002).

Factors such as plant species, genotype, growth conditions, and different organs and tissues of the same plant species affect the micronutrient content of a plant. Genetic makeup together with physiological and environmental factors changes the concentrations of micronutrients inside the plants, deficiencies, or toxicities (Table 21.1). Environmental variables and differences between plant species, as well as genetic variation within a plant species, can affect micronutrient concentrations in higher plants (Luber and Taureau 1990; Benton-Jones 1991; Benton-Jones et al. 1992). Micronutrients play a significant role in both general and specific physiological processes in plants. Fe, Cu, Mn, and Cl are involved in different metabolic processes like photosynthesis acting as cofactors. Fe, Mn, Zn, Cu, Ni, Mo, and Cl contribute to the activity of different enzymes such as DNA/RNA polymerases, N-metabolizing enzymes, dismutases, catalases, superoxide, dehydrogenases, oxidases, ATPases, and enzymes involved in redox processes (Broadley et al. 2012). Zn specifically plays a role in the enzymatic biosynthesis of auxin (Hossain et al. 1997; Fageria 2002) which enhances root growth. Ni is involved in N metabolism of plants by transforming urea to ammonia (Polaccao et al. 1999; Sirko and Brodzik 2000). Mo is used by both symbiotic and free-living N-fixing bacteria for N fixation since it is a component of the nitrogenase enzyme system (Barron et al. 2009). The role of micronutrients as cofactors is crucial for enzyme and nonenzyme activities in plant metabolism depending on the environment especially in abiotic and biotic stress mitigation by plants. Zn is shown to modulate the activity of membrane-bound NADPH oxidase (Cakmak 2008) in homeostasis of reactive oxygen species which regulates defense and signaling by the host during drought or other abiotic stresses (Bagci et al. 2007; Golldack et al. 2014). Cu is an essential

Table 21.1 Summary of major functions, deficiency, and toxicity levels and symptoms of micronutrient defficiency and toxicity

Element	Major functions	Deficient concentration in plant	Deficiency symptoms	Toxic concentration in plants	Toxicity symptoms	References
Fe	Biological redox systems (i.e., electron transport chains in photosynthesis and respiration), enzyme activation, carrier in N fixation (i.e., leghemoglobin in bacteroids of legume roots)	Less than 50–100μg g^{-1}	Interveinal chlorosis in young leaves (caused by decreased chlorophyll synthesis), retarded/stunted growth and reduced activity of hill reaction	Above 50–100μg g^{-1}	Growth inhibition, reduced chlorophyll synthesis, inhibition of photosynthesis	Welch and Shuman (1995), Tripathi et al. (2015)
Mn	Enzyme activation, biological redox systems (e.g., electron transport reactions in photosynthesis), detoxification of oxygen free radicals, secondary plant metabolite synthesis, structural constituent of ribosomes, disease resistance	Less than 10–100μg g^{-1}	Interveinal chlorosis in young tissue, appearance of greenish-gray specks at the lower base of monocots, development of brown necrotic spots on the cotyledons of legume plants, premature leaf fall, white-gray spots of leaf and delayed maturity	Above 10–100μg g^{-1}	Interferes with absorption and utilization of other mineral elements, affects the energy metabolism, decreases photosynthetic rates, causes oxidative stress	Welch and Shuman (1995), Tripathi et al. (2015)
Zn	Membrane integrity, enzyme activation, gene expression and regulation, carbohydrate metabolism, anaerobic root respiration, protein synthesis, structural integrity of ribosomes, detoxification of	Below 15 mg Zn kg^{-1} dry weight	Impaired stem elongation in tomato, root apex necrosis ("dieback"), interveinal chlorosis ("mottled leaf"), development of reddish-brown or bronze "bronzing," internode shortening	Above 20 mg Zn kg^{-1} dry weight	Reduced yields and stunted growth. Leafy vegetable crops are sensitive to Zn toxicity. Soybean and rice have been recognized as Zn sensitivity crops in which Zn toxicity instigates genetic variation,	Welch and Shuman (1995), Tripathi et al. (2015)

(continued)

Table 21.1 (continued)

Element	Major functions	Deficient concentration in plant	Deficiency symptoms	Toxic concentration in plants	Toxicity symptoms	References
	superoxide radicals, phytohormone activity (e.g., IAA and gibberellic acid), gene structure (Zn finger motif), disease resistance		("rosetting"), epinasty, inward curling of leaf lamina "goblet" leaves and reductions in leaf size (little leaf)		increased lipoxygenase activity and lipid peroxidation enhancing antioxidative activity in plants	
Cu	Physiological redox processes (e.g., photosynthetic electron transport, respiration), detoxification of superoxide radicals, lignification, disease resistance, pollen viability	Below 5μg g^{-1} dry weight	Improper growth rate and distortion or whitening (chlorosis) of young leaves, decrease in cell wall formation lignification in several tissues and curling of leaf margins, damages apical meristem, pollen development, the fruit and seed production, wood production	Above 20μg g^{-1} dry weight	Chlorosis and necrosis, stunting, and inhibition of root and shoot growth inhibit enzyme activity and protein function, which later produces highly toxic hydroxyl radicals leading to oxidative damage of plant cell	Welch and Shuman (1995), Tripathi et al. (2015)
Ni	Urea and ureide metabolism, iron absorption, seed viability, nitrogen fixation, reproductive growth	0.05–10 mg kg^{-1} of dry weight	Leaf tip necrosis (legumes), chlorosis and patchy necrosis (*Gramineae*)	>10 mg kg^{-1} dry weight (DW) in sensitive species like barley, water spinach, and wheat >50 mg kg^{-1} DW in moderately tolerant species and >1000 mg kg^{-1} DW in	Inhibition of mitotic activities, reductions in plant growth and adverse effects on fruit yield and quality. Extremely high soil Ni concentrations have left some farm land unsuitable for growing	Chen et al. (2009), Alloway (2008)

				Ni hyper-accumulator plants	crops, fruits, and vegetables	
Cl	Osmoregulation, charge compensation (i.e., counter ion in cation uptake), reactivity of enzymes, photosynthesis, disease resistance, stomatal regulation	Can vary from 0.1 to 6 mg g^{-1} (dry matter) or between 0.03 and 0.17 mmol L^{-1} of the plant tissue water content for different species	Wilting of leaves, especially at margins, shriveling and necrosis of leaves, frond fracture and stem cracking in coconut, sub-apical swelling in roots	Crops with high Cl endurance, such as corn, sugar beet, grain sorghum, cotton, and spinach, can endure the Cl of >600 mg kg^{-1} with no visible negative effects The crops with mid-Cl endurance, such as wheat, rice, cucumber, tomato, cabbage, peanut, and grape seedling, can endure the Cl of 300–600 mg kg^{-1} The crops with low Cl endurance, such as soybean, lettuce, sweet potato, strawberry, and apple seedling, can't endure the environmental Cl when it exceeds 300 mg kg^{-1}	Reduced yields	Chen et al. (2009), Alloway (2008)
B	Cell wall formation and stabilization, lignification, xylem differentiation, membrane	Less than 0.3 ppm and 0.14 mg kg^{-1} (generally required in greater amounts by	Stunted growth inhibition of cell expansion, cracking or rotting of fruits, wilted or curled	Above 0.3–1 ppm and 3–100μg g^{-1} dry weight	Yellowing of the leaf tips and distorted shoot growth, chlorotic and necrotic patches in the	Welch and Shuman (1995), Tripathi

(continued)

Table 21.1 (continued)

Element	Major functions	Deficient concentration in plant	Deficiency symptoms	Toxic concentration in plants	Toxicity symptoms	References
	integrity, carbohydrate utilization, pentose phosphate metabolism, phenol metabolism and auxin activity, pollen germination and growth, inhibition of callose formation, stomatal regulation	dicotyledonous plant species than by monocotyledons)	leaves, water-soaked petiole		margin/older leaves, spots on fruits	et al. (2015)
Mo	Electron transfer reactions, nitrate reduction, nitrogen fixation, ureide metabolism, sulfate oxidation, pollen formation, protein synthesis	0.2–2 ppm in forage legumes	Chlorosis of leaf margins, "whiptail" of leaves and distorted curding of cauliflower; "fired" margin and deformation of leaves due to NO3 excess and destruction of embryonic tissues	100–1000 mg kg^{-1}	Stunted growth with yellow-brown leaf discolorations	Maathuis (2013), Baligar et al. (2001)

element for lignin synthesis needed for cell wall strengthening (Yruela 2009; Ryan et al. 2013) which facilitates withstanding abiotic stresses such as wilting, wind, and rain. B facilitates cross-linking of pectic polysaccharides serving cell wall functioning and maintaining the structural support of the cytoskeleton (Miwa et al. 2008). Chloride plays a role in stomatal regulation and protects plants from wilting and death (Broadley et al. 2012).

When considering crop plants, the role of micronutrients in combating abiotic and biotic stresses, improving nutritional quality, increasing yield, and enhancing uptake of essential macronutrients, NPK, is substantial in addition to their physiological role. Micronutrients assist plants to mitigate biotic stresses by means of developing resistance to plant diseases either directly affecting the pathogens in the rhizosphere or inducing different types of physiological responses in the plant during pathogen attack through mechanisms such as siderophore production (Kloepper et al. 1980; Lim and Kim 1997; Fernandez et al. 2005; Vansuyt et al. 2007; Dimkpa et al. 2009, 2015a; Radzki et al. 2013), inducing cellular activity, disease resistance (Shirasu et al. 1999; Datnoff et al. 2007) and microbial biocontrol agents to produce antimicrobials (Dimkpa et al. 2012, 2015a; Duffy and Defago 1999) and acidification of soil (Dimkpa et al. 2013a, b). Zn and Mn were reported to suppress diseases (Huber and Wilhelm 1988), and Cu, Ni, Mn, Mo, and B have been reported through mechanisms such as inducing the production of antioxidants, strengthening cell walls through the production of lignin and suberin, and controlling N metabolism in the plant (Huber and Wilhelm 1988; Römheld and Marschner 1991; Boyd et al. 1994; Bai et al. 2006; Evans et al. 2007; Stangoulis and Graham 2007; Taran et al. 2014; Servin et al. 2015) which make the crop more resistant to diseases and drought conditions.

Even though there should be a remarkable contribution of micronutrients for the nutritional value of crop produce with a number of physiological roles governed by them in plants, their relationship has been inconsistent (Dimkpa and Bindraban 2016). Some studies demonstrated a positive outcome with the addition of micronutrients (Rietra et al. 2015; Dimkpa et al. 2015a, b; Kumar et al. 2009), especially through positively modulating the uptake of other micronutrients. Some studies reported that the levels of other micronutrients can be reduced by the addition of a specific micronutrient perhaps because of the competition for uptake among micronutrients (Dimkpa and Bindraban 2016). The positive effects of micronutrients in the yield, quality, earliness, fruit setting, postharvest life, and biotic and abiotic stresses in vegetable crops have been documented by Sidhu et al. (2019). In addition to crop nutritional quality, they have been reported to enhance seed vitality and thereby good seed emergence and vigorous seedling growth (Brodrick et al. 1995; Nestel et al. 2006; Eggert and von Wirén 2013; Velu et al. 2014). The role of micronutrients in improving the agronomic quality including yield has been overserved with the supplement of individual element (Dimkpa and Bindraban 2016) as well as in combined applications (Yaseen et al. 2013; Vanlauwe et al. 2014). The significance of fortification of micronutrients into the soil through different means in order to improve the quality and quantity of food crops, has been discussed only recently and its real value will be persuaded in the future.

21.4 Micronutrients, Ecosystems, and Environment

The major ecological role micronutrients play in soils is increasing the NPK fertilizer efficiency use by plants that would otherwise be lost via leaching, fixation, and/or volatilization. Additionally micronutrients in soils may improve water use efficiency of crops under water-deficient conditions (Movahhedy-Dehnavy et al. 2009; Molden et al. 2010; Ashraf et al. 2014), and when coupled with organic matter in soils, they have been observed to enhance the ion exchange capacity, soil structure, and water storage capacity, improve drainage and aeration, and decrease soil salinity (Dhaliwal et al. 2019). These benefits together with the suppression of plant diseases will enhance the sustainable agricultural production systems. However, most of the micronutrients are heavy metals; therefore, nonstrategic applications would exert ecological and environmental challenges. Further, contaminations with high levels of micronutrients can also occur through water irrigation, carrying pesticides and heavy metal-containing wastewaters, as biosolid accumulates in soil (Alloway and Jackson 1991; Wuana and Okieimen 2011) which ultimately can cause micronutrient toxicity to soil flora and fauna. Despite both beneficial and detrimental agronomical, environmental, ecological, and health effects of micronutrients to living beings, increasing their availability in soil strategically will offer great potential in mitigating some challenges related to food security and "hidden hunger" through enhancing the quality and quantity of food produce.

21.5 Addressing Micronutrient Deficiencies

Since micronutrient fertilization is not a prevailing cultural farming practice, mitigating their deficiencies in soil would require vigilant fortification intervened after prior evaluations of crop and soil conditions (Joy et al. 2014; Kumssa et al. 2015) with systematically determined nutrient ratios and antagonistic interactions among the micronutrients, as well as between micronutrients and macronutrients. This will lead to the need of vigorous experimentation on the micronutrient applications in combination with fertilizer regimes to ensure more plant-specific and balanced ratios of micronutrient in fertilizer formulations (Dimkpa and Bindraban 2016; Rietra et al. 2015; Kaura et al. 2020). Micronutrient can be fortified through a number of methods which have their own limitations. The most common and direct method is the agronomic fortification through soil applications (Cakmak 2008; Velu et al. 2014; Duffner et al. 2014), as foliar sprays or as seed treatments (Farooq et al. 2012; Mondal and Bose 2019) in which foliar sprays have been more effective in yield improvement and grain enrichment, but are restricted by high cost (Johnson et al. 2005). Soil application would require higher doses due to low nutrient-use efficiency since 60–90% of the total applied fertilizer is lost and only the remaining 10–40% is taken up by plants. Seed treatment with its easiness to apply would be a better option both economically and environmentally as less micronutrient is needed while improving seedling growth (Farooq et al. 2012; Mondal and Bose 2019; Singh et al. 2003). However, this is under thorough investigation in terms of optimization of formulation, application protocols, and storage methods (Farooq et al. 2012). As alternatives to agronomic fortification, methods such as plant breeding, genetic engineering (biofortification), and postharvest biofortification of food are used in different countries. These methods are usually time consuming, and several tedious optimization trials such as screening of germplasm, crossing between varieties, molecular marker-assisted selection, and new crop breed phenotyping and high technical skills are needed (Velu et al. 2014; Waters and Sankaran 2011). In addition, transformation of micronutrients to available forms through building up of soil organic matter content has gained attention recently (Dhaliwal et al. 2019); however, it is yet to be further investigated with field trials done in large scale and stimulation models to better understand the relationship and to formulate management strategies (Dhaliwal et al. 2019). With these practical issues encountered with other alternatives, soil applications as chemical fertilizers have been the most common method of micronutrient fortification; however, they would result in chemical residues in soil due to their low use efficiency which will lead to severe environmental problems and toxicities to plants if not duly addressed. Further, imbalanced application can enhance the micronutrient deficiency levels in soils (Dhaliwal et al. 2019). The use of chemical fertilizers causes soil acidification (Chun-Li et al. 2014) and groundwater and air pollution (Youssef and Eissa 2014). More importantly, concerns over the contribution of chemical fertilizer for global warming and climate change have led the world toward sustainable fortification strategies in the recent past. Using indigenous microorganisms which are capable of promoting growth (plant growth-promoting rhizosphere microorganisms, PGPMs) and disease resistance (biocontrol agents) in crops through different mechanisms including converting nutritionally significant elements from unavailable to available form (mineralization/solubilization) has gained attention in the recent past as a substitute to chemical fertilizer in sustainable farming (Kaura et al. 2020; White and Broadley 2009; Bhardwaj et al. 2014; Fomina et al. 2005b). It has also been identified as an eco-friendly and cheaper approach in maintaining a sustainable environment with their potential to conserve and increase the soil biodiversity (Vessey 2003; Raja 2013) and to reduce environmental pollution including heavy metal contamination. When considering a sustainable environment and increasing bioavailability of micronutrients, the role of indigenous microbes in bioleaching and bioremediation process should also be considered since they exist as ores mainly in insoluble forms, while some (Fe, Cu, Zn, Ni) are categorized as heavy metals. Therefore, there is a recent advancement of research on micronutrient-solubilizing indigenous microorganisms toward developing sustainable environments (Cai et al. 2013; Kumar and Gopal 2015).

21.6 Indigenous Microorganisms (IMOs)

Indigenous microorganisms refer to a group of beneficial microorganisms that are native to a given area which are different from effective microorganisms that are laboratory-cultured mixture of microorganisms (Kumar and Gopal 2015). Ideally indigenous microorganisms are a mixture of a variety of beneficial microorganisms yet can also be considered as organized microbial communities. Their ability of microbial biofilm formation and their microbiome networks in various activities in the soil have been discussed for the last few years (Mandakovic et al. 2018; Horton et al. 2019; Akkaya et al. 2020). Their potentiality in plant growth promotion through processes such as biodegradation, nitrogen fixation, soil fertility improvement, and mineral solubilization has been observed for decades (Umi and Sariah 2006). In addition, their role in bio-composting, biodegradation, bioremediation, bioleaching, and natural farming has gained attention in the recent past and has been the focus of many researchers (Dhaliwal et al. 2019; Kumar and Gopal 2015; Gadd 2010; Sangeetha et al. 2020; Saravanabhavan et al. 2020; Sarker and Rahman 2020; Sharma et al. 2020). These have made them potential tools in developing sustainable approaches in agriculture, environmental restoration, and safeguarding targets since they are composed of a natural microbiome (Kumar and Gopal 2015). Depending on the purpose, a variety of terms are being used to refer to indigenous microorganisms. In agriculture, they are mostly being termed as "biofertilizers" referring to the products containing a combination of different types of microorganisms which are applied to crops in order to increase their quality and quantity. In some other context, they are named as "plant growth-promoting microorganisms (PGPMs)" which inhabit in root rhizosphere and considered as bioprotectants of plants (Akkaya et al. 2020; Yang et al. 2009; Ahmad et al. 2018; Pitiwittayakul and Tanasupawat 2020). As biofertilizers and PGPMs, they lead to crop productivity through decomposition of organic matter, nutrient acquisition, absorption of water, nutrient recycling, weed control and bio-control of plant pathogens (Bhardwaj et al. 2014; Vessey 2003; Sangeetha et al. 2020; Ahmad et al. 2018; Berg et al. 2013) improving the soil structure and function. Mineralization and solubilization have been identified as two main methods by which IMOs increase the bioavailability of nutrients by increasing solubilization which ultimately promote growth and yield of plants (Vessey 2003). Further, the role of IMOs in bioremediation and bioleaching of metals has also been investigated and might have an impact on regulating nutrient contents in soil. Solubilization of nutrients, mainly NPK and to some extent other macro- and micronutrients, by solubilizing IMOs has widely been researched and reviewed recently (Djajadi and Hidayati 2020).

21.7 Nutrient-Solubilizing IMOs

Both macro- and micronutrients are originated from minerals deposited under the Earth's crust as ores. Many nutrients are metals from which some are considered as potentially hazardous metals or heavy metals when in high concentrations in soil, water, and biological tissues. The majority of metals exist as minerals in soil with a number of mineral forms for each metal element having varying distribution in the environment with different physicochemical properties (Gadd 2010; Ehrlich and Newman 2009). The minerals in soil are subjected to various geological processes such as chemical cycling of elements including mineral formation (mineralization), mineral deterioration, and chemical transformations of metals, metalloids, and radionuclides (solubilization/mobilization). Solubilization refers to "the preparation of a thermodynamically stable isotropic solution of a substance normally insoluble or very slightly soluble in a given solvent by the introduction of an additional amphiphilic component or components" (Yadav et al. 2018) so that its availability is increased. Mineralization refers to the conversion of organic compounds (metals) into inorganic compounds through various decomposition procedures. Microorganisms significantly contribute to all of these geological processes. Microbes are in continuous interaction with metals and minerals under natural and artificial environments. Their interactions alter the physical and chemical state of metals and minerals, while microbial growth, activity, and survival are in return affected by the characteristics of metals and minerals (Gadd 2010). As a result many minerals are biogenic in origin (biomineralization) and some make structural components for many organisms such as diatoms (Ehrlich 1996; Gadd and Raven 2010). Most biominerals are in the form of silicates, calcium carbonates, and iron oxides or sulfides (Baeuerlein 2000; Bazylinski 2001). All kinds of microbes (bacterial, fungi, protists) and their symbiotic associations such as lichens and mycorrhizae contribute actively to the above geological processes (Macalady and Banfield 2003; Bottjer 2005; Chorover et al. 2007; Konhauser 2007; Gleeson et al. 2007; Gadd 2008), especially metal and mineral transformations (Ehrlich 1996). Specific groups of microbes that are directly involved in geochemical transformations include both pro- and eukaryotes such as manganese-oxidizing and manganese-reducing bacteria, iron-oxidizing and iron-reducing bacteria, sulfate-reducing bacteria, and sulfur-oxidizing and sulfur-reducing bacteria that can form or degrade silicates, carbonates, phosphates, and other minerals (Gadd 2007, 2010; Ehrlich 1996; Kim and Gadd 2008). In addition, soil microorganisms, especially mycorrhizal fungi (Tao et al. 2008), are solely responsible for nutrient cycling through decomposition of soil organic matter and also by making chemically fixed nutrients such as phosphorus (P), zinc (Zn), potassium (K), and iron (Fe) available (Ahmad et al. 2018). In addition, early stages of soil formation are supported by the activity of microbes such as lichens through weathering process (Purvis and Pawlik-Skowronska 2008; Gilmour and Riedel 2009; Uroz et al. 2009). General metabolic activities of all microbes affect metal distribution and bioavailability through cellular accumulation, decomposition, or biodeterioration of organic and inorganic substrates (Gadd 2007;

Warren and Haack 2001; Huang et al. 2004). However, mineral and metal solubilization in other terms may have negative contribution when they are potentially hazardous/heavy metals in certain context such as contaminated soil including solid wastes (Sayer et al. 1999; Fomina et al. 2004, 2005a, b).

21.8 Role of Nutrients in Microorganisms

Microbial growth, metabolism and differentiation require nutrients (Gadd 1992). Microbes interact with minerals containing nutrients in several ways depending on the type of nutrient, organism, and environment. All microbes use nutrients for structural functions and/or catalytic functions (Ehrlich 1997). The structure and the function of microbes also affect metal speciation and thereby solubility, mobility, bioavailability, and toxicity of nutrients (Gadd 2010). When these elements are metals, they particularly interact with microbes in different ways. Firstly, microbes incorporate trace metals into metalloenzymes or utilize enzyme activation (Wackett et al. 1989) such as nitrogenase (Mo/Fe or sometimes V/Fe or Fe only), cytochromes (Fe) and cytochrome oxidase aa3 (Fe, Cu), superoxide dismutases (Fe, Mn, Cu, or Zn), bacteriochlorophyll (Mg), iron-sulfur proteins, CO dehydrogenase with Ni in anaerobic bacteria and Mo in aerobic bacteria, NADP-dependent formate dehydrogenase (W/Se/Fe), and formate dehydrogenase H (Mo/Se/Fe) (Wackett et al. 1989; Fridovich 1978; Yamamoto et al. 1983; Robson et al. 1986; Scheer 1991; Orme-Johnson 1992; Boyington et al. 1997). Some, especially by eubacteria and archaea, use certain metals/metalloids as electron donors or acceptors in energy metabolism (Ehrlich 1996). The entire energy demand of chemolithotrophs like eubacteria *Thiobacillus ferrooxidans* and *Leptospirillum ferrooxidans* and the archaea *Acidianus brierleyi* and *Sulfolobus acidocaldarius* can be satisfied by oxidizable metals or metalloids, particularly through oxidation (FeII) to Fe(III) (Ehrlich 1996, 1997). Thirdly, microbes can enzymatically detoxify harmful metals or metalloids by oxidation or reduction or, when cannot be detoxified, removed from cell interior using efflux systems (molecular pumps) (Gadd 2010). Anaerobes such as sulfate-reducing bacteria enzymatically catalyze biocorrosion through cathodic depolarization. Anaerobic biocorrosion is thought to be regulated by biofilms consisting of a consortium of a variety of bacteria, often including aerobic, facultative, and anaerobic bacteria, each with specific locations in the biofilm (Ehrlich 1997). Non-enzymatic usage of metals by microbes occurs with accumulating and even with dead cells, binding them as cations to the cell surface with a passive process (Gadd 1993).

Despite the positive interactions of microbes and nutrients, their toxicities for microorganisms can occur through natural geochemical events and anthropogenic contamination in aquatic and terrestrial ecosystems by domestic, agricultural, and industrial activities. However, many microbes grow and survive in locations polluted with metals by different mechanisms which contribute to resistance (Avery 2001; Holden and Adams 2003; Verma and Kuila 2019; Tarekegn et al. 2020).

Changes in metal speciation affect the survival of microbes. These changes can be redox transformations, efflux and intracellular compartmentalization with cell walls, production of metal-binding peptides and proteins (e.g., metallothioneins, phytochelatins), active transport, organic inorganic precipitation, and other constituents with metal-binding abilities (Gadd 2010). They also can convert the pollutants into metabolic intermediates and be utilized as primary substrates for cell growth (Verma and Kuila 2019). In addition, the presence of plasmids containing resistance genes (Rosen et al. 2005; Silver and Phung 1996) also affects the activity of bacteria and fungi toward certain nutrients (Van Ho et al. 2002). Many microbial processes such as energy generation, cell adhesion, biofilm formation, and nutrient acquisition (Hochella 2002; Brown et al. 2008) are influenced by minerals and nutrients. Further, some mineral surface properties such as surface composition, microtopography, surface charge, and hydrophobicity affect thigmotropism, microbial attachment and detachment, and thereby colonization and biofilm formation (Brown et al. 2008; Vaughan et al. 2002; Bowen et al. 2007; Gleeson et al. 2010). Oxides of some micronutrients such as Fe significantly influence microbial activity by altering soil behavior through soil physical, chemical, and biological processes (Huang et al. 2005).

21.9 Mechanisms of Solubilization/Mobilization Nutrients

Nutrient solubilization/mobilization from different substrates such as rocks, minerals, soil, and others can occur by different processes and can result in volatilization through protonolysis, complexation by excreted metabolites and Fe(III)-binding siderophores, chemical oxidation or reduction, indirect Fe(III) attack, and methylation (Gadd 2010). Other metabolites that are excreted with metal-complexing properties such as amino acids, phenolic compounds, and organic acids may also play a role. For example, oxalic acid can form soluble oxalate complexes with nutrients such as Al and Fe (Strasser et al. 1994). Microbes play a major role in all of these mechanisms which affect their bioavailability and toxicity. Extracellular compounds such as enzymes and other metabolic products such as gluconic acid and its derivatives (Gadd and Sayer 2000; Saravanan et al. 2007; Khan et al. 2013), H_2S, formate, or other secondary metabolites produced by microbes regulate redox processes (Fe) (Ehrlich and Newman 2009). Metal chelators of microbial origin related to Fe(III) solubilization include oxalate, citrate, humic acids, and tannins. Methylation is another mechanism by which some microbes such as methanogens, clostridia, and sulfate-reducing bacteria under anaerobic conditions and fungi (*Penicillium* and *Alternaria*), under aerobic conditions, solubilize nutrients (Gadd 2010). For example, the production of siderophores is the key mechanism by which Fe assimilation occurs in fungi and bacteria (Kalinowski et al. 2000; Glasauer et al. 2004). Since this chapter is focused on micronutrients, a detailed review of their importance to plants, deficiency and toxicity symptoms, availability, experimental records on solubilizing microorganisms, their role in microbial growth and function, and mechanisms of solubilization is presented in the following section and summarized in Tables 21.1 and 21.2.

21.10 Iron-Solubilizing IMOs

With a significant role in some life-sustaining processes of microbes and plants, iron is considered to be an essential, multifunctional micronutrient. It is required for the different physiochemical processes in plants and plays a vital role in the activation of chlorophyll, photosynthesis, structural component of the chloroplast membrane, respiration, and synthesis of many iron-sulfur (Fe-S) clusters and heme proteins as cofactors of proteins that function in the life of plants.

Iron mostly occurs in two oxidation states (+2 and +3) in nature. Plants absorb iron as Fe^{2+} and must be in the general range of >7.7-10 mol L^{-1} to avoid any deficiency (Lindsay and Schwab 1982). The functions of iron are mainly based on the reversible redox reaction of Fe^{2+} (ferrous) and Fe^{3+} (ferric) ions. The biosynthesis ALA, which is a precursor in the formation of chlorophyll, might need an intermediate that contains iron in the electron transfer chain, ferredoxin. This could control the reduction and activation of one or more enzymes responsible for ALA formation (Miller et al. 1984). Deficiency symptoms in plants include interveinal chlorosis in young leaves and stunted growth, while toxicity causes growth inhibition, reduced chlorophyll synthesis, and inhibition of photosynthesis (Table 21.1).

Iron is the fourth most prevailing element after O, Si, and Al in the crust and soils. The forms of Fe in the soil can be categorized into four types as Fe^{II} in primary minerals, Fe^{III} in secondary minerals, Fe crystalline minerals and poorly ordered crystalline (hydro) oxides, soluble and exchangeable Fe, and organic matter-bounded Fe in soluble or insoluble forms (Colombo et al. 2014).

Iron release by weathering of soil mineral deposits is a very slow process and it is regulated by pH value, O_2 concentration, and the dissolution-precipitation process of both crystalline and poorly ordered Fe-hydroxide minerals (Lindsay 1988). Once mobilized in weathering processes, occurrence of redox reaction and pH conditions of the soil environment affect the destiny of Fe^{II}. Although there is a more than enough iron (Fe) content in soils for plant requirement, especially in calcareous soils, bioavailability of Fe is often severely limited. Those types of soils are mainly found in dry areas of the earth. Plants grown in calcareous soils usually show iron deficiencies. These plant species have evolved various strategies to enhance their uptake of iron. However, usually these strategies are not sufficient to avoid Fe deficiency completely. Hence, soil microbial community plays a significant role in influencing plant Fe uptake.

The close relationship between microbes and oxides of Fe coexists in soils, and they provide adequate opportunities for mutual interactions. Primary minerals may provide Fe as well as many other important nutrients such as K, P, and S while accommodating microbes in mineral cycling (Lowenstam 1981; Lower et al.

Table 21.2 Summary of micronutrient-solubilizing microorganisms and the mechanisms of solubilization

Element	Solubilizing microorganism	Mechanism(s) of solubilization	References
Mo	*Escherichia coli*, *Enterobacter cloacae* strain, *Pseudomonas* sp., *Serratia* spp., *Enterobacter* sp., *Acinetobacter calcoaceticus*	Production of siderosphores	Halmi et al. (2013)
	Thiobacillus ferrooxidans (now *Acidithiobacillus ferrooxidans*)	Bioleaching by changing the soil pH	Frascoli and Hudson-Edwards (2018)
	Klebsiella, *Bacillus*, *Rhodobacter*	Possess NADH-dependent nitrate reductase caterlizers	Schaechter (2009)
	Azotobacter vinelandii	Siderophore production, protochelin and azotochelin	Hänsch and Mendel (2009)
	Rhizobium bacteria	Cofactor for the enzyme nitrate reductase which is involved in nitrogen assimilation	Hänsch and Mendel (2009)
	Rhodobacter capsulatus	Possess Mo and Fe nitrogenases. Mo-nitrogenases exhibit higher specific activities than the alternative nitrogenases	
Cu	*Aspergillus niger*	Production of organic acids such as oxalic, citric, malic, and tartaric acids	Mulligan et al. (2004)
	Pseudomonas lurida	Promote Cu uptake by roots and leaves in plants	Kumar et al. (2020)
	Bacillus toyonensis	Exhibited a considerable capacity for $Cu_2(OH)_2CO_3$ solubilization, increased the soluble Cu concentration in the soil	Sheng et al. (2012)
	Penicillium bilaji	Chelating mechanisms by lowering the solution/soil pH to 4.0	Asea et al. (1988)
	Herbaspirillum sp.	By metabolic products of the strain	Govarthanan et al. (2014)
	Phosphorus-solubilizing bacteria (PSB)	Production of low-molecular-weight organic acids	Li and Ramakrishna (2011)
	Mycorrhizal colonization	Redoxolysis, acidolysis, and complexolysis	Nouren et al. (2011)
	PGPR (plant growth-promoting bacteria)	Secreting siderophores and organic acid and by	Ke et al. (2020)

(continued)

Table 21.2 (continued)

Element	Solubilizing microorganism	Mechanism(s) of solubilization	References
		increasing soil organic carbon content	
Zn	*Aspergillus niger*, *A. oryzae*, *A. nomius*	Secretion of gluconic acid and its 2- and 2,5-keto-derivatives during growth via decreasing soil pH	White et al. (1997)
	Aspergillus niger	Production of citric and oxalic acid	White et al. (1997)
	Aspergillus terreus	Decrease in pH through the production of gluconic acid	Anitha et al. (2015)
	Trichoderma harzianum Rifai	Releasing Zn^{2+} ion via oxidative dissolution process	Altomare et al. (1999)
	Beauveria caledonica	Process of acidolysis, complex lysis, and metal accumulation	Fomina et al. (2004)
	Ericoid mycorrhizal fungus *Oidiodendron maius* Arbuscular mycorrhizae	Production of organic acids (solubilize ZnO and $Zn_3(PO_4)_2$)	Martino et al. (2003), Subramanian et al. (2009)
	Bacillus sp. alone or in combination *Bacillus pumilus* *Bacillus* sp. AZ6	Organic acid production Production of amino acids, plant hormones, chelating ligands, and organic acids via oxido-reductive systems and proton extrusion	Yadav et al. (2018), Monreal et al. (2015), Mahdi et al. (2010), Jha (2019), Hussain et al. (2015), Saravanan et al. (2004)
	Bacillus aryabhattai	Production of organic acids	Vidyashree et al. (2018)
	Pseudomonas sp. *Pseudomonas pseudoalcaligenes* *P. putida*	Organic compound production, keto-D-glutarate, propionic acid, formic acid, lactic acid, gluconic acid acetic acid, glycolic acid, citric acid, fumaric acid, succinic acid, malic acid, and oxalic acid. Auxin production Soluble carbohydrate production	Yadav et al. (2018), Jha (2019), Patten and Glick (2002), Vazquez et al. (2000)
	P. fragi	Production of siderophores	Kamran et al. (2017)
	Pseudomonas taiwanensis	Production of gluconic and 2-keto-gluconic acid	Fasim et al. (2002)
	Gluconacetobacter diazotrophicus	Solubilize insoluble zinc source especially ZnO, $ZnCO_3$, and $Zn_3(PO_4)_2$	Saravanan et al. (2007)
	Azotobacter, *Azospirillum*	Production of chelating agents	Biari et al. (2008)

(continued)

Table 21.2 (continued)

Element	Solubilizing microorganism	Mechanism(s) of solubilization	References
	Burkholderia, Acinetobacter	Production of organic acids	Vaid et al. (2014)
	Consortia of *Azospirillum lipoferum Pseudomonas* sp., *Agrobacterium sp.*	Production of organic acids	Mengel et al. (2001)
	Pantoea agglomerans	Auxin production Extracellular enzyme production	Kamran et al. (2017)
	Enterobacter cloacae	Extracellular enzyme production	Kamran et al. (2017)
	Providencia sp. *Anabaena* sp. *Calothrix* sp. *Anabaena* sp.		Rana et al. (2011)
Ni	*Sphingomonas macrogoltabidus Microbacterium liquefaciens Microbacterium arabinogalactanolyticum*	Those rhizobacteria are shown to play an important role in increasing the availability of Ni in soil, thus enhancing Ni accumulation by *Alyssum murale*	Abou-Shanab et al. (2003)
	PGPR (*Pseudomonas* sp.)	Siderophore production	Tank and Saraf (2009)
	Aspergillus niger, Aspergillus fumigatus, Acidithiobacillus ferrooxidans	Solubilize nickel at room temperature 30–37 °C, whereas organism unable to solubilize nickel at higher temperatures 45 °C	Mohapatra et al. (2007)
	Pseudomonas sp. SRI2, *Psychrobacter* sp. SRS8, *Bacillus* sp. SN9	Production of indole-3-acetic acid (IAA), siderophores, utilization of 1-aminocyclopropane-1-carboxylic acid (ACC)	Ma et al. (2009)
	Azotobacter chroococcum (N-fixing bacteria), *Bacillus megaterium* (P-solubilizer), *Bacillus mucilaginosus* (K-solubilizer), and *Bacillus sp.* RJ16	pH reduction by producing acids	Arunakumara et al. (2015)
Cl	No available records	Most of the micronutrients are present in the form of chloride complexes of their cationic forms. Most of the soil microorganisms produce acids and siderophores and release into the soil. These	

(continued)

Table 21.2 (continued)

Element	Solubilizing microorganism	Mechanism(s) of solubilization	References
		chemical components reduce the soil pH and cause the changes of the pH which facilitate the breaking down metal-Cl complexes and release of chloride in to the soil	
Fe	Fe-reducing bacteria *Gallionella* spp. *Leptothrix* spp. (microaerophilic bacteria) *Shewanella alga* *Shewanella putrefaciens*	Release of low-molecular-weight Fe-binding molecules – siderophores	Colombo et al. (2014)
	Fe-oxidizing bacteria *Acidithiobacillus ferrooxidans*	Fe-oxide formation onto extracellular polymers in order to enhance metabolic energy generation The oxidation of Fe (II) increases the pH gradient across the cell membrane, which in return increases the proton motive force and the energy-generating potential of the cells	Graham and Webb (1991), Alloway (2009)
	Leptospirillum ferrooxidans	Catalysis of sulfide oxidation by ferric iron at very low pH (0.7–1.0)	
	Pseudomonas and *Trichoderma* genera	By the synthesis and release of siderophores	Singh (2020)
Mn	Acidophilic Mn solubilizers *Enterobacter* sp. *Bacillus cereus* *Bacillus nealsonii* *Staphylococcus hominis*	Enzymatic conversion, metal effluxing, and reduction in sensitivity of cellular targets, intra- or extracellular sequestration, and permeability barrier exclusion Direct solubilization by utilization of MnO_2 as a final electron acceptor in the bacterial respiratory chain, instead of oxygen Indirect solubilization by the formation of metabolic reductive compounds Metal anion protonation, soluble Mn ligand complex formation, and production	Samourgiannidis and Matsi (2013)

(continued)

Table 21.2 (continued)

Element	Solubilizing microorganism	Mechanism(s) of solubilization	References
		of bio-generated organic acids	
	Mn oxidation *Leptothrix* sp. *Pedomicrobium* sp. *Hyphomicrobium*, *Caulobacter*, or common Gram-positive or Gram-negative bacteria, e.g., *Arthrobacter*, *Micrococcus*, *Bacillus*, *Chromobacterium*, *Pseudomonas*, *Vibrio*, *Oceanospirillum*	Indirect oxidation by the production of hydrogen peroxide, free radical, or oxidant Direct oxidation (an enzymatic reaction) catalyzed by Mn binding and oxidizing proteins found in crude or purified extracts	Gounot (1994)
	Mn reduction *Pseudomonas* spp. *Bacillus* spp. *Corynebacterium* *Acinetobacter johnsonii* *Pseudomonas fluorescens*	Mn oxide reduction through a drop of pH and/or redox potential due to bacterial metabolism Reduction through direct or indirect processes Mn (IV) be reduced by inorganically or organic reductants produced by microorganisms Enzymatic Mn (IV) reduction (e.g., *Acinetobacter calcoaceticus*)	Maathuis and Diatloff (2013)
	Bacillus polymyxa	By coupling of metal reduction with oxidation of a non-fermentative carbon source like lactate	Marschner (2012)
	Geobacter metallireducens, *Shewanella* (formerly *Alteromonas*) *putrefaciens*, a facultative anaerobe and obligate respire *Shewanella putrefaciens*		Lovley et al. (1993)
B	Boron-tolerant microorganisms *Lysinibacillus boronitolerans* *Chimaereicella boritolerans* *Gracilibacillus boraciitolerans*	High B efflux and exclusions which actively pump boric acid from the cells and are thus able to maintain a lower B concentration in the cell than in the external medium	Ahmed and Fujiwara (2010)

(continued)

Table 21.2 (continued)

Element	Solubilizing microorganism	Mechanism(s) of solubilization	References
	Bacillus boroniphilus *Arthrobacter* sp. *Rhodococcus* sp. *Lysinibacillus* sp. *Algoriphagus* sp.		
	Boron accumulators *Variovorax boronicumulans*		Miwa et al. (2008)
	Boron uptake promoters *Bacillus pumilus*		Masood et al. (2019)

2001). Microbes especially by bacteria like *Thiobacillium* and *Metallogenium* sp. dissolve primary minerals which contain iron through various processes which are termed as solubilization or chelation, sorption, accumulation, transformation, and precipitation (Colombo et al. 2014; Mengel 1994). These mechanisms are even more complex within the rhizosphere, due to the activity of the plants. The activity of plant roots can affect the abundance, diversity, and activity of microbes as well as Fe availability, and the interactions between Fe minerals and microbes (Colombo et al. 2014).

Iron solubilization mediated by PGPR was reported by Kloepper et al. (1980). Many microbes belonging to bacterial genus *Pseudomonas* and fungal genus *Trichoderma* are found to possess the ability to solubilize iron (Singh 2020). In line with Jin et al. (2010), an isolated *Pseudomonas* sp. could grow and produce siderosphores under Fe-deficient medium (Jin et al. 2010). Their work also showed that phenolic compounds exuded from plant (red clover) roots under Fe-deficient conditions favor the rhizosphere microbes to secrete more siderosphores which help to improve plant iron uptake.

Reduction or oxidation of iron minerals provides energy for anaerobic ferric-reducing and ferrous-oxidizing bacteria. This apparently plays an important role in catalyzing iron transformations in anoxic environments. Lithotrophic acidophilic and neutrophilic bacteria oxidize ferrous iron aerobically (Harrison Jr 1984). The acidophile, *Acidithiobacillus ferrooxidans*, is the most widely studied of all iron oxidizers. Phototrophic purple, non-sulfur bacteria were found capable of anaerobic ferrous iron oxidation by utilizing ferrous iron as electron donor in the light (Widdel et al. 1993). *Desulfuromusa kysingii*, *Geospirillum barnesii*, *Rhodobacter capsulatus*, *Desulfofrigus oceanense*, *Desulfotalea psychrophila*, *Geobacter metallireducens*, *G. sulfurreducens*, and *Shewanella putrefaciens* (now *S. oneidensis*) are some of the reported iron-reducing bacteria (Straub et al. 2001).

21.11 Mechanisms of Fe Metabolism

The most accepted mechanism for iron solubilization by microbes is by production of siderosphores under iron-deficient growth conditions. Siderophores are chelating agents that are secreted by bacteria and fungi with formation constants for ferric iron in the range of 10^{25} to 10^{35} and in exceptional cases as high as 10^{51} (Hider 1984). Iron availability in the surrounding environment highly regulates the siderophore production of microorganisms (Kalinowski et al. 2000). Siderosphores differ in structure and are low in molecular mass. The main groups of siderophores are catecholates, hydroxamates, and carboxylates. The catecholate is a main siderophore which is produced by bacteria, whereas fungi produce hydroxamate (Miethke and Marahiel 2007; Hider and Kong 2010). Stable soluble complex, made by iron with siderophore in soil solution and at the mineral surface, makes them available for uptake by the cell membrane of plant roots. The siderophore is either destroyed or recycled during this reduction in some cases.

There are three different mechanisms in transporting siderophore combined with Fe across the cell membranes in microorganisms. Membrane-spanning proteins may involve in binding these to the substrate which are transported into the cell after undergoing a conformational change. The location of Fe release mechanisms differs. Alternatively, Fe may also be removed by hydrolytic destruction of the chelate. The second mechanism is termed as the direct shuttle (Crowley et al. 1991). In that mechanism, ferric siderophore binds to a cell surface receptor where Fe is cleaved and simultaneously transported without concomitant transport of the desferri siderophore. In the third mechanism, an indirect shuttle (extracellular dissociation) acquires Fe in which Fe is removed through reduction at a site some distance to the carrier protein (Crowley et al. 1991).

21.12 Manganese-Solubilizing IMOs

Manganese is an essential plant trace element that plays a significant role in plant metabolism and development but could be toxic at high concentrations. Mn occurs in many oxidation states like II, III, and IV in approximately 35 enzymes of a plant cell (Hebbern et al. 2009). In plant proteins, manganese acts either as a catalytically active metal or as an activator of enzymes. Superoxide dismutase which contains manganese protects the cells from the damaging effect of free radicals, oxalate oxidase, and manganese-containing water splitting system of photosystem II. Manganese activates PEP carboxykinase, malic enzyme, isocitrate dehydrogenase, and phenylalanine ammonia-lyase enzymes (Hänsch and Mendel 2009). Mn plays an important role in the synthesis of lignin along with Cu and provides resistance in root tissues to pathogens. Symptoms of Mn deficiency in plants include chlorosis, premature leaf fall, and delayed maturity, while toxicity causes reduced yields and stunted growth (Table 21.1).

Manganese is the fifth most abundant metal found on the earth's surface. In the Earth's crust, Mn is mainly found as minor components of rock-forming silicate minerals such as olivine, pyroxenes, and amphiboles along with Fe. Manganese oxide (pyrolusite) and Mn carbonate (rhodochrosite) minerals are among the most important Mn ore resources in the world. Mn is extensively available in deposits of complex ores, nodules on ocean floors (Patrick 2010), wastewater sludge (Wang et al. 2011), and municipal solid wastes (Abdulsalam et al. 2011). Reduced soluble or adsorbed Mn (II) and insoluble Mn (III) and Mn (IV) oxides are the naturally found forms of Mn in soil. Solubility and availability of Mn in soil are increased with increasing state of reduction. Oxidation of Mn in soil is basically a biological process, while reduction of Mn may be either chemical or biological. Mn availability in the rhizosphere soil depends on the redox condition and the pH ranges (Gounot 1994).

Both oxidation and reduction of manganese in natural environments is dominantly promoted by microbial catalysis, but abiotic converters are often important too and it may compete with the biological processes (Gounot 1994). Oxidation of Mn has been reported by many types of microorganisms such as fungi, bacteria, and algae (Ghiorse 1984). Most of them are bacteria that belong to common Gram-positive or Gram-negative bacteria or group of sheathed bacteria, *Leptothrix*, and budding and appendaged bacteria: *Pedomicrobium*, *Hyphomicrobium*, and *Caulobacter* (Gounot 1994). The demosponge *Suberites domuncula* was found to have a Mn-oxidizing bacterium by Wang et al. (2011) which belongs to *Bacillus* strain BAC-SubDo-03. Most Mn-oxidizing bacteria are heterotrophic aerobic bacteria that use organic substances as the substrate. Some rhizosphere bacteria like *Bacillus*, *Pseudomonas*, and *Geobacter* can reduce oxidized Mn^{+4} into Mn^{+2} which is the plant metabolite form of Mn.

Effective rhizosphere Mn-reducing bacteria (*Pseudomonas* sp.) have been reported by Marschner and Dell (1994). Most of Mn-oxidizing bacteria are heterotrophic aerobic bacteria that grow on organic substances. The bacterial isolates *Bacillus anthracis*, *Acinetobacter* sp., *Lysinibacillus* sp., and *Bacillus* sp. are capable of solubilizing Mn in a range of pH (Ghosh et al. 2016). *Bacillus thuringiensis* has been found to have the capability to tolerate high concentrations as 4000 mg L^{-1} of Mn (II) and the highest removal rate of Mn (II). Hence, *Bacillus thuringiensis* plays a significant role in detoxifying and immobilizing excessive Mn in soil (Huang et al. 2020). Many fungi like *Acremonium* spp. can also take part in manganese oxidation (Miyata et al. 2004). The white rot fungus *Ganoderma lucidum* has possessed a good potential in solubilizing Mn under shaking and non-shaking conditions (Nouren et al. 2011). The soil fungi *Aspergillus niger* and *Serpula himantioides* have shown the ability to tolerate and solubilize manganese oxides (Wei et al. 2012).

21.13 Mechanisms of Mn Solubilization

The possible mechanisms of Mn oxidation by microorganisms can be described as direct or indirect. Production of hydrogen peroxide, free radical, or oxidant indicates the indirect oxidation of Mn which is due to the change of the surrounding environment. *Arthrobacter and Leptothrix* like bacterial groups are found to oxidize Mn by producing hydrogen peroxide as a mechanism of protecting the cells from the harmful effects of hydrogen. Direct oxidation is an enzymatic reaction which is facilitated through Mn binding and oxidizing enzymes which are found in crude or purified extracts. Examples can be found as a spore protein of *Bacillus* SG-1 (de Vrind et al. 1986) and an intracellular protein of a *Pseudomonas* sp. (Jung et al. 1998). *G. lucidum* was able to solubilize Mn by production of organic acids such as citric acid, tartaric acid, and oxalic acid (Nouren et al. 2011). In addition, roots and rhizosphere bacteria produce chelating agents like phenolic compounds and organic acids and other elements and hence avoid precipitation of Mn (Marschner and Dell 1994).

21.14 Zinc-Solubilizing IMOs

Zinc is another vital micronutrient for normal growth and development of plants. The normal concentration range for zinc in plant tissue is 15–60 ppm. Zinc requirement in plants is 30–100 mg kg^{-1}, below which would result in its deficiency. Plants require Zn for optimum fruit size, crop production, and yield. It is also used in the carbonic anhydrase activity involved in photosynthetic tissues for biosynthesis of chlorophyll (Xi-Wen et al. 2013). Further, Zn plays a key role in the synthesis of protein, activation of enzymes, RNA and DNA synthesis and regulations, and oxidation and metabolism of carbohydrates and prevents the peroxidation of lipids and proteins due to reactive oxygen species. Zinc is important for auxin production and for normal fruit and flower development. In plants, genes responsible for environmental stress tolerance are Zn dependent (Hafeez et al. 2013). Several studies reported that the use of zinc-containing fertilizers and micronutrients improves crop quality (Hussain et al. 2018).

Soil contains considerable amounts of Zn but in insoluble forms. Zinc is easily mobile in oxidizing acid soils, whereas it is immobile in poorly reducing neutral or alkaline soils. In the soil solution, Zn is a divalent cation or complexes with ligand via different transporter systems which is carried by mass flow, diffusion, and root extension in the direction of roots. A majority of the Zn absorption happens via active transport system, and it is transported from the root to the shoots via both xylem and phloem tissues. A little amount of Zn is retained at the basal node which governs the distribution of Zn in plants (White et al. 1997). The deficiency symptoms appear in the new leaves due to its immobility. Symptoms depend on the crop. Deficiency symptoms are expressed as preliminary in young leaves and could be visualized as patterns of chlorosis of the new leaves (often interveinal) and necrotic spots on the margins or leaf tips (Table 21.1). These affected leaves are smaller in size and form leaf rosette. The shortened internodes give the plant a rosette appearance and poor bud development which result in reduced branching and flowering. Crops with Zn deficiency may have susceptibility to injury or infection (Ghosh et al. 2014; Gandhi and Muralidharan 2016).

Several studies have found that inoculations of potent strain of Zn mobilizer rhizobacteria have been found to increase the yield of field crop such as rice wheat barley and maize (Hussain et al. 2018; Kutman et al. 2010; Tariq and Ashraf 2016). For example, Zn-mobilizing PGPR inoculation had a significant impact on root weight (74%), root length (54%), root area (75%), root volume (62%), shoot weight (23%), and panicle emergence index (96%) (Kutman et al. 2010) which exhibits potential in mitigating Zn deficiency in soils and crops.

The composition of Zn in polluted soils is dependent on both soil location and sources of pollution (Kabata-Pendias and Pendias 2001). Zinc distribution in agricultural soils ranges from 10 to 300 mg kg^{-2}. Under anaerobic conditions, higher concentration of iron reduces the zinc bioavailability in soil (Hussain et al. 2018). Geochemical composition and weathering of the parent rock will determine the Zn content in the developing soil. Environmental pollution or Zn-rich products can add up and alter the parent rock composition. Zn composition in the Earth's crust is 78 mg kg^{-1} which varies in different parent rocks. In soil, active Zn occurs either in the divalent form (Zn^{2+}) or in complex form like $ZnOH^+$, $ZnHCO_3^+$, $Zn(OH)_3^-$, and ZnO_2^-. Zinc exists in five different pools within the soil, namely, water-soluble, organically bound, exchangeable, chelated, and adsorbed. The strength of these forms will determine their ability to plant uptake and leaching.

The bioavailability of Zn to plants is influenced by total Zn contents in the soil, soil pH, elevated concentration of cations (Na, Ca, and Mg), phosphate soluble forms, anion bicarbonate, soil organic matter, and $CaCO_3$ content. Zn is strongly adsorbed on calcium carbonate ($CaCO_3$), magnesium carbonates ($MgCO_3$), iron oxide (FeO), or manganese oxide (MnO) (Alloway 2009). Chemical fertilizers indirectly affect the conversion of soluble Zn into different insoluble Zn fractions. Therefore, several studies have suggested the use of biofertilizers containing Zn-solubilizing microbes to increase the soluble Zn concentration in the soil (Kamran et al. 2017).

Several microorganisms play a significant role in solubilization and mobilization Zn (Kamran et al. 2017; Fasim et al. 2002; Javed et al. 2018). Several studies have reported the effectiveness of rhizospheric fungi in solubilizing insoluble Zn compounds both in vitro and in vivo. The production of organic acids by microorganisms has been found to increase solubilization and release of Zn compounds (Agusto da Costa and Duta 2001). Some filamentous non-mycorrhizal fungi, namely, *Aspergillus niger*, *A. oryzae*, and *A. nomius* isolated from a Zn mining site at Tak Province, Thailand, showed that the solubilization of insoluble Zn compounds ZnO, $Zn_3(PO_4)_2$, and $ZnCO_3$ occurs through the secretion of gluconic acid and its 2- and 2,5-keto-derivatives during growth (White et al. 1997). Release of Zn from organic complexes and calcium carbonate is facilitated by microbes through mineralization and solubilization, respectively. It has been reported that *Aspergillus niger* solubilizes insoluble ZnO, $Zn_3(PO_4)_2$, and $Ca_3(PO_4)_2$ to soluble form through the production of citric and oxalic acid (White et al. 1997). *Aspergillus terreus* (ZSF-9) isolated from Tiruppur District, India, was found to solubilize ZnO, $ZnCO_3$, and $Zn_3(PO_4)_2$ through the production of gluconic acid (Anitha et al. 2015). *Trichoderma harzianum Rifai* 1295-22 (T-22) converts insoluble Zn present in the soil into soluble form by releasing Zn^{2+} ion via oxidative dissolution process. During the process, fungus releases a complex compound which segregates Zn^{2+}, resulting in the enhancement of dissolution of metallic Zn in the soil. Fungus *Beauveria caledonica* converts insoluble $Zn_3(PO_4)_2$ into soluble Zn through the process of acidolysis, complex lysis, and metal accumulation (Fomina et al. 2004). Similarly, ZnO and $Zn_3(PO_4)_2$ can be solubilized by the ericoid mycorrhizal fungus *Oidiodendron maius* (Martino et al. 2003).

Bacterial species such as *Pseudomonas striata*, *Gluconacetobacter diazotrophicus*, *Thiobacillus thiooxidans*, *Burkholderia cenocepacia*, *Pseudomonas pseudoalcaligenes*, *P. fluorescens*, *P. japonica*, *P. fragi*, *Acinetobacter*, *Serratia marcescens*, *S. liquefaciens*, *Enterobacter cloacae*, and *Pantoea agglomerans* and several cyanobacterial species have been reported to solubilize insoluble Zn (Kamran et al. 2017; Zaheer et al. 2019; Altomare et al. 1999; Bapiri et al. 2012; Abaid-Ullah et al. 2015; Hussain et al. 2015). However, some *Bacillus* sp. (*Bacillus subtilis*, *Bacillus pumilus*, *Bacillus thuringiensis*, *Bacillus aryabhattai*) alone or in combination with cheaper insoluble Zn such as ZnO, $ZnCO_3$, and ZnS has been suggested as an effective alternative to costly $ZnSO_4$ and found to be more effective than other Zn solubilizers (Mahdi et al. 2010; Pawar et al. 2015; Mumtaz et al. 2017; Jha 2019; Zaheer et al. 2019).

21.15 Mechanisms of Zn Solubilization

Microorganisms can solubilize Zn by either a single mechanism or multiple mechanisms. As for other micronutrients, soil pH affects the availability of Zn where a 100 times increase in solubility can be achieved by decreasing one unit in pH (Mumtaz et al. 2017; Havlin et al. 2005). In addition, *Pseudomonas, Bacillus* spp. (Saravanan et al. 2004) and arbuscular mycorrhizae (Subramanian et al. 2009) were observed to reduce pH in the solubilization of ZnS, ZnO, and $ZnCO_3$. The main mechanism by which plant growth-promoting bacteria improve the Zn availability is by releasing organic acids like gluconate (Saravanan et al. 2011) or the derivatives of gluconic acids, e.g., 2-ketogluconic acid (Fasim et al. 2002), 5-ketogluconic acid (Saravanan et al. 2007), and various other organic acids (Tariq et al. 2007) and extrude protons (Fasim et al. 2002; Wu et al. 2006). *Bacillus* sp. AZ6 was found to secrete organic acids like cinnamic acid, ferulic acid, caffeic acid, chlorogenic acid, syringic acid, and gallic acid in a liquid medium (Hussain et al. 2004). Mycorrhizal fungi were also observed to secrete organic acids (Martino et al. 2003) to solubilize Zn from insoluble Zn_3 $(PO_4)_2$ and ZnO. Organic acids produced by some *Bacillus* sp. and *Pseudomonas* sp. include keto-D-glutarate, propionic acid, formic acid, lactic acid, gluconic acid acetic acid, glycolic acid, citric acid, fumaric acid, succinic acid, malic acid, and oxalic acid. Three ZSB isolates were found to produce 11 organic acids against ZnO, $ZnCO_3$, and $Zn_3(PO_4)_2$ including lactic acid, malonic acid, malic acid, citric acid, and succinic acid being the major acids. *Bacillus aryabhattai* produced many organic acids during Zn solubilization process as compared to *Pseudomonas taiwanensis* and other *Bacillus* sp. Organic acid secretions by *Bacillus* and *Pseudomonas* were dependent on the substrate of Zn minerals (Vidyashree et al. 2018). Some *Bacillus* strains were found to produce some other compounds in relation to Zn solubilization such as amino acids and plant hormones (Saravanan et al. 2004) and glucose or sucrose (*Gluconacetobacter diazotrophicus*) (Saravanan et al. 2007). Enhancement of chlorophyll, carotenoid, and antioxidant enzymes catalase (CAT)- and peroxidase (PO)-related functions by *Bacillus pumilus* and *Pseudomonas pseudoalcaligenes* had been reported to protect plants from salinity injuries. Further plants inoculated with the above ZMB also accumulated soluble carbohydrates in leaves, helping plants to overcome osmotic stress under salinity. Further, both bacterial isolates were positive for auxin production, *P. pseudoalcaligenes* showing more than *B. pumilus* in the presence of Zn in the medium compared to the control (Jha 2019). Auxin production in response to Zn has also been observed by Patten and Glick (2002) in *P. putida* which increased the length of canola seedling roots.

Zn-chelating compounds released by plant roots facilitate Zn solubilization by microbes in the rhizosphere (Obrador et al. 2003; Velazquez et al. 2016). Bacterial metabolites reduce reaction of Zn in the soil by forming complexes with Zn^{2+} (Tarkalson et al. 1998). At the root surface, Zn chelates the ligand (Zn^{2+}). *Pseudomonas monteilii*, *Microbacterium saperdae*, and *Enterobacter cancerogenesis* are thought to manufacture Zn-chelating metallophores (Whiting et al. 2001). A biofertilizer containing *Pseudomonas* sp. (96-51), *Azospirillum lipoferum* (JCM-1270, ER-20), and *Agrobacterium* sp. (Ca-18) produces chelating agent ethylenediaminetetraacetic acid (Tariq et al. 2007), and *Penicillium bilaji* was reported to increase Zn bioavailability to plants through chelating mechanism (Kucey 1988).

Zinc bioavailability can also be increased by facilitating nutrient uptake from long distance through improving root growth and surface area. Mycorrhizal fungi can change the root architecture which enables plant to obtain Zn from a distance of 40 mm from the root surface (Burkert and Robson 1994). An increase in Zn concentration up to 4% in cereal grains and increased root length by mycorrhizal fungus were reported (Subramanian et al. 2009). Tariq et al. (2007) reported significant increase in weight, length, and volume of root with biofortification of rice with Zn-solubilizing bacteria.

21.16 Copper-Solubilizing IMOs

Copper is another micronutrient needed for the growth of plants. Among many roles, it involves several enzyme processes and is the key to the formation of chlorophyll. Its normal range in most tissues is between 3 and 10 ppm. Cu requirement in plants is 3–20 mg kg^{-1} and below this range would result in deficiency. Although copper deficiencies or toxicities rarely occur, deficiencies have been reported in several parts of the world and need to be addressed since either extremes can have a negative influence on crop growth and quality. Cu ions act as a cofactor in enzymes such as Cu/Zn-superoxide dismutase (Cu/ZnSOD), cytochrome oxidase, ascorbate oxidase, amino oxidase, laccase, plastocyanin, biogenesis of molybdenum cofactor, and polyphenol oxidase (Krämer and Clemens 2006). Cu plays a vital role in the signaling of the transcription protein trafficking machinery, cell wall metabolism, iron mobilization, and oxidative phosphorylation and oxidative stress protection at cellular level (Yruela 2009; Puig et al. 2007). Cu is also required in photosynthesis and plant respiration electron transport chains and plant metabolism of carbohydrates and proteins. Its ethylene-sensing ability supports to intensify flavor and color of vegetables and flowers. Copper also acts as a structural element in certain metalloproteins (Pilon et al. 2006).

The deficiency symptoms of copper occur in the newer leaves. Symptoms vary depending on the crop. The symptoms are slight chlorosis of either in the whole leaf or between the veins of the new leaves. Small necrotic spots may form within the chlorotic areas on the leaf margins. The newest leaves are smaller in size, lose their sheen, and ultimately may wilt. Necrosis occurs on the apical meristems, which leads to death, inhibiting the growth of lateral branches. Lighter colored flowers than normal are produced by the plant (Welch and Shuman 1995; Tripathi et al. 2015).

Copper stress condition of the plant can cause burning of the root tips and thereby causes excess lateral root growth (Franco et al. 2011). When the copper concentration is higher in the soil, iron and sometimes molybdenum or zinc nutrients have to compete with copper for micronutrient availability in the soil and plant uptake (Tyler and Olsson 2001). Affected plants can exhibit symptoms of iron deficiency or other micronutrient deficiencies. Copper toxicity ultimately can reduce branching. Legumes have a tendency to be the most sensitive plants to copper toxicity (Carruthers 2016a, b).

The composition of Cu in polluted soils is dependent on both sources of pollution and soil location. Cu level in the environment has been increased by the mining of Cu-containing ores and industrial activities (Engelhardt et al. 2020). In soil, active Cu occurs mostly in the divalent form (Cu^{2+}) or in complex form with soil organic matter. The largest portion of Cu is usually present in the crystal lattices of primary and secondary minerals. The Cu ion is adsorbed to organic and inorganic negatively charged groups and is dissolved as Cu^{2+} and organic Cu complexes in the soil solution. Also it is specifically adsorbed to soil organic matter, carbonates, phyllosilicates, and hydrous oxides of Fe, Mn, and Al. The strength of these forms will determine their ability to plant uptake and leaching (Engelhardt et al. 2020).

The bioavailability of Cu in soil is influenced by physical, chemical, and biological properties at soil-root interface in rhizosphere. It is affected by different properties such as organic matter, soil type, pH, soil moisture, clay particles, temperature, retention, permeability, and different metal ions and their oxides (Hinsinger et al. 2009). In general, the bioavailable form of Cu^{2+} in the soil solution is decreased dramatically with an increasing soil pH. However, organic Cu complexes may dissolve at higher pH (Kumar et al. 2020). Calcareous or alkaline soils can limit the phytoavailability of Cu. Hence, the ability of plants to efficiently uptake Cu from soil solution, and distribution of this among different organs and tissues can strongly affect the crop growth and yield under Cu-limiting conditions (Migocka and Malas 2018).

Some micronutrients including Cu have limited mobility in soils which are transported to roots by slow diffusion. Even though Cu is usually present in large quantities in the bulk soil, the plant-available fraction in the rhizopheric soil solution can be insufficient to satisfy plant requirements. Copper allocation in the soils was found to be in the order of strong organic > residual > water soluble > ion exchangeable > carbonate > reducible > weak organic fractions, indicating that Cu is more distributed in organic fraction (84.67%). High affinity of Cu to organic matters make organic bound Cu distributed in the soil in large quantities. Cu in the crystalline lattice of the residual fraction cannot be easily released (Govarthanan et al. 2014). As one of the strategies to overcome this problem, microorganisms have been investigated in many studies.

Both bacterial and fungi have shown Cu-solubilizing ability (Table 21.2). Some filamentous non-mycorrhizal fungi, namely, *Penicillium* and *Aspergillus*, have been the most active metal-leaching fungi (Burgstaller and Schinner 1993). Several studies have shown the potential of *Aspergillus niger* to generate organic acids such as oxalic, citric, malic, and tartaric acids which resulted in maximum solubilization of Cu, Zn, and Ni (Mulligan et al. 2004). Bioleaching of Cu from ores has been done with *Penicillium simplicissimum* (Sukla and Panchanadikar 1993).

Several studies have reported the effectiveness of rhizosphere microorganisms in solubilizing insoluble Cu compounds both in vitro and in vivo. They mobilize and degrade organic pollutants. Cu-resistant bacterium, *Pseudomonas* sp. *DGS6* isolated from a natural Cu-contaminated soil, stimulated root elongation of maize and sunflower (Yang et al. 2013). *Pseudomonas lurida* strain EOO26 was found to increase Cu uptake by 8.6-fold by roots and 1.9-fold by leaves in inoculated plants (Kumar et al. 2020). The bacteria isolated from the rhizosphere, *Elsholtzia splendens*, significantly increased the bioavailability of Cu while stimulating the other heavy metals like Zn in the soil (Chen et al. 2005). Increasing Cu bioavailability and bioaccumulation with inoculations was observed to be species-specific in certain cases. The influence of *Bacillus* spp. on Cu bioavailability and the bioaccumulation for ryegrass and fescue were different to each other (Ke et al. 2020). Similarly, Liu et al. (2014) found that inoculations of 11 PGPR strains increased the Cu concentration in *Oenothera erythrosepala* and *Medicago sativa*, but decreased that in *Pennisetum purpureum*, which suggests the complex interactions between plants, microbes, and the soil (Liu et al. 2014). The addition of

Bacillus toyonensis alone had the maximum effect on $Cu_2(OH)_2CO_3$ solubilization (Sheng et al. 2012). *Penicillium bilaji* was able to solubilize cuprous and cupric oxide, cupric carbonate. This is mainly by chelating mechanisms under a low pH value as 4.0 (Asea et al. 1988). This mechanism involves the use of organic acids which have been reported to have phosphorus-solubilizing abilities (Khan and Bhatnagar 1977). Another study reported Cu leaching ability of *Herbaspirillum* sp. from ion-exchangeable, reducible, strong organic, and residual fractions (Govarthanan et al. 2014). The maximum solubilization (40%) in strong organic fractions was suggested to be resulted by the metabolic products of the microbe (Deng et al. 2012). The metabolites of microorganisms can act as indirect reactive species and solubilize metal sulfides and oxides during the bioleaching process (Mishra et al. 2008).

21.17 Mechanisms of Cu Solubilization

There are no clear mechanisms defined for Cu solubilization by microbes. Most of the literature suggest that microorganisms release chemical compound siderosphores that have ability to oxidize Cu ore, making them available to plants. However, the principal mechanisms of bioleaching of metal by fungi are documented to be redoxolysis, acidolysis, and complexolysis. The fungi are also found to produce organic acids such as citric, oxalic, malic, and gluconic acids during bioleaching which might contribute to solubilization (Mulligan et al. 2004; Johnson 2006). In addition, siderophores, organic acids and soil organic carbon content increased by the activity of PGPR improve soil Cu bioavailability (Ke et al. 2020). Phosphorus-solubilizing bacteria (PSB) also enhance Cu availability (Li and Ramakrishna 2011) by secreting low-molecular-weight organic acids. Many organic acids such as malic, lactic, 2-ketogluconic, citric, oxalic, glycolic, malonic, valeric, piscidic, tartaric, formic, and succinic have been identified as chemical compounds secreted by PSB, which have chelating properties.

21.18 Nickel-Solubilizing IMOs

Nickel is considered as an essential micro-element for plant growth since the late 1980s (Brown et al. 1987). The usual range for nickel in most plant tissue is between 0.05 and 5 ppm (Chen et al. 2009). Nickel is required to the plant as a component of certain plant enzymes like urease, superoxide dismutase, and hydrogenase. Legume plants use nickel as a catalyst in nitrogen fixation enzymes (Ahmad et al. 2012). It is required for urease enzyme which metabolizes urea nitrogen into usable ammonia in the plant. It prevents the accumulation of toxic levels of urea in plant tissues forming necrotic legions on the leaf tips. Therefore, the deficiency of nickel in plant can cause urea toxicity (Krämer 2005), reduced leaf size, disruption of amino acid metabolism, and urea accumulation in leaf (Bai et al. 2006). Nickel provides tolerance to plant diseases; however, the mechanism is unclear.

Nickel deficiency is unusual and minor and is often difficult to identify due to less symptom development. In certain cases, it can reduce yield and growth of plants. As nickel is a mobile element, its deficiency symptoms first appear typically in the mature leaves of the plants (Chen et al. 2009). In legume plants, deficiency causes whole leaf chlorosis along with necrotic leaf tips due to the increased levels of urea. In woody ornamentals, deficiency causes shortened internodes and it gives a rosette appearance to the plant, weak shoot growth, death of terminal buds, and eventual death of shoots and branches. The symptoms in pecans include decreased expansion of the leaf blade and necrosis of the leaf tips (Bai et al. 2006). Nickel turns out to be less available for plant uptake at higher pH of the soil or in growing media. Some other micronutrients like magnesium, zinc, iron, copper, cobalt, or cadmium in high amounts in the growth medium can result in nickel deficiency to the plant. Legumes (beans and alfalfa), barley, pecans, peach, plum, wheat, citrus, and certain wetland plants are some plants which are most sensitive to nickel deficiency (Merlot 2020).

The Earth's crust composition is comprised of approximately 3% of Ni and it is the 24th most abundant element. Its concentration in plant leaves ranges from 0.05 to 5 mg kg^{-1}, which is equal to 0.05–5 ppm on a dry weight basis. The required content of Ni in vegetative tissues of plants is between 2 and 4 ng g^{-1} dry biomass (Dalton et al. 1988) and up to 90 ng g^{-1} dry biomass in barley. Nickel concentrations $\geq$10 ppm are generally considered to be toxic to sensitive species. Ni^{2+} is the available form of Ni for plant. Rapid oxidation of Ni ion (Ni^{2+}) to unavailable forms in the soil makes total Ni concentration not a useful measure for Ni bioavailability. Thus, plants grown in high pH soils are vulnerable to Ni deficiency (Brown et al. 1987). In soil, active Ni occurs almost exclusively in divalent form (Ni^{2+}) or in complex form with soil organic matter. High pH soils can cause Ni deficiency. Other than that, excessive use of Cu and Zn can result in Ni deficiency in soil because Ni, Cu, and Zn share a common nutrient uptake system in the plant.

Rhizosphere microorganisms play a major role in Ni solubilization. Three bacteria, *Sphingomonas macrogoltabidus*, *Microbacterium liquefaciens*, and *Microbacterium arabinogalactanolyticum*, isolated from the rhizosphere of *Alyssum murale*, were observed to increase Ni uptake into the shoot by 17%, 24%, and 32.4%, respectively, compared to the non-inoculated control (Abou-Shanab et al. 2003). Tank and Saraf (2009) suggested that PGPR (*Pseudomonas* sp.) positively influence the growth of plants and also facilitate plant growth in Ni-contaminated soils. Mohapatra et al. (2007) reported that *Aspergillus niger*, *Aspergillus fumigatus*, and *Acidithiobacillus ferroxidans* solubilized nickel at room temperature, 30–37 °C, whereas organisms were unable to solubilize nickel at higher temperatures as 45 °C. In a pot experiment, inoculation of plants (*Brassica juncea* and *B. oxyrrhina*) with Ni-mobilizing strains of *Pseudomonas* sp. SRI2, *Psychrobacter* sp. SRS8, and *Bacillus* sp. SN9 maximized the biomass of the plants. In addition, strain SN9 was observed to increase Ni concentration in the root and shoot tissues of *B. juncea* and *B. oxyrrhina* (Ma et al. 2009).

21.19 Mechanisms of Ni Solubilization

The possible mechanisms of Ni solubilization by microbes include pH changes in the soil, siderophore production, and phosphate solubilization (Burd et al. 2000). Siderophore production in relation to Ni has significantly increased the size and chlorophyll content of leaf (Tank and Saraf 2009). Bacteria such as *Azotobacter chroococcum* (N-fixing bacteria), *Bacillus megaterium* (P-solubilizer) and *Bacillus mucilaginosus* (K-solubilizer), and *Bacillus* sp. RJ16 were reported to decrease the pH by producing acids which enhance the bioavailability of Ni in the soil (Arunakumara et al. 2015). Zaidi et al. (2006) reported a reduction in pH from 7.5 to 4.8 with the solubilizing *Bacillus subtilis* SJ-101, resulting in increased Ni availability. In addition, acidic soil conditions created by phosphate solubilization have shown to increase Ni accumulation in the presence of some bacteria (Rajkumar et al. 2008). Ma et al. (2009) reported the production of indole-3-acetic acid (IAA) and siderophores and utilization of 1-aminocyclopropane-1-carboxylic acid (ACC) by Ni-mobilizing bacterial strains.

21.20 Chlorine-Solubilizing IMOs

Chlorine is another micronutrient which is needed for the proper growth and processes in plants including osmotic and stomatal regulation, evolution of oxygen in photosynthesis, and disease resistance. Plant uses the ion (Cl^-) rather than the gas (Cl_2). In chloroplast, chloride is a structural constitute of photosystem II in the oxygen-evolving complex which is one of the three important cofactors (Kusunoki 2007). Chloride stimulates the ATPase at the tonoplast. However, excess chloride is accumulated in certain tissues such as guard cells, and their opening and closing is regulated by the flux of potassium and anions such as malate and chloride and therefore important for plant photosynthesis. It also maintains the rigidity of leaves (Chen et al. 2010).

Reports of Cl deficiency are rare in agriculture (Dordas 2008). Fairly a larger amount of Cl application was reported to enhance disease resistance in plants. These amounts are much higher than those required as a micronutrient but far less than those required to induce toxicity (Mann et al. 2004). Cl has been shown to be effective on a number of diseases such as stalk rot in corn, stripe rust in wheat, take-all in wheat, northern corn leaf blight and downy mildew of millet, and septoria in wheat (Graham and Webb 1991; Mann et al. 2004). The mechanism by which Cl increases resistance is not well understood. However, it appears to be nontoxic in vitro and does not stimulate lignin synthesis in wounded wheat leaves. It was suggested that Cl can compete with NO^{-3} absorption and influences the rhizosphere pH by suppressing nitrification and increasing the availability of Mn mediating reduction of $Mn^{III,IV}$ oxides which increases tolerance to pathogens.

As a beneficial micronutrient, Cl^- regulates increased fresh and dry biomass, greater leaf expansion, elongation of leaf and root cells, improved water relations, higher mesophyll diffusion to CO_2, and better water- and nitrogen-use efficiency (Colmenero-Flores et al. 2019). In most cases, deficient leaves exhibit distinct characteristic and continuous boundaries between the affected and healthy tissue and appear as blotchy leaf chlorosis and necrosis. In such cases, Cl deficiency may result in wilting and bronzing of leaves. Chlorine toxicity can occur naturally in plants grown in coastal soils due to the excess Cl. Chlorine toxicity usually results in necrosis along the leaf margins. Leaves are smaller than usual. They may be yellow and drop early. The symptoms first appear on mature leaves. In some species, chlorosis may also occur. Chlorine toxicity can result from air pollution, in the form of chlorine gas, or from excess chloride in the soil (Table 21.1).

Generally, soil contains sufficient amount of chloride in the soil. The plant available form of chlorine is an anionic form which is chloride (Cl^-). Anionic form is the dominant form of chlorine in soils. Chloride is thought to pass through the root by a symplastic pathway and is mobile within the plant (White and Broadley 2001). The content of Cl^- fluctuates greatly in soils. Most soils contain sufficient levels of chlorine. However, Cl may become deficient in inland soil under frequent high rainfall and irrigation. Plants may be able to absorb some metal-Cl complex such as $CdCl^+$, but with a minimal percentage (Weggler et al. 2004). Negatively charged chloride ion tends to be repelled from the surfaces of soil particles, making it difficult to form complexes readily with negatively charged mineral soils. Therefore, chloride in the bulk solution contains a higher concentration than in the diffuse layers surrounding soil particles. Water fluxes, relationship between precipitation, and evapotranspiration determine the movement of chloride ion within the soil (Chen et al. 2010).

There is no available literature on Cl-solubilizing microorganisms in the soil. However, the soil microbes make changes in soil pHs which will ultimately release chloride into the soils in plant available forms.

21.21 Mechanisms of Cl Solubilization

Most of the micronutrients are present as forms of chloride complexes in their cationic forms. Soil microorganisms produce acids which reduce the soil pH. Reduced pHs facilitate breaking down of metal-Cl complexes making them available for plants.

21.22 Boron-Solubilizing IMOs

Boron is a non-metal micronutrient which essentially optimizes plant growth and development. The critical concentration of B in plant tissues is 20–25 mg kg^{-1} (usually 35 mg kg^{-1}) on a dry mass basis (Ahmad et al. 2012). It plays an important role in cell wall synthesis and structural integration as well as in protein and enzymatic functioning of the cell membrane, providing improved membrane integrity (Brown et al. 2002). B is cross-linked with pectin assembly, glycosylinositol phosphorylceramides (GIPCs), and rhamnogalacturonan-II (RG-II) that control the tensile strength and porosity of the cell wall (Shireen et al. 2018). Optimum B concentration in cells enhances the plasma membrane hyperpolarization, while its deficiency alters the membrane potential and reduces H^+-ATPase activity. In young growing tissues, B acts primarily in cell division and elongation, and starvation leads to the inhibition in root elongation with deformed flower and fruit formation. Boron is also involved in phenolic metabolism and nitrogen metabolism in plants. The role of B in rhizobial N fixation, actinomycete symbiosis, and cyanophyceae heterocyst formation in leguminous crops has been highlighted in previous studies. B deficiency is thought to affect photosynthesis indirectly by weakening the vascular tissues responsible for ion transport (Rasheed 2009).

Boron affects the availability and uptake of other plant nutrients from the soil. B application increased the uptake and translocation of P, N, K, Zn, Fe, and Cu in leaves, buds, and seeds of cotton (Ahmed and Fujiwara 2010). Boron deficiency has occurred in over 132 crops and 80 countries during last 60 years. After Zn, B is the second most deficient micronutrient severely affecting the growth of crops on global scale (Alloway 2008). Deficiency symptoms depend on the age of the plant and include stunted root growth, restricted apical meristem growth, stunted root growth, reduced chlorophyll content, brittle leaves, and photosynthetic activity, disruption in ion transport, increased phenolic and lignin contents, and reduced crop yield (Shireen et al. 2018).

In the soil, the presence of boron is common as boric acid or borate. Boron is percolated in the form of uncharged molecules rather than as ions. It is extremely deficient in soils which are developed from calcareous, loessial, or alluvial deposits and also in highly leached soils (Borkakati and Takkar 2000). There are various other factors, including sandy/coarse texture, drought, alkalinity, liming, and intensive cultivation with more nutrient uptake and less fertilizer application, which affect the availability of B to plants (Ahmad et al. 2012). In many regions of the world such as Brazil, the USA, China, Japan, and Korea, B availability is limited which is resulted by its high solubility and leaching off by irrigation water or rainfall in shallow or coarse-textured soils. In drought conditions and in soils with less organic matter, the availability of B is low due to alkalization and breakdown of organic matter (Shireen et al. 2018). Optimal B availability in soils can be achieved by using several beneficial and eco-friendly techniques.

Boric acid uptake is affected by the transpiration stream. Enhancement of transpiration-driven water flow can be affected by plant growth-promoting bacteria which can increase B accumulation in plant. This may also cause B toxicity. Inoculation of these bacteria under low pH into soil increased the growth of rapeseed. Further, addition of P enhanced the uptake of B by rapeseed, while *B. pumilus* inoculation inhibited the growth of rapeseed under B supply (Masood et al. 2019). There have not been many studies done on increasing the bioavailability of B especially with indigenous microorganisms despite it being the second most deficient micronutrient affecting crops worldwide, therefore warranting investigations.

21.23 Molybdenum-Solubilizing IMOs

Molybdenum is most common in agricultural soils which can exist in several oxidation states ranging from zero to VI (Kaiser et al. 2005). In a plant, Mo performs various physiological and metabolic functions. Despite its requirement in small amounts for normal plant development, it plays a critical role in the regulation of various plant functions. The required concentration range in plant tissue for its normal function is between 0.3 and 1.5 ppm. Mo has been utilized by specific plant enzymes to participate in reduction and oxidative reactions (Thomas et al. 2017).

Molybdenum is an essential component in nitrogenase enzyme used by symbiotic nitrogen-fixing bacteria in legumes to fix atmospheric nitrogen. Plants also use Mo to convert inorganic phosphorus into organic forms in the plant (Beevers and Hageman 1969). Molybdenum deficiency can affect nitrogen deficiency in plant, since it is closely linked to nitrogen fixation process. Molybdenum is the only mobile micronutrient in the plants. Therefore, older and middle leaves show Mo deficiency symptoms early, but it spreads up to the stem and affects the new leaves (Table 21.1). Some plants such as poinsettias show thin chlorotic, leaf margins around the leaf perimeter followed by necrosis (Carruthers 2016a), which restricts plant growth and flower formation. Molybdenum deficiency or toxicity is uncommon in many plants. However, crops that are most sensitive to molybdenum deficiency are crucifers (broccoli, cauliflower, cabbage), legumes (beans, peas, clovers), poinsettias, and primula (Carruthers 2016a). Research has shown that high sulfates can reduce plant uptake of molybdenum (Kaiser et al. 2005).

The average concentration of Mo in the lithosphere is 2–3 mg kg^{-1} but can increase to a concentration like 300 mg kg^{-1} with significant content of organic matter (Kaiser et al. 2005; Reddy et al. 1997; Fortescue 1992). Different environmental factors such as soil pH, extent of water drainage, concentration of adsorbing oxides (e.g., Fe oxides), and organic compounds found in the soil influence the availability of molybdenum for plant growth. Molybdenum becomes more soluble and is accessible to plants mainly in its anion form MoO_4^- in alkaline soils, while it decreases in acidic soils ($pH < 5.5$) (Reddy et al. 1997). In agricultural soils, the complex that molybdenum is present depends on the chemical speciation of the soil zone. Molybdenite (MoS_2), wulfenite ($PbMoO_4$), and ferrimolybdenite [$Fe_2(MoO_4)$]

are the mineral forms of molybdenum found in rocks (Kaiser et al. 2005). Weathering releases Mo from solid mineral forms (Kaiser et al. 2005). Molybdenum is typically added to the soil by fertilization or by the addition of other chemicals such as sodium or ammonium molybdate.

Molybdate reduction by microbes has been reported from 100 years ago and includes mainly bacteria (Table 21.2). Potential mechanisms of reduction of molybdenum were first reported in *Escherichia coli*, *Thiobacillus ferrooxidans* (now *Acidithiobacillus ferrooxidans*), *Enterobacter cloacae* strain, *Pseudomonas* sp., *Serratia* spp., *Enterobacter* sp., *Acinetobacter calcoaceticus*, and *Klebsiella* sp. (Halmi et al. 2013). Mo-reducing soil bacteria are reported from Pakistan, Sudan (*Enterobacter* sp. strain Zeid-6), Indonesia, and Antarctica (*Pseudomonas* sp. strain DRY1). Except *Bacillus* sp., molybdenum-reducing bacteria are gram negative (Frascoli and Hudson-Edwards 2018). Many Mo-reducing bacteria isolated from Pakistani soils were resistant to high Mo concentrations (up to 50 mM) (Khan et al. 2014). Some plants eliminate Mo from their roots and shoots (e.g., *Cistus*, *Quercus* species), while some take up Mo without any harmful effects (e.g., *Baccharis* species) (Frascoli and Hudson-Edwards 2018). Autotrophic bacteria *Acidithiobacillus ferrooxidans* and *Thiobacillus thiooxidans* isolated from drainage from Kennecott's open-pit mine in Bingham Canyon, Utah, USA, were capable of bioleaching molybdenite (Frascoli and Hudson-Edwards 2018; Bryner and Anderson 1957).

Molybdenum is an essential component in nitrogenase enzyme and thereby in nitrogen fixation process (Hänsch and Mendel 2009). Molybdenum is the cofactor for the enzyme nitrate reductase during nitrogen assimilation (Hänsch and Mendel 2009). Also Mo is the key regulatory component for nodule initialization and maintenance of nitrogen fixation in legumes (Franco and Munns 1981), and the enzyme activity was elevated with a high Mo content. Several microorganisms are associated with the biofertilization process and they enhanced the activity of the Mo in plant growth and development. *Bradyrhizobium* inoculation and Mo fertilization with at least 50 g ha^{-1} increased the yield of peanut pods and kernels (Crusciol et al. 2019). Chatterjee and Bandyopadhyay (2017) found that the application of biofertilizers together with boron and molybdenum enhanced the growth, nodulation, and pod yield of vegetable cowpea in acid soil of eastern Himalayan region.

21.24 Mechanisms of Mo Solubilization

Molybdenum seems to induce the production of iron-chelating compounds such as dihydroxybenzoic acid (DHBA) and tris(catechol) protochelin and bis(catechol) azotochelin. Protochelin and azotochelin production also increases at lower concentrations of Mo and vanadium (V). Protochelin and azotochelin act as strong complexing agents for Fe(III), molybdate, and vanadate. Azotochelin (LH5) reacts with molybdate to form a 1:1 complex with Mo (VI) ($LH_4^- + MoO_4^{2-} \rightarrow MoO_2L^{3-} + 2H_2O$). Essential metals (Fe, Mo, and V) are acquired by these compounds while excluding toxic ones (such as W). At low concentrations, these catechol compounds form siderophore complexes with essential metals (Fe, Mo, V) and are taken up by the bacteria through specialized transport systems (McRose et al. 2017).

Microorganisms, those who have a capability of fixing atmospheric nitrogen and form ammonia, have Mo-nitrogenase or three forms of nitrogenase enzymes. Diazotrophic organisms such as *Klebsiella* and *Rhizobium* have Mo-nitrogenase. Mo in nitrogenase enzyme can also be combined with other metals depending on the microbial species. For example, *Azotobacter chroococcum* possesses Mo and V nitrogenases, *Rhodobacter capsulatus* has Mo and Fe nitrogenases, and *Azotobacter vinelandii* contains the three enzymes. The most commonly occurring nitrogenases have Mo in their active center to form the iron-molybdenum cofactor. Mo-nitrogenases exhibit higher efficacy than the alternative nitrogenases with respect to N_2 reduction rates. The metal cluster called FeMoco, an abbreviation for the iron-molybdenum cofactor, is the site of conversion of N_2 into ammonia (Hänsch and Mendel 2009).

Assimilatory nitrate reductases (Nas) that catalyze the first reaction in nitrate assimilation are molybdoenzymes. Molybdenum acts as the cofactor in nitrate reductase. Nitrate reductase in higher plants is proposed to be a homodimer, with two identical subunits joined and held together by the Mo cofactor. In bacteria, there are two types of nitrate reductases, first the ferredoxin- or flavodoxin-dependent enzyme found in cyanobacteria, *Azotobacter*, and the archaeon *Haloferax mediterranei*, and second, the NADH-dependent enzyme present in heterotrophic bacteria and *R. capsulatus*. The cyanobacterial nitrate reductase enzyme is an 80 kDa monomer, encoded by narB. Electrons from ferredoxin or flavodoxin are transferred to the cluster and the Mo-bis-molybdopterin guanine dinucleotide (Mo-bis-MGD) cofactor, the nitrate reduction site. NADH-dependent nitrate reductase catalizers are present in *Klebsiella*, *Bacillus*, and *Rhodobacter*. In *Klebsiella*, a catalytic subunit and a small electron transfer subunit are present in the enzyme, while the large subunit (nasA gene product) binds to Mo-bis-MGD (Schaechter 2009).

21.25 Conclusions and Future Prospects

This review discusses the significance of micronutrients for plants with special focus on increasing the productivity of crop plants and the role of microorganisms in sustainable agriculture while maintaining a sustainable environment. Furthermore, it presents developments in research and their applications in agriculture and environmental management and highlights their potential applications in achieving sustainable environments by taking into account their dimensions mainly in processes such as bio-composting, biodegradation, and other processes such as bioremediation and bioleaching. This review also highlights the fact that research has not given enough attention to microbes in terms of micronutrient solubilization and when did, the focuse has mainly been on bacteria with fungi to some extent. Other groups of microorganisms such as cyanobacteria and algae have not been investigated much for micronutrient solubility. Research especially on field trials/application of micronutrient-solubilizing microbes are currently being restricted to certain regions of the world mainly to India, Pakistan, and Africa and therefore should be expanded. With already successful stories and extensive future research, biofortification/ biofertilization of crops with micronutrient-solubilizing microorganisms will open up new avenues in addressing the "hidden hunger" in a sustainable environment in years to come while creating a clean and efficient environment for sustainable developmental goals to be achieved.

References

Abaid-Ullah M, Nadeem M, Hassan M et al (2015) Plant growth promoting rhizobacteria: an alternate way to improve yield and quality of wheat (*Triticum aestivum*). Int J Agric Biol 17 (1):51–60

Abdulsalam S, Bugaje IM, Adefila SS, Ibrahim S (2011) Comparison of bio-stimulation and bio augmentation for remediation of soil contaminated with spent motor oil. Int J Environ Sci Technol 8(1):187–194

Abou-Shanab RA, Angle JS, Delorme TA et al (2003) Rhizobacterial effects on nickel extraction from soil and uptake by *Alyssum murale*. New Phytol 158(1):219–224

Agusto da Costa AC, Duta FP (2001) Bioaccumulation of copper, zinc, cadmium and lead by *Bacillus* sp., *Bacillus cereus*, *Bacillus sphaericus*, and *Bacillus subtilis*. Braz J Microbiol 32:1–5

Ahmad W, Zia MH, Malhi SS, Niaz A, Ullah S (2012) Boron deficiency in soils and crops: a review. In: Crop plant. InTechOpen, London, pp 65–97

Ahmad M, Pataczek L, Hilger TH et al (2018) Perspectives of microbial inoculation for sustainable development and environmental management. Front Microbiol. https://doi.org/10.3389/fmicb.2018.02992

Ahmed I, Fujiwara T (2010) Mechanism of boron tolerance in soil bacteria. Can J Microbiol 56 (1):22–26

Akkaya Ö, Şeker MG, Çiftçi YÖ (2020) Plant growth promoting microbiome networks. In: Sangeetha J, Thangadurai D, Islam S (eds) Beneficial microbes for sustainable agriculture and environmental management. Apple Academic Press, Boca Raton

Allen HE (2002) Bioavailability of metals in terrestrial ecosystems: importance of partitioning for bioavailability to invertebrates, microbes, and plants. SETAC Foundation, Pensacola

Alloway BJ (2008) Micronutrients and crop production: an introduction. In: Micronutrient deficiencies in global crop production. Springer, Dordrecht, pp 1–39

Alloway BJ (2009) Soil factors associated with zinc deficiency in crops and humans. Environ Geochem Health 31(5):537–548

Alloway BJ, Jackson AP (1991) The behavior of heavy-metals in sewage sludge-amended soils. Sci Total Environ 100:151–176. https://doi.org/10.1016/0048-9697(91)90377-Q

Altomare C, Norvell WA, Björkman T, Harman GE (1999) Solubilization of phosphates and micronutrients by the plant growth-promoting and biocontrol fungus *Trichoderma harzianum* Rifai 1295-22. Appl Environ Microbiol 65(7):2926–2933

Anitha S, Padma Devi SN, Sunitha Kumari K (2015) Isolation and identification of zinc solubilizing fungal isolates from agricultural fields. Indian J Agr Sci 85(12):638–1642

Arunakumara KKIU, Walpola BC, Yoon MH (2015) Bioaugmentation-assisted phytoextraction of Co, Pb and Zn: an assessment with a phosphate-solubilizing bacterium isolated from metal-contaminated mines of Boryeong Area in South Korea. BASE

Asea PEA, Kucey RMN, Stewart JWB (1988) Inorganic phosphate solubilization by two *Penicillium* species in solution culture and soil. Soil Biol Biochem 20:459–464

Ashraf MY, Iqbal N, Ashraf M, Akhter J (2014) Modulation of physiological and biochemical metabolites in salt stressed rice by foliar application of zinc. J Plant Nutr 37:447–457. https://doi.org/10.1080/01904167.2013.864309

Avery SV (2001) Metal toxicity in yeast and the role of oxidative stress. Adv Appl Microbiol 49:111–142

Baeuerlein E (2000) Biomineralization. Wiley-VCH, Weinheim

Bagci SA, Ekiz H, Yilmaz A, Cakmak I (2007) Effects of zinc deficiency and drought on grain yield of field-grown wheat cultivars in Central Anatolia. J Agron Crop Sci 193:198–206. https://doi.org/10.1111/j.1439-037X.2007.00256.x

Bai C, Reilly CC, Wood BW (2006) Nickel deficiency disrupts metabolism of ureides, amino acids, and organic acids of young pecan foliage. Plant Physiol 140:433–443. https://doi.org/10.1104/pp.104.900185

Baligar VC, Fageria NK, He ZL (2001) Nutrient use efficiency in plants. Commun Soil Sci Plant Anal 32:921–950. https://doi.org/10.1081/CSS100104098

Bapiri A, Asgharzadeh A, Mujallali H et al (2012) Evaluation of Zinc solubilization potential by different strains of Fluorescent Pseudomonads. J Environ Manage 16(3):295298

Barker AV, Pilbeam DJ (2015) Handbook of plant nutrition, 2nd edn. CRC Press Taylor & Francis Group, Boca Raton

Barron AR, Wurzburger N, Bellenger JP et al (2009) Molybdenum limitation of asymbiotic nitrogen fixation in tropical forest soils. Nat Geosci 2:42–45. https://doi.org/10.1038/ngeo366

Bazylinski DA (2001) Bacterial mineralization. In: Encyclopedia of materials: science and technology. Elsevier, Amsterdam, pp 441–448

Beevers L, Hageman RH (1969) Nitrate reduction in higher plants. Annu Rev Plant Physiol 20 (1):495–522

Berg G, Zachow C, Müller H, Phillips J, Tilcher R (2013) Next-generation bio-products sowing the seeds of success for sustainable agriculture. Agronomy 3:648–656

Bhardwaj D, Ansari MW, Sahoo R, Tuteja N (2014) Biofertilizers function as key player in sustainable agriculture by improving soil fertility, plant tolerance and crop productivity. Microb Cell Fact 13(1):66. https://doi.org/10.1186/1475-2859-13-66

Biari A, Gholami A, Rahmani H (2008) Growth promotion and enhanced nutrient uptake of maize (*Zea mays* L.) by application of plant growth promoting rhizobacteria in arid region of Iran. J Biol Sci 8. https://doi.org/10.3923/jbs.2008.1015.1020

Borkakati K, Takkar PN (2000) Forms of boron in acid alluvial and lateritic soils in relation to ecosystem and rainfall distribution. In: Proceedings of international conference on managing resources for sustainable agricultural production in the 21st century. Better Crops, vol 2, pp 127–128

Bottjer DJ (2005) Geobiology and the fossil record: eukaryotic, microbes, and their interactions. Palaeogeogr Palaeoclimatol Palaeoecol 219:5–21

Bowen AD, Davidson FA, Keatch R, Gadd GM (2007) Induction of contour sensing in *Aspergillus niger* by stress and its relevance to fungal growth mechanics and hyphal tip structure. Fungal Genet Biol 44:484–491

Boyd RS, Shaw JJ, Martens SN (1994) Nickel hyperaccumulation defends *Streptanthus polygaloides* (Brassicaceae) against pathogens. Am J Bot 81:294–300. https://doi.org/10.1007/s00374-015-1039-7

Boyington JC, Gladyshev VN, Kangulov SV et al (1997) Crystal structure of formate dehydrogenase H: catalysis involving Mo, molybdopterin, selenocysteine, and an Fe4S4 cluster. Science 275:1305–1308

Broadley M, Brown P, Cakmak I, Rengel Z, Zhao F (2012) Functions of nutrient: micronutrients. In: Marschner P (ed) Marschner's mineral nutrition of higher plants, 3rd edn. Elsevier, Oxford, pp 243–248

Brodrick SJ, Amijee F, KipeNolt JA, Giller KE (1995) Seed analysis as a means of identifying micronutrient deficiencies of *Phaseolus vulgaris* L in the tropics. Trop Agric 72:277–284
Brown PH, Welch RM, Cary EE (1987) Nickel: a micronutrient essential for all higher plants. Plant Physiol 85:801–803
Brown PH, Bellaloui N, Wimmer MA et al (2002) Boron in plant biology. Plant Biol 4(2):205–223
Brown GE, Trainor TP, Chaka AM (2008) Geochemistry of mineral surfaces and factors affecting their chemical reactivity. In: Nilsson A, Pettersson LGM, Norskov JK (eds) Chemical bonding at surfaces and interfaces. Elsevier, Amsterdam, pp 457–509
Bryner LC, Anderson R (1957) Microorganisms in leaching sulfide minerals. Ind Eng Chem 49:1721–1724
Burd GI, Dixon DG, Glick BR (2000) Plant growth promoting bacteria that decrease heavy metal toxicity in plants. Can J Microbiol 46:237–245
Burgstaller W, Schinner F (1993) Leaching of metals with fungi. J Biotechnol 27:91–116
Burkert B, Robson A (1994) Zn uptake in subterranean clover (*Trifolium subterraneum* L.) by three vesicular–arbuscular mycorrhizal fungi in a root–free sandy soil. Soil Biol Biochem 26:1117–1124
Cai M, Yao J, Yang H, Wang R, Masakorala K (2013) Aerobic biodegradation process of petroleum and pathway of main compounds in water flooding well of Dagang oil field. Bioresour Technol 144:100–106
Cakmak I (2008) Enrichment of cereal grains with zinc: agronomic or genetic biofortification? Plant Soil 302:1–17. https://doi.org/10.1007/s11104-007-9466-3
Carruthers S (2016a) Copper: deficiency and toxicity. Practical hydroponics and greenhouses 173:50–53
Carruthers S (2016b) Molybdenum: deficiency and toxicity. Practical hydroponics and greenhouses 170:40–42
Chatterjee R, Bandyopadhyay S (2017) Effect of boron, molybdenum and biofertilizers on growth and yield of cowpea (*Vigna unguiculata* L. Walp.) in acid soil of eastern Himalayan region. J Saudi Soc Agric Sci 16(4):332–336
Chen YX, Wang YP, Lin Q, Luo YM (2005) Effect of copper-tolerant rhizosphere bacteria on mobility of copper in soil and copper accumulation by *Elsholtzia splendens*. Enviro Int 31 (6):861–866
Chen C, Huang D, Liu J (2009) Functions and toxicity of nickel in plants: recent advances and future prospects. Clean Soil Air Water 37(4-5):304–313. https://doi.org/10.1002/clen.200800199
Chen W, He ZL, Yang XE, Mishra S, Stoffella PJ (2010) Chlorine nutrition of higher plants: progress and perspectives. J Plant Nutr 33(7):943–952
Chorover J, Kretzschmar R, Garcia-Pichel F, Sparks DL (2007) Soil biogeochemical processes within the critical zone. Elements 3:321–326
Chun-Li W, Shiuan-Yuh C, Chiu-Chung Y (2014) Present situation and future perspective of bio-fertilizer for environmentally friendly agriculture. Annu Rep 21:1–5
Colmenero-Flores JM, Franco-Navarro JD, Cubero-Font P et al (2019) Chloride as a beneficial macronutrient in higher plants: new roles and regulation. Int J Mol Sci 20(19):4686
Colombo C, Palumbo G, He JZ et al (2014) Review on iron availability in soil: interaction of Fe minerals, plants, and microbes. J Soils Sediments 14(3):538–548
Crowley DE, Wang YC, Reid CPP, Szaniszlo PJ (1991) Mechanisms of iron acquisition from siderophores by microorganisms and plants. In: Iron nutrition and interactions in plants. Springer, Dordrecht, pp 213–232
Crusciol CAC, Ferrari Neto J, Mui TS et al (2019) Rhizobial inoculation and molybdenum fertilization in peanut crops grown in a no tillage system after 20 years of pasture. Rev Bras Ciên Solo 43:e0170399
Dalton AD, Russell SA, Evans HJ (1988) Nickel as a micronutrient element for plants. Biofactors 1:11–16

Datnoff LE, Elmer WH, Huber DM (2007) Mineral nutrition and plant disease. The American Phytopathological Society, St. Paul

Deng X, Chai L, Yang Z et al (2012) Bioleaching of heavy metals from a contaminated soil using indigenous *Penicillium chrysogenum* strain F1. J Hazard Mater 233:25–32

Dhaliwal SS, Naresh RK, Mandal A et al (2019) Dynamics and transformations of micronutrients in agricultural soils as influenced by organic matter build-up: a review. Environ Sustain Indicators 1–2:10007

Dimkpa CO, Bindraban PS (2016) Fortification of micronutrients for efficient agronomic production: a review. Agron Sustain Dev 36:7

Dimkpa CO, Merten D, Svatoš A et al (2009) Metal-induced oxidative stress impacting plant growth in contaminated soil is alleviated by microbial siderophores. Soil Biol Biochem 41:154–162. https://doi.org/10.1016/j.soilbio.2008.10.010

Dimkpa CO, McLean JE, Britt DW, Anderson AJ (2012) Bioactivity and biomodification of Ag, ZnO and CuO nanoparticles with relevance to plant performance in agriculture. Ind Biotechnol 8:344–357. https://doi.org/10.1089/ind.2012.0028

Dimkpa CO, Latta DE, McLean JE et al (2013a) Fate of CuO and ZnO nano- and microparticles in the plant environment. Environ Sci Technol 47:4734–4742. https://doi.org/10.1021/es304736y

Dimkpa CO, McLean JE, Britt DW, Anderson AJ (2013b) Antifungal activity of ZnO nanoparticles and their interactive effect with a biocontrol bacterium on growth antagonism of the plant pathogen, *Fusarium graminearum*. BioMetals 26:913–924. https://doi.org/10.1007/s10534-013-9667-

Dimkpa CO, Hansen T, Stewart J et al (2015a) ZnO nanoparticles and root colonization by a beneficial pseudomonad influence metal responses in bean (*Phaseolus vulgaris*). Nanotoxicology 9:271–278. https://doi.org/10.3109/17435390.2014.900583

Dimkpa CO, McLean JE, Britt DW, Anderson AJ (2015b) Nano-CuO and interaction with nano-ZnO or soil bacterium provide evidence for the interference of nanoparticles in metal nutrition of plants. Ecotoxicology 24:119–129. https://doi.org/10.1007/s10646-014-1364-x

Djajadi RS, Hidayati SN (2020) Effect of NPK fertilizer, biofertilizer containing N fixer and P solubilizer, and green manure of *C. juncea* on nutrients uptake and growth of sugarcane. Earth Environ Science 418. https://doi.org/10.1088/1755-1315/418/1/012068

Dordas C (2008) Role of nutrients in controlling plant diseases in sustainable agriculture. A review. Agron Sustain Dev 28(1):33–46

Duffner A, Hoffland E, Stomph TJ et al (2014) Eliminating zinc deficiency in rice-based systems. VFRC report 2014/2. Virtual Fertilizer Research Center, Washington, DC, p 35

Duffy BK, Defago G (1999) Environmental factors modulating antibiotic and siderophore biosynthesis by *Pseudomonas fluorescens* biocontrol strains. Appl Environ Microbiol 65:2429–2438

Eggert K, von Wirén N (2013) Dynamics and partitioning of the ionome in seeds and germinating seedlings of winter oilseed rape. Metallomics 5:1316. https://doi.org/10.1039/C3MT00109A

Ehrlich HL (1996) How microbes influence mineral growth and dissolution. Chem Geol 132:5–9

Ehrlich HL (1997) Microbes and metals. Appl Microbiol Biotechnol 48:687–692

Ehrlich HL, Newman DK (2009) Geomicrobiology, 5th edn. CRC Press/Taylor & Francis, Boca Raton

Engelhardt MM, de Lima FRD, Martins GC et al (2020) Copper phytotoxicity in agricultural crops cultivated in tropical soils. Semina Ciênc Agrár 41(6 suppl 2):2883–2898

Evans I, Solberg E, Huber DM (2007) Copper and plant diseases. In: Datnoff LE, Elmer WH, Huber DM (eds) Mineral nutrition and plant disease. American Phytopathological Society, St. Paul, pp 177–188

Fageria NK (2002) Influence of micronutrients on dry matter yield and interaction with other nutrients in annual crops. Pesq agropec bras Brasília 37:1765–1772

Farooq M, Wahid A, Siddique KHM (2012) Micronutrient application through seed treatments – a review. J Soil Sci Plant Nutr 12(1):125–142

Fasim F, Ahmed N, Parsons R, Gadd GM (2002) Solubilization of zinc salts by a bacterium isolated from the air environment of a tannery. FEMS Microbiol Lett 213:1–6

Fernandez V, Ebert G, Winkelmann G (2005) The use of microbial siderophores for foliar iron application studies. Plant Soil 272:245–252. https://doi.org/10.1007/s11104-5212-2
Fomina MA, Alexander IJ, Hillier S, Gadd GM (2004) Zinc phosphate and pyromorphite solubilization by soil plant-symbiotic fungi. Geomicrobiol J 21:351–366
Fomina MA, Alexander IJ, Colpaert JV, Gadd GM (2005a) Solubilization of toxic metal minerals and metal tolerance of mycorrhizal fungi. Soil Biol Biochem 37:851–866
Fomina MA, Hillier S, Charnock JM et al (2005b) Role of oxalic acid over-excretion in toxic metal mineral transformations by *Beauveria caledonica*. Appl Environ Microbiol 71:371–381
Fortescue JAC (1992) Landscape geochemistry: retrospect and prospect. Appl Geochem 7:1–53
Franco AA, Munns DN (1981) Response of *Phaseolus vulgaris* L. to molybdenum under acid conditions [Nitrogen fixation]. Soil Science Soc Am J 45(6):1144–1148
Franco JA, Bañón S, Vicente MJ et al (2011) Root development in horticultural plants grown under abiotic stress conditions–a review. J Hortic Sci Biotechnol 86(6):543–556
Frascoli F, Hudson-Edwards KA (2018) Geochemistry, mineralogy and microbiology of molybdenum in mining-affected environments. Minerals 8(2):42
Fridovich I (1978) The biology of oxygen radicals. Science 201:875–880
Gadd GM (1992) Metals and microorganisms: a problem of definition. FEMS Microbiol Lett 100:197–204
Gadd GM (1993) Interactions of fungi with toxic metals. New Phytol 124:25–60
Gadd GM (2007) Geomycology: biogeochemical transformations of rocks, minerals, metals and radionuclides by fungi, bioweathering and bioremediation. Mycol Res 111:3–49
Gadd GM (2008) Bacterial and fungal geomicrobiology: a problem with communities? Geobiology 6:278–284
Gadd GM (2010) Metals, minerals and microbes: geomicrobiology and bioremediation. Microbiology 156:609–643
Gadd GM, Raven JA (2010) Geomicrobiology of eukaryotic microorganisms. Geomicrobiol J 27 (6–7):491
Gadd GM, Sayer JA (2000) Fungal transformations of metals and metalloids. In: Lovley DR (ed) Environmental microbe–metal interactions. American Society for Microbiology, Washington, DC, pp 237–256
Gandhi A, Muralidharan G (2016) Assessment of zinc solubilizing potentiality of *Acinetobacter* sp. isolated from rice rhizosphere. Eur J Soil Biol 76:1–8
Ghiorse WC (1984) Biology of iron- and manganese-depositing bacteria. Annu Rev Microbiol 38:515–550
Ghosh PK, Hazra KK, Kumar N et al (2014) Resource conservation technology in pulses. In: Zinc in crops and human health - an overview. Scientific Publishers, India, pp 213–223
Ghosh S, Mohanty S, Nayak S et al (2016) Molecular identification of indigenous manganese solubilising bacterial biodiversity from manganese mining deposits. J Basic Microbiol 56 (3):254–262
Gilmour C, Riedel G (2009) Biogeochemistry of trace metals and metalloids. In: Likens GE (ed) Encyclopedia of inland waters. Elsevier, Amsterdam, pp 7–15
Glasauer S, Beveridge TJ, Burford EP et al (2004) Metals and metalloids, transformations by microorganisms. In: Hillel D, Rosenzweig C, Powlson DS, Scow KM, Singer MJ, Sparks DL, Hatfield J (eds) Encyclopedia of soils in the environment. Elsevier, Amsterdam, pp 438–447
Gleeson D, McDermott F, Clipson N (2007) Understanding microbially active biogeochemical environments. Adv Appl Microbiol 62:81–104
Gleeson DB, Melville K, McDermott FP et al (2010) Molecular characterization of fungal communities in sandstone. Geomicrobiol J 27(6–7):559–571
Golldack D, Li C, Mohan H, Probst N (2014) Tolerance to drought and salt stress in plants: unraveling the signaling networks. Front Plant Sci 5:151. https://doi.org/10.3389/fpls.2014.00151
Gounot AM (1994) Microbial oxidation and reduction of manganese: consequences in groundwater and applications. FEMS Microbiol Rev 14(4):339–349

Govarthanan M, Lee GW, Park JH et al (2014) Bioleaching characteristics, influencing factors of Cu solubilization and survival of *Herbaspirillum* sp. GW103 in Cu contaminated mine soil. Chemosphere 109:42–48

Graham DR, Webb MJ (1991) Micronutrients and disease resistance and tolerance in plants. In: Mortvedt JJ, Cox FR, Shuman LM, Welch RM (eds) Micronutrients in agriculture, 2nd edn. Soil Science Society of America, Madison, pp 329–370

Hafeez B, Khanif YM, Saleem M (2013) Role of zinc in plant nutrition-a review. Am J Exp Agric 3 (2):374

Halmi MIE, Zuhainis SW, Yusof MT et al (2013) Hexavalent molybdenum reduction to Mo-blue by a sodium-dodecyl-sulfate-degrading *Klebsiella oxytoca* strain DRY14. Biomed Res Int 1:384541. https://doi.org/10.1155/2013/384541

Hänsch R, Mendel RR (2009) Physiological functions of mineral micronutrients (Cu, Zn, Mn, Fe, Ni, Mo, B, Cl). Currt Opin Plant Biol 12(3):259–266

Harrison AP Jr (1984) The acidophilic Thiobacilli and other acidophilic bacteria that share their habitat. Annu Rev Microbiol 38(1):265–292

Havlin J, Beaton JD, Tisdale SL, Nelson WL (2005) Soil fertility and fertilizers: an introduction to nutrient management. Pearson Educational, Inc, Upper Saddle River, New Jersey

Hebbern CA, Laursen KH, Ladegaard AH et al (2009) Latent manganese deficiency increases transpiration in barley (*Hordeum vulgare*). Physiol Plants 135:307–316

Hider RC (1984) Siderophore mediated absorption of iron. Struct Bonding 58:25–87

Hider RC, Kong X (2010) Chemistry and biology of siderophores. Nat Prod Rep 27(5):637–657

Hinsinger P, Bengough AG, Vetterlein D, Young IM (2009) Rhizosphere: biophysics, biogeochemistry and ecological relevance. Plant Soil 321(1–2):117–152

Hochella MF (2002) Sustaining Earth: thoughts on the present and future roles in mineralogy in environmental science. Mineral Mag 66:627–652

Holden JF, Adams MWW (2003) Microbe–metal interactions in marine hydrothermal vents. Curr Opin Chem Biol 7:160–165

Horton DJ, Theis KR, Uzarski DG, Learman DR (2019) Microbial community structure and microbial networks correspond to nutrient gradients within coastal wetlands of the Laurentian Great Lakes. FEMS Microbiol Ecol 95(4). https://doi.org/10.1093/femsec/fiz033

Hossain B, Hirata N, Nagatomo Y et al (1997) Internal zinc accumulation is correlated with increased growth in rice suspension culture. J Plant Growth Regul 16:239–243. https://doi.org/10.1007/PL00007003

Huang PM, Wang MC, Wang MK (2004) Mineral–organic–microbial interactions. In: Hillel D, Rosenzweig C, Powlson DS, Scow KM, Singer MJ, Sparks DL, Hatfield J (eds) Encyclopedia of soils in the environment. Elsevier, Amsterdam, pp 486–499

Huang PM, Wang MK, Chiu CY (2005) Soil mineral–organic matter–microbe interactions: impacts on biogeochemical processes and biodiversity in soils. Pedobiologia (Jena) 49:609–635

Huang H, Zhao Y, Xu Z et al (2020) A high Mn (II)-tolerance HM7, isolated from manganese ore and its bio sorption characteristics. Peer J 8:e8589. https://doi.org/10.7717/peerJ.8589

Huber DM, Wilhelm NS (1988) The role of manganese in disease resistance. In: Graham RD, Hannam RJ, Uren NC (eds) Manganese in soils and plants. Kluwer Academic, London, pp 155–173

Hussain D, Haydon MJ, Wang Y et al (2004) P-type ATPase heavy metal transporters with roles in essential zinc homeostasis in *Arabidopsis*. Plant Cell 16:1327–1339

Hussain A, Arshad M, Zahir ZA, Asghar M (2015) Prospects of zinc solubilizing bacteria for enhancing growth of maize. Pak J Agric Sci 52(4):915–922

Hussain A, Zahir ZA, Asghar HN et al (2018) Zinc solubilizing bacteria for zinc biofortification in cereals: a step toward sustainable nutritional security. In: Role of rhizospheric microbes in soil. Springer, Singapore, pp 203–227

Itelima JU, Bang WJ, Sila MD et al (2018) A review: biofertilizer - a key player in enhancing soil fertility and crop productivity. Microbial Biotechnol Rep 1(1):22–29

Javed H, Akhtar MJ, Asghar HN, Jamil A (2018) Screening of zinc solubilizing bacteria and their potential to increase grain concentration in wheat (*Triticum aestivum*). Int J Agric Biol 20:547–553

Jha Y (2019) The importance of zinc-mobilizing rhizosphere bacteria to the enhancement of physiology and growth parameters for paddy under salt-stress conditions. Jordan J Biol Sci 12 (2):167–173

Jin CW, Li GX, Yu XH, Zheng SJ (2010) Plant Fe status affects the composition of siderophore-secreting microbes in the rhizosphere. Ann Bot 105:835–841. https://doi.org/10.1093/aob/mcq071

Johnson DB (2006) Biohydrometallurgy and the environment: intimate and important interplay. Hydrometallurgy 83:153–166

Johnson SE, Lauren JG, Welch RM, Duxbury JM (2005) A comparison of the effects of micronutrient seed priming and soil fertilization on the mineral nutrition of chickpea (*Cicer arietinum*), lentil (*Lens culinaris*), rice (*Oryza sativa*) and wheat (*Triticum aestivum*) in Nepal. Expert Agric 41:427–448

Joy EJM, Ander EL, Young SD et al (2014) Dietary mineral supplies in Africa. Physiol Plantarum 151:208–229. https://doi.org/10.1111/ppl.12144

Joy EJM, Stein AJ, Young SD et al (2015) Zinc-enriched fertilisers as a potential public health intervention in Africa. Plant Soil 389:1–24. https://doi.org/10.1007/s11104-015-2430-8

Benton-Jones J Jr (1991) Plant tissue analysis of micronutrients. In: Mortvedt JJ, Cox FR, Shuman LM, Welch RM (eds) Micronutrients in agriculture, 2nd edn. Soil Science Society of America, Madison, pp 477–521

Benton-Jones J Jr, Wolff JB, Mills HA (1992) Plant analysis handbook, a practical sampling, preparation, analysis and interpretation guide. Micro-Macro Publishing, Athens

Jung HC, Lebeault JM, Pan JG (1998) Surface display of *Zymomonas mobilis* levansucrase by using the ice-nucleation protein of *Pseudomonas syringae*. Nat Biotechnol 16:576–580

Kabata-Pendias A, Pendias H (2001) Trace elements in soils and plants, 3rd edn. CRC Press, Boca Raton

Kaiser B, Gridley KL, Ngaire BJ et al (2005) The role of molybdenum in agricultural plant production. Ann Bot 96(5):745–754

Kalinowski BE, Liermann LJ, Givens S, Brantley SL (2000) Rates of bacteria-promoted solubilization of Fe from minerals: a review of problems and approaches. Chem Geol 169(3-4):357–370

Kamran S, Shahid I, Baig DN et al (2017) Contribution of zinc solubilizing bacteria in growth promotion and zinc content of wheat. Front Microbiol 8:2593

Kanwar JS, Youngdahl JJ (1985) Micronutrient needs of tropical food crops. In: Vlek PLG (eds) Micronutrients in tropical food crop production. Developments in plant and soil sciences, vol 14. Springer, Dordrecht, pp 43–67. https://doi.org/10.1007/978-94-009-5055-9_2

Karthick R, Rajalingam GV, Praneetha S et al (2018) Effect of micronutrients on growth, flowering and yield of bitter gourd (*Momordica charantia*) cv. CO1. Int J Chem Stud 6(1):845–848

Katyal JC, Ponamperuma FN (1974) Zinc deficiency: a widespread nutritional disorder in rice in Agusan del Norte. Philipp Agric 58:79–89

Kaura T, Ranaa KL, Koura D et al (2020) Microbe-mediated biofortification for micronutrients: present status and future challenges. In: Rastegari AA, Yadav AN, Yadav N (eds) New and future developments in microbial biotechnology and bioengineering. Elsevier, Amsterdam, pp 1–18

Ke T, Zhang J, Tao Y et al (2020) Individual and combined application of Cu-tolerant *Bacillus* spp. enhance the Cu phytoextraction efficiency of perennial ryegrass. Chemosphere 263:127952

Keuskamp DH, Kimber R, Bindraban PS et al (2015) Plant exudates for nutrient uptake. VFRC Report 2015/4. Virtual Fertilizer Research Center, Washington, DC, p 53

Khan JA, Bhatnagar RM (1977) Studies on solubilization of insoluble phosphates by microorganisms. Fertil Technol 14:329–333

Khan M, Noushina I, Masood A et al (2013) Salicylic acid alleviates adverse effects of heat stress on photosynthesis through changes in proline production and ethylene formation. Plant Signal Behav 8:e26374

Khan A, Halmi MIE, Shukor MY (2014) Isolation of Mo-reducing bacterium in soils from Pakistan. J Environ Microbiol Toxicol 2:38–41

Kim BH, Gadd GM (2008) Bacterial physiology and metabolism. Cambridge University Press, Cambridge

Kloepper JW, Leong J, Teintze M, Schroth MN (1980) Pseudomonas siderophores—a mechanism explaining disease-suppressive soils. Curr Microbiol 4:317–320. https://doi.org/10.1007/BF02602840

Kochian LV (1991) Mechanisms of micronutrient uptake and translocation in plants. In: Mortvedt JJ, Cox FR, Shuman LM, Welch RM (eds) Micronutrients in agriculture, 2nd edn. Soil Science Society of America, Madison, pp 229–296

Konhauser K (2007) Introduction to geomicrobiology. Blackwell, Oxford

Krämer U (2005) Phytoremediation: novel approaches to cleaning up polluted soils. Curr Opin Biotechnol 16:133–141

Krämer U, Clemens S (2006) Functions and homeostasis of zinc, copper, and nickel in plants. In: Tamás M, Martinoia E (eds) Molecular biology of metal homeostasis and detoxification from microbes to man. Springer, Berlin, pp 214–272

Kucey RMN (1988) Effect of *Penicillium bilaji* on the solubility and uptake of P and micronutrients from soil by wheat. Can J Soil Sci 68(2):261–270

Kumar BL, Gopal DS (2015) Effective role of indigenous microorganisms for sustainable environment. 3 Biotech 5(6):867–876. https://doi.org/10.1007/s13205-015-0293-6

Kumar R, Mehrotra NK, Nautiyal BD et al (2009) Effect of copper on growth, yield and concentration of Fe, Mn, Zn and Cu in wheat plants (*Triticum aestivum* L.). J Environ Biol 30:485–488

Kumar V, Pandita S, Sidhu GPS et al (2020) Copper bioavailability, uptake, toxicity and tolerance in plants: a comprehensive review. Chemosphere 2:127810

Kumssa DB, Joy EJM, Ander EL et al (2015) Dietary calcium and zinc deficiency risks are decreasing but remain prevalent. Sci Rep 5:10974. https://doi.org/10.1038/srep10974

Kusunoki M (2007) Mono-manganese mechanism of the photosystem II water splitting reaction by a unique Mn4Ca cluster. Biochim Biophys Acta Bioenergetics 1767(6):484–492

Kutman UB, Yildiz B, Ozturk L, Cakmak I (2010) Biofortification of durum wheat with zinc through soil and foliar applications of nitrogen. Cereal Chem 87(1):1–9

Li K, Ramakrishna W (2011) Effect of multiple metal resistant bacteria from contaminated lake sediments on metal accumulation and plant growth. J Hazard Mater 189:531–539

Lim HS, Kim SD (1997) Role of siderophores in biocontrol of *Fusarium solani* and enhanced growth response of bean by *Pseudomonas fluorescens* GL20. J Microbiol Biotechnol 7:12–20

Lindsay WL (1988) Solubility and redox equilibria of iron compounds in soils. In: Iron in soils and clay minerals. Springer, Dordrecht, pp 37–62

Lindsay WL (1991) Inorganic equilibria affecting micronutrients in soils. In: Mortvedt JJ, Cox FR, Shuman LM, Welch RM (eds) Micronutrients in Agriculture, 2nd edn. Soil Science Society of America, Madison, pp 89–112

Lindsay WL, Schwab AP (1982) The chemistry of iron in soils and its availability to plants. J Plant Nutr 5(4-7):821–840

Liu W, Yang C, Shi S, Shu W (2014) Effects of plant growth-promoting bacteria isolated from copper tailings on plants in sterilized and non-sterilized tailings. Chemosphere 97:47–53

Lovley DR, Giovannoni SJ, White DC et al (1993) *Geobacter metallireducens* gen. nov. sp. nov., a microorganism capable of coupling the complete oxidation of organic compounds to the reduction of iron and other metals. Arch Microbiol 159:336–344

Lowenstam HA (1981) Minerals formed by organisms. Science 211(4487):1126–1131

Lower SK, Hochella MF, Beveridge TJ (2001) Bacterial recognition of mineral surfaces: nanoscale interactions between Shewanella and α-FeOOH. Science 292(5520):1360–1363

Luber E, Taureau JC (1990) Plant analysis as diagnosis of mineral nutrition. C R Acad Agric France 76:151–166
Ma Y, Rajkumar M, Freitas H (2009) Isolation and characterization of Ni mobilizing PGPB from serpentine soils and their potential in promoting plant growth and Ni accumulation by *Brassica* spp. Chemosphere 75(6):719–725
Maathuis FJM (2013) Plant mineral nutrients: methods and protocols. Methods in molecular biology series. Springer, London. https://doi.org/10.1007/978-1-62703-152-3
Maathuis F, Diatloff E (2013) Roles and functions of plant mineral nutrients. Methods Mol Biol 953:1–21. https://doi.org/10.1007/978-1-62703-152-3_1
Macalady J, Banfield JF (2003) Molecular geomicrobiology: genes and geochemical cycling. Earth Planet Sci Lett 209:1–17
Mahdi S, Hassan G, Samoon S et al (2010) Bio-fertilizers in organic agriculture. J Phytol 2:42–54
Mallarino AP, Oltmans RR, Prater JR, Villavicencio CX (2011) Nutrient uptake by corn and soybean, removal and recycling with crop residue. In Integrated crop management conference. Iowa State University No. 103
Mandakovic D, Maldonado J, Pulgar R et al (2018) Microbiome analysis and bacterial isolation from Lejía Lake soil in Atacama Desert. Extremophiles 22:665–673. https://doi.org/10.1007/s00792-018-1027-6
Mann RL, Kettlewell PS, Jenkinson P (2004) Effect of foliar-applied potassium chloride on septoria leaf blotch of winter wheat. Plant Pathol 53:653–659
Marschner P (2012) Marschner's mineral nutrition of higher plants, 3rd edn. Elsevier, Oxford
Marschner H, Dell B (1994) Nutrient uptake in mycorrhizal symbiosis. In: Robson AD, Abott LK, Malaccjuk N (eds) Management of mycorrhizas in agriculture, horticulture and forestry. Academic Publishers, Netherlands, pp 89–102
Martino E, Perotto S, Parsons R, Gadd GM (2003) Solubilization of insoluble inorganic zinc compounds by ericoid mycorrhizal fungi derived from heavy metal polluted sites. Soil Biol Biochem 35(1):133–141
Masood S, Zhao XQ, Shen RF (2019) *Bacillus pumilus* increases boron uptake and inhibits rapeseed growth under boron supply irrespective of phosphorus fertilization. AoB Plants 11 (4). https://doi.org/10.1093/aobpla/plz036
Maurya PK, Yadav LM, Patel P, Thakur G (2018) Effect of micronutrients application on quality and shelf life of kharif onion (*Allium cepa* L.). Int J Chem Studies 6(2):1121–1124
McRose DL, Baars O, Morel FMM, Kraepiel AML (2017) Siderophore production in *Azotobacter vinelandii* in response to Fe-, Mo- and V-limitation. Environ Microbiol 19(9). https://doi.org/10.1111/1462-2920.13857
Mengel K (1994) Iron availability in plant tissues-iron chlorosis on calcareous soils. Plant Soil 165 (2):275–283
Mengel K, Kosegarten H, Kirkby EA, Appel T (eds) (2001) Principles of plant nutrition. Springer, London
Merlot S (2020) Understanding nickel responses in plants: more than just an interaction with iron homeostasis. Plant Cell Physiol 61(3):443–444
Miethke M, Marahiel MA (2007) Siderophore-based iron acquisition and pathogen control. Microbiol Mol Biol Rev 71(3):413–451
Migocka M, Malas K (2018) Plant responses to copper: molecular and regulatory mechanisms of copper uptake, distribution and accumulation in plants. In: Plant micronutrient use efficiency. Academic, Boston, pp 71–86
Miller GW, Pushnik JC, Welkie GW (1984) Iron chlorosis, a world wide problem, the relation of chlorophyll biosynthesis to iron. J Plant Nutr 7(1-5):1–22
Miller ER, Lei X, Ullrey DE (1991) Trace elements in animal nutrition. In: Mortvedt JJ, Cox FR, Shuman LM, Welch RM (eds) Micronutrients in agriculture, 2nd edn. Soil Science Society of America, Madison, pp 593–662
Mishra D, Kim DJ, Ralph DE et al (2008) Bioleaching of spent hydro-processing catalyst using acidophilic bacteria and its kinetics aspect. J Hazard Mater 152:1082–1091

Miwa H, Ahmed I, Yoon J, Yokota A, Fujiwara T (2008) *Variovorax boronicumulans* sp. nov., a boron-accumulating bacterium isolated from soil. Int J Syst Evol Microbiol 58(1):286–289
Miyata N, Tani Y, Iwahori K, Soma M (2004) Enzymatic formation of manganese oxides by an Acremonium-like hyphomycete fungus, strain KR21-2. FEMS Microb Ecol 47:101–109
Mohapatra S, Bohidar S, Pradhan N et al (2007) Microbial extraction of nickel from Sukinda chromite overburden by *Acidithiobacillus ferrooxidans* and *Aspergillus* strains. Hydrometallurgy 85(1):1–8
Molden D, Oweis T, Steduto P et al (2010) Improving agricultural water productivity: between optimism and caution. Agric Water Manage 97:528–535. https://doi.org/10.1016/j.agwat.2009.03.023
Mondal S, Bose B (2019) Impact of micronutrient seed priming on germination, growth, development, nutritional status and yield aspects of plants. J Plnt Nutr 42(19):2577
Monreal CM, DeRosa M, Mallubhotla SC et al (2015) Nanotechnologies for increasing the crop use efficiency of fertilizer-micronutrients. Biol Fertil Soils. https://doi.org/10.1007/s00374-015-1073-5
Movahhedy-Dehnavy M, Modarres-Sanavy SAM, Mokhtassi-Bidgoli A (2009) Foliar application of zinc and manganese improves seed yield and quality of safflower (*Carthamus tinctorius* L) grown under water deficit stress. Ind Crops Prod 30:82–92. https://doi.org/10.1016/j.indcrop.2009.02.004
Mulligan C, Kamali M, Gibbs B (2004) Bioleaching of heavy metals from a low-grade mining ore using *Aspergillus niger*. J Hazard Mater 110:77–84
Mumtaz MZ, Ahmad M, Jamil M, Hussain T (2017) Zinc solubilizing *Bacillus* spp. potential candidates for biofortification in maize. Microbiol Res 202:51–60
Nestel P, Bouis HE, Meenakshi JV, Pfeiffer W (2006) Biofortification of staple food. J Nutr 136:1064–1067
Nielsen FH (1992) Nutritional requirements for boron, silicon, vanadium, nickel, and arsenic: current knowledge and speculation. FASEB J 5:2661–2667
Nouren S, Bhatti HN, Ilyas S (2011) Bioleaching of copper, aluminum, magnesium and manganese from brown shale by *Ganoderma lucidum*. Afr J Biotechnol 10(52):10664–10673
Obrador A, Novillo J, Alvarez JM (2003) Mobility and availability to plants of two zinc sources applied to a calcareous soil. Soil Sci Soc Am J 67:564–572
Oliver MA, Gregory PJ (2015) Soil, food security and human health: a review. Eur J Soil Sci 66:257–276. https://doi.org/10.1111/ejss.12216
Orme-Johnson WH (1992) Nitrogenase structure: where to now? Science 257:1639–1640
Patrick K (2010) Manganese nodules and the age of the ocean floor. J Creat 24(3):82–86
Patten C, Glick B (2002) The role of bacterial indoleacetic acid in the development of the host plant root system. Appl Environ Microbiol 68:3795–3801
Pawar A, Ismail S, Mundhe S, Patil VD (2015) Solubilization of insoluble zinc compounds by different microbial isolates *in vitro* condition. J Trop Agric 33:865–869
Pilon M, Abdel-Ghany SE, Cohu CM et al (2006) Copper cofactor delivery in plant cells. Curr Opin Plant Biol 9:256–263. https://doi.org/10.1016/j.pbi.2006.03.007
Pitiwittayakul N, Tanasupawat S (2020) Plant growth-promoting endophytic bacteria and their potential benefits in asian countries. In: Sangeetha J, Thangadurai D, Islam S (eds) Beneficial microbes for sustainable agriculture and environmental management. Apple Academic Press, Boca Raton, pp 81–114
Polaccao J, Freyermuth S, Gerendas J, Cianzio S (1999) Soybean genes involved in nickel insertion into urease. J Exp Bot 50:1149–1156. https://doi.org/10.1093/jxb/50.336.1149
Puig S, Andrés-Colás N, García-Molina A, Peñarrubia L (2007) Copper and iron homeostasis in *Arabidopsis*: responses to metal deficiencies, interactions and biotechnological applications. Plant Cell Environ 30:271–290. https://doi.org/10.1111/j.1365-3040.2007.01642.x
Purvis OW, Pawlik-Skowronska B (2008) Lichens and metals. In: Stress in yeasts and filamentous fungi. Elsevier, Amsterdam, pp 175–200

Radzki W, Gutierrez Mañero FJ, Algar E et al (2013) Bacterial siderophores efficiently provide iron to iron-starved tomato plants in hydroponics culture. Antonie van Leeuwenhoek 104:321–330. https://doi.org/10.1007/s10482-013-9954-9

Raja N (2013) Biopesticides and biofertilizers: ecofriendly sources for sustainable agriculture. J Biofertil Biopestic 4:112

Rajkumar M, Ma Y, Freitas H (2008) Characterization of metal-resistant plant-growth promoting *Bacillus weihenstephanensis* isolated from serpentine soil in Portugal. J Basic Microbiol 48:500–508

Rana A, Saharan B, Kabi S et al (2011) Providencia, a PGPR with biocontrol potential elicits defense enzymes in wheat. Ann Pl Protec Sci 19(1):138–141

Rasheed MK (2009) Role of boron in plant growth: a review. J Agric Res 47(3):329–338

Reddy KJ, Munn LC, Wang L (1997) Chemistry and mineralogy of molybdenum in soils. In: Gupta UC (ed) Molybdenum in agriculture. Cambridge University Press, Cambridge, pp 4–22

Riaz U, Mehdi SM, Iqbal S et al (2020) Bio-fertilizers: eco-friendly approach for plant and soil environment. In: Hakeem K, Bhat R, Qadri H (eds) Bioremediation and biotechnology. Springer, Cham, pp 189–213. https://doi.org/10.1007/978-3-030-35691-0_9

Rietra RPJJ, Heinen M, Dimkpa CO, Bindraban PS (2015) Effects of nutrient antagonism and synergism on fertilizer use. VFRC report 2015/5. Virtual Fertilizer Research Center, Washington, DC, p 42

Robson RL, Eady RR, Richardson TH, Miller RW, Hawkins M, Postgate JR (1986) The alternative nitrogenase of *Azotobacter chroococcum* is a vanadium enzyme. Nature 322:388–390

Römheld V, Marschner H (1991) Function of micronutrients in plants. In: Mortvedt JJ, Cox FR, Shuman LM, Welch RM (eds) Micronutrients in agriculture. Soil Science Society of America, Madison, pp 297–328

Rosen K, Zhong WL, Martensson A (2005) Arbuscular mycorrhizal fungi mediated uptake of Cs-137 in leek and ryegrass. Sci Total Environ 338:283–290

Ryan J, Rashid A, Torrent J et al (2013) Micronutrient constraints to crop production in the Middle East-West Asia region: significance, research and management. Adv Agron 122:1–84. https://doi.org/10.1016/b978-0-12-417187-9.00001-2

Samourgiannidis G, Matsi T (2013) Comparison of two sequential extraction methods and the DTPA method for the extraction of micronutrients from acidic soils. Commun Soil Sci Plant Anal 44:38–49. https://doi.org/10.1080/00103624.2012.734030

Sangeetha J, Thimmappa SC, Thangadurai D et al (2020) Dynamic potential of indigenous and effective microbes in wastewater treatment processes. In: Sangeetha J, Thangadurai D, Islam S (eds) Beneficial microbes for sustainable agriculture and environmental management. Apple Academic Press, Boca Raton, pp 319–351

Saravanabhavan S, Govindasamy M, Natesan S, Gopal S (2020) Bacteriological removal of azo dyes: an eco-friendly approach. In: Sangeetha J, Thangadurai D, Islam S (eds) Beneficial microbes for sustainable agriculture and environmental management. Apple Academic Press, Boca Raton, pp 219–258

Saravanan VS, Subramoniam SR, Raj SA (2004) Assessing in vitro solubilization potential of different zinc solubilizing bacterial (ZSB) isolates. Braz J Microbiol 35(1–2):121–125

Saravanan VS, Kalaiarasan P, Madhaiyan M, Thangaraju M (2007) Solubilization of insoluble zinc compounds by *Gluconacetobacter diazotrophicus* and the detrimental action of zinc ion (Zn^{2+}) and zinc chelates on root knot nematode *Meloidogyne incognita*. Lett Appl Microbiol 44 (3):235–241

Saravanan VS, Malavath RK, Sa T (2011) Microbial zinc solubilization and their role on plants. https://doi.org/10.1007/978-3-642-21061-7_3

Sarker M, Rahman MM (2020) Microbial degradation of wastes for environmental protection. In: Sangeetha J, Thangadurai D, Islam S (eds) Beneficial microbes for sustainable agriculture and environmental management. Apple Academic Press, Boca Raton

Sayer JA, Cotter-Howells JD, Watson C et al (1999) Lead mineral transformation by fungi. Curr Biol 9:691–694

Schaechter M (2009) Encyclopedia of microbiology. Academic, Boston
Scheer H (ed) (1991) Chlorophylls. CRC Press, Boca Raton
Servin A, Elmer W, Mukherjee A et al (2015) A review of the use of engineered nanomaterials to suppress plant disease and enhance crop yield. J Nanopart Res 17:92. https://doi.org/10.1007/s11051-015-2907-7
Sharma CP (2006) Plant micronutrients. Science Publishers, Enfield
Sharma N, Singh A, Batra N (2020) Impact of soil-plant microbe interaction in metal-contaminated soils. In: Sangeetha J, Thangadurai D, Islam S (eds) Beneficial microbes for sustainable agriculture and environmental management. Apple Academic Press, Boca Raton
Sheng X, Sun L, Huang Z et al (2012) Promotion of growth and Cu accumulation of bio-energy crop (*Zea mays*) by bacteria: implications for energy plant biomass production and phytoremediation. J Environ Manage 103:58–64
Shirasu K, Lahaye T, Tan MW, Zhou FS, Azevedo C, Schulze-Lefert P (1999) A novel class of eukaryotic zinc-binding proteins is required for disease resistance signaling in barley and development in *C. elegans*. Cell 99:355–366. https://doi.org/10.1016/S0092-8674(00)81522-6
Shireen F, Nawaz MA, Chen C et al (2018) Boron: functions and approaches to enhance its availability in plants for sustainable agriculture. Int J Mol Sci 19(7):1856
Sidhu MK, Raturi HC, Kachwaya ES, Sharma A (2019) Role of micronutrients in vegetable production: a review. J Pharmacogn Phytochem SP1:332–340
Silver S, Phung LT (1996) Bacterial heavy metal resistance: new surprises. Annu Rev Microbiol 50:753–789
Singh SK (2020) Microbes for iron chlorosis remediation in peach. In: Biostimulants in plant science. IntechOpen, London
Singh B, Natesan SKA, Singh BK, Usha K (2003) Improving zinc efficiency of cereals under zinc deficiency. Curr Sci 88:36–44
Sirko A, Brodzik R (2000) Plant ureases: roles and regulation. Acta Biochim Polon 47:1189–1195
Sobral LF, Smyth JT, Fageria NK, Stone LF (2013) Comparison of copper, manganese and zinc extraction with Mehlich 1, Mehlich 3 and DTPA solutions for soils of the Brazilian coastal tablelands. Commun Soil Sci Plant Anal 44:2507–2513. https://doi.org/10.1080/00103624.2013.812731
Stangoulis JCR, Graham RD (2007) Boron and plant disease. In: Datnoff LE, Elmer WH, Huber DM (eds) Mineral nutrition and plant disease. American Phytopathological Society, St. Paul, pp 139–153
Strasser H, Burgstaller W, Schinner F (1994) High yield production of oxalic acid for metal leaching purposes by *Aspergillus niger*. FEMS Microbiol Lett 119:365–370
Straub KL, Benz M, Schink B (2001) Iron metabolism in anoxic environments at near neutral pH. FEMS Microbiol Ecol 34(3):181–186
Subramanian KS, Tenshia V, Jayalakshmi K, Ramachandran V (2009) Role of arbuscular mycorrhizal fungus (*Glomus intraradices*) (fungus aided). Agric Biotechnol Sustain Dev 1:29–38
Sukla LB, Panchanadikar V (1993) Bioleaching of lateritic nickel ore using a heterotrophic microorganism. Hydrometallurgy 32:373–379
Tank N, Saraf M (2009) Enhancement of plant growth and decontamination of nickel-spiked soil using PGPR. J Basic Microbiol 49(2):195–204
Tao GC, Tian SJ, Cai MY, Xie GH (2008) Phosphate-solubilizing and -mineralizing abilities of bacteria isolated from soils. Pedosphere 18:515–523
Taran NY, Gonchar OM, Lopatko KG et al (2014) The effect of colloidal solution of molybdenum nanoparticles on the microbial composition in rhizosphere of *Cicer arietinum* L. Nanoscale Res Lett 9:289. https://doi.org/10.1186/1556-276X-9-289
Tarekegn MM, Salilih FZ, Osman IAI (2020) Microbes used as a tool for bioremediation of heavy metal from the environment. Cogent Food Agric 6:1. https://doi.org/10.1080/23311932.2020.1783174
Tariq SR, Ashraf A (2016) Comparative evaluation of phytoremediation of metal contaminated soil of firing range by four different plant species. Arabian J Chem 9(6):806–814

Tariq M, Hameed S, Malik KA, Hafeez FY (2007) Plant root associated bacteria for zinc mobilization in rice. Pak J Bot 39:245–253

Tarkalson DD, Jolley VD, Robbins CW, Terry RE (1998) Mycorrhizal colonization and nutrient uptake of dry bean in manure and composted manure treated subsoil and untreated topsoil and subsoil. J Plant Nutr 21:1867–1878

Thomas E, Rathore I, Tarafdar JC (2017) Bioinspired production of molybdenum nanoparticles and its effect on chickpea (*Cicer arietinum* L). J Bionanosci 11(2):153–159

Tripathi DK, Singh S, Singh S et al (2015) Micronutrients and their diverse role in agricultural crops: advances and future prospective. Acta Physiol Plant 37(7):139

Tyler G, Olsson T (2001) Plant uptake of major and minor mineral elements as influenced by soil acidity and liming. Plant Soil 230(2):307–321

Umi KMS, Sariah M (2006) Utilization of microbes for sustainable agriculture in Malaysia: current status. Bio prospecting and management of microorganisms. In National conference on agro biodiversity conservation and sustainable utilization, pp 27–29

Uroz S, Calvaruso C, Turpault MP, Frey-Klett P (2009) Mineral weathering by bacteria: ecology, actors and mechanisms. Trends Microbiol 17:378–387

Vaid S, Kumar B, Sharma A et al (2014) Effect of zinc solubilizing bacteria on growth promotion and zinc nutrition of rice. J Soil Sci Plant Nutr 14:889–910

Van Campen DR (1991) Trace elements in human nutrition. In: Mortvedt JJ, Cox FR, Shuman LM, Welch RM (eds) Micronutrients in agriculture, 2nd edn. Soil Science Society of America, Madison, pp 663–701

Van Ho A, Ward DM, Kaplan J (2002) Transition metal transport in yeast. Annu Rev Microbiol 56:237–261

Vanlauwe B, Descheemaeker K, Giller K et al (2014) Integrated soil fertility management in Sub-Saharan Africa: unravelling local adaptation. Soil Discuss 1:1239–1286. https://doi.org/10.5194/soild-1-1239-2014

Vansuyt G, Robin A, Briat JF et al (2007) Iron acquisition from Fe-pyoverdine by *Arabidopsis thaliana*. Mol Plant Microbe Interact 20:441–447. https://doi.org/10.1094/MPMI-20-4-0441

Vatansever R, Ozyigit II, Filiz E (2017) Essential and beneficial trace elements in plants, and their transport in roots: a review. Appl Biochem Biotechnol 181:464–482. https://doi.org/10.1007/s12010-016-2224-3

Vaughan DJ, Pattrick RAD, Wogelius RA (2002) Minerals, metals and molecules: ore and environmental mineralogy in the new millennium. Mineral Mag 66:653–676

Vazquez P, Holguin G, Puente ME et al (2000) Phosphate-solubilizing microorganisms associated with the rhizosphere of mangroves in a semiarid coastal lagoon. Biol Fertil Soils 30 (5–6):460–468

Velazquez E, Silva LR, Ramírez-Bahena MH, Peix A (2016) Diversity of potassium-solubilizing microorganisms and their interactions with plants. In: Meena VS, Maurya BR, Verma JP, Meena RS (eds) Potassium solubilizing microorganisms for sustainable agriculture. Springer, New Delhi, pp 99–110. https://doi.org/10.1007/978-81-322-2776-2_7

Velu G, Ortiz-Monasterio I, Cakmak I, Hao Y, Singh RP (2014) Biofortification strategies to increase grain zinc and iron concentrations in wheat. J Cereal Sci 59:365–372. https://doi.org/10.1016/j.jcs.2013.09.001

Verma S, Kuila A (2019) Bioremediation of heavy metals by microbial process. Environ Technol Innov 14:100369

Vessey JK (2003) Plant growth promoting rhizobacteria as biofertilizers. Plant Soil 255:571–586

Vidyashree DN, Muthuraju N, Panneerselvam P, Mitra D (2018) Organic acids production by zinc solubilizing bacterial isolates. Int J Curr Microbiol Appl Sci 7(10):626–633

Voortman R, Bindraban PS (2015) Beyond N and P: toward a land resource ecology perspective and impactful fertilizer interventions in Sub-Saharan Africa. VFRC Report 2015/1. Virtual Fertilizer Research Center, Washington, DC, p 49

de Vrind JPM, de Vrind-de Jong EW, de Voogt JHH, Westbrook P, Boogerd FC, Rosson RA (1986) Manganese oxidation by spores and spore coats of a marine *Bacillus* species. Appl Environ Microbiol 52:1091–1100
Wackett L, Brusseau GA, Householder SR, Hanson RS (1989) Survey of microbial oxygenases: trichloroethylene degradation by propane-oxidizing bacteria. Appl Environ Microbiol 55 (11):2960–2964. https://doi.org/10.1128/AEM.55.11.2960-2964.1989
Wang X, Wiens M, Divekar M, Grebenjuk VA (2011) Isolation and characterization of a Mn (II) oxidizing *Bacillus* strain from the demosponge suberites domuncula. Mar Drugs 9:1–28. https://doi.org/10.3390/md9010001
Warren LA, Haack EA (2001) Biogeochemical controls on metal behaviour in freshwater environments. Earth Sci Rev 54:261–320
Waters BM, Sankaran RP (2011) Moving micronutrients from the soil to the seeds: genes and physiological processes from a biofortification perspective. Plant Sci 180:562–574. https://doi.org/10.1016/j.plantsci.2010.12.003
Weggler K, McLaughlin MJ, Graham RD (2004) Effect of chloride in soil solution on the plant availability of biosolid-borne cadmium. J Environ Qual 33:496–504
Wei Z, Hillier S, Gadd GM (2012) Biotransformation of manganese oxides by fungi: solubilization and production of manganese oxalate biominerals. Environ Microbiol 14(7):1744–1753
Welch RM, Shuman L (1995) Micronutrient nutrition of plants. Crit Rev Plant Sci 14(1):49–82
White PJ, Broadley MR (2001) Chloride in soils and its uptake and movement within the plant: a review. Ann Bot 88(6):967–988
White PJ, Broadley MR (2009) Biofortification of crops with seven mineral elements often lacking in human diets—iron, zinc, copper, calcium, magnesium, selenium and iodine. New Phytol 182:49–84. https://doi.org/10.1111/j.1469-8137.2008.02738.x
White C, Sayer JA, Gadd GM (1997) Microbial solubilization and immobilization of toxic metals: key biogeochemical processes for treatment of contamination. FEMS Microbiol Rev 20 (3–4):503–516
White PJ, George TS, Dupuy LX et al (2013) Root traits for infertile soils. Front Plant Sci 4:193. https://doi.org/10.3389/fpls.2013.00193
Whiting SN, Souza MD, Terry N (2001) Rhizosphere bacteria mobilize Zn for hyper accumulator by *Thlaspi caerulescens*. Environ Sci Technol 35:3144–3150
Widdel F, Schnell S, Heising S et al (1993) Ferrous iron oxidation by anoxygenic phototrophic bacteria. Nature 362(6423):834–836
Wu SC, Cheung KC, Luo YM (2006) Wong effects of inoculation of plant growth promoting rhizobacteria on metal uptake by *Brassica juncea*. Environ Pollut 140:124–135
Wuana A, Okieimen FE (2011) Heavy metals in contaminated soils: a review of sources, chemistry, risks and best available strategies for remediation. ISRN Ecol 10:5402–402647. https://doi.org/10.5402/2011/402647
Xi-Wen Y, Xiao-Hong T, Xin-Chun L, William G, Yu-Xian C (2013) Foliar zinc fertilization improves the zinc nutritional value of wheat *(Triticum aestivum* L.) grain. Afr J Biotechnol 10:14778–14785
Yadav LM, Singh YP, Kumar J et al (2018) Response of zinc and boron application on yield, yield parameters and storage quality of garlic (*Allium sativum* L.) var. G-282. J Pharmacogn and Phytochem 7(1):1768–1770
Yamamoto I, Takashi S, Liu S-M, Ljungdahl LG (1983) Purification and properties of NADP-dependent dehydrogenase from *Clostridium thermoaceticum*, a tungsten-selenium-iron protein. J Biol Chem 258:1826–1832
Yang J, Kloepper JW, Ryu CM (2009) Rhizosphere bacteria help plants tolerate abiotic stress. Trends Plant Sci 14:1–4. https://doi.org/10.1016/j.tplants.2008.10.004
Yang R, Luo C, Chen Y et al (2013) Copper-resistant bacteria enhance plant growth and copper phytoextraction. Int J Phytoremediat 15(6):573–584

Yaseen M, Ahmed W, Shahbaz M (2013) Role of foliar feeding of micronutrients in yield maximization of cotton in Punjab. Turk J Agric For 37:420–426. https://doi.org/10.3906/tar-1206-56

Youssef MMA, Eissa MFM (2014) Biofertilizers and their role in management of plant parasitic nematodes: a review. Biotechnol Pharm Res 5(1):1–6

Yruela I (2009) Copper in plants: acquisition, transport and interactions. Funct Plant Biol 26:409–430. https://doi.org/10.1071/FP08288

Zaheer A, Malik A, Sher A (2019) Isolation, characterization, and effect of phosphate-zinc-solubilizing bacterial strains on chickpea (*Cicer arietinum* L.) growth. Saudi J Biol Sci 26 (5):1061–1067

Zaidi S, Usmani S, Singh BR, Musarrat J (2006) Significance of *Bacillus subtilis* strain SJ-101 as a bioinoculant for concurrent plant growth promotion and nickel accumulation in *Brassica juncea*. Chemosphere 64:991–997

Chapter 22
Synergistic Interaction of Methanotrophs and Methylotrophs in Regulating Methane Emission

Vijaya Rani, Rajeev Kaushik, Sujan Majumder, A. T. Rani, Asha Arambam Devi, Pratap Divekar, Priyanka Khati, K. K. Pandey, and Jagdish Singh

Abstract The atmospheric methane concentration is increasing rapidly at the rate of around 10 ppb/year. A concerted effort is required to reduce methane emission. Methanotrophs possess methane monooxygenase enzyme system and can consume a major portion of the methane produced in the environment. These microbes play a major role in the single-carbon-driven microbial food web. Microbial interaction is an important component of microbial ecology studies, and its role in community functioning and various biogeochemical cycles still remains unclear. A synergistic interaction occurs between the methanotrophs and non-methane-utilizing methylotrophs (NUM) in the natural ecosystem. The intermediates produced by the methanotrophs can be used as a carbon source by the NUM and support its existence. On the other hand, NUM consumes toxic intermediates like methanol and formaldehyde of the methanotrophs and prolongs their growth. The consumption of the intermediates (methanol, formaldehyde and formate) of the methane utilization pathway by methylotrophs as a result of cross-feeding enhances the methane utilization rate of that ecosystem. Co-inoculation of methanotrophs and NUM in the natural habitat particularly paddy ecosystem can aid in the reduction of net methane emission. This chapter highlights the role of microbial interactions, particularly between methanotrophs and methylotrophs, that can be harnessed to mitigate methane emission from the methane-rich environment.

Keywords Methylotrophs · Methanotrophs · Cross-feeding · Methane · Methanol

---

V. Rani (✉) · S. Majumder · A. T. Rani · P. Divekar · K. K. Pandey
Division of Crop Protection, ICAR-Indian Institute of Vegetable Research, Varanasi, India

R. Kaushik · A. A. Devi
Division of Microbiology, ICAR-Indian Agricultural Research Institute, New Delhi, India

P. Khati
ICAR-Vivekananda Parvatiya Krishi Anusandhan Sansthan, Almora, Uttarakhand, India

J. Singh
ICAR-Indian Institute of Vegetable Research, Varanasi, India

P. Bhatt et al. (eds.), *Microbial Technology for Sustainable Environment*,
https://doi.org/10.1007/978-981-16-3840-4_22

22.1 Introduction

Methane, the principal component of natural gas, is a colourless, odourless greenhouse gas (GHG) and contributes around 14% to the total greenhouse gas emission. Methane-rich environments like paddy fields, wetlands, sewage, landfills and digestive system of ruminants and termites possess huge diversity of methanogen, methanotrophs and other methylotrophs (Kirschke et al. 2013; Lee et al. 2014). Methanotrophs are those bacteria that can consume methane to meet their carbon and energy requirements before it gets released into the atmosphere and plays a major role in reducing net methane emission, thereby maintaining global carbon balance. On the other hand, methylotrophs are a diverse group of bacteria, yeast, fungi and archaea that can utilize C1 compounds like methanol, monomethylamine, dimethylamine, trimethylamine, methanesulfonate and dimethylsulfonate including methane as the sole source of carbon and energy. Traditional methanotrophs of the group Alpha- and Gammaproteobacteria widespread in Types I, II and X of methanotrophs with the prefix 'methylo' are well studied and investigated. Members of Betaproteobacteria of the genera *Methylophilus* (Madhaiyan et al. 2009), *Methylovorus* (Govorukhina and Trotsenko 1991) and *Methylibium* (Nakatsu et al. 2006) are recently recognized as methane oxidizers. Besides Proteobacteria, few members of the phylum Verrucomicrobia belonging to genera *Methylacidimicrobium* and *Methylacidiphilum* can also utilize methane (Op den Camp et al. 2009; Sharp et al. 2013; van Teeseling et al. 2014). Methylotrophs, on the other hand, cover all the three domains of microorganisms, viz. Bacteria, Archaea and Eukarya. Methylotrophs are microorganisms with a diverse group that besides utilizing methane (methanotrophs) also include those that can utilize other carbon substrates with no C–C bonds like methanol and other methylated compounds like methylamine (Chistoserdova et al. 2009). Since all methylotrophs cannot utilize methane, it can be said that all methanotrophs are methylotrophs but all methylotrophs are not methanotrophs. The ability to oxidize methanol has been reported in both prokaryotes and eukaryotes. Eukaryotic yeast belonging to genera *Candida*, *Pichia*, *Hansenula* and *Torulopsis* can utilize methanol as the sole carbon source (Negruţa et al. 2010). The prokaryotic members capable of oxidizing methanol are spread across Alphaproteobacteria (*Methylobacteria*, *Hyphomicrobium*), Betaproteobacteria (*Burkholderia*, *Methylibium*, *Methyloversatilis*) and Gammaproteobacteria (*Clonothrix fusca*, *Beggiatoa*, *Pseudomonas*), Verrucomicrobia, Cytophagales, Bacteroidetes (*Flavobacterium*), Firmicutes (*Bacillus methanolicus*, *Paenibacillus*) and Actinobacteria (*Microbacterium*, *Gordonia*, *Arthrobacter* and *Mycobacterium*) (Rani et al. 2021b; Kolb 2009; Madhaiyan et al. 2010; Waturangi et al. 2011; Jhala et al. 2014; McTaggart et al. 2015; Macey et al. 2020).

Non-methane-utilizing methylotrophs (NUM) are known to co-occur with methanotrophs in the natural ecosystem and affect methane utilization rate. Modern-day techniques like stable-isotope probing have indicated that a synergistic interaction occurs between the methanotrophic and non-methane-utilizing methylotrophic community (Shiau et al. 2020; van Grinsven et al. 2020). NUM is known to survive on methane-derived carbon particularly methanol and enhance the methane oxidation rate (Krause et al. 2017). Moreover, emergent properties like interaction-induced production of metabolites may arise when microorganisms interact leading to altered community functions otherwise not possible in the individual cells (Watrous et al. 2012; Abrudan et al. 2015). The transfer of metabolites from methanotrophs is not only restricted to NUM but to a wide range of microbial taxa as evident from the DNA-SIP study (Beck et al. 2013). These findings suggest that the assimilation of methane by methanotrophs in the methane-rich environment provides carbon to a diverse group of microbes (NUM and other heterotrophs) and sometimes to other life forms as well (Sanseverino et al. 2012; Oshkin et al. 2015; Yu et al. 2017).

22.2 Pathway for Methane Utilization

The unique ability of the methanotrophs to metabolize methane comes from the presence of methane monooxygenase (MMO) enzyme system. It's the first enzyme in the metabolic pathway of methanotrophs. MMO enzyme can be either housed in an intracytoplasmic membrane known as particulate MMO (pMMO) or suspended freely in the cytoplasm known as soluble MMO (sMMO). pMMO, a copper-containing, membrane-associated enzyme, is found in all the methanotrophs except for the genera *Methylocella* and *Methyloferula* (Theisen et al. 2005) but is less studied as it is membrane-associated when compared to sMMO. Both sMMO and pMMO enzyme can act on a wide range of substrates ranging from single carbon substrate, methane to as long as eight carbon compounds. They can act on alkane, alkenes, cycloalkanes and even halogenated derivatives (McDonald et al. 2006). Alkanes can be oxidized by a group of enzymes like cytochrome P450, alkane hydroxylases, sMMO and pMMO (Beilen and Funhoff 2005). However, among these, only sMMO and pMMO can act on methane. Some methanotrophs can produce both pMMO and sMMO, and their expression is regulated by copper concentration in the environment. pMMO is expressed under high copper-to-biomass ratios, whereas sMMO is expressed when the copper-to-biomass ratio is low (Murrell et al. 2000).

Methanol produced by the action of MMO is further acted upon by methanol dehydrogenase to produce formaldehyde. Methanol dehydrogenase (Mdh) is pyrroloquinoline quinone (PQQ)-containing NAD^+-dependent oxidoreductase enzyme (Anthony and Williams 2003). Formaldehyde produced by methylotrophs can be assimilated either by RuMP pathway (Type I) or by serine pathway (Type II). RuMP (ribulose monophosphate pathway) was earlier thought to be restricted to methylotrophic bacteria. However, they are now reported in various prokaryotic microorganisms and their role in formaldehyde fixation and detoxification has been established (Nobuo et al. 2006). Anaerobic methane oxidation by archaea differs in their mechanism to utilize methane. They utilize methane via reverse and modified methanogenesis pathway. Various intermediates of the methane oxidation pathway,

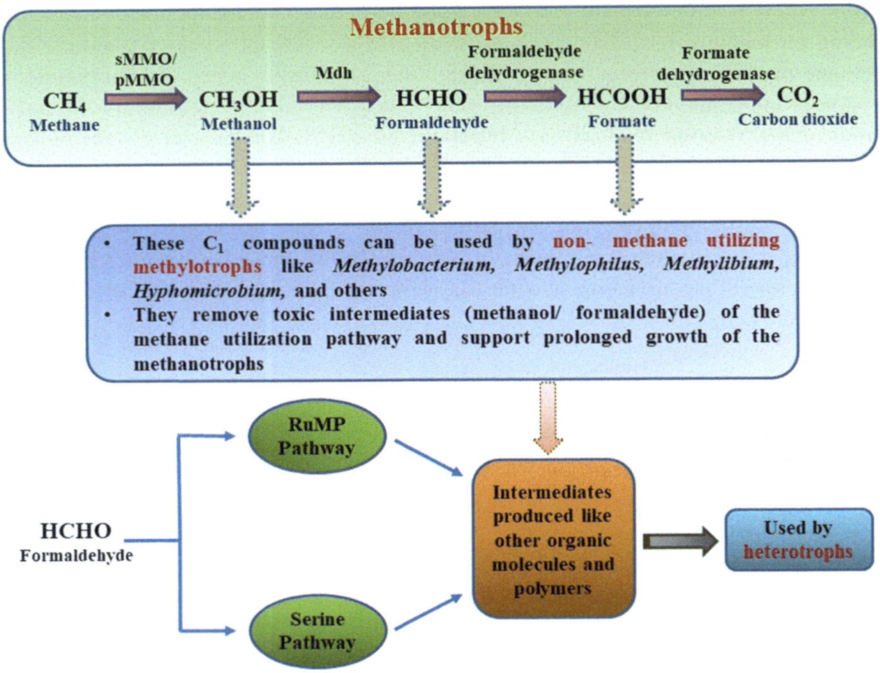

Fig. 22.1 Cross-feeding of metabolites produced by methanotrophs. *sMMO* soluble methane monooxygenases, *pMMO* particulate methane monooxygenases, *Mdh* methanol dehydrogenases, *RuMP* ribulose monophosphate pathway

viz. methanol, formaldehyde, formate, acetate and other organic acids when secreted by the methanotrophs, can be used as a growth substrate by both non-methane-utilizing methylotrophs and other heterotrophs as shown in Fig. 22.1. Methane-fed microbial microcosm study showed the abundance of methanotrophs of the family *Methylococcaceae* particularly *Methylobacter* along with other methylotrophs (*Methylotenera*) and heterotrophs, suggesting that there is a distribution of carbon from methane among diverse bacterial populations rather than a single type of microbe and thus methanotrophs play an important role in methane cycling (Oswald et al. 2016).

22.3 Cross-Feeding of Methane by NUM

Non-methanotrophs, particularly methylotrophs, play a major role in combating climate change in a methane-rich environment. Other heterotrophic forms may affect the growth of methanotrophic bacteria as a result of its various metabolic activities (secretion of growth factors or removal of toxic intermediates) (Hrsak and Begonja

2000). NUM is often known to coexist with methane-utilizing bacteria by cross-feeding on methane-derived carbon, particularly methanol (Takeuchi et al. 2019). Methane-oxidizing microorganisms possess monooxygenases that catalyze the conversion of methane to methanol. Methanol produced in the periplasmic space by the action of MMO enzyme system can easily diffuse out and serve as an alternative carbon source for other groups of microorganisms (Corder et al. 1986). Methanol-dependent cross-feeding between methanotrophs and other methylotrophs is largely dependent upon methanol excreted by the methanotrophic bacteria. Microflora residing in the rhizosphere, phyllosphere and non-rhizosphere or as endophytes of plants can utilize methanol and consume a major proportion of it (Kolb 2009; Iguchi et al. 2015; Chistoserdova and Kalyuzhnaya 2018). Methane-derived carbons (methanol, formaldehyde and formate) particularly methanol from methanotrophs can be utilized by NUM and enhance methane utilization rate by cross-feeding (Hanson and Hanson 1996; Qiu et al. 2008).

The findings of various researchers confirm the abundance of NUM along with methanogens and methanotrophs in the environment as shown in Table 22.1. Synergistic associations of methane and methanol oxidizers have been reported that favours the utilization of methane due to the removal of its intermediate methanol by the other partner (Krause et al. 2017; Jeong and Kim 2019). The coordinated response of *Methylococcaceae* (methanotroph) and *Methylophilaceae* (NUM) to changing methane and nitrate levels suggests that the two different functional groups of microbes may be involved in some type of cooperative behaviour (Beck et al. 2013). Similarly, methane oxidation by *Methylocystis* was found to increase in the presence of helper organism *Hyphomicrobium* according to the experiments carried out by (Jeong and Kim 2019). The transfer of methanol from the methanotrophic partner *Methylobacter tundripaludum* to the non-methanotroph methylotrophic partner *Methylotenera mobilis* has been confirmed in a microcosm model by (Krause et al. 2017). Their findings indicate that the non- methanotrophic partner induces a change in the gene expression of the methanotrophic partner causing the synthesis of less efficient methanol dehydrogenase enzyme (MxaF-type catalysing the conversion of methanol to formaldehyde) resulting in methanol excretion.

In the natural ecosystem, a complex interaction occurs between methanotroph, NUM and other heterotrophs. The success of single carbon-based microbial food web is determined by the effective transfer of intermediates from one microbial group to the other, allowing them to survive in methane-rich environments. A successful example of cross-feeding is the experiment carried out by Yu et al. (2017). They made a synthetic community of 50 bacterial cultures comprising of 10 methanotrophs (*Methylomonas*, *Methylobacter*, *Methylosarcina* and *Methylosinus*), 28 methanol-utilizing methylotrophs (*Methylotenera*, *Methylovorus*, *Methylophilus*, *Ancylobacter*, *Hyphomicrobium*, *Labrys*, *Methylobacteria*, *Methylopila*, *Paracoccus*, *Xanthobacter* and *Methyloversatilis*), 8 non-methanol-utilizing methylotrophs (*Aminobacter*, *Arthrobacter*, *Mycobacterium* and *Bacillus)* and 4 heterotrophs (*Pseudomonas*, *Janthinobacterium* and *Flavobacterium)* to study syntrophy in the aerobic methane-oxidizing environment (Yu et al. 2017). The metatranscriptomics analysis showed that across all the treatment with varying

Table 22.1 Studies showing cross-feeding of metabolites from methanotrophic to non-methane-utilizing methylotrophic partner

Methanotroph	Non-methanotroph	Substrate transferred	Salient findings	References
Methylobacter tundripaludum	*Methylotenera mobilis*	Methanol	Transcriptome analysis showed high expression of genes involved in methanol oxidation in the methylotrophic partner *Methylotenera mobilis* causes a change in the expression of methanotrophic partner causing it to secrete methanol	Krause et al. (2017)
Methylocaldum marinum	*Methyloceanibacter caenitepidi* (facultative methylotroph)	Acetate	Observed syntrophic association between *M. caenitepidi* and *M. marinum* Under co-culture condition, genes involved in serine pathway were downregulated in *M. caenitepidi* Organic compound probably acetate might be the major carbon source for the methylotrophic partner *M. caenitepidi*	Takeuchi et al. (2019)
Members of *Methylococcaceae* and others	*Methylophaga*	Not studied (may be methanol)	DNA-SIP experiment identified members of *Methylococcaceae* as major $^{13}CH_4$ consumers Microbial mats showed diverse assemblage of bacteria, protozoa with *Methylophaga* as key consumers of methane-derived organic matter	Paul et al. (2017)
Methylomicrobium	*Methylophaga*, *Hyphomicrobium* and other unrecognized methylotrophs	Not studied (may be methanol)	DNA-SIP study indicates that methane-derived carbon particularly methanol produced by methanotrophs may be consumed by *Methylophaga* and other related	Jensen et al. (2008)

(continued)

Table 22.1 (continued)

Methanotroph	Non-methanotroph	Substrate transferred	Salient findings	References
			uncultivated Gammaproteobacteria	
Methylococcaceae, particularly *Methylobacter*	*Methylophilaceae*, particularly *Methylotenera*	Not studied (may be methanol)	Observed coordinated response of both methanotroph and methylotroph to changing methane and nitrate levels, suggesting cooperative behaviour	Beck et al. (2013)
Methylobacter sp.	*Methylotenera*	Not studied (may be methanol)	No physical contact was required between the partners for the transfer of carbon as was confirmed by stable-isotope probing (SIP) and nanoscale secondary ion mass spectrometry (NanoSIMS) Requires nitrate for carbon transfer as it is potentially used by *Methylotenera* sp. and its deficiency may affect the methane oxidation rate of *Methylobacter* sp.	van Grinsven et al. (2020)
Methylococcaceae	*Methylophilaceae*	Methanol	Made synthetic bacterial communities of 50 isolates including methanotrophs, methylotrophs and heterotrophs with varying oxygen and methane levels Observed predominance of the methanotrophs of the family *Methylococcaceae* and non-methanotrophic methylotrophs of the family *Methylophilaceae* across all the treatments Vitamin B_{12} produced by *Methyloversatilis*	Yu et al. (2017)

(continued)

Table 22.1 (continued)

Methanotroph	Non-methanotroph	Substrate transferred	Salient findings	References
			may be shared among other community members	

nitrogen, oxygen and methane concentration, methanotroph of the family *Methylococcaceae* and methylotroph of the family *Methylophilaceae* did outcompete other species. Heterotrophs of the genera *Janthinobacterium* and *Pseudomonas* were detected in only a few treatments. Their research shows that methane-utilizing bacteria support the growth of other NUM and heterotrophs through the transfer of metabolites.

The bacterial community structure in a methane-rich environment is influenced by various factors like the existing concentration of methane, oxygen, nitrogen and other nutrients. The eutrophic lakes have high nitrate concentration and are one of the major sources of aquatic methane production. The nitrate in the aquatic ecosystem does influence the growth of microbial species and affect the cross-feeding of metabolites. The transfer of methane-derived carbon between *Methylobacter* (methanotroph) and *Methylotenera* (NUM) is based on the nitrate levels as it is required by the methylotrophic partner (van Grinsven et al. 2020). It has been observed that nitrate can cause stimulation in methane oxidation resulting in increased transfer of associated carbon compounds. Similarly, oxygen level selects the population of methanotrophs and methylotrophs, thereby determining their microbial diversity in a particular niche. The effect of oxygen on the conversion of methane-derived carbon has been studied (Wei et al. 2016). They observed greater transfer of methane-derived carbon at high O_2 concentration (21%) as compared to that observed at 2.5 and 5% O_2 concentration. They even reported higher microbial diversity index at 2.5% O_2 concentration and concluded that more methane-derived carbon was exuded into the environment and available for the growth of non-methanotrophs in O_2-limiting environments. Similar findings were reported where speciation within *Methylococcaceae* and *Methylophilaceae* family at different oxygen gradient with an abundance of *Methylosarcina* (methanotroph) and *Methylophilus* (NUM) at high O_2 tension (150–225 μM) and *Methylobacter* (methanotroph) and *Methylotenera* (NUM) at low initial O_2 tension (15–75 μM) was observed (Hernandez et al. 2015). The specific species differentiation observed within the methanotrophic and methylotrophic members of the *Methylococcaceae* and *Methylophilaceae* family is driven towards niche adaptation to specific oxygen gradient. The change in the population of methanotrophs and NUM to varying oxygen and methane concentration has been observed, suggesting that the relative concentration of methane and oxygen selects microbial community that can thrive under such situations. A synthetic community model comprising 50 bacterial species (methanotrophs, methylotrophs and heterotrophs) showed a change in the species composition with the abundance of methanotrophs of the family *Methylococcaceae* and methylotrophs of the family *Methylophilaceae* at varying methane and oxygen concentration (Yu et al. 2017). Lanthanum (Ln), a rare earth metal, also affects the transfer of methane-derived carbon as it is an important co-factor of XoxF-type methanol dehydrogenases (MDHs) present in Gram-negative methylotrophs (Vu et al. 2016; Yanpirat et al. 2020). A shift in the expression of methanol dehydrogenases from lanthanide-dependent MDH (XoxF) type to the more efficient calcium-dependent MDH (MxaF) type occurs when non-methanotrophs are cultured along with methanotrophs, allowing an excess of methanol production that can be used by the methylotrophic partner (Krause et al. 2017). The presence of lanthanides allows a partner-induced change in gene expression and influences microbial interactions in the environment. The above finding suggests that the existing concentration of methane, oxygen, nitrate and other nutrients in the natural ecosystem plays a major role in determining the community composition of methanotrophs and methylotrophs, thereby influencing the transfer of methane-derived carbon and methane oxidation capacity of that particular ecosystem.

22.4 Approaches Used to Study the Interaction of Methanotrophs and NUM

Techniques involving the cultivation of different microbial groups cannot be very useful for interaction studies as it is difficult to simulate natural conditions under laboratory and most of the microorganisms still remain un-culturable due to their specific growth requirement. A useful approach is to simulate the natural environment under controlled condition through a microcosm or mesocosm experiment depending upon the scale of the model ecosystem and use molecular tools to determine community composition. Microcosms are artificial, controlled, simplified ecosystem used to simulate natural ecosystems mostly done under laboratory conditions, whereas mesocosms are bounded and partially enclosed outdoor experiment used to bridge the gap between the laboratory and the real world in environmental science (Bruckner et al. 1995). Microcosm and mesocosm experiment reduces the credibility gap and helps us to provide a solution to large-scale environmental problems. They provide a better understanding of the ecological problems by bringing them to spatial and temporal scale convenient enough to carry out the study (Benton et al. 2007). Microcosm experiments have widely been designed to study the diversity and dynamics of both methanotrophs and methylotrophs in soil and sediment samples collected from the natural environment (Shiau et al. 2020; Oshkin et al. 2015; Morris et al. 2002). Research shows that the activity pattern of methane-oxidizing bacteria and the population structure of methylotrophs follow the same pattern under field and microcosm condition (Eller et al. 2005). It can be concluded that the findings of the microcosm study can be extrapolated to field scale keeping in mind the concerned quantitative changes. Various molecular tools and techniques are commonly being used to study the interaction of methanotrophs with NUM. Some of them are mentioned below.

DNA-Based Stable Isotope Probing (DNA-SIP): It is a powerful means to study the flow of intermediates from microbes with one functional group to the other. In DNA-SIP study, environmental samples are fed with substrate labelled with a heavy isotope (^{13}C). The labelled isotope then gets incorporated into the cell biomass including DNA, which can be processed and analysed to determine phylogenetic affiliations of species with labelled DNA. Isotope labelled $^{13}CH_4$ is used to study the cross-feeding of intermediates produced by methanotrophs determining the association of methanotrophs with methylotrophs and other heterotrophs in the natural environment. DNA-SIP helps us to establish a direct link between CH_4 oxidation and taxonomic identity for active methanotrophs and methylotrophs in complex environments (Shiau et al. 2020). It has been widely used to study metabolic interactions in methane-fed communities (van Grinsven et al. 2020; Paul et al. 2017; Jensen et al. 2008). DNA-SIP experiments are widely used to uncover the participants involved in the C1 cycle and give a clear picture of the transfer of metabolites from one microbe to the other. It provides confirmatory evidence of the associations of actively interacting microorganisms, sharing carbon derived from a single-key biogeochemical process.

PCR-Based Method: Functional marker genes unique to the physiology and metabolism of methanotrophs and methylotrophs can be targeted to study the diversity of microbes involved in the metabolism of single carbon compound. Functional genes commonly targeted to study the diversity of methanotrophs and methylotrophs are those of methane monooxygenases (*pmoA* and *mmoX*), methanol dehydrogenase (*mxaF*), 16S rRNA region targeting serine pathway and RuMP, dinitrogen reductase (*nifH*) and formyltransferase/hydrolase complex (*fhcD*) (McDonald et al. 2008). PCR product can be run on denaturing gradient gel to separate amplicons even with a single-nucleotide difference. PCR followed by denaturing gradient gel electrophoresis (PCR-DGGE) will help us to determine the degree of genetic polymorphism in the target regions within the community (Bodelier et al. 2005; Piterina and Pembroke 2013). One major limitation of DGGE methodology is that the size of the amplicon should be between 100 and 500 bp, and therefore, primer set should be carefully designed (Marzorati et al. 2008). Eller et al. (2005) used three universal eubacterial primers set targeting methylotrophs with RuMP (533F/907R and 197F/533R) and serine pathway (142F/533R) followed by DGGE to study the community composition of methylotrophic bacteria in soil samples collected from the paddy field. The advantage of PCR-DGGE over DNA-SIP technique is that it does not require a closed controlled environment and can be used to determine community composition of samples directly collected from the natural environment.

Next-Generation Sequencing (NGS): Metagenomic and transcriptomic approach to study microbial diversity requires sequencing of a large amount of DNA and transcripts. Next-generation sequencing methods are more sensitive and can detect low-frequency variants. It is a high-throughput process that handles hundreds and thousands of genes simultaneously and provides a comprehensive gene coverage (Krishna et al. 2019). Storage, analysis and interpretation of NGS data are the major rate-limiting steps of NGS technology. A large number of online bioinformatics tools are available that can process original raw sequencing data to functional biology (Kulski 2016). Techniques involving the use of NGS technology are widely used to study the interaction between methanotrophs and NUM (Krause et al. 2017; Beck et al. 2013; Takeuchi et al. 2019). Whole-genome sequencing and transcriptomic approach were used to study the interaction between the *Methylocaldum marinum* (methanotroph) and *Methyloceanibacter caenitepidi* (NUM) and observed that there is non-methanol-based cross-feeding (particularly acetate) of metabolites between the partners (Takeuchi et al. 2019). Pyrosequencing of 16S rRNA gene (27F/519R) was done to study the community dynamics in methane-fed microbial microcosms (Oshkin et al. 2015). The result showed low species diversity with the predominance of *Methylococcaceae* species, closely related to *Methylobacter tundripaludum* with few members of *Methylotenera*, *Flavobacterium*, *Pseudomonas*, *Janthinobacterium*, *Achromobacter* and *Methylophilus*. They also studied the community dynamics through Illumina sequencing of prepared DNA libraries and observed the predominance of methanotroph (*Methylobacter*) followed by NUM of the family *Methylophilaceae* (*Methylobacter tundripaludum*, *Methylophilus methylotrophus*, *Methylotenera versatilis* and *Methylotenera mobilis*). Both these techniques confirmed the strong correlation of the population of methanotrophs to that of NUM, suggesting that there may be the flow of intermediates between the two partners.

22.5 Interaction of Methanotrophs with Microbes of Different Functional Group

Besides methylotrophs, intermediates of the methanotrophic bacteria also support the growth of few heterotrophic bacteria. Synergistic interactions occur between the methanotrophs and heterotrophs where one provides the other with carbon source and the other produces growth factor or remove toxic intermediates from the environment (Stock et al. 2013; Ho et al. 2014; Veraart et al. 2018; Singh et al. 2019). Growth stimulation of methane-utilizing *Methylovulum miyakonense* in the presence of *Rhizobium* has been documented (Iguchi et al. 2011). They identified cobalamin secreted by *Rhizobium* as the key factor responsible for stimulating the growth of the methanotroph. Removal of toxic intermediates like organic acids can also support the growth and proliferation of methanotrophic partners (Singh et al. 2019). The effect of the interaction of methanotrophs with non-methanotrophs (heterotrophs/ autotrophs) has been summarized in Table 22.2.

Methanotrophic bacteria can grow with other organisms and aid in the removal of other greenhouse gas (Singh et al. 2019). Co-culture of alkaliphilic methanotrophic

Table 22.2 Beneficial effect of the interaction of methanotrophs with non-methanotrophs

Methanotroph	Non-methanotroph	Effect of interaction	References
Gammaproteobacteria (*Methylosarcina* and *Methylocaldum*)	Algae (autotroph)	Autotrophs provide O_2 to the methanotrophs and increase methane oxidation rate, whereas methanotrophs provide them CO_2 for photosynthesis	Yoshida et al. (2014)
Methylocystis	*Sphagnum* mosses (autotroph)	The autotrophs provide O_2 to the methanotrophs and increase methane oxidation rate, whereas methanotrophs provide them CO_2 for photosynthesis	Kip et al. (2011)
Methylobacter luteus	*Pseudomonas mandelii* (heterotroph)	Growth stimulation and increased methane oxidation	Veraart et al. (2018)
Methylovulum, *Methyloparacoccus*, *Methylomonas*	*Rhizobium* sp., *Mesorhizobium* sp. and *Sinorhizobium* sp. (heterotroph)	Heterotrophs produce vitamin B_{12} and support the growth of methanotrophs	Hoefman et al. (2014)
Methylomonas methanica	*Rhizobium/Ochrobactrum/Pseudomonas/Escherichia coli* (heterotroph)	Growth promotion	Ho et al. (2014)
Methylovulum miyakonense	*Rhizobium* sp. (heterotroph)	Growth stimulation	Iguchi et al. (2011)
Methylomonas	*Cupriavidus taiwanensis* (heterotroph)	Heterotroph could synthesize quinone, pyridoxine and vitamin B_{12} and supported the growth of methanotroph	Stock et al. (2013)

bacteria with microalga *Scenedesmus obtusiusculus* in the ratio 3:1, 4:1 and 5:1 can lead to complete CH_4 and CO_2 uptake and thus is a promising strategy for greenhouse gas mitigation in a single step (Ruiz-Ruiz et al. 2020). Methanol-independent cross-feeding occurs in the natural ecosystem and supports the existence of non-methylotrophic heterotrophic bacteria. A recent study shows that methane-oxidizing bacteria can undergo mixed acid fermentation under the anoxic condition and release other products like acetate, succinate and H_2 (Kalyuzhnaya et al. 2013; Xin et al. 2004). These fermentation products can be used as a substrate by a diverse group of heterotrophic bacteria. The complex interaction of methanotrophs with other microbes occurs in the natural environment and thus can greatly influence net methane emission from these areas.

22.6 Importance of Interaction of Methane Utilizers with Non-methanotrophs in the Natural Ecosystem

Methanotrophs allow microbial food web to work at locations where it is difficult for other microbes to survive and consume methane which is the most reduced form of carbon. At the oxic-anoxic interface, aerobic methanotrophs survive that consume methane produced by methanogenic archaea and support the growth of other methylotrophs as well as heterotrophs. The type of interaction between these microbial functional groups in a methane-rich environment has been shown in Fig. 22.2.

Methylotrophic partner removes toxic intermediates of the methane utilizers like methanol and formaldehyde and allows sustained growth of the methanotrophs. Reports on excretion of methanol (up to 100 μM) in the culture medium are available that suggests a mismatch between the methanol produced and methanol that can be further assimilated into the cell biomass (Xin et al. 2004; Tavormina et al. 2017). The release of methanol will decrease the methane oxidation rate and inhibit methanol production by the methanotrophic culture. The presence of methanol-utilizing methylotrophs will allow removal of the released methanol and allow the sustained activity of methane monooxygenase enzyme system. Low methanol concentration in the environment is associated with low ozone concentration in the atmosphere and thus plays an important role in atmospheric chemistry (Warneke et al. 1999; Galbally and Kirstine 2002). Methanol-utilizing methylotrophs thereby play a key role and consume both plant-derived methanol and those obtained from methanotrophs

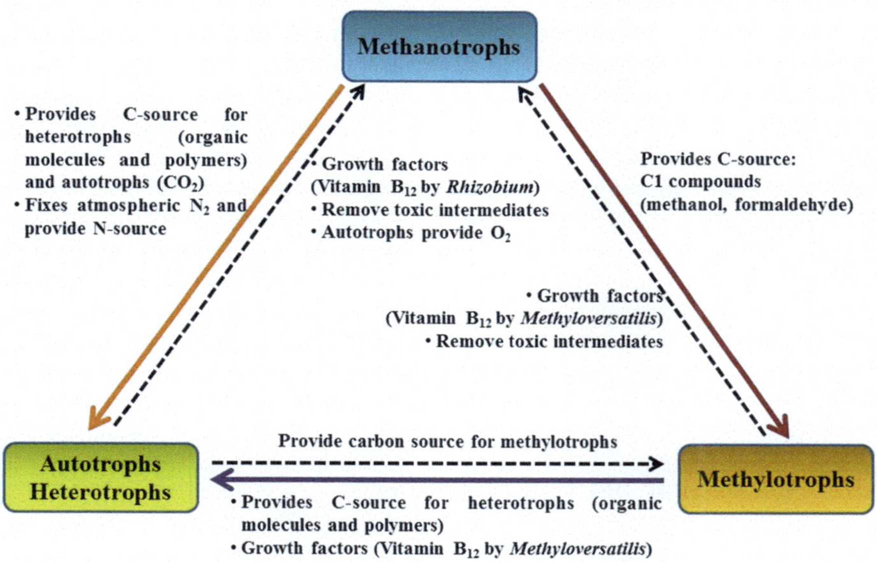

Fig. 22.2 The effect of the interaction of methanotrophs with non-methanotrophs (methylotrophs/heterotroph/autotroph) in a methane-rich environment

before it gets released into the atmosphere. Isolation of proteobacterial methanotroph requiring lanthanides by enrichment culture technique led to co-isolation of non-methanotrophic community, belonging to the genera *Methylophilus*, *Methyloversatilis*, *Hyphomicrobium*, *Methylobacteria*, *Pseudomonas* and *Thiobacillus*, as they can utilize intermediate compounds of the methane oxidation like methanol, acetate, and formate (Kato et al. 2020). Mesocosm experiments showed that there is a relative abundance of methanotrophs and NUM, indicating that a large part of methane-derived product (methanol, acetate and others) was being transferred from methanotrophs to non-methane-utilizing methylotrophs (Kuloyo et al. 2020).

In a natural environment, methanotrophs are found along with other methylotrophs, heterotrophs and autotrophs. Metabolites produced by each one of them may support or suppress the growth of other bacteria. Besides methylotrophs, heterotrophs and autotrophs also affect the activity of methane-utilizing bacteria. Growth factors (quinone, pyridoxine and vitamin B_{12}) produced by these organisms may support the activity of methanotrophs (Stock et al. 2013; Ho et al. 2014; Hoefman et al. 2014). Research shows that synergistic interaction exists between methanotrophs, methylotrophs and heterotrophs. A methane-utilizing mixed culture composed of a methanotroph, methanol-utilizing methylotroph (*Methyloceanibacter caenitepidi*) and a heterotroph was successfully established from the sample collected from marine sediments in Japan (Takeuchi et al. 2014). The stable association of these three functional groups on a medium with methane as carbon source shows that the methanotrophs via providing its metabolic intermediates (methanol, formaldehyde, acetate and formate) support the growth of methylotrophs as well as other heterotrophs in the environment. The close association of methane-oxidizing bacteria with autotrophs (macrophytic algae/*Sphagnum* mosses) suggests that their photosynthetic activity may provide O_2 to the methanotrophs and support its growth and proliferation (Yoshida et al. 2014; Kip et al. 2011). In turn, the methanotrophs may provide fixed nitrogen (NH_4^+) to the *Sphagnum* mosses by its N_2 fixation activity and exert beneficial effect (Larmola et al. 2014). Research suggests that the flow of methane-derived carbon does not stop at the microbial level but sometimes extend to the whole aquatic food web, up to the fish level (Sanseverino et al. 2012). Their findings indicate the importance of methanotrophs in the C1 cycle (particularly methane) and the role it plays in the food web of aquatic systems. Natural methane-rich environments possess a diverse group of microflora right from methanogens to methylotrophs, heterotrophs and autotrophs in close association, thereby allowing the microbial community to thrive.

22.7 Conclusion and Future Prospects

Studies have emphasized the importance of biotic interactions, particularly microbial interactions, as key modulators of biogeochemical processes. Methanotrophs along with other microbes allow methane-based food web to function in various anaerobic ecosystems. Mitigation of methane emission through the use of methane-utilizing bacteria from various anthropogenic sources (paddy fields, wastewater treatment and landfills) has gained impetus in recent years (Oswald et al. 2016; Strong et al. 2017; Davamani et al. 2020). With the increase in anthropogenic methane emissions, the importance of these bacteria is set to increase as they play an important role in reducing global methane sink. Artificial inoculation of methanotrophs with plant growth-promoting traits in paddy field can cause a substantial reduction in methane emission and an increase in grain yield (Rani et al. 2021a; Davamani et al. 2020). Removal of methane from anoxic lake waters upon inoculation with γ-proteobacterial methanotrophs has been reported (Oswald et al. 2016). However, efforts to harness the synergistic interaction of methanotrophs with other microbial groups have not been undertaken. We propose that co-inoculation of NUM with methanotrophs may expedite the methane removal process due to their synergistic interaction. Studies in this area have still not gained impetus, and the effect of microbial co-inoculation on the removal of methane has still not been explored much. This chapter provides enough evidence and confirms the transfer of metabolites from methanotrophs to the other microbial groups. This microbial synergistic interaction can be tapped for reducing methane emission from various anoxic habitats.

References

Abrudan MI, Smakman F, Grimbergen AJ, Westhoff S, Miller EL, Van Wezel GP, Rozen DE (2015) Socially mediated induction and suppression of antibiosis during bacterial coexistence. Proc Natl Acad Sci 112(35):11054–11059

Anthony C, Williams P (2003) The structure and mechanism of methanol dehydrogenase. Biochim Biophys Acta 1647:18–23

Beck DAC, Kalyuzhnaya MG, Malfatti S, Tringe SG, Rio TG, Ivanova N, Lidstrom ME, Chistoserdova L (2013) A metagenomic insight into freshwater methane-utilizing communities and evidence for cooperation between the *Methylococcaceae* and the *Methylophilaceae*. PeerJ 1:e23

Beilen JB, Funhoff EG (2005) Expanding the alkane oxygenase toolbox: new enzymes and applications. Curr Opin Biotechnol 16:308–314

Benton TG, Solan M, Travis JM, Sait SM (2007) Microcosm experiments can inform global ecological problems. Trends Ecol Evol 22(10):516–521

Bodelier PL, Meima-Franke M, Zwart G, Laanbroek HJ (2005) New DGGE strategies for the analyses of methanotrophic microbial communities using different combinations of existing 16S rRNA-based primers. FEMS Microbiol Ecol 52(2):163–174

Bruckner A, Wright J, Kampichler C, Bauer R, Kandeler E (1995) A method of preparing mesocosms for assessing complex biotic processes in soils. Biol Fertil Soils 19(2):257–262

Chistoserdova L, Kalyuzhnaya MG (2018) Current trends in methylotrophy. Trends Microbiol 26 (8):703–714

Chistoserdova L, Kalyuzhnaya MG, Lidstrom ME (2009) The expanding world of methylotrophic metabolism. Annu Rev Microbiol 63:477–499

Corder RE, Johnson ER, Vega JL, Clausen EC, Gaddy JL (1986) Biological production of methanol from methane. Am Chem Soc Energy Fuels Div 33(3):469–478

Davamani V, Parameswari E, Arulmani S (2020) Mitigation of methane gas emissions in flooded paddy soil through the utilization of methanotrophs. Sci Total Environ 726:138570

Eller G, Krüger M, Frenzel P (2005) Comparing field and microcosm experiments: a case study on methano- and methylotrophic bacteria in paddy soil. FEMS Microbiol Ecol 51(2):279–291

Galbally IE, Kirstine W (2002) The production of methanol by flowering plants and the global cycle of methanol. J Atmos Chem 43(3):195–229

Govorukhina NI, Trotsenko YA (1991) *Methylovorus*, a new genus of restricted facultatively methylotrophic bacteria. Int J Syst Bacteriol 41(1):158–162

van Grinsven S, SinningheDamsté JS, Harrison J, Polerecky L, Villanueva L (2020) Nitrate promotes the transfer of methane-derived carbon from the methanotroph *Methylobacter* sp. to the methylotrophs *Methylotenera* sp. in eutrophic lake water. Limnol Oceanogr. https://doi.org/10.1002/lno.11648

Hanson RS, Hanson TE (1996) Methanotrophic bacteria. Microbiol Rev 60:439–471

Hernandez ME, Beck DA, Lidstrom ME, Chistoserdova L (2015) Oxygen availability is a major factor in determining the composition of microbial communities involved in methane oxidation. PeerJ 3:e801

Ho A, De Roy K, Thas O, De Neve J, Hoefman S, Vandamme P, Heylen K, Boon N (2014) The more, the merrier: heterotroph richness stimulates methanotrophic activity. ISME J 8:1945–1948

Hoefman S, van der Ha D, Iguchi H, Yurimoto H, Sakai Y, Boon N, Vandamme P, Heylen K, De Vos P (2014) *Methyloparacoccus murrellii* gen. nov., sp. nov., a methanotroph isolated from pond water. Int J Syst Evol Microbiol 64(6):2100–2107

Hrsak D, Begonja A (2000) Possible interactions within a methanotrophic-heterotrophic groundwater community able to transform linear alkylbenzenesulfonates. Appl Environ Microbiol 66 (10):4433–4439

Iguchi H, Yurimoto H, Sakai Y (2011) Stimulation of methanotrophic growth in cocultures by cobalamin excreted by rhizobia. Appl Environ Microbiol 77:8509–8515

Iguchi H, Yurimoto H, Sakai Y (2015) Interactions of methylotrophs with plants and other heterotrophic bacteria. Microorganisms 3(2):137–151

Jensen S, Neufeld JD, Birkeland NK, Hovland M, Murrell JC (2008) Methane assimilation and trophic interactions with marine *Methylomicrobium* in deep-water coral reef sediment off the coast of Norway. FEMS Microbiol Ecol 66(2):320–330

Jeong S, Kim TG (2019) Development of a novel methanotrophic process with the helper microorganism *Hyphomicrobium* sp. NM3. J Appl Microbiol 126(2):533–544

Jhala YK, Vyas RV, Shelat HN, Patel HK, Patel KT (2014) Isolation and characterization of methane utilizing bacteria from wetland paddy ecosystem. World J Microbiol Biotechnol 30 (6):1845–1860

Kalyuzhnaya MG, Yang S, Rozova ON, Smalley NE, Clubb J, Lamb A, Gowda GN, Raftery D, Fu Y, Bringel F, Vuilleumier S (2013) Highly efficient methane biocatalysis revealed in a methanotrophic bacterium. Nat Commun 4(1):1–7

Kato S, Takashino M, Igarashi K, Kitagawa W (2020) Isolation and genomic characterization of a Proteobacterial methanotroph requiring lanthanides. Microbes Environ 35(1):ME19128

Kip N, Ouyang W, van Winden J, Raghoebarsing A, van Niftrik L, Pol A, Pan Y, Bodrossy L, van Donselaar EG, Reichart GJ, Jetten MS (2011) Detection, isolation, and characterization of acidophilic methanotrophs from Sphagnum mosses. Appl Environ Microbiol 77(16):5643–5654

Kirschke S, Bousquet P, Ciais P, Saunois M, Canadell JG, Dlugokencky EJ, Bergamaschi P, Bergmann D, Blake DR, Bruhwiler L, Cameron-Smith P (2013) Three decades of global methane sources and sinks. Nat Geosci 6:813–823

Kolb S (2009) The quest for atmospheric methane oxidizers in forest soils. Environ Microbiol Rep 1 (5):336–346

Krause SM, Johnson T, Karunaratne YS, Fu Y, Beck DA, Chistoserdova L, Lidstrom ME (2017) Lanthanide-dependent cross-feeding of methane-derived carbon is linked by microbial community interactions. Proc Natl Acad Sci 114(2):358–363

Krishna BM, Khan MA, Khan ST (2019) Next-generation sequencing (NGS) platforms: an exciting era of genome sequence analysis. In: Microbial genomics in sustainable agroecosystems. Springer, Singapore, pp 89–109

Kuloyo O, Ruff SE, Cahill A, Connors L, Zorz JK, Hrabe de Angelis I, Nightingale M, Mayer B, Strous M (2020) Methane oxidation and methylotroph population dynamics in groundwater mesocosms. Environ Microbiol 22:1222–1237

Kulski JK (2016) Next-generation sequencing—an overview of the history, tools, and "omic" applications. In: Next generation sequencing-advances, applications and challenges. InTechOpen, London, pp 3–60

Larmola T, Leppänen SM, Tuittila ES, Aarva M, Merilä P, Fritze H, Tiirola M (2014) Methanotrophy induces nitrogen fixation during peatland development. Proc Natl Acad Sci 111(2):734–739

Lee HJ, Kim SY, Kim PJ, Madsen EL, Jeon CO (2014) Methane emission and dynamics of methanotrophic and methanogenic communities in a flooded rice field ecosystem. FEMS Microbiol Ecol 88(1):195–212

Macey MC, Pratscher J, Crombie AT, Murrell JC (2020) Impact of plants on the diversity and activity of methylotrophs in soil. Microbiome 8(1):1–17

Madhaiyan M, Poonguzhali S, Kwon ASAT (2009) *Methylophilus rhizosphaerae* sp. nov., a restricted facultative methylotroph isolated from rice rhizosphere soil. Int J Syst Evol Microbiol 59:2904–2908

Madhaiyan M, Poonguzhali S, Lee JS, Lee KC, Sundaram S (2010) *Flavobacterium glycines* sp. nov., a facultative methylotroph isolated from the rhizosphere of soybean. Int J Syst Evol Microbiol 60(9):2187–2192

Marzorati M, Wittebolle L, Boon N, Daffonchio D, Verstraete W (2008) How to get more out of molecular fingerprints: practical tools for microbial ecology. Environ Microbiol 10 (6):1571–1581

McDonald IR, Miguez CB, Rogge G, Bourque D, Wendlandt KD, Groleau D, Murrell JC (2006) Diversity of soluble methane monooxygenase-containing methanotrophs isolated from polluted environments. FEMS Microbiol Lett 255(2):225–232

McDonald IR, Bodrossy L, Chen Y, Murrell JC (2008) Molecular ecology techniques for the study of aerobic methanotrophs. Appl Environ Microbiol 74(5):1305–1315

McTaggart TL, Beck DA, Setboonsarng U, Shapiro N, Woyke T, Lidstrom ME, Kalyuzhnaya MG, Chistoserdova L (2015) Genomics of methylotrophy in Gram-positive methylamine-utilizing bacteria. Microorganisms 3(1):94–112

Morris SA, Radajewski S, Willison TW, Murrell JC (2002) Identification of the functionally active methanotroph population in a peat soil microcosm by stable-isotope probing. Appl Environ Microbiol 68(3):1446–1453

Murrell JC, McDonald IR, Gilbert B (2000) Regulation of expression of methane monooxygenases by copper ions. Trends Microbiol 8:221–225

Nakatsu CH, Hristova K, Hanada S, Meng X, Hanson JR, Scow KM, Kamagata Y (2006) *Methylibium petroleiphilum* gen. nov., sp. nov., a novel methyl tert-butyl ether-degrading methylotroph of the Betaproteobacteria. Int J Syst Evol Microbiol 56:983–989

Negruţa OANA, Csutak O, Stoica I, Rusu E, Vassu T (2010) Methylotrophic yeasts: diversity and methanol metabolism. Rom Biotechnol Lett 15:5369–5375

Nobuo K, Hiroya Y, Rudolf KT (2006) The physiological role of the ribulose monophosphate pathway in bacteria and archaea. Biosci Biotechnol Biochem 70(1):10–21

Op den Camp HJ, Islam T, Stott MB, Harhangi HR, Hynes A, Schouten S, Jetten MS, Birkeland NK, Pol A, Dunfield PF (2009) Environmental, genomic and taxonomic perspectives on methanotrophic Verrucomicrobia. Environ Microbiol Rep 1(5):293–306

Oshkin IY, Beck DA, Lamb AE, Tchesnokova V, Benuska G, McTaggart TL, Kalyuzhnaya MG, Dedysh SN, Lidstrom ME, Chistoserdova L (2015) Methane-fed microbial microcosms show differential community dynamics and pinpoint taxa involved in communal response. ISME J 9 (5):1119–1129. https://doi.org/10.1038/ismej.2014.203

Oswald K, Milucka J, Brand A, Hach P, Littmann S, Wehrli B, Kuypers MM, Schubert CJ (2016) Aerobic gammaproteobacterial methanotrophs mitigate methane emissions from oxic and anoxic lake waters. Limnol Oceanogr 61(S1):S101–S118

Paul BG, Ding H, Bagby SC, Kellermann MY, Redmond MC, Andersen GL, Valentine DL (2017) Methane-oxidizing bacteria shunt carbon to microbial mats at a marine hydrocarbon seep. Front Microbiol 8:186

Piterina AV, Pembroke JT (2013) Use of PCR-DGGE based molecular methods to analyse microbial community diversity and stability during the thermophilic stages of an ATAD wastewater sludge treatment process as an aid to performance monitoring. Int Scholarly Res Notices 2013:13

Qiu Q, Noll M, Abraham WR, Lu Y, Conrad R (2008) Applying stable isotope probing of phospholipid fatty acids and rRNA in a Chinese rice field to study activity and composition of the methanotrophic bacterial communities in situ. ISME J 2:602–614

Rani V, Bhatia A, Kaushik R (2021a) Inoculation of plant growth promoting-methane utilizing bacteria in different N-fertilizer regime influences methane emission and crop growth of flooded paddy. Sci Total Environ 775:145826

Rani V, Bhatia A, Nain L, Tomar GS, Kaushik R (2021b) Methane utilizing plant growth-promoting microbial diversity analysis of flooded paddy ecosystem of India. World J Microbiol Biotechnol 37(4):56

Ruiz-Ruiz P, Gomez-Borraz TL, Revah S, Morales M (2020) Methanotroph-microalgae co-culture for greenhouse gas mitigation: effect of initial biomass ratio and methane concentration. Chemosphere 259:127418

Sanseverino AM, Bastviken D, Sundh I, Pickova J, Enrich-Prast A (2012) Methane carbon supports aquatic food webs to the fish level. PLoS One 7(8):e42723

Sharp CE, den Camp HJO, Tamas I, Dunfield PF (2013) Unusual members of the PVC superphylum: the methanotrophic verrucomicrobia genus "*Methylacidiphilum*". In: Planctomycetes: cell structure, origins and biology. Humana Press, Totowa, pp 211–227

Shiau YJ, Lin CW, Cai Y, Jia Z, Lin YT, Chiu CY (2020) Niche differentiation of active methane-oxidizing bacteria in estuarine mangrove forest soils in Taiwan. Microorganisms 8(8):1248

Singh R, Ryu J, Kim SW (2019) Microbial consortia including methanotrophs: some benefits of living together. J Microbiol 57(11):939–952

Stock M, Hoefman S, Kerckhof FM, Boon N, De Vos P, De Baets B, Heylen K, Waegeman W (2013) Exploration and prediction of interactions between methanotrophs and heterotrophs. Res Microbiol 164(10):1045–1054

Strong PJ, Karthikeyan OP, Zhu J, Clarke W, Wu W (2017) Methanotrophs: methane mitigation, denitrification and bioremediation. In: Agro-environmental sustainability. Springer, Cham, pp 19–40

Takeuchi M, Katayama T, Yamagishi T, Hanada S, Tamaki H, Kamagata Y, Oshima K, Hattori M, Marumo K, Nedachi M, Maeda H (2014) *Methyloceanibacter caenitepidi* gen. nov., sp. nov., a facultatively methylotrophic bacterium isolated from marine sediments near a hydrothermal vent. Int J Syst Evol Microbiol 64(2):462–468

Takeuchi M, Ozaki H, Hiraoka S, Kamagata Y, Sakata S, Yoshioka H, Iwasaki W (2019) Possible cross-feeding pathway of facultative methylotrophs *Methyloceanibacter caenitepidi* Gela4 on methanotroph *Methylocaldum marinum* S8. PLoS One 14(3):e0213535

Tavormina PL, Kellermann MY, Antony CP, Tocheva EI, Dalleska NF, Jensen AJ, Valentine DL, Hinrichs KU, Jensen GJ, Dubilier N, Orphan VJ (2017) Starvation and recovery in the deep-sea methanotroph *Methyloprofundus sedimenti*. Mol Microbiol 103(2):242–252

van Teeseling MC, Pol A, Harhangi HR, van der Zwart S, Jetten MS, den Camp HJO, van Niftrik L (2014) Expanding the verrucomicrobial methanotrophic world: description of three novel species of *Methylacidimicrobium* gen. Nov. Appl Environ Microbiol 80(21):6782–6791

Theisen AR, Ali MH, Radajewski S, Dumont MG, Dunfield PF, McDonald IR, Dedysh SN, Miguez CB, Murrell JC (2005) Regulation of methane oxidation in the facultative methanotroph *Methylocella silvestris* BL2. Mol Microbiol 58:682–692

Veraart AJ, Garbeva P, Van Beersum F, Ho A, Hordijk CA, Meima-Franke M, Zweers AJ, Bodelier PLE (2018) Living apart together-bacterial volatiles influence methanotrophic growth and activity. ISME J 12:1163–1166

Vu HN, Subuyuj GA, Vijayakumar S, Good NM, Martinez-Gomez NC, Skovran E (2016) Lanthanide-dependent regulation of methanol oxidation systems in *Methylobacterium extorquens* AM1 and their contribution to methanol growth. J Bacteriol 198(8):1250–1259

Warneke C, Karl T, Judmaier H, Hansel A, Jordan A, Lindinger W, Crutzen PJ (1999) Acetone, methanol, and other partially oxidized volatile organic emissions from dead plant matter by abiological processes: significance for atmospheric HOx chemistry. Glob Biogeochem Cycles 13(1):9–17

Watrous J, Roach P, Alexandrov T, Heath BS, Yang JY, Kersten RD, van der Voort M, Pogliano K, Gross H, Raaijmakers JM, Moore BS (2012) Mass spectral molecular networking of living microbial colonies. Proc Natl Acad Sci 109(26):E1743–E1752

Waturangi DE, Francisca I, Susanto CO (2011) Genetic diversity of methylotrophic bacteria from human mouth based on amplified ribosomal DNA restriction analysis (ARDRA). Hayati J Biosci 18(2):77–81

Wei XM, He R, Chen M, Su Y, Ma RC (2016) Conversion of methane-derived carbon and microbial community in enrichment cultures in response to O_2 availability. Environ Sci Pollut Res 23(8):7517–7528

Xin JY, Cui JR, Niu JZ, Hua SF, Xia CG, Li SB, Zhu LM (2004) Production of methanol from methane by methanotrophic bacteria. Biocatal Biotransformation 22(3):225–229

Yanpirat P, Nakatsuji Y, Hiraga S, Fujitani Y, Izumi T, Masuda S, Mitsui R, Nakagawa T, Tani A (2020) Lanthanide-dependent methanol and formaldehyde oxidation in *Methylobacterium aquaticum* strain 22A. Microorganisms 8(6):822

Yoshida N, Iguchi H, Yurimoto H, Murakami A, Sakai Y (2014) Aquatic plant surface as a niche for methanotrophs. Front Microbiol 5:30

Yu Z, Beck DA, Chistoserdova L (2017) Natural selection in synthetic communities highlights the roles of *Methylococcaceae* and *Methylophilaceae* and suggests differential roles for alternative methanol dehydrogenases in methane consumption. Front Microbiol 8:2392

Chapter 23
Biopesticides: An Alternative to Synthetic Insecticides

A. T. Rani, Vasudev Kammar, M. C. Keerthi, Vijaya Rani, Sujan Majumder, K. K. Pandey, and Jagdish Singh

Abstract The modern agriculture is negatively affected due to the rapid exploitation of natural resources, indiscriminate use of pesticide application, and frequent weather events influenced by climate change. Biocontrol action is a significant tool for IPM, offers alternative management techniques that are safer for human and environmental health. It is also worth noting that biological control has worked in a versatile manner in different agricultural management systems and with different types of disease causing organisms existing in the nature. Several key pests and invasive pests were successfully controlled with the application of biological agents. Although biopesticides very slowly replace the use of pesticide that may be due to biopesticide exploration and application range. In India the main challenge for biopesticide is related to their shelf life, narrow host range for pathogens, variation in the lab to land performances, economic regulation, etc. Integrated approach will be beneficial for biopesticide application for this private and government sectors that come together with farmers to the village level and to build confidence in the use of biopesticides.

A. T. Rani (✉)
Agricultural Entomology, ICAR-Indian Institute of Vegetable Research, Varanasi, Uttar Pradesh, India

V. Kammar
Quality Control Cell, Pune, Department of Food and Public distribution, M/o Consumer Affairs and Food and Public distribution, Government of India, Pune, India

M. C. Keerthi
Agricultural Entomology, ICAR-Indian Grassland and Fodder Research Institute, Jhansi, Uttar Pradesh, India

V. Rani
Microbiology, ICAR-Indian Institute of Vegetable Research, Varanasi, Uttar Pradesh, India

S. Majumder
Agricultural Chemicals, ICAR-Indian Institute of Vegetable Research, Varanasi, Uttar Pradesh, India

K. K. Pandey · J. Singh
ICAR-Indian Institute of Vegetable Research, Varanasi, Uttar Pradesh, India

P. Bhatt et al. (eds.), *Microbial Technology for Sustainable Environment*,
https://doi.org/10.1007/978-981-16-3840-4_23

Keywords Biopesticides · Entomopathogens · Success stories · Semiochemicals · Bacteria · Neem

23.1 Introduction

Agriculture is critical in fostering people's food and nutrition welfare, supporting farmers' livelihoods, and ensuring the long-term growth of countries like India (Pawlak and Kołodziejczak 2020). Sustainable development requires the maintenance or enhancement of environmental quality through conservation of natural resources simultaneously (Ansari et al. 2012). Sustainable agriculture therefore necessitates effective agricultural resource management, with the goal of reducing pest and disease issues to the point that they do not significantly harm crops by not disrupting nature's equilibrium (Kogan 1998). India grows into self-sufficient in food production after the introduction of modern technologies such as the use of high yielding crop varieties, irrigation facilities, and chemical inputs including fertilizers and synthetic pesticides (Rena 2004). Overuse of chemical pesticides resulted in pesticide resistance, resurgence of pests, and accumulation of residues on non-target areas, resulting in the eradication of natural predators, threats to farmworkers, and negative environmental consequences (Gill and Garg 2014).

Considering the negative effect of pesticides, the EU and USA have already banned few chemical insecticides and also many conventional products have been withdrawn from use because of raising concern about environment and human health related issues (Pesticides Safety Directorate 2008; Damalas and Eleftherohorinos 2011). In India, 32 pesticide active ingredients were banned from import, manufacture, and use, eight were withdrawn from the market and 13 pesticides were restricted for use as of October 2015 (CIBRC 2017). Remaining insecticides were appeared to be ineffective in controlling insect pests due to development of insecticide resistance (Kranthi et al. 2002; Mishra et al. 2015). Several other limitations of synthetic insecticides including low biodegradability associated with high persistence (Tayade et al. 2013), detrimental effects on soil microbiota lead to loss of biodiversity and recycling of nutrients (Su et al. 2014), environmental contamination (soil and water), and harmful effect of insecticide residues on beneficial organisms (bees, spiders, earthworms, plants) (Singh et al. 2014). In such conditions integrated approach offers different management actions that are eco-friendly and sustainable for human and environmental health. In integrated approach, biological control is an important tool that includes utilization of efficient biocontrol agent to reduce the damage caused by harmful organisms (DeBach and Rosen 1991). Biopesticide can work in a diversified manner for different agricultural management systems, landscapes with potential actions.

The basic concept is to use biological species and natural products to control the properties of an ecological environment and its elements in order to reduce insect pests' biotic and reproductive capacity (Ruiu 2018). Biopesticides offer solutions to challenges such as insecticide tolerance, environmental, and public health concerns in the age of sustainable agriculture (Chandler et al. 2011). Biopesticides are well-known for being much cleaner for the environment than natural pesticides, but this long-term solution is competing for a place in the current synthetic pesticide industry. To popularize or to promote the biopesticides, immediate attention is required to address the major issues related to technological challenges and long-standing sustainability for more adoptability.

23.2 Biopesticides and Its Classification

Biopesticides or biological pesticides are derived from naturally occurring living organisms including plants, animals, and microbes (fungi, bacteria, virus, nematodes, etc.) used as such or as their products or by-products that can control serious pest and diseases of plants by their non-toxic eco-friendly mechanism. Biopesticides are described as mass-produced agents obtained from living organisms or a natural material manufactured and marketed for the control of plant pests, according to the Organization for Economic Co-operation and Development (OECD). The target specific biopesticides gaining importance all over the world because as they offer an effective and ecologically sound solution to the pest problems.

Biopesticides are broadly classified into four different categories based on the origin of the active substance or the living organism used in the control of target pest. It includes (1) microbial pesticides; (2) biochemical pesticides (botanical pesticides and semiochemicals); (3) transgenics (plant-incorporated products); (4) natural enemies (insect predators and parasitoids) (Fig. 23.1).

23.2.1 Microbial Pesticides

Microbial pesticides are the pest control products comprising of several pathogenic microorganisms such as bacteria, fungi, baculoviruses, protozoa, nematodes, and nematode-associated bacteria as their active ingredient (Fig. 23.1). Entomopathogenic viruses, bacteria, and fungi are the most commonly used among farming community used as alternatives to traditional insecticides. Some of the examples of commonly used microbial pesticides and their target pests are listed in Table 23.1. Among various microbial pesticides, *B. thuringiensis* (*Bt*) is the most widely used entomopathogenic bacterial biopesticide. Nearly 90% of the biopesticide market was covered by the different strains/serotypes and subspecies of *Bt* (Chattopadhyay et al. 2004; Romeis et al. 2006). Over 6000 isolates are stored in many repositories distributed around the world (Ansari et al. 2012). These biopesticides can control diverse range of pests and each separate active ingredient is specific to a target pest providing safety to the non-target organisms. Most often these pesticides are applied to crops in a similar way to chemical pesticides and they may be delivered either in the form of spores, as living organisms or dead organisms.

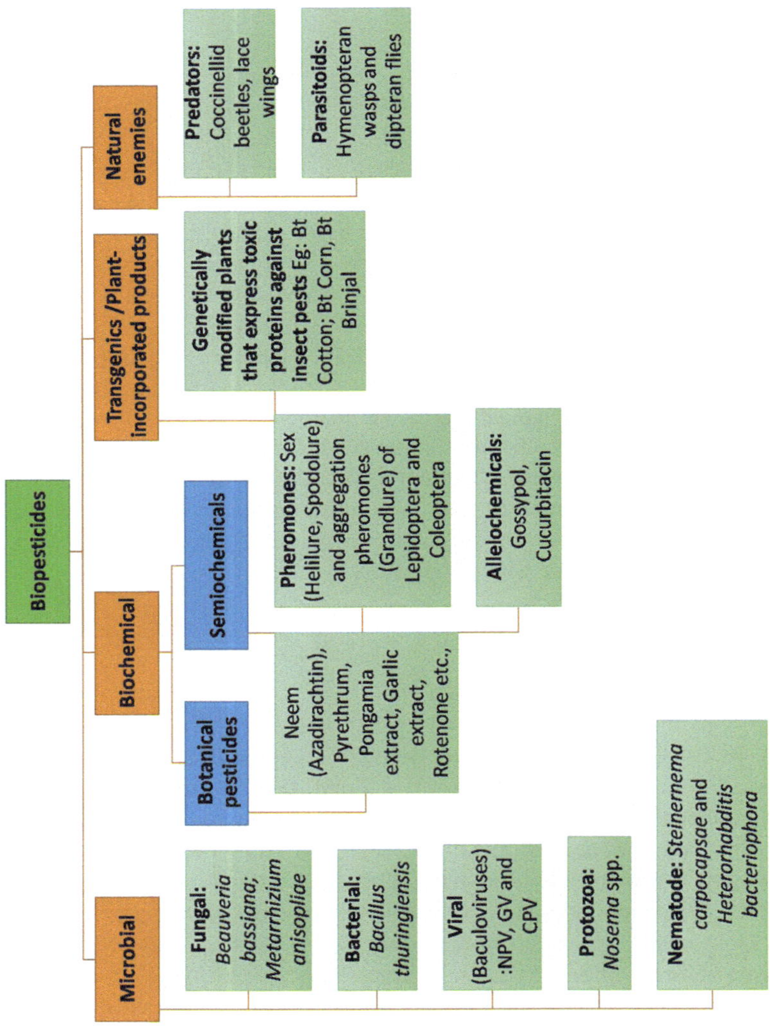

Fig. 23.1 Classification of biopesticides

Table 23.1 Some of the examples of microbial biopesticides for insect pest management

Microbial species/ variety	Target pest	Selected references
Bacterial		
Bacillus thuringiensis var. *kurstaki,* var. *aizawai*, var. *galleriae;*	Lepidoptera (armyworms, diamondback moth, *Helicoverpa armigera, Spodoptera litura,* etc.)	Pawar and Singh (1993), Mohan et al. (2009)
Bacillus thuringiensis var. *tenebrionis*	Coleoptera (Colorado potato beetle, Japanese beetle, etc.)	Pawar and Singh (1993), Bravo et al. (2007)
Bacillus popilliae		
Bacillus thuringiensis var. *isralensis*; *Bacillus sphaericus*	Diptera (mosquitoes and black flies, etc.)	Pawar and Singh (1993), Bravo et al. (2007)
Bacillus moritai		
Fungal		
Beauveria bassiana	Broad spectrum of mites and insects	Bhattacharya et al. (2003), Mohammad Beigi and Port (2015), Tadele and Emana (2017)
Beauveria brongniartii	Root grubs, B. borer, *H. armigera*	Ruiu (2018), Zimmermann (2007)
Metarhizium anisopliae	Coleoptera, Diptera, Hemiptera, Isoptera	Mohammad Beigi and Port (2015), Erler and Ates (2015)
Lecanicillium lecanii	Leafminers, mealybugs, scale insects, aphids, whiteflies, thrips.	Kim et al. (2008), Nithya and Rani (2019), Bouhous and Larous (2012)
Nomuraea rileyi	Lepidoptera	Ruiu (2018), Pekrul and Grula (1979)
Hirsutella thompsonii	Spider mites	Ruiu (2018)
Isaria fumosorosea	Whitefly	Ruiu (2018), Zimmermann (2007)
Paecilomyces fumosoroseus	Insects, mites, nematodes, thrips	Ruiu (2018), Siddiqui and Akhtar (2008)
Baculoviruses		
Granulosis virus (GV)	*Adoxophyes orana, Plutella xylostella, Cydia pomonella, Phthorimaea operculella*	Ansari et al. (2012), Ruiu (2018)
Nuclear polyhedrosis virus (NPV)	*Helicoverpa armigera, Heliothis virescens, H. zea, Spodoptera litura, S. littoralis, S. exigua, Lymantria dispar*	Ansari et al. (2012), Ruiu (2018), Ramakrishnan (1993)
Entomopathogenic nematode		
Steinernema carpocapsae	Caterpillars, *R. ferrugineus*, B. beetles, moth larvae crane fly, Tipulidae	Ruiu (2018), Shapiro-Ilan et al. (2012)
Steinernema feltiae	*Bradysia* spp., *C. syngenesiae*, codling moth larvae, *P.vitalbae*, soil-dwelling pests, sciarids, thrips	Ruiu (2018)

(continued)

Table 23.1 (continued)

Microbial species/ variety	Target pest	Selected references
Heterorhabditis bacteriophora	*M. melolontha*, *Otiorhynchus* spp., chestnut moths, black vine weevil, leaf miners, caterpillars, cutworms, and soil-dwelling beetle larvae	Ruiu (2018), Shapiro-Ilan et al. (2012)
Protozoa		
Nosema and *Vairimorpha*	Lepidopteran and orthopteran insects	Lewis (2002)

Since the active ingredients are living substances they offer them potentially higher benefit over chemical pesticides, because of their target specificity, reproducibility, and hence provide continuous pest control (SP-IPM 2006). The detailed mode of action of many of these microbial pesticides has already been reported by several workers (Ansari et al. 2012; Sahayaraj et al. 2011; Senthil-Nathan et al. 2009). These pesticides suppress insect pests by causing disease or by producing specific toxins against the target pest or through competition by preventing the establishment of other organisms (Clemson 2007). The specific and diverse range of properties of microbial pesticides make them desirable component for integrated pest management.

23.2.2 *Natural Biochemical Pesticides*

Naturally occurring biochemical substances are retrieved from plants or insects and used in pest management by eco-friendly mode of action. Biochemical pesticides are broadly classified into three categories that include botanical pesticides and semiochemicals (Fig. 23.1). It is estimated that approximately 2500 plant species belonging to 235 different families have showed measurable biocidal properties against various pests (Ansari et al. 2012). The botanical pesticides include plant extracts and essential oils, natural plant growth regulators, and secondary metabolites (Table 23.2). The natural compounds derived from plant extracts have biologically active compounds that are employed in pest control. Among different botanical pesticides neem tree, *Azadirachta indica* (Meliaceae) is one of the best known and most effective plant that contains azadirachtin as active ingredient, which has various effects on insects such as antifeedant, repellent, deterrent (Saxena 1989), and insect growth regulator because of its activity same as an insect hormone and can be used in pest control. Semiochemicals are organic compounds produced by an organism (either plant or insect) that mediate interactions between individuals of same species (intraspecific communication caused by pheromones) or individuals of a different species (interspecific communication caused by allelochemicals).

Semiochemicals include pheromones and allelochemicals. They can be used for behavioral manipulation of insects and can be used in the control of insect pest. The

Table 23.2 Some of the examples of biochemical pesticides for insect pest management

Botanical pesticides			
Plant	Active compound(s)	Target pests	Selected references
Neem tree, *Azadirachta indica*	Azadirachtin	Most of the herbivorous insects	–
Chrysanthemum	Pyrethrum	Most of the herbivorous insects	Casida and Quistad (1995)
Thyme, *Thymus vulgaris*	Thymol	Varroa mite (*Varroa destructor*)	Floris et al. (2004)
Ageratum houstonianum	Precocene I and II	Heteroptera, Homoptera, and Orthoptera	Bowers et al. (1976), Bowers (1991), Grafton-Cardwell et al. (2006)
Saponin	IGR	Mites, lepidopterans, beetles, and many other insects	Ishaaya et al. (1969)
Basil, *Ocimum basilicum*	Juvocimenes	Several field crop pests	Bowers and Nishida (1980), Lopez et al. (2005)
Garlic, *Allium sativum*	Diallyl disulfide	*B. brassicae, Sitotroga cerealella, S. littorals, T. molitor, C. maculatus, P. xylostella*	Plata-Rueda et al. (2017), Lengai and Muthomi (2018)
Insect pheromones			
Insect pest	Compound(s)	Type of pheromone	Selected references
Helicoverpa armigera	(*Z*)-11-hexadecenal + (*Z*)-9-hexadecenal and (*Z*)-9-tetradecenal	Sex pheromone	Zhang et al. (2012)
Spodoptera litura	(*Z,E*)-9,11- and (*Z,E*)-9,12-tetradecadienyl acetate	Sex pheromone	Chen et al. (2018)
Leucinodes orbonalis	(*E*)-11-hexadecenyl acetate, (*E*)-11-hexadecen-1-ol	Sex pheromone	Cork et al. (2001)
Fall armyworm (*Spodoptera frugiperda*)	7-dodecen-1-ol acetate, dodecan-1-ol acetate, (*Z*)-11-dodecen-1-ol acetate, (*Z*)-9-tetradecen-1-ol acetate, (*Z*)-11-hexadecenal, (*Z*)-9-tetradecenal, and (*Z*)-11-hexadecen-1-ol acetate	Sex pheromone	Tumlinson et al. (1986)

(continued)

Table 23.2 (continued)

Pink bollworm (*Pectinophora gossypiella*)	(*Z*,*Z*)-7,11- and (*Z*,*E*)-7,11-hexadecadienyl acetates	Sex pheromone	Foster and Roelofs (1988)
Japanese beetle, *Popillia japonica*	(*R*,*S*)-5-(1-decenyl)-dihydr0-2-(3H)-furanone	Sex pheromone	Klein et al. (1981)
Cotton boll weevil, *Anthonomus grandis*	Grandlure (cyclobutane alcohol (3 parts), a cyclohexane alcohol (4 parts), and *Z*-and *E*-pair of cyclohexane and acetaldehydes)	Aggregation pheromone	Ansari et al. (2012)
Colorado potato beetle, *Leptinotarsa decemlineata* (Say)	(*S*)-3,7-dimethyl-2-oxo-oct-6-ene-1,3-diol	Aggregation pheromone	Dickens et al. (2002)
Southern pine beetle, *Dendroctonus frontalis*	1,5-Dimethyl-6,8-dioxabicyclo[3.2.1]octane	Aggregation pheromone	Renwick and Vite (1969)
California five-spined ips, *Ips confusus*	(+)-2-Methyl-6-methylene-2,7-octadien-4-ol	Aggregation pheromone	Silverstein et al. (1966)
Green peach aphid, Myzus persicae,	(*E*)-β-farnesene	Alarm pheromone	Kunert et al. (2010), De-Vos et al. (2010)
Allelochemicals			
Plant species	Compound(s)	Target pest	Selected references
Neem tree, *Azadirachta indica*	Azadirachtin (OR, AF, GR)	Most of the herbivorous insects	–
Lemon grass oil, *Cymbopogon nardus*	Methyl eugenol (4-allyl-1,2-dimethoxybenzenecarboxylate) (A)	Several *Bacterocera species*	Cunningham (1989), Wee et al. (2002), Khrimian et al. (2006)
Raspberry	Cue lure [4-(*p* acetoxyphenyl)-2-butanone] (A)	*B. cucurbitae*	Khrimian et al. (2006)
	Geraniol, eugenol, and phenethyl alcohol (A)	Japanese beetle, *P. japonica*	Hamilton et al. (1971), Schwartz (1975)

Cotton, *Gossypium hirsutum*	Gossypol (Ab)	Tobacco bud worm, *Heliothes virescens* and cotton leaf worm, *Spodoptera littoralis* a number of herbivorous insects	Parrott (1990), Guo et al. (2013)
Cucurbits	Cucurbitacins (A, Ab)	Cucumber leaf beetles (*Phyllotreta* spp., *Phaedon* spp., and *Ceratoma trifurcate); stem borer (Margonia hyalinata);* and *red spider mite (Phyllotreta* spp., *Phaedon* spp., and *Ceratoma trifurcate);* and other *arthropods (Phyllotreta* spp., *Phaedon* spp.*)*	Da-Costa and Jones (1971)

OR oviposition repellent, *AF* antifeedant, *GR* growth regulation, *A* attractant, *Ab* Antibiosis most commonly used semiochemicals in insect pest management are sex pheromones and aggregation pheromones (Table 23.2). Insect sex pheromones are an essential component of insect pest control because of their high species specificity and low toxicity (Goldansaz et al. 2004). Some of the examples of biochemical pesticides are listed in Table 23.2.

Biochemical pesticides' active ingredients typically disrupt insect pests' growth, development, reproduction, and overall biology. A single molecule or a combination of molecules, such as plant secondary metabolites and/or essential oils, or a mixture of structurally related molecules (isomers), such as insect pheromones, may be used as the active ingredient (Ansari et al. 2012). While these active ingredients are normally only in nature, a synthetic analog might be possible. Allelochemicals can be further grouped into attractants, repellants, deterrents, stimulants, growth regulators, or mating disruptors based on the behavior elicited by them. Methyl eugenol and cue lure are the most commonly used attractants in the male annihilation technique (MAT) for the control mango fruit fly, *Bactrocera dorsalis* and melon fruit fly, *Zeugodacus cucurbitae* (Table 23.2; insect pheromones are often used for tracking or mass trapping (Reddy et al. 2009), lure-and-kill Schemes (El-Sayed et al. 2009), and mating destruction (Witzgall et al. 2010). There are about 122 registered active ingredients of biochemical pesticides that include 6 insect growth regulators, 20 plant growth regulators, 18 floral attractants, 19 repellents, and 36 insect pheromones (Mandula 2008).

23.2.3 Transgenics/Plant-Incorporated Products

Transgenic crops are referred as plant-incorporated protectants (PIPs), which are grouped under biopesticides as an alternative to chemical insecticides. Plants themselves produce protecting substance (toxins) after the introduction of genetic material coding for that toxic substance. Plants inserted with such transgene are called genetically modified crops, plant pesticides, or plant-incorporated protectants. The best-known example for the PIPs is Bt transgenic plants inserted with gene coding for the Bt toxin into the chromosome of the crop plants, thus the plants become toxic to the pest. Examples of some of the transgenic plants developed for management of insect pests are listed in Table 23.3. In this situation, the Environmental Protection Agency (EPA) regulates the Bt pesticidal protein and its genetic content, but not the plant itself (Mazid et al. 2011). Transgenic crops are low cost and eco-friendly technology for the resource poor farmers to manage pests and diseases as well as other constraints such as abiotic stress and human vitamin A deficiency (SP-IPM 2006).

Table 23.3 Some of the examples of transgenics/plant-incorporated products developed for insect pest management

Plant	Gene	Target pest	Selected references
Tomato	*cry 1Ab, cry 1Ac*	*Heliothis virescens, M. Sexta, H. armigera*	Mandaokar et al. (2000), Leng et al. (2011), Arora and Shera (2014)
Potato	*cry 3, cry 3a, cry 3b, cry 2a5, cry 1Ab, cry 1Ac9, cry 5*	*Leptinotarsa decemlineata, Phthorimaea operculella*	Arora and Shera (2014), Rico et al. (1998), Coombs et al. (2002)
Soybean	*cry1A(c)*	*H. zea, H. virescens*, cigarette beetle *Pseudoplusia includens*	Leng et al. (2011), Stewart et al. (1996a)
Maize/corn	*cry 1Ab, cry 9C, cry 3Bb, cry 1F, cry 34Ab1/cry 35 Ab1, cry 1 Ab + cry 3Bb, cry 1F + cry 34Ab1/cry 35 Ab1*	*Ostrinia nubilalis, Chilo partellus, Busseola fusca, H. zea, Diatraea grandiosella, D. saccharalis, S. frugiperda, Diabrotica undecimpunctata howardi, D. virgifera virgifera*	Leng et al. (2011), Arora and Shera (2014), Jansen et al. (1997)
Rice	*cry 1 Ab, cry 1 Ac, cry 1Ab/cry1Ac, cry2a*	*Chilo suppressalis, Cnaphalocrocis medinalis, Scirpophaga incertulas*	Leng et al. (2011), Arora and Shera (2014)
Cotton	*cry 1Ac, cry 1Ab/cry 1Ac, cry 1Ac + cry2Ab, cry1C*	*Helicoverpa armigera, Pectinophora gossypiella, Earias* spp., *Heliothis virescens, H. zea, Trichoplusia ni, Spodoptera* spp.	Arora and Shera (2014), Adamczyk and Hardee (2001)
Canola	*cry1A(c), cry 1C*	Diamondback moth, *H. zea*, cabbage looper, *Susumia exigua, H. zea, S. exigua*	Arora and Shera (2014), Stewart et al. (1996b)
Poplar	*cry 1Aa, cry3Aa*	*Lymantria dispar, Chrysomela tremulae* F.	Arora and Shera (2014), Cornu et al. (1996)
Sorghum	*cry 1Ac*	*Chilo partellus*	Arora and Shera (2014)
Sugar cane	*cry 1Ab*	*D. saccharalis*	Arora and Shera (2014)
Chickpea	*cry 1Ac*	*H. armigera*	Arora and Shera (2014)
Tobacco	*cry 3, cry 2a5, cry 1Aa, cry 1Ab, cry 1Ac*	*H. virescens, Manduca sexta, H. armigera, H. zea, Leptinotarsa decemlineata*	Arora and Shera (2014), McBride et al. (1995)
Brinjal	*cry 1Ac, cry 3b*	*Leucinodes orbonalis, Leptinotarsa decemlineata*	Arora and Shera (2014)
Chinese cabbage	*cry 1Ab, cry 1Ac*	*Plutella xylostella*	Arora and Shera (2014)

(continued)

Table 23.3 (continued)

Plant	Gene	Target pest	Selected references
Broccoli	*cry 1C*	*P. xylostella, T. ni, Pieris rapae*	Zhao et al. (2001)
Groundnut	*cry 1Ac*	*Elasmopalpus lignosellus*	Arora and Shera (2014)
Alfalfa	*cry 1C*	*S. littoralis*	Arora and Shera (2014)

23.2.4 Natural Enemies (Predators and Parasitoids of Insects)

Plants are protected or vector colonies are reduced by natural enemies such as insect predators (Coccinellid beetle and lacewings) and parasitoids (hymenopteran wasps and dipteran flies). Natural enemies are responsible for natural suppression of pest population. A few of the most common insect predators and parasitoids, respective target pests and their host plants are provided in Table 23.4.

23.3 Success Stories of Insect Pest Management Using Biopesticides

In India, several good cases of traditional biological management of insect pests and weeds have been analyzed in-depth (Singh 2004). Some of the success stories of agricultural pests management in India including weeds by using biological control agents are given in brief.

23.3.1 Controlling the Papaya Mealy Virus

Paracoccus marginatus Williams and Granara de Willink (Hemiptera: Pseudococcidae), the papaya mealybug (Hemiptera: Pseudococcidae) is a dreaded insect pest native to Mexico and/or Central America (Miller et al. 1999) that was first described in 1992. It never attained the status of a pest in the native country due to the existence of endemic natural enemies' complex, however posed a serious threat to horticultural industry following its invasion into the Caribbean region, south and southeast Asian countries. Natural enemies from native areas, such as *Acerophagus papayae* (Noyes and Schauff), *Anagyrus loecki* (Noyes and Menezes), *Anagyrus californicus Compere, Pseudleptomastix mexicana,* and *Pseudaphycus* sp., were used to monitor the pest in invaded countries (U.S. Department of Agriculture, Animal and Plant Health Inspection Service 1999, 2000; Meyerdirk and Kauffman 2001; Noyes and Schauff 2003). In the Dominican Republic, Guam, and Puerto

Table 23.4 Some of the examples of natural enemies used in insect pest management

Natural enemy	Target pest	Host	Selected references
Predators			
Cryptolaemus montrouzieri	Coffee green scale, *Coccus viridis* (Green); *Planococcus citri* (Risso), *P. lilacinus* (Cockerell), *Ferrisia virgata* (Cockerell), *Maconellicoccus hirsutus* (Green)	Several plantation and horticultural crops	Mayne (1953)
Rodolia cardinalis	Cottony cushion scale, *Icerya purchasi*	Citrus, wattle, and other Acacia spp. are among the crops affected	Subramanyam (1955)
Platymeris laevicollis	Oryctes rhinoceros	Coconut	Singh (1994)
Phytoseiulus spp.	*Tetranychus urticae* *Tetranychus evansi*	*Phaseolus vulgaris; Solanum lycopersicon; Fragaria ananassa*	Rhodes and Liburd (2006), Silva et al. (2010), Hoque et al. (2010)
Neoseiulus spp.	*Tetranychus urticae* *Oligonychus perseae*	*Vicia faba; Capsicum annuum; Persea americana*	Elmoghazy et al. (2012), Takano-Lee and Hoddle (2002)
Amblyseius swirskii	*Scirtothrips dorsalis*	*Capsicum* sp.	Arthurs et al. (2009)
Parasitoids			
Trichogramma spp.	Many lepidopteran pests	Several field and horticultural crops	Sankaran (1974)
Copidosoma koehleri	*Phthorimaea operculella*	Potato	Noyes and Hayat (1994)
Cotesia sp.	*Chilo infuscatellus, C. sacchariphagus indicus, Scirpophaga excerptalis*	Sugarcane	Rao et al. (1971)
Chelonus blackburni Cameron	*Pectinophora gossypiella, Earias* spp., *Phthorimaea operculella*	Cotton, okra, hollyhock, potato	Singh (1994)

(continued)

Table 23.4 (continued)

Natural enemy	Target pest	Host	Selected references
Diadegma semiclausum	*Plutella xylostella*	Cruciferous vegetables	Singh (1994)
Encarsia guadeloupae	Spiraling whitefly, *Aleurodicus dispersus*	Several agricultural and horticultural crops and ornamentals	Ramani et al. (2002)
Aphidius colemani	*Aphis gossypii*	*Dendranthema grandiflora*	Vasquez et al. (2006)
Encarsia formosa	*Bemisia tabaci*	*Solanum lycopersicon*	Dai et al. (2014), Moreno-Ripoll et al. (2012)
Tetrastichus sp.	*Chilo infuscatellus*, *C. sacchariphagus indicus*, *Scirpophaga excerptalis*	Sugarcane	Sankaran (1974)
Tachinid flies	Caterpillars, beetles, bugs	Sugarcane	–

Rico, respectively, the release of parasitoid wasps resulted in a 99.7%, 99.5%, and 97.5% decline in the density of mealybug species (Meyerdirk and Kauffman 2001; Kauffman et al. 2001). Later during 2008 the pest was first detected in Tamil Nadu, India subsequently spread to the rest of the country caused serious damage to papaya. The pest was successfully brought under control through the importation of three natural enemies, i.e. *A. papayae*, *A. loecki,* and *P. mexicana.* Even though the aforesaid bio-agents able to parasitize the mealybug, the establishment of the parasitoids was not uniform over the varied geographic region. In Puerto Rico, the Dominican Republic, and the Indian subcontinent, *Acerophagus* sp. has arisen as the dominant parasitoid group (Meyerdirk and Kauffman 2001).

23.3.2 Management of Sugarcane Wooly Aphid: **Ceratovacuna lanigera** *Zehnt-ner*

The aphid is native to India and was first reported from West Bengal during 1958 and limited distribution to northern and northeastern region of India. The aphid invaded the tropical Indian states of Maharashtra and Karnataka in 2002 and later spread to southern Indian states (Patil et al. 2004; Joshi and Viraktamath 2004; Thirumurugan et al. 2004; Srikanth 2007). A year after its first report the pest was observed in 3.13 lakh ha of sugarcane-growing areas in Maharashtra and Karnataka alone. Following its damage the sugarcane growers resorted to spraying chemical pesticides to control the pest, but in vain. The chemical gave only temporary relief and often farmers had to spray repeatedly. This resulted in the destruction of the natural enemy's complex as well as environmental hazards. Hence effective bio-agents were selected and released for the management of wooly aphids and farmers were advised not to apply the chemical pesticides.

Two effective parasitoids, namely *Micromus igorotus* Banks (Neuroptera: Hemerobiidae) and *Dipha aphidivora* (Lepidoptera: Pyralidae), were mass multiplied and released under field condition. In addition, farmers have been advised to conserve existing natural enemies in the sugarcane ecosystem. Within a few years of release, bio-agents have been successful in reducing the aphid population. An estimated benefit of approximately 398.23 crores was realized through this biocontrol intervention. Continuous surveys and surveillance of wooly aphid in sugarcane-growing areas in southern states revealed the scanty presence of wooly aphid (up to 5% incidence), indicating the effectiveness of the biocontrol strategy.

23.3.3 Biological Control Cassava Mealybug in Sub-Saharan Africa

Phenacoccus manihoti, a mealybug, was inadvertently introduced to Africa from South America in the 1970s and soon became a major threat to cassava, one of Africa's most valuable staple crops. In order to manage the *P. manihoti*, the parasitic wasp, *Anagyrus lopezi* was imported from South America, which is the native place of Cassava and cassava mealybug. The parasitic wasp reared in the laboratories and released into the field which began controlling the mealybug almost immediately after release that caused measurable reduction in the pest damage. The mealybug population was reduced to 10% of peak numbers within 2–4 years and that continues till today. This is one of the most active traditional biological control programs in history (SP-IPM 2006).

23.3.4 Classical Biological Control of Terrestrial and Aquatic Weeds Using Insect Biocontrol Agents

Via trade and transportation, many weed species were either intentionally or unintentionally introduced into India. Prickly Pear, *Opuntia elatior Miller, Opuntia Stricta* (Haworth), and *Opuntia vulgaris Miller* (Cactaceae), which entered India through cochineal trade, have become a serious problem in agriculture in South India. This weed species was successfully controlled by the introduction of cochineal insect, *Dactylopius ceylonicus* during 1795. *O. Stricta* was spectacularly controlled by the introduction of North American species *D. Opuntiae* in 1926 from Sri Lanka. Presently *D. ceylonicus* continues to successfully control *O. vulgaris* in southern India and northern Sri Lanka reducing it to the state of virtual extinction (Singh 2004).

Similarly, in 1982, the arrival of the weevil, *Cyrtobagous salviniae,* from Australia, which was endemic to Brazil, was used to biologically kill the water fern, Salvinia molesta. Water hyacinth, *Eichhornia crassipes*, a free-floating aquatic grass, was successfully regulated in India in 1982 by introducing three exotic natural enemies, namely hydrophilic weevils *Neochetina bruchi* (Ex. Argentina) and *N. eichhorniae* (Ex. Argentina) and galumnid mite *Orthogalumna terebrantis* (Ex. South America), and galumnid mite (Singh 2004).

23.3.5 Entomopathogenic Microorganisms: An Unsung Warrior of Biological Control

Entomopathogenic organisms including bacteria, fungi, and viruses are the environment friendly and best alternative to traditional insecticides (Usta 2013). Its usage is not generalized because of high specificity each pest has its own strain. They are specific to target insects and highly safe to mammals and the environment. Despite their vulnerability to environmental changes, microbial insecticides are preferable to conventional insecticides in that they destroy pest colonies while maintaining natural predator and parasite species under certain conditions (Chattopadhyay et al. 2017).

Rather the bio-agents are self-perpetuating and only control tactic that increases in numbers and spread, with increasing pest population.

The bio-agents are the very important component of IPM that helps in successful control of key insect pest under different ecosystem. Among the biopesticides, the products based on entomopathogenic bacteria (EPB) are most commonly used among the farming community (Chattopadhyay et al. 2017). Many public and private laboratories are producing biopesticides and distributing to farmers.

23.4 Status and Market Scenario of Biopesticides in India

For a very long time, the idea of biological control of insect pests and pathogens has been in use (Schmutterer 1985). For a long time, seed grain protectants have been made from the derivatives of the neem tree, Azadirachta indica, such as leaf extract, fat, and seed cake (Isman 1997; Brahmachari 2004). When chemical insecticides failed to combat *Helicoverpa armigera, Spodoptera litura*, and other cotton pests in India, the value of biocontrol became clear (Kranthi et al. 2002). It was recognized that biocontrol is the only alternative management approach that can be used to combat chemical pesticide resistance in insect pests and diseases.

In India, there are 361 biocontrol laboratories and/or units, according to the Directorate of Plant Protection, Quarantine, and Storage (DPPQS) that includes private sector laboratories and also those private sector laboratories aided by GOI grants, state as well as ICAR/SAUs/DBT laboratories and central IPM centers. Although only a handful of them are engaged in processing (Fig. 23.2). According

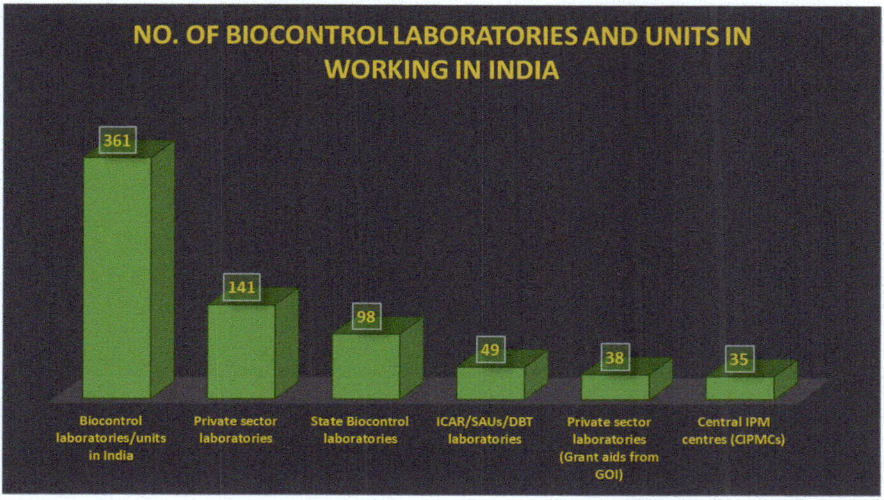

Fig. 23.2 The total number of biocontrol laboratories and units currently operating in India. (Source: Mishra et al. 2020)

to the data biopesticide use seems to have risen in the last two decades. The share of biopesticides, however, is only two percent of the overall market for pesticides. There are currently 970 biopesticide products registered with the Central Insecticides Board and Registration Committee (CIBRC), which is India's top governing body for all forms of biopesticides use. The total number of registered biopesticides and their formulation types is listed in Table 23.5. Totally 51 biopesticides formulations

Table 23.5 Under section 9(3) of the Insecticides Act, 1968, a list of approved biopesticides and their formulations for use in the country is available. (As on 01.01.2021)

Name of bio-insecticide	Formulation type	No. of formulation
I. A. Microbial–bacterial biopesticides		
Ampelomyces quisqualis	2.00%WP	1
Pseudomonas fluorescens	0.5%WP,1.75%WP, 1%WP,1.5%WP, 2.0% AS, 1.5% LF	6
Bacillus sphaericus	1.3% FC	1
Bacillus subtilis	2.0% AS	1
Bacillus thuringiensis var. galleriae	1.3% FC	1
Bacillus thuringiensis var. israelensis	5% AS, 12% AS, 5.0% WP	3
Bacillus thuringiensis var. kurstaki	5% WP, 2.5% As, 0.5% WP, 3.5% AS	4
I. B. Microbial–fungal biopesticides		
Beauveria bassiana	1.15% WP, 1.0% WP,1.15%SC, 10% SC, 5.0% WP	5
Metarhizium anisopliae		
Verticillium lecanii	1.15% WP, 1.5% liquid	2
Verticillium chlamydosporium	1.00 WP	1
Trichoderma harzianum	0.50% WS, 1.0% WP, 2.0% WP, 2.0% AS	4
Trichoderma viride	1.0% WP, 1.15% WP, 5.0% SC, 1.0% AS, 5.0% L. F.	5
I. C. Microbial–viral biopesticides		
Nuclear polyhyderosis virus of *Helicoverpa armigera*	0.43% AS, 2.0% AS. 0.5% AS	3
Nuclear polyhyderosis virus of *Spodoptera litura*	0.5% AS	1
II. A. Biochemical–botanical pesticides		
Azadirachtin (neem products)	25%, 10%, 0.03% EC, 0.1% EC, 0.15% EC, 5% EC, 0.3%EC, 15% extract concentrâtes, 1% EC, 0.1% Gr, 0.15%Gr, 0.03% (300 PPM) w/w min	12
III. B. Biochemical pesticides–pheromones		
Gossyplure (PB-RopeL)	Dispenser	1
	Total	51

Data collected from the DPPQS of the Ministry of Agriculture and Farmers Welfare of the Government of India.; ppqs.gov.in/divisions/cib-rc/major-uses-of-pesticides

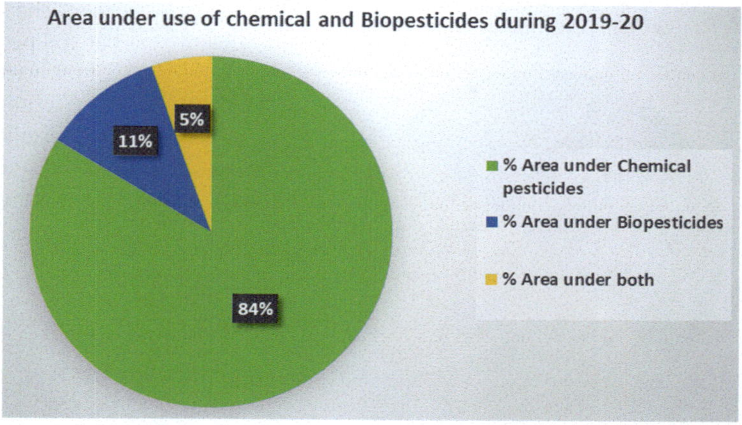

Fig. 23.3 Area under chemical and biopesticides in India during 2019–2020. (DPPQS, Ministry of Agriculture and Farmers Welfare, Government of India data; ppqs.gov.in/statistical-database)

have been registered for use in the country under the Insecticide Act 1968. Azadirachtin or Neem products have registered maximum number of formulations compared to other biopesticide products. The industries are producing bacterial, fungal, microbial, and other (plant-based, pheromones) biopesticides that account for 29, 66, 4, and 1% of total production, respectively.

As per the DPPQS data, the area under use of biopesticides is 11%, whereas, maximum area comes under the use of chemical pesticides with a share of 84% and only 5% area comes under both chemical and biopesticides (Fig. 23.3). Among different types of biopesticides, the fungal biopesticides consumption is maximum with a 50% share as compared to bacterial (21%), viral (15%), and botanical pesticides (14%), respectively (Fig. 23.4). The data suggests that the consumption of various biopesticide formulations in different states has showed steady increase from 5152 MT during 2014–2015 to 7804 MT of technical grade material during 2019–2020 (Fig. 23.5).

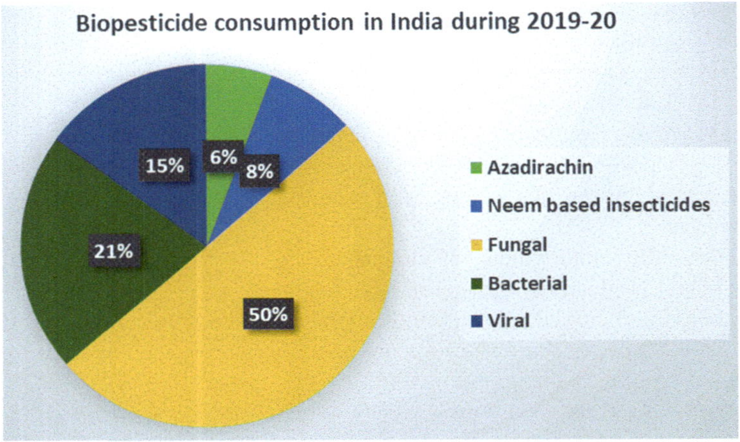

Fig. 23.4 Pesticide wise consumption of indigenous biopesticides in India during 2019–2020. (Data collected from the DPPQS of the Ministry of Agriculture and Farmers Welfare of the Government of India; ppqs.gov.in/statistical-database)

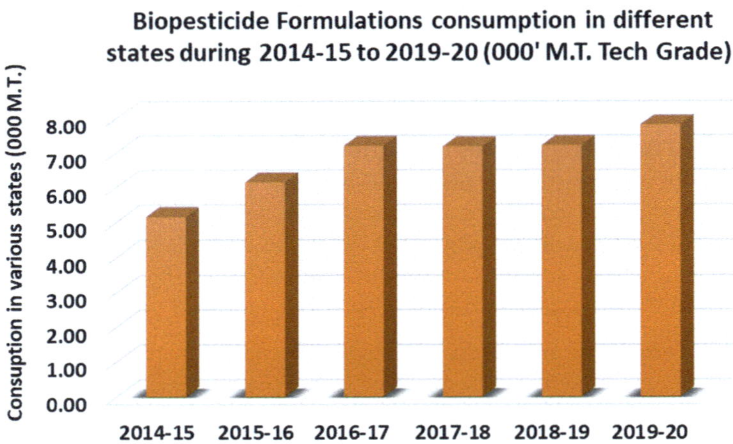

Fig. 23.5 Consumption of biopesticides formulations in various states during 2014–2015 to 2019–2020. (Data collected from the DPPQS of the Ministry of Agriculture and Farmers Welfare of the Government of India; ppqs.gov.in/statistical-database)

23.5 Synthetic Pesticides Versus Biological Pesticides (Table 23.6)

Table 23.6 Comparison between synthetic and biological pesticides

Synthetic or chemical pesticides	Biological pesticides or biopesticides
Uses non-living substance, synthetic products in the form of different type of formulations	Derived from naturally occurring living organisms including plants, animals, and microbes, used as such or as their products or by-products
Quick and easy to use, highly effective as they are fast-acting	Most of them are slower acting, often less toxic than conventional pesticides
Broadspectrum of action	Very specific to the target pest
Relatively expensive	Relatively cheaper
Easily stored for long duration, i.e. more shelf life	Cannot be stored for long duration, i.e. less shelf life
Readily available through a long-established market	Not readily available
Causes serious environmental pollution	Eco-friendly; safe to environment
Hazardous to natural predators, pollinators, and non-target species are all negatively affected	Pest natural predators, pollinators, and non-target species are all safer
They work against nature disrupt ecosystems	They work with nature, maintain ecological balance
Pests eventually become resistant	Less likely to have resistance issues
Diminishing market	Growing market preference
Longer residual activity, provide greater persistent control under field conditions	Effective in small quantities, resulting in lower operator exposure
The cost effectiveness is low, but the cost of spraying is higher	Expenses are higher, but the number of applications is lower
High persistence and long-term impact	Low persistence and decompose quickly
Easy handling and bulkiness but danger and harmful	Bulk: Carrier based, easy: Liquid formulation
Curative nature of control	Preventive nature of control

23.6 Limitations/Constraints in the Promotion and Consumption of Biopesticides

The need for new synthetic insecticide substitutes that are good for the environment and human health has become a major focus of science research and may lead to the production of safe foods. While biopesticides will never be able to completely offset the use of conventional insecticides, they will help to solve some of the problems associated with their use. The major limitation of use of biopesticides is their limited shelf life and timely unavailability. In addition, the efficacy of biopesticides shows high variability under different environmental condition. The users prefer consistency and reliability of pest control methods. In the era of commercial agriculture, the farmers prefer to use conventional insecticides having quick knockdown effect on the insect pests.

23.7 Future Directions

The modern agriculture is increasingly affected by degradation and overexploitation of natural resources, frequent occurrence of inclement weather events influenced by climate change, and excessive application of synthetic inputs on farms including injudicious use of pesticide or even use of banned or spurious ones. The government should set stringent safety criteria on conventional insecticides, which result in the fewer molecules in the market which promote biopesticides. The use of biopesticides must be a mass movement rather than an individual's effort. Biopesticides have been used to protect seeds from rodents and diseases for decades. However, in contrast to traditional chemicals, their demand and position among agrochemicals are still far behind.

The effectiveness, shelf life, processing processes, limited range of host or target pathogens/pests, low field efficiency, distribution system issues, economics, and regulations are all obstacles to widespread biopesticide use in India. To create trust in the system, both the private and public sectors will need to collaborate and work with farmers at the grassroots level.

23.8 Conclusions

The farmers around the globe are majorly using synthetic insecticides for the management of a pest in their agricultural ecosystem. Even though the application of chemicals helps in bringing down the insect population, it unfortunately resulted in posing multiple side effects to the surrounding environment. Hence, there is an immediate need to develop an appropriate pest management approach like use of biopesticides. These pesticides are promising alternatives for use in pest management tactics. The biopesticides like entomopathogens (bacteria, fungi, viruses, protozoa, and nematodes), insect growth regulators, semiochemicals (pheromones), botanicals, plant-incorporated protectants (alkaloids, steroids, terpenoids, essential oils) have been proposed as safer and ecologically alternative to conventional synthetic pesticides for sustainable agricultural production.

Acknowledgements The first author is thankful to the editor for inviting to contribute this chapter. This work was supported by the ICAR-Indian Institute of Vegetable Research, Varanasi.

References

Adamczyk LC, Hardee DD (2001) Field efficacy and seasonal expression profiles for terminal leaves of single and double *Bacillus thuringiensis* toxin cotton genotypes. J Econ Entomol 94:1589–1593

Ansari MS, Ahmad N, Hasan F (2012) Potential of biopesticides in sustainable agriculture. In: Malik A, Grohmann E (eds) Environmental protection strategies for sustainable development. Springer, Dordrecht, pp 529–595. https://doi.org/10.1007/978-94-007-1591-2_17

Arora R, Shera PS (2014) Genetic improvement of biocontrol agents for sustainable pest management. In: Sahayaraj K (ed) Basic and applied aspects of biopesticides. Springer, New Delhi, pp 255–285. https://doi.org/10.1007/978-81-322-1877-7

Arthurs S, McKenzie CL, Chen J, Dogramaci M, Brennan M, Houben, K. And Osborne, L. (2009) Evaluation of *Neoseiulus cucumeris* and *Amblyseius swirskii* (Acari: Phytoseiidae) as biological control agents of chilli thrips, *Scirtothrips dorsalis* (Thysanoptera: Thripidae) on pepper. Biol Control 49:91–96

Bhattacharya AK, Mondal P, Ramamurthy VV, Srivastava RP (2003) Beauveriabassiana: a potential bioagent for innovative integrated pest management programme. In: Srivastava RP (ed) Biopesticides and bioagents in integrated pest management of agricultural crops. International Book Distributing Co., Lucknow, pp 381–492

Bouhous M, Larous L (2012) Efficiency of the entomopathogenic fungus *Verticillium lecanii* in the biological control of *Trialeurodes vaporariorum*, (Homoptera: Aleyrodidae), a greenhouse culture pest. Afr J Microbiol Res 6(10):2435–2442

Bowers WS (1991) Insect hormones and antihormones in plants. In: Rosenthal GA, Berenbaum MR (eds) Herbivores: their interactions with secondary plant metabolites, vol 1. Academic, New York, pp 431–456

Bowers WS, Nishida R (1980) Juvocimenes; potent juvenile hormone mimics from sweet basil. Science 209:1030–1032

Bowers WS, Ohta T, Cleere JS, Marsella PA (1976) Discovery of insect anti-juvenile hormones in plants. Science 193:542–547

Brahmachari G (2004) Neem—an omnipotent plant: a retrospection. Chembiochem 5(4):408–421. https://doi.org/10.1002/cbic.200300749

Bravo A, Gill S, Soberón M (2007) Mode of action of *Bacillus thuringiensis* Cry and Cyt toxins and their potential for insect control. Toxicology 49:423–435

Casida JF, Quistad GB (1995) Metabolism and synergism of pyrethrins. In: Casida JE, Quistad GB (eds) Pyrethrum flowers: production, chemistry, toxicology, and uses. Oxford University Press, New York, pp 259–276

Chandler D, Bailey AS, Tatchell GM, Davidson G, Greaves J, Grant WP (2011) The development, regulation and use of biopesticides for integrated pest management. Philos Trans R Soc B Biol Sci 366(1573):1987–1998

Chattopadhyay A, Bhatnagar N, Bhatnagar R (2004) Bacterial insecticidal toxins. Crit Rev Microbiol 30:33

Chattopadhyay P, Banerjee G, Mukherjee S (2017) Recent trends of modern bacterial insecticides for pest control practice in integrated crop management system. 3 Biotech 7(1):60

Chen Y, Chen X, Chen Y, Wei H, Lin S, Tian H et al (2018) Preparation, characterisation, and controlled release of sex pheromone-loaded MPEG-PCL diblock copolymer micelles for Spodoptera litura (Lepidoptera: Noctuidae). PLoS One 13(9):e0203062. https://doi.org/10.1371/journal.pone.0203062

CIBRC (2017) CIBRC, 2017. Central Insecticides Board and Registration Committee. Ministry of Agriculture and Farmers Welfare, Government of India. http://cibrc.nic.in. Accessed 25 June 2017

Clemson HG (2007) Organic pesticides and biopesticides, Clemson extension, home and garden information center. Clemson University, Clemson

Coombs JJ, Douches DS, Li WB, Grafius EJE, Pett WL (2002) Combining engineered (Bt-cry3A) and natural resistance mechanisms in potato for control of Colorado potato beetle. J Am Soc Hortic Sci 127:62–68

Cork A, Alam SN, Das A, Das CS, Ghosh GC, Farman DI, Hall DR, Maslen NR, Vedham K, Phythian SJ, Rouf FMA, Srinivasan K (2001) Female sex pheromone of brinjal fruit and shoot borer, *Leucinodes orbonalis* blend optimization. J Chem Ecol 27(9):1867–1877

Cornu D, Leplé JC. Bonadé-Bottino M, Ross A, Augustin S, Delplanque A, Jouanin L, Pilate G (1996) Expression of proteinase inhibitor and a *Bacillus thuringiensis* delta–endotoxin in transgenic poplars.. In: Proceedings IUFRO meeting on somatic cell genetics and molecular genetic of trees. Kluwer, Dordrecht, pp 131–136, 287p

Cunningham RT (1989) Male annihilation. In: Robinson AS, Hooper G (eds) World crop pests. Elsevier, Amsterdam, pp 78–81

Da-Costa CP, Jones CM (1971) Cucumber beetle resistance and mite susceptibility controlled by the bitter gene in Cucumis sativus L. Science 172(3988):1145–1146

Dai P, Ruan CC, Zang LS, Wan FH, Liu LZ (2014) Effects of rearing host species on the host-feeding capacity and parasitism of the whitefly parasitoid *Encarsia formosa*. J Insect Sci 14:118. https://doi.org/10.1093/jis/14.1.118.Article118

Damalas CA, Eleftherohorinos IG (2011) Pesticide exposure, safety issues, and risk assessment indicators. Int J Environ Res. Public Health 8:1402–1419. https://doi.org/10.3390/ijerph8051402

DeBach P, Rosen D (1991) Biological control by natural enemies. CUP archive, 2nd edn. Cambridge University Press, Cambridge

De-Vos M, Cheng WY, Summers HE, Raguso RA, Jander G (2010) Alarm pheromone habituation in *Myzus persicae* has fitness consequences and causes extensive gene expression changes. PNAS 107(33):14673–14678. https://doi.org/10.1073/pnas.1001539107

Dickens JC, Oliver JE, Hollister B, Davis JC, Klun JA (2002) Breaking a paradigm: male-produced aggregation pheromone for the Colorado potato beetle. J Exp Biol 205:1925–1933

Elmoghazy MME, El-Saiedy EMA, Romeih AHM (2012) Integrated control of the two-spotted spider mite *Tetranychus urticae* Koch (Acari: Tetranychidae) on Faba bean *Vicia faba* (L.) in an open field at Behaira gGovernorate, Egypt. Int J Environ Sci Eng 2:93–100

El-Sayed AM, Suckling DM, Byers JA, Jang EB, Wearing CH (2009) Potential of "lure and kill" in long-term pest management and eradication of invasive species. J Econ Entomol 102:815–835. https://doi.org/10.1603/029.102.0301

Erler F, Ates AO (2015) Potential of two entomopathogenic fungi, *Beauveria bassiana* and *Metarhizium anisopliae* (Coleoptera: Scarabaeidae), as biological control agents against the June beetle. J Insect Sci 15(1):44

Floris I, Satta A, Cabras P, Garau VL, Angioni A (2004) Comparison between two thymol formulations in the control of Varroa destructor: effectiveness, persistence, and residues. J Econ Entomol 97(2):187–191

Foster SP, Roelofs WL (1988) Pink bollworm sex pheromone biosynthesis from oleic acid. Insect Biochem 18(3):281–286

Gill HK, Garg H (2014) Pesticide: environmental impacts and management strategies. Pestic Toxic Asp 8:187. https://doi.org/10.5772/57399

Goldansaz SH, Dewhirst S, Birkett MA, Hooper AM, Smiley DWM, Pickett JA, Wadhams L, McNeil JN (2004) Identification of two sex pheromone components of the potato aphid, Macrosiphum euphorbiae (Thomas). J Chem Ecol 30:819–834

Grafton-Cardwell EE, Leea JE, Stewartb JR, Olsenc KD (2006) Role of two insect growth regulators in integrated pest management of citrus scales. J Econ Entomol 99(3):733–744

Guo JY, Wu G, Wan FH (2013) Effects of high-gossypol cotton on the development and reproduction of *Bemisia tabaci* (Hemiptera: Aleyrodidae) MEAM1 cryptic species. J Econ Entomol 106(3):1379–1385. https://doi.org/10.1603/ec12401

Hamilton DW, Schwartz PH, Townshend BG, Jester CW (1971) Traps reduce an isolated infestation of Japanese beetle. J Econ Entomol 64:150

Hoque FM, Islam MW, Khalequzzaman M (2010) Functional Eesponse of *Phytoseiulus persimilis* Athias-Henriot to *Tetranychus urticae* Koch: effects of prey life stages and temperature. Univ J Zool Rajshahi Univ 29:1–8

Ishaaya I, Birk Y, Bondi A, Tencer Y (1969) Soybean saponins IX. Studies of their effect on birds, mammals and coldblooded organisms. J Sci Food Agric 20:433–436

Isman MB (1997) Neem and other botanical insecticides: barriers to commercialization. Phytoparasitica 25(4):339. https://doi.org/10.1007/BF02981099

Jansen SV, Dickburt A, Buysse C, Piens L, Saey C, De Wulf B, Gossele A, Paez VA, Gobel E (1997) Transgenic corn expressing a cry9C insecticidal protein from *Bacillus thuringiensis* protected from European corn borer damage. Crop Sci 37:1616–1624

Joshi S, Viraktamath CA (2004) The sugarcane woolly aphid, *Ceratovacuna lanigera*. Curr Sci 87 (3):307–316

Kauffman WC, Meyerdirk DE, Miller D, Schauff M, Hernandez HG, Villanueva Jimenez JA (2001) Papaya mealy bug biological control in Puerto Rico and Dominican Republic. Presented 11 December 2001 at the ESA annual meeting, San Diego

Khrimian A, Jang EB, Nagata J, Carvalho L (2006) Consumption and metabolism of 1,2 dimethoxy-4-(3-fluoro-2-propenyl)benzene, a fluorine analog of methyl eugenol, in the oriental fruit fly, Bactrocera dorsalis (Hendel). J Chem Ecol 32:1513–1526

Kim JJ, Goettel MS, Gillespie DR (2008) Evaluation of *Lecanicillium longisporum*, Vertalec® for simultaneous suppression of cotton aphid, *Aphis gossypii*, and cucumber powdery mildew, *Sphaerotheca fuliginea*, on potted cucumbers. Biol Control 45:404–409

Klein MG, Tumlinson JH, Ladd TL Jr, Doolittle RE (1981) Japanese beetle (Coleoptera: Scarabaeidae):response to synthetic sex attractant plus phenethyl propionate Eugenol. J Chem Ecol 7:1–7

Kogan M (1998) Integrated pest management: historical perspectives and contemporary developments. Annu Rev Entomol 43:243–270

Kranthi KR, Jadhav DR, Kranthi S, Wanjari RR, Ali SS, Russell DA (2002) Insecticide resistance in five major insect pests of cotton in India. Crop Prot 21:449–460

Kunert G, Reinhold C, Gershenzon J (2010) Constitutive emission of the aphid alarm pheromone, (E)-b-farnesene, from plants does not serve as a direct defense against aphids. BMC Ecol 10:23

Leng P, Zhang Z, Pan G, Zhao M (2011) Applications and development trends in biopesticides. Afr J Biotechnol 10(86):19864–19873

Lengai GMW, Muthomi JW (2018) Biopesticides and their role in sustainable agricultural production. J Biosci Med 6:7–41. https://doi.org/10.4236/jbm.2018.66002

Lewis LC (2002) Protozoan control of pests. In: Pimental D (ed) Encyclopedia of pest management. Taylor & Francis, New York, pp 673–676

Lopez O, Fernandez-Bolanos JG, Gil MV (2005) New trends in pest control: the search for greener insecticides. Green Chem 7:431–442

Mandaokar AD, Goyalb RK, Shukla A, Bisaria S, Bhalla R, Reddy VS, Chaurasia A, Sharma RP, Altosaar I, Kumar PA (2000) Transgenic tomato plants resistant to fruit borer (*Helicoverpa armigera* Hu¨bner). Crop Prot 19:307–312

Mandula B (2008) Personal communication. US EPA Office of Pesticide Programs, Biopesticide and Pollution Prevention Division, Washington DC

Mayne WW (1953) Cryptolaemus montrouzieri Mulsant in South India. Nature 172:85

Mazid S, Kalita JC, Rajkhowa RC (2011) A review on the use of biopesticides in insect pest management. Int J Sci Adv Technol 1:169–178

McBride KE, Svab Z, Schaaf DJ, Hogan PS, Stalker DM, Maliga P (1995) Amplification of a chimeric Bacillus gene in chloroplasts leads to an extraordinary level of an insecticidal protein in tobacco. Biotechnology 13:362–365

Meyerdirk DE, Kauffman WC (2001) Status on the development of a biological control program for *Paracoccus marginatus* Williams, papaya mealybug. Internal USDA, APHIS, PPQ Report

Miller DR, Williams DJ, Hamon AB (1999) Notes on a new mealybug (Hemiptera: Coccoidea: Pseudococcidae) pest in Florida and the Caribbean: the papaya mealybug, *Paracoccus marginatus* Williams and Granara de Willink. Insecta Mundi 13:3–4

Mishra J, Tewari S, Singh S, Arora NK (2015) Biopesticides: where we stand? In: Plant microbes symbiosis: applied facets. Springer, New Delhi, pp 37–75

Mishra J, Dutta V, Arora NK (2020) Biopesticides in India: technology and sustainability linkages. 3 Biotech 10:1–2

Mohammad Beigi A, Port G (2015) Effect of infection by *Beauveria bassiana* and *Metarhizium anisopliae* on the feeding of *Uvarovistia zebra*. J Insect Sci 15(1):88

Mohan M, Sushil SN, Selvakumar G, Bhatt JC, Gujar G, Gupta HS (2009) Differential toxicity of bacillus thuringiensis strains and their crystal toxins against high-altitude Himalayan populations of diamondback moth, *Plutella xylostella* L. Pest Manag Sci 65(1):27–33

Moreno-Ripoll R, Gabarra R, Symondson WOC, King RA, Agustí N (2012) Trophic relationships between predators, whiteflies and their parasitoids in tomato greenhouses: a molecular approach. Bull Entomol Res 102:415–423. https://doi.org/10.1017/S0007485311000836

Nithya PR, Rani R (2019) Enriched bioformulations of *Lecanicillium lecanii* (Zimmermann) Zare and Gams against sucking pests of yard long bean, *Vigna ungiculata* L. Walp *sesquipedalis*. J Entomol Research 43(4):445

Noyes JS, Hayat M (1994) Oriental mealybug parasitoids of the Anagyrini (Hymenoptera: Encyrtidae). CAB International/The Natural History Museum, London, 554p

Noyes JS, Schauff ME (2003) New encyrtidae (Hymenoptera) from papaya mealybug (*Paracoccus marginatus* Williams and Granara de Willink) (Hemiptera: Sternorrhyncha: Pseudococcidae). Proc Entomol Soc Wash 105:180–185

Parrott WL (1990) Plant resistance to insects in cotton. Florida Entomol 73(3):392–396. https://doi.org/10.2307/3495457

Patil AS, Shinde VD, Magar SB, Yadav RG, Nerkar YS (2004) Sugarcane woolly aphid (*Ceratovacuna lanigera* Zehnt) its history and control measures. Coop Sug 36(1):37–48

Pawar AD, Singh B (1993) Prospects of botanicals and biopesticides. In: Parmar BS, Devakumar C (eds) Botanicals and biopesticides. SPS Publishing no.4, Society of Pesticides Science, India and Westville Publishing House, New Delhi, pp 188–196

Pawlak K, Kołodziejczak M (2020) The role of agriculture in ensuring food security in developing countries: considerations in the context of the problem of sustainable food production. Sustainability 12(13):5488. https://doi.org/10.3390/su12135488

Pekrul S, Grula EA (1979) Mode of infection of the corn earworm (Heliothiszea) by Beauveriabassiana as revealed by scanning electron microscopy. J Invertebr Pathol 34:238–247

Pesticides Safety Directorate (2008) Assessment of the impact on crop protection in the UK of the 'cut-off criteria' and substitution provisions in the proposed regulation of the European Parliament and of the Council concerning the placing of plant protection products in the market. Pesticides Safety Directorate, York, p 46

Plata-Rueda A, Martínez LC, Santos MHD, Fernandes FL, Wilcken CF, Soares MA, Zanuncio JC (2017) Insecticidal activity of garlic essential oil and their constituents against the mealworm beetle, *Tenebrio molitor* Linnaeus (Coleoptera: Tenebrionidae). Sci Rep 7:46406. https://doi.org/10.1038/srep46406

Ramakrishnan N (1993) Baculovirus pesticides. In: de Parmar BS, Devakumar C (eds) Botanicals and biopesticis. SPS Publishing no.4, Society of Pesticides Science, India and Westville Publishing House, New Delhi, pp 178–187

Ramani S, Poorani J, Bhumannavar BS (2002) Spiralling whitefly, *Aleurodicus dispersus* Russell in India. Biocontrol News Inf 23:55–62

Rao VP, Ghani AM, Sankaran T, Mathur KC (1971) A review of the biological control of insects and other pests in South East Asia and the Pacific region. Commonwealth Institute of Biological Control Technical Communication, vol 6, 149p. CAB, Slough, UK

Reddy GV, Cruz ZT, Guerrero A (2009) Development of an efficient pheromone-based trapping method for the banana root borer *Cosmopolites sordidus*. J Chem Ecol 35:111–117. https://doi.org/10.1007/s10886-008-9580-6

Rena R (2004) Green revolution: Indian agricultural experience-a paradigm for Eritrea. Eritrean Stud Rev 4(1):103–130

Renwick JAA, Vite JP (1969) Bark beetle attractants: mechanism of colonization by *Dendroctonus frontalis*. Nature 224:1222–1122

Rhodes ME, Liburd OE (2006) Evaluation of predatory mites and acramite for control of two-spotted spider mites in strawberries in north Central Florida. J Econ Entomol 4:1291–1298. https://doi.org/10.1093/jee/99.4.1291

Rico E, Ballester V, Mensua JL (1998) Survival of two strains of *Phthorimaea operculella* (Lepidoptera: Gelechiidae) reared on transgenic potatoes expressing a *Bacillus thuringiensis* crystal protein. Agronomie 18:151–155

Romeis J, Meissle M, Bigler F (2006) Transgenic crops expressing bacillus thuringiensis toxins and biological control. Nat Biotechnol 24:63–71

Ruiu L (2018) Microbial biopesticides in agroecosystems. Agronomy 8(11):235

Sahayaraj K, Namasivayam SK, Rathi JM (2011) Compatibility of entomopathogenic fungi with extracts of plants and commercial botanicals. Afr J Biotechnol 10:933–938

Sankaran T (1974) Natural enemies introduced in recent years for biological control of agricultural pests in India. Indian J Agric Sci 44(7):425–433

Saxena RC (1989) Insecticides from neem. In: Arnason JT, BJR P, Morand P (eds) Insecticides of plant origin. ACS symposium series 387. American Chemical Society, Washington, DC, pp 110–135

Schmutterer H (ed) (1985) The neem tree *Azadirachtia indica* A. Juss and other meliaceous plants: sources of unique natural products for integrated pest management, medicine, industry and other purposes. VCH, Weinheim, p 696

Schwartz PH Jr (1975) Control of insects on deciduous fruit and tree nuts in the home orchard without insecticides. USDA Home Garden Bull 211:36

Senthil-Nathan S, Choi MY, Seo HY, Paik CH, Kalaivani K (2009) Toxicity and behavioral effect of 3β, 24, 25-trihydroxycycloartane and beddomei lactone on the rice leaf folder Cnaphalocrocis medinalis (Guenée) (Lepidoptera: Pyralidae). Ecotoxicol Environ Saf 72:1156–1162

Shapiro-Ilan DI, Han R, Dolinksi C (2012) Entomopathogenic nematode production and application technology. J Nematol 44:206–217

Siddiqui ZA, Akhtar MS (2008) Synergistic effects of antagonistic fungi and a plant growth promoting rhizobacterium, an arbuscular mycorrhizal fungus, or composted cow manure on populations of *Meloidogyne incognita* and growth of tomato. Biocontrol Sci Tech 18:279–290

Silva FR, Moraes GJ, Gondim MGC, Knapp M, Rouam SL, Paes JLA, Oliveira GM (2010) Efficiency of *Phytoseiulus longipes* Evans as a control agent of *Tetranychus evansi* Baker & Pritchard (Acari: Phytoseiidae: Tetranychidae) on screenhouse tomatoes. Neotrop Entomol 6:991–995. https://doi.org/10.1590/S1519-566X2010000600022

Silverstein RM, Rodin JO, Wood DL (1966) Sex attractants in frass produced by male *Ips confusus* in ponderosa pine. Science 154:509–510

Singh SP (1994) Fifteen years of AICRP on biological control. Technical bulletin no. 8. Project Directorate of Biological Control (ICAR), Bangalore, India, 323p

Singh SP (2004) Some success stories in classical biological control of agricultural pests in India. APAARI publication 2004/2. FAO-Regional office for Asia and the Pacific, Bangkok, Thailand, 50p

Singh R, Singh P, Sharma R (2014) Microorganism as a tool of bioremediation technology for cleaning environment: a review. Proc Int Acad Ecol Environ Sci 4:1–6

SP-IPM (2006) Biological alternatives to harmful chemical pesticides. IPM Research brief No. 4. SP-IPM Secretariat, International Institute of Tropical Agriculture (IITA), Cotonou, Benin

Srikanth J (2007) World and Indian scenario of sugarcanewooly aphid. In: Mukunthan N, Srikanth J, Singaravelu B, Rajula Shanthy T, Thiagarajan R, Puthira Prathap D (eds) Woolly aphid management in Sugarcane. Extn Publ no 154. ICAR-Sugarcane Breeding Institute, Coimbatore, pp 1–12

Stewart CN Jr, AdangMJ AJN, Boerma HR, Cardineaux G, Tucker D, Parrot WA (1996a) Genetic transformation recovery, and characterization of fertile soybean transgenic for a synthetic *Bacillus thuringiensis* cryIA (c) gene. Plant Physiol 112:121–129

Stewart CN Jr, Adang MJ, All JN, Raymer PL, Ramachandran S, Parrott WA (1996b) Insect control and dosage effects in transgenic canola containing a synthetic *Bacillus thuringiensis* cryIAc gene. Plant Physiol 112:115–120

Su C, Jiang L, Zhang W (2014) A review on heavy metal contamination in the soil world wide: situation, impact and remediation techniques. Environ Skeptics Crit 3:24–38

Subramanyam VK (1955) Control of the fluted scale in Peninsular India. Indian J Entomol 11 (1):61–70

Tadele S, Emana G (2017) Entomopathogenic effect of *Beauveria bassiana* (Bals.) and *Metarrhizium anisopliae* (Metschn.) on Tuta absoluta (Meyrick) (Lepidoptera: Gelechiidae) larvae under laboratory and glasshouse conditions in Ethiopia. J Plant Pathol Microbiol 8:411

Takano-Lee M, Hoddle M (2002) Predatory behaviors of Neoseiulus californicus and Galendromus helveolus (Acari: Phytoseiidae) attacking Oligonychus perseae (Acari: Tetranychidae). Exp Appl Acarol 26:13–26. https://doi.org/10.1023/A:1020949531639

Tayade S, Patel ZP, Mutkule DS, Kakde AM (2013) Pesticide contamination in food: a review. IOSR J Agric Vet Sci 6:7–11

Thirumurugan A, Koodalingam K, Baskaran TL (2004) Status of woolly aphid, *Ceratovacuna lanigera* Zehntner on sugarcane in northern districts of Tamil Nadu. In Agenda notes of 36th meeting of sugarcane research and development workers of Tamil Nadu. Sugarcane Breeding Institute, Coimbatore, pp 57–63

Tumlinson JH, Mitchell ER, Teal PEA, Heath RR, Mengelkoch LJ (1986) Sex pheromone of fall armyworm, *Spodoptera frugiperda* (J.E. smith): identification of components critical to attraction in the field. J Chem Ecol 12:1909–1926

U.S. Department of Agriculture, Animal and Plant Health Inspection Service (1999) Control of the papaya mealybug, Paracoccus marginatus (Homoptera: Pseudococcidae). Environmental assessment, Riverdale

U.S. Department of Agriculture, Animal and Plant Health Inspection Service (2000) Control of the papaya mealybug, Paracoccus marginatus (Homoptera: Pseudococcidae). Environmental assessment (Supplement), Riverdale

Usta C (2013) Microorganisms in biological pest control—a review (bacterial toxin application and effect of environmental factors). Curr Prog Biol Res 24:287–317

Vasquez GM, Orr DB, Baker JR (2006) Efficacy assessment of *Aphidius colemani* (Hymenoptera : Braconidae) for suppression of *Aphis gossypii* (Homoptera : Aphididae) in greenhouse-grown chrysanthemum. J Econ Entomol 99:1104–1111

Wee SL, Hee AKW, Tan KH (2002) Comparative sensitivity to and consumption of methyl eugenol in three *Bactrocera dorsalis* (Diptera: Tephritidae) complex sibling species. Chemoecology 12:193–197

Witzgall P, Kirsch P, Cork A (2010) Sex pheromones and their impact on pest management. J Chem Ecol 36:80–100

Zhang JP, Salcedo C, Fang YL, Zhang RJ, Zhang ZN (2012) An overlooked component: (Z)-9-tetradecenal as a sex pheromonein *Helicoverpa armigera*. J Insect Physiol 58(9):1209–1216. https://doi.org/10.1016/j.jinsphys.2012.05.018

Zhao JZ, Li YX, Collins HL, Cao J, Earle ED, Shelton AM (2001) Different cross-resistance patterns in the diamondback moth (Lepidoptera: Plutellidae) resistant to *Bacillus thuringiensis* toxin crylC. J Econ Entomol 94:1547–1552

Zimmermann G (2007) Review on safety of the entomopathogenic fungus, Metarhizium anisopliae. Biocont Sci Technol 17(9):879–920

Chapter 24
Impact of Pesticides on Microbial Population

Sujan Majumder, Anindita Paul, Anup Kumar, Chandan K. Verma, Pratap A. Divekar, Vijaya Rani, A. T. Rani, Jaydeep Halder, K. K. Pandey, and Jagdish Singh

Abstract Microbes are constituting elements of the soil environment and their abundance, enzymatic activity, degradation process, and biodiversity indicate the balance in the agro-ecological system. It is necessary to keep strengthening the scientific basis of modern agriculture because pesticides may be purposefully used only if their persistence, bioaccumulation, and toxicity in agro-ecosystems are strictly controlled. The use of agrochemicals, such as chemical fertilisers and pesticides, is important in modern crop management strategies (mainly insecticides and herbicides). Pesticide poisoning affects three million people worldwide, according to the WHO. Long-term and indiscriminate pesticide use has serious negative consequences for soil microbes, the nutrient cycle, the decomposition process, and the atmosphere, resulting in long-term negative consequences for food stability, human health, and the environment. Pesticide application can alter microbial diversity, which can be detrimental to plant growth and development by decreasing nutrient availability or disrupting the nutrient cycle. Therefore, the qualitative, innovative, and demand-driven pest management is the need of the hour. Hence, this chapter covers the positive and negative consequences of pesticides on microbes and their environment and current issues about the extensive use of pesticides.

Keywords Microbes · Pesticides · Decomposition processes · Nutrient cycle

---

S. Majumder (✉) · C. K. Verma · P. A. Divekar · V. Rani · A. T. Rani · J. Halder · K. K. Pandey · J. Singh
ICAR-Indian Institute of Vegetable Research, Varanasi, India

A. Paul
ICAR-Central Tobacco Research Institute, Rajahmundry, India

A. Kumar
ICAR-Indian Grassland and Fodder Research Institute, Jhansi, India

P. Bhatt et al. (eds.), *Microbial Technology for Sustainable Environment*,
https://doi.org/10.1007/978-981-16-3840-4_24

24.1 Introduction

Pesticides are effective agrochemicals that are used to protect crops from pests in the agricultural system. Pesticides are chemicals used to deter, eliminate, or control pests, unwanted species of plants, or animals that are causing harm or interfering with the cultivation, manufacturing, storage, transportation, or selling of food, agricultural crops, wood, wood products, or animal feedstuffs or which can be given to animals to keep flies, mites/spider mites, and other pests out of or on their bodies (FAO 1989).

Pesticide residue is described as substances that remain in or on a feed or food product, soil, air, or water after a pesticide has been applied. It contains the parent compound as well as any toxicologically relevant degradation products, metabolites, or impurities. Pesticides are often applied in crop to manage the pest during cropping seasons and non-cropping seasons to regulate the weed in fallow land and those pesticides reach the soil, and water. The non-judicious use of pesticides caused hazard and toxicity to non-target organism (birds) and environmental pollution. The effect of pesticides on soil microorganisms is determined by the physical and chemical characteristics of the soil, as well as the chemical composition and concentration of the pesticides. Microbes have been shown to be able to thrive in the presence of pesticides and use the pesticide molecules as carbon and energy sources in numerous studies. Pollen and nectar poisoned by systemic insecticides (pesticide moves inside the plant via xylem) can kill bees and other pollinators. Pesticide usage over time can lead to bioaccumulation and biomagnification in plants and other species. Pesticides that disrupt microorganisms' activities can have an effect on the nutrient cycle and soil nutritional quality, resulting in severe ecological imbalance (Handa et al. 1999).

The microbial population of 1 g of soil during a counting is a barometer of a country's agricultural prosperity. The total mass of microflora and fauna underneath the soil is 20 times that of the entire world's human population. In one gramme of healthy soil, there are one million to 100 million bacteria involved in organic matter breakdown, 0.15–0.5 mg of fungal hyphae, 10,000–100,000 protozoa, a few to several hundred microarthropods, 15–500 nematodes, and a couple of earthworms (Lavelle et al. 2001). Microbial activity (primarily bacteria and fungi) aids in the breakdown of soil organic matter and the management of soil aggregates, while other soil components help to maintain proper number of bacteria and fungi through prey–predator encounters, assisting in the recycling and preservation of basic nutrients.

These soil food cycle components are in harmony and are masterfully crafted in interdependent relationships. The structure and performance of the soil food cycle, i.e. the number, operation, and community structure, serve as a key indicator of ecosystem health. Directly monitoring the active and total biomass of each organism group can aid in detecting the dynamics of change that lead to ecosystem harm.

Once pesticides are applied, they dissipate in the soil, water, and environment and persist for a long time or even year after year and have negative impacts on microbial populations. As a result of the decrease in microbe numbers, the food chain conducted by a group is disrupted, as are the components that depend on it. Owing to the loss of soil organic matter, most microbial groups undergo a series of modifications, resulting in altered predator–prey modules, which cause changes in soil aggregation, soil chemistry, pH, and structure. Over the course of a few decades, the soil becomes barren due to a lack of organic matter to support microorganism growth and development (Chowdhury et al. 2008).

24.2 Recent Trend of Pesticides in India

Chemicals such as insecticides, herbicides, or fungicides are commonly used for the control of various pests in agriculture. There are thousands of pesticides of both biological and chemical origins which are used commonly all over the world to minimise losses of crop production. In India insecticides contribute a higher share in total consumption of pesticides. Both per hectare consumption and total consumption of pesticides increase significantly after 2009–2010 (Fig. 24.1).

Pesticide intake per hectare in 2014–2015 was 0.29 kg ha^{-1}, which is approximately 50% higher than the consumption in 2009–2010. Due to the rising cost of manual labour for weed control, herbicides play an important role in the increased use of pesticides (FICCI 2015). Punjab had the highest per hectare intake in 2016–2017 (0.74 kg), followed by Haryana (0.62 kg) and Maharashtra (0.62 kg) (0.57 kg). Pesticide consumption per hectare in India is 0.69 kg on average. When compared to other countries such as China (13.06 kg ha^{-1}), Japan (11.85 kg ha^{-1}),

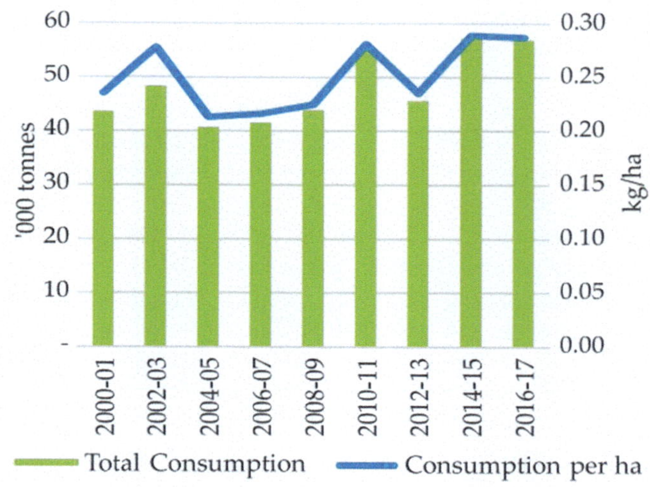

Fig. 24.1 Pesticide (technical grade) consumption in India. (Source*:* Data from Ministry of Chemicals and Fertilizers, Govt. of India)

Brazil (4.57 kg ha^{-1}), and other Latin American countries, this is significantly lower (FAOSTAT 2017).

24.3 Pesticide Production Scenario

In India, insecticides are preceded by fungicides, herbicides, and rodenticides in order of development (Fig. 24.2).

The production share of insecticides lower down from more than 70% in the year 2003–2004 to 39% in the year 2016–2017. The shares of other groups of pesticides such as herbicides, fungicides, and rodenticides grow over a period of time. The increase in the production of fungicides is mainly due to increased use in vegetables and fruits.

24.4 Trade in Agro-chemicals

In the year 2016–2017, 377.76 thousand tonnes of pesticides were exported from India in which fungicide contributes the largest share of 45.94%, followed by herbicides with a share of 28.19% (Fig. 24.3).

Mancozeb, cypermethrin, sulphur, acephate, and chlorpyrifos were the top five pesticides exported in 2016–2017, according to the Central Board of Excise and Customs (CBEC), while glyphosate and atrazine were the top two imported items.

Brazil, the USA, and France are the major countries where pesticides were exported from India (Table 24.1), while China and Germany were major exporters to India.

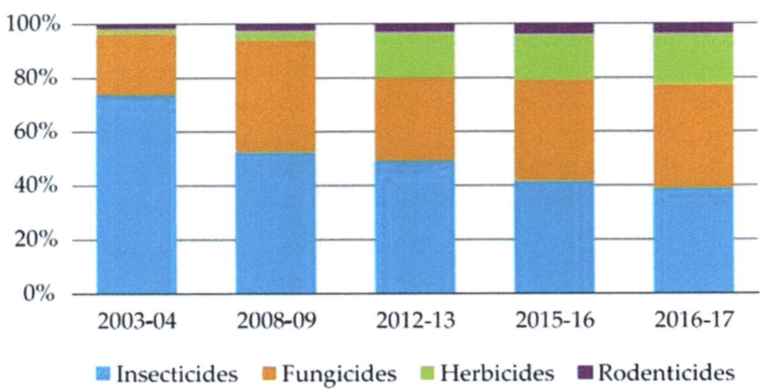

Fig. 24.2 Share of different groups of pesticides (technical grade) in terms of production. (Source: Ministry of Chemicals and Fertilizers, Govt. of India)

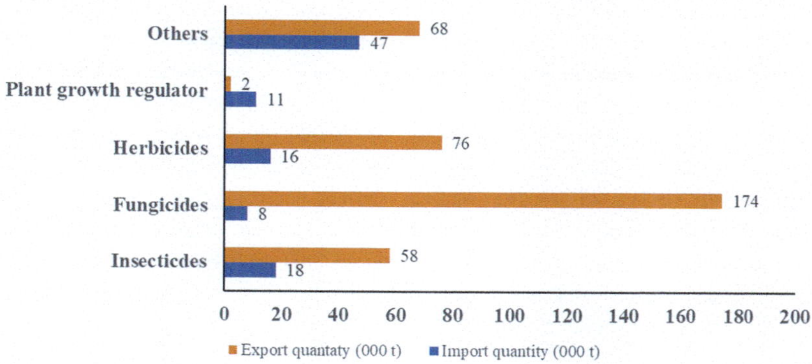

Fig. 24.3 Export and import of major pesticides in India for the year 2016–2017. (Source: DGCI&S, Ministry of Commerce and Industry taken from Subash et al. 2017)

Table 24.1 Major import and export countries for pesticide trade in 2016–2017 (tonnes)

	Country	Insecticides	Fungicides	Herbicides
Export	Brazil	9437.61	42,898.27	20,457.02
	USA	3275.35	8307.82	6095.06
	France	–	7954.77	–
Import	China	11,095.18	2220.28	15,243.93
	Germany	1065.93	1523.81	–
	Japan	–	–	2428.02
	Israel	–	–	4732.66

Source: DGCI&S, Ministry of Commerce and Industry, Govt. of India, taken from Subash et al. (2017)

24.5 Effect of Insecticides on Microbes

In plant protection perspective, insecticides are widely used in agriculture. Due to their xenobiotic properties, the growth of soil microbes and their related soil bioremediation can be negatively impacted by pesticides. Insecticide-contaminated soils are found to inhibit nitrogen-fixing and phosphorus-solubilising microorganisms. Recent studies show that certain pesticides impair plant-to-plant molecular interactions with N-fixing rhizobacteria, thereby retarding the essential biological nitrogen fixation cycle. Similarly, several studies have shown that insecticides suppress soil enzyme activity, which is a key indicator of soil health. Several biochemical reactions can also be affected by pesticides such as mineralisation of organic matter, nitrification, denitrification, ammonification, methanogenesis, etc. On the other hand, a few studies show some positive effects of chemicals applied on soil health.

Pesticides undergo a sequence of degradation, transport, and adsorption/desorption processes in soil, which are influenced by the pesticide's chemical composition (Laabs et al. 2007) and soil properties (Shixian et al. 2018). Interaction of pesticides with soil microbes and their metabolic activities influence the physiological and biochemical behaviour of the microbes (Singh et al. 2006). Microbial biomass is a key indicator of microbial activity and provides a direct measure of the association between microbial activities and the transformation of nutrients and further ecological processes (Schultz et al. 2008). In general, a decline in soil respiration represents a decrease in microbial biomass (Klose et al. 2004) or an increase in respiration (Haney et al. 2000). Some microbial groups are able to replicate using applied pesticides as a source of energy and nutrients, whereas the pesticide may be detrimental to other species (Johnson et al. 2001). The use of pesticides can indeed suppress or destroy certain groups of microorganisms and outnumber other groups by eliminating them from competition. Various positive and negative effects of insecticides on microbes are summarised in Table 24.2.

24.6 Effect of Herbicides on Microbes

Use of herbicides is a common practice all over the world to control unwanted plants in cropped as well as non-cropped areas. Herbicides that were used onto the surface of soil are more likely to influence the growth of microflora as well as micro-fauna. Due to excessive addition of herbicides in the soil, qualitative as well as quantitative alteration in terms of microbial population as well as their enzymatic activities may occur (Min et al. 2002; Saeki et al. 2004). Application of herbicides may also kill the sensitive species of bacteria, fungi, and protozoa that compete with the disease-causing microorganisms, leading to upsetting balance between harmful and beneficial microorganisms. This might lead to a rise of a problem and give opportunities to the pathogens to infect the main crops (Kalia et al. 2004). It is well known in the literature that some microorganisms degrade herbicides leading to stimulated growth of microorganism, whereas some of the other microorganisms were adversely affected in terms of their growth and population depending on the rate of application and type of herbicides used in the field and also on microorganism species and environmental conditions (Sebiomo et al. 2011; Zain et al. 2013).

Many researchers looked into the impact of various types of herbicides on different types of microbes (Adhikary et al. 2014). The effects of three widely used herbicides (pendimethalin, oxyfluorfen, and propaquizafop) on soil microbial species in chilli (total bacteria, fungi, and actinomycetes) were studies and found that herbicide treatments inhibited the production of all three microbial populations in the soil, with the degree of inhibition varying depending on the herbicide used during chilli growth. From the start of the effect until 15 days after application, there was a growing pattern of inhibition on microbial population growth.

Herbicides (paraquat, glyphosate, glufosinate-ammonium, and metsulfuron-methyl) have an impact on microbial population growth, with the degree of inhibition being strongly linked to the rates at which they are applied in oil palm plantations. To bacteria and actinomycetes, paraquat had the greatest inhibitory effect, while glyphosate had the greatest effect on fungi, and metsulfuron-methyl

Table 24.2 Effect of insecticides on microbes

Sl. no.	Insecticides	Microbes	Impact	References
1.	Chlorpyrifos, methylpyrimifos, profenofos	Nitrogen-fixing bacteria, denitrifying bacteria, and nitrifying bacteria	Decreased beneficial microorganism populations	Martinez-Toledo et al. (1992)
2.	Fenvalerate	*Azotobacter* *Azospirillum brasilense* *Azospirillum lipoferum*	Decreased respiration rate and protein contents of diazotrophs	Omar et al. (1992)
3.	Profenofos	Soil fungi, *Penicillium chrysogenum*	Reduction in total-N in fungi *P. chrysogenum*	Moharram et al. (1994)
4.	Carbofuran	Methanotrophs	Carbofuran stimulated the proliferation of methanotrophs	Kumaraswamy et al. (1998)
5.	Endosulfan, monocrotophos, and deltamethrin	Entomopathogenic microorganisms *Beauveria bassiana*, *Metarhizium anisopliae*, and *Sporothrix insectorum*	Reduced the production of conidia and vegetative growth of entomopathogenic fungi	Filho et al. (2001)
6.	Carbofuran	Soil microflora	Adversely affected soil microorganisms	Kalam et al. (2001)
7.	Lindane and dieldrin	*Nitrosomonas*, *Nitrobacter*, and *Thiobacillus*	Toxic	Odokuma et al. (2004)
8.	DDT	Soil algae	A decrease in the diversity of soil algal forms was observed, as well as a decrease in the amount of viable soil algae	Mallavarapu (2002)
9.	Monocrotophos, lindane, dichlorvos, endosulfan, chlorpyrifos, malathion	*Gluconacetobacter diazotrophicus*	May affect the cell morphology and produced pleomorphic cells in large number	Madhaiyan et al. (2006)
10.	Acetamiprid	*Escherichia coli*, *Pseudomonas*, and *Bacillus subtilis*	Stress enzymes in microbes: *Superoxide dismutase*, *catalase*, and *ATPase* were all negatively affected	Yao et al. (2006)
11.	Methamidophos	Soil microorganism	Increase in population of some microbes but decrease in total biomass	Wang et al. (2006)
12.	DDT, methyl parathion	Rhizobium	Interfere with legume–rhizobium chemical	Rockets (2007)

(continued)

Table 24.2 (continued)

Sl. no.	Insecticides	Microbes	Impact	References
			signalling which results in reduced nitrogen fixation	
13.	Cypermethrin	Bacteria and fungi	In the cucumber phyllosphere, there was a rise in bacterial biomass and a decline in fungal biomass, as well as a decrease in the ratio of gram-positive to gram-negative bacteria	Zhang et al. (2008)

had the least inhibitory effect. Glufosinate ammonium and metsulfuron methyl also had similar effects (Zain et al. 2013).

Another study showed the effect of 2,4-dichlorophenoxyaceticacid (2,4-D) and 2,4,5-trichlorophenoxyaceticacid (2,4,5-T) on respiration of *Azotobacter* and concluded that both inhibited the respiration of *Azotobacter* but inhibition was more by 2,4,5-T in comparison to 2,4-D (Magee et al. 1955).

24.7 Pesticidal Impact on Processes of Decomposition

The decomposition of organic matter in soil is an integral portion of nutrient cycling method in soil. During spraying pesticides get in contact with crop residues in the soil. The nonselective preplant herbicides like, glyphosate and paraquat are extensively used, which contribute to slow decomposition of various plant materials. There is inhibition of decomposition of paraquat-treated cellulose thread, but very negligible decomposition of paraquat-treated soil was examined (Grossbard et al. 1978). Further studies on barley and wheat straw also revealed that when straws were kept on soil surface, greater inhibition occurs with application of paraquat (Grossbard et al. 1981). Although in some experiments glyphosate has shown its inhibitory effect, its effect are inconsistent too (Grossbard et al. 1981); (Hendrix et al. 1985; House GJ et al. 1987). Collectively, these studies indicated that the recommended doses of paraquat and glyphosate can hinder the crop residue decomposition. When paraquat and glyphosate were sprayed on crop residues, they appeared to improve decomposition, but their overall results are mixed. Decomposition was reduced more when residues were left on the soil surface than when they were added. The effects of those herbicides on crop residue decomposition were apparently due to high herbicide concentrations remaining after spraying. Paraquat and glyphosate concentrations that inhibit microorganisms in pure culture were typically higher than those present in soil after field application (Grossbard et al. 1979; Grossbard et al. 1985), and both herbicides used in the soil have negligible harmful effects on microbial species (Roslycky 1982). Both paraquat and glyphosate are highly adsorbed in soil, which may explain why herbicides applied to soil are ineffective. Other herbicides, such as 2,4-D and 2,4,5-T (Gottschalk et al. 1979; Fletcher et al. 1986; Sikka et al. 1982), trifluralin, and its metabolites (Boyette et al. 1988), tend to have little effect on cellulose or plant residue decomposition, at least at high concentrations. However, the effects of other recently established herbicide classes, such as sulfonylurea herbicides, which inhibit amino acid synthesis in microorganisms and are similar to glyphosate, have yet to be evaluated.

Though the reduction rates of crop residue decomposition had not yet been assessed fundamentally, some of the effects can be drawn from this. The effects of paraquat and glyphosate will be more in without-tilled crop residue than in other different residue management systems because of surface retention of different crop residues which increases the use of those herbicides. Greater amount of surface pesticide residues may interfere with plant growth as well as weed control, and these lead to increased inhibition of decomposition process. The residence time of straw or other crop residues on the soil surface is determined by additional crop residue coverage in cropping systems. While herbicides can temporarily delay the decomposition process, there is no evidence that they have any long-term effects. Furthermore, there is no evidence that the use of these herbicides affects long-term nutrient turnover in crop residues.

24.8 Pesticidal Impact on Nutrient Cycling

24.8.1 Effects on Nitrogen Transformations

One of the most crucial criteria for crop production is the management of soil N. The conversion of plant nitrogen found in crop residues and soil organic matter to NH_4 and then to NO_3 is known as mineralisation, and it accounts for 40–60% of crop nitrogen, with fertiliser N accounting for the remainder. The residual nitrogen in crop residues was returned to the soil, where it was gradually re-mineralised as the residues decomposed. A wide variety of soil microorganisms participate in the mineralisation process, but only a few species transform NH_4 to NO_3^-. Many studies have been conducted to determine the effects of pesticides on nitrogen mineralisation, and it is clear that the contrast of NO_3^- and NH_4 nitrogen produced in pesticide-treated soils to that produced in untreated soils is the most important factor. The capacity for N production in a soil is influenced by the C/N ratio of soil organic matter and crop residues, as well as the size of the microbial biomass.

Goring and Laskowski (1982) examined the impact of insecticides, herbicides, fungicides, and nematicides on N transformations in soil and found that the majority of pesticides have a marginal effect, inhibiting nitrification and mineralisation by less than 25%. It was observed that most of the pesticides inhibit those natural processes only above the recommended rates for field application. Soil fumigants, e.g. chloropicrin or methyl bromide, particularly have important effects on nitrogen mineralisation and nitrification (Martin 1972). Initially, decreased populations of all microorganisms and N mineralisation were observed. Since fumigated soils get recolonised, nitrogen mineralisation process enhances and progressively may exceed than that of non-fumigated soil. However, long after fumigation nitrification may get suppressed, thus newer systemic nematicides and insecticides replaced soil fumigants to a large extent. There was no declination in microbial biomass or soil nitrogen mineralised in laboratory and field experiments using nematicides such as fenamiphos and oxamyl (Tu 1980; Ross et al. 1984, 1985). Fensulfothion, but not carbofuran, inhibited the populations of various fungi and bacteria. Moreover Tu (1972) reported that neither chemical inhibits mineralisation or nitrification. Dithiocarbamate fungicides may inhibit nitrogen mineralisation under certain conditions. Single pesticide applications have varying effects on nitrogen availability, but repeated pesticide applications or single applications above the recommended levels will reduce both ammonification and nitrification (Jaques et al. 1959; Dubey et al. 1970; Mazur et al. 1975). Maximum 10 ppm of those fungicides can cause reductions in fungal and bacterial populations, and subsequently increasing rates inhibit soil respiration (Tu 1980) and glucose metabolism (Boyette et al. 1988), indicating that those compounds are relatively nonselective towards microbes. Since certain fungicides are relatively nonpersistent in soil and microbial activity, soil nitrogen mineralisation ability will easily recover following their degradation. Corke et al. (1970) discovered that certain pesticide-degraded products had slight effects on nitrogen transformations, such as a 2- to 4-day lag for nitrification when low concentrations of 3,4-dichloroaniline were present. This is a popular phenylamide herbicide and substituted urea degradation product.

When applied at the same concentrations as the metabolites of the herbicides, the parent herbicides had no effect. The metabolites of insecticides including terbufos and phorate, such as sulfoxide and sulfone, reduced nitrification in soil to a lesser degree than the parent compounds (Tu 1980). Other pesticide metabolites have little attention towards mineralisation. The effect of pesticides on microbial fertiliser transformations has yet to be thoroughly investigated. Marsh (1985) investigated the impact of seven herbicides on nitrogen and phosphorus transformations in both fertilised (triple superphosphate and NH_4NO_3) and unfertilised soil. Asulam not only avoided nitrification in fertilised soil, but it also greatly reduced nitrogen mineralisation. While glyphosate, chloridazon, paraquat, and isoproturon decreased phosphorus availability, the effects were very subtle and not agronomically important. In general, the above pesticides tend to have a much lower inhibitory effect on the nitrification process than commercially available nitrification inhibitors (Bundy et al. 1973; Turner 1979), although some direct comparisons have been made. The soil enzyme urease converts urea to NH4+, which is a common nitrogen fertiliser. Monuron, fenuron, diuron, linuron, and neburon, among other substituted urea herbicides, inhibited soil urease production (Cervelli et al. 1976). According to one study, conversion of urea to NH4+ was reduced by 8–39% depending on herbicide concentration (2–10 ppm) and soil type (Maria et al. 2013). Mancozeb, a dithiocarbamate fungicide, inhibited urease activity as well. Pesticides that inhibit urease activity could theoretically reduce the conversion of this fertiliser to NO_3^- over time, particularly if used in combination or close proximity to urea treated in the soil. Mild inhibition of the urease enzyme will avoid NH_4^+ toxicity issues on occasion and limit nitrogen losses from NH_4 volatilisation and nitrate leaching. Denitrification has been shown to be inhibited by pesticides when used at higher rates. At low concentrations, the insecticide carbaryl and the herbicide dalapon were found to be inhibitory (Grant et al. 1982; Weeraratna 1980). Later studies by Yeomans et al. (1985, 1987) found that although dalapon's inhibitory effects were not verified, higher rates of metribuzin or dinoseb induced denitrification. Although inhibiting denitrification can help with efficient nitrogen management, these findings showed that herbicides used at recommended rates have a fairly consistent impact on denitrification.

24.8.2 *Transformation of Sulphur*

The oxidation of elemental sulphur to sulphate and the reduction of sulphite are the most important transformations since sulphate is the main plant-available nutrient. Despite previous assumptions that heterotrophic microorganisms are more essential for nitrification, the oxidation process is carried out by advanced chemoautotrophic bacteria. Due to sulphur deficiency in different parts of the world, pesticides have a significant impact on the sulphur oxidation mechanism (Coleman et al. 1966). Sulphur, which must be oxidised before it can be used by plants, can also be used to address such shortages. Given the importance of sulphur in crop production, the literature includes only a few studies on pesticide effects on sulphur oxidation. The organophosphate group of insecticides had little impact on soil sulphate, while nematicides such as DDVP, carbofuran, and Vorlex reduced sulphur oxidation marginally (Tu 1972; Aristeidis et al. 2020). Paraquat, on the other hand, slows down the process significantly (Tu et al. 1968). Audus (Audus et al. 1970) discovered that when insecticides (DDT, BHC, aldrin, dieldrin, etc.) are applied at field concentrations, they have no effect on sulphur-oxidising bacteria.

24.8.3 *Availability of Trace Elements*

The solubility of trace elements in soils has been shown to be impaired by soil fumigants (Warcup 1957). Fumigants appear to affect manganese in particular, so steam sterilisation increases toxic levels (Sonneveld et al. 1973). Some pesticides' effects on trace elements have received much less publicity. Smith and Weeraratna (Smith et al. 1974) found that the herbicides simazine and ioxynil increased manganese solubility as well as Mg, Ca, Fe, and Cu solubility in acidic and alkaline soil media.

Wainwright and Pugh (Wainwright et al. 1974) discovered a similar pattern in their research after fungicide treatment in both laboratory and field treated soils. The mechanism of the process is unknown; it may be caused by lysis of dead microorganism cells or by microorganisms solubilising the components.

24.9 Conclusions

The primary aim of agricultural growth is to feed and supply enough food, nutrition, and surplus to the increasing human population while mitigating environmental and ecological harm. Pesticides are regarded as one of the most effective crop protection tools in developing countries. Since pesticides and their derivatives stay in the soil system for such a long time, they pose significant risks to soil health, the soil microbial ecosystem, and human health. Plant defence chemicals have been shown to minimise soil bacteria, fungi, and almost all flora and fauna populations, as well as soil microbial activity, biomass carbon, and nitrogen mineralisation. The use of natural pesticides and bio-pesticides, as well as judicious application of agrochemicals, should be encouraged.

Pesticide use must be limited in order to reduce the negative effects of pesticides on humans and the environment, which necessitates public awareness campaigns among farmers and other stakeholders. Long-term impact of pesticides on soil microbial populations and the soil ecosystem should be researched in detail. Organic pesticides can aid in the preservation of our environment's microbial niche at this early stage of organic agriculture.

References

Adhikary P, Shil S, Patra PS et al (2014) Effect of herbicides on soil microorganisms in transplanted chilli. Glob J Biol Agric and Health Sci 3:236–238

Aristeidis S, Tsagkaris O, Leos U, Jana P, Jana H et al (2020) Screening of carbamate and organophosphate pesticides in food matrices using an affordable and simple spectrophotometric acetylcholinesterase assay. Appl Sci 10(2):565

Audus LJ et al (1970) The action of herbicides and pesticides on the microflora. Meded Rijksfac LandbWet Gent 35:465–492

Boyette KD, Moorman TB, Koskinen WC et al (1988) Effects of trifluralin and metabolites on the decomposition of selected substrates by soil microorganisms. Biol Fertil Soils 6(2):100–105

Bundy LG, Bremner JM et al (1973) Inhibition of nitrification in soils. Soil Sci Soc Am J 37 (3):396–398

Cervelli S, Nannipier IP, Giovannini G, Perna A et al (1976) Relationships between substituted urea herbicides and soil urease activity. Weed Res 16(6):365–368

Chowdhury A, Pradhan S, Saha M, Sanyal N, Indian J et al (2008) Impact of pesticides on soil microbiological parameters and possible bioremediation strategies. Microbiology 48:114–127

Coleman R et al (1966) The importance of sulfur as a plant nutrient in world crop production. Soil Sci 101(4):230–239

Corke CT, Thompson FR et al (1970) Effects of some phenylamide herbicides and their degradation products on soil nitrification. Can J Microbiol 16(7):567–571

Dubey HD, Rodriguez RL et al (1970) Effect of dyrene and maneb on nitrification and ammonification, and their degradation in tropical soils. Soil Sci Soc Am J 34(3):435–439

FAO (1989) Guidelines for legislation on the control of pesticides. Food and Agriculture Organization of the United Nation, Rome

FAOSTAT (2017) Pesticides. Food and Agriculture Organization, Rome

FICCI (2015) Ushering in the 2nd green revolution: role of crop protection chemicals. Federation of Indian Chambers of Commerce and Industry, New Delhi

Filho AB, Almeida JEM, Clovis Lamas C et al (2001) Effect of thiamethoxam on entomopathogenic microorganisms. Neotrop Entomol 30(3):437–447

Fletcher K, Freedman B et al (1986) Effects of the herbicides glyphosate 2 4 5-trichlorophenoxyacetic acid, 2 4-dichlorophenoxyacetic acid on forest litter decomposition. Can J For Res 16(1):6–9

Goring CAI, Laskowski DA et al (1982) The effects of pesticides on nitrogen transformations in soils. Nitrogen Agric Soils 22:689–720

Gottschalk MR, Shure DJ et al (1979) Herbicide effects on leaf litter decomposition processes in an oak-hickory forest. Ecology 60(1):143–151

Grant MA, Payne WJ et al (1982) Effects of pesticides on denitrifying activity in salt marsh sediments. J Environ Qual 11(3):369–372

Grossbard E, Wingfield GI et al (1978) Effects of paraquat, aminotriazole and glyphosate on cellulose decomposition. Weed Res 18(6):347–353

Grossbard E, Harris D (1979) Effects of herbicides on the decay of straw. In: Grossbard E (ed) Straw decay and its effect on disposal and utilization. John Wiley, Chichester, pp 167–176

Grossbard E, Harris D et al (1981) Effects on straw decay of the herbicides paraquat and glyphosate in combination with nitrogen amendments. Ann Appl Biol 98(2):277–288

Grossbard E, Atkinson D et al (1985) Effects of glyphosate on the microflora with reference to the decomposition of treated vegetation and interaction with some plant pathogens. In: Herbicide glyphosate. Butterworths, London

Handa SK, Agnihotri NP, Kulshrestha G (1999) Maximum residue limits of pesticides. In: Handa SK, Agnihotri NP, Kulshrestha G (eds) Pesticide residues significance, management and analysis. Research Periodicals and Book Publishing House, Houston, pp 184–198

Haney RL, Senseman SA, Hons FM, Zuberer DA et al (2000) Effect of glyphosate on soil microbial activity and biomass. Weed Sci 48:89–93

Hendrix PF, Parmelee RW et al (1985) Decomposition, nutrient loss and microarthropod densities in herbicide-treated grass litter in a Georgia piedmont agroecosystem. Soil Biol Biochem 17 (4):421–428

House GJ, Worsham AD, Sheets TJ, Stinner RE et al (1987) Herbicide effects on soil arthropod dynamics and wheat straw decomposition in a North Carolina no-tillage agroecosystem. Biol Fertil Soils 4(3):109–114

Jaques RP, Robinson JB, Chase FE et al (1959) Effects of thiourea, ethyl urethane and some dithiocarbamate fungicides on nitrification in Fox sandy loam. Can J Soil Sci 39(2):237–245

Johnson K, Jacobsen CS, Torsvik V et al (2001) Pesticides effects on bacterial diversity in agricultural soils—a review. Biol Fertil Soils 33:443–453

Kalam A, Mukherjee AK et al (2001) Influence of hexaconazole, carbofuran and ethion on soil microflora and dehydrogenase activities in soil and intact cell. Indian J Exp Biol 39:90–94

Kalia A, Gupta RP et al (2004) Disruption of food web by pesticides. Ind J Ecol 31:85–92

Klose S, Ajwa HA et al (2004) Enzymes activities in agricultural soils fumigated with methyl bromide alternatives. Soil Biol Biochem 36:1625–1635

Kumaraswamy S, Rath AK, Satpathy SN, Ramakrishnan B, Adhya TK, Sethunathan N et al (1998) Influence of the carbofuran on the production and oxidation of methane in a flooded rice soil. Biol Fertil Soil 25:362–366

Laabs V, Wehrhan A, Pinto A, Dores E, Amelung W et al (2007) Pesticide fate in tropical wetlands of Brazil: an aquatic microcosm study under semi-field conditions. Chemosphere 67:975–989
Lavelle P, Spain AV et al (2001) Soil organisms. Soil Ecol 78:201–356
Madhaiyan A, Poonguzhali S, Hari K, Saravanan VS, Sa T et al (2006) Influence of pesticides on the growth rate and plant-growth promoting traits of Gluconacetobacter diazotrophicus. Pestic Biochem Physiol 84:143–154
Magee LA, Colmer AR et al (1955) The effect of herbicides on soil microorganisms: III. The effect of some herbicides on the respiration of Azotobacter. Appl Microbiol 3(5):288
Mallavarapu M (2002) Heavy pesticide use lowers soil health. Farm Head 121:37–38
Maria AMM, Bruna CVC, Márcia MH et al (2013) Toxicity of herbicides: impact on aquatic and soil biota and human health. https://doi.org/10.5772/55851
Marsh JA (1985) Effects of herbicide-fertiliser interactions on nitrogen and phosphorus transformations and herbicide persistence in soil. Pestic Sci 16(1):93–100
Martin JP (1972) Side effects of organic chemicals on soil properties and plant growth. Chapter 12. In: Goring CAI, Hamaker JW (eds) Organic chemicals in the soil environment. Marcel Deckker, New York, pp 733–792
Martinez-Toledo MV, Salermon V, Gonzalez-Lopez J et al (1992) Effect of an organophosphorus insecticide, profenofos, on agricultural soil microflora. Chemosphere 24:71–80
Mazur AR, Hughes TD et al (1975) Nitrogen transformations in soil as affected by the fungicides benomyl, dyrene, and Maneb 1. Agron J 67(6):755–758
Min HYF, Ye ZY, Chen WX, Wu Du YF et al (2002) Effects of butachlor on microbial enzyme activities in paddy soil. J Environ Sci 14:413–417
Moharram AM, Abdel-Malek AY, Abdel-Kader MIA, Omar SO et al (1994) Effect of three pesticides on nitrogen content of some soil fungi. J Islamic Acad Sci 7(2):82–87
Odokuma LO, Osuagwu C et al (2004) Tolerance of chemolithotrophic bacteria to organochlorine, organophosphate and carbamate pesticides. J Agric Environ Eng Technol 1(1):7–15
Omar SA, Abd-Alla MH et al (1992) Effect of pesticides on growth, respiration and nitrogenase activity of Azotobacter and Azospirillum. World J Microbiol Biotechnol 8:326–328
Rockets R (2007) Down of the farm yields, nutrients and soil quality. Take action! How to eliminate pesticides use. National Audubon Society, pp 1–3
Roslycky EB (1982) Glyphosate and the response of the soil microbiota. Soil Biol Biochem 14 (2):87–92
Ross DJ, Speir TW, Cowling JC, Whale KN et al (1984) Influence of field applications of oxamyl and fenamiphos on biochemical activities of soil under pasture. New Zeal J Sci 27:247–254
Ross DJ, Speir TW et al (1985) Changes in biochemical activities of soil incubated with the nematicides oxamyl and fenamiphos. Soil Biol Biochem 17(1):123–125
Saeki M, Toyota K et al (2004) Effect of bensulfuron-methyl (a sulfonylurea herbicide) on the soil bacterial community of a paddy soil microcosm. Biol Fertil Soils 40:110–118
Schultz P, Urban NR et al (2008) Effects of bacterial dynamics on organic matter decomposition and nutrient release from sediments: a modeling study. Ecol Model 210:1–14
Sebiomo A, Ogundero VW, Bankole SA et al (2011) Effects of four herbicides on microbial population, soil organic matter and dehydrogenase activity. Afr J Biotechnol 10(5):770–778
Shixian S, Virinder S, Yuhong R, Zheng Y et al (2018) Pesticide pollution in agricultural soils and sustainable remediation methods: a review. Curr Poll 4(3):1–11
Sikka HC, Pack EJ et al (1982) Environmental fate, effects, and health hazards of Fenac. Syracuse Research Corp, New York
Singh BK, Walker A et al (2006) Microbial degradation of organophosphorus compounds. FEMS Microbiol Rev 30:428–471
Smith MS, Weeraratna CS et al (1974) The influence of some biologically active compounds on microbial activity and on the availability of plant nutrients in soils. I. simazine, ioxynil and 2, 3 6-tba. Pestic Sci 5(6):721–729
Sonneveld C, Voogt S et al (1973) The effects of soil sterilisation with steam-air mixtures on the development of some glasshouse crops. Plant Soil 38(2):415–423

Tu CM (1972) Effect of four nematocides on activities of microorganisms in soil. Appl Microbiol 23(2):398–401
Tu CM (1980) Influence of pesticides and some of the oxidized analogues on microbial populations, nitrification and respiration activities in soil. Bull Environ Contam Toxicol 24(1):13–19
Tu CM, Bollen WB et al (1968) Effect of paraquat on microbial activities in soils. Weed Res 8 (1):28–37
Turner FT (1979) Soil nitrification retardation by rice pesticides. Soil Sci Soc Am J 43(5):955–957
Wainwright M, Pugh GJF et al (1974) The effects of fungicides on certain chemical and microbial properties of soils. Soil Biol Biochem 6(4):263–267
Wang MC, Gong M, Zang HB, Hua XM, Yao J, Pang YJ, Yang YH et al (2006) Effect of methamidophos and urea application on microbial communities in soils as determined by microbial biomass and community level physiological profiles. J Environ Sci Health B 41:399–413
Warcup JH (1957) Chemical and biological aspects of soil sterilization. Soils Fertil 20:1–5
Weeraratna CS (1980) Effect of dalapon-sodium on nitrification and denitrification in a tropical loam soil. Weed Res 20(5):291–293
Yao XH, Min H, Lv ZM et al (2006) Response of superoxide dismutase, catalase, and ATPase activity in bacteria exposed to acetamiprid. Biomed Environ Sci 19(4):309–314
Yeomans JC, Bremner JM et al (1985) Denitrification in soil: effects of herbicides. Soil Biol Biochem 17(4):447–452
Yeomans JC, Bremner JM et al (1987) Effects of dalapon, atrazine and simazine on denitrification in soil. Soil Biol Biochem 19(1):31–34
Zain NMM, Mohamad RB, Sijam K, Morshed MM, Awang Y et al (2013) Effects of selected herbicides on soil microbial populations in oil palm plantation of Malaysia: a microcosm experiment. Afr J Microbiol Res 7(5):367–374
Zhang B, Zhang H, Jin B, Tang L, Yang J, Li B, Zhuang G, Bai Z et al (2008) Effect of cypermethrin insecticide on the microbial community in cucumber phyllosphere. J Environ Sci 20(11):1365–1362

Chapter 25
Microbe-Mediated Removal of Xenobiotics for Sustainable Environment

Helina Patel, Sneha Trivedi, Deepesh Bhatt, Manoj Nath, and Naresh Butani

Abstract Xenobiotics are man-made chemical compounds widely used in pesticides, dyes, drugs, explosives and other industrial chemicals. Poor waste disposal practices, intensive agricultural practices and fossil fuel combustion are some of the reasons that lead to the release of these compounds into natural ecosystem such as soil and water. These cause serious damage to aquatic and terrestrial ecosystems due to noxious nature of chemical compounds. Xenobiotics when build up in soil kill beneficial microbes of soil that are accountable for soil fertility. Degradation of xenobiotic compounds is being done using physical and chemical means; however, these methods consequently result in the formation of lethal intermediates and end products. Therefore, microbial remediation is adopted as a sustainable emerging technique to eliminate these pollutants from nature. The present study highlights on the involvement of several bacterial and fungal genera in catabolism of recalcitrant xenobiotics. Moreover, phytoremediation approach employing plants for treating chemically contaminated soil is also discussed.

Keywords Xenobiotics · Microbial remediation · Phytoremediation · Mycoremediation · Bioremediation

H. Patel · S. Trivedi · N. Butani (✉)
Department of Microbiology, Shree Ramkrishna Institute of Computer Education and Applied Sciences, Sarvajanik University, Surat, Gujarat, India

D. Bhatt
Department of Biotechnology, Shree Ramkrishna Institute of Computer Education and Applied Sciences, Sarvajanik University, Surat, Gujarat, India

M. Nath
ICAR—Directorate of Mushroom Research, Solan, Himachal Pradesh, India

P. Bhatt et al. (eds.), *Microbial Technology for Sustainable Environment*,
https://doi.org/10.1007/978-981-16-3840-4_25

25.1 Introduction

Xenobiotic chemicals are distant to the biosphere, are effectively lethal and are the cause of undesirable physiological and/or ecological effects, consequently leading to disease conditions in humans, other living creatures and pollution, respectively (Olicón-Hernández et al. 2017). Environmentally predominant synthetic xenobiotic compounds include hazardous pollutants such as pesticides, fuels, solvents, alkanes, hydrocarbon derivatives, synthetic polymers, dyes, plastics, etc. (Sharma et al. 2018). Antibiotics, steroids and biomedical waste mixtures also are an example of man-made xenobiotic compounds. Xenobiotic substances are becoming progressively a major problem as they are relatively new substances that resist degradation, mainly because of their recalcitrant nature. Being foreign to the organisms, xenobiotics are not easily recognized by the microbes in nature, and hence these compounds do not enter common metabolic pathways, thereby remaining persistent in nature. Such chemical contaminants are referred as persistent organic pollutants (Bhatt et al. 2019). The accumulation of persistent pollutants affects the environment and survival of higher as well as lower eukaryotes. Moreover, their recalcitrant nature is a major risk factor that poses threat to human health as it disrupts normal cellular pathways that play a significant role in their development and reproduction (Gangola et al. 2018; Baker et al. 2019). Nevertheless, prolong exposure with these xenobiotics may even cause neurological damage, immunosuppression and cancer (Kalyoncu et al. 2009).

Thousands of persistent organic pollutants exist in the atmosphere and hold an extended half-life of several days in air, or about a decade in soil and in living organisms. For instance, xenobiotic organic pesticides DDT, BHC, polychlorinated biphenyls, halogenated aromatic compounds and other such compounds that mainly partake agricultural and industrial routines are discarded in the environment in huge amounts globally. The chemical pesticides are being used to eliminate insect-borne diseases for healthier harvest, so as to meet the growing demand of food. Although these substances are advantageous to humans, their prolonged presence in the biosphere poses perilous effects, as such chemical compounds are not easily biodegraded and hence their concentration gradually upsurges with time (Varsha et al. 2011). These chemical compounds build up in natural reserves (soil, air and water) and also found to be accumulated in plants, animals and humans; thus, they remain present in the ecosystem for decades (Breivik et al. 2004).

25.2 Points of Xenobiotic Discharge

The prime or direct sources of xenobiotics entering into the environment are effluents from (1) chemical and pharmaceutical industries which include alkylphenols, octylphenol and biphenyls such as bisphenol A, stilbene, genistein, estrogens, etc. These pharmaceutically active complexes are controversial endocrine disruptors which disturb the physiological stability in animals and are bothersome to human health (Asgher et al. 2008). (2) Paper and pulp bleaching are major sources of release of harmful polychlorinated phenols (PCPs) and other organic compounds and include dyes like azo dye and crystal violet (D'Souza et al. 2006). Effluent discharge from textile industries and paper printing where synthetic dyes are majorly being used, adversely affects the aquatic life (Kumari et al. 2014). (3) Mining releases heavy metals into biogeochemical cycles. (4) Fossil fuel combustion and gas plants produce polycyclic aromatic hydrocarbons like naphthalene and benzo (a)pyrene, which are noxious and possibly carcinogenic xenobiotics (Gojgic-Cvijovic et al. 2012). (5) Intensive agriculture practices use enormous amount of chemical fertilizers, herbicides and pesticides such as chlorinated aromatic compounds and their derivatives: DDT, chlordane and lindane. Bioaccumulation and biomagnification of pesticides lead to toxic behavioural effects on animals and the human race.

25.3 Classes of Xenobiotic Compounds

On the basis of chemical composition, the recalcitrant xenobiotic compounds that are released from different industrial residual wastes, viz. paper and pulp remnants, dye effluents, chemicals, plastics and pharma waste, are categorized into the following types.

25.3.1 *Halocarbons*

These are volatile compounds comprised of different numbers of halogens (Cl, F, Br, I) in place of hydrogen atoms. These compounds are mainly used in the preparation of organochlorine pesticides, i.e. insecticides (DDTs and its metabolites [toxaphene, chlordane, etc.], BHC, lindane, etc.), herbicides (dalapon, 2,4-dichlorophenoxyacetic acid, 2,4,5-trichlorophenoxyacetic acid, etc.) and fungicides (Qadir et al. 2017). Volatile compounds when escaped into the environment cause damage to the ozone layer, and when deposited in soil and leach into waterbodies, they lead to biomagnification (Bharadwaj 2018).

25.3.2 *Polychlorinated Biphenyls (PCBs)*

PCBs (2-chlorobiphenyl, 4,4′-dichlorobiphenyl, 2,2′,5,5′-tetrachlorobiphenyl, etc.) are prevalent toxic pollutants, inert in nature and highly stable mixtures resistant to extreme temperature and pressure (Tigini et al. 2009a). These compounds are covalently linked with two benzene rings having chlorine in place of hydrogen atoms. These are broadly used in plasticisers, in electrical equipment like capacitors and insulator coolants in transformers, etc.

25.3.3 Synthetic Polymer

Synthetic polymers are high molecular weight compounds also known as plastic polymers which include polystyrene, polypropylene, polyethylene, polyvinyl chloride, etc. Polyamide such as nylon, is expansively being used for wrapping materials, in garments, etc. (Shrivastava 2018).

25.3.4 Alkylbenzylsulfonates

These are anionic surfactants (branched alkylbenzene sulfonates and linear alkylbenzene sulfonates) broadly used in formulation of detergents. They have hydrophilic sulfonate group present at one end, as a result of which they are resistant to degradation by microorganisms, while at another end hydrophobic alkylbenzene tail is present, making it recalcitrant if it is branched (Bharadwaj 2018).

25.3.5 Oil Mixture

Oil is a natural product, is insoluble in water, possesses some toxic constituents and thus is a recalcitrant. Microbial degradation of oil has varying rates of degradation based on the complexes present in it. However, biological means of degradation of large oil spills over the water surface is ineffective, resulting in severe pollution problems (Qadir et al. 2017).

25.4 Hazardous Effect of Xenobiotic Compound

Xenobiotic compounds are potentially perilous to both lower and higher eukaryotes and even to humans. Exposure to xenobiotic pesticides increases the risk of diabetes, neurological disorder and several skin diseases and extended exposure may even cause cancer (Qadir et al. 2017). As described in the reports by Kelce et al. (1995), scientific inferences from environmental impact assessment studies depict that persistent organic pollutants are a major hazard that causes impairment of brain function, reproductive dysfunction and endocrine disruption. Besides, their recalcitrant nature leads to their gradual accumulation in the environment with time, thereby entering into the food chain and hence upsetting the ecosystem (Bharadwaj 2018).

25.5 Microbial Remediation of Xenobiotics

The use of chemicals that are noxious to human beings and that damage the wilderness of nature is prevalent. Industries manufacture different chemical compounds to satisfy the need of people for better living of life. Although the usage of pesticides, paints, plastics, pharmaceuticals and textiles that contribute to xenobiotics cannot be neglected from our daily needs, steps should be taken to eliminate these xenobiotic chemicals from the environment. Degradation of organic pollutants using physical and/or chemical route is economically not feasible; besides, chances of undesirable toxic intermediates and end products being formed are high. For this purpose, exploitation of microorganisms is the most competent, sustainable and feasible way to achieve efficient degradation of xenobiotic pollutants (Fig. 25.1). Microbes are likely used because of their rapid growth rate and that they possess complex enzymatic machinery that helps to degrade complex toxic compounds to innocuous or less harmful degraded products. Microorganisms also have evolved excellent biochemical control mechanisms so as to utilize pollutants as a source of carbon and energy and degrade them. However, microorganisms fail to degrade all chemical compounds for the reason that they are unable to break certain chemical bonds present in them (Gangola et al. 2019).

25.5.1 Bacterial Remediation

Successful bioremediation requires potent microbial strains which can withstand and degrade hazardous pollutants. Bioremediation mechanisms include both isolation of naturally occurring xenobiotic-degrading microbes from heterogenous microbial population and genetically engineered microorganisms. Biodegradation is affected by various factors like hydrophobic nature of hydrocarbons, bioavailability and predominant environmental conditions. Microorganisms can enhance the hydrophobicity of the cell surface by inherently changing their outer membrane so as to facilitate the uptake of hydrocarbons (Shukla and Singh 2020). Microbial communities get colonized at contaminated sites as they metabolize recalcitrant xenobiotic compounds (Galvão et al. 2005). Several aerobic (*Pseudomonas*, *Bacillus*, *Escherichia*, *Serratia*, *Gordonia*, *Moraxella*, *Micrococcus*, *Sphingobium*, *Pandoraea*, *Rhodococcus*), anaerobic (*Desulfovibrio*, *Desulfotomaculum*, *Methanospirillum*, *Methanosaeta*, *Pelatomaculum*, *Syntrophobacter*, *Syntrophus*) bacteria, methanotrophic and methanogenic bacteria, cyanobacteria and sphingomonads possess xenobiotic degradative potential (Varsha et al. 2011; Sinha et al. 2009).

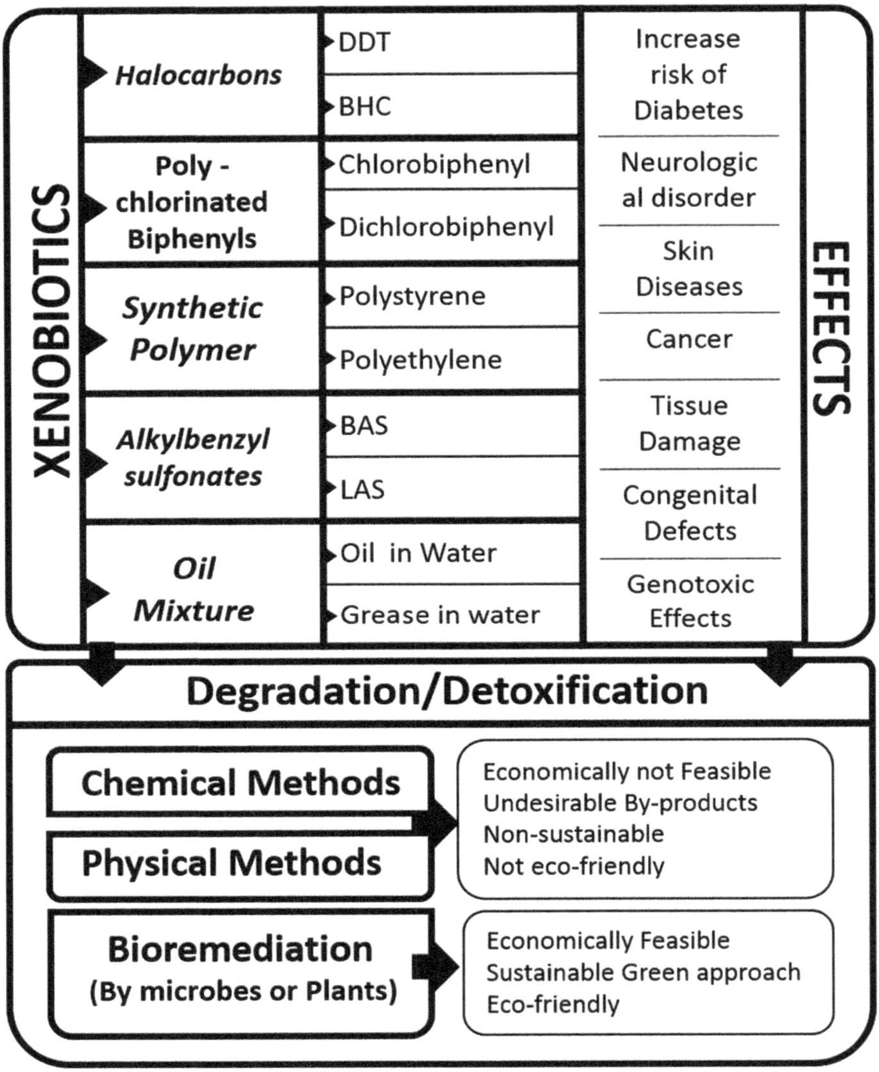

Fig. 25.1 Xenobiotic compounds, its types, effects and various strategies for degradation and detoxification (*BAS* branched alkylbenzene sulfonates, *LAS* linear alkylbenzene sulfonates)

A soil bacterium *Serratia* sp. strain DS001 utilized methyl parathion (an organophosphate insecticide), 4-nitrocatechol, *p*-nitrophenol and 1,2,4-benzenetriol as sole carbon and energy source and effectively metabolized methyl parathion, signifying the role of enzyme parathion hydrolase in degradation (Pakala et al. 2007). Yao et al. (2006) reported efficient degradation of phenolic compounds using H_2O_2 as oxidizer and an enzyme extracted from *Serratia* sp. AB 90027 as a catalyst. Using this enzyme/H_2O_2 treatment, high COD removal efficiency was also achieved. Kafilzadeh et al. (2015) in their studies collected samples from the sediments of Kor River, Iran, from the area with high agricultural activity and isolated five bacterial genera (*Klebsiella*, *Acinetobacter*, *Alcaligenes*, *Flavobacterium* and *Bacillus*) that were able to metabolize endosulfan, a lipophilic insecticide. Degraded metabolites of endosulfan produced depicted less toxicity when compared with endosulfan itself.

Extremophiles such as halotolerant and thermotolerant *Bacillus* sp. strain DHT also possess the capability of utilizing fuels, crude oil, several pure alkanes and polycyclic aromatic hydrocarbons as a sole carbon and energy source (Kumar et al. 2007). Thermophilic bacterium *Brevibacillus borstelensis* degraded polyethylene xenobiotic compounds for the sole source of carbon (Hadad et al. 2005). Efficient catabolism of phenanthrene and crude oil was attained by M19F and M16K strains of *Bacillus subtilis* on day 28 and day 18 post-inoculation, respectively (Oyetibo et al. 2017). Furthermore, *Bacillus drentensis* strain S1 isolated from sewage sample proved its potential in biodegradation of drug acetaminophen (paracetamol) (Chopra and Kumar 2020). A pure strain of *Pseudomonas* sp. YATO411 in an immobilized and a freely suspended system exhibited biodegradation of benzene and toluene, indicating its potential to catabolize high concentrations of these xenobiotics (Tsai et al. 2013). Additionally, bacteria of *Pseudomonas* species have been testified for partial and complete decomposition of fungicides, pesticides, aliphatic or polycyclic aromatic hydrocarbons, recalcitrant dyes, phenolic compounds and hexavalent toxic heavy metals (Poornima et al. 2010; Wasi et al. 2010; Joe et al. 2011). *Rhodococcus erythropolis* bacteria presented its potentiality in bio-desulfurization of crude oil (Amin 2011). An idea of "Super Bugs" also has been advanced to catabolize an extensive array of xenobiotic pollutants (Furukawa 2003). A metabolic by-product, known as "microbial concrete", is produced by urease-producing bacterial strains, namely, *P. aeruginosa*, *P. mirabilis* and *Micrococcus*. Microbial concrete is employed for remediation and re-establishing the buildings (Reddy and Yang 2011). Cyanobacterial mats also were exploited for cleaning up contamination of oil spinoffs (Bordenave et al. 2009; Raeid 2011). Sarkar et al. (2017) reported 98% removal of total petroleum hydrocarbons with the aid of microcosms bioaugmented with *Enterobacter*, *Pandoraea* and *Burkholderia* strains.

Owing to the refractory properties of xenobiotics, their bioavailability is very low and thus their accessibility might be difficult under subsurface environment, and therefore they are persistent in the environment, escaping the metabolism by microorganisms. Consequently, organisms having degradative potential can be genetically manipulated to enhance mobility so as to access these pollutants (Díaz 2004). A major restraint of bioremediation process is optimal physico-chemical conditions that are obligatory for accurate metabolic working of microorganisms, as a prerequisite for degradation of xenobiotics, which may be tough to achieve in natural environment (Singh and Ward 2004). In several studies, syntrophic bacterial consortia are being used for degradation of xenobiotics, as single microorganism in some cases may not be able to perform all metabolic activities required (Díaz 2004). Therefore, when mixed bunch of bacteria works in combination, the dead-end products from one organism will be then broken down by another bacterium (Singh and Ward 2004). Biodegradation of chrysene, a persistent polycyclic aromatic hydrocarbon, was achieved by employing bacterial consortium consisting of *Bacillus* sp., *Rhodococcus* sp. and *Burkholderia* sp. Chrysene was utilized as a sole source of carbon and energy by the consortium (Vaidya et al. 2018). Researches also reported for possibility of application of a microaerophilic mixed bacterial consortium for complete mineralization of azo dyes methyl orange and Congo red (Dissanayake et al. 2021). Furthermore, anaerobic microbial consortia also are reported for anaerobic degradation of levofloxacin (Shu et al. 2021).

25.5.2 Myco-Remediation

Fungi are characterized as one of the most diverse collections of microorganisms that exhibit a significant part in nature as decomposers, mutualists or pathogens (Schmit and Mueller 2007). Mineralization exploiting fungus may be accomplished by direct metabolism, where a xenobiotic component may be totally degraded by fungi to non-toxic end products like inorganic chemical compounds and CO_2 (Mougin et al. 2009). Amongst diverse groups of fungi, white rot fungi can degrade several xenobiotics, including polycyclic aromatic hydrocarbons, persistent pesticides, dyes, chlorinated phenols and pharmaceuticals (Vanhulle et al. 2008). Besides, a wide variety of fungal species that assist in degradation of recalcitrant xenobiotic compounds are listed in Table 25.1.

25.5.3 Phytoremediation

Phytoremediation, termed as green remediation, agro- or botano-remediation, is a technique that employs plants for treating chemically contaminated soils (Wenzel et al. 1999), thereby reducing the concentration of hazardous compounds (Utmazian and Wenzel 2006). Plants are used because they can endure reasonably elevated amounts of xenobiotic chemicals deprived of noxious effects (Briggs et al. 1982). Moreover, plants have a potential to take up these chemicals and convert them into less toxic compounds (Bock et al. 2002). Therefore, it is one of best green routes to target chemical pollutants present in the environment for their removal. Enzyme secreted within plants aids in degradation of chlorinated compounds, herbicides and other organic pollutants; this method of removal of pollutants is termed as phytotransformation (Shukla et al. 2010). Likewise, rhizodegradation also leads to degradation and detoxification of recalcitrant pollutants present in soil with the aid of plant roots. The process is employed for catabolism of chlorinated solvents, surfactants, PCBs, petroleum hydrocarbons and various pesticides (Goyal and Basniwal 2017). For elimination of toxic metals from the soil, a method known as phytomining is employed, in which metal ions are absorbed by plant roots. If

Table 25.1 List of fungal species that possess xenobiotic degradation potential

Xenobiotic compounds	Fungi degrading xenobiotics	References
Pesticides		
Diazinon	*Aspergillus niger* MK640786	Hamad (2020)
Chlorpyrifos, profenofos and methyl parathion	*Aspergillus sydowii* CBMAI 935	Soares et al. (2021)
DDT	Consortium *of fungus Fomitopsis pinicola* + bacterium *Ralstonia pickettii*	Purnomo et al. (2020a)
	Xerocomus chrysenteron	Huang and Wang (2013)
	P. aeruginosa + *P. ostreatus*	Purnomo et al. (2017)
Allethrin	*Fusarium proliferatum* CF2	Bhatt et al. (2020a)
Chlorfenvinphos	*Penicillium citrinum*, *Aspergillus fumigatus*, *Aspergillus terreus* and *Trichoderma harzianum*	Oliveira et al. (2015)
Endosulfan and chlorpyrifos	*Cladosporium cladosporioides*, *Phanerochaete chrysosporium*, *Trichoderma harzianum*, *Trichoderma virens*, *Trametes hirsuta* and *Trametes versicolor*	Bisht et al. (2019)
PCBs		
Aroclor 1254	*Phanerochaete chrysosporium*	Eaton (1985)
3,3′,4,4′-Tetrachlorobiphenyl, 2,3,3′,4,4′-pentachlorobiphenyl, 2,3′,4,4′,5-pentachlorobiphenyl, 3,3′,4,4′,5-pentachlorobiphenyl and 2,3′,4,4′,5,5′-hexachlorobiphenyl	*Phlebia brevispora*	Kamei et al. (2006)
2-Chlorobiphenyl, 4,4′-dichlorobiphenyl and 2,2′,5,5′-tetrachlorobiphenyl	*Aspergillus fumigatus* MUT 4026, *Penicillium chrysogenum* MUT 4021, *Fusarium solani* MUT 4020, *Penicillium digitatum* MUT 4079, *Scedosporium apiospermum* MUT 641 and *Scedosporium apiospermum* MUT 631	Tigini et al. (2009b)
PCBs	*Pleurotus sajor-caju* LBM 105	Sadañoski et al. (2019)
Dye		
Methyl orange	*Gloeophyllum trabeum*	Purnomo et al. (2020b)
Triphenylmethane dyes	*Penicillium simplicissimum*	Chen et al. (2019)

(continued)

Table 25.1 (continued)

Xenobiotic compounds	Fungi degrading xenobiotics	References
Acid red 88	*Achaetomium strumarium*	Bankole et al. (2018a)
Reactive blue 4, Remazol brilliant blue and acid blue 129 (AB129)	*Trametes hirsuta* D7	Alam et al. (2021)
Scarlet RR dye	*Peyronellaea prosopidis*	Bankole et al. (2018b)
Polycyclic aromatic hydrocarbons		
Mixture of four polycyclic aromatic hydrocarbons	*Fusarium oxysporum*	Marchand et al. (2017)
Phenanthrene, anthracene and pyrene	*Trematophoma genus*	Moghimi et al. (2017)
Anthracene	*Trichoderma harzianum*, *Cladosporium* sp., *Aspergillus sydowii*, *Penicillium citrinum* and *Mucor racemosus*	Birolli et al. (2018)
Anthracene, anthrone, anthraquinone, acenaphthene, fluorene, phenanthrene, fluoranthene, pyrene and nitropyrene	*Cladosporium* sp. CBMAI 1237	
Phenanthrene, anthracene and pyrene	*Coriolopsis caperata*, *Fomes fomentarius*, *Pluteus chysophaeus*	Hadibarata and Yuniarto (2020)
Benzo[a]pyrene	*Penicillium canescens*, *Cladosporium cladosporioides*, *Fusarium solani* and *Talaromyces helicus*	Fayeulle et al. (2019)
Anthracene and dibenzothiophene	*Penicillium oxalicum*	Aranda et al. (2017)
Estrogenic xenobiotics		
Bisphenol A, estrone, 17β-estradiol, estriol, 17α-ethinylestradiol, triclosan and 4-*n*-nonylphenol	*Pleurotus ostreatus* HK 35	Křesinová et al. (2018)
4-*t*-Octylphenol	*Fusarium falciforme*	Rajendran et al. (2017)
Testosterone and 17α-ethynylestradiol	*Lentinula edodes*	Muszyńska et al. (2018)

required, metals can be extracted by incineration of plants to yield ash and assimilation of plant incinerations will help metal reuse (Shukla et al. 2010). *Thlaspi caerulescens*, *Alyssum murale*, *Alyssum markgrafii*, *Bornmuellera baldacii* subsp. *markgrafii* and *Leptoplax emarginata* are examples of some plants involved in phytomining (Aboudrar et al. 2007; Bani et al. 2007, 2009). Trends in phytoremediation also use transgenic or genetically engineered plants that help to elevate tolerance and metabolism of xenobiotic chemicals for remediation by plants (Sonoki et al. 2011). In contrast to conventional mechanical methods being used, phytoremediation is an eco-friendly and economically feasible approach to clean up polluted groundwater (Bhatt et al. 2020b). Nevertheless, it is a time-consuming process as it is reliant on the growth of the plants. Additionally, for successful remediation by means of this green technique, it is necessary to look for right plant for right pollutant (Goyal and Basniwal 2017; Bhandari and Bhatt 2021).

25.6 Conclusion and Future Prospects

Environmental pollution is escalating due to increased global industrialization, leading to generation of hefty portions of xenobiotic wastes that present potent health hazard to mankind. Degradation of recalcitrant chemicals is thus the need of the hour so as to lessen contamination of these environmental pollutants. In this context, remediation strategies employing microorganisms, plants and their enzymes have gained substantial attention. As cited in this chapter, microbial remediation using bacteria and fungi and phytoremediation are capable of mineralization of xenobiotics to innocuous or less toxic end products. Though microbial-based systems for removal of pollutants are slow, it offers several pros over physico-chemical methods of remediation, as it is an economically feasible and eco-friendly method. Furthermore, comprehensive studies on the utility of these microorganisms and their enzyme machineries and studies on cloning and expression of genes appear to be compelling tools to decrease the levels of toxic chemicals and to understand the mechanism of biodegradation, respectively. Thus, it can be concluded that elimination of toxic pollutants from the environment through green approach is only possible with the assistance of microorganisms and plants.

References

Aboudrar W, Schwartz C, Benizri E, Morel JL, Boularbah A (2007) Soil microbial diversity as affected by the rhizosphere of the hyperaccumulator Thlaspi caerulescens under natural conditions. Int J Phytoremediation 9(1):41–52

Alam R, Ardiati FC, Solihat NN, Alam MB, Lee SH, Yanto DHY et al (2021) Biodegradation and metabolic pathway of anthraquinone dyes by Trametes hirsuta D7 immobilized in light expanded clay aggregate and cytotoxicity assessment. J Hazard Mater 405:124176

Amin GA (2011) Integrated two-stage process for biodesulfurization of model oil by vertical rotating immobilized cell reactor with the bacterium Rhodococcus erythropolis. J Pet Environ Biotechnol 2(107):2

Aranda E, Godoy P, Reina R, Badia-Fabregat M, Rosell M, Marco-Urrea E, García-Romera I (2017) Isolation of Ascomycota fungi with capability to transform PAHs: insights into the biodegradation mechanisms of Penicillium oxalicum. Int Biodeter Biodegr 122:141–150

Asgher M, Bhatti HN, Ashraf M, Legge RL (2008) Recent developments in biodegradation of industrial pollutants by white rot fungi and their enzyme system. Biodegradation 19(6):771

Baker P, Tiroumalechetty A, Mohan R (2019) Fungal enzymes for bioremediation of xenobiotic compounds. In: Recent advancement in white biotechnology through fungi. Springer, Cham, pp 463–489

Bani A, Echevarria G, Sulçe S, Morel JL, Mullai A (2007) In-situ phytoextraction of Ni by a native population of Alyssum murale on an ultramafic site (Albania). Plant and Soil 293(1):79–89

Bani A, Echevarria G, Mullaj A, Reeves R, Morel JL, Sulçe S (2009) Nickel hyperaccumulation by Brassicaceae in serpentine soils of Albania and northwestern Greece. Northeast Nat 16 (sp5):385–404

Bankole PO, Adekunle AA, Govindwar SP (2018a) Enhanced decolorization and biodegradation of acid red 88 dye by newly isolated fungus, Achaetomium strumarium. J Environ Chem Eng 6 (2):1589–1600

Bankole PO, Adekunle AA, Obidi OF, Chandanshive VV, Govindwar SP (2018b) Biodegradation and detoxification of Scarlet RR dye by a newly isolated filamentous fungus, Peyronellaea prosopidis. Sustain Environ Res 28(5):214–222

Bhandari G, Bhatt P (2021) Concepts and application of plant–microbe interaction in remediation of heavy metals. In: Microbes and signaling biomolecules against plant stress. Springer, Singapore, pp 55–77

Bharadwaj A (2018) Bioremediation of xenobiotics: an eco-friendly cleanup approach. In: Green chemistry in environmental sustainability and chemical education. Springer, Singapore, pp 1–13

Bhatt P, Pathak VM, Joshi S, Bisht TS, Singh K, Chandra D (2019) Major metabolites after degradation of xenobiotics and enzymes involved in these pathways. In: Smart bioremediation technologies. Academic, Boston, pp 205–215

Bhatt P, Zhang W, Lin Z, Pang S, Huang Y, Chen S (2020a) Biodegradation of allethrin by a novel fungus Fusarium proliferatum strain CF2, isolated from contaminated soils. Microorganisms 8 (4):593

Bhatt P, Gangola S, Bhandari G, Zhang W, Maithani D, Mishra S, Chen S (2020b) New insights into the degradation of synthetic pollutants in contaminated environments. Chemosphere 268:128827

Birolli WG, Santos DDA, Alvarenga N, Garcia AC, Romão LP, Porto AL (2018) Biodegradation of anthracene and several PAHs by the marine-derived fungus Cladosporium sp. CBMAI 1237. Mar Pollut Bull 129(2):525–533

Bisht J, Harsh NSK, Palni LMS, Agnihotri V, Kumar A (2019) Biodegradation of chlorinated organic pesticides endosulfan and chlorpyrifos in soil extract broth using fungi. Remediat J 29 (3):63–77

Bock C, Kolb M, Bokern M, Harms H, Mackova M, Chroma L et al (2002) Advances in phytoremediation: phytotransformation. In: Innovative approaches to the on-site assessment and remediation of contaminated sites. Springer, Dordrecht, pp 115–140

Bordenave S, Goni-Urriza M, Caumette P, Duran R (2009) Differential display analysis of cDNA involved in microbial mats response after heavy fuel oil contamination. J Microbial Biochem Technol 1:1–4

Breivik K, Alcock R, Li YF, Bailey RE, Fiedler H, Pacyna JM (2004) Primary sources of selected POPs: regional and global scale emission inventories. Environ Pollut 128(1–2):3–16

Briggs GG, Bromilow RH, Evans AA (1982) Relationships between lipophilicity and root uptake and translocation of non-ionised chemicals by barley. Pestic Sci 13(5):495–504

Chen SH, Cheow YL, Ng SL, Ting ASY (2019) Biodegradation of triphenylmethane dyes by non-white rot fungus Penicillium simplicissimum: enzymatic and toxicity studies. Int J Environ Res 13(2):273–282

Chopra S, Kumar D (2020) Characterization, optimization and kinetics study of acetaminophen degradation by Bacillus drentensis strain S1 and waste water degradation analysis. Bioresour Bioprocess 7(1):1–18

D'Souza DT, Tiwari R, Sah AK, Raghukumar C (2006) Enhanced production of laccase by a marine fungus during treatment of colored effluents and synthetic dyes. Enzyme Microb Technol 38(3-4):504–511

Díaz E (2004) Bacterial degradation of aromatic pollutants: a paradigm of metabolic versatility. Int J Microbiol 7:173–180

Dissanayake M, Liyanage N, Herath C, Rathnayake S, Fernando EY (2021) Mineralization of persistent azo dye pollutants by a microaerophilic tropical lake sediment mixed bacterial consortium. Environ Adv 3:100038

Eaton DC (1985) Mineralization of polychlorinated biphenyls by Phanerochaete chrysosporium: a ligninolytic fungus. Enzyme Microb Technol 7(5):194–196

Fayeulle A, Veignie E, Schroll R, Munch JC, Rafin C (2019) PAH biodegradation by telluric saprotrophic fungi isolated from aged PAH-contaminated soils in mineral medium and historically contaminated soil microcosms. J Soil Sediment 19(7):3056–3067

Furukawa K (2003) 'Super bugs' for bioremediation. Trends Biotechnol 21(5):187–190

Galvão TC, Mohn WW, de Lorenzo V (2005) Exploring the microbial biodegradation and biotransformation gene pool. Trends Biotechnol 23(10):497–506

Gangola S, Sharma A, Bhatt P, Khati P, Chaudhary P (2018) Presence of esterase and laccase in *Bacillus subtilis* facilitates biodegradation and detoxification of cypermethrin. Sci Rep 8 (1):1–11

Gangola S, Joshi S, Kumar S, Pandey SC (2019) Comparative analysis of fungal and bacterial enzymes in biodegradation of xenobiotic compounds. In: Smart bioremediation technologies. Academic, Boston, pp 169–189

Gojgic-Cvijovic GD, Milic JS, Solevic TM, Beskoski VP, Ilic MV, Djokic LS, Vrvic MM (2012) Biodegradation of petroleum sludge and petroleum polluted soil by a bacterial consortium: a laboratory study. Biodegradation 23(1):1–14

Goyal P, Basniwal RK (2017) Environmental bioremediation: biodegradation of xenobiotic compounds. In: Xenobiotics in the soil environment. Springer, Cham, pp 347–371

Hadad D, Geresh S, Sivan A (2005) Biodegradation of polyethylene by the thermophilic bacterium Brevibacillus borstelensis. J Appl Microbiol 98(5):1093–1100

Hadibarata T, Yuniarto A (2020) Biodegradation of polycyclic aromatic hydrocarbons by high-laccase basidiomycetes fungi isolated from tropical forest of Borneo. Biocatal Agric Biotechnol 28:101717

Hamad MTMH (2020) Biodegradation of diazinon by fungal strain Apergillus niger MK640786 using response surface methodology. Environ Technol Innov 18:100691

Huang Y, Wang J (2013) Degradation and mineralization of DDT by the ectomycorrhizal fungi, Xerocomus chrysenteron. Chemosphere 92(7):760–764

Joe J, Kothari RK, Raval CM, Kothari CR, Akbari VG, Singh SP (2011) Decolorization of textile dye Remazol Black B by Pseudomonas aeruginosa CR-25 isolated from the common effluent treatment plant. J Bioremed Biodegr 2(2):118

Kafilzadeh F, Ebrahimnezhad M, Tahery Y (2015) Isolation and identification of endosulfan-degrading bacteria and evaluation of their bioremediation in Kor River, Iran. Osong Public Health Res Perspect 6(1):39–46

Kalyoncu L, Agca I, Aktumsek A (2009) Some organochlorine pesticide residues in fish species in Konya, Turkey. Chemosphere 74(7):885–889

Kamei I, Sonoki S, Haraguchi K, Kondo R (2006) Fungal bioconversion of toxic polychlorinated biphenyls by white-rot fungus, Phlebia brevispora. Appl Microbiol Biotechnol 73(4):932–940

Kelce WR, Stone CR, Laws SC, Gray LE, Kemppainen JA, Wilson EM (1995) Persistent DDT metabolite p, p′–DDE is a potent androgen receptor antagonist. Nature 375(6532):581–585

Křesinová Z, Linhartová L, Filipová A, Ezechiáš M, Mašín P, Cajthaml T (2018) Biodegradation of endocrine disruptors in urban wastewater using Pleurotus ostreatus bioreactor. Nat Biotechnol 43:53–61

Kumar M, León V, Materano ADS, Ilzins OA (2007) A halotolerant and thermotolerant Bacillus sp. degrades hydrocarbons and produces tensio-active emulsifying agent. World J Microbiol Biotechnol 23(2):211–220

Kumari M, Ghosh P, Thakur IS (2014) Microcosmic study of endosulfan degradation by Paenibacillus sp. ISTP10 and its toxicological evaluation using mammalian cell line. Int Biodeter Biodegrad 96:33–40

Marchand C, St-Arnaud M, Hogland W, Bell TH, Hijri M (2017) Petroleum biodegradation capacity of bacteria and fungi isolated from petroleum-contaminated soil. Int Biodeter Biodegr 116:48–57

Moghimi H, Tabar RH, Hamedi J (2017) Assessing the biodegradation of polycyclic aromatic hydrocarbons and laccase production by new fungus Trematophoma sp. UTMC 5003. World J Microbiol Biotechnol 33(7):1–10

Mougin C, Boukcim H, Jolivalt C (2009) Soil bioremediation strategies based on the use of fungal enzymes. In: Advances in applied bioremediation. Springer, Berlin, pp 123–149

Muszyńska B, Żmudzki P, Lazur J, Kała K, Sułkowska-Ziaja K, Opoka W (2018) Analysis of the biodegradation of synthetic testosterone and 17α-ethynylestradiol using the edible mushroom Lentinula edodes. 3 Biotech 8(10):1–15

Olicón-Hernández DR, González-López J, Aranda E (2017) Overview on the biochemical potential of filamentous fungi to degrade pharmaceutical compounds. Front Microbiol 8:1792

Oliveira BR, Penetra A, Cardoso VV, Benoliel MJ, Crespo MB, Samson RA, Pereira VJ (2015) Biodegradation of pesticides using fungi species found in the aquatic environment. Environ Sci Pollut Res 22(15):11781–11791

Oyetibo GO, Chien MF, Ikeda-Ohtsubo W, Suzuki H, Obayori OS, Adebusoye SA et al (2017) Biodegradation of crude oil and phenanthrene by heavy metal resistant Bacillus subtilis isolated from a multi-polluted industrial wastewater creek. Int Biodeter Biodegr 120:143–151

Pakala SB, Gorla P, Pinjari AB, Krovidi RK, Baru R, Yanamandra M et al (2007) Biodegradation of methyl parathion and p-nitrophenol: evidence for the presence of ap-nitrophenol 2-hydroxylase in a Gram-negative Serratia sp. strain DS001. Appl Microbiol Biotechnol 73(6):1452–1462

Poornima K, Karthik L, Swadhini SP, Mythili S, Sathiavelu A (2010) Degradation of chromium by using a novel strains of Pseudomonas species. J Microbial Biochem Technol 2(4):95

Purnomo AS, Ashari K, Hermansyah FT (2017) Evaluation of the synergistic effect of mixed cultures of white-rot fungus Pleurotus ostreatus and biosurfactant-producing bacteria on DDT biodegradation. J Microbiol Biotechnol 27(7):1306–1315

Purnomo AS, Sariwati A, Kamei I (2020a) Synergistic interaction of a consortium of the brown-rot fungus Fomitopsis pinicola and the bacterium Ralstonia pickettii for DDT biodegradation. Heliyon 6(6):e04027

Purnomo AS, Andyani NEA, Nawfa R, Putra SR (2020b) Fenton reaction involvement on methyl orange biodegradation by brown-rot fungus Gloeophyllum trabeum. In AIP conference proceedings, vol 2237, no 1, p. 020002. AIP Publishing

Qadir A, Hashmi MZ, Mahmood A (2017) Xenobiotics, types, and mode of action. In: Xenobiotics in the soil environment. Springer, Cham, pp 1–7

Raeid MMA (2011) Unraveling the role of cyanobacterial mats in the cleanup of oil pollutants using modern molecular and microsensor tools. In: World congress on biotechnology, New Delhi.

Rajendran RK, Lin CC, Huang SL, Kirschner R (2017) Enrichment, isolation, and biodegradation potential of long-branched chain alkylphenol degrading non-ligninolytic fungi from wastewater. Mar Pollut Bull 125(1–2):416–425

Reddy N, Yang Y (2011) Potential of plant proteins for medical applications. Trends Biotechnol 29 (10):490–498

Sadañoski MA, Benítez SF, Fonseca MI, Velázquez JE, Zapata PD, Levin LN, Villalba LL (2019) Mycoremediation of high concentrations of polychlorinated biphenyls with Pleurotus sajor-caju LBM 105 as an effective and cheap treatment. J Environ Chem Eng 7(6):103453

Sarkar P, Roy A, Pal S, Mohapatra B, Kazy SK, Maiti MK, Sar P (2017) Enrichment and characterization of hydrocarbon-degrading bacteria from petroleum refinery waste as potent bioaugmentation agent for in situ bioremediation. Bioresour Technol 242:15–27

Schmit JP, Mueller GM (2007) An estimate of the lower limit of global fungal diversity. Biodivers Conserv 16(1):99–111

Sharma B, Dangi AK, Shukla P (2018) Contemporary enzyme based technologies for bioremediation: a review. J Environ Manage 210:10–22

Shrivastava A (2018) Introduction to plastics engineering. William Andrew, Burlington

Shu W, Zhang Y, Wen D, Wu Q, Liu H, Cui MH et al (2021) Anaerobic biodegradation of levofloxacin by enriched microbial consortia: effect of electron acceptors and carbon source. J Hazard Mater 414:125520

Shukla AK, Singh AK (2020) Exploitation of potential extremophiles for bioremediation of xenobiotics compounds: a biotechnological approach. Curr Genomics 21(3):161–167

Shukla KP, Singh NK, Sharma S (2010) Bioremediation: developments, current practices and perspectives. Genet Eng Biotechnol J 3:1–20

Singh A, Ward OP (eds) (2004) Applied bioremediation and phytoremediation, vol 1. Springer, New York

Sinha S, Chattopadhyay P, Pan I, Chatterjee S, Chanda P, Bandyopadhyay D et al (2009) Microbial transformation of xenobiotics for environmental bioremediation. Afr J Biotechnol 8(22)

Soares PRS, Birolli WG, Ferreira IM, Porto ALM (2021) Biodegradation pathway of the organophosphate pesticides chlorpyrifos, methyl parathion and profenofos by the marine-derived fungus Aspergillus sydowii CBMAI 935 and its potential for methylation reactions of phenolic compounds. Mar Pollut Bull 166:112185

Sonoki T, Kajita S, Uesugi M, Katayama Y, Iimura Y (2011) Effective removal of bisphenol a from contaminated areas by recombinant plant producing lignin peroxidase. J Pet Environ Biotechnol 2(105):2

Tigini V, Prigione V, Di Toro S, Fava F, Varese GC (2009a) Isolation and characterisation of polychlorinated biphenyl (PCB) degrading fungi from a historically contaminated soil. Microb Cell Fact 8(1):5

Tigini V, Prigione V, Di Toro S, Fava F, Varese GC (2009b) Isolation and characterisation of polychlorinated biphenyl (PCB) degrading fungi from a historically contaminated soil. Microb Cell Fact 8(1):1–14

Tsai SL, Lin CW, Wu CH, Shen CM (2013) Kinetics of xenobiotic biodegradation by the Pseudomonas sp. YATO411 strain in suspension and cell-immobilized beads. J Taiwan Inst Chem Eng 44(2):303–309

Utmazian MN, Wenzel WW (2006) Phytoextraction of metal polluted soils in Latin America. In: Environmental applications poplar and willow working party, pp 16–20

Vaidya S, Devpura N, Jain K, Madamwar D (2018) Degradation of chrysene by enriched bacterial consortium. Front Microbiol 9:1333

Vanhulle S, Enaud E, Trovaslet M, Billottet L, Kneipe L, Jiwan JLH et al (2008) Coupling occurs before breakdown during biotransformation of Acid Blue 62 by white rot fungi. Chemosphere 70(6):1097–1107

Varsha YM, Naga Deepthi CH, Chenna S (2011) An emphasis on xenobiotic degradation in environmental clean up. J Bioremed Biodegr 11:1–10

Wasi S, Tabrez S, Ahmad M (2010) Isolation and characterization of a Pseudomonas fluorescens strain tolerant to major Indian water pollutants. J Bioremed Biodegr 1:101. https://doi.org/10.4172/2155-6199.1000101

Wenzel WW, Lombi E, Adriano DC (1999) Biogeochemical processes in the rhizosphere: role in phytoremediation of metal-polluted soils. In: Heavy metal stress in plants. Springer, Berlin, pp 273–303

Yao RS, Sun M, Wang CL, Deng SS (2006) Degradation of phenolic compounds with hydrogen peroxide catalyzed by enzyme from Serratia marcescens AB 90027. Water Res 40 (16):3091–3098

Chapter 26
Harnessing the Rhizomicrobiome Interactions for Plant Growth Promotion and Sustainable Agriculture: Mechanisms, Applications and Recent Advances

Geeta Bhandari and Niki Nautiyal

Abstract Rhizomicrobiome is of utmost significance in agricultural sciences on account of highly diverse, distinct rhizospheric microflora present in direct or indirect association with plants. It is vital for plant growth promotion due to its involvement in processes such as nutrient uptake and incorporation; enhancement of soil characteristics; regulation of the production of plant growth hormones, secondary metabolites and antibiotics; regulation of stress tolerance; and rhizoremediation. The diverse rhizomicrobiomes are significant for maintaining sustainability in agricultural practices. Fulfilment of increasing food demand with employment of minimal chemical fertilizers has become a great challenge for both researchers and farmers. PGPR and their secretions play a pivotal role in efficiently stimulating plant growth and enhance stress tolerance for the biotic and abiotic stresses. The versatile PGPR-based biofertilizers have been formulated for use in agricultural practices, thus minimizing chemical fertilizers and agrochemical input. Therefore, a deep and comprehensive insight into the plant rhizomicrobiome and their mechanism is critical for exploring the aforesaid sustainability of agricultural practices.

Keywords Rhizomicrobiome · PGPR · Biofertilizer · Sustainable agriculture

26.1 Introduction

Soil is a vital, living matrix and is a significant reservoir crucial for agriculture productivity, food security and sustenance of living beings. It is regarded as a repository of bacterial metabolism even though microbes occupy not more than 5% of the entire soil space. The area of soil around the plant root is commonly termed as rhizosphere, which is under chemical, physical and biological influence of

G. Bhandari (✉) · N. Nautiyal
Department of Biotechnology, Sardar Bhagwan Singh University, Dehradun, Uttarakhand, India

P. Bhatt et al. (eds.), *Microbial Technology for Sustainable Environment*,
https://doi.org/10.1007/978-981-16-3840-4_26

plant roots. The rhizospheric microbial communities comprise of several organisms such as bacteria, archaea, fungi, algae and viruses. Rhizospheric microbes are constituent of microflora that are associated with the plant roots and are regarded as highly intricate ecosystems on the Earth (Wagg et al. 2014). The plant, the soil and the microbes in the rhizosphere are all interconnected to one another forming "microbiome" which allows the occurrence of several processes beneficial for enhanced plant production (Ahmed et al. 2019). The microbiomes of the rhizospheric region encompass potential microbes significant for mineral exchange, recycling, improving soil fecundity and plant protection and production (Jacoby et al. 2017). The microbes present in the rhizosphere are usually known as plant growth-promoting rhizobacteria—"PGPR". These are the useful microbes that inhabit the rhizospheric zone and augment the plant development directly or indirectly, such as inhibition of plant pathogens, degradation of hazardous pollutants, stimulation of induced systemic resistance (ISR), release of phytostimulating compounds and furnishing of beneficial nutrients, such as N and P (Glick 2012). Root exudates produced by the plant roots alter soil physiology (Mukherjee et al. 2018) and eventually modulate heterogeneity and structure of rhizospheric microbiome (Huang et al. 2014) and thus selectively stimulate beneficial microbes required for plant growth and development (Chaparro et al. 2014).

26.2 Plant Growth-Promoting Rhizospheric Microbiome

The valuable results of several traditional sustainable farming practices on plant development have been reported 100 years ago by the ancient Greeks and Romans. They found that combining various soil types or adding organic amendments enhances soil health and fecundity and thus subsequently improves plant yield (Tisdale and Nelson 1975). Thereafter, Kloepper and Schroth then coined the word "plant growth-promoting rhizobacteria" (PGPR) for the rhizospheric microbial population (Kloepper and Schroth 1981). PGPR is a broad term that especially refers to such microbial strains inhabiting the rhizosphere and they considerably enhance plant growth, productivity, soil health, pest resistance and synthesis of plant growth hormones (Prasad et al. 2019; Compant et al. 2019), and thus the PGPR are given prominence in farming practices. The PGPR are divided into two groups, i.e. extracellular PGPR (ePGPR) and intracellular PGPR (iPGPR) (Viveros et al. 2010). The ePGPR are usually present in the rhizospheric soil or inhabit the outermost root cortical region and belong to genera *Flavobacterium*, *Azotobacter*, *Caulobacter*, *Pseudomonas*, *Serratia*, *Burkholderia*, *Erwinia*, *Chromobacterium*, *Azospirillum*, *Arthrobacter*, *Agrobacterium*, *Bacillus* and *Micrococcus*. In contrast, the iPGPR form symbiotic association with the plant root and reside in the root nodule and include genera such as *Rhizobium*, *Bradyrhizobium*, *Allorhizobium*, *Mesorhizobium*, etc. (Viveros et al. 2010; Bhattacharyya and Jha 2012). The PGPR enhance plant growth and development in direct and indirect mechanisms and thus render soil fecundity and enhance plant productivity (Gupta et al. 2015). Direct mechanism includes phosphate solubilization, biological nitrogen fixation, siderophore synthesis and iron acquisition and synthesis of plant growth protomory compounds. Indirect methods involve biocontrol through antibiotic, antifungal and volatile organic acid production (Glick 2012; Ahemad and Khan 2012; Jahanian et al. 2012; Liu et al. 2016). The PGPR should first inhabit and thrive in the rhizospheric zone so as to work as plant growth promoters. Rhizosphere colonization by PGPR is dependent on several aspects including soil characteristics, occurrence of protozoans, synthesis of antibacterial metabolites of plants or microflora and root exudate usage.

Currently, the employment of nonhazardous substitutes for accomplishing increased plant productivity in a sustainable manner is of utmost significance since the use of environment-friendly agricultural practices is the fundamental principle of sustainable agriculture. Various efforts are being done for screening and exploiting PGPR to be employed for improving soil fertility and health in place of synthetic agrochemicals. Thus, various symbiotic (*Bradyrhizobium*, *Rhizobium*, *Mesorhizobium*) and nonsymbiotic (*Flavobacterium*, *Chromobacterium*, *Klebsiella*, *Pseudomonas*, *Agrobacterium*, *Azomonas*, *Enterobacter*, *Micrococcus*, *Burkholderia*, *Variovorax*, *Azotobacter*, *Bacillus*, *Serratia*, *Azospirillum*, *Caulobacter*, *Erwinia* and *Arthrobacter*) rhizobacteria are globally employed as biofertilizers for improving crop production in various agroclimatic zones (Glick 2012; Ahemad and Khan 2012; Egamberdieva and Lugtenberg 2014). Several PGPR employed for plant growth promotion are listed in Table 26.1.

26.3 Nutrient Acquisition by Plant Growth-Promoting Rhizospheric Microbiome

26.3.1 Phosphate Solubilization

Phosphorus (P) is a vital nutrient necessary for optimum development of plants. It plays a significant role in plant developmental activities such as biomolecule synthesis, cellular respiration, photosynthesis and signalling pathways (Anand et al. 2016). The plants are able to utilize only monobasic ($H_2PO_4^-$) and dibasic ($H_2PO_4^{2-}$) forms of phosphate (Bhattacharyya and Jha 2012). Although phosphate exists abundantly in the soil, its plant-utilizable form is generally very less since greater than 95% of phosphorus is insoluble and/or immobile and thus is not accessible to the plants (Ahemad and Kibret 2014). Phosphorus is rapidly converted into insoluble metal oxide complexes in soils, thus making it deficit of the available phosphate (Bhattacharyya and Jha 2012). PGPR have evolved several mechanisms for converting insoluble soil P into plant-utilizable soluble forms and such microbes are known as phosphate-solubilizing bacteria (PSMs) (Alori et al. 2017). Solubilization of phosphorus by microbes occurs via (a) production of organic acids (gluconic and citric acid) which drop the soil pH resulting in the release of insoluble P; (b) synthesis of extracellular phosphatases and phytases which break down

Table 26.1 Plant growth-promoting mechanisms of plant rhizomicrobiomes

Rhizobacteria	PGP traits	Crops/plant rhizosphere	References
Mesorhizobium ciceri	Phosphate solubilization, ammonia production	*Cicer arietinum*	Ahmad et al. (2008)
Bacillus sp.	Phosphate solubilizer	*Zea mays*	Oliveira et al. (2009)
Azospirillum brasilense Az39, *Bradyrhizobium japonicum* E109	Phytostimulation	*Glycine max*	Cassan et al. (2009)
P. fluorescens Aur6, *Chryseobacterium balustinum* Aur9	Biocontrol agents	*Oryza sativa*	Lucas et al. (2009)
Chryseobacterium palustre, *Chryseobacterium humi*, *Sphingobacterium*, *Bacillus*, *Achromobacter*	IAA, HCN, NH_3, siderophore production, ACC deaminase activity	*Zea mays*	Marques et al. (2010)
Bacillus sp. PSB10	IAA, HCN, siderophore production	*Cicer arietinum*	Wani and Khan (2010)
Rhizobium MRPI	Nitrogen fixation and phosphate solubilization	*Pisum sativum*	Ahemad and Khan (2011)
Rhodococcus sp. EC35, *Pseudomonas* sp. EAV, *Arthrobacter nicotinovorans* EAPAA	Phosphate solubilizer	*Zea mays*	Sofia et al. (2014)
Pseudomonas fluorescens strain Psd	Zinc solubilizer, nitrate reducer	*Triticum aestivum*	Sirohi et al. (2015)
Bacillus circulans	Potassium solubilization	*Solanum lycopersicum*	Mehta et al. (2015)
Bacillus sp. strain WG4	Antifungal metabolite production	*Zingiber officinale*	Jimtha et al. (2016)
Burkholderia sp. *Pseudomonas aeruginosa* strain MAJ PIA 03 *Bacillus firmus* strain MAJ PSB12	IAA, GA3, ACC deaminase activity, HCN production, NH_3 production, siderophore production, antagonistic activity, phosphate solubilizer	*Ricinus communis*	Sandilya et al. (2016, 2017)
Bacillus, *Azotobacter*, *Pseudomonas* and *Acinetobacter*	IAA, NH_3, HCN, siderophore production, phosphate solubilizer, antagonistic activity, nitrate reducer	*Momordica charantia*	Singh et al. (2017)
Bacillus megaterium PSB12	Phosphate solubilization	*Triticum aestivum*	Mukhtar et al. (2017)

(continued)

Table 26.1 (continued)

Rhizobacteria	PGP traits	Crops/plant rhizosphere	References
Bacillus megaterium	Phosphate solubilization	*Vigna radiata*	Biswas et al. (2018)
Azotobacter chroococcum	Siderophore production	Cereals	Zhang et al. (2019)
Azospirillum brasilense	Nitrogen fixation	*Oryza sativa*	Thomas et al. (2019)
Bacillus siamensis	Gibberellin production	*Arabidopsis mutants*	Hossain et al. (2019)
Burkholderia cenocepacia	Phosphate solubilization	*Nicotiana tabacum L.*	Liu et al. (2019)
Enterobacter ludwigii	Zn solubilization	*Triticum aestivum L.*	Singh et al. (2018)

phosphoric esters, resulting in mineralization of P; (c) biological P mineralization by substrate degradation; and (d) production of chelating or phosphorus-solubilizing molecules, e.g. organic acid anions, protons, hydroxyl ions and CO_2 (Glick 2012; Zaidi et al. 2009). Moreover, these microbes in the existence of unstable C immobilize even very minute concentrations of P, and thus these PSMs act as P source upon their starvation, predation or death (Butterly et al. 2009).

PSMs generally belong to genera *Azospirillum*, *Achromobacter*, *Acetobacter*, *Acinetobacter*, *Serratia*, *Microbacterium*, *Azotobacter*, *Rhodococcus*, *Klebsiella*, *Erwinia*, *Arthrobacter*, *Rhizobium*, *Enterobacter*, *Burkholderia*, *Pseudomonas*, *Beijerinckia*, *Mesorhizobium*, *Bacillus* and *Flavobacterium* (Bhattacharyya and Jha 2012; Kumar and Dubey 2012). *Bacillus megaterium* is marketed as Bio Phos (Bio Power Lanka, Sri Lanka) and employed as "P" biofertilizer (Samina 2016). Other reported P-solubilizing strains by AgriLife (India) are *Pseudomonas striata*, *B. polymyxa* and *B. megaterium* (Samina 2016).

26.3.2 Nitrogen Fixation

Nitrogen is another vital mineral required for several developmental activities in plants governing their growth. Nitrogen being present in limiting amount, its biological fixation is a significant process carried out by several PGPR in the symbiotic and free-living forms. Biological nitrogen fixation (BNF) is defined as enzymatic reduction of atmospheric dinitrogen to ammonium using a nitrogenase enzyme complex (Masson-Boivin and Sachs 2018). Since the utilizable form of nitrogen (nitrate or ammonium) is available in limiting amounts in comparison to biological demand, nitrogen fixation is an ecologically significant process for providing fixed nitrogen in several terrestrial and aquatic ecosystems (Vitousek and Howarth 1991; Arp 2000). The BNF microbes include symbiotic nitrogen fixers of genera *Rhizobium*, *Bradyrhizobium*, *Azorhizobium*, *Sinorhizobium*, *Allorhizobium*, *Mesorhizobium*, *Azoarcus*, *Beijerinckia*, *Frankia*, *Pantoea*, *Burkholderia*, *Klebsiella*, *Achromobacter* and *Herbaspirillum* and free-living nitrogen fixers of genera such as *Arthrobacter*, *Acetobacter*, *Clostridium*, *Azotobacter*, *Azospirillum*, *Bacillus*, *Pseudomona*, *Enterobacter*, *Burkholderia*, *Gluconacetobacter*, *Cyanobacterium* and *Diazotrophicus* (Dinnage et al. 2019; Babalola 2010; Pérez-Montaño 2014; Turan et al. 2016). The symbiotic N_2-fixing bacteria inhabit inside the root cells of various plants and cause formation of root nodule, which are the site of transformation of atmospheric nitrogen to fixed forms of nitrogen that can be utilized by the plants (Ahemad and Kibret 2014). In exchange, bacteria obtain fixed form of carbon such as dicarboxylates from the plants which enable them to thrive and carry out the highly energy-demanding process of nitrogen fixation (Udvardi and Poole 2013). All of the nitrogen-fixing microbes contain *nif* genes responsible for the production of the enzyme "nitrogenase". Nitrogenase consists of two multisubunit metalloproteins, i.e. dinitrogen reductase (Fe) and dinitrogenase (MoFe) (Howard and Rees 1996). The *nifHDK* genes code for the enzyme nitrogenase and are generally arranged contiguously. The component I of the enzyme complex is translated from *nifD* and *nifK* genes and has a molecular weight of approximately 250 kDa and is responsible for reduction of N_2. The component II having a molecular weight of 70 kDa is involved in coupling ATP hydrolysis to electron transfer chain and is coded by the *nifH* gene. Both these components work in a coordinated way for reduction of N_2 to NH_3 (Santi et al. 2013). Numerous authors have described various kinds of nitrogenase complexes with differing metal ions found in association with the nitrogenase enzyme complex (Ahemad and Kibret 2014).

Nitrogen-fixing microbes find diverse applications in agriculture such as growth enhancement, pathogen inhibition and supply of fixed nitrogen (Damam et al. 2016). *Azospirillum*, a nonsymbiotic nitrogen-fixing microbe, has been reported to improve growth development of various plants (Lin et al. 2015). Fukami et al. have suggested that *Azospirillum* can be employed for improving yield and productivity of sunflower by supplying nitrogen (Fukami et al. 2018). Yadav and Verma observed the impact of native symbiotic N_2-fixing strain, *Rhizobium leguminosarum* BHURC04, and *Pseudomonas aeruginosa* on the development and production of *Cicer arietinum* L. var. C-235 and reported enhanced nitrogen-fixing capacities and improved plant growth (Yadav and Verma 2014).

26.3.3 Potassium Solubilization

Potassium is another vital nutrient required for enhancement of plant development. Greater than 90% of potassium occurs as insoluble forms and thus the plant-utilizable form is often limited in the soil (Parmar and Sindhu 2013). Potassium deficiency affects root growth and complete plant development and thus results in poor crop productivity. Thus, there is a need of exploring alternative resources of potassium for sustaining plant development without imposing adverse impact on the environment (Kumar and Dubey 2012). Potassium-solubilizing bacteria are considered as highly definitive and effective due to their capability to synthesize organic acids for solubilization of insoluble potassium in rocks and minerals. K solubilization occurs due to synthesis of organic and inorganic acids and protons (Meena et al. 2015) which cause the conversion of insoluble K found in biotite feldspar, mica and muscovite to soluble K by lowering the soil pH and releasing K ions by chelation of Si^{4+}, Al^{3+}, Fe^{2+} and Ca^{2+} ions found in K minerals (Meena et al. 2014; Verma et al. 2017; Yadav et al. 2019). Several acids including oxalic, fumaric, citric, malonic, succinic, tartaric, glycolic, lactic, gluconic, propionic, malic and 2-keto-gluconic are responsible for K solubilization (Saiyad et al. 2015). Several potassium-solubilizing PGPR such as *Bacillus mucilaginosus*, *Pseudomonas* sp., *Bacillus edaphicus*, *Burkholderia* sp., *Acidithiobacillus ferrooxidans* and *Paenibacillus* sp., are well known (Liu et al. 2012). Enhanced growth and potassium uptake of vital crops, e.g. cotton, maize, cucumber, rape, peanut and pepper in the presence of potassium-solubilizing microbes, have been reported (Ashley et al. 2006). Carboxylic acid synthesis for potassium solubilization by *Bacillus* species has been well documented for enhancing soil fecundity and crop production (Majeed et al. 2018; Saha et al. 2016a).

26.3.4 Siderophore Production

Iron is a beneficial micronutrient for all living beings and works as a cofactor of many oxidoreductases. Majority of iron occurs in insoluble form of ferric hydroxide and oxyhydroxides in soil, thus making it a limiting factor for uptake and assimilation by plant and bacteria even in iron-rich soils (Rajkumar et al. 2010). The available form of iron is usually limiting owing to conversion of Fe^{++} to Fe^{+++}. In iron-limiting soil, bacteria synthesize iron-complexing molecules known as siderophores, which form complexes with Fe^{3+}. Siderophores exists in extracellular and intracellular form and are manifested to be significant iron-solubilizing agents. In intracellular chelation, Fe^{3+} is converted to Fe^{2+} having lesser affinity for siderophores and thus released from the chelation complex inside the cell and is employed for bacterial growth and development (Boukhalfa and Crumbliss 2002). Siderophores secreted by plant growth-promoting bacteria (PGPB) possess maximum affinity for iron in comparison to siderophores synthesized by either plants or fungi (Saha et al. 2016b). Pseudomonads are the chief secretors of siderophores and thus play a crucial role in the development of plants (Sandilya et al. 2017). Siderophores contain three Fe-binding domains linked by a flexible backbone with and two oxygen atoms being joined to each functional group. The functional groups present are generally hydroxamates or catecholates, carboxylate, citrate or ethylenediamine (Laschat et al. 2017). Siderophore-producing PGPR such as *Aeromonas*, *Streptomyces* sp., *Bacillus*, *Azotobacter*, *Serratia*, *Rhizobium*, *Burkholderia* and *Pseudomonas* have been reported (Sujatha and Ammani 2013).

Siderophores are also significant for the existence of both plants and microbes due to inhibition of pathogenic fungal strains and other bacterial competitors in the soil (Shen et al. 2013). Siderophores chelate Fe in the rhizosphere and thus limit the pathogenic organisms, e.g. *Fusarium oxysporum and Pythium ultimum*, from iron required for their growth and work as biocidal agent for these pathogens. Moreover, siderophores are documented to chelate several heavy metals (Cd, Zn, Ga, Al, Pb and Cu), thus increasing their amount, and assist in the alleviation of heavy metal stress in plants (Rajkumar et al. 2010; Neubauer et al. 2000). Kloepper et al. have reported siderophore production by *Pseudomonas putida* B10 and assistance in biocontrol against *Fusarium oxysporum* in iron-limited conditions, but this inhibition was however terminated on amending iron in the soil (Kloepper et al. 1980). Siderophores have also been documented in PGPR *Chryseobacterium* spp. C138 and were found to be efficient in supplying iron to plants (Radzki et al. 2013).

26.3.5 Phytohormone Production

Phytohormones are organic compounds which greatly regulate the biochemical, physiological and morphological activities of plants or may act as chemical messengers even in very low concentrations (Fuentes-Ramírez and Caballero-Mellado 2006). Phytohormone production capability is broadly found in various microorganisms, and several PGPR capable of producing plant growth-promoting hormones such as indole-3-acetic acid, gibberellins, cytokinins and ethylene are well known (Spaepen et al. 2007). Several PGPR belonging to genera such as *Proteus mirabilis*, *Azotobacter chroococcum*, *Escherichia coli*, *Stenotrophomonas maltophilia*, *Pseudomonas vulgaris*, *Pseudomonas aeruginosa*, *Rhizobium leguminosarum*, *Klebsiella pneumoniae*, *Mesorhizobium ciceri*, *Bacillus cereus*, *Paenibacillus polymyxa*, *Klebsiella oxytoca*, *Pseudomonas putida* and *Enterobacter asburiae* are documented to synthesize an extensive variety of phytohormones (Ahemad and Kibret 2014) (Table 26.2).

26.3.5.1 Indole Acetic Acid (IAA)

IAA is a naturally existing auxin and significantly impacts the plant growth promotion activities such as cell division, differentiation and root extension (Miransari and

Table 26.2 Phytohormone production by plant growth-promoting rhizomicrobiomes

Phytohormone	PGPR	Crop	References
IAA	*Enterobacter* sp. C1D	*Vigna radiata* L.	Subrahmanyam and Archana (2011)
	Enterobacter sp.	*Cicer arietinum* L.	Fierro-Coronado et al. (2014)
	Proteus vulgaris JBLS202	*Arabidopsis thaliana*	Bhattacharyya et al. (2015)
	Pseudomonas sp., *Bacillus* sp.	*Sulla carnosa*	Hidri et al. (2016)
	Bacillus licheniformis	*Triticum aestivum* L.	Singh and Jha (2016)
	Bacillus subtilis	*Acacia gerrardii* Benth	Hashem et al. (2016)
	Pseudomonas sp.	*Zea mays*	Mishra et al. (2017)
	Azospirillum brasilense	Cereals	Schillaci et al. (2019)
Cytokinin	*Micrococcus luteus*	*Zea mays*	Raza and Faisal (2013)
	Bacillus subtilis	*Platycladus orientalis*	Liu et al. (2013)
	Paenibacillus polymyxa	*Lens culinaris*	Gupta et al. (2015)
	Pseudomonas fluorescens	*Arabidopsis thaliana*	Großkinsky et al. (2016)
	Bacillus subtilis	*Solanum lycopersicum L*	Tahir et al. (2017)
	Advenella kashmirensis	–	Ibal et al. (2018)
	Frankia casuarinae, F. inefficax, F. irregularis and F. saprophytica	–	Nouioui et al. (2019)
	Proteus vulgaris JBLS202	*Arabidopsis thaliana*	Bhattacharyya et al. (2015)
Gibberellin	*Phoma glomerata*, *Penicillium* sp.	*Cucumis sativus*	Waqas et al. (2012)
	Pseudomonas putida	Brassica campestris L.	Kang et al. (2014)
	Bacillus amyloliquefaciens	*Oryza sativa*	Shahzad et al. (2016)
	Proteus vulgaris JBLS202	*Arabidopsis thaliana*	Bhattacharyya et al. (2015)
ACC deaminase	*Methylobacterium fujisawaense*	*Brassica* sp.	Madhaiyan et al. (2006)
	Ralstonia sp. J1-22-2, *P. agglomerans* Jp3-3, *Pseudomonas thivervalensis* Y1-3-9	*B. napus/B. juncea*	Zhang et al. (2011)

(continued)

Table 26.2 (continued)

Phytohormone	PGPR	Crop	References
	P. putida (N21), *P. aeruginosa* (N39), *Serratia proteamaculans* (M35)	*Triticum aestivum*	Zahir et al. (2009)
	Bacillus, *Microbacterium*, *Methylophaga*, *Agromyces*	*Oryza sativa*	Bal et al. (2013)
	Enterobacter sp. C1D	*Vigna radiata* L.	Subrahmanyam et al. (2018)

Smith 2014; Bhatt et al. 2020g). Patten and Glick have documented that about 80% of PGPR produce and secrete auxins as a secondary metabolite (Glick 2012; Spaepen et al. 2007; Patten and Glick 1996). IAA synthesized by the microbes affects various metabolic processes such as vegetative development, geotropism and phototropism, apical dominance and lateral root growth, cell division and differentiation, xylem development, photosynthesis, germination, pigment production, resistance against biotic and abiotic stress and establishment of rhizomicrobiome (Spaepen and Vanderleyden 2011; Grobelak et al. 2015). IAA is also responsible for microbial signalling and affects bacterial metabolism (Spaepen et al. 2007; Spaepen and Vanderleyden 2011). The precursor molecule for microbial IAA biosynthesis has been found to be tryptophan (Etesami et al. 2009) and four tryptophan-dependent pathways have been reported. The pathways are referred on the basis of the intermediates formed, e.g. indole-3-acetamide, indole-3-pyruvic acid, indole-3-acetonitrile and the indole-3-tryptamine (Imada et al. 2017). Innumerous microbes including *Erwinia herbicola*, *Klebsiella*, *Pseudomonas*, *Agrobacterium*, *Enterobacter*, *Bradyrhizobium*, *Azospirillum* and *Rhizobium* have been well documented for synthesizing IAA (Patten and Glick 1996; Spaepen and Vanderleyden 2011). The IAA produced by rhizobacteria has been conferred to magnify the root surface area and length, making it easier for mineral acquisition by the crops from the soil (Ahemad and Khan 2012). IAA synthesized by PGPR has been documented to cause induction of transcriptional changes in the hormone and cell wall genes which result in improved root length and root biomass and reduced stomata size and density (Backer et al. 2018; Vacheron et al. 2013).

26.3.5.2 Cytokinins and Gibberellins

Cytokinins and gibberellins are essential phytohormones required for regulating vital metabolic processes of plant development. Several PGPR such as *Rhizobium*, *Paenibacillus polymyxa*, *Pantoea agglomerans*, *Azotobacter*, *Gluconacetobacter diazotrophicus*, *Acinetobacter calcoaceticus*, *Pseudomonas fluorescens*, *Azospirillum*, *Rhodospirillum rubrum*, *Bacillus subtilis*, *Achromobacter xylosoxidans* and *Bradyrhizobium* have been documented to produce these phytohormones (Glick 2012; Singh 2015; Deka et al. 2015). Cytokinins are generally responsible for regulating cell division and lateral and adventitious root growth

(Salamone et al. 2005). *Azospirillum* was reported to metabolize and absorb gibberellins under different conditions while being associated with higher plants (Bottini et al. 2004). Dobert et al. reported significant internode elongation on inoculation of *Phaseolus lunatus* with *Bradyrhizobium* sp. (Dobert et al. 1992). In a similar study by Kucey, increased N uptake and root growth were reported on inoculation of wheat and maize with gibberellin-synthesizing *Azospirillum* sp. (Kucey 1988). Lucangeli and Bottini documented the reversion of the dwarf phenotype in rice and maize on inoculation with *Azospirillum* sp. (Lucangeli and Bottini 1996). A genetically modified *Sinorhizobium meliloti* strain was capable of cytokinin overproduction and improved tolerance of alfalfa plants towards water stress (Xu et al. 2012).

26.3.5.3 Ethylene

Ethylene is another essential plant growth hormone involved in regulating plant growth, development and survival (Bhattacharyya and Jha 2012). It plays an indispensible role in promoting fruit maturation, defoliation, flower wilt and induction of some other phytohormones (Glick et al. 2007). Additionally, ethylene synthesis increases endogenously in abiotic and biotic stress and thus it is implicated to play an essential role in stress tolerance. Ethylene biosynthesis employs 1-aminocyclopropane-1-carboxylate (ACC) as a precursor and is under tight regulation of several transcription and post-transcription factors controlled by biotic and abiotic stresses (Hardoim et al. 2008). During stress state, ethylene endogenously controls plant physiology and causes reduction in root and shoot development. Diverse PGPR genera such as *Achromobacter*, *Serratia*, *Azospirillum*, *Alcaligenes*, *Burkholderia*, *Ralstonia*, *Enterobacter*, *Acinetobacter*, *Rhizobium*, *Bacillus*, *Agrobacterium* and *Pseudomonas* synthesize an enzyme ACC deaminase that inhibits ACC by transforming it into 2-oxobutanoate and NH_3, thus restraining the deleterious impact of ethylene and augmenting plant development by allowing resistance against drought and salt (Glick et al. 2007; Nadeem et al. 2009). Iqbal et al. documented the inoculation of *Pseudomonas* sp. resulting in lowered ethylene production, thus enhancing root nodule development and straw and grain production (Iqbal et al. 2012). Actinomycetes *Microbispora* sp. and *Streptomyces* sp. were found to synthesize ACC deaminase and IAA (Glick 2014). A recent study has reported the effect of ACC deaminase on regulating ethylene content by the ACC deaminase-synthesizing *Pseudomonas fluorescens* YsS6 on assisting the nodulation by rhizobia (Nascimento et al. 2019).

26.4 Plant Growth-Promoting Rhizospheric Microbiome as Biocontrol Agents

The agricultural sector experiences severe economic loss every year owing to various plant diseases causedby over 60 pathogens resulting in major damage in plant productivity. However, the utilization of agrochemicals has improved the crop production but the agroecosystems have become unbalanced (Kremer et al. 2006). The surge for sustainable development has invigorated interest in search of novel biocontrol methods as a significant part of Integrated Pest Management (IPM) or Integrated Plant Disease Management (IPDM) for plant disease control (Sayyed and Chincholkar 2010). Since the last two decades, PGPR have been employed for their immense potential to perform both as biocidal and growth promotion agents. The biocontrol PGPR confer a resistance to plants against various plant pathogens by synthesizing several allelochemicals, e.g. volatile organic compounds, iron-complexing siderophores, lytic and detoxification enzymes, antibiotics and exopolysaccharides (Table 26.3). PGPR work as biocontrol agents by employing two vital processes which can be divided according to elicitation and regulation methods employed: (1) systemic acquired resistance and (2) induced systemic resistance.

26.4.1 Systemic Acquired Resistance (SAR)

SAR is induced or acquired through exposure of the plants to virulent and nonpathogenic bacterial strains or organic compounds such as salicylic acid, benzo-(1,2,3)-thiadiazole-7-carbothioic acid *S*-methyl ester (BTH) and 2,6-dichloro-isonicotinic acid (INA) (Sticher et al. 1997). Exposure of plants with "necrotizing" pathogens (eliciting a hypersensitive response) produces enhanced resistance towards several other pathogens due to a biological phenomenon referred to as systemic acquired resistance. Activation of group of SAR genes results in the synthesis of proteins called pathogenesis-related proteins (PR proteins). PR-1 is one of most documented PR genes and thus is generally employed as a biomarker for SAR (Van Loon et al. 2006). Several reports on the production of salicylic acid by PGPB due to activation of SAR are available in literature (Chen et al. 1999).

26.4.2 Induced Systemic Resistance (ISR)

Induced systemic resistance (ISR) is a defined as phenomena involving nonpathogenic microbes, including PGPB, which reduce the detrimental impact of pathogenic organisms by inducing resistance in the plants through production of antimicrobial agents (Van Loon et al. 1998). The presence of bacterial antagonist in the

Table 26.3 Plant growth-promoting rhizomicrobiomes as biocontrol agents

Biocontrol agent	Crop	Pathogen	Mode of action	References
Burkholderia cepacia BAM-6, *B. cepacia BAM-12* and *Pseudomonas fluorescens BAM-4*	*Vigna radiata*	–	IAA production, siderophore and chitinase production	Minaxi and Saxena (2011)
B. amyloliquefaciens LJ02	Cucurbits	Powdery mildew disease	ISR-SA mediated	Li et al. (2015)
Pantoea agglomerans strain ENA1	*Glycine max* (L.)	*Macrophomina phaseolina* Merrill	Antibiosis by pyrrolnitrin	Vasebi et al. (2015)
Bacillus anthracis	*Triticum aestivum L.*	–	Protease production	Verma et al. (2016)
Bacillus endophyticus	*Curcuma longa*	*Fusarium solani*	Chitinase production	Chauhan et al. (2016)
Bacillus subtilis RB14	*Solanum lycopersicum*	*Rhizoctonia solani*	Antimicrobial Inturin A	Zohora et al. (2016)
P. aeruginosa strain LV	*C. sinensis* cv. *Valencia*	*Xanthomonas citri* subsp. *Citri*	Antibiosis by organocopper compound	De Oliveira et al. (2016)
Aneurinibacillus aneurinilyticus CKMV1	*Solanum lycopersicum*	*Sclerotium rolfsii*, *Fusarium oxysporum*, *Dematophora necatrix*, *Rhizoctonia solani*, *Alternaria* sp. and *Phytophthora* sp.	Antifungal activity, IAA, HCN and siderophore production	Chauhan et al. (2017)
Burkholderia cepacia JBK9	*Piper nigrum*	*F. oxysporum*, *R. solani*, *P. capsici*	Pyrrolnitrin	Jung et al. (2018)
Bacillus sonorensis	*Capsicum annuum L.*	–	Chitinase production	Thilagar et al. (2018)
Bacillus amyloliquefaciens	*Solanum lycopersicum*	*Agrobacterium tumefaciens*	Antibiotics	Abdallah et al. (2018)
Bacillus aerius	*Piper nigrum*	*Phytophthora capsici*	HCN production	San Fulgencio et al. (2018)
Bacillus sp. *strain* B25	*Zea mays*	*Fusarium verticillioides*	Antibiosis by chitinases, glycoside hydrolases, siderophores and antibiotics	Douriet-Gámez et al. (2018)

rhizosphere allows the plant to develop induced systemic resistance for a wide range of plant pathogenic organisms (Lugtenberg and Kamilova 2009). The mechanism of "ISR" involves fungal cell wall lysis (Maksimov et al. 2011), root inhabitation (Kamilova et al. 2005), decreased ethylene content (Van Loon 2007) and synthesis of siderophore and antibiotics (Beneduzi et al. 2012). PGPR induce ISR in plants resulting in the activation of pathogenetic genes, mediation of phytohormone signalling mechanisms and production of regulatory proteins which protect plants from hazardous impact of pathogenic organisms (Pieterse et al. 2014).

26.4.3 Antibiosis

One of the major mechanisms employed by PGPB for countering the detrimental effects of phytopathogens is the production of antimicrobial agents (Couillerot et al. 2009; Raaijmakers and Mazzola 2012). Several different types of antimicrobial, antiviral, antihelminthic, cytotoxic, antioxidant and antitumor agents have been documented, e.g. phenazine, kanosamine, pyoluteorin, cyclic lipopeptides, karalicin, pyrrolnitrin, pseudomonic acid, cepafungins, tensin, phenazine-1-carboxylic acid (PCA), tropolone, rhamnolipids, viscosinamide, amphisin, 2,4-diacetylphloroglucinol (DAPG), azomycin, ecomycins, aerugine, hydrogen cyanide, oomycin A, cepaciamide A and butyrolactones synthesized by *Pseudomonas*, and oligomycin A, xanthobaccin, subtilosin, Tas A, bacillaene, zwittermicin A, fengycin, kanosamine, chlorotetain, sublancin, iturin, subtilin, bacilysin and surfactin synthesized by *Bacillus*, *Streptomyces* and *Stenotrophomonas* spp. (Goswami et al. 2016). Soil actinomycetes have also been found to synthesize an array of antibiotics and thus are biocidal for a wide range of pathogenic organisms (Verma et al. 2009). Antibiotic-producing genes such as *srf*, *dfn*, *fen*, *bmy*, *mln*, *nrs*, *bac*, *dhb* and *bae* have been located in *Bacillus subtilis* 168 and *Bacillus amyloliquefaciens* FZB42 and are involved in the production of peptides and polyketides through NRPSs and PKS enzymes (Chang et al. 2007).

26.4.4 Production of Enzymes

PGPB also enhance plant development by synthesizing various metabolites, contributing to the antibiosis and antifungal properties. PGPR synthesize various fungal cell wall hydrolytic enzymes including chitinase (Husson et al. 2017), β-1,3-glucanase (Vaddepalli et al. 2017), protease and lipase (Friedrich et al. 2012). The cell wall of fungi is generally composed of chitin and β-glucan; therefore, chitinases and β-glucanase-synthesizing microbes can assist in inhibiting fungal growth. *Serratia plymuthica* synthesizes the enzyme chitinase which results in the inhibition of spore germination and germ tube expansion in *Botrytis cinerea* (Frankowski et al. 2001). *Serratia marcescens* also works as a biocontrol agent against *Sclerotium rolfsii* by producing extracellular chitinases (Ordentlich et al. 1988). Lim et al. reported the production of extracellular enzymes such as laminarinase and chitinase by *Pseudomonas stutzeri* capable of degrading *Fusarium solani* mycelia (Lim et al. 1991). *Pseudomonas* spp. are capable of inhibiting two of the most detrimental crop pathogens *Rhizoctonia solani* and *Phytophthora capsici* by releasing extracellular enzymes chitinase and beta-glucanases. The β-1,3-glucanase produced by *Burkholderia cepacia* is capable of degrading the cell wall of several pathogenic microbes, e.g. *Rhizoctonia solani*, *Pythium ultimum* and *Sclerotium rolfsii.* The synthesis of these hydrolytic enzymes is under tight regulation of several regulatory systems such as GrrA/GrrS or GacA/GacS (Ovadis et al. 2004). *Enterobacter asburiae* BQ9 mediates tolerance against tomato yellow leaf curl virus by activation of defence-related genes and production of phenylalanine, ammonia lyase, superoxide dismutase, peroxidase and catalase (Yan et al. 2018). The bacteria also synthesize ACC deaminase and contribute to tolerance in tomatoes caused by *Scelerotium rolfsii* to southern blight disease. The inoculated plants demonstrated modulated ethylene metabolism and antioxidant enzyme activity; the pathogen-related gene expression analysis corroborated systemic tolerance (Dixit et al. 2016).

26.4.5 HCN Production

Cyanide is an extremely hazardous compound well known for its toxicity and thus can be employed for inhibition of pathogenic organisms damaging agricultural crops. HCN is a secondary metabolite synthesized by several PGPB, is generally employed for weed and pest control and simultaneously does not impose any harmful impact on the host plants (Aarab et al. 2019). HCN toxicity is mediated by inhibition of cytochrome C oxidase along with other vital metalloenzymes (Nandi et al. 2017). Several PGPR such as *Rhizobium*, *Aeromonas*, *Alcaligenes*, *Pseudomonas* and *Bacillus* are reported to produce HCN and assist in biocontrol of dreadful pathogenic strains including *Pythium ultimum*, *Fusarium oxysporum* and *Agrobacterium* (Ahmad et al. 2008; Das et al. 2017; Zachow et al. 2017). Elimination of *Meloidogyne javanica* and *Odontotermes obesus* crop pests in India has been conferred due to HCN production (Kumar et al. 2015; Siddiqui et al. 2006). HCN production by PGPB is effective for elimination of crop pests and is also vital for metal-chelate complex formation necessary for biogeochemical cycling (Kumari et al. 2018).

26.4.6 Production of Volatile Organic Compounds (VOCs)

The production of VOCs by various microorganisms in the rhizospheric zone is an additional method of influencing plant development indirectly. *Arthrobacter agilis*,

Azospirillum brasilense, *Chromobacterium violaceum*, *Bacillus* spp., *Pseudomonas fluorescens* and *Burkholderia pyrrocinia* are documented for synthesizing several kinds of VOCs (Yadav and Yadav 2018; Santoro et al. 2011) such as 2-pentylfuran, tetradecane, jasmonates, 2,3-butanediol, acetonin, methyl decane, 1-chlorooctadecane and cyclohexanes (Zou et al. 2010; Ryu et al. 2003; Kanchiswamy et al. 2015). VOCs released in the rhizosphere function as a signalling molecule for triggering the resistance mechanisms in plant towards plant pathogens and induce systemic resistance (ISR) (Santoro et al. 2011). Cotton (*Gossypium hirsutum*) plants on inoculation with Bacillus spp. were reported to secrete gossypol and jasmonic acid, thus lowering larval feeding by *Spodoptera exigua* (Zebelo et al. 2016). Higher transcript level of genes responsible for synthesizing alleochemicals and jasmonates was reported in comparison to pest control in PGPR-inoculated plants (Zebelo et al. 2016). Khan et al. studied the efficacy of endophytic fungal strain *Penicillium janthinellum* LK5 (PjLK5) for production of defence-related endogenous phytohormone salicylic acid to counter the deleterious impact of metal stress in *Solanum lycopersicum* (Khan et al. 2015). Several VOCs such as dimethyl disulphide, acetaldehyde, β-caryophyllene, 2,3-butanediol, methanol, dimethylhexadecylamine, geosmin, furfural, camphor, 5-hydroxy-methylfurfural, camphene, 2-methyl isoborneol, propanoic acid, 1-octen-3-ol, butanoic acid and α-pinene are secreted by PGPR species including *Agrobacterium*, *Serratia*, *Burkholderia*, *Arthrobacter*, *Paenibacillus*, *Bacillus*, *Rahnella*, *Enterobacter*, *Pseudomonas*, *Collimonas* and *Pedobacter*, thus causing inhibition of various fungal pathogens and resulting in improved soil fertility and health (Kanchiswamy et al. 2015; Chenniappan et al. 2019). These VOCs have significant applications for biocontrolling plant pathogens, stimulating of immune system in plants and modulating root expansion (Kanchiswamy et al. 2015; Rojas-Solís et al. 2018).

26.4.7 Production of Antifungal Agents

Several PGPB are capable of synthesizing a wide range of antifungal compounds including phenazines, pyoluteorin, 2, 4-diacetylphloroglucinol (DAPG), HCN, viscosinamide, tensin and pyrrolnitrin (Bhattacharyya and Jha 2012). Majority of identified *Pseudomonas* biocontrol strains synthesizes antifungal metabolites (phenazine, DAPG, pyoluterin, tensin, pyrrolnitrin and viscosinamide) able to inhibit growth of phytopathogenic fungi (Bloemberg and Lugtenberg 2001; Thrane et al. 2000). A pigment prodigiosin reported from rhizospheric microbe *Serratia* sp. possesses the potential to act as an antifungal agent (John et al. 2017). John and Radhakrishnan have also documented the synthesis of an antifungal compound pyrrolnitrin by *Serratia plymuthica* isolated from *Curcuma amada* (John and Radhakrishnan 2018). Pyrrolnitrin was found to provide tolerance against soft rot due to *Pythium myriotylum* to ginger rhizome (John and Radhakrishnan 2018).

26.5 Rhizoremediation

Bioremediation is defined as implementation of biological agents/processes for the removal of toxic xenobiotics in the contaminated environment and emphasizes on combining phytoremediation and bioaugmentation to result into rhizoremediation (Bhatt 2019; Bhatt et al. 2019a, b; Negi et al. 2014). Rhizoremediation relies on the application of root exudates derived from the plants for stimulation, survival and degradation activities of soil microbes, which ultimately allow efficient degradation of toxic pollutants (Bhatt et al. 2020a). Several PGPB have been reported to be vital in the geochemical recycling of minerals, thus causing cleanup of the contaminated ecosystem (Gangola et al. 2018; Bhatt et al. 2020b, 2019c). Many PGPR strains possessing the ability to degrade a wide range of pollutants are well documented and generally belong to genera such as *Pseudomonas*, *Flavobacterium*, *Ralstonia*, *Bacillus*, *Enterobacter*, *Corynebacterium*, *Achromobacter*, *Agrobacterium*, *Rhodococcus*, *Sphingomonas* and *Azospirillum* (Bhatt et al. 2020c, d, e, f; Gangola et al. 2016) (Table 26.4). Rylott et al. have described the conversion of

Table 26.4 Bioremediation of xenobiotics by plant growth-promoting rhizomicrobiomes

PGPR	Pollutant	References
Pseudomonas putida (PML2)	Polychlorinated biphenyls	Narasimhan et al. (2003)
Azospirillum lipoferum strains 15	Crude oil	Muratova et al. (2005)
*Achromobacter xylosoxidans*_Ax10	Copper	Ma et al. (2009b)
Pseudomonas sp. *SRA 2, SRA 1, B. cereus SRA 10, B. juncea, Brassica oxyrrhina*	Nickel	Ma et al. (2009a)
Serratia sp. *SY5*	Cadmium and copper	Koo and Cho (2009)
Bacillus sp. *PSB10*	Chromium	Wani and Khan (2010)
Bacillus pumilus	Chlorpyrifos	Ahmad et al. (2012)
Pseudomonas fluorescens	Polychlorinated biphenyls	Toussaint et al. (2012)
Burkholderia metalliresistens	Cadmium, copper, lead, zinc	Guo et al. (2015)
Chryseobacterium sp. *PYR2*	Organochlorine pesticides: hexachlorocyclohexane (HCH)	Qu et al. (2015)
Brevundimonas diminuta	Arsenic	Singh et al. (2016)
Pseudomonas putida	Tannery effluent: lead and chromium	Nokman et al. (2019)
Bacillus megaterium	Palladium	Chen et al. (2019)

2,4,6-trinitrotoluene (TNT) by *Enterobacter cloacae* PB2 and *Pseudomonas fluorescens* IC (Rylott et al. 2011). *Ceratobasidium stevensii*, an endophyte of members of Euphorbiaceae, is capable of metabolizing 89.51% of phenanthrene (Dai et al. 2010). PGPR such as *Azospirillum lipoferum*, *P. fluorescens*, *Enterobacter cloacae* and *Pseudomonas putida* possess the ability to degrade several petroleum hydrocarbons, PAHs and trichloroethylene (Glick 2012). *Pseudomonas putida* KT2440, a PGPR, has the potential of degrading naphthalene (Fernandez et al. 2012). Uhlik et al. have analysed the activated modification of root exudates in the rhizomicrobiome composition and its bioremediation ability in polychlorinated biphenyl (PCB)-contaminated soil (Uhlik et al. 2013). Bisht et al. have reported several PGPR, *Kurthia* sp., *Bacillus circulans*, *Micrococcus varians* and *Deinococcus radiodurans*, obtained from the rhizospheric soil of *Populus deltoides* capable of degrading naphthalene and anthracene (Bisht et al. 2010). *Acinetobacter calcoaceticus* P23, obtained from the rhizospheric zone of duckweeds, has the potential to degrade phenol (Yamaga et al. 2010). Yang et al. have investigated the effect of inoculation of VAM in legumes, *Robinia pseudoacacia*, for the remediation of lead-contaminated environments (Yang et al. 2016). PGPR possessing metal tolerance ability are capable of immobilizing various toxic heavy metals, thus reducing their detrimental impact on plants (Kong and Glick 2017). Prapagdee et al. have documented cadmium-tolerant microbial strains: *Micrococcus* sp. MU1 and *Klebsiella* sp. BAM1 capable of enhancing cadmium mobilization and plant growth promotion (Prapagdee et al. 2013). Jing et al. have obtained metal-tolerant strains of *Enterobacter* sp. and *Klebsiella* sp. from the rhizosphere of *Polygonum pubescens* cultivated in metal-contaminated soil (Jing et al. 2014). On inoculating these strains in *Brassica napus*, hyperaccumulation of heavy metals (Zn, Pb and Cd) was observed along with improved plant growth. Hassen et al. reported a biosurfactant-producing PGPR strain *Pseudomonas rhizophila* S211, capable of biofertilization, biocontrol and bioremediation (Hassen et al. 2018).

26.6 Conclusions and Perspectives

Unravelling the prospective PGPR mechanisms of growth promotion and relationship with plants is indispensible for augmenting plant growth and production. PGPR possess enormous potential to contribute towards sustainable agricultural practices with minimal harm to the environment. Microbial metabolites affect the plant physiology in complex manner, governing plant development, nutrition and resistance to biotic and abiotic challenges consequently. Plant-microbe relationships have played a significant impact in development of several biofertilizer, biocontrol and bioremediation agents. Despite this, there exists a vast difference in progression from the in vitro conditions to field due to variable environmental conditions, deficiency in microbial inhabitation and constraints in persistence in the rhizosphere, thus reducing the possibilities of implication of PGPR for development of sustainable agricultural practices. PGPR research should shift focus towards understanding the genetic mechanisms regulating growth promontory processes. Concomitant investigations regarding genetic modification of the plant, chemical genomics strategy, rhizospheric engineering and colonizing with large subpopulations of rhizomicrobiomes may assist in overcoming these constraints and thus can be significant for evaluating vital microbial molecular components regulating plant development and executing PGPR in the field. By recognizing and interpreting the particular processes of plant-microbe interaction, safer, more efficient and sustainable PGPR applications may be designed for sustainable agricultural practices.

References

Aarab S, Ollero J, Megías M, Laglaoui A, Bakkali M, Arakrak A (2019) Some characteristics of phosphate solubilizing rhizobacteria as an ecological strategy for sustainable agriculture. Mater Today Proc 13:1224–1228. https://doi.org/10.1016/j.matpr.2019.04.091

Abdallah DB, Frikha-Gargouri O, Tounsi S (2018) Rizhospheric competence, plant growth promotion and biocontrol efficacy of *Bacillus amyloliquefaciens* subsp. plantarum strain 32a. Biol Control 124:6167. https://doi.org/10.1016/j.biocontrol.2018.01.013

Ahemad M, Khan MS (2011) Effect of tebuconazole-tolerant and plant growth promoting *Rhizobium* isolate MRP1 on pea-Rhizobium symbiosis. Sci Hortic 129:266–272

Ahemad M, Khan MS (2012) Evaluation of plant-growth promoting activities of rhizobacterium *Pseudomonas putida* under herbicide stress. Ann Microbiol 62:1531–1540

Ahemad M, Kibret M (2014) Mechanisms and applications of plant growth promoting rhizobacteria: current perspective. J King Saud Univ Sci 26:1–20

Ahmad F, Ahmad I, Khan MS (2008) Screening of free-living rhizospheric bacteria for their multiple plant growth promoting activities. Microbiol Res 163(2):173–181. https://doi.org/10.1016/j.micres.2006.04.001

Ahmad F, Iqbal S, Anwar S, Afzal M, Islam E, Mustafa T, Khan QM (2012) Enhanced remediation of chlorpyrifos from soil using ryegrass (*Lollium multiflorum*) and chlorpyrifos-degrading bacterium *Bacillus pumilus* C2A1. J Hazard Mater 237:110–115. https://doi.org/10.1016/j.jhazmat.2012.08.006

Ahmed T, Shahid M, Noman M, Hussain S, Khan MA, Zubair M, Mahmood F (2019) Plant growth-promoting rhizobacteria as biological tools for nutrient management and soil sustainability. In: Plant growth promoting rhizobacteria for agricultural sustainability. Springer, Singapore, pp 95–110. https://doi.org/10.1007/978-981-13-7553-8_5

Alori ET, Glick BR, Babalola OO (2017) Microbial phosphorus solubilization and its potential for use in sustainable agriculture. Front Microbiol 8:971

Anand K, Kumari B, Mallick MA (2016) Phosphate solubilizing microbes: an effective and alternative approach as bio-fertilizers. Int J Pharm Sci 8:37–40

Arp DJ (2000) The nitrogen cycle. In: Triplett EW (ed) Prokaryotic nitrogen fixation. Horizon Scientific Press, Wymondham, pp 1–14

Ashley MK, Grant M, Grabov A (2006) Plant responses to potassium deficiencies: a role for potassium transport proteins. J Exp Bot 57(2):425–436. https://doi.org/10.1093/jxb/erj034

Babalola OO (2010) Beneficial bacteria of agricultural importance. Biotechnol Lett 32:1559–1570

Backer R, Stefan Rokem J, Ilangumaran G, Lamont J, Praslickova D, Ricci E, Subramanian S, Smith DL (2018) Plant growth-promoting rhizobacteria: context, mechanisms of action, and roadmap to commercialization of biostimulants for sustainable agriculture. Front Plant Sci 9:1473

Bal HB, Das S, Dangar TK, Adhya TK (2013) ACC deaminase andIAAproducing growth promoting bacteria from the rhizosphere soil of tropical rice plants. J Basic Microbiol 53(12):972–984

Beneduzi A, Ambrosini A, Passaglia LMP (2012) Plant growth-promoting rhizobacteria (PGPR): their potential as antagonists and biocontrol agents. Genet Mol Biol 35(4):1044–1051

Bhatt P (2019) Smart bioremediation technologies: microbial enzymes. Elsevier Science, Amsterdam

Bhatt P, Pal K, Bhandari G, Barh A (2019a) Modelling of the methyl halide biodegradation in bacteria and its effect on environmental systems. Pestic Biochem Physiol 158:88–100

Bhatt P, Huang Y, Zhan H, Chen S (2019b) Insight into microbial applications for the biodegradation of pyrethroid insecticides. Front Microbiol 10:1778. https://doi.org/10.3389/fmicb.2019.01778

Bhatt P, Gangola S, Chaudhary P, Khati P, Kumar G, Sharma A (2019c) Pesticide induced up-regulation of esterase and aldehyde dehydrogenase in indigenous *Bacillus* spp. Bioremediat J 23(1):42–52. https://doi.org/10.1080/10889868.2019.1569586

Bhatt P, Gangola S, Bhandari G, Zhang W, Maithani D, Mishra S, Chen S (2020a) New insights into the degradation of synthetic pollutants in contaminated environments. Chemosphere. https://doi.org/10.1016/j.chemosphere.2020.128827

Bhatt P, Sethi K, Gangola S, Bhandari G, Verma A, Adnan M, Singh Y, Chaube S (2020b) Modeling and simulation of atrazine biodegradation in bacteria and its effect in other living systems. J Biomol Struct Dyn. https://doi.org/10.1016/j.chemosphere.2020.128827

Bhatt P, Huang Y, Zhang W, Sharma A, Chen S (2020c) Enhanced cypermethrin degradation kinetics and metabolic pathway in *Bacillus thuringiensis* Strain SG4. Microorganisms 8(2):223. https://doi.org/10.3390/microorganisms8020223

Bhatt P, Huang Y, Rene ER, Kumar AJ, Chen S (2020d) Mechanism of allethrin biodegradation by a newly isolated *Sphingomonas trueperi* strain CW3 from wastewater sludge. Bioresour Technol 305:123074. https://doi.org/10.1016/j.biortech.2020.123074

Bhatt P, Rene ER, Kumar AJ, Zhang W, Chen S (2020e) Binding interaction of allethrin with esterase: bioremediation potential and mechanism. Bioresour Technol 315:123845. https://doi.org/10.1016/j.biortech.2020.123845

Bhatt P, Bhatt K, Huang Y, Lin Z, Chen S (2020f) Esterase is a powerful tool for the biodegradation of pyrethroid insecticides. Chemosphere 244:125507. https://doi.org/10.1016/j.chemospheree.2019.125507

Bhatt P, Verma A, Verma S, Anwar MS, Prasher P, Mudila H, Chen S (2020g) Understanding phytomicrobiome: a potential reservoir for better crop management. Sustainability 12(13):5446. https://doi.org/10.3390/su12135446

Bhattacharyya PN, Jha DK (2012) Plant growth-promoting rhizobacteria (PGPR): emergence in agriculture. World J Microbiol Biotechnol 28:1327–1350

Bhattacharyya D, Garladinne M, Lee Y (2015) Volatile indole produced by rhizobacterium *Proteus vulgaris* JBLS202 stimulates growth of Arabidopsis thaliana through auxin, cytokinin, and brassinosteroid pathways. J Plant Growth Regul 34:158–168

Bisht S, Pandey P, Sood A, Sharma S, Bisht NS (2010) Biodegradation of naphthalene and anthracene by chemo-tactically active rhizobacteria of *Populus deltoides*. Braz J Microbiol 41 (4):922–930

Biswas JK, Banerjee A, Rai M, Naidu R, Biswas B, Vithanage M, Meers E (2018) Potential application of selected metal resistant phosphate solubilizing bacteria isolated from the gut of earthworm (*Metaphire posthuma*) in plant growth promotion. Geoderma 330:117–124. https://doi.org/10.1016/j.geoderma.2018.05.034

Bloemberg GV, Lugtenberg BJJ (2001) Molecular basis of plant growth promotion and biocontrol by rhizobacteria. Curr Opin Plant Biol. 4(4):343–350

Bottini R, Cassán F, Piccoli P (2004) Gibberellin production by bacteria and its involvement in plant growth promotion and yield increase. Appl Microbiol Biotechnol 65:497–503

Boukhalfa H, Crumbliss AL (2002) Chemical aspects of siderophore mediated iron transport. Biometals 15:325–339

Butterly CR, Bunemann EK, McNeill AM, Baldock JA, Marschner P (2009) Carbon pulses but not phosphorus pulses are related to decrease in microbial biomass during repeated drying and rewetting of soils. Soil Biol Biochem 41:1406–1416

Cassan F, Maiale S, Masciarellia O, Vidal A, Luna V, Ruiz O (2009) Cadaverine production by Azospirillum brasilense and its possible role in plant growth promotion and osmotic stress mitigation. Eur J Soil Biol 45:12–19

Chang WT, Chen YC, Jao CL (2007) Antifungal activity and enhancement of plant growth by *Bacillus cereus* grown on shellfish chitin wastes. Bioresour Technol 98:1224–1230

Chaparro JM, Badri DV, Vivanco JM (2014) Rhizosphere microbiome assemblage is affected by plant development. ISME J 8(4):790

Chauhan AK, Maheshwari DK, Kim K, Bajpai VK (2016) Termitarium-inhabiting *Bacillus endophyticus* TSH42 and *Bacillus cereus* TSH77 colonizing *Curcuma longa* L.: isolation, characterization, and evaluation of their biocontrol and plant-growth-promoting activities. Can J Microbiol 62(10):880–892. https://doi.org/10.1139/cjm-2016-0249

Chauhan A, Guleria S, Balgir PB, Walia A, Mahajan R, Mehta P, Shirkot CK (2017) Tricalcium phosphate solubilization and nitrogen fixation by newly isolated *Aneurinibacillus aneurinilyticus* CKMV1 from rhizosphere of *Valeriana jatamansi* and its growth promotional effect. Braz J Microbiol 48(2):294–304

Chen C, Bélanger RR, Benhamou N, Paulitz TC (1999) Role of salicylic acid in systemic resistance induced by *Pseudomonas* spp. against *Pythium aphanidermatum* in cucumber roots. Eur J Plant Pathol 105:477–486

Chen Y, Hu K, Chen Y (2019) The effect of biotic and abiotic environmental factors on Pd (II) adsorption and reduction by *Bacillus megaterium* Y-4. Chemosphere 220:1058–1066. https://doi.org/10.1016/j.chemosphere.2019.01.011

Chenniappan C, Narayanasamy M, Daniel GM, Ramaraj GB, Ponnusamy P, Sekar J, Ramalingam PV (2019) Biocontrol efficiency of native plant growth promoting rhizobacteria against rhizome rot disease of turmeric. Biol Control 129:55–64. https://doi.org/10.1016/j.biocontrol.2018.07.002

Compant S, Samad A, Faist H, Sessitsch A (2019) A review on the plant microbiome: ecology, functions, and emerging trends in microbial application. J Adv Res 19:29–37. https://doi.org/10.1016/j.jare.2019.03.004

Couillerot O, Prigent-Combaret C, Caballero-Mellado J, Moënne-Loccoz Y (2009) *Pseudomonas fluorescens* and closely-related fluorescent pseudomonads as biocontrol agents of soil-borne phytopathogens. Lett Appl Microbiol 48:505–512

Dai CC, Tian LS, Zhao YT, Chen Y, Xie H (2010) Degradation of phenanthrene by endophytic fungus *Ceratobasidium stevensii* found in *Bischofia polycarpa*. Biodegradation 21(2):245–255

Damam M, Kaloori K, Gaddam B, Kausar R (2016) Plant growth promoting substances (phytohormones) produced by rhizobacterial strains isolated from the rhizosphere of medicinal plants. Int J Pharm Sci Rev 37:130–136

Das K, Prasanna R, Saxena AK (2017) Rhizobia: a potential biocontrol agent for soilborne fungal pathogens. Folia Microbiol 62:425–435. https://doi.org/10.1007/s12223-017-0513-z

De Oliveira AG, Spago FR, Simionato AS, Navarro MOP, da Silva CS, Barazetti AR, Cely MVT, Tischer CA, San Martin JAB, de Jesus Andrade CGT, Novello CR, Mello JCP, Andarde G (2016) Bioactive organocopper compound from *Pseudomonas aeruginosa* inhibits the growth of *Xanthomonas citri* subsp. citri. Front Microbiol 7:1–12. https://doi.org/10.3389/fmicb.2016.00113

Deka H, Deka S, Baruah C (2015) Plant growth promoting rhizobacteria for value addition: mechanism of action. In: Plant-growth promoting rhizobacteria (PGPR) and medicinal plants. Springer, New York, pp 305–321

Dinnage R, Simonsen AK, Barrett LG, Cardillo M, Raisbeck-Brown N, Thrall PH, Prober SM (2019) Larger plants promote a greater diversity of symbiotic nitrogen-fixing soil bacteria associated with an Australian endemic legume. J Ecol 107(2):977–991. https://doi.org/10.1111/1365-2745.13083

Dixit R, Agrawal L, Gupta S, Kumar M, Yadav S, Chauhan PS, Nautiyal SC (2016) Southern blight disease of tomato control by 1-aminocyclopropane-1-carboxylate (ACC) deaminase producing Paenibacillus lentimorbus B-30488. Plant Signal Behav 11(2):e1113363

Dobert RC, Rood SB, Blevins DG (1992) Gibberellins and the legume-Rhizobium symbiosis. I. Endogenous gibberellins of lima bean (*Phaseolus lunatus* L.) stems and nodules. Plant Physiol 98:221–224

Douriet-Gámez NR, Maldonado-Mendoza IE, Ibarra-Laclette E et al (2018) Genomic analysis of *Bacillus* sp. Strain B25, a biocontrol agent of maize pathogen *Fusarium verticillioides*. Curr Microbiol 75(3):247–255. https://doi.org/10.1007/s00284-017-1372-1

Egamberdieva D, Lugtenberg B (2014) Use of plant growth-promoting rhizobacteria to alleviate salinity stress in plants. In: Miransari M (ed) Use of microbes for the alleviation of soil stresses. Springer, New York, pp 73–96

Etesami HA, Alikhani HA, Akbari A (2009) Evaluation of plant growth hormones production (IAA) ability by Iranian soils rhizobial strains and effects of superior strains application on wheat growth indexes. World Appl Sci J 6:1576–1584

Fernandez M, Niqui-Arroyo JL, Conde S, Ramos JL, Duque E (2012) Enhanced tolerance to naphthalene and enhanced rhizoremediation performance of *Pseudomonas putida* KT2440 via the NAH7 catabolic plasmid. Appl Environ Microbiol 78(15):5104–5110

Fierro-Coronado RA, Quiroz-Figueroa FR, García-Pérez LM, Ramírez-Chávez E, Molina-Torres J, Maldonado-Mendoza IE (2014) IAA-producing rhizobacteria from chickpea (*Cicer arietinum* L.) induce changes in root architecture and increase root biomass. Can J Microbiol 60 (10):639–648

Frankowski J, Lorito M, Scala F, Schmid R, Berg G, Bahl H (2001) Purification and properties of two chitinolytic enzymes of Serratia plymuthica HRO-C48. Arch Microbiol 176(6):421–426. https://doi.org/10.1007/s002030100347

Friedrich N, Hagedorn M, Soldati-Favre D, Soldati T (2012) Prison break: pathogens' strategies to egress from host cells. Microbiol Mol Biol Rev 76:707–720

Fuentes-Ramírez LE, Caballero-Mellado J (2006) Bacterial biofertilizers. In: Siddiqui ZA (ed) PGPR: biocontrol and biofertilization. Springer, Netherlands, pp 143–172

Fukami J, Cerezini P, Hungria M (2018) *Azospirillum*: benefits that go far beyond biological nitrogen fixation. AMB Express 8(1):73. https://doi.org/10.1186/s13568-018-0608-1

Gangola S, Negi G, Srivastava A, Sharma A (2016) Enhanced biodegradation of endosulfan by *Aspergillus* and *Trichoderma* spp. isolated from an agricultural field of tarai region of Uttarakhand. Pest Res J 27:223–230

Gangola S, Sharma A, Bhatt P, Khati P, Chaudhary P (2018) Presence of esterase and laccase in *Bacillus subtilis* facilitates biodegradation and detoxification of cypermethrin. Sci Rep 8 (1):12755. https://doi.org/10.1038/s41598-018-31082-5

Glick BR (2012) Plant growth-promoting bacteria: mechanisms and applications. Scientifica 2012:963401. https://doi.org/10.6064/2012/963401

Glick BR (2014) Bacteria with ACC deaminase can promote plant growth and help to feed the world. Microbiol Res 169:30–39

Glick BR, Todorovic B, Czarny J, Cheng Z, Duan J, McConkey B (2007) Promotion of plant growth by bacterial ACC deaminase. Crit Rev Plant Sci 26:227–242

Goswami D, Thakker JN, Dhandhukia PC, Tejada Moral M (2016) Portraying mechanics of plant growth promoting rhizobacteria (PGPR): a review. Cogent Food Agric 2:1127500

Grobelak A, Napora A, Kacprzak M (2015) Using plant growth-promoting rhizobacteria (PGPR) to improve plant growth. Ecol Eng 84:22–28

Großkinsky DK, Tafner R, Moreno MV, Stenglein SA, De Salamone IEG, Nelson LM, Roitsch T (2016) Cytokinin production by *Pseudomonas fluorescens* G20-18 determines biocontrol activity against *Pseudomonas syringae* in *Arabidopsis*. Sci Rep 6:23310. https://doi.org/10.1038/srep23310

Guo JK, Ding YZ, Feng RW, Wang RG, Xu YM, Chen C, Chen WM (2015) *Burkholderia metalliresistens* sp. nov., a multiple metal-resistant and phosphate-solubilising species isolated from heavy metal-polluted soil in Southeast China. Antonie Leeuwenhoek 107(6):1591–1598. https://doi.org/10.1007/s10482-015-0453-z

Gupta G, Parihar SS, Ahirwar NK, Snehi SK, Singh V (2015) Plant growth promoting rhizobacteria (PGPR): current and future prospects for development of sustainable agriculture. J Microb Biochem Technol 7(2):96–102. https://doi.org/10.4172/1948-5948.1000188

Hardoim PR, Van Overbeek LS, Elsas JD (2008) Properties of bacterial endophytes and their proposed role in plant growth. Trends Microbiol 16:463–471

Hashem A, Abd Allah EF, Alqarawi A, Al-Huqail AA, Wirth S, Egamberdieva D (2016) The interaction between arbuscular mycorrhizal fungi and endophytic bacteria enhances plant growth of Acacia gerrardii under salt stress. Front Plant Sci 7:1089

Hassen W, Nefiar M, Cherif H, Najjari A, Chouchane H, Drioucih RC, Salah A, Naili F, Mosbah A, Souissi Y, Raddadi N, Ouzari HL, Fava F, Cherif A (2018) *Pseudomonas rhizophila* S211, a new plant growth-promoting rhizobacterium with potential in pesticide-bioremediation. Front Microbiol 9:34. https://doi.org/10.3389/fmicb.2018.00034

Hidri R, Barea JM, Mahmoud OM, Abdelly C, Azcón R (2016) Impact of microbial inoculation on biomass accumulation by Sulla carnosa provenances, and in regulating nutrition, physiological and antioxidant activities of this species under non-saline and saline conditions. J Plant Physiol 201:28–41

Hossain MT, Khan A, Harun-Or-Rashid M, Chung YR (2019) A volatile producing endophytic *Bacillus siamensis* YC7012 promotes root development independent on auxin or ethylene/ jasmonic acid pathway. Plant Soil 439(1–2):309–324. https://doi.org/10.1007/s11104-019-04015-y

Howard JB, Rees DC (1996) Structural basis of biological nitrogen fixation. Chem Rev 96:2965–2982

Huang XF, Chaparro JM, Reardon KF, Zhang R, Shen Q, Vivanco JM (2014) Rhizosphere interactions: root exudates, microbes, and microbial communities. Botany 92(4):267–275

Husson E, Hadad C, Huet G, Laclef S, Lesur D, Lambertyn V, Jamali A, Gottis S, Sarazin C, Nhien ANV (2017) The effect of room temperature ionic liquids on the selective biocatalytic hydrolysis of chitin via sequential or simultaneous strategies. Green Chem 19:4122–4131. https://doi.org/10.1039/C7GC01471F

Ibal JC, Jung BK, Park CE, Shin JH (2018) Plant growth-promoting rhizobacteria used in South Korea. Appl Biol Chem 61(6):709–716. https://doi.org/10.1007/s13765-018-0406-0

Imada EL, de Oliveira ALM, Hungria M, Rodrigues EP (2017) Indole-3-acetic acid production via the indole-3-pyruvate pathway by plant growth promoter *Rhizobium tropici* CIAT 899 is strongly inhibited by ammonium. Res Microbiol 168(3):283–292. https://doi.org/10.1016/j.resmic.2016.10.010

Iqbal MA, Khalid M, Shahzad SM, Ahmad M, Soleman N (2012) Integrated use of *Rhizobium leguminosarum*, plant growth promoting rhizobacteria and enriched compost for improving growth, nodulation and yield of lentil (*Lens culinaris* Medik). Chilean J Agric Res 72:104–110

Jacoby R, Peukert M, Succurro A, Koprivova A, Kopriva S (2017) The role of soil microorganisms in plant mineral nutrition – current knowledge and future directions. Front Plant Sci 8 (1617):1–17

Jahanian A, Chaichi MR, Rezaei K, Rezayazdi K, Khavazi K (2012) The effect of plant growth promoting rhizobacteria (PGPR) on germination and primary growth of artichoke (*Cynara scolymus*). Int J Agric Crop Sci 4:923–929

Jimtha JC, Jishma P, Arathy GB, Anisha C, Radhakrishnan EK (2016) Identification of plant growth promoting Rhizosphere Bacillus sp. WG4 antagonistic to Pythium myriotylum and its enhanced antifungal effect in association with Trichoderma. J Soil Sci Plant Nutr 16(3):578–590

Jing Y, Yan J, He H, Yang D, Xiao L (2014) Characterization of bacteria in the rhizosphere soils of *Polygonum pubescens* and their potential in promoting growth and Cd, Pb, Zn uptake by *Brassica napus*. Int J Phytoremediat 16:321–333

John CJ, Radhakrishnan EK (2018) Chemicobiological insight into anti-phytopathogenic properties of rhizospheric serratia plymuthica R51. Proc Natl Acad Sci India Sect B Biol Sci 88 (4):1629–1635

John CJ, Jishma P, Sreelekha S, Chithra S, Radhakrishnan EK (2017) Antifungal properties of prodigiosin producing rhizospheric Serratia sp. Rhizosphere. 3:105–108

Jung BK, Hong SJ, Park GS, Kim MC, Shin JH (2018) Isolation of *Burkholderia cepacia* JBK9 with plant growth-promoting activity while producing pyrrolnitrin antagonistic to plant fungal diseases. Appl Biol Chem 61(2):173–180. https://doi.org/10.1007/s13765-018-0345-9

Kamilova F, Validov S, Azarova T, Mulders I, Lugtenberg B (2005) Enrichment for enhanced competitive plant root tip colonizers selects for a new class of biocontrol bacteria. Environ Microbiol 7:1809–1817

Kanchiswamy CN, Malnoy M, Maffei ME (2015) Chemical diversity of microbial volatiles and their potential for plant growth and productivity. Front Plant Sci 6:151. https://doi.org/10.3389/fpls.2015.00151

Kang SM, Radhakrishnan R, Khan AL, Kim MJ, Park JM, Kim BR, Lee IJ (2014) Gibberellin secreting rhizobacterium, *Pseudomonas putida* H-2-3 modulates the hormonal and stress physiology of soybean to improve the plant growth under saline and drought conditions. Plant Physiol Biochem 84:115–124. https://doi.org/10.1016/j.plaphy.2014.09.001

Khan AL, Waqas M, Hussain J, Al-Harrasi A, Hamayun M, Lee IJ (2015) Phytohormones enabled endophytic fungal symbiosis improve aluminum phytoextraction in tolerant *Solanum lycopersicum*: an examples of *Penicillium janthinellum* LK5 and comparison with exogenous GA 3. J Hazard Mater 295:70–78

Kloepper JW, Schroth MN (1981) Relationship of in vitro antibiosis of plant growth promoting rhizobacteria to plant growth and the displacement of root microflora. Phytopathology 71:1020–1024

Kloepper JW, Leong J, Teintze M, Schroth MN (1980) *Pseudomonas* siderophores: a mechanism explaining disease-suppressive soils. Curr Microbiol 4:317–320

Kong Z, Glick BR (2017) The role of plant growth-promoting bacteria in metal phytoremediation. Adv Microbial Physiol 71:97–132

Koo SY, Cho KS (2009) Isolation and characterization of a plant growth-promoting rhizobacterium, Serratia sp. SY5. J Microbiol Biotechnol 19:1431–1438

Kremer RJ, Caesar AJ, Souissi T (2006) Soilborne microorganisms of Euphorbia are potential biological control agents of the invasive weed leafy spurge. Appl Soil Ecol 32(1):27–37

Kucey RMN (1988) Plant growth-altering effects of *Azospirillum brasilense* and *Bacillus* C-11-25 on two wheat cultivars. J Appl Bacteriol 64:187–196

Kumar P, Dubey RC (2012) Plant growth promoting rhizobacteria for biocontrol of phytopathogens and yield enhancement of *Phaseolus vulgaris*. J Curr Perspect Appl Microbiol 1:6–38

Kumar A, Bahadur I, Maurya B, Raghuwanshi R, Meena V, Singh D, Dixit J (2015) Does a plant growth promoting rhizobacteria enhance agricultural sustainability. J Pure Appl Microbiol 9:715–724

Kumari P, Meena M, Gupta P, Dubey MK, Nath G, Upadhyay RS (2018) Plant growth promoting rhizobacteria and their biopriming for growth promotion in mung bean (*Vigna radiata* (L.) R. Wilczek). Biocatal Agric Biotechnol 16:163–171. https://doi.org/10.1016/j.bcab.2018.07.030

Laschat S, Bilitewski U, Blodgett J, Duhme-Klair AK, Dallavalle S, Routledge A, Schobert R (2017) Chemical and biological aspects of nutritional immunity-perspectives for new anti-infectives targeting iron uptake systems. Angew Chem Int Ed 56:14360

Li Y, Gu Y, Li J, Xu M, Wei Q, Wang Y (2015) Biocontrol agent *Bacillus amyloliquefaciens* LJ02 induces systemic resistance against cucurbits powdery mildew. Front Microbiol 6:883. https://doi.org/10.3389/fmicb.2015.00883

Lim HS, Kim YS, Kim SD (1991) Pseudomonas stutzeri YPL-1 genetic transformation and antifungal mechanism against *Fusarium solani*, an agent of plant root rot. Appl Environ Microbiol 57(2):510–516

Lin SY, Hameed A, Liu YC, Hsu YH, Lai WA, Shen FT, Young CC (2015) *Azospirillum* soli sp. nov., a nitrogen-fixing species isolated from agricultural soil. Int J Syst Evol Microbiol 65:4601–4607

Liu D, Lian B, Dong H (2012) Isolation of *Paenibacillus* sp. and assessment of its potential for enhancing mineral weathering. J Geomicrobiol 29:413–421

Liu F, Xing S, Ma H, Du Z, Ma B (2013) Cytokinin-producing, plant growth-promoting rhizobacteria that confer resistance to drought stress in Platycladus orientalis container seedlings. Appl Microbiol Biotechnol 97:9155–9164

Liu W, Wang Q, Hou J, Tu C, Luo Y, Christie P (2016) Whole genome analysis of halotolerant and alkalotolerant plant growth-promoting rhizobacterium *Klebsiella* sp. D5A. Sci Rep 6:26710

Liu C, Mou L, Yi J, Wang J, Liu A, Yu J (2019) The Eno Gene of *Burkholderia cenocepacia* strain 71-2 is involved in phosphate solubilization. Curr. Microbiol. 76(4):495–502. https://doi.org/10.1007/s00284-019-01642-7

Lucangeli C, Bottini R (1996) Reversion of dwarfism in dwarf-1 maize (Zea mays L.) and dwarf-x rice (Oryza sativa L.) mutants by endophytic Azospirillum sp. Biocell 20:223–228

Lucas JA, Solano BR, Montes F, Ojeda J, Megias M, Gutierrez Ma-nero FJ (2009) Use of two PGPR strains in the integrated management of blast disease in rice (Oryza sativa) in Southern Spain. Field Crop Res 114:404–410

Lugtenberg B, Kamilova F (2009) Plant-growth-promoting rhizo-bacteria. Annu Rev Microbiol 63:541–556

Ma Y, Rajkumar M, Freitas H (2009a) Inoculation of plant growth promoting bacteria *Achromobacter xylosoxidans* strain Ax10 for improvement of copper phytoextraction by Brassica juncea. J Environ Manage 90:831–837. https://doi.org/10.1016/j.jenvman.2008.01.014

Ma Y, Rajkumar M, Freitas H (2009b) Improvement of plant growth and nickel uptake by nickel resistant-plant-growth promoting bacteria. J Hazard Mater 166:1154–1161

Madhaiyan M, Poonguzhali S, Ryu J, Sa T (2006) Regulation of ethylene levels in canola (Brassica campestris) by 1-aminocyclopropane-1-carboxylate deaminase-containing Methylobacterium fujisawaense. Planta 224:268–278

Majeed A, Muhammad Z, Ahmad H (2018) Plant growth promoting bacteria: role in soil improvement, abiotic and biotic stress management of crops. Plant Cell Rep. 37(12):1599–1609. https://doi.org/10.1007/s00299-018-2341-2

Maksimov IV, Abizgil'dina RR, Pusenkova LI (2011) Plant growth promoting rhizobacteria as alternative to chemical crop protectors from pathogens (Review). Appl Biochem Microbiol 47:333–345

Marques APGC, Pires C, Moreira H, Rangel AOSS, Castro PML (2010) Assessment of the plant growth promotion abilities using Zea mays as indicator plant. Soil Biol Biochem 42:1229–1235

Masson-Boivin C, Sachs JL (2018) Symbiotic nitrogen fixation by rhizobia—the roots of a success story. Curr Opin Plant Biol 44:7–15. https://doi.org/10.1016/j.pbi.2017.12.001

Meena VS, Maurya BR, Verma JP (2014) Does a rhizospheric microorganism enhance K+ availability in agricultural soils? Microbiol Res 169(5–6):337–347

Meena VS, Maurya BR, Verma JP, Aeron A, Kumar A, Kim K, Bajpai VK (2015) Potassium solubilizing rhizobacteria (KSR): isolation, identification, and K-release dynamics from waste mica. Ecol Eng 81:340–347

Mehta P, Walia A, Kulshrestha S, Chauhan A, Shirkot CK (2015) Efficiency of plant growth-promoting P-solubilizing *Bacillus circulans* CB7 for enhancement of tomato growth under net house conditions. J Basic Microbiol 55(1):33–44. https://doi.org/10.1002/jobm.201300562

Minaxi, Saxena J (2011) Efficacy of rhizobacterial strains encapsulated in nontoxic biodegradable gel matrices to promote growth and yield of wheat plants. Appl Soil Ecol 48(3):301–308

Miransari M, Smith DL (2014) Plant hormones and seed germination. Environ Exp Bot 99:110–121

Mishra SK, Khan MH, Misra S, Dixit KV, Khare P, Srivastava S, Chauhan PS (2017) Characterisation of Pseudomonas spp. and Ochrobactrum sp. isolated from volcanic soil. Antonie Van Leeuwenhoek 110:253–270

Mukherjee J, Mridha N, Mondal S, Chakraborty D, Kumar A (2018) Identifying suitable soil health indicators under variable climate scenarios: a ready reckoner for soil management. In: Bal S, Mukherjee J, Choudhury B, Dhawan A (eds) Advances in crop environment interaction. Springer, Singapore

Mukhtar S, Shahid I, Mehnaz S, Malik KA (2017) Assessment of two carrier materials for phosphate solubilizing biofertilizers and their effect on growth of wheat (*Triticum aestivum L.*). Microbiol Res 205:107–117

Muratova AY, Turkovskaya OV, Antonyuk LP, Makarov OE, Pozdnyakova LI, Ignatov VV (2005) Oil-oxidizing potential of associative rhizobacteria of the genus Azospirillum. Microbiology 74:210–215

Nadeem SM, Zahir ZA, Naveed M (2009) Rhizobacteria containing ACC-deaminase confer salt tolerance in maize grown on salt-affected fields. Can J Microbiol 55:1302–1309

Nandi M, Selin C, Brawerman G, de Kievit T, Dilantha Fernando WG (2017) Hydrogen cyanide, which contributes to Pseudomonas chlororaphis strain PA23 biocontrol, is upregulated in the presence of glycine. In: Biological control: theory and applications in pest management, vol 108. Acedemic, Boston, pp 47–54. https://doi.org/10.1016/j.biocontrol.2017.02.008. AGR: IND605652605

Narasimhan K, Basheer C, Bajic VB, Swarup S (2003) Enhancement of plant–microbe interactions using a rhizosphere metabolomics-driven approach and its application in the removal of polychlorinated biphenyls. Plant Physiol 2003:146–153. https://doi.org/10.1104/pp.102.016295

Nascimento FX, Tavares MJ, Franck J, Ali S, Glick BR, Rossi MJ (2019) ACC deaminase plays a major role in *Pseudomonas fluorescens* YsS6 ability to promote the nodulation of Alpha-and Betaproteobacteria rhizobial strains. Arch Microbiol 201(6):817–822. https://doi.org/10.1007/s00203-019-01649-5

Negi G, Pankaj, Srivastava A, Sharma A (2014) *In situ* biodegradation of endosulfan, imidacloprid, and carbendazim using indigenous bacterial cultures of agriculture fields of Uttarakhand, India. World Acad Sci Eng Technol 8:894–902

Neubauer U, Furrer G, Kayser A, Schulin R (2000) Siderophores, NTA, and citrate: potential soil amendments to enhance heavy metal mobility in phytoremediation. Int J Phytoremediat 2:353–368

Nokman W, Benluvankar V, Packiam SM, Vincent S (2019) Screening and molecular identification of heavy metal resistant *Pseudomonas putida* S4 in tannery effluent wastewater. Biocatal Agric Biotechnol 18:101052. https://doi.org/10.1016/j.bcab.2019.101052

Nouioui I, Cortes CJ, Carro L, Castro JF, Gtari M, Ghodhbane-Gtari F, Goodfellow M (2019) Genomic insights into plant-growth-promoting potentialities of the genus *Frankia*. Front Microbiol 10:1457. https://doi.org/10.3389/fmicb.2019.01457

Oliveira CA, Alves VMC, Marriel IE, Gomes EA, Scotti MR, Carneiro NP, Guimaraes CT, Schaffert RE, So NMH (2009) Phosphate solubilizing microorganisms isolated from rhizosphere of maize cultivated in an oxisol of the Brazilian Cerrado Biome. Soil Biol Biochem 41:1782–1787

Ordentlich A, Elad Y, Chet I (1988) The role of chitinase of *Serratia marcescens* in biocontrol of *Sclerotium rolfsii*. Phytopathology 78:84–88

Ovadis M, Liu X, Gavriel S, Ismailov Z, Chet I, Chernin L (2004) The global regulator genes from biocontrol strain *Serratia plymuthica* IC1270: cloning, sequencing, and functional studies. J Bacteriol 186:4986–4993

Parmar P, Sindhu SS (2013) Potassium solubilisation by rhizosphere bacteria: influence of nutritional and environmental conditions. J Microbial Res 3:25–31

Patten CL, Glick BR (1996) Bacterial biosynthesis of indole-3-acetic acid. Can J Microbiol 42:207–220

Pérez-Montaño F (2014) Plant growth promotion in cereal and leguminous agricultural important plants: from microorganism capacities to crop production. Microbiol Res 169:325–336

Pieterse CM, Zamioudis C, Berendsen RL, Weller DM, Van Wees SC, Bakker PA (2014) Induced systemic resistance by beneficial microbes. Annu Rev Phytopathol. 52:347–375

Prapagdee B, Chanprasert M, Mongkolsuk S (2013) Bioaugmentation with cadmium-resistant plant growth-promoting rhizobacteria to assist cadmium phytoextraction by *Helianthus annuus*. Chemosphere 92:659–666

Prasad M, Srinivasan R, Chaudhary M, Choudhary M, Jat LK (2019) Plant growth promoting rhizobacteria (PGPR) for sustainable agriculture: perspectives and challenges. In: PGPR amelioration in sustainable agriculture. Woodhead Publishing, Sawston, pp 129–157. https://doi.org/10.1016/B978-0-12-815879-1.00007-0

Qu J, Xu Y, Ai GM, Liu Y, Liu ZP (2015) Novel Chryseobacterium sp. PYR2 degrades various organochlorine pesticides (OCPs) and achieves enhancing removal and complete degradation of DDT in highly contaminated soil. J Environ Manage 161:350–357. https://doi.org/10.1016/j.jenvman.2015.07.025

Raaijmakers JM, Mazzola M (2012) Diversity and natural functions of antibiotics produced by beneficial and plant pathogenic bacteria. Annu Rev Phytopathol 50:403–424

Radzki W, Gutierrez Mañero FJ, Algar E, Lucas García JA, García-Villaraco A, Ramos Solano B (2013) Bacterial siderophores efficiently provide iron to iron-starved tomato plants in hydroponics culture. Antonie Van Leeuwenhoek 104:321–330

Rajkumar M, Ae N, Prasad MNV, Freitas H (2010) Potential of siderophore-producing bacteria for improving heavy metal phytoextraction. Trends Biotechnol 28:142–149

Raza A, Faisal M (2013) Growth promotion of maize by desiccation tolerant Micrococcus luteuschp37 isolated from Cholistan desert, Pakistan. Aust J Crop Sci 7:1693–1698

Rojas-Solís D, Zetter-Salmon E, Contreras-Perez M, del Carmen Rocha-Granados M, Macías-Rodríguez L, Santoyo G (2018) *Pseudomonas stutzeri* E25 and *Stenotrophomonas maltophilia* CR71 endophytes produce antifungal volatile organic compounds and exhibit additive plant growth-promoting effects. Biocatal Agric Biotechnol 13:46–52. https://doi.org/10.1016/j.bcab.2017.11.007

Rylott EL, Lorenz A, Bruce NC (2011) Biodegradation and transformation of explosives. Curr Opin Biotechnol 22:434–440

Ryu CM, Farag MA, Hu CH, Reddy M, Wei HX, Pare PW, Kloepper JW (2003) Bacterial volatiles promote growth in *Arabidopsis*. Proc Natl Acad Sci U S A 100:4927–4932

Saha M, Maurya BR, Meena VS, Bahadur I, Kumar A (2016a) Identification and characterization of potassium solubilizing bacteria (KSB) from Indo-Gangetic Plains of India. Biocatal Agric Biotechnol 7:202–209. https://doi.org/10.1016/j.bcab.2016.06.007

Saha M, Sarkar S, Sarkar B, Sharma BK, Bhattacharjee S, Tribedi P (2016b) Microbial siderophores and their potential applications: a review. Environ Sci Pollut Res 23:3984–3999

Saiyad SA, Jhala YK, Vyas RV (2015) Comparative efficiency of five potash and phosphate solubilizing bacteria and their key enzymes useful for enhancing and improvement of soil fertility. Int J Sci Res Publ 5(2):1–6

Salamone IEG, Hynes RK, Nelson LM (2005) Role of cytokinins in plant growth promotion by rhizosphere bacteria. In: Siddiqui ZA (ed) PGPR: biocontrol and biofertilization. Springer, Dordrecht, pp 173–195

Samina M (2016) An overview of globally available bioformulations. In: Bioformulations: For sustainable agriculture. Ch-15, pp 267–281

San Fulgencio NS, Suarez-Estrella F, Lopez MJ, Jurado MM, Lopez-Gonzalez JA, Moreno J (2018) Biotic aspects involved in the control of damping-off producing agents: the role of the thermotolerant microbiota isolated from composting of plant waste. Biol Control 124:82–91. https://doi.org/10.1016/j.biocontrol.2018.04.015

Sandilya SP, Bhuyan PM, Gogoi DK, Kardong D (2016) Phosphorus solubilization and plant growth promotion ability of rhizobacteria of R. communis L. growing in Assam, India. Proc Natl Acad Sci India Sect B Biol Sci 88(3):959–966

Sandilya SP, Bhuyan PM, Vijay N, Gogoi DK, Kardong D (2017) Impact of *Pseudomonas aeruginosa* strain MAJ PIA03 affecting the growth and phytonutrient production of castor, a primary host-plant of Samia ricini. J Soil Sci Plant Nutr 17(2):499–515

Santi C, Bogusz D, Franche C (2013) Biological nitrogen fixation in non-legume plants. Ann Bot 111:743–767
Santoro MV, Zygadlo J, Giordano W, Banchio E (2011) Volatile organic compounds from rhizobacteria increase biosynthesis of essential oils and growth parameters in peppermint (*Mentha piperita*). Plant Physiol Biochem 49:1177–1182
Sayyed RZ, Chincholkar SB (2010) Growth and siderophores production in *Alcaligenes faecalis* is regulated by metal ions. Indian J Microbiol 50(2):179–182
Schillaci M, Gupta S, Walker R, Roessner U (2019) The role of plant growth-promoting bacteria in the growth of cereals under abiotic stresses. Root Biol. https://doi.org/10.5772/intechopen.87083
Shahzad R, Waqas M, Khan AL, Asaf S, Khan MA, Kang SM, Lee IJ (2016) Seed-borne endophytic *Bacillus amyloliquefaciens* RWL-1 produces gibberellins and regulates endogenous phytohormones of *Oryza sativa*. Plant Physiol Biochem 106:236–243. https://doi.org/10.1016/j.plaphy.2016.05.006
Shen X, Hu H, Peng H, Wang W, Zhang X (2013) Comparative genomic analysis of four representative plant growth-promoting rhizobacteria in Pseudomonas. BMC Genomics 14:1471–2164
Siddiqui IA, Shaukat SS, Sheikh IH, Khan A (2006) Role of cyanide production by *Pseudomonas fluorescens* CHA0 in the suppression of root-knot nematode, *Meloidogyne javanica* in tomato. World J Microbiol Biotechnol 22:641–650
Singh SGG (2015) Plant growth promoting rhizobacteria (PGPR): current and future prospects for development of sustainable agriculture. J Microb Biochem Technol 07(02):96–102
Singh RP, Jha PN (2016) A halotolerant bacterium Bacillus licheniformis HSW-16 augments induced systemic tolerance to salt stress in wheat plant (Triticum aestivum). Front Plant Sci 7:1890
Singh N, Marwa N, Mishra J, Verma PC, Rathaur S, Singh N (2016) *Brevundimonas diminuta* mediated alleviation of arsenic toxicity and plant growth promotion in *Oryza sativa* L. Ecotoxicol Environ Safe 125:25–34. https://doi.org/10.1016/j.ecoenv.2015.11.020
Singh R, Pandey DK, Kumar A, Singh M (2017) PGPR isolates from the rhizosphere of vegetable crop Momordica charantia: characterization and application as biofertilizer. Int J Curr Microbiol App Sci 6(3):1789–1802
Singh RP, Mishra S, Jha P, Raghuvanshi S, Jha PN (2018) Effect of inoculation of zinc-resistant bacterium *Enterobacter ludwigii* CDP-14 on growth, biochemical parameters and zinc uptake in wheat (*Triticum aestivum* L.) plant. Ecol Eng 116:163–173. https://doi.org/10.1016/j.ecoleng.2017.12.033
Sirohi MH, Jackson J, Edwards M, Ollerton J (2015) Diversity and abundance of solitary and primitively eusocial bees in an urban centre: a case study from Northampton (England). J Insect Conserv 19(3):123–136
Sofia IA, Paula P, Castro ML (2014) Phosphate solubilizing rhizobacteria enhance Zea mays growth in agricultural P-deficient soils. Ecol Eng 73:526–535
Spaepen S, Vanderleyden J (2011) Auxin and plant-microbe interactions. Cold Spring Harb Perspect Biol 3:a001438
Spaepen S, Vanderleyden J, Remans R (2007) Indole-3-acetic acid in microbial and microorganism-plant signaling. FEMS Microbiol Rev 31:425–448
Sticher L, Mauch-Mani B, Métraux JP (1997) Systemic acquired resistance. Annu Rev Phytopathol 35:235–270
Subrahmanyam G, Archana GK (2011) Plant growth promoting activity of Enterobacter sp. C1D in heavy metal contaminated soils. In: Plant growth-promoting rhizobacteria (PGPR) for sustainable agriculture, pp 440–446
Subrahmanyam G, Sharma RK, Kumar GN, Archana G (2018) Vigna radiata var.GM4plant growth enhancement and root colonization by a multi-metal-resistant plant growth-promoting bacterium Enterobacter sp. C1D in Cr (VI)-amended soils. Pedosphere 28(1):144–156

Sujatha N, Ammani K (2013) Siderophore production by the isolates of fluorescent Pseudomonads. Int J Cur Res Rev 5:1–7

Tahir HA, Gu Q, Wu H, Raza W, Hanif A, Wu L, Gao X (2017) Plant growth promotion by volatile organic compounds produced by *Bacillus subtilis* SYST2. Front Microbiol 8:171. https://doi.org/10.3389/fmicb.2017.00171

Thilagar G, Bagyaraj DJ, Podile AR, Vaikuntapu PR (2018) *Bacillus sonorensis*, a novel plant growth promoting rhizobacterium in improving growth, nutrition and yield of chilly (*Capsicum annuum* L.). Proc Natl Acad Sci India B 88(2):813–818. https://doi.org/10.1007/s40011-016-0822-z

Thomas J, Kim HR, Rahmatallah Y, Wiggins G, Yang Q, Singh R, Mukherjee A (2019) RNA-seq reveals differentially expressed genes in rice (*Oryza sativa*) roots during interactions with plant-growth promoting bacteria, *Azospirillum brasilense*. PLoS One 14:e0217309. https://doi.org/10.1371/journal.pone.0217309

Thrane T, Harder Nielsen T, Neiendam Nielsen M, Rensen JSA, Olsson S (2000) Viscosinamide-producing Pseudomonas fluorescens DR54 exerts a biocontrol effect on Pythium ultimum in sugar beet rhizosphere. FEMS Microbiol Ecol 33(2):139–146

Tisdale SL, Nelson WL (1975) Soil fertility and fertilizers, 3rd edn. Macmillan, New York, p 694

Toussaint JP, Pham T, Barriault D, Sylvestre M (2012) Plant exudates promote PCB degradation by a rhodococcal rhizobacteria. Appl Microbiol Biotechnol 95:1589–1603. https://doi.org/10.1007/s00253-011-3824-z

Turan M, Kıtır N, Alkaya Ü, Günes A, Tüfenkçi Ş, Yıldırım E, Nikerel E (2016) Making soil more accessible to plants: the case of plant growth promoting rhizobacteria. In: Plant growth. InTech, Rijeka. https://doi.org/10.5772/64826

Udvardi M, Poole PS (2013) Transport and metabolism in legume-rhizobia symbioses. Annu Rev Plant Biol 64:781–805

Uhlik O, Musilova L, Ridl J, Hroudova M, Vleck C, Koubek J, Holeckova M, Mackova M, Macek T (2013) Plant secondary metabolite-induces shifts in bacterial community structure and degradative ability in contaminated soil. Appl Microbiol Biotechnol 97(20):9245–9256

Vacheron J, Desbrosses G, Bouffaud ML, Touraine B, Moenne-Loccoz Y, Muller D, Legendre L, Wisniewski-Dye F, Prigent-Combaret C (2013) Plant growth-promoting rhizobacteria and root system functioning. Front Plant Sci 4:356

Vaddepalli P, Fulton L, Wieland J, Wassmer K, Schaeffer M, Ranf S, Schneitz K (2017) The cell wall-localized atypical β-1, 3 glucanase ZERZAUST controls tissue morphogenesis in *Arabidopsis thaliana*. Development 144:2259–2269

Van Loon LC (2007) Plant responses to plant growth-promoting rhizobacteria. Eur J Plant Pathol 119:243–254

Van Loon L, Bakker P, Pieterse C (1998) Systemic resistance induced by rhizosphere bacteria. Annu Rev Phytopathol 36:453–483

Van Loon LC, Rep M, Pieterse CM (2006) Significance of inducible defense-related proteins in infected plants. Annu Rev Phytopathol 44:135–162

Vasebi Y, Alizadeh A, Safaie N (2015) *Pantoea agglomerans* ENA1 as a biocontrol agent of *Macrophomia phaseolina* and growth enhancer of soybean. J Crop Prot 4:43–57

Verma VC, Gond SK, Kumar A, Mishra A, Kharwar RN, Gange AC (2009) Endophytic actinomycetes from *Azadirachta indica* A. Juss.: isolation, diversity, and antimicrobial activity. Microb Ecol 57:749–756. https://doi.org/10.1007/s00248-008-9450-3

Verma P, Yadav AN, Khannam KS, Kumar S, Saxena AK, Suman A (2016) Molecular diversity and multifarious plant growth promoting attributes of *Bacilli* associated with wheat (*Triticum aestivum* L.) rhizosphere from six diverse agro-ecological zones of India. J Basic Microbiol. 56 (1):44–58. https://doi.org/10.1002/jobm.201500459

Verma P, Yadav AN, Khannam KS, Saxena AK, Suman A (2017) Potassium-solubilizing microbes: diversity, distribution, and role in plant growth promotion. In: Panpatte DG, Jhala YK, Vyas RV, Shelat HN (eds) Microorganisms for green revolution, Microbes for sustainable crop production, vol 1. Springer, Singapore, pp 125–149. https://doi.org/10.1007/978-981-10-6241-4_7

Vitousek PM, Howarth RW (1991) Nitrogen limitation on land and in the sea: How can it occur? Biogeochemistry 13:87–115

Viveros OM, Jorquera MA, Crowley DE, Gajardo G, Mora ML (2010) Mechanisms and practical considerations involved in plant growth promotion by rhizobacteria. J Soil Sci Plant Nutr 10:293–319

Wagg C, Bender SF, Widmer F, van der Heijden MGA (2014) Soil biodiversity and soil community composition determine ecosystem multifunctionality. Proc Natl Acad Sci 111:5266–5270

Wani PA, Khan MS (2010) Bacillus species enhance growth parameters of chickpea (*Cicer arietinum* L.) in chromium stressed soils. Food Chem Toxicol 48:3262–3267

Waqas M, Khan AL, Kamran M, Hamayun M, Kang SM, Kim YH, Lee IJ (2012) Endophytic fungi produce gibberellins and indoleacetic acid and promotes host-plant growth during stress. Molecules 17:10754–10773

Xu J, Li XL, Luo L (2012) Effects of engineered *Sinorhizobium meliloti* on cytokinin synthesis and tolerance of alfalfa to extreme drought stress. Appl Environ Microbiol 78:8056–8061

Yadav J, Verma JP (2014) Effect of seed inoculation with indigenous *Rhizobium* and plant growth promoting rhizobacteria on nutrients uptake and yields of chickpea (*Cicer arietinum* L.). Eur J Soil Biol 63:70–77

Yadav N, Yadav A (2018) Biodiversity and biotechnological applications of novel plant growth promoting methylotrophs. J Appl Biotechnol Bioeng 5:342–344

Yadav AN, Yadav N, Sachan SG, Saxena AK (2019) Biodiversity of psychrotrophic microbes and their biotechnological applications. J Appl Biol Biotechnol 7:99–108

Yamaga F, Washio K, Morikawa M (2010) Sustainable biodegradation of phenol by *Acinetobacter calcoaceticus* P23 isolated from the rhizosphere of duckweed *Lemna aoukikusa*. Environ Sci Technol 44(16):6470–6474

Yan Z, Perez-de-Castro A, Dıez MJ, Hutton SF, Visser RGF, Wolters AMA, Bai Y, Li J (2018) Resistance to tomato yellow leaf curl virus in tomato germplasm. Front Plant Sci 9:1198

Yang Y, Liang Y, Han X, Chiu TY, Ghosh A, Chen H, Tang K (2016) The roles of arbuscular mycorrhizal fungi (AMF) in phytoremediation and tree-herb interactions in Pb contaminated soil. Sci Rep 6:20469

Zachow C, Müller H, Monk J, Berg G (2017) Complete genome sequence of *Pseudomonas brassicacearum* strain L13-6-12, a biological control agent from the rhizosphere of potato. Stand Genomic Sci 12:6

Zahir ZA, Ghani U, Naveed M, Nadeem SM, Asghar HN (2009) Comparative effectiveness of *Pseudomonas* and *Serratia* sp containing ACC-deaminase for improving growth and yield of wheat (*Triticum aestivum* L) under salt-stressed conditions. Arch Microbiol 191:415–424

Zaidi A, Khan MS, Ahemad M, Oves M (2009) Plant growth promotion by phosphate solubilizing bacteria. Acta Microbiol Immunol Hung 56:263–284

Zebelo S, Song Y, Kloepper JW, Fadamiro H (2016) Rhizobacteria activates (C)-delta-cadinene synthase genes and induces systemic resistance in cotton against beet armyworm (Spodoptera exigua). Plant Cell Environ. 39:935–943. https://doi.org/10.1111/pce.12704

Zhang Y, He L, Chen Z, Wang Q, Qian M, Sheng X (2011) Characterization of ACC deaminase-producing endophytic bacteria isolated from copper-tolerant plants and their potential in promoting the growth and copper accumulation of *Brassica napus*. Chemosphere 83:57–62

Zhang X, Baars O, Morel FM (2019) Genetic, structural, and functional diversity of low and high-affinity siderophores in strains of nitrogen fixing *Azotobacter chroococcum*. Metallomics 11 (1):201–212. https://doi.org/10.1039/C8MT00236C

Zohora US, Ano T, Rahman MS (2016) Biocontrol of *Rhizoctonia solani* K1 by iturin A producer *Bacillus subtilis* RB14 seed treatment in tomato plants. Adv Microbiol 6(6):424. https://doi.org/10.4236/aim.2016.66042

Zou C, Li Z, Yu D (2010) *Bacillus megaterium* strain XTBG34 promotes plant growth by producing 2-pentylfuran. J Microbiol 48:460–466

Chapter 27
Fungal Mycelium-Based Biocomposites: An Emerging Source of Renewable Materials

Dhanushka Udayanga and Shaneya Devmini Miriyagalla

Abstract Fungi are efficiently used to produce a variety of medicinal compounds, functional foods, and environmentally sustainable raw materials for a wide range of consumer goods due to their distinctive biological properties. Mycelium, the vegetative structure of filamentous fungi, acts as a natural, self-assembling adhesive as it grows, binding the fragments of organic substrates, leading to the production of fungal mycelium-based biocomposites (MBCs). These biocomposites are biodegradable alternatives for many synthetic polymers, such as polystyrene, and are therefore considered as a widely applicable, emerging class of renewable materials. MBCs are excellent examples of circular materials, ensuring a cradle-to-cradle (C2C) design, in which biodegradable products can be returned to the ecosystem after its use. Diverse species of fungi can be used to produce MBCs together with a range of agricultural and other plant-based lignocellulosic substrates. Several business start-ups, by innovative investors, are globally leading in mycelium-based product manufacturing. MBCs, including both mycelium-based foams (MBFs) and mycelium-based sandwich composites (MBSCs), are known for their potential industrial applications, such as packaging materials, architectural design, construction, fashion, and automotive insulation products. Both the mycelium binder and substrate type have an immense impact on the significant material properties of MBCs, including their hydrophobicity, acoustic nature, thermal insulation, and fire resistance. This chapter summarizes the diversity of the fungi used to produce MBCs as well as their potential feeding substrates, manufacturing process, physical and mechanical properties, innovative applications, and future directions for related research endeavours.

Keywords Biodegradation · Biofabrication · Circular economy · Mushrooms

D. Udayanga (✉) · S. D. Miriyagalla
Department of Biosystems Technology, Faculty of Technology, University of Sri Jayewardenepura, Pitipana, Homagama, Sri Lanka
e-mail: dudayanga@sjp.ac.lk

P. Bhatt et al. (eds.), *Microbial Technology for Sustainable Environment*,
https://doi.org/10.1007/978-981-16-3840-4_27

27.1 Introduction

Fungi are important microorganisms due to their vital role as a major decomposer in natural ecosystems as well as their industrial, food, and medicinal applications (Blackwell 2011; Hyde et al. 2019). Mycelium, consisting of bundles of thread-like hyphae, is the vegetative structure of filamentous fungi. Hyphae spreading through the substratum or an artificial medium absorb nutrients required for the growth of the mycelium of a fungus. Therefore, mycelium represents the major component of fungal biomass. Vegetative mycelia of the filamentous fungi and mushrooms are traditionally well-known for their capability to produce antibiotics, enzymes, organic acids, nutritious functional foods, high-value health supplements, and natural food flavouring agents (Rathore et al. 2019). However, recent advances in environmental engineering and biomaterial research have revealed the potential of filamentous fungal mycelium as a promising raw material for biocomposites (Attias et al. 2020; Ghazvinian et al. 2019; Jones et al. 2020). This is due to their possible use as a substitute for a variety of non-biodegradable, inert synthetic materials, such as polystyrene and other plastics, including moulded or fabricated expanded polystyrene (EPS), expanded polypropylene (EPP), and expanded polyethylene (EPE). Therefore, MBCs have now become an eco-friendly alternative in various commercial applications, such as architecture, textile, fashion accessories, footwear, automotive insulation padding, electronics, healthcare, and packaging, where EPS-like synthetic polymers were traditionally used.

Mycelium is primarily comprised of polysaccharide chitin, which makes up the cell wall of fungi. Additionally, lipids, glycoproteins, minerals, fibre, and other polysaccharides, like mannans and beta-glucans, are also found in the mycelium (Bowman and Free 2006). The unique structure and composition of fungal mycelium have made it a convenient source of natural composite materials with tuneable and well-ordered structural and mechanical properties. In some cases, the properties of mycelium grown on a natural substrate can be engineered by changing the type of medium as well as other growth conditions. Therefore, fungal mycelium grown in a large scale can be used as a raw material for the manufacture of various sustainable products.

All fungi are heterotrophs, which means they require organic substrates for their growth. MBCs are produced by ceasing the growth of mycelia before its organic substrate or medium is fully degraded. Among the diverse groups of fungi, mushroom-forming basidiomycetes are the primary choice of microorganisms used in MBC production (Cerimi et al. 2019). Therefore, MBCs are also popularly known as "mushroom-composites", "mushroom-material", "mycomaterial", and "mycocomposites" by commercial producers and innovators. In MBCs, when the fibrous substrate material is used, fungal hyphae bind organic fibres or particles together while colonizing the substrate (Appels and Wösten 2020). The ability of fungi to bind and digest lignocellulosic plant material produces an inherent bonding that can be developed into a natural, unique, and lightweight biocomposite. These MBCs are usually free of synthetic adhesives; therefore, they represent an eco-friendly, biodegradable alternative for many industrial applications. The morphology and mechanical properties of the composite material are largely defined by the feeding substrate and biological nature of the fungus. Based on the composition of the feeding substrate of fungi, the final fibrous structures exhibit varying structural compositions.

The global environmental crisis, which stems from the accumulation of synthetic plastic debris and the depletion of fossil fuels, has caused the global scientific community to re-consider renewable sources of material in industrial applications. The major environmental issues in the world have arisen due to the continued practice of a linear economy based on the "cradle-to-grave" concept in which raw natural resources are taken, transformed into products, and disposed into the natural ecosystems after their use. In contrast, a circular economy model, which is based on the "cradle-to-cradle" design concept, aims to close the gap between production and the disposal of related waste, turning the waste of a given process into the raw material for another product. Therefore, biodegradable and compostable MBCs are ideal solutions to the global demand towards sustainable living through a circular economy. In this review, we discuss the fungi used in the production, the manufacturing process, the physical and mechanical properties of MBCs, and the recent developments of advanced applications of similar materials.

27.2 Fungi Used in the Production of MBCs

The saprotrophic fungal species belong in the phyla Basidiomycota and Ascomycota have generally been used as binding components in MBCs. Mushroom-forming Basidiomycota represent the widely used species of fungi in MBCs. Their prevalence in nature, ability to colonize a large area by making extensive three-dimensional hyphal networks, and desirable physicochemical properties are invaluable traits in MBC design, which make them the prime candidates for this up-and-coming sustainable material design concept (Appels and Wösten 2020).

Fungi are ideal for the partial degradation of the majority of plant-derived filler materials of MBCs consisting of fibrous or particulate matter with a rich lignocellulose composition. For instance, fungal species belonging to the genera *Trametes*, *Ganoderma*, and *Pleurotus* are widely used in MBCs, since those fungi can be easily collected and have a high abundance of extracellular enzymes, which allows them to specialize in partially degrading more complex nutrients, such as cellulose and lignin in substrates (Jones et al. 2020).

When selecting a suitable fungal species for composite manufacturing, growth predictors and the biology of the organism, such as the growth rate, mycelium density, hyphae nature, biochemical composition, and pathogenicity, must be considered. Non-obligate pathogens can become versatile colonizers of filler materials due to their innate ability to aggressively compete with living hosts (Jones et al. 2018a). The mitic system, comprised of mono-mitic, di-mitic, and tri-mitic networks, defines the different types of hyphal networks found within the mycelia of basidiomycete fungi (Pegler 1996). In MBC design, tri-mitic strains, especially *Trametes versicolor* and *T. multicolor*, which are composed of all three hyphal types, generative, binding, and skeletal hyphae, are favoured. Those fungi with tri-mitic strains are known to have a better mechanical performance within a composite structure provided by their often solid, thick-walled hyphal filaments than mono-mitic strains, such as *Pleurotus*, that are composed of less complex generative hyphae (Jones et al. 2020). *Trametes versicolor* had been used in binding hemp and hardwood chips in a novel MBC design (Zimele et al. 2020). In a similar study, *T. multicolor* was experimented on for beech sawdust, rapeseed straw, and cotton waste (Appels et al. 2019). Studies on the fire-resistant properties of high silica agricultural and industrial waste-based composites have been conducted with *T. versicolor* as a binder by Jones et al. (2018b).

Fungi belong in the genus *Ganoderma* form slow-growing, coarse, densely matted, woody mushrooms. Therefore, the incorporation of the mycelium of *Ganoderma* in MBCs may result in mechanically resilient end-products. Two species of mushroom-forming fungi, *G. lucidum* and *Pleurotus ostreatus*, with sawdust, ground walnut shells, rice hulls, millet hulls, and cardboard have been evaluated for a potential application as thermal insulators for buildings (Yang 2020). In a related study, researchers investigated the possibility of using *G. lucidum* as a binder for a composite board made from palm sugar fibre and cassava bagasse fibre (Agustina et al. 2019). Moreover, spent mushroom substrate generated by sawdust, food waste, and diaper waste has been used along with *G. lucidum*, *P. ostreatus*, and *Auricularia polytricha* in strong fire- and water-resistant bio-boards (Khoo et al. 2020). The San Francisco-based start-up, MycoWorks, dedicated to introducing new mycelium materials to the world, primarily uses *G. lucidum* and *P. ostreatus* as their binder fungi.

With a global distribution, the genus *Pleurotus* has been studied at a fundamental level in a variety of MBC designs. For instance, *P. ostreatus*, *P. djamor*, *P. eryngii*, *P. ostreatus* var. *columbinus* (Kalisz and Rocco 2011), *P. pulmonarius*, *P. salmoneostramineus* (Attias et al. 2017), and *P. citrinopileatus* (Silverman 2018) have been explored in many similar studies. *Grifola frondosa*, *Lentinula edodes*, *Agrocybe aegerita*, *Coprinus comatus* (Kalisz and Rocco 2011), *Pycnoporus sanguineus*, *Lentinus velutinus* (Bruscato et al. 2019), *Fomitopsis pinicola*, *Gloeophyllum sepiarium*, *Laetiporus sulphureus*, *Phaeolus schweinitzii*, *Piptoporus betulinus*, *Polyporus arcularius*, *Schizophyllum commune* (Lakk et al. 2018), *Trametes pubescens*, *T. suaveolens*, *Trichaptum abietinum* (Wimmers et al. 2019), *Polyporus brumalis* (Jones et al. 2018a), and *Agaricus bisporus* (Tacer-Caba et al. 2020) include some other mushroom species that have been investigated in MBC design.

When compared to basidiomycete fungi, members of the phylum Ascomycota are relatively less studied as binding agents for biocomposites, since the reliance on fungi already limits the rate at which biocomposites can be manufactured. The characteristics of the fungi such as high hyphal extension rates to effectively colonize new areas of the substrate and density of mycelium are some contributing factors when selecting a species to produce MBCs. In ascomycete fungi, the hyphal extension rate has been found to be decreased with an increase of branching. When the colonies branch with a high mycelial density, inhibitory compounds are produced with the increased utilization of substrates, thus reducing the hyphal extension rates. This could potentially explain the reason for ascomycetes being an unpopular option for MBC design (Jones et al. 2018a). However, due to the availability of a range of enzymes, such as cellulases, pectinases, proteases, and xylanases, as well as fast growth and colonization rates of the mycelium, *Fusarium oxysporum* has been tested for biocomposite production with waste paper and spent coffee substrates (Iordache et al. 2018). Tacer-caba et al. (2020) evaluated the potential of using ascomycete fungus, *Trichoderma asperellum* in oat husk and rapeseed cake substrate fillers. In addition, a recent patent review highlights a MBC based on ascomycete, *Morchella angusticeps* that has been granted a patent (Cerimi et al. 2019). However, the same review indicated four applications based on a few wood-associated ascomycete fungi-based (*Xylaria polymorpha*, *X. hypoxylon*, *X. filiformis*, and *X. longipes*) MBCs, which have not been granted the patents or awaiting consideration (Cerimi et al. 2019).

27.3 Feeding Substrates for MBCs

The feeding substrate for MBC formation usually depends on the availability of excess plant materials from the agricultural waste, food processing, or wood and landscape industries. In addition to a carbon source, effective mycelium growth requires micro-nutrients, moisture, an ambient temperature, and oxygen. Therefore, any cellulose-rich or fibre material in sufficient supply is ideal for the production of biocomposites if the resulting composite meets the required mechanical property standards (Lelivelt 2015). Cellulose-rich substrates hinder mushroom-forming fungi from producing fruiting bodies, and the substrate provides optimum mechanical properties of the final product (Jones et al. 2017a). Therefore, wood-containing substrates, such as sawdust or wood chips of common landscape trees (i.e. spruce, pine, and fir), are commonly used in prototype preparation and industrial applications. Agricultural waste materials, such as rice straw, rice husk, wheat straw, wheat bran, corn straw, corn cobs, coconut fibre, sorghum stubbles, flax shive, kenaf fibre, cotton bur fibre, hemp hurd, and sugarcane bagasse, are also ideal sources of organic fibrous material, that have been used in various studies (Lelivelt 2015; Arifin and Yusuf 2013; Jones et al. 2017b; Pelletier et al. 2017). Therefore, typically, any non-toxic cellulose-rich plant material is an ideal source for the biofabrication of MBCs, as mycelium can be grown with or without supplements of artificial substrates or other chemicals.

In addition to its wood and fungal components, essential nutrients and artificial fungal growth media can be added to the substrate to be used in MBC production. For instance, Haneef et al. used pure cellulose and a cellulose–Potato Dextrose Broth (PDB) mixture as a fibre-nutrient source to develop MBCs (Haneef et al. 2017). Moreover, attempts have been made to produce hybrid composites via the incorporation of cellulose nano-fibrils (CNFs) in novel panel composites in addition to mycelium and wood components (Sun et al. 2019). Producing a hybrid composite requires a mycelium-treated wood produced by the growth of mycelia on soft wood, which is then hybridized with various levels of CNF as a binder (Sun et al. 2019).

Although MBCs are self-grown materials on a substrate, pure mycelium can be separately grown in a broth culture, or an artificial media can be used to make a composite by mixing the harvested mycelia with a sterile fibre substrate. Similarly, an organic substrate can be amended with an appropriate amount of inert material, such as metals, plastics, or ceramics, in order to produce composites for various purposes (Appels and Wösten 2020).

27.4 Manufacturing Process of MBCs

The positive environmental impact of MBCs is partially accomplished through their eco-friendly, sustainable, and circular manufacturing processes (Fig. 27.1). This is highly beneficial when compared to how their conventional petroleum-based, non-renewable composite counterparts are manufactured. In MBC production, relatively less energy-consuming, natural, and low-cost production processes are facilitated by the selection of usually mesophilic, indigenous, or commonly available fungi, which prefer ambient temperatures. Therefore, unlike in the industrial processing of synthetic polymers, no additional energy is required for maintenance or incubation under specific high or low temperatures.

The typical process of manufacturing a MBC starts with the selection of the substrate material and the fungal strain that is biocompatible with the substrate of choice. The significant factors that determine the growth performance (growth rate, mycelial density, and composition) of the fungal mycelium are the nutritional properties of the substrate, environmental conditions, and the genetic factors of the species used (Jones et al. 2018a; Appels et al. 2018). Low-cost agricultural wastes are the popular choice in many available MBC products and prototypes, but sometimes higher-grade, expensive, more nutritious substrates, such as wheat grains, have also been used (Jones et al. 2018b, c).

The second step in the production is the pre-treatment of the substrate, which involves soaking (to supply the optimum moisture content for the mycelium that is to be inoculated) followed by the homogenization of the substrates in order to increase the surface area for mycelial colonization. The chopping or cutting of feedstock, grinding, milling, or blending can be employed to macerate the substrates (Elsacker et al. 2019). Heat treatment of wood fibres to open and expand the air cavities is suggested to increase the porosity of the MBCs produced (Attias et al. 2017). Following the pre-treatments, sterilization of the substrates is a crucial step in the process, preferably achieved through wet heat methods, such as autoclave or pressure cooking, to ensure a contaminant-free medium without drying out of the substrate (Lelivelt 2015).

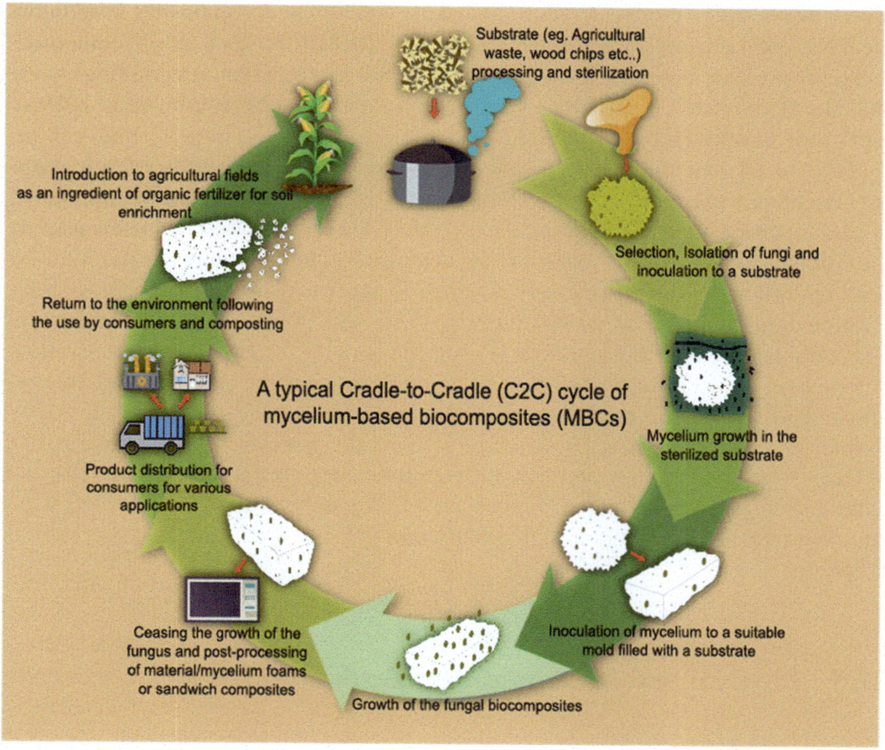

Fig. 27.1 A typical cradle-to-cradle (C2C) cycle of mycelium-based biocomposites (MBCs). Both the fungal mycelium and the natural cellulosic substrate used in biocomposite manufacturing are returned to the environment as a nutrient-rich ingredient for soil enrichment. Therefore, in the C2C life cycle, materials flow cyclically in an appropriate, continuous biological or technical nutrient cycle. All the waste materials are productively re-incorporated into a new sustainable product within the cycle

Once the substrate is pretreated and sterilized, the introduction of the fungal culture into the sterilized substrate should be done via inoculation. The fungal cultures can be introduced in the form of a spore suspension or through an intermediate nutrient-rich substrate, such as grain or sawdust ramified with mycelia, focusing on the even distribution of the inoculum within the substrate. The inoculated substrate is then filled into moulds to provide a defined shape for the MBCs (Jiang et al. 2013; Bayer et al. 2008).

During the incubation period, followed by the substrate inoculation, the mycelium initiates growth from the points of inoculation and colonizes the filler material, adhering the particles/fibres together into a dense networked structure while maintaining the mould shape, a process that can take anywhere from days to months depending on the strain (Jones et al. 2020). Temperatures around 25–35 °C are preferred, since temperatures above 40 °C may induce fruiting bodies. The humidity levels of 60–65%, proper aeration, and dark conditions are generally considered ideal for optimum mycelial growth (Silverman 2018; Butu et al. 2020). Although common practice involves incubation in ambient environmental conditions, controlled environments (e.g. higher temperatures) may also be employed to achieve rapid growth rates (Jones et al. 2020).

To ensure a contamination-free and properly ventilated incubation, growth chambers are often used in most studies (Silverman 2018). The final and most critical step in MBC manufacturing is stopping the growth of the fungus on the substrate prior to the complete degradation of the fillers by neutralizing/dehydrating the mycelium via heating/drying, thus simultaneously making the material denser (Jones et al. 2018b). Heating of the fully colonized structures from 2 to 48 h in 60–125 °C temperatures in convectional ovens or hot pressing is used industrially for mycelium deactivation. Alternatively, air drying for 48 h can be considered a more practical and energy-efficient method to be used in large-scale, industrial settings (Jiang et al. 2019). Further processing steps, such as hot pressing, cold pressing (Appels et al. 2019), hybridization with natural polymers (i.e. cellulose nano-fibrils) (Sun et al. 2019), and natural resin/glue impregnation (Jiang et al. 2013), can be carried out if desired at the end of production to obtain a high-quality end-product with enhanced mechanical properties.

27.5 Physical and Mechanical Properties of MBCs

Apart from the key advantage of being biodegradable, MBCs have several other desirable physical and mechanical material properties that encourage their use in various applications. These properties considered here include density, tensile strength, compressive strength, rigidity and material's strain response parameters, water absorption propensities, fire resistance, heat-insulating capability, and acoustic properties. According to recent studies, the density of MBCs falls within the range of 60–300 kg/m^3. According to an analysis by Jones et al., MBCs made with lignin-rich woody substrates are relatively denser than those made with more fibrous, cellulose-rich agricultural waste (Jones et al. 2020). Compared with similar building materials, such as plywood (680 kg/m^3) and particle board (750 kg/m^3), the apparent disparity in the density values between these conventional construction materials and MBCs is remarkable.

Tensile strength, or the maximum load a MBC material can bear once stretched, is primarily dependent on how strongly the binder fungus adheres the fillers together with adequately strong interfacial bonding, which is directly influenced by the nutrient richness of the substrate that can facilitate a rich network of mycelium (Jones et al. 2020). The compressive strength of the MBC is affected by the porosity of the substrate, which explains why more porous, fibrous composites have lesser compressive strengths than particulate ones. Compressive strength values range from 29 to 567 kPa for MBFs, which is considered relatively lower than that of standard categories of EPS foams (Girometta et al. 2019). The flexural strengths/

bending strengths of the structure are determined by the tensile and compressive stress limits the two opposite surfaces of the composite structure can withstand when flexed. The formation of a thick, continuous mycelial mat on air-exposed surfaces is paramount to withstand these conditions, and this is achieved through the utilization of nutrient-rich substrates that can support the rich growth of mycelium via proper aeration.

In materials, elastic deformation is usually expressed by means of the shear modulus (G) and Young's modulus (*E*), which are types of elastic moduli that can be used to determine the rigidity and strain response parameters, respectively. Young's modulus, which measures the ability of a material to withstand changes in length when under lengthwise tension or compression, can be calculated by taking the slope of the linear part of a stress/strain curve (Appels et al. 2018; Jastrzebski 1959). The shear modulus is defined as the ratio of the shear stress to strain (IUPAC 1997). Both E and G are known to be relatively higher in fibrous MBCs, which are mechanically more rigid (Haneef et al. 2017) compared to some synthetic counterparts, like large lightweight EPS blocks (geoforms). For instance, the Young's moduli of MBCs have been reported in the range of 3–11 and 17–27 MPa for *Ganoderma lucidum* and *Pleurotus ostreatus* pure mycelia, respectively (Haneef et al. 2017), with the average value of the initial elastic modulus of EPS ranging from 5.08 to 3.31 MPa (with a standard deviation of 1.89–1.48 MPa) (Bathurst et al. 2007). However, these numbers might not be directly comparable as they are highly density-dependent as well as influenced by the specimen size, rate of loading, and loading path (Zarnani and Bathurst 2007).

27.6 Water Absorption Propensities

The relatively high water absorption propensity of MBCs is one drawback that may limit their application, except when used as acoustic and thermal insulators within door cores with dry indoor environments that prevent moisture contact. The increasing of weight when in contact with water for 48–192 h are ranged from 30 to 580% as reported in several studies experimenting on this quality (Zimele et al. 2020; Appels et al. 2019; Agustina et al. 2019; Sun et al. 2019; Elsacker et al. 2019; Ziegler et al. 2016). The hydrophilic nature of the lignocellulosic filler substrates, the hydrophobic nature of the mycelium binder, and the porosity of the composites govern the overall extent of the water absorbed by the material. The cellulose-rich filler materials in general contain ample hydrophilic, hydroxyl (–OH) groups that readily bind with water. Meanwhile, certain fungal mycelia contain the protein hydrophobin, which makes them hydrophobic (Appels et al. 2018; Girometta et al. 2019). For instance, the absence of the sc3 gene in *Schizophyllum commune* affects the cell wall composition, where the deletion of the sc3 gene increases the total amount of cell wall polysaccharide schizophyllan as well as decreases the amount of glucan cross-linked to chitin (Sietsma and Wessels 1977, 1981; van Wetter et al. 2000). Haneef et al. reported a high sensitivity to water in mycelia with low chitin content in the cell walls (Haneef et al. 2017). According to results obtained by Khoo et al., high water-resistant nature of their tested bio-boards was attributed to a thick impenetrable skin formed outside the structure by *Pleurotus* and *Ganoderma* binders (Khoo et al. 2020). During swelling, water is absorbed into the pores of the MBCs through capillary action. Therefore, the lower void volume in particulate fillers like sawdust and the higher density of the final MBC product can reduce the absorbance of the water into the material (Jones et al. 2020; Zimele et al. 2020; Elsacker et al. 2019).

27.7 Fire Resistance, Insulating Capability, and Acoustic Properties

When using MBCs as an innovative, novel material for application in building construction and packaging, it is paramount to consider their fire-retardant property, since these composites are unequivocally rich in combustible organic matter that can easily catch fire. Therefore, research on fire-resistant properties of MBCs needs to be encouraged to promote their usage in the above applications. Wheat- and rice husk-based MBCs start degradation at around 250 °C, react to fire in three degradation stages, and form a stable carbonaceous char at 600 °C (around 23% by weight), producing a relatively greater residual yield when compared to other polymers; this indicates lower amounts of toxic volatiles and smoke being released upon combustion (Jones et al. 2018b). This significant amount of char formation hinders the oxygen migration in the core and acts as a thermal insulator, thereby limiting combustion (Mouritz and Gibson 2007). Phosphorus, a commonly used flame retardant in polymers, had been found extensively within char and acts as a dehydration catalyst in the char generation reaction by forming phosphoric and polyphosphoric acids. As evident from scanning electron microscopic (SEM) images of pyrolyzed MBCs, their fibrous structures were preserved post-combustion due to the presence of mycelial chitin, which is an excellent flame retardant, thermostable compound (Jones et al. 2018c). The heat release rate, one of the most important features to be considered in the fire safety of a material, was found to be between 33 and 107 kW/m^2 (Jones et al. 2018b) in MBCs. Therefore, the heat release rate of MBCs is also lower than that of synthetic building materials (EPS foam, 114 kW/m^2; particleboard, 134 kW/m^2). Thus, MBCs can be used in the products in order to significantly ensure the fire security. Utilizing substrates like rice husks, which are rich in silicon and lignin, can further enhance the fire-related properties of MBCs. This is because plenty of more stable, silicon-rich residual char can be generated by the combustion of such material that can act as a better thermal and oxygen insulator (Jones et al. 2018b).

One of the practical applications of MBCs thus far has been their use in thermal insulators and acoustic materials, and, according to recent research, they can emulate commonly used synthetic insulators, such as polystyrene and polyurethane foam, in terms of performance. They can be jointly used within walls (intra wall panels) as thermal and acoustic insulations (Butu et al. 2020). The thermal conductivity (the ease with which heat can travel across a material) values of MBCs fall within the range of 0.05–0.07 W/m*K, which is a competitive value when compared with conventional insulation foams (XPS foam, 0.03º0.04 W/m*K; polyurethane, 0.006–0.18 W/m*K). While there is an optimal density for high insulator performance (Yang 2020), the low-density straw and hemp-based substrates also have shown promising results (Elsacker et al. 2019; Xing et al. 2018). The characteristic porous nature and the presence of air cavities within the MBCs that can trap large volumes of air can be noted as a reason limiting the heat conduction across the material. Low-moisture content is also associated with better thermal insulators, and completely dry conditions are required for MBCs to perform optimally (Yang et al. 2017). A study by Wimmers et al. discovered that sawdust-based MBCs have remarkable insulation properties (0.051–0.055 W/m*K) that are better than those of conventional wood panels (0.12 W/m*K–0.19 W/m*K) (Wimmers et al. 2019). Ecovative Design's prototype product Greensulate™ is a MBC insulation panel designed and tested at both residential and commercial levels (US Environmental Protection Agency 2019).

Pelletier et al. proposed the use of MBCs as acoustic insulators in automotive panels as well as in installations for acoustic damping (Pelletier et al. 2013). Acoustic absorbers work by converting sound waves into heat, which prevents sound accumulation and reflection. An acoustic absorption of over 70–75% at 1000 Hz was achieved in this study, with a 50–50% composition of switchgrass and sorghum fibres showing the highest absorption levels (Pelletier et al. 2013). A combination of surface porosity and the fibrous quality of the MBCs determines the degree of sound absorption. Surface porosity of MBCs causes the sound waves to enter the material while the fibres present inside generate a resistance to the air flow, thus damping the sound waves by converting them to heat (Jones et al. 2020).

27.8 Improvement of Material Properties

There are limitations of applications of MBCs partially due to the lack of documented information available regarding the physical and mechanical properties (Jones et al. 2020). Physically, after colonization, the surface of the MBC material appears white due to a mycelial mat formation. Several studies have reported the formation of a more smooth mycelial skin on top of the material, particularly with longer periods of colonization. Thus, the incubation time can be used to optimize physical and mechanical properties during the colonization step.

The overall physical and mechanical properties of MBCs can be improved through post-processing steps, such as cold pressing, hot pressing, and the addition of various chemicals. During cold pressing, the structure undergoes compaction due to the reduction in its porosity, and, when heat-pressed, the moisture inside the composite evaporates, causing steam to diffuse through the cells and the voids in the mycelium to increase the plasticity and density of the material. A non-enzymatic browning reaction that causes cross-linking between amino acids and reducing sugars is another factor that strengthens MBCs when heat-pressed (Martins et al. 2000).

27.9 Innovative Applications of MBCs

MBCs are increasingly growing in prevalence globally, with primary interest moving towards renewable alternatives in industrial applications. Many of the recent successful and potential uses of MBCs are discussed in recent scientific publications, patent applications, and various news sources regarding science and technology. Among them, only a limited number of successful business start-ups and industrial-scale applications are available worldwide, which may not fully satisfy the global demand. For instance, a few leading companies based in Europe and the USA, such as Ecovative Design, Mogu, the Magical Mushroom Company, MycoWorks, and Officina Corpuscoli, provide a range of innovative novel products based on MBCs. The applications of these composites include artwork, packaging, shipment padding, construction, textiles, shoe soles, automotive interior padding, model making, interior design, vertical green walls, raised garden beds, electrical circuit boards, and many more innovative alternatives replacing petroleum-based synthetic material.

27.10 Architectural Designs Using MBCs

MBCs are known for their potential use as a renewable substitute for architectural construction constituents. These composites are ideal for specific use as architectural raw materials due to their properties of insulation, fire resistance, and plasticity that can be exploited to produce desired shapes and patterns compared to synthetic petroleum-based materials, such as EPS, XPS (extruded polystyrene), and polyurethane (Xing et al. 2018). Mogu, a leading European company based in Italy, has developed mycelium-based wall panels, ceiling panels, and acoustic panels with remarkable sound proof properties that can be effectively used in the interior of the modern architectural designs (Fig. 27.2).

As an initiative of potential construction projects based on mycocomposites, the Hi-Fi tower, a mycelium brick structure opened in the courtyard of MoMA's PS1 (Museum of Modern Arts, Public School-1 building) space in New York and designed by Architect David Benjamin, has been popular in the last few years (Scott 2014). MycoTree, a spatial branching structure, which was featured at the Seoul Biennale of Architecture and Urbanism 2017 exhibition in Seoul, Korea, had been built using *Ganoderma lucidum* to bind bamboo plates in an innovative, geometrically weight-bearing architectural design (Heisel et al. 2017). Furthermore, Mycotecture Alpha, an architectural design by Phil Ross, was grown from the

Fig. 27.2 Mycelium-based interior products by Mogu (https://mogu.bio/). (**a**) Mycelium resilient—oyster tile and samples. (**b**) Mycelium-resilient floors. (**c**) Flooring mycelium panels—interiors, kitchen. (**d**) Acoustic mycelium panel—plain (interiors). (**e**) Acoustic mycelium panel. (**f**) Acoustic mycelium panel—interiors. (Courtesy: Images published with the copyright permission from Mogu Press and Communication team. MOGU S.r.l. Via San Francesco d'Assisi 62, 21020 Inarzo (VA), Italy)

mushroom *G. lucidum* (Reishi), made into MBC bricks, and stacked into an arch (Stone 2015). It was also confirmed that *Schizophyllum commune* mycelium can potentially be used in designing biocomposite building materials for habitats on Mars and the Moon (Lakk et al. 2018; Wösten et al. 2018). The unique properties of MBCs have made it possible to use them as thermal insulation panels, acoustic architectural designs, and other construction materials. Compared to the many man-made, synthetic materials used to make building materials, MBCs represent an excellent non-toxic, cleaner alternative for the growing needs in sustainable alternatives for construction and interior designs.

27.11 MBCs as Packaging Material

The majority of the retail products bought by consumers around the world come in plastic, polyethylene, or polystyrene packaging. Once the consumer receives the product, the packaging material is eventually disposed of in the environment, leading to the global crisis of plastic pollution. MBCs have been developed over time as various types of packaging material. These products are an ideal substitute for polystyrene and other synthetic hydrocarbon-based packaging materials (Abhijith et al. 2018). One of the leading start-up companies based on mycelium-based products, Ecovative Design, produced MBCs to be used as a wide range of packaging materials as an alternative to polystyrene and other synthetic packaging materials. Ikea, a leading multinational company founded in Sweden that designs and sells ready-to-assemble furniture, kitchen appliances, and home accessories around the world, recently announced their intention to use MBCs in their packaging materials (Hammon 2020). Apart from commercial providers of packaging material, various studies have been conducted to develop packaging solutions based on MBCs. For instance, six blends of processed cotton plant biomass materials were evaluated as a substrate for the colonization of selected fungi in the manufacture of moulded packaging materials (Holt et al. 2012). The same study revealed that the resulted MBCs had more favourable properties when compared to extruded polystyrene foam. Ecovative Design (Bayer et al. 2008) and Shenzhen Tech (Lacourse and Altieri 1989) have received several patents for mushroom packaging solutions and their production protocols (Cerimi et al. 2019).

Packaging solutions based on MBCs are an excellent example of a circular economy. These are bio-based materials resulting from a bioprocess similar to that of solid-state fermentation followed by a few processing steps. Therefore, the production process requires less energy. Once the synthetic packaging material is used, the remains are dumped into the environment directly unless it is not properly recycled. However, the biocomposite-based packaging material will be compostable, thus closing the loop, in accordance with the C2C model for materials, by reducing waste and promoting the continual use of resources. The hydrophilic nature of some mycelium materials gives them the advantage of superabsorbent materials. In contrast, the properties of MBCs, including hydrophobicity, low density, insulation, and fire resistance, have made it more promising sustainable future material in packaging applications.

27.12 Mycelium-Based Sandwich Composites and Their Applications

One of the most recently identified applications of mycelium materials is their use in making sandwich composites (Bayer et al. 2008). These composites comprise of a core made up of a lightweight mycelium-based material sandwiched within two thin, but relatively stiff, laminates. The uniqueness in its design provides MBSCs with advanced material properties, like greater shear strength, flexural strength, and tensile strength, over conventional synthetic materials, in addition to the advantage of being lightweight overall.

In contrast to the MBSCs, polymer matrix composites (PMCs) are the materials made up of fibres, embedded in an organic polymer matrix. The PMCs consist of three constituents including a matrix, reinforcement, and interface, which affect its overall mechanical properties. The matrices of PMCs are made up of a petroleum-based polymer component. The reinforcement can be either glass or carbon fibres with plastic resin, epoxy, polyester, and nylon metal sheets as laminate skins. The use of biodegradable materials like sawdust and agricultural waste to reinforce PMCs could seal them in an inert matrix, preventing the natural degradation process of those cellulosic components. Therefore, the PMCs reinforced with wood or agricultural waste are also considered to be less eco-friendly compared to MBSCs (Jiang et al. 2013, 2014a). Attempts have been made in the material technologies to overcome the disadvantage of PMCs, by seeking a bio-based approach to build MBSCs with natural fibres as reinforcements and natural glues as matrices in the laminate skins with MBCs in the cores, which would render the whole structure biodegradable. Ecovative Design, along with the Center for Automation Technologies and Systems (CATS) at Rensselaer Polytechnic Institute, pioneered research on this tech-front (Jiang et al. 2013, 2014a). In a series of studies by Jiang et al. (2013, 2014a, b, 2016, 2017a, b, 2019), a seven-step protocol was proposed to manufacture MBC laminate and MBSC structures with several process optimization stages.

For mycelium-based cores, different textile fabrics with natural fibres are being tested as a natural fibre reinforcement for the laminate (Jiang et al. 2013). The fabric can first be cut into the required ply shape using cutting dies, an industrial clicker press, laser cutters, or hand cutting. The laminate can be made with a single ply of the fabric or with multiple plies. The cut fabric plies should then be impregnated with a natural glue; for this, Jiang et al. used a pinch roller impregnator method (Jiang et al. 2013, 2014a, 2017b, 2019), and dip-and-soak, spray coating, and curtain coating were also tested (Jiang et al. 2013). Corn starch mixed with maltodextrin was used as the natural reinforcement glue, which acts as a matrix for binding the fabric fibres (Jiang et al. 2014a). This skin preform is next moulded to give it a defined shape and then sterilized before the core material is introduced. Moulding and sterilization were carried out simultaneously, using a matched mould-forming method in which the mould was heated, thus activating the glue and stiffening the preform at the same time while creating a gap between the two skin preforms to fill in the core material (Jiang et al. 2013, 2014a).

The selection of the lightweight substrate/filler for the core and the fungal strain as the binder takes place in the same way as in manufacturing basic MBCs. Loose, pre-colonized core materials are introduced between the laminates, and the incubation and mycelium inactivation steps are similar to basic MBC production protocols. Table 27.1 depicts the main components of the sandwich structures used in a few selected studies. During incubation, the mycelium colonizes the reinforcement fibres and core fillers, firmly binding the two skins to the core and forming a unitized sandwich composite within 5–7 days. In addition to the filler material, the impregnated natural glues in the skins also act as a substrate for the colonizing mycelium (Jiang et al. 2014a).

Jiang et al. (2014b, 2019), further experimented upon infusing bio-resins (soy-based polyols) into the material, thus forming MBSCs, in order to further increase their strength, stiffness, and other material properties. Compared with a resin-less MBSC beam, the resin-infused beams had shown better core shear ultimate stress, core shear yield stress, skin ultimate stress, and flexural strengths.

Ziegler et al. (2016) suggest packaging, shipping, marine flotation, and other non-structural applications for these sandwich structures due to their comparatively

Table 27.1 Examples of laminate and sandwich biocomposite structures with mycelium-based cores

Skin/laminate	Core substrate	Reference(s)
100% jute burlap (loose-weave) 100% linen cloth (tight-weave)	Unspecified agricultural waste	Jiang et al. (2013, 2014a, 2016)
Biotex jute Biotex flax BioMid cellulose	Kenaf/hemp mix (50/50% by weight) and corn Stover/hemp mix (50/50% by weight)	Jiang et al. (2014b, 2017a, 2019)
Recycled jute Recycled cotton Cornstarch	Sawdust spawn block mixture with Flour, feathers, and psyllium husk	Silverman (2018)
Unspecified woven/ non-woven mat	Cotton by-product (ginning waste) and hemp	Ziegler et al. (2016)
Biotex jute (plain weave) Biotex flax twill	Kenaf/hemp	Jiang et al. (2017b)

lower resistance to compression. Moreover, doors, cubicle walls, basement panelling, conventional insulating applications, roof insulation, and table tops were also suggested as potential applications (Bayer et al. 2008).

27.13 MBCs in the Fashion Industry

Textile and fashion industries have rapidly evolved from traditional plant- and animal-based natural products to synthetic products over las few dacades. The annual global production of all apparel and textile fibres is estimated at more than 110 million tons (Ütebay et al. 2020). The accumulation of piles of textile waste and the release of hazardous effluents from fabric industries have caused a major environmental threat globally. Additionally, growing concerns regarding consumers rejecting animal leather have also triggered the fashion industries to seek biological alternatives for their raw materials.

The project MycoTEX/Mycelium Textile by an artist, Aniela Hoitink from the Netherlands, investigates the use of pure fungal mycelia as material for dresses (NEFFA 2004–2020). MycoTEX team has extended the project to create biodegradable MycoTEX fabric grown from mycelium which has gained the attention of the scientific community, fashion industry, as well as the field of contemporary arts and crafts (Cairns et al. 2019; Goncu-Berk 2019; Nai and Meyer 2016; Meyer 2021). A flexible composite product called MycoTEX developed by Aniela Hoitink comprised of disc-shaped mycelium and placed them on a mannequin to shape the dress, retaining the flexibility needed without using traditional textile material (Vasquez and Vega 2019; Nayak et al. 2020).

The specific application tested by Jiang et al. was a benchmark product of Ecovative, i.e. a shoe sole for an outdoor sandal (Jiang et al. 2014a, 2017b). Cost modelling was conducted as well to minimize the cost and maximize the efficiency in the optimization of the manufacturing system of this product (Jiang et al. 2016). Mycelium-based cores were also tested for another shoe sole application, in which the design was an incomplete sandwich with one skin and the core (Silverman 2018). Its applicability was suggested for the sole of a casual shoe/sandal with less compressive force given by the wearer. Mushroom MBCs derived from reishi, oyster, king oyster, and yellow oyster with or without a natural fabric mat have been tested as a potential source for footwear products (Silverman et al. 2020). Among them, the material made from king oyster mycelium provided higher compressive strength when compared to other mushroom species tested in the same study.

27.14 Future Directions of MBC Research

Although both tropical and temperate species of mushroom-forming fungi have been used in research and start-ups related to MBC production, only a few species of the fungi kingdom have been tested and developed into viable products thus far. Therefore, new fungi from different ecological niches can be explored for the potential use of mycelium-based product discovery and development (Attias et al. 2019). Despite the approximately similar chemical composition of mycelium compared to diverse groups of fungi, the physical and mechanical properties of mycelium can vary among species due to their growth characteristics; nature of substratum; growth conditions such as temperature, relative humidity, etc.; and processing techniques (Jones et al. 2019). The genetic and environmental factors affecting the chemical composition and mechanical properties of mycelium have to be investigated in detail in order to improve the productivity and quality of the MBCs (Jones et al. 2018a). Fungi growing under mesophilic conditions which also produce dense mycelium are ideal for high-quality MBC-based products. Improvements in the mechanical performance of the mycelium have been challenging in most MBC research in the past decade. The deletion of the hydrophobin gene from a strain of *Schizophyllum commune* has resulted in a remarkable increase in the mechanical properties of mycelium material, such as the Young's modulus and maximum tensile strength (Appels et al. 2018; van Wetter et al. 1996; Momenteller 2017). Therefore, an approach based on genetic manipulations as well as rigorous optimization of environmental factors could be used to improve the properties of mycelium, thus enhancing the efficiency and cost of production. In addition, the incorporation of material engineering and nanotechnology in MBC research could result into major tech breakthroughs in developing healthy, renewable materials that may eventually resolve many prevailing environmental issues.

27.15 Concluding Remarks

Globally, the "fungal mycelium-based materials" is a topic that experiences a surge in popularity among researchers working on biocomposites and environmental sustainability. MBCs are significant, in terms of their capability to attract scientists, investors, and policy-makers looking forward to overcome challenges in global environmental issues. This is because MBCs and their variants are eco-friendly alternatives for various polystyrene foams and similar products that cause a serious environmental crisis upon large-scale disposal to the environment. Furthermore, possible rewards of using renewable sources of materials are highly complementary with the goals of the 2030 agenda for sustainable development adopted by the United Nations. Therefore, we encourage more research and innovative projects based on MBCs worldwide, through multidisciplinary collaborations among scientists, technologists, and prospective investors from various industries.

References

Abhijith R, Ashok A, Rejeesh CR (2018) Sustainable packaging applications from mycelium to substitute polystyrene: a review. Mater Today Proc 5(1):2139–2145. https://doi.org/10.1016/j.matpr.2017.09.211

Agustina W, Aditiawati P, Kusumah SS, Dungani R (2019) Physical and mechanical properties of composite boards from the mixture of palm sugar fiber and cassava bagasse using mycelium of *Ganoderma lucidum* as a biological adhesive. In: The eighth international symposium for sustainable Humanosphere, IOP conference series: earth and environmental science, Medan, Indonesia, 18–19 Oct 2018, vol 374. IOP Publishing, p 012012

Appels FVW, Wösten HAB (2020) Mycelium materials. In: Reference module in life sciences. Elsevier, Amsterdam. https://doi.org/10.1016/B978-0-12-809633-8.21131-X

Appels FV, Dijksterhuis J, Lukasiewicz CE, Jansen KM, Wösten HA, Krijgsheld P (2018) Hydrophobin gene deletion and environmental growth conditions impact mechanical properties of mycelium by affecting the density of the material. Sci Rep 8(1):1–7. https://doi.org/10.1038/s41598-018-23171-2

Appels FV, Camere S, Montalti M, Karana E, Jansen KM, Dijksterhuis J, Krijgsheld P, Wösten HA (2019) Fabrication factors influencing mechanical, moisture-and water-related properties of mycelium-based composites. Mater Des 161:64–71. https://doi.org/10.1016/j.matdes.2018.11.027

Arifin YH, Yusuf Y (2013) Mycelium fibers as new resource for environmental sustainability. Proc Eng 53:504–508. https://doi.org/10.1016/j.proeng.2013.02.065

Attias N, Danai O, Ezov N, Tarazi E, Grobman YJ (2017) Developing novel applications of mycelium based bio-composite materials for design and architecture. In: Proceedings of building with biobased materials: best practice and performance specification, 6–7 September 2017, pp 76–77

Attias N, Danai O, Tarazi E, Pereman I, Grobman YJ (2019) Implementing bio-design tools to develop mycelium-based products. Des J 22:1647–1657. https://doi.org/10.1080/14606925.2019.1594997

Attias N, Danai O, Abitbol T, Tarazi E, Ezov N, Pereman I, Grobman YJ (2020) Mycelium bio-composites in industrial design and architecture: comparative review and experimental analysis. J Clean Prod 246:119037. https://doi.org/10.1016/j.jclepro.2019.119037

Bathurst RJ, Keshavarz A, Zarnani S, Take WA (2007) A simple displacement model for response analysis of EPS geofoam seismic buffers. Soil Dyn Earthq Eng 27(4):344–353. https://doi.org/10.1016/j.soildyn.2006.07.004

Bayer E, McIntyre G, Swersey BL (2008) Method for producing grown materials and products made thereby. US Patent 12/001,556, 19 Jun 2008

Blackwell M (2011) The fungi: 1, 2, 3... 5.1 million species? Am J Bot 98(3):426–438. https://doi.org/10.3732/ajb.1000298

Bowman SM, Free SJ (2006) The structure and synthesis of the fungal cell wall. BioEssays 28 (8):799–808. https://doi.org/10.1002/bies.20441

Bruscato C, Malvessi E, Brandalise RN, Camassola M (2019) High performance of macrofungi in the production of mycelium-based biofoams using sawdust—sustainable technology for waste reduction. J Clean Prod 234:225–232. https://doi.org/10.1016/j.jclepro.2019.06.150

Butu A, Rodino S, Miu B, Butu M (2020) Mycelium-based material for the ecodesign of bioeconomy. Dig J Nanomater Biostruct 15(4):1129–1140

Cairns TC, Zheng X, Zheng P, Sun J, Meyer V (2019) Moulding the mould: understanding and reprogramming filamentous fungal growth and morphogenesis for next generation cell factories. Biotechnol Biofuels 12(1):77. https://doi.org/10.1186/s13068-019-1400-4

Cerimi K, Akkaya KC, Pohl C, Schmidt B, Neubauer P (2019) Fungi as source for new bio-based materials: a patent review. Fungal Biol Biotechnol 6(1):1–10. https://doi.org/10.1186/s40694-019-0080-y

Elsacker E, Vandelook S, Brancart J, Peeters E, De Laet L (2019) Mechanical, physical and chemical characterisation of mycelium-based composites with different types of lignocellulosic substrates. PLoS One 14(7):e0213954. https://doi.org/10.1371/journal.pone.0213954

Ghazvinian A, Farrokhsiar P, Vieira F, Pecchia J, Gursoy B (2019) Mycelium-based bio-composites for architecture: assessing the effects of cultivation factors on compressive strength. In: Sousa JP, Henriques GC, Xavier JP (eds) The eCAADe and SIGraDi conference. University of Porto, Portugal, pp 11–13

Girometta C, Picco AM, Baiguera RM, Dondi D, Babbini S, Cartabia M, Pellegrini M, Savino E (2019) Physico-mechanical and thermodynamic properties of mycelium-based biocomposites: a review. Sustainability 11(1):281. https://doi.org/10.3390/su11010281

Goncu-Berk G (2019) Smart textiles and clothing: an opportunity or a threat for sustainability. In: Proceedings of the textile intersections, London, 12–14 Sept 2019

Hammon D (2020) IKEA commits to biodegradable mushroom packaging. Yahoo News. news.yahoo.com. Accessed 30 Jan 2021

Haneef M, Ceseracciu L, Canale C, Bayer IS, Heredia-Guerrero JA, Athanassiou A (2017) Advanced materials from fungal mycelium: fabrication and tuning of physical properties. Sci Rep 7(1):1–11. https://doi.org/10.1038/srep41292

Heisel F, Schlesier K, Lee J, Rippmann M, Saeidi N, Javadian A, Nugroho AR, Hebel D, Block P (2017) Design of a load-bearing mycelium structure through informed structural engineering. In: World congress on sustainable technologies (WCST-2017), University of Cambridge, 11--14 December 2017, pp 45–49

Holt GA, Mcintyre G, Flagg D, Bayer E, Wanjura JD, Pelletier MG (2012) Fungal mycelium and cotton plant materials in the manufacture of biodegradable molded packaging material: evaluation study of select blends of cotton byproducts. J Biobaased Mater Bioenergy 6(4):431–439. https://doi.org/10.1166/jbmb.2012.1241

Hyde KD, Xu J, Rapior S et al (2019) The amazing potential of fungi: 50 ways we can exploit fungi industrially. Fungal Divers 97:1–136. https://doi.org/10.1007/s13225-019-00430-9

Iordache O, Perdum E, Mitran EC, Chivu A, Dumitrescu I, Irina-Mariana MF (2018) Novel myco-composite material obtained with *Fusarium oxysporum*. In: International conference on advanced materials and systems (ICAMS), Bucharest, 18–20 Oct 2018, pp 111–116

IUPAC (1997) Compendium of chemical terminology, 2nd edn (the "Gold Book"). Compiled by McNaught AD, Wilkinson A. Blackwell Scientific Publications, Oxford. Online version (2019-) created by Chalk SJ. ISBN 0-9678550-9-8. https://doi.org/10.1351/goldbook

Jastrzebski ZD (1959) Nature and properties of engineering materials, International edn. Wiley, New York

Jiang L, Walczyk D, Mooney L, Putney S (2013) Manufacturing of mycelium-based biocomposites. In: Proceedings of the international SAMPE technical conference, pp 1944–1955

Jiang L, Walczyk D, McIntyre G (2014a) A new process for manufacturing biocomposite laminate and sandwich parts using mycelium as a binder. ASC 2014 proceedings, pp 8–10

Jiang L, Walczyk DF, McIntyre G (2014b) Vacuum infusion of mycelium-bound biocomposite preforms with natural resins. In: CAMX conference proceedings, Orlando, 13–16 Oct 2014, pp 13–16

Jiang L, Walczyk D, McIntyre G, Chan WK (2016) Cost modeling and optimization of a manufacturing system for mycelium-based biocomposite parts. J Manuf Syst 41:8–20. https://doi.org/10.1016/j.jmsy.2016.07.004

Jiang L, Walczyk D, McIntyre G, Bucinell R, Tudryn G (2017a) Manufacturing of biocomposite sandwich structures using mycelium-bound cores and preforms. J Manuf Process 28:50–59. https://doi.org/10.1016/j.jmapro.2017.04.029

Jiang L, Walczyk D, McIntyre G (2017b) A new approach to manufacturing biocomposite sandwich structures: investigation of preform shell behavior. J Manuf Sci Eng 139(2):021014. https://doi.org/10.1115/1.4034278

Jiang L, Walczyk D, McIntyre G, Bucinell R, Li B (2019) Bioresin infused then cured mycelium-based sandwich-structure biocomposites: resin transfer molding (RTM) process, flexural properties, and simulation. J Clean Prod 207:123–135. https://doi.org/10.1016/j.jclepro.2018.09.255

Jones M, Huynh T, Dekiwadia C, Daver F, John S (2017a) Mycelium composites: a review of engineering characteristics and growth kinetics. J Bionanosci 11(4):241–257. https://doi.org/10.1166/jbns.2017.1440

Jones M, Bhat T, Wang CH, Moinuddin K., John S (2017b) Thermal degradation and fire reaction properties of mycelium composites. In: Proceedings of the 21st international conference on composite materials, Xi'an, China, pp 20–25

Jones M, Huynh T, John S (2018a) Inherent species characteristic influence and growth performance assessment for mycelium composite applications. Adv Mater Lett 9(1):71–80. https://doi.org/10.5185/amlett.2018.1977

Jones M, Bhat T, Huynh T, Kandare E, Yuen R, Wang CH, John S (2018b) Waste-derived low-cost mycelium composite construction materials with improved fire safety. Fire Mater 42 (7):816–825. https://doi.org/10.1002/fam.2637

Jones M, Bhat T, Kandare E, Thomas A, Joseph P, Dekiwadia C, Yuen R, John S, Ma J, Wang CH (2018c) Thermal degradation and fire properties of fungal mycelium and mycelium-biomass composite materials. Sci Rep 8(1):1–10. https://doi.org/10.1038/s41598-018-36032-9

Jones MP, Lawrie AC, Huynh TT, Morrison PD, Mautner A, Bismarck A, John S (2019) Agricultural by-product suitability for the production of chitinous composites and nanofibers utilising *Trametes versicolor* and *Polyporus brumalis* mycelial growth. Process Biochem 80:95–102. https://doi.org/10.1016/j.procbio.2019.01.018

Jones M, Mautner A, Luenco S, Bismarck A, John S (2020) Engineered mycelium composite construction materials from fungal biorefineries: a critical review. Mater Des 187:108397. https://doi.org/10.1016/j.matdes.2019.108397

Kalisz RE, Rocco CA (2011) Hardened mycelium structure and method. US Patent 12/796,876, 15 Dec 2011

Khoo SC, Peng WX, Yang Y, Ge SB, Soon CF, Ma NL, Sonne C (2020) Development of formaldehyde-free bio-board produced from mushroom mycelium and substrate waste. J Hazard Mater 400:123296. https://doi.org/10.1016/j.jhazmat.2020.123296

Lacourse NL, Altieri PA (1989) Biodegradable packaging material and the method of preparation thereof. US Patent 4,863,655, 5 Sept 1989

Lakk H, Krijgsheld P, Montalti M, Woesten H (2018) Fungal based biocomposite for habitat structures on the Moon and Mars. In: 69th International astronautical congress, Berlin, 1–5 Oct 2018

Lelivelt RJJ (2015) The mechanical possibilities of mycelium materials. Dissertation, Eindhoven University of Technology (TU/e)

Martins SI, Jongen WM, Van Boekel MA (2000) A review of Maillard reaction in food and implications to kinetic modelling. Trends Food Sci Technol 11(9–10):364–373. https://doi.org/10.1016/S0924-2244(01)00022-X

Meyer V (2021) The beauty and the morbid: fungi as source of inspiration in contemporary art. w/k-Zwischen Wissenschaft & Kunst

Momenteller R (2017) Comparing two *Schizophyllum* species' developmental strategies through their hydrophobins. Dissertation, Southern Illinois University at Edwardsville

Mouritz AP, Gibson AG (2007) Fire properties of polymer composite materials (volume 143 of solid mechanics and its applications). Springer, New York

Nai C, Meyer V (2016) The beauty and the morbid: fungi as source of inspiration in contemporary art. Fungal Biol Biotechnol 3(1):1–5. https://doi.org/10.1186/s40694-016-0028-4

Nayak R, Nguyen LVT, Panwar T, Jajpura L (2020) Sustainable technologies and processes adapted by fashion brands. In: Nayak R (ed) Sustainable technologies for fashion and textiles. Woodhead Publishing, Sawston, pp 233–248

NEFFA (2004–2020). https://neffa.nl/portfolio/mycotex/. Accessed 30 Jan 2021

Pegler DN (1996) Hyphal analysis of basidiomata. Mycol Res 100(2):129–142. https://doi.org/10.1016/S0953-7562(96)80111-0

Pelletier MG, Holt GA, Wanjura JD, Bayer E, McIntyre G (2013) An evaluation study of mycelium based acoustic absorbers grown on agricultural by-product substrates. Ind Crop Prod 51:480–485. https://doi.org/10.1016/j.indcrop.2013.09.008

Pelletier MG, Holt GA, Wanjura JD, Lara AJ, Tapia-Carillo A, McIntyre G, Bayer E (2017) An evaluation study of pressure-compressed acoustic absorbers grown on agricultural by-products. Ind Crop Prod 95:342–347. https://doi.org/10.1016/j.indcrop.2016.10.042

Rathore H, Prasad S, Kapri M, Tiwari A, Sharma S (2019) Medicinal importance of mushroom mycelium: mechanisms and applications. J Funct Foods 56:182–193. https://doi.org/10.1016/j.jff.2019.03.016

Scott R (2014) Hy-Fi, the organic mushroom-brick tower opens at MoMA's PS1 courtyard. ArchDaily. https://www.archdaily.com/521266/hy-fi-the-organic-mushroom-brick-tower-opens-at-moma-s-ps1-courtyard. Accessed 30 Jan 2021

Sietsma JH, Wessels JGH (1977) Chemical analysis of the hyphal wall of *Schizophyllum commune*. Biochim Biophys Acta 496:225–239. https://doi.org/10.1016/0304-4165(77)90131-3

Sietsma JH, Wessels JGH (1981) Solubility of (1–3)-β-D/(1–6)-β-D-glucan in fungal walls: importance of presumed linkage between glucan and chitin. J Gen Microbiol 125:209–212. https://doi.org/10.1099/00221287-125-1-209

Silverman J (2018) Development and testing of mycelium-based composite materials for shoe sole applications. Dissertation, University of Delaware

Silverman J, Cao H, Cobb K (2020) Development of mushroom mycelium composites for footwear products. Cloth Text Res J 38(2):119–133. https://doi.org/10.1177/2F0887302X19890006

Stone M (2015) The technology that will build our future may be found in mushrooms. Gizmodo. https://www.gizmodo.com.au/2015/04/the-technology-that-will-build-our-future-may-be-found-in-mushrooms/. Accessed 30 Jan 2021

Sun W, Tajvidi M, Hunt CG, McIntyre G, Gardner DJ (2019) Fully bio-based hybrid composites made of wood, fungal mycelium and cellulose nanofibrils. Sci Rep 9(1):1–12. https://doi.org/10.1038/s41598-019-40442-8

Tacer-Caba Z, Varis JJ, Lankinen P, Mikkonen KS (2020) Comparison of novel fungal mycelia strains and sustainable growth substrates to produce humidity-resistant biocomposites. Mater Des 192:108728

US Environmental Protection Agency (2019) Science in action- fact sheet, National Service Center for environmental publications (NSCEP). https://nepis.epa.gov. Accessed 7 Jan 2021

Ütebay B, Çelik P, Çay A (2020) Textile wastes: status and perspectives. In: Körlü A (ed) Waste in textile and leather sectors. IntechOpen, London

Vasquez ESL, Vega K (2019) Myco-accessories: sustainable wearables with biodegradable materials. In: Proceedings of the 23rd international symposium on wearable computers, London, 11–13 Sept 2019, pp 306–311

van Wetter MA, Schuren FHJ, Schuurs TA, Wessels GJ (1996) Targeted mutation of the SC3 hydrophobin gene of *Schizophyllum commune* affects formation of aerial hyphae. FEMS Microbiol Lett 140(2–3):265–269. https://doi.org/10.1111/j.1574-6968.1996.tb08347.x

van Wetter MA, Wösten HAB, Sietsma JH, Wessels JGH (2000) Hydrophobin gene expression affects hyphal wall composition in *Schizophyllum commune*. Fungal Genet Biol 31:99–104. https://doi.org/10.1006/fgbi.2000.1231

Wimmers G, Klick J, Tackaberry L, Zwiesigk C, Egger K, Massicotte H (2019) Fundamental studies for designing insulation panels from wood shavings and filamentous fungi. Bioresources 14(3):5506–5520

Wösten H, Krijgsheld P, Montalti M, Läkk H, Summerer L (2018) Growing fungi structures in space. European Space Agency

Xing Y, Brewer M, El-Gharabawy H, Griffith G, Jones P (2018) Growing and testing mycelium bricks as building insulation materials. In: IOP conference series: earth and environmental science, vol 121. IOP Publishing, Bristol, p 022032

Yang K (2020) Investigations of mycelium as a low-carbon building material. Dissertation, Thayer School of Engineering Dartmouth College Hanover, New Hampshire

Yang Z, Zhang F, Still B, White M, Amstislavski P (2017) Physical and mechanical properties of fungal mycelium-based biofoam. J Mater Civ Eng 29(7):04017030

Zarnani S, Bathurst RJ (2007) Experimental investigation of EPS geofoam seismic buffers using shaking table tests. Geosynth Int 14(3):165–177. https://doi.org/10.1680/gein.2007.14.3.165

Ziegler AR, Bajwa SG, Holt GA, McIntyre G, Bajwa DS (2016) Evaluation of physico-mechanical properties of mycelium reinforced green biocomposites made from cellulosic fibers. Appl Eng Agric 32(6):931–938

Zimele Z, Irbe I, Grinins J, Bikovens O, Verovkins A, Bajare D (2020) Novel mycelium-based biocomposites (MBB) as building materials. J Renew Mater 8(9):1067–1076. https://doi.org/10.32604/jrm.2020.09646

Chapter 28
An Endophytic Bacterial Approach: A Key Regulator of Drought Stress Tolerance in Plants

Sudha Bind, Sandhya Bind, and Dinesh Chandra

Abstract Drought is the chief limiting factor for crop productivity around the world. The continuous rise in temperature due to global climate change leads to scarcity of water which directly enhances drought frequency. One eco-friendly and safer approach is to utilize the endophytic bacteria as biofertilizers for agriculture productivity and sustainability. Plant-endophytic bacteria improve plant health through three types of interaction; they may be symbiotic, mutualistic and commensalistic. Endophytic bacteria have direct and indirect working mechanism for plant growth promotion. Direct mechanism involves making nutrient accessible to the plants via fixing nitrogen and solubilizing the phosphate, zinc and potassium, and siderophore, phytohormone and 1-aminocyclopropane-1-carboxylate (ACC) deaminase production, whereas indirect mechanism involves production of antibiotics, exopolysaccharide and hydrolytic enzymes and competition for nutrient and space to inhibit the pathogen. Thus, the aim of this chapter is to provide a better understanding of plant-endophytic bacterial interaction in amelioration of drought stress and determinants produced by endophytic bacteria beneficial in plant growth promotion under drought stress.

Keywords Endophytes · ACC deaminase · Drought · Sustainability · Nutrient uptake

---

S. Bind · S. Bind
Department of Biological Sciences, CBS&H, G.B. Pant University of Agriculture and Technology, Pantnagar, Uttarakhand, India

D. Chandra (✉)
Department of Biological Sciences, CBS&H, G.B. Pant University of Agriculture and Technology, Pantnagar, Uttarakhand, India

GIC Chamtola, Almora, Uttarakhand, India

P. Bhatt et al. (eds.), *Microbial Technology for Sustainable Environment*,
https://doi.org/10.1007/978-981-16-3840-4_28

28.1 Introduction

Drought is a critical abiotic stress throughout the world and its intensity is increasing due to climate change and global warming. Drought negatively affects the crop yield and productivity. According to the studies from 1980 to 2015, drought reduced 40% yield in maize and 21% in wheat (Daryanto et al. 2016). To provide food to the huge population of the world, productivity of crop must be enhanced under drought conditions. Drought can be divided into metrological, agricultural and hydrological drought (Dai 2011). A period of months to years with below normal precipitation is called metrological drought. In metrological drought, the temperature remains above the normal temperature, also caused by high temperature and pressure. Hydrological drought occurs because of low water content in rivers, streams, lakes and ground water after the long persistence of metrological drought (Van Loon 2015). Drought disturbs the water potential and turgor of the plant and thus interferes with normal physiological functioning (Hsiao 2000).

28.2 Responses of Drought in Plant

Drought stress causes reduction in diffusion and supply of nutrients such as nitrate, calcium, sulphate, magnesium and silicon (Selvakumar et al. 2012). Drought also increases oxidative stress in plant by generating high quantity of reactive oxygen species (ROS) such as superoxide radical, hydrogen peroxide and hydroxyl radical (Chandra et al. 2019a, b). A high concentration of ROS induces destruction of the lipid, nucleic acid and protein in cells (Sgherri et al. 2000). Under drought stress, chlorophyll concentration decreases due to photo-oxidation (Rahdari et al. 2012; Chandra et al. 2018a, b). Drought negatively affects the nitrate reductase functionality and thus leads to lower uptake of nitrate from the soil (Caravaca et al. 2005). Ethylene concentration also increased under drought condition which inhibits plant growth (Ali et al. 2014).

28.3 Mechanism of Drought Tolerance in Plants

Plants resist drought by four mechanisms: recovery, tolerance, avoidance and escape. Drought tolerance is the capability of plants to withstand severe dehydration via osmotic regulation and osmo-protectants. Various methods are being used to increase the plant tolerance towards drought stress for higher crop production. Methods being currently used in agriculture are non-renewable and expensive (Ullah et al. 2019; Chandra et al. 2020). Excessive usage of chemicals as fertilizers and pesticides creates environmental pollution and generation of greenhouse gases that negatively affected the soil property, thus causing a hazardous effect on human health (Sulbhi et al. 2021; Bhandari et al. 2021; Bhatt et al. 2020a, b, 2021a, b, c, d, e, f). These xenobiotics enter into the food chain via the process known as biomagnification (Bhatt et al. 2019a, 2020c, d, e, f). Conventional breeding and genetic engineering were used for the production of drought-tolerant and high-yielding varieties of various plants (Ngumbi and Kloepper 2016). Conventional breeding has some drawbacks; it requires a lot of labour and time and loses other useful traits and breeding can be granted one crop at one time (Philippot et al. 2013). On the other hand, genetic engineering is faster but the development of genetically modified crops brings about time and labour challenges until commercialization (Ullah et al. 2019; Eisenstein 2013). Moreover, due to response of consumers to the genetically modified crop which varies among various countries, the success of transgenic crops on the market cannot be guaranteed (Fedoroff et al. 2010). Recently, plant-endophytic bacterial interaction receives much attention because of their plant growth enhancement ability under drought condition (Finkel et al. 2017).

28.4 Endophytic Bacteria: A Sustainable Approach to Reduce Drought Stress

Endophytic bacteria are considered as beneficial bacteria which live inside the plant tissue without any harmful effect. Plant-endophytic bacterial interaction may be symbiotic, mutualistic and commensalistic. Endophytic bacteria are present in almost all the parts of plants; they are isolated and characterized from different types of plant and plant parts (such as seeds, roots, leaves, stems, fruits, etc.). Roots contain the highest population of endophytic bacteria as compared to the other aboveground plant parts (Rosenblueth and Martínez-Romero 2006). Endophytic bacteria have a broad host range, making them a powerful tool for sustainable agriculture. Endophytic bacteria present in all plants are studied till now, and plants without endophytic bacteria are rare exception in nature (Partida-Martinez and Heil 2011). Plants without endophytic bacteria are more susceptible against phytopathogens and stress conditions (Timmusk et al. 2011). Endophytic bacteria having ACC deaminase confer resistance against various types of abiotic stress (high temperature, drought, salinity, flood, acidity) by reducing the level of ethylene inside the plant tissues. They also break down ACC, which is a precursor of ethylene, into α-ketobutyrate and ammonia, thus eliminating the negative impact of high ethylene concentration on plant growth. Numerous studies on drought tolerance in plants mediated by endophytic bacteria are summarized in Table 28.1.

Table 28.1 Drought tolerance in plants by endophytic bacteria

Endophytic bacteria	Benefitted plant	Source of plant species	Effect of bacterial endophytes on plant	References
Bacillus sp. (Acb9), *Providencia* sp. (Acb11), *Staphylococcus* sp. (Acb12) *Staphylococcus* sp. (Acb13) and *Staphylococcus* sp. (Acb14)	*Ananas comosus*	Leaf	IAA, ACC production, nitrogen fixation	Jayakumar et al. (2020)
Staphylococcus sp. Ceb1	*Curcuma longa*	Rhizome	IAA production	
Bacillus megaterium HX-2	*Agastache rugosa*	Leaf	IAA production, phosphate solubilization	Li et al. (2019)
Bacillus cereus, *B. amyloliquefaciens*	*Fagonia mollis*	Leaf	IAA production, phosphate solubilization, cellulase, gelatinase, pectinase, xylanase	ALKahtani et al. (2020)
Brevibacillus agri	*Achillea fragrantissima*	Leaf	Ammonia, IAA production	
Achromobacter xylosoxidans	*Helianthus annuus*	Root	IAA, jasmonic acid, 12-oxo-phytodienoic acid, ABA, phosphate solubilization	Forchetti et al. (2007)
Enterobacter cloacae EB-48, *Microbacterium* sp. EB-65	*Sorghum bicolor*	Root	P-solubilization, siderophores, IAA production, high expression of sbP5CS 2 gene, N-fixation, ACC deaminase production	Govindasamy et al. (2020)
Bacillus subtilis, *Paenibacillus illinoinensis*	*Capsicum annuum*	Root	Phosphate solubilization, siderophore production, overexpression of root vacuolar proton pump H+-PPase (V-PPase)	Vigani et al. (2019)
Paraburkholderia megapolitana MGT9	*Ricinus communis*		Auxin production, phosphate solubilization, siderophore production enhanced proline, CAT, SOD, POD and GR antioxidant activity	Trivedi et al. (2019)
Pseudomonas azotoformans	*Alyssum serpyllifolium*	Leaf	Increased enzymatic activity of POD, SOD, CAT and proline content	Timmusk et al. (2014)
Sphingomonas sp. LK11	*Tephrosia apollinea*	Leaf	ABA, jasmonic acid production, enhanced proline, glycine, glutathione and glutamine content	Khan et al. (2014)

Bacillus amyloliquefaciens SB-9	*Vitis vinifera*	Root	Melatonin, tryptamine, 5-hydroxytryptophan, serotonin and *N*-acetylserotonin production	Jiao et al. (2016)
Bacillus subtilis B26	*Panicum virgatum*	Leaf	Phosphate solubilization, cellulase production	Gagne-Bourgue et al. (2013)
Pseudomonas frederickbergensis	*Phoenix dactylifera*	Root	Siderophore, auxin production and phosphate solubilization	Cherif et al. (2015)
Pantoea alhagi	*Alhagi sparsifolia*	Leaf	IAA, siderophore, EPS, protease, ammonia production, phosphate solubilization	Chen et al. (2017)
B. amyloliquefaciens	*Euphorbia trigona*	Stem, root	EPS production, phosphate solubilization	Eke et al. (2019)
Bacillus licheniformis	*Lepidium perfoliatum*	Root	Catalase, protease, esterase, amylase, biofilm formation	Li et al. (2017)
Staphylococcus pasteuri strain MBL_B3, *Bacillus* sp. strain MBL_B15, *Kocuria* sp. strain MBL_B19	*Corchorus olitorius*	Leaf, root, seed and seedling	Phosphate solubilization, siderophores, IAA and ACC deaminase production	Haidar et al. (2018)
Arthrobacter sp. *Bacillus* sp. TW4	*Capsicum annum*	Stem	ACC deaminase, IAA production	Sziderics et al. (2007)
Klebsiella sp.	*Alhagi sparsifolia*	Root	IAA, protease, ammonia, acetoin, 2,3-butanediol production, nitrogen fixation	Zhang et al. (2017)
Pseudomonas azotoformans ASS1	*Trifolium arvense*	Leaves of *Alyssum serpyllifolium*	Enhanced chlorophyll content; increased Cu, Zn and Ni uptake; and also enhanced proline, antioxidant enzymes, decreased MDA content	Ma et al. (2017)
Pseudomonas plecoglossicida	*Vitis vinifera*	Root	1AA, siderophore, ammonia production, phosphate solubilization	Rolli et al. (2015)
Pseudomonas poae	*Astragalus mongholicus*	Root	IAA production, drought tolerant	Sun et al. (2019)
Bacillus pseudomycoides, *B. subtilis* subsp. *inaquosorum*	*Zea mays*	Seed	IAA, ammonia, protease, lipase, esterase, amylase, cellulase, pectinase	Bodhankar et al. (2017)

(continued)

Table 28.1 (continued)

Endophytic bacteria	Benefitted plant	Source of plant species	Effect of bacterial endophytes on plant	References
Bacillus subtilis WL19, *B. thuringiensis* WS11	*Triticum aestivum*	Leaf, stem	IAA, catalase, oxidase, chitinase, HCN and EPS production, phosphate solubilization, carbonic anhydrase activity	Aslam et al. (2018)
Xanthomonas sacchari, *Bacillus idriensis*, *Pseudomonas geniculate*	*Dodonaea viscosa*	Root	Ammonia, HCN, IAA production phosphate solubilization, protease, pectinase, cellulase, chitinase	Afzal et al. (2017)
Bacillus subtilis LK14	*Moringa peregrina*	Bark	IAA production, ACC deaminase, phosphate solubilization	Latif Khan et al. (2016)
Pseudomonas putida	*Zingiber officinale*	Rhizome	Siderophore, IAA, and ACC deaminase, production	Jasim et al. (2014)

28.5 Endophytic Bacteria Against Drought Stress

Drought stress is most critical among all abiotic stresses. The interaction of plant and endophytic bacteria increased the tolerance or resistance towards drought stress (Paul and Lade 2014). Plant-microbe interaction is very effective in plant tolerance towards drought stress. Endophytic bacteria represent the subclass of rhizospheric bacteria having multiple traits of plant growth promotion (Bhatt et al. 2015a, b, 2016a, b, 2019b, c). These bacteria are better over rhizospheric bacteria in plant growth promotion because they are having a close association with plant tissue and exert a direct beneficial effect to plant growth promotion and development under drought condition (Cooper et al. 2014). Plant shows various morphological, physiological, biochemical and molecular responses towards drought stress, and endophytic bacteria help in the modulation of these responses to improve their tolerance capacity under drought stress. Endophytes increase stress-responsive gene expression, osmolytes and antistress metabolite during stress condition (Lata et al. 2018), thus enhancing plant tolerance and plant growth under drought stress. Endophytic bacteria increase plant growth and development by direct and indirect mechanism. Direct mechanism includes making nutrients available to plant by nitrogen fixation; phosphatesolubilization, zinc solubilization and potassium solubilization; and production of phytohormones, siderophore and 1-aminocyclopropane-1-carboxylate (ACC) deaminase, and indirect mechanism includes the production of antibiotics, exopolysaccharide, hydrolytic enzymes, and competition for nutrient and space to inhibit the pathogen. These mechanisms enhance the plant growth and thus help to cope with different kinds of biotic and abiotic stresses.

28.5.1 Direct Mechanism of Plant Growth Promotion by Endophytic Bacteria

Endophytic bacteria improve plant growth with direct mechanism by making the nutrients available (phosphate, potassium and zinc solubilization, nitrogen fixation, siderophore production), production of phytohormone (IAA, gibberellin, cytokinin, etc.) and ACC deaminase activity.

28.5.1.1 Nutrient Availability

Soil usually contains insufficient amount of nutrients which are crucial for plant growth. Some of the nutrients present in the soil are not easily available to the plants. Endophytic bacteria increase the availability of the limiting nutrients (N, P, Fe) for better growth of plants.

28.5.1.2 Nitrogen

Most of nitrogen in the atmosphere is present in unobtainable form. Endophytic bacteria help in increasing the nitrogen availability for the plants. They fix the atmospheric nitrogen with the help of nitrogenase enzyme. Nitrogenase is a complex enzyme consisting of dinitrogenase and dinitrogenase reductase, which helps in N_2 fixation. Some N_2-fixing endophytic bacteria are *Azospirillum brasilense*, *Burkholderia* spp., *Klebsiella oxytoca*, *Azoarcus* spp., *Gluconobacter*, *Enterobacter cloacae*, *Klebsiella pneumoniae* and *Pantoea* sp. and *Herbaspirillum*.

28.5.1.3 Phosphorus

Phosphorus is the chief micronutrient for various enzymatic reactions that is responsible for many physiological processes of the plant. It is present in large quantity but in insoluble form which is unavailable for the plants. Seventy-five per cent of phosphorus applied as fertilizers makes the complex with soil and becomes inaccessible to the plant (Ezawa et al. 2002). Endophytic bacteria solubilize the insoluble phosphate into soluble form by chelation, organic acid production, acidification and ion exchange (Nautiyal et al. 2000) and also produce the acid phosphatase for mineralization of organic phosphate. Some phosphate-solubilizing endophytic bacteria are *Bacillus methylotrophicus*, *B. megaterium*, *B.* amyloliquefaciens, *B. subtilis*, *Acinetobacter* sp., *Pseudomonas* spp., *Enterobacter asburiae*, and *Pantoea* spp.

28.5.1.4 Iron

Iron is a crucial element in the entire living organism and is involved in various important functions such as DNA synthesis, respiration and photosynthesis. It is a chief content of chlorophyll and essential for maintenance and function of chloroplast. In normal state iron is found in Fe^{3+} state which is insoluble and cannot be absorbed by plant. Endophytic bacteria produce siderophore (low-molecular-weight iron-chelating compound). Siderophore provides iron to the plant under iron-limiting condition (Ma et al. 2011). Some endophytic bacteria for siderophore production are *Pseudomonas*, *Bacillus*, *Azotobacter*, *Enterobacter*, *Methylobacteria*, *Rhizobium*, *Burkholderia*, etc.

28.5.1.5 Phytohormone Production

Endophytic bacteria increase the growth of plant by accumulating nutrient and metabolism of the host plant by producing various phytohormones. Five main phytohormones produced by endophytic bacteria are auxin, gibberellin, abscisic acid, cytokinin and ethylene. IAA is the chief plant hormone produced by endophytic bacteria.

28.5.1.5.1 IAA Production

IAA is a significant plant hormone having a key role in various plant physiological processes. It plays a chief role in growth and development of plant that includes cell-cell signalling, cell division and elongation and tissue differentiation. Other roles of IAA are initiation of adventitious and lateral root formation, response to light and gravity, apical dominance, mediation of resistance to plant stress and induction of plant defence. IAA can also modulate the synthesis of ethylene (Woodward and Bartel 2005; Glick 2012). Production of IAA by endophytic bacteria resulted in increased root biomass and lateral root in the host plant (Tsavkelova et al. 2007; Taghavi et al. 2009). According to the study of Lata et al. (2006), *Pseudomonas stutzeri* P3 strain can produce IAA. Etminani et al. (2018) found that endophytic bacteria *Bacillus pumilus and Pseudomonas protegens* isolated from wild pistachio tree have potential of IAA production. According to the study of Patten and Glick (2002), *Pseudomonas putida* GR12-2 defective in IAA synthesis is not able to increase the root growth and lateral root formation, indicating the role of IAA in lateral root formation.

28.5.1.5.2 Cytokinin

Cytokinin has the main role in plant growth, physiology and development. It is a plant growth regulator playing an important role in apical dominance, seed germination, leaf senescence and flower and fruit development. Cytokinin also enhances the resistance against various plant pathogens (fungi, bacteria, pest and insect). Higher concentration of cytokinin in plants increases the resistance towards the pathogen. Endophytic bacteria *Pseudomonas resinovorans* and *Paenibacillus polymyxa* isolated from *Gynura procumbens* have the ability to produce cytokinin (Bhore et al. 2010).

28.5.1.5.3 Gibberellin

Gibberellins are diterpenoid compound, having an important role in growth and development of plant which includes stem elongation, seed germination, flowering fruiting and seed dormancy. According to Gutierrez-Manero et al. (2001), GA1, GA3, GA4 and GA20 were detected in liquid media of *Bacillus pumilus* and *Bacillus licheniformis*. Forchetti et al. (2007) and Eke et al. (2019) studied the production of GA in the culture media of *Bacillus pumilus* and *Achromobacter xylosoxidans*. According to Cohen et al. (2009), *Azospirillum lipoferum*-treated maize plant increased the GA under drought condition. According to the study of Khan et al.

(2014) *Sphingomonas* sp. LK11 isolated from *Tephrosia apollinea* leaves produced the GA4, GA9 and GA20. Piccoli et al. (2011) isolated *Arthrobacter* from *Prosopis strombulifera* root and they showed GA1 and GA3 production.

28.5.1.5.4 ABA Production

ABA is an important plant hormone regulating plant growth, development and stress responses. ABA plays a significant role in various physiological processes of plants including stomatal closure, leaf senescence, bud dormancy, seed germination, cuticular wax deposition and osmotic regulation. According to a study of Shahzad et al. (2017), endophytic bacteria *Bacillus amyloliquefaciens* RWL-1 isolated from the seed of rice produced ABA under saline condition. *Bacillus pumilus* isolated from sunflower produced ABA under drought condition (Eke et al. 2019). According to Cohen et al. (Bhore et al. 2010) inoculation of maize with *Azospirillum lipoferum* increased the ABA content under drought condition. Endopyhtic bacteria *Curtobacterium* spp. SAK1 when inoculated to soybean plant under salt stress increased the ABA content (Khan et al. 2019).

28.5.1.6 Control of Ethylene Level via ACC Deaminase

Ethylene is a small organic molecule with biological activity and acts as a plant growth regulator at very low concentration. It is associated with various biological and developmental processes, e.g. fruit ripening, senescence of leaf, root initiation and nodulation, abscission, auxin transport and cell elongation (Sun et al. 2016). During plant growth and development, the concentration of ethylene remains low, but it enhanced during senescence and fruit ripening. Ethylene concentration increased due to various biotic (mechanical wounding, disease, pest) and abiotic (temperature extreme, water stress, UV light) stress and inhibits the elongation of root, nodulation of legumes, lateral root and root hair formation, defoliation, leaf senescence, destruction of chlorophyll, leaf abscission and epinasty. Regulation of ethylene concentration is necessary for proper plant growth.

Although endophytic bacteria use different mechanisms of plant growth, the possession of ACC deaminase activity is a key trait of endophytic bacteria for enhancement of plant growth in drought stress. ACC deaminase reduces the ethylene level by breaking the ACC, the precursor of ethylene. ACC deaminase-possessing bacteria use ACC as a nitrogen source, bind to the surface of the root and degrade the ACC by cleavage of ACC into ammonia and α-ketobutyrate (Woodward and Bartel 2005). Thus, by utilizing the ACC deaminase activity, endophytic bacteria alleviate the stress effect and improve plant growth and development. ACC deaminase is a pyridoxal phosphate-dependent enzyme, induced in the presence of its substrate ACC. ACC deaminase induction is complex and slow process. Activity of ACC deaminase gradually decreases after few hours of induction (Jacobson et al. 1994). According to the model for functioning of bacterial ACC deaminase proposed by

Glick et al. (1998), it has been revealed that an appropriate amount of ACC is released by the plant's root and taken up by the bacteria, and ACC deaminase activity of microbes breaks this ACC into ammonia and α-ketobutyrate. Uptake and breakdown of ACC by bacteria decreases the ACC level outside the plant root, thus maintaining the equilibrium between external and internal environments through exudation of ACC from inside to the rhizosphere. Thus, the maintenance of low ACC level, to reduce the biosynthesis of ethylene, is important for plant growth and development.

28.5.1.7 Indirect Mechanism of Plant Growth by Endophytic Bacteria

Endophytic bacteria by producing various compounds, e.g., antibiotics, volatile compound, HCN, siderophore, hydrolytic enzymes and toxins, inhibit the growth of phytopathogen, thus increasing the growth of plant indirectly. Endophytic bacteria increase the resistance of plant against pathogen by producing secondary metabolites including alkaloids, phenolics, terpenoids, steroids, quinines and peptides (Yu et al. 2010). Endophytic bacteria inhibit growth and activity of both fungal and bacterial plant pathogens. *Pseudomonas*, *Bacillus*, *Enterobacter, Paenibacillus* and *Serratia* are common genera of endophytic bacteria having antimicrobial activity. According to the study of Hong-Thao et al. (2016), endophytic bacteria *Streptomyces* from citrus fruit exhibited antimicrobial activity against *Colletotrichum truncatum*, *F. oxysporum*, *F. udum* and *Geotrichum candidum*. Ramesh et al. (2009) reported in their study that endophytic bacteria *Burkholderia cepacia*, *Pseudomonas* sp. and *Enterobacter cloacae* decreased the eggplant wilt by 70% by inhibiting *Ralstonia solanacearum*. Liu et al. (2020) studied *Bacillus megaterium* isolated from potato tuber and showed its inhibitory effect against *Streptomyces scabies* and *Erwinia carotovora* subsp. *Atroseptica*. Endophytic bacteria *Staphylococcus warneri* and *Bacillus velezensis* isolated from *Gnetum gnemon* showed inhibition against *Ralstonia solanacearum* (Agarwal et al. 2020). *Bacillus subtilis* isolated from sugarcane inhibited the growth of red rot causing *Colletotrichum falcatum* (Shastri et al. 2020; Bhatt et al. 2019d, e; Sharma and Bhatt 2016; Sharma et al. 2016; Bhatt and Nailwal 2018; Khati et al. 2018a; Gangola et al. 2018a; Bhatt 2018; Bhatt and Barh 2018; Bhandari and Bhatt 2020; Bhatt and Bhatt 2021). Jayakumar et al. (2020) isolated *Bacillus axarquiensis* and *Bacillus licheniformis* from sugarcane and showed their antagonistic effect against *Colletotrichum falcatum*. Thus, there is an immense need of eco-friendly strategies in agriculture for crop production under abiotic stresses (Singh et al. 2021; Zhang et al. 2020a, b; Mishra et al. 2020; Feng et al. 2020; Lin et al. 2020; Zhan et al. 2020; Ye et al. 2019; Huang et al. 2019, 2020). Utilization of endophytes as biofertilizer is an eco-friendly approach for sustainable agriculture (Huang et al. 2020; Fan et al. 2020; Pang et al. 2020; Gangola et al. 2018b; Gupta et al. 2018; Khati et al. 2018b). Indigenous microbial strains are able to make the environment sustainable. Both bacteria and fungi played a direct role in resource recovery of the agricultural fields via various mechanisms (Khati et al. 2017a, b, 2018b; Kumar et al. 2017).

28.6 Conclusion and Future Prospects

Drought stress affects the growth and productivity in all crop varieties. However, tolerance towards drought varies according to different species. Endophytic bacteria play a major role to adopt and resist drought stress in plants and thus can be a future tool for solving the future food security. Plant and endophytic bacteria interaction increases the growth and development of plant via direct and indirect mechanisms (Fig. 28.1). Direct mechanism includes enhancing nutrient availability, production of siderophore, regulation of phytohormones and accumulation of several compatible solutes/osmolytes, whereas indirect mechanism increases plant growth by acting against the phytopathogen by producing antibiotic, lytic enzymes and HCN and making nutrients unavailable for phytopathogen. Various approaches have been used for enhancement of drought tolerance in different varieties of plants. The future research needs to develop an effective microbial formulation for improving plant productivity under drought conditions. Understanding the molecular level of interaction between plant and endophytic bacteria is also necessary for better crop yield.

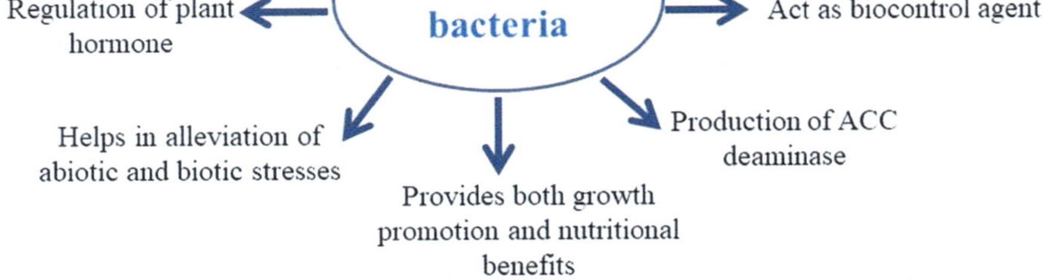

Fig. 28.1 Conceptual diagram showing different mechanisms of plant growth promotion by endophytic bacteria

References

Afzal I, Iqrar I, Shinwari ZK, Yasmin A (2017) Plant growth-promoting potential of endophytic bacteria isolated from roots of wild *Dodonaea viscosa* L. Plant Growth Regul 81:399–408

Agarwal H, Dowarah B, Baruah PM, Bordoloi KS, Krishnatreya DB, Agarwala N (2020) Endophytes from *Gnetum gnemon* L. can protect seedlings against infection of the phytopathogenic bacterium *R. solanacearum* as well as promote plant growth in tomato. Microbiol Res 238:126503

Ali SZ, Sandhya V, Rao LV (2014) Isolation and characterization of drought-tolerant ACC deaminase and exopolysaccharide-producing fluorescent *Pseudomonas* sp. Ann Microbiol 64:493–502

ALKahtani MD, Fouda A, Attia KA, Al-Otaibi F, Eid AM, EED E, Abdelaal KA (2020) Isolation and characterization of plant growth promoting endophytic bacteria from desert plants and their application as bioinoculants for sustainable agriculture. Agronomy 10:1325

Aslam A, Ahmad Zahir Z, Asghar HN, Shahid M (2018) Effect of carbonic anhydrase-containing endophytic bacteria on growth and physiological attributes of wheat under water-deficit conditions. Plant Prod Sci 21:244–255

Bhandari G, Bhatt P (2020) Concepts and application of plant microbe interaction in remediation of heavy metals. In: Sharma A (ed) Microbes and signaling biomolecules against plant stress. Rhizosphere biology. Springer, Singapore. https://doi.org/10.1007/978-981-15-7094-0_4

Bhandari G, Bagheri AR, Bhatt P, Bilal M (2021) Occurrence, potential ecological risks, and degradation of endocrine disruptor, nonylphenol from the aqueous environment. Chemosphere. https://doi.org/10.1016/j.chemosphere.2021.130013

Bhatt P (2018) *Insilico* tools to study the bioremediation in microorganisms. In: Pathak V, Navneet (eds) Handbook of research on microbial tools for environmental waste management. Hershey, IGI Global, pp 389–395. https://doi.org/10.4018/978-1-5225-3540-9.ch018

Bhatt P, Barh A (2018) Bioinformatic tools to study the soil microorganisms: an *in silico* approach for sustainable agriculture. In: Choudhary D, Kumar M, Prasad R, Kumar V (eds) In silico approach for sustainable agriculture. Springer, Singapore. https://doi.org/10.1007/978-981-13-0347-0_10

Bhatt K, Bhatt P (2021) Rhizospheric biology: alternate tactics for enhancing sustainable agriculture. In: Verma A, Jitendra Kumar JS, AEL H, Bahadur Singh HB (eds) Phytomicrobiome interactions and sustainable agriculture, pp 164–186. https://doi.org/10.1002/9781119644798.ch9

Bhatt P, Nailwal TK (2018) Crop improvement through microbial technology: a step toward sustainable agriculture. In: Prasad R, Singh Gill S, Tuteja N (eds) Crop improvement through microbial biotechnology. Elsevier, pp 245–253

Bhatt P, Bisht TS, Pathak VM, Barh A, Chandra D (2015a) Optimization of amylase production from the fungal isolates of Himalayan region Uttarakhand. Ecol Environ Conserv 21 (3):1517–1521

Bhatt P, Negi G, Gangola S, Khati P, Srivastava A, Sharma A (2015b) Optimization of sulfosulfuron biodegradation through response surface methodology using indigenous bacterial strain isolated from contaminated agriculture field. Int J Curr Microbiol App Sci 4(8):105–112

Bhatt P, Negi G, Gangola S, Khati P, Kumar G, Srivastava A, Sharma A (2016a) Differential expression and characterization of cypermethrin degrading potential proteins in *Bacillus thuringiensis* strain, SG4. 3 Biotech 6:225

Bhatt P, Sharma A, Gangola S, Khati P, Kumar G, Srivastava A (2016b) Novel pathway of cypermethrin biodegradation in *a Bacillus* sp. strain SG2 isolated from cypermethrin-contaminated agriculture field. 3 Biotech 6:45

Bhatt P, Huang Y, Zhan H, Chen S (2019a) Insight into microbial applications for the biodegradation of pyrethroid insecticides. Front Microbiol 10:1778

Bhatt P, Pal K, Bhandari G, Barh A (2019b) Modeling of methyl halide biodegradation on bacteria and its effect on other environmental systems. Pestic Biochem Physiol 158:88–100

Bhatt P, Gangola S, Chaudhary P, Khati P, Kumar G, Sharma A, Srivastava A (2019c) Pesticide induced up-regulation of esterase and aldehyde dehydrogenase in indigenous *Bacillus* spp. Biorem J 23(1):42–52

Bhatt P, Joshi D, Kumar N, Kumar N (2019d) Recent trends to study the functional analysis of mycorrhizosphere. In: Varma A, Choudhary D (eds) Mycorrhizosphere and pedogenesis. Springer, Singapore. https://doi.org/10.1007/978-981-13-6480-8_11

Bhatt P, Pathak VM, Joshi S, Bisht TS, Singh K, Chandra D (2019e) Major metabolites after degradation of xenobiotics and enzymes involved in these pathways. In: Smart bioremediation technologies: microbial enzymes, chap 12. Academic, Boston, pp 205–215. https://doi.org/10.1016/B978-0-12-818307-6.00012-3

Bhatt P, Rene ER, Kumar AJ, Kumar AJ, Zhang W, Chen S (2020a) Binding interaction of allethrin with esterase: bioremediation potential and mechanism. Bioresour Technol 315:13845. https://doi.org/10.1016/j.biortech.2020.123845

Bhatt P, Verma A, Verma S, Anwar MS, Prasher P, Mudila H, Chen S (2020b) Understanding phytomicrobiome: a potential reservoir for better crop management. Sustainability 12:5446

Bhatt P, Huang Y, Rene ER, Kumar AJ, Chen S (2020c) Mechanism of allethrin biodegradation by a newly isolated *Sphingomonas trueperi* strain CW3 from wastewater sludge. Bioresour Technol 305:123074

Bhatt P, Zhang W, Lin Z, Pang S, Huang Y, Chen S (2020d) Biodegradation of allethrin by a novel fungus *Fusarium proliferatum* strain CF2, isolated from contaminated. Soils Microorg 8:593

Bhatt P, Huang Y, Zhang W, Sharma A, Chen S (2020e) Enhanced cypermethrin degradation kinetics and metabolic pathway in *Bacillus thuringiensis* strain, SG4. Microorganisms. 8:223

Bhatt P, Bhatt K, Huang Y, Ziqiu L, Chen S (2020f) Esterase is a powerful tool for the biodegradation of pyrethroid insecticides. Chemosphere 244:125507

Bhatt P, Zhou X, Huang Y, Zhang W, Chen S (2021a) Characterization of the role of esterases in the biodegradation of organophosphate, carbamate and pyrethroid group pesticides. J Hazard Mater 411:125026

Bhatt P, Joshi T, Bhatt K, Zhang W, Huang Y, Chen S (2021b) Binding interaction of glyphosate oxidoreductase and C-P lyase: molecular docking and molecular dynamics simulation studies. J Hazard Mater 409:124927

Bhatt P, Bhatt K, Sharma A, Zhang W, Mishra S, Chen S (2021c) Biotechnological basis of microbial consortia for the removal of pesticides from the environment. Crit Rev Biotechnol. https://doi.org/10.1080/07388551.2020.1853032

Bhatt P, Sethi K, Gangola S, Bhandari G, Verma A, Adnan M, Singh Y, Chaube S (2021d) Modeling and simulation of atrazine biodegradation in bacteria and its effect in other living systems. J Biomol Struct Dyn. https://doi.org/10.1080/07391102.2020.1846623

Bhatt P, Gangola S, Bhandari G, Zhang W, Maithani D, Mishra S, Chen S (2021e) New insights into the degradation of synthetic pollutants in contaminated environments. Chemosphere. https://doi.org/10.1016/j.chemosphere.2020.128827

Bhatt P, Sharma A, Rene ER, Kumar AJ, Zhang W, Chen S (2021f) Bioremediation mechanism, kinetics of fipronil degradation using *Bacillus* sp. FA3 and resource recovery potential from contaminated environments. J Water Process Eng 39:101712

Bhore SJ, Ravichantar N, Loh CY (2010) Screening of endophytic bacteria isolated from leaves of Sambung Nyawa [*Gynura procumbens* (Lour.) Merr.] for cytokinin-like compounds. Bioinformation 5:191

Bodhankar S, Grover M, Hemanth S, Reddy G, Rasul S, Yadav SK, Srinivasarao C (2017) Maize seed endophytic bacteria: dominance of antagonistic, lytic enzyme-producing Bacillus spp. 3 Biotech 7:232

Caravaca F, Alguacil MM, Hernández JA, Roldán A (2005) Involvement of antioxidant enzyme and nitrate reductase activities during water stress and recovery of mycorrhizal *Myrtus communis* and *Phillyrea angustifolia* plants. Plant Sci 169:191–197

Chandra D, Srivastava R, Glick BR, Sharma AK (2018a) Drought-tolerant *Pseudomonas* spp. improve the growth performance of finger millet (*Eleusine coracana* (L.) Gaertn.) under non-stressed and drought-stressed conditions. Pedosphere 28:227–240

Chandra D, Srivastava R, Sharma AK (2018b) Influence of IAA and ACC deaminase producing fluorescent pseudomonads improves the growth of wheat (*Triticum aestivum* L.) under non-stressed and drought stressed conditions. Agric Res 7:290–299

Chandra D, Srivastava R, Gupta VV, Franco CM, Sharma AK (2019a) Evaluation of ACC-deaminase-producing rhizobacteria to alleviate water-stress impacts in wheat (*Triticum aestivum* L.) plants. Can J Microbiol 65:387–403

Chandra D, Srivastava R, Gupta VV, Franco CM, Paasricha N, Saifi SK, Tuteja N, Sharma AK (2019b) Field performance of bacterial inoculants to alleviate water stress effects in wheat (*Triticum aestivum* L.). Plant Soil 441:261–281

Chandra D, Srivastava R, Glick BR, Sharma AK (2020) Rhizobacteria producing ACC deaminase mitigate water-stress response in finger millet (*Eleusine coracana* (L.) Gaertn.). 3 Biotech 10:65

Chen C, Xin K, Liu H, Cheng J, Shen X, Wang Y, Zhang L (2017) *Pantoea alhagi*, a novel endophytic bacterium with ability to improve growth and drought tolerance in wheat. Sci Rep 7:41564

Cherif H, Marasco R, Rolli E, Ferjani R, Fusi M, Soussi A, Cherif A (2015) Oasis desert farming selects environment-specific date palm root endophytic communities and cultivable bacteria that promote resistance to drought. Environ Microbiol Rep 7:668–678

Cohen AC, Travaglia CN, Bottini R, Piccoli PN (2009) Participation of abscisic acid and gibberellins produced by endophytic *Azospirillum* in the alleviation of drought effects in maize. Botany 87:455–462

Cooper M, Gho C, Leafgren R, Tang T, Messina C (2014) Breeding drought tolerant maize hybrids for the US corn-belt: discovery to product. J Exp Bot 65:6191–6204

Dai A (2011) Characteristics and trends in various forms of the Palmer Drought Severity Index during 1900–2008. J Geophys Res Atmos 116(D12)

Daryanto S, Wang L, Jacinthe PA (2016) Drought effects on root and tuber production: a meta-analysis. Agric Water Manag 176:122–131

Eisenstein M (2013) Discovery in a dry spell. Nature 501:S7–S9

Eke P, Kumar A, Sahu KP, Wakam LN, Sheoran N, Ashajyothi M, Fekam FB (2019) Endophytic bacteria of desert cactus (*Euphorbia trigonas* Mill) confer drought tolerance and induce growth promotion in tomato (*Solanum lycopersicum* L.). Microbiol Res 228:126302

Etminani F, Harighi B (2018) Isolation and identification of endophytic bacteria with plant growth promoting activity and biocontrol potential from wild pistachio trees. Plant Pathol J 34:208

Ezawa T, Smith SE, Smith FA (2002) P metabolism and transport in AM fungi. Plant Soil 244:221–230

Fan X, Ye T, Li Q, Bhatt P, Zhang L, Chen S (2020) Potential of a quorum quenching bacteria isolate *Ochrobactrum intermedium* D-2 against soft rot pathogen *Pectobacterium carotovorum* subsp. Carotovora Front Microbiol 11:898

Fedoroff NV, Battisti DS, Beachy RN, Cooper PJM, Fischhoff DA, Hodges CN, Zhu JK (2010) Radically rethinking agriculture for the 21st century. Science 327:833–834

Feng Y, Huang Y, Zhan H, Bhatt P, Chen S (2020) An overview of strobilurin fungicide degradation: current status and future perspective. Front Microbiol 11:389

Finkel OM, Castrillo G, Paredes SH, González IS, Dangl JL (2017) Understanding and exploiting plant beneficial microbes. Curr Opin Plant Biol 38:155–163

Forchetti G, Masciarelli O, Alemano S, Alvarez D, Abdala G (2007) Endophytic bacteria in sunflower (*Helianthus annuus* L.): isolation, characterization, and production of jasmonates and abscisic acid in culture medium. Appl Microbiol Biotechnol 76:1145–1152

Gagne-Bourgue F, Aliferis KA, Seguin P, Rani M, Samson R, Jabaji S (2013) Isolation and characterization of indigenous endophytic bacteria associated with leaves of switchgrass (*Panicum virgatum* L.) cultivars. J Appl Microbiol 114:836–853

Gangola S, Bhatt P, Chaudhary P, Khati P, Kumar N, Sharma A. Bioremediation of industrial waste using microbial metabolic diversity. In Pankaj, & A. Sharma (Eds.), Microbial biotechnology in environmental monitoring and cleanup (pp. 1–27). Hershey: IGI Global. 2018a; doi:https://doi.org/10.4018/978-1-5225-3126-5.ch001

Gangola S, Sharma A, Bhatt P, Khati P, Chaudhary P (2018b) Presence of esterase and laccase in *Bacillus subtilis* facilitates biodegradation and detoxification of cypermethrin. Sci Rep 8:12755. https://doi.org/10.1038/s41598-018-31082-5

Glick BR (2012) Plant growth-promoting bacteria: mechanisms and applications. Scientifica 2012:963401

Glick BR, Penrose DM, Li J (1998) A model for lowering of plant ethylene concentrations by plant growth promoting bacteria. J Theor Biol 190:62–68

Govindasamy V, George P, Kumar M, Aher L, Raina SK, Rane J, Minhas PS (2020) Multi-trait PGP rhizobacterial endophytes alleviate drought stress in a senescent genotype of sorghum [*Sorghum bicolor* (L.) Moench]. 3 Biotech 10:13

Gupta S, Bhatt P, Chaturvedi P (2018) Determination and quantification of asiaticoside in endophytic fungus from *Centella asiatica* (L.) urban. World J Microbiol Biotechnol 34:111

Gutiérrez-Mañero FJ, Ramos-Solano B, Probanza AN, Mehouachi JR, Tadeo F, Talon M (2001) The plant-growth-promoting rhizobacteria *Bacillus pumilus* and *Bacillus licheniformis* produce high amounts of physiologically active gibberellins. Physiol Plant 111:206–211

Haidar B, Ferdous M, Fatema B, Ferdous AS, Islam MR, Khan H (2018) Population diversity of bacterial endophytes from jute (*Corchorus olitorius*) and evaluation of their potential role as bioinoculants. Microbiol Res 208:43–53

Hong-Thao PH, Mai-Linh NV, Hong-Lien NT, Hieu NV (2016) Biological characteristics and antimicrobial activity of endophytic Streptomyces sp. TQR12-4 isolated from elite *Citrus nobilis* cultivar Ham Yen of Vietnam. Int J Microbiol 2016:7207818

Hsiao A (2000) Effect of water deficit on morphological and physiological characterizes in Rice (*Oryza sativa*). J Agric For 3:93–97

Huang Y, Zhan H, Bhatt P, Chen S (2019) Paraquat degradation from contaminated environments: current achievements and perspectives. Front Microbiol 10:1754

Huang Y, Lin Z, Zhang W, Pang S, Bhatt P, Rene ER, Kumar AJ, Chen S (2020) New insights into the microbial degradation of D-cyphenothrin in contaminated water/soil environments. Microorganisms 8:473

Jacobson CB, Pasternak JJ, Glick BR (1994) Partial purification and characterization of 1-aminocyclopropane-1-carboxylate deaminase from the plant growth promoting rhizobacterium *Pseudomonas putida* GR12-2. Can J Microbiol 40:1019–1025

Jasim B, Joseph AA, John CJ, Mathew J, Radhakrishnan EK (2014) Isolation and characterization of plant growth promoting endophytic bacteria from rhizome of *Zingiber officinale*. 3 Biotech 4:197–204

Jayakumar A, Krishna A, Nair IC, Radhakrishnan EK (2020) Drought-tolerant and plant growth-promoting endophytic *Staphylococcus* sp. having synergistic effect with silicate supplementation. Arch Microbiol 202:1899–1906

Jiao J, Ma Y, Chen S, Liu C, Song Y, Qin Y, Liu Y (2016) Melatonin-producing endophytic bacteria from grapevine roots promote the abiotic stress-induced production of endogenous melatonin in their hosts. Front Plant Sci 7:1387

Khan AL, Waqas M, Kang SM, Al-Harrasi A, Hussain J, Al-Rawahi A, Al Khiziri S, Ullah I, Ali L, Jung HY, Lee IJ (2014) Bacterial endophyte *Sphingomonas* sp LK11 produces gibberellins and IAA and promotes tomato plant growth. J Microbiol 52:689–695

Khan MA, Asaf S, Khan AL, Ullah I, Ali S, Kang SM, Lee IJ (2019) Alleviation of salt stress response in soybean plants with the endophytic bacterial isolate *Curtobacterium* sp. SAK1. Ann Microbiol 69:797–808

Khati P, Sharma A, Gangola S, Kumar R, Bhatt P, Kumar G (2017a) Impact of agriusable nanocompounds on soil microbial activity: an indicator of soil health. Soil Air Water 45 (5):1600458

Khati P, Parul, Gangola S, Bhatt P, Sharma A (2017b) Nanochitosan induced growth of Zea mays with soil health maintenance. 3 Biotech 7(1):81

Khati P, Gangola S, Bhatt P, Kumar R, Sharma A (2018a) Application of nanocompounds for sustainable agriculture system. In: Pankaj, Sharma A (eds) Microbial biotechnology in environmental monitoring and cleanup. Hershey, IGI Global, pp 194–211. https://doi.org/10.4018/978-1-5225-3126-5.ch012

Khati P, Parul, Bhatt P, Nisha Kumar R, Sharma A (2018b) Effect of nanozeolite and plant growth promoting rhizobacteria on maize. 3 Biotech 8:141

Kumar G, Arya K, Verma A, Pankaj, Khati P, Gangola S, Kumar R, Sharma A, Singh H (2017) Bioremediation of petrol engine oil polluted soil using microbial consortium and wheat crop. J Pure Appl Microbiol 11(3):1583–1588

Lata H, Li XC, Silva B, Moraes RM, Halda-Alija L (2006) Identification of IAA-producing endophytic bacteria from micropropagated Echinacea plants using 16S rRNA sequencing. Plant Cell Tissue Organ Cult 85:353–359

Lata R, Chowdhury S, Gond SK, White JF Jr (2018) Induction of abiotic stress tolerance in plants by endophytic microbes. Lett Appl Microbiol 66:268–276

Latif Khan A, Ahmed Halo B, Elyassi A, Ali S, Al-Hosni K, Hussain J, Al-Harrasi A, Lee IJ (2016) Indole acetic acid and ACC deaminase from endophytic bacteria improves the growth of *Solarium lycopersicum*. Electron J Biotechnol 19:58–64

Li Y, Cheng C, An D (2017) Characterisation of endophytic bacteria from a desert plant *Lepidium perfoliatum* L. Plant Prot Sci 53:32–43

Li H, Zhao Y, Jiang X (2019) Seed soaking with Bacillus sp. strain HX-2 alleviates negative effects of drought stress on maize seedlings. Chil J Agric Res 79:396–404

Lin Z, Zhang W, Pang S, Huang Y, Mishra S, Bhatt P, Chen S (2020) Current approaches to and future perspectives on methomyl degradation in contaminated soil/water environments. Molecules 25:738

Liu JM, Wang SS, Zheng X, Jin N, Lu J, Huang YT, Wang FZ (2020) Antimicrobial activity against phytopathogens and inhibitory activity on solanine in potatoes of the endophytic bacteria isolated from potato tubers. Front Microbiol 11:570926

Ma Y, Prasad MNV, Rajkumar M, Freitas H (2011) Plant growth promoting rhizobacteria and endophytes accelerate phytoremediation of metalliferous soils. Biotechnol Adv 29:248–258

Ma Y, Rajkumar M, Moreno A, Zhang C, Freitas H (2017) Serpentine endophytic bacterium *Pseudomonas azotoformans* ASS1 accelerates phytoremediation of soil metals under drought stress. Chemosphere 185:75–85

Mishra S, Zhang W, Lin Z, Pang S, Huang Y, Bhatt P, Chen S (2020) Carbofuran toxicity and its microbial degradation in contaminated environments. Chemosphere 259:127429

Nautiyal CS, Bhadauria S, Kumar P, Lal H, Mondal R, Verma D (2000) Stress induced phosphate solubilization in bacteria isolated from alkaline soils. FEMS Microbiol Lett 182:291–296

Ngumbi E, Kloepper J (2016) Bacterial-mediated drought tolerance: current and future prospects. Appl Soil Ecol 105:109–125

Pang S, Lin Z, Zhang W, Mishra S, Bhatt P, Chen S (2020) Insights into the microbial degradation and biochemical mechanisms of neonicotinoids. Front Microbiol 11:868

Partida-Martinez LPP, Heil M (2011) The microbe-free plant: fact or artifact? Front. Plant Sci 2:100

Patten CL, Glick BR (2002) Regulation of indoleacetic acid production in *Pseudomonas putida* GR12-2 by tryptophan and the stationary-phase sigma factor RpoS. Can J Microbiol 48:635–642

Paul D, Lade H (2014) Plant-growth-promoting rhizobacteria to improve crop growth in saline soils: a review. Agron Sustain Dev 34:737–752

Philippot L, Raaijmakers JM, Lemanceau P, van der Putten WH (2013) Going back to the roots: the microbial ecology of the rhizosphere. Nat Rev Microbiol 11:789–799

Piccoli P, Travaglia C, Cohen A, Sosa L, Cornejo P, Masuelli R, Bottini R (2011) An endophytic bacterium isolated from roots of the halophyte *Prosopis strombulifera* produces ABA, IAA, gibberellins A 1 and A 3 and jasmonic acid in chemically-defined culture medium. Plant Growth Regul 64:207–210

Rahdari P, Hoseini SM, Tavakoli S (2012) The studying effect of drought stress on germination, proline, sugar, lipid, protein and chlorophyll content in Purslane (*Portulaca oleraceae* L.) leaves. J Med Plants Res 6:1539–1547

Ramesh R, Joshi AA, Ghanekar MP (2009) Pseudomonads: major antagonistic endophytic bacteria to suppress bacterial wilt pathogen, *Ralstonia solanacearum* in the eggplant (*Solanum melongena* L.). World J Microbiol Biotechnol 25:47–55

Rolli E, Marasco R, Vigani G, Ettoumi B, Mapelli F, Deangelis ML, Pierotti Cei F (2015) Improved plant resistance to drought is promoted by the root-associated microbiome as a water stress-dependent trait. Environ Microbiol 17:316–331

Rosenblueth M, Martínez-Romero E (2006) Bacterial endophytes and their interactions with hosts. Mol Plant-Microbe Interact 19:827–837

Selvakumar G, Panneerselvam P, Ganeshamurthy AN (2012) Bacterial mediated alleviation of abiotic stress in crops. In: Maheshwari DK (ed) Bacteria in agrobiology: stress management. Springer, Berlin, pp 205–224

Sgherri CLM, Maffei M, Navari-Izzo F (2000) Antioxidative enzymes in wheat subjected to increasing water deficit and rewatering. J Plant Physiol 157:273–279

Shahzad R, Khan AL, Bilal S, Waqas M, Kang SM, Lee IJ (2017) Inoculation of abscisic acid-producing endophytic bacteria enhances salinity stress tolerance in *Oryza sativa*. Environ Exp Bot 136:68–77

Sharma A, Bhatt P (2016) Bioremediation: a microbial technology for improvising wildlife. In: Saxena A (ed) Wildlife management concept analysis and conservation, pp 29–40. ISBN 10: 9351247929

Sharma A, Pankaj, Khati P, Gangola S, Kumar G (2016) Microbial degradation of pesticides for environmental cleanup. In: Bioremediation of industrial pollutants, chap 6. Write and Print Publication, pp 178–205

Shastri B, Kumar R, Lal RJ (2020) Isolation and identification of antifungal metabolite producing endophytic Bacillus subtilis (S17) and its in vitro effect on *Colletotrichum falcatum* causing red rot in sugarcane. Vegetos 33:493–503

Singh K, Gera R, Sharma R, Maithani D, Chandra D, Bhat AM, Kumar R, Bhatt P (2021) Mechanism and application of *Sesbania* root nodulating bacteria: an alternative for chemical fertilizers and sustainable development. Arch Microbiol 203:1259–1270. https://doi.org/10.1007/s00203-020-02137-x

Sulbhi V, Bhatt P, Verma A, Mudila H, Prasher P, Rene ER (2021) Microbial technologies for heavy metal remediation: effect of process conditions and current practices. Clean Technol Environ Policy. https://doi.org/10.1007/s10098-021-02029-8

Sun L, Wang X, Li Y (2016) Increased plant growth and copper uptake of host and non-host plants by metal-resistant and plant growth-promoting endophytic bacteria. Int J Phytoremediation 18:494–501

Sun H, Kong L, Du H, Chai Z, Gao J, Cao Q (2019) Benefits of *Pseudomonas poae* s61 on *Astragalus mongholicus* growth and bioactive compound accumulation under drought stress. J Plant Interact 14:205–212

Sziderics AH, Rasche F, Trognitz F, Sessitsch A, Wilhelm E (2007) Bacterial endophytes contribute to abiotic stress adaptation in pepper plants (*Capsicum annuum* L.). Can J Microbiol 53:1195–1202

Taghavi S, Garafola C, Monchy S, Newman L, Hoffman A, Weyens N, van der Lelie D (2009) Genome survey and characterization of endophytic bacteria exhibiting a beneficial effect on growth and development of poplar trees. Appl Environ Microbiol 75:748–757

Timmusk S, Paalme V, Pavlicek T, Bergquist J, Vangala A, Danilas T, Nevo E (2011) Bacterial distribution in the rhizosphere of wild barley under contrasting microclimates. PLoS One 6: e17968

Timmusk S, Abd El-Daim IA, Copolovici L, Tanilas T, Kannaste A, Behers L, Niinemets U (2014) Drought-tolerance of wheat improved by rhizosphere bacteria from harsh environments: enhanced biomass production and reduced emissions of stress volatiles. PLoS One 9:e96086

Trivedi G, Patel P, Saraf M (2019) Synergistic effect of endophytic selenobacteria on biofortification and growth of Glycine max under drought stress. S Afr J Bot 134:27–35

Tsavkelova EA, Cherdyntseva TA, Botina SG, Netrusov AI (2007) Bacteria associated with orchid roots and microbial production of auxin. Microbiol Res 162:69–76

Ullah A, Nisar M, Ali H, Hazrat A, Hayat K, Keerio AA, Khan A (2019) Drought tolerance improvement in plants. An endophytic bacterial approach. Appl Microbiol Biotechnol 103:7385–7397

Van Loon AF (2015) Hydrological drought explained. Wiley Interdiscip Rev Water 2:359–392

Vigani G, Rolli E, Marasco R, Dell'Orto M, Michoud G, Soussi A, Daffonchio D (2019) Root bacterial endophytes confer drought resistance and enhance expression and activity of a vacuolar H^+-pumping pyrophosphatase in pepper plants. Environ Microbiol 9:3212–3228

Woodward AW, Bartel B (2005) Auxin: regulation, action, and interaction. Ann Bot 95:707–735

Ye T, Zhou T, Fan X, Bhatt P, Zhang L, Chen S (2019) *Acinetobacter lactucae* strain QL-1, a novel quorum quenching candidate against bacterial pathogen *Xanthomonas campestris* pv. Campestris. Front Microbiol 10:2867

Yu H, Zhang L, Li L, Zheng C, Guo L, Li W, Sun P, Qin L (2010) Recent developments and future prospects of antimicrobial metabolites produced by endophytes. Microbiol Res 165:437–449

Zhan H, Huang Y, Lin Z, Bhatt P, Chen S (2020) New insights into the microbial degradation and catalytic mechanism of synthetic pyrethroids. Environ Res 182:109138

Zhang L, Zhong J, Liu H, Xin K, Chen C, Li Q, Shen X (2017) Complete genome sequence of the drought resistance-promoting endophyte Klebsiella sp. LTGPAF-6F. J Biotechnol 246:36–39

Zhang W, Pang S, Lin Z, Mishra S, Bhatt P, Chen S (2020a) Biotransformation of perfluoroalkyl acid precursors from various environmental system: advances and perspectives. Environ Pollut 272:115908

Zhang W, Lin Z, Pang S, Bhatt P, Chen S (2020b) Insights into the biodegradation of lindane (γ-hexachlorocyclohexane) using a microbial system. Front Microbiol. 11:522

The manufacturer's authorised representative in the EU is Springer
Nature Customer Service Centre GmbH, Europaplatz 3, 69115 Heidelberg,
Germany. If you have any concerns regarding our products, please
contact ProductSafety@springernature.com

Printed and bound by CPI Group (UK) Ltd, Croydon, CR0 4YY
30/09/2026
02225240-0014